Quantity	Symbol	SI Unit	Symb	
length	l	metre	m	
magnetizing force	H	ampere-turns	—	
magnetic moment	m	Wb m	—	
magnification, linear	m	*a ratio*	—	
mass	m	kilogramme	kg	fundamental unit
mass number	A	*a number*	—	number of neutrons + protons
molar volume	V_m	(dm^3)	—	volume of 1 mole
molar solution	M	*a ratio*	—	moles/dm^3
moment of force	—	N m	—	force × perp. distance
neutron number	N	*a number*	—	number of neutrons
number	n	—	—	—
number of molecules	N	—	—	—
number turns on coil	n	*a number*	—	—
number order spectrum	p	*a number*	—	—
object distance	u	metre	m	fundamental unit
peak current	I_o	ampere	A	see current
peak e.m.f.	E_o	volt	V	see e.m.f.
period	T	second	s	fundamental unit
permeability	μ	H m^{-1}	—	henry/metre
permeability, vacuum	μ_o	H m^{-1}	—	
permeability, relative	μ_r	*a ratio*	—	$\mu = \dfrac{\mu}{\mu_o}$
permittivity	ε	F m^{-1}	—	farad/metre
permittivity, vacuum	ε_o	F m^{-1}	—	farad/metre
permittivity, relative	ε_r	*a ratio*	—	$\varepsilon_r = \dfrac{\varepsilon}{\varepsilon_o}$
potential, electric	V	volt	V	energy/charge
potential difference	V	volt	V	energy/charge
power	P	watt	W	J s^{-1}
pressure	p	pascal	Pa	N m^{-2}: force/area
radius	r	metre	m	fundamental unit
reactance	X	ohm	Ω	E_o/I_o
refractive index	n	*a ratio*	—	—
resistance	R	ohm	Ω	p.d./current
resistivity, electrical	ρ	ohm-metre	—	resistance × length
relative density	d	*a ratio*	—	ρ_{sub}/ρ_{water}
r.m.s. current	$I_{r.m.s}$	ampere	A	see current
r.m.s. voltage	$V_{r.m.s}$	volt	V	see e.m.f.
slit separation	s	metre	m	fundamental unit
tension	T	newton	N	see force
temperature, Celsius	θ	degree C	°C	from kelvin
temp. interval	θ	degree	° or K	—
temp. absolute	T	kelvin	K	fundamental unit
thickness	d	metre	m	fundamental unit
time	t	second	s	fundamental unit
torque	T	Nm	—	see moment
turns ratio	T	*a ratio*	—	n_{sec}/n_{prim}
(unit of electricity)	—	kW h	—	kilowatt × hour
velocity	u, υ	ms^{-1}	—	distance/time
velocity, angular	ω	second^{-1}	s^{-1}	angle/time
velocity, e.m.waves	c	m s^{-1}	—	—
velocity of sound	υ	m s^{-1}	—	—
volume	V	metre cubed	m^3	l × b × h
wavelength	λ	metre	m	fundamental unit
work	w	joule	J	force x distance (Nm)
weight	W	newton	N	kg m s^{-2} or mg

Industrial Electronics

4th Edition

James T. Humphries
Santa Fe Community College
Gainesville, Florida

Leslie P. Sheets
College of Technical Careers
Southern Illinois University at Carbondale

Delmar
Publishers Inc.

NOTICE TO THE READER

Cover photo: Chuck O'Rear/West Light
Cover design: Lisa L. Pauly

DELMAR STAFF
Administrative Editor: Wendy J. Welch
Project Editor: Melissa A. Conan
Production Coordinator: Barbara A. Bullock
Design Coordinator: Lisa L. Pauly
Art Coordinator: Brian Yacur

For information, address Delmar Publishers Inc.
3 Columbia Circle, Box 15015
Albany, NY 12212-5015

Printed in the United States of America
Published simultaneously in Canada
by Nelson Canada,
a division of The Thomson Corporation

1 2 3 4 5 6 7 8 9 10 XXX 99 98 97 96 95 94 93

Library of Congress Cataloging–in–Publication Data

Humphries, James T., 1946–
    Industrial electronics/James T. Humphries, Leslie P. Sheets—4th ed.
        p. cm.
    Includes index.
    ISBN 0–8273–5825–3
    1. Industrial electronics. I. Sheets, Leslie P., 1941–
II. Title.
TK7881.H85 1993
621.381—dc20
92–29767
CIP

To our wives, Marg and Joyce

She is far more precious than jewels—Proverbs 31:10

Contents

15:   Introduction to Robotics   591

Data Sheets   609

Appendixes   655

Bibliography   690

Answers to Odd-Numbered Problems   696

Index   705

Preface

Few areas of electronics have changed as much as that of industrial electronics. Faced with pressures from manufacturers overseas, managers in industry have used the latest developments in electronics to make their businesses more competitive. The growth in this area places an increasing demand on instructors to prepare students to work in the industrial electronics area. Most managers and instructors would agree that a thorough knowledge of electrical theory is not enough to be successful in today's industrial environments. Therefore, the fourth edition of *Industrial Electronics* covers not only the theory but also the applications in industrial systems necessary to survive as a technician in industry today.

FEATURES

- All the chapters present in the third edition have been retained in this fourth edition. A survey of users of the text feel that the coverage in most areas is adequate. A chapter covering *robotics* and one covering *optoelectronic devices* (lasers and light-emitting diodes) have been added. These two areas are seldom covered in other courses. On the recommendation of several users, an appendix on Operational Amplifiers has been added for reference.

- *Bibliographies* are included at the end of the book for each chapter. These references may be used as supplemental reading or study assignments for topics that are to be covered in depth.

- *Data sheets* have also been included at the back of the book for reference. They present detailed information about some of the IC chips used in the text. We hope that students will use this information to gain greater understanding of the IC chips and circuits we have used as examples. These same IC chips can also be used in student design projects, and the data sheets should help.

- A book of *laboratory experiments* is available as an accompaniment to the fourth edition. We have tried to include experiments in the lab manual that will reinforce the concepts taught in the text material. Every effort has been made to use inexpensive, generic components and circuits for experiments. In many cases, equipment already on hand can be adapted for use in these experiments. The laboratory manual is available as a separate publication and is keyed to reading assignments in the text.

- An ancillary *instructor's guide* is available without charge from the publisher for the convenience of instructors who adopt this book for classroom use. The instructor's guide will include the solutions to problems at the end of the chapters, as well as solutions to the laboratory experiments.

GOALS

The primary goal of the fourth edition of *Industrial Electronics* is the same as all other editions: to provide the student with an understanding of the basic components and systems used in industrial electronics in an interesting and easy-to-understand style. We feel that students need a course that introduces a systems approach to many of the devices they have studied earlier in their coursework. They also need a text that covers devices they may not have been exposed to in other courses. The explosion of knowledge in the electronics field—especially in the area of digital electronics and microprocessors—has caused many electronics programs to decrease or eliminate coverage of many devices that are important in industrial electronics.

Success in any facet of electronics depends on a knowledge of fundamentals. This text presents the basic facts, concepts, and principles of industrial electronics. Since an industrial electronics course tries to prepare students for entry into the workplace, any text should reflect those changes that are occurring in industry. In this fourth edition, we have made every effort to bring this text up-to-date to reflect what industry is demanding of our graduates and what instructors are now teaching and should be teaching in their courses.

- *The text avoids design questions.* Instead, it focuses on the underlying concepts and the operation of electronic devices, circuits, and systems. We feel that if concepts are understood, designing circuits, in most cases, is not a problem. We definitely do not subscribe to the notion that the best way to understand electronic circuits is to design them.
- *The text is comprehensive.* Experience has shown that a course in industrial electronics requires the coverage of a large number of topics. But how can these topics be covered in one course? One solution is to use several textbooks for the course. Another is to supplement one text extensively with instructor-prepared materials. A third approach is to modify the course and teach only those topics covered in the textbook. However, we feel that all of these alternatives are unacceptable. Thus, we have purposely written a comprehensive text that includes most of the required topics.
- *The text presentation is flexible.* We have written all chapters so that certain topics can be easily omitted. Although we cover most of the topics in the book in a one-semester course, some of the topics can be treated in other courses. Thus, the material in this book could easily be expanded to two semesters or two quarters, depending on the depth of treatment for each topic.

- *The text presentation is not overly mathematical.* We expect students to have a mathematics background no higher than algebra and trigonometry, and we have avoided any higher level of mathematics in our presentation. We realize, however, that mathematics is a concise way of representing concepts. Thus, we have included it where we feel that an adequate grasp of the concept demands a mathematical treatment.

LEVEL

This book is intended for use in electronics programs offered in two- or four-year colleges. All electronics students at the College of Technical Careers at Southern Illinois University, Carbondale, are required to take an industrial electronics course. We feel that every graduate of a two-year or four-year electronics program should have a basic understanding of both digital and analog circuitry. Many of our graduates report that their preparation in analog circuits and systems has been invaluable. They typically find themselves acting as liaisons between analog applications and people who have a predominantly digital background.

We also expect students to have some background in basic digital gates and logic gained from an introductory course in digital electronics earlier in their electronics education. It may also be helpful, but not essential, for the student to have completed a technical physics course.

ACKNOWLEDGMENTS

Publishing a textbook is the result of a coordinated effort that involves many people. We would like to express our thanks to the following individuals who reviewed the manuscript at different levels in the production process:

James Allenbrand, Mike Burch, William A. Campbell, John Clemons, Donald Crawford, Ron Hessman, Clarke Homoly, Charles L. R. Klingensmith, Thomas C. Labonté, Scotty D. Richardson, and Ralph E. Scogin. We would especially like to thank Mark E. Kirkegaard, who helped us to ensure the accuracy of the text; our colleagues at Santa Fe Community College and at Southern Illinois University; and, finally, our wives and children.

James T. Humphries
Leslie P. Sheets

Figure Credits

Chapter 1: Figure 1.27, National Semiconductor Corporation; Figures 1.49 and 1.53, RCA Solid State Division.

Chapter 2: Figure 2.16, Reprinted with permission of National Semiconductor Corporation. National holds no responsibility for any circuitry described. National reserves the right at any time to change without notice said circuitry.

Chapter 3: Figures 3.1A and 3.4, Reliance Electric Company; Figure 3.5, reprinted from EDN, 1978 © Cahners Publishing Company; Table 3.6, reprinted from EDN, 1978 © Cahners Publishing Company.

Chapter 4: Figure 4.6, Bodine Electric Company, Chicago, IL 60618; Figures 4.12, 4.15, and 4.19, Pacific Scientific; Figure 4.21, Pittman, a division of Penn Engineering & Manufacturing Corp.; Figures 4.23, 4.26, 4.27, 4.28, 4.29, 4.30, 4.31, 4.32, 4.33, 4.34, 4.35, 4.36, 4.37, and 4.38, Western Division IMC Magnetics Corporation; portions of the section on brushless DC motors are drawn from Chapter 6 of *DC Motors, Speed Controls, and Servo Systems*, 6th edition, 1982 Electro-Craft Corporation, Hopkins, MN; portions of the section on stepper motors are drawn from the *TORMAX*® *Steppers and Control Manual*, Western Division IMC Magnetics Corporation, Maywood, CA.

Chapter 5: Figure 5.1, Reliance Electric Company; Figure 5.1, G.M. Ferree, STC Photography; Figures 5.7 and 5.11, Bob White, STC Photography.

Chapter 6: Figures 6.2, 6.3, 6.4, and 6.5, Micro Switch, a Honeywell Division; Figure 6.6, Allen-Bradley Company; Figures 6.8 and 6.11, Micro Switch, a Honeywell Division; Figure 6.13, Chilton Company, a division of ABC, Inc.; Figure 6.14,

Ledex, Inc.; Figure 6.19, Square D Company; Figures 6.21 and 6.22, reprinted from *Machine Design*, May 19, 1983, copyright 1983, by Penton/IPC, Inc., Cleveland, OH; Figure 6.23, Sprecher & Schuh, Inc.; Figure 6.31, E G & G Wakefield Engineering; Figure 6.32, reprinted with permission, Signetics Corporation, copyright Signetics Corporation; Figure 6.55, copyrighted by and reprinted with permission of General Electric Company; Figure 6.56, Motorola Semiconductor Products, Inc.; Figure 6.57, Hamlin, Inc.; Figure 6.60, Simpson Electric Company; Figures 6.62 and 6.64, copyrighted by and reprinted with permission of General Electric Company.

Chapter 7: Figure 7.1B, Motorola; Figures 7.4, 7.5, and 7.6, copyrighted by and reprinted with permission of General Electric Company; Figure 7.8, RCA Solid State Division; Figures 7.11 and 7.12, Furnas Electric Company; Figure 7.17, Reliance Electric Company; Figures 7.19 and 7.20, copyrighted by and reprinted with permission of General Electric Company; Figures 7.21, 7.28, and 7.29, Sprague Electric Company; Figure 7.23, National Semiconductor Corporation; Figure 7.31, Airpax Corporation, a North American Philips Company; Figures 7.32, 7.39, and 7.40, copyrighted by and reprinted with permission of General Electric Company; Figure 7.41, International Rectifier; Tables 7.2 and 7.3, Westinghouse Electric Corporation.

Chapter 8: Figure 8.7, Hewlett-Packard Company; Figure 8.10, Omega Engineering, Inc.; Figure 8.11, Nanmac Corporation; Figures 8.12, 8.14, and 8.15, Fenwal Electronics, Division of Kidde, Inc.; Figure 8.19, Omega Engineering, Inc.; Figure 8.24, Hewlett-Packard Company; Figures 8.33 and 8.36, Schaevitz Engineering; Figures 8.37, 8.38, and 8.39, Chilton Company, a division of ABC, Inc.; Figure 8.47, Micro Switch, a Honeywell Division; Figure 8.53, National Semiconductor Corporation; Figure 8.60, Hersey Products, Inc., Industrial Measurement Group; Figure 8.67, Metritape, Inc.; Table 8.2, *Temperature Measurement Handbook*, 1983, p. T-15, Omega Engineering, Inc., Stamford, CT; Table 8.3, © 1981 *Instrument and Control Systems*, Chilton Company.

Chapter 9: Figures 9.63 and 9.72A, Clairex Electronics; Figure 9.68, Siemens Optoelectronics Division; Figure 9.80, Micro Switch, a Honeywell Division.

Chapter 10: Figures 10.12 and 10.19, copyrighted by and reprinted with permission of General Electric Company; Figure 10.33, adapted by Analog Devices by permission of Moore-Reed, Ltd.; Figure 10.36, Analog Devices, Inc.; Figures 10.39, and 10.40, C.A. Norgren Company.

Chapter 11: Figures 11.35, 11.36, 11.37, and 11.38, photos by Leslie P. Sheets.

Chapter 12: Figures 12.1, 12.4, 12.5, 12.7, 12.8, 12.9, 12.10, and 12.11, adapted from information contained in IRIG 106-80, *Telemetry Standards*, May, 1973, written by the Telemetry Group, Range Commander's Council and published by Range Commander's Council, Secretariat, White Sands Missile Range, NM; Figure 12.13, reprinted with permission of National Semiconductor Corporation. National holds no responsibility for any circuitry described. National reserves the right at any time to change without notice said circuitry. Figure 12.20, photo by Leslie P. Sheets; Table 12.1, adapted from information contained in IRIG 106-80, *Telemetry Standards*, May, 1973, written by the Telemetry

Group, Range Commander's Council and published by Range Commander's Council, Secretariat, White Sands Missile Range, NM.

Chapter 13: Figure 13.12, photo by Leslie P. Sheets; Figures 13.16 and 13.20, Allen-Bradley Company.

Chapter 14: Figure 14.2, Instruments and Control Systems, Chilton Company, a division of ABC, Inc.; Figures 14.3 and 14.4, Allen-Bradley Company (PLC is the registered trademark of Allen-Bradley Company).

Chapter 15: Figure 15.8 from Malcolm, Robotics: An Introduction, 1988, Delmar Publishers Inc.

The data sheets have been reproduced courtesy of National Semiconductor Corporation (© 1982) and Texas Instruments, Incorporated.

Appendix A, Table A.8, has been reproduced from *Electronic Engineers Master Catalog*, courtesy of Hearst Business Communications, Inc.

Appendix E has been reprinted, with permission, from the November 1987 issue of *Personal Engineering & Instrumentation News*, Box 300, Brookline, MA 02146.

Appendix F has been reproduced courtesy of *Control Engineering*.

Introduction

Industrial electronics can be defined as the control of industrial machinery and processes through the use of electronic circuits and systems. Each of the topics in this text has been carefully chosen to help you, the future technician, survive in such an environment. We feel that knowledge gained by studying these topics will make you better prepared for entry-level employment as an electronics technician.

Although many topics have been included in this text, there are many additional topics you will need to know to function successfully in industry. As a starting point, we assume that your previous electronics courses have given you a firm grounding in alternating and direct current theory, the functions of electrical and electronic components, and mathematics through algebra and trigonometry.

The topics in this text are built around a very general process control system since we believe the electronics technician should be a generalist. Furthermore, it is not possible to cover in one text all the circuits and systems you are likely to encounter in industry. The range of industrial applications is simply too broad. Therefore, we have concentrated on some basic circuit and component concepts with appropriate examples. Once you have mastered the basic circuit and component functions, you will be ready to put these parts together into a functional system. That is, your knowledge of the functions of subparts and subsystems will help you understand how the overall system functions.

An example of a very basic process control system is illustrated in Figure I.1. Each block in the

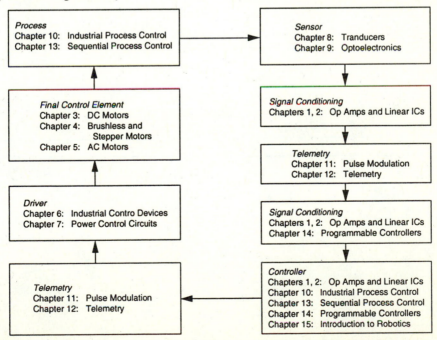

FIGURE I.1 General Process Control System

diagram represents a division of the elements within a process control system. Note that it is sometimes difficult to separate physically one block or topic from another in a real system. And, of course, not all blocks will be used in every control system; however, all are likely to be encountered in industry.

Each block in Figure I.1 contains the associated chapter in which that topic is discussed. The chapters by themselves may seem disconnected from one another, but they obviously are related when you consider the complete system.

The process in Figure I.2 is a hypothetical example. It shows a conveyor belt and a motor driver that must keep a constant speed on the belt regardless of the load. This example is related to the system of Figure I.1 in the following manner: The process in Figure I.1 is the transportation of the load from one point to another at a constant speed. As the load on the belt changes, some type of speed sensor is used to detect changes in speed and convert this change into an electric output. If the controlling system is some distance from the belt, a communications circuit can be used to transport the information from the sensor to the controller.

Before the controller can operate on this sensor output, it may require conditioning of some type to get the electric signal in the correct range or form. The controller can then make a decision, on the basis of the results of the incoming data, as to whether the belt should have more or less power than it had previously or the same power. This decision is then transmitted by telemetry back to the belt drive circuitry, which will apply the appropriate amount of power, depending on the results of the controller's decision. The final control element, the motor, will then be provided with the appropriate power necessary to keep the belt operating at a constant speed. This simple process illustrates the concept of process control and the basis for the inclusion of the topics in this text.

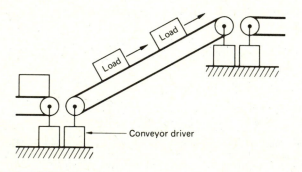

FIGURE I.2 Conveyor Belt System Used to Illustrate a Simple Process and Its Control

Operational Amplifiers for Industrial Applications

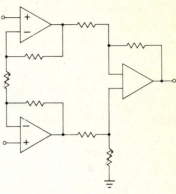

OBJECTIVES

On completion of this chapter, you should be able to:

- Describe the operation of an instrumentation amplifier.
- Calculate the output voltage and gain, given an instrumentation amplifier circuit and an input voltage.
- Calculate the output voltage of a logarithmic amplifier with a given input voltage.
- Explain the operation and calculate the output frequency of a Wien-bridge oscillator.
- Compare the current-differencing amplifier (CDA) to the op amp.
- Calculate the output voltages and gains of the noninverting and inverting CDAs.
- Calculate the trigger voltages and currents of both inverting and noninverting CDA comparators with and without hysteresis.
- Explain the operation and applications of the CDA window comparator.
- Describe the operation of the operational transconductance amplifier (OTA) and draw its schematic symbol.
- Describe an application for the op amp current-to-voltage converter.
- List the general procedures for troubleshooting op amps.

INTRODUCTION

The op amp is the basic building block for many industrial control circuits. This chapter will discuss some simple applications of op amps. These applications should help you understand more about the versatility of the devices. That op amps are versatile can be seen in the variety of their applications.

Other amplifiers have come on the scene since the op amp made its appearance in the early 1970s. One such device we will study is the current-differencing amplifier, a device almost as versatile as the op amp. Finally, we will discuss the operational transconductance amplifier. This linear-integrated circuit has many applications, especially in telecommunications circuits. All of these devices enjoy wide application in industry. The applications we discuss have been specifically chosen to demonstrate their simplicity and their versatility in design and use.

INSTRUMENTATION AMPLIFIER

An *instrumentation amp* is an amplifier that is generally used to amplify DC and low-frequency signals produced by a transducer. (Transducers are discussed in Chapter 8.) These signals are typically only a few millivolts in amplitude, while common-mode noise may be several volts. The instrumentation amp has good common-mode rejection characteristics. Common-mode rejection is the ability of the amplifier to reject noise common to both inputs. Since noise will be common to both inputs, the difference amplifier of Figure 1.1 is frequently used when a signal may be masked by noise or when a signal is very small.

One problem inherent in the differential or difference amplifier of Figure 1.1 is its input impedances. The input impedances to this amplifier may be unequal. And they are relatively low for use in instrumentation applications when the inverting and noninverting gains are not equal. This problem is reduced by making the gains large and equal; that is, $R_F = R_B$ and $R_i = R_A$, with R_F and R_B large with respect to R_i and R_A. The input impedances then become much higher but may still be unequal. The amplifier can be improved by adding a voltage follower to each input. Such an amplifier is then

called an instrumentation amp. Desirable characteristics of very high input impedance and excellent common-mode rejection combine in the instrumentation amp of Figure 1.2.

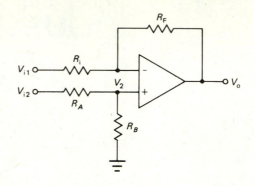

FIGURE 1.1 Difference Amp

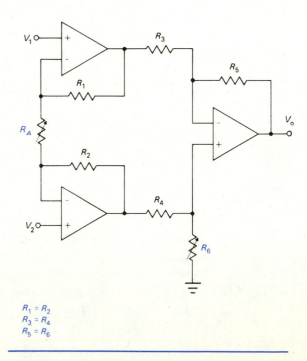

$R_1 = R_2$
$R_3 = R_4$
$R_5 = R_6$

FIGURE 1.2 Instrumentation Amp

The gain (A_V) of this circuit is expressed by the following equation:

$$A_V = \left(\frac{R_5}{R_3}\right)\left(\frac{2R_1}{R_A} + 1\right)$$ (1.1)

And since $V_i = V_2 - V_1$, we have

$$V_o = A_V (V_2 - V_1)$$

or

$$V_o = \left(\frac{R_5}{R_3}\right)\left(\frac{2R_1}{R_A} + 1\right)(V_2 - V_1)$$ (1.2)

where

R_3 = resistance between voltage follower and output op amp

R_5 = feedback resistance to inverting terminal of output op amp and resistance to ground at noninverting terminal of output op amp

V_o = DC output voltage

$R_1 = R_2$, $\quad R_3 = R_4$, $\quad R_5 = R_6$

• EXAMPLE 1.1

Given the instrumentation amplifier shown in Figure 1.2, with component values of $R_1 = R_2 = R_3 = R_4 = 10$ kΩ, $R_A = 2$ kΩ, and $R_5 = R_6 = 20$ kΩ, find the output voltage if the top input (V_1) is +0.15 V and the bottom input (V_2) is +0.1 V.

Solution

Using Equation 1.2 and substituting values, we get

$$V_o = \left(\frac{20 \text{ k}\Omega}{10 \text{ k}\Omega}\right)\left[\frac{2(10 \text{ k}\Omega)}{2 \text{ k}\Omega} + 1\right](0.1 \text{ V} - 0.15 \text{ V})$$

$$= -1.1\text{V}$$

• EXAMPLE 1.2

Given the instrumentation amplifier of Example 1.1 with a variable resistance for R_A and all other resistances as in Example 1.1, find the required resistance of R_A for an output voltage of 5 V and $V_i = V_2 - V_1 = 0.075$ V.

Solution

Using Equation 1.1, we have

$$A_V = \left(\frac{R_5}{R_3}\right)\left(\frac{2R_1}{R_A} + 1\right) = \frac{V_o}{V_i}$$

Solving this equation for R_A, we get

$$R_A = \frac{2R_1}{\left(\frac{V_o R_3}{V_i R_5}\right) - 1}$$ (1.3)

Substituting values, we obtain

$$R_A = \frac{2(10 \text{ k}\Omega)}{\left[\frac{(5\text{V})(10 \text{ k}\Omega)}{(0.075 \text{ V})(20 \text{ k}\Omega)}\right] - 1} = 618.6 \ \Omega$$

Instrumentation Amplifiers with Bridges

Many industrial circuits must detect the small change in resistance of a sensor to a voltage or current. For example, a strain gauge—a device used to measure stress or strain—may have a resistance of 100 Ω. Ten milliamperes of current through that device will produce a 1 V drop across it. Let us say that this gauge increases by 1 mΩ (0.001 Ω). With the same current flowing through the gauge, the change in resistance will produce a 10 µV change in voltage. If we amplify that change by 1000, we will have an output change of 10 mV. The entire resistance, however, will produce close to 100 V, a voltage too high for most low-voltage circuits.

One solution to this dilemma is to use a bridge, as shown in Figure 1.3. Note that the bridge is connected to an instrumentation amplifier. When the bridge produces voltages that are equal at points A and B, the bridge is said to be in a "nulled" state. The differential input to the instrumentation amplifier ($V_2 - V_1$) is zero, so the output of the differential amplifier is 0 V. If we assume that resistor R_9 is adjusted to exactly 120 Ω and the other resistances are as shown, the voltages at A and B will be equal, each at +4 V.

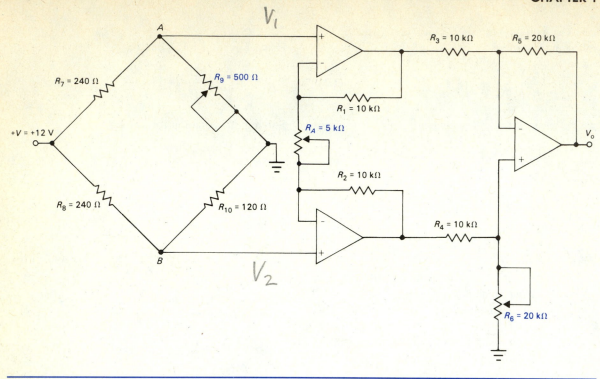

FIGURE 1.3 Instrumentation Amplifier Driven by a Bridge

If the resistance of R_9 were to increase to 121 Ω, what would the output voltage of the instrumentation amplifier be? We need to calculate the voltage at point A. Using the following voltage divider equation,

$$V_{R_9} = \left(\frac{R_9}{R_7 + R_9}\right)(V) \qquad (1.4)$$

we obtain

$$V_{R_9} = \left(\frac{121\ \Omega}{240\ \Omega + 121\ \Omega}\right)(+12\ V)$$

$$= 4.022\ V$$

The input voltage is $V_2 - V_1$, or 4.022 V − 4.0 V = 0.022 V. With a gain of 22, the output voltage is 22 times the input voltage, or (22)(0.022 V) = 0.484 V.

• EXAMPLE 1.3

Given the instrumentation amplifier shown in Figure 1.3 with $R_1 = R_2 = R_3 = R_4 = 10$ kΩ, $R_A = 2$ kΩ, $R_5 = R_6 = 20$ kΩ, and resistor R_9 adjusted to 122.5 Ω, find the output voltage.

Solution

The voltage at point A must be calculated, using Equation 1.4.

$$V_{R_9} = \left(\frac{R_9}{R_7 + R_9}\right)(V)$$

$$= \left(\frac{122.5\ \Omega}{240\ \Omega + 122.5\ \Omega}\right)(+12\ V)$$

$$= 4.055\ V$$

The gain of the amplifier is the same as that calculated in Example 1.1, $A_V = 22$. The output voltage will equal the input times 22, so

$$V_o = (A_V)(V_i) = (22)(0.055 \text{ V}) = 1.21 \text{ V}$$

The instrumentation amplifier finds use in industry where signals are masked by noise and where it is desirable not to load the sensor unduly. Some manufacturers make an instrumentation amp within a single chip. The LM725 is an example. A single chip is desirable because, in that case, the amplifiers and resistors of the instrumentation amp can be more closely matched. Close matching improves instrumentation amp performance.

LOGARITHMIC AMPLIFIER

A circuit frequently seen in analog computers is the logarithmic amplifier (log amp). In a *log amp*, the output is proportional to the logarithm of the input. The log amp takes advantage of the logarithmic relationship between the current through a semiconductor device and the voltage across it. Two such nonlinear devices are the semiconductor diode and the bipolar junction transistor (BJT). The circuit configurations for these log amps are illustrated in Figures 1.4A and 1.4B.

In the circuit shown in Figure 1.4A, the voltage across the diode (V_D) is logarithmically related to the current through the diode (I_D) by the following relationship:

$$V_D = A \log_B \frac{I_D}{I_r} \qquad (1.5)$$

where

$\log_B$ = logarithm to a specified base (B)
A = constant to proportionality, which depends on the base of the logarithm
I_r = theoretical reverse leakage current of diode

Likewise, in Figure 1.4B the collector current (I_C) divided by its reverse leakage current (I_{Cr}) is logarithmically related to the voltage from base to emitter (V_{BE}) by the following equation:

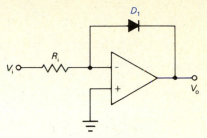

A. With Semiconductor Diode

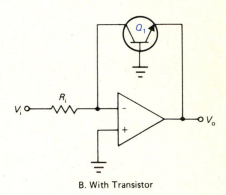

B. With Transistor

FIGURE 1.4 Ideal Log Amp

$$V_{BE} = A \log_B \frac{I_C}{I_{Cr}} \qquad (1.6)$$

Practical Log Amp

In actual use of the log amp, a more practical circuit such as the one in Figure 1.5A would be utilized. Here, the more practical equation is as follows:

$$V_{o1} = -V_{BE} = A \log_B (V_i) + K \qquad (1.7)$$

where

V_{o1} = output of log amp
V_{BE} = voltage from base to emitter of transistor
A = constant of proportionality
K = offset value

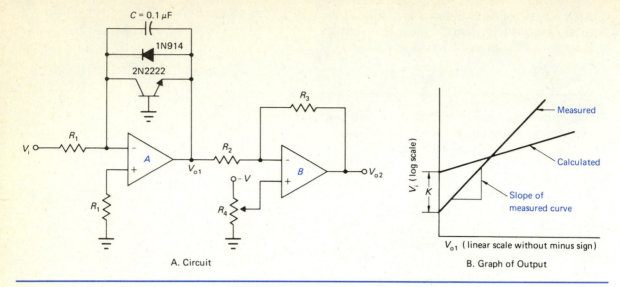

A. Circuit   B. Graph of Output

FIGURE 1.5 Practical Log Amp

The value of resistance for both R_1's in Figure 1.5A is given by the following expression:

$$\frac{V_{i(max)}}{I_{C(max)}} \leq R_1 \leq \frac{V_{i(min)}}{I_b} \qquad (1.8)$$

where

$V_{i(max)}$ = maximum input voltage
$V_{i(min)}$ = minimum input voltage
$I_{C(max)}$ = maximum collector current of transistor used
I_b = input bias current of op amp

For the 741 op amp, I_b is approximately 80 nA.

For evaluation of A and K in Equation 1.7, the following equations may be used:

$$A = \frac{V_{o2} - V_{o1}}{\log (V_{i2}) - \log (V_{i1})} \qquad (1.9)$$

$$K = V_{o1} - A \log (V_{i1}) \qquad (1.10)$$

where

V_{i1} = first applied input voltage
V_{i2} = second applied input voltage
V_{o1} = output voltage of log amp due to V_{i1}
V_{o2} = output voltage of amp due to V_{i2}

An example may be helpful at this point to demonstrate the evaluation of A and K. Assume the input voltage applied will range over three different values: 1 V, 2 V, and 3 V. For these three inputs, three output voltages (V_{o1} in Figure 1.5A) result: 0.60 V, 0.62 V, and 0.64 V (hypothetically).

In Equation 1.9, V_{o1} is 0.60 V, V_{o2} is 0.62 V, V_{i1} is 1 V, and V_{i2} is 2 V. (Notice that the negative signs were dropped from the output voltage measurements.) Substituting these values into Equation 1.9 results in a value of 0.0664 for A. In Equation 1.10, V_{o1} is 0.60 V, A is 0.0664, and V_{i1} is 1 V. Substituting these values into Equation 1.10 results in a value of 0.60 for K.

Now, the procedure is repeated for the second group of values for V_i and V_o. For Equation 1.9 and the second group of values, V_{o1} is now 0.62 V, V_{o2} is 0.64 V, V_{i1} is 2 V, and V_{i2} is 3 V. Substituting these values into Equation 1.9 results in a value of 0.114 for A. In Equation 1.10, V_{o1} is now 0.62 V, A is 0.114, and V_{i1} is 2 V. Substituting these values into Equation 1.10 results in a value of 0.586 for K.

After all values for A and K have been calculated, they can be averaged in order to obtain values that can be used in the amplifier circuit following the log amp in Figure 1.5A.

After A and K are determined, they are used with the second amplifier (B amplifier) in Figure 1.5A. The noninverting terminal of amplifier B is set to the value of K, and the gain of amplifier B is set to the reciprocal of A. When a voltage is input at V_1, the logarithm of that voltage is now output at V_{o2}.

As a precaution for the circuit of Figure 1.5A, be sure to use only positive input voltages in the range for which you determined the R_1 value in Equation 1.8. In addition, the transistor in the feedback circuit of the log amp (2N2222 in Figure 1.5A) can be any small-signal, high-speed transistor. If the transistor's response is a straight line on the graph in Figure 1.5B, the circuit of Figure 1.5A should produce an accurate logarithmic output. Whichever transistor you choose, though, will be heat-sensitive. For best results, either keep the transistor at a constant temperature or use a temperature-stabilized circuit. The bibliography for this chapter given at the back of the text lists good sources for such designs.

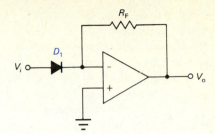

A. With Semiconductor Diode

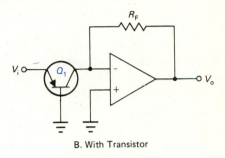

B. With Transistor

FIGURE 1.6 Ideal Antilog Amp

Antilog Amp

The *antilog amplifier* performs the inverse function of the log amp. This operation is done by simply reversing the position of the semiconductor and the input resistor. The antilog amp is shown in Figures 1.6A and 1.6B.

Designing and working with log amps can be a tedious and exacting job. For this reason, several manufacturers have designed and built log amps within a single integrated circuit (IC) chip.

Applications of Log Amps

Applications of log amps are numerous. For example, they are used for compressing large voltage ranges into smaller ones. Also, since many transducers have a logarithmic transfer function, the log amp is used to compensate for this nonlinear response.

Probably the most frequent use of the log amp is in analog computers. Recall that multiplying and dividing may be accomplished by adding and subtracting the logarithms of numbers and then obtaining the antilogarithm. Since log amps perform these

operations, they can be used in multipliers and dividers in computers. The block diagram of a multiplier is shown in Figure 1.7.

ACTIVE FILTER

The operation of many electronic systems requires that certain bands of frequencies be passed and others attenuated. A device that does this operation is called a *filter*.

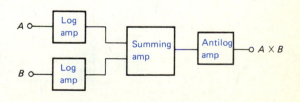

FIGURE 1.7 Multiplying with Log and Antilog Amps

Classification of filters depends on point of view. If the components from which the filter is made are the basis of classification, then we say filters are either passive or active. A *passive filter* uses capacitors, resistors, and inductors. *Active filters* employ amplifiers along with passive components. Inductors usually are not used at audio frequencies since they can be simulated with amplifiers, capacitors, and resistors. Also, inductors are relatively expensive, large, and heavy. Figure 1.8 shows this type of classification of filters and the frequencies at which they are used.

Since an in-depth treatment of active filters would be quite lengthy, only an introduction is given in this text. For more information on active filters, consult the bibliography for this chapter (at the end of the book) or other reference material.

As we will see, the frequency behavior of any given amplifier can be controlled by adding reactive components to it in different configurations. Such active filters are reliable, low in cost, and easy to tune. In addition, when op amps are used, they give not only voltage gain but also high input impedance and low output impedance characteristics.

Filters are also classified according to the shapes of their response curves. Figure 1.9 shows some of these curves. Table 1.1 compares these filter types.

The most common classification of filters is related to how they treat frequencies. These classifications are low-pass, high-pass, bandpass, and band-reject (or notch).

Low-Pass Filter

A *low-pass filter* is defined as "a circuit that passes all frequencies below a certain frequency while attenuating all higher frequencies." A simple low-pass Butterworth filter with its bandpass characteristic is shown in Figure 1.10.

As shown in Figure 1.10B, the output of the filter is constant until the *cutoff frequency* (f_c) is approached. After f_c is passed, the output voltage decreases, or rolls off, at a rate of –6 dB per octave, or –20 dB per decade. (Recall that a decade is a tenfold increase or decrease in frequency, and an octave is a twofold increase or decrease in frequency.) This filter is called a *first-order filter* because its transfer function is a linear equation and because it has only one *RC* filter combination.

Figure 1.11 shows a second-order Butterworth filter. Notice that there are two *RC* networks here. This filter exhibits a roll-off of –12 dB per octave, or –40 dB per decade. A third-order Butterworth filter has three *RC* networks and roll-offs of –18 dB

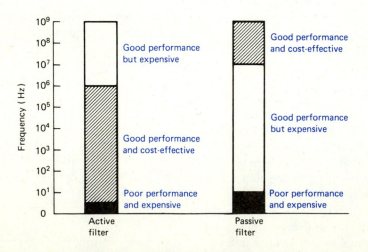

FIGURE 1.8 Performance of Active and Passive Filters

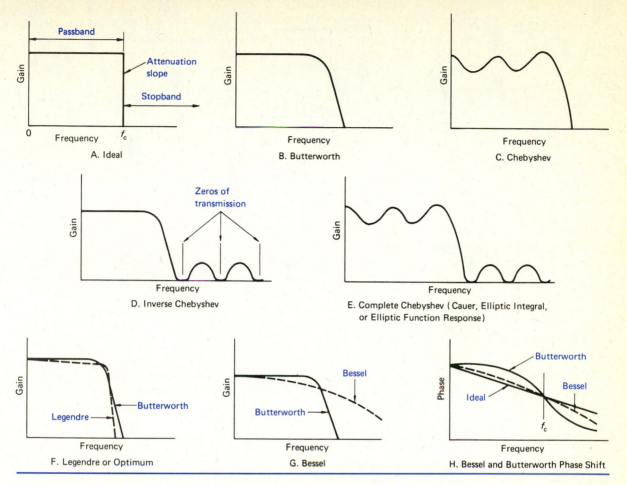

FIGURE 1.9 Classification of Filters by Shape of Response Curves

TABLE 1.1 Comparison of Filter Types

Name of Filter Type	Distinguishing Characteristic	Remarks
Butterworth	Maximally flat amplitude response	Most popular general-purpose filter
Chebyshev	Equal-amplitude ripples in passband	Attenuation slope steeper than in Butterworth near cutoff
Inverse Chebyshev	Equal-amplitude ripples in stopband	No passband ripple; zeros of transmission in stopband
Complete Chebyshev (also called Cauer, elliptic function, elliptic integral, or Zolatarev)	Equal-amplitude ripples in both passband and stopband	Zeros of transmission in stopband
Legendre	No passband ripple, but steeper attenuation slope than in Butterworth	Not maximally flat
Bessel (also called Thomson)	Phase characteristic nearly linear in pass region, giving maximally flat group delay	Good for pulse circuits because ringing and overshoot minimized; poor attenuation slope

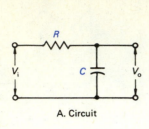

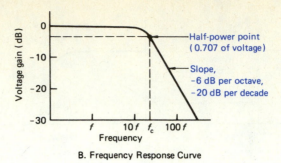

A. Circuit

B. Frequency Response Curve

FIGURE 1.10 Low-Pass Filter

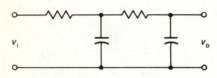

FIGURE 1.11 Second-Order, Low-Pass Butterworth Filter

per octave, or –60 dB per decade; and so on. Table 1.2 shows the phase shifts at the half-power (0.707) points for these filters.

TABLE 1.2 Phase Shifts at Half-Power Point

Order of Filter	Phase Shift at f_o or f_c*	Roll-Off (dB per Decade)
1	–45°	–20
2	–90°	–40
3	–135°	–60
4	–210°	–80

*f_o = oscillation frequency for oscillators; f_c = cutoff frequency for filters.

The diagram in Figure 1.12 shows a second-order, low-pass filter. The roll-off is –12 dB per octave, or –40 dB per decade. The cutoff frequency is found from Equation 1.11.

$$f_c = \frac{1}{2\pi\sqrt{R_1 R_2 C_1 C_2}}$$ **(1.11)**

For the component values shown in Figure 1.12, the cutoff frequency is

$$f_c = \frac{1}{2\pi\sqrt{R_1 R_2 C_1 C_2}}$$

$$= \frac{1}{2\pi\sqrt{(10\ k\Omega)\,(10\ k\Omega)\,(0.02\ \mu F)\,(0.01\ \mu F)}}$$

$$= \frac{1}{2\pi\,(0.0001414)}$$

$$= 1125\ Hz$$

• EXAMPLE 1.4

Given the low-pass filter illustrated in Figure 1.12, with $C_1 = 0.1\ \mu F$, $C_2 = 0.05\ \mu F$, and $R_1 = R_2 = 7.5\ k\Omega$, find the filter's cutoff frequency.

Solution

From Equation 1.11, we find

$$f_c = \frac{1}{2\pi\sqrt{R_1 R_2 C_1 C_2}}$$

$$= \frac{1}{2\pi\sqrt{(7.5\ k\Omega)\,(7.5\ k\Omega)\,(0.1\ \mu F)\,(0.05\ \mu F)}}$$

$$= 300\ Hz$$

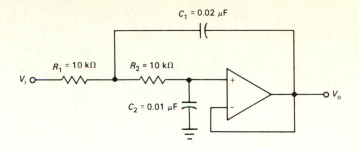

FIGURE 1.12   Second-Order, Low-Pass Filter

GYRATOR

While not strictly a filter circuit, the gyrator may be classified along with filters because of its unique reactive properties. The *gyrator* acts as a simulated inductor. We have already discussed the disadvantages of inductors at audio frequencies. At frequencies below radio frequencies (RF), the gyrator can simulate the behavior of an inductor. The gyrator is basically a two-terminal (or single-port) device. When a capacitor is connected to one port, the device rotates electrically—or gyrates—the capacitor so that it looks like an inductor at the other port. Two examples of gyrators are shown in Figure 1.13.

The gyrator shown in Figure 1.13A produces an inductance (L) equal to the product of R_1, R_2, and C. The equation is

$$L = (R_1)(R_2)(C) \qquad (1.12)$$

If $R_1 = 1000 \ \Omega$, $R_2 = 100{,}000 \ \Omega$, and $C = 1 \ \mu F$, the inductance is

$$L = (1000 \ \Omega)(100 \ k\Omega)(1 \ \mu F) = 100 \ H$$

In the gyrator shown in Figure 1.13B, the inductance is equal to the resistance squared times the capacitance, or

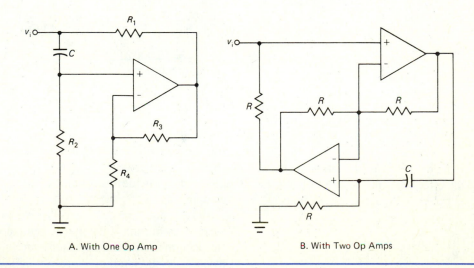

A. With One Op Amp   B. With Two Op Amps

FIGURE 1.13   Gyrator (Simulated Inductor)

$$L = R^2 C \qquad (1.13)$$

For $R = 10$ kΩ and $C = 1$ μF, the inductance is

$$L = (10,000 \ \Omega)^2 \ (1 \ \mu F) = 100 \ H$$

• EXAMPLE 1.5

Given the gyrator in Figure 1.13A, where $R_1 = R_3 = 1500 \ \Omega$, $R_2 = 100,000 \ \Omega$, and $C = 1.5$ μF, find the inductance of the gyrator (in henrys).

Solution

From Equation 1.12, we find

$$L = (R_1)(R_2)(C)$$

$$= (1500 \ \Omega)(100 \ k\Omega)(1.5 \ \mu F) = 225 \ H$$

Gyrators may be used to replace inductors in low-frequency circuits. However, one terminal of the replaced inductor must be grounded for these circuits to work properly.

CAPACITANCE MULTIPLIER

Capacitance may be multiplied in a similar fashion to the creation of inductance. The circuit shown in Figure 1.14 will multiply the capacitance of C_1 according to the following relationship:

$$C_m = C_1 \left(\frac{R_1}{R_3} \right) \qquad (1.14)$$

The resistance of R_1 should be as large as possible since the impedance appears in series with the effective capacitance C_m.

By substituting values into Equation 1.14, we can calculate the effective capacitance of C_m for the capacitance multiplier in Figure 1.14. We find

$$C_m = C_1 \left(\frac{R_1}{R_3} \right) = (10 \ \mu F) \left(\frac{10 \ M\Omega}{1 \ k\Omega} \right)$$

$$= 100,000 \ \mu F$$

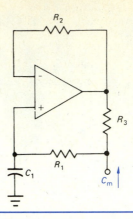

FIGURE 1.14 Capacitance Multiplier

The effective capacitance is 100,000 μF, a value many times larger than the 10 μF capacitance without multiplication.

• EXAMPLE 1.6

Given the capacitance multiplier in Figure 1.14, with $C_1 = 1.5$ μF, find the effective multiplied capacitance.

Solution

From Equation 1.14, we find

$$C_m = C_1 \left(\frac{R_1}{R_3} \right) = (1.5 \ \mu F) \left(\frac{10 \ M\Omega}{1 \ k\Omega} \right)$$

$$= 15,000 \ \mu F$$

OSCILLATOR

An *oscillator* is a circuit that generates an alternating current (AC) waveform with only a direct current (DC) input. Commonly, the frequency of alternation is determined by the oscillator's components.

Oscillators are quite widely used in industrial electronics for both analog and digital applications. Because most oscillators require an amplification device of some kind, op amps can be used to form basic oscillator circuits.

Sine Wave Oscillator

Sine wave oscillators, whether they use op amps or other amplification devices, must meet certain criteria to sustain oscillations. These criteria are sometimes called the Barkhausen criteria. The *Barkhausen criteria* are as follows:

1. The overall gain of the circuit (called *loop gain*) must be 1 or greater. Thus, losses around the circuit must be compensated for by an amplifying device.
2. The phase shift around the circuit (input to output and back to the input) must be 0° (or 360°).

Wien-Bridge Oscillator. A common sine wave oscillator is the Wien-bridge oscillator, shown in Figure 1.15. The *Wien-bridge oscillator* operates on the balanced bridge principle. When the impedance of the R_1C_1 branch equals the impedance of the R_2C_2 branch, the feedback voltage is in phase with the output voltage.

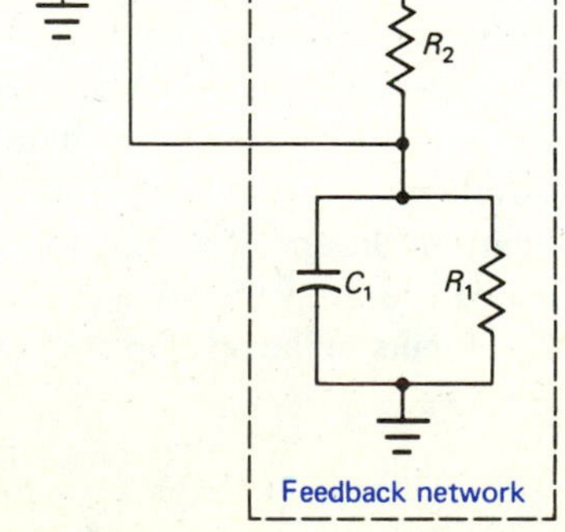

Again, when the gain of the amplifier is sufficient to replace losses around the circuit, both Barkhausen criteria are met. In the Wien-bridge oscillator, the criteria are met when the gain is 3. If the gain of the amplifier is less than 3, the loop gain is less than 1. The oscillator will not oscillate. If the gain is exactly 3, the loop gain is 1 and the circuit will oscillate. If the loop gain is greater than 1, however, the amplifier will be driven into saturation. A saturated amplifier will produce a distorted output and may stop oscillation.

In a practical sense, it is difficult to adjust an amplifier to a gain of exactly 3. A more practical Wien-bridge oscillator is shown in Figure 1.16. When the gain is low and the signal is small, the zener diodes will be off. When the gain is set to greater than 3, the output amplitude will increase. When the output amplitude is sufficient to turn on the zener diodes, part of the feedback resistance is shunted, reducing the gain. The reduced gain should be sufficient to keep the amplifier from saturating. The output voltage amplitude will equal the zener breakdown voltage.

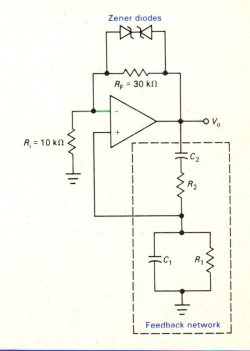

FIGURE 1.15 Wien-Bridge Oscillator

FIGURE 1.16 Practical Wien-Bridge Oscillator

For the resistors and capacitors as labeled in Figure 1.16, the frequency of oscillation (f_o) of the Wien-bridge oscillator is as follows:

$$f_o = \frac{1}{2\pi\sqrt{R_1 C_1 R_2 C_2}} \qquad (1.15)$$

MISCELLANEOUS OP AMP APPLICATIONS

Throughout the text we show applications of op amps in conjunction with other components and devices. We have chosen six applications of op amps to discuss here to further show the flexibility and versatility of op amps.

Op Amp Supply

Figure 1.17 illustrates a method by which a dual power supply can be made from a single supply input, one op amp, two transistors, a few resistors, and a few capacitors. The circuit is basically a high-gain amplifier with a very high input impedance. Resistors R_1 and R_2 form a voltage divider that references the noninverting terminal at about half the input voltage. Feedback current from the junction of the transistor Q_1 and Q_2 emitters to the input of the op amp keeps the output voltage evenly balanced.

An example will clarify how this device works. First, we will consider the condition when the loads, R_{L1} and R_{L2}, are equal to 500 Ω with $V_{i(DC)}$ equal to 10 V. Note that the input resistors evenly split the supply voltage. Since each load has the same voltage across it (5 V), each draws exactly the same amount of current. The exact amount of current can be found by Ohm's law:

$$I_L = \frac{+V/2}{R_L} \qquad (1.16)$$

So we obtain

$$I_{L1} = I_{L2} = \frac{5\text{ V}}{500\text{ Ω}} = 10\text{ mA}$$

Each load will draw 10 mA. The op amp output is +5 V, and both transistors are off.

If the load resistance of R_{L2} were to increase to 750 Ω, the voltage across the load would try to go to

$$V_L = \left(\frac{R_{L2}}{R_{L2} + R_{L1}}\right)(V_{i(DC)})$$

$$= \left(\frac{750\text{ Ω}}{750\text{ Ω} + 500\text{ Ω}}\right)(10\text{ V}) = 6\text{ V} \qquad (1.17)$$

This increased potential is felt at the inverting input terminal of the op amp, driving it 1 V above the 5 V potential on the noninverting terminal. The op amp responds by driving its output in a negative or less positive direction. When the op amp output reaches 4.4 V, transistor Q_2 is turned on. Transistor Q_2 diverts some current around R_{L2}. The transistor is driven hard enough to drop the voltage across R_{L2} back to 5 V. The inputs to the op amp are now equal again. Load 2 draws

$$\frac{5\text{ V}}{750\text{ Ω}} = 6.67\text{ mA}$$

and load 1 draws 10 mA.

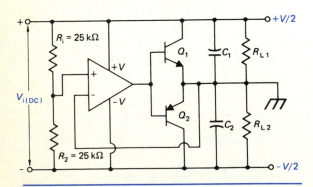

FIGURE 1.17 Op Amp Power Supply Circuit

• EXAMPLE 1.7

Given the op amp circuit in Figure 1.17, with $R_{L1} = 200 \ \Omega$, $R_{L2} = 400 \ \Omega$, and $V_{i(DC)} = +15$ V, find the load currents for each load resistance.

Solution

From Equation 1.16, we find

$$I_{L1} = \frac{V/2}{R_{L1}} = \frac{(15 \ V)/2}{200 \ \Omega} = 37.5 \ mA$$

$$I_{L2} = \frac{V/2}{R_{L2}} = \frac{(15 \ V)/2}{400 \ \Omega} = 18.75 \ mA$$

• EXAMPLE 1.8

Given an LED overvoltage indicator like the one in Figure 1.18, with $R_F = 30$ kΩ, $R_i = 10$ kΩ, and zener voltages of 5 V, find the input voltage that will cause the overvoltage indicators to switch.

Solution

The output LEDs will switch when the output voltage goes above and below 5 V. The amplifier has a gain of 3. The input voltage will then equal V_o/A_V, or 5 V/3 = 1.67 V. If the input goes above +1.67 V, the output will be lower (more negative) than −5 V, turning on diodes D_3 and D_1. If the input goes more negative than −1.67 V, diodes D_4 and D_2 will turn on.

LED Overvoltage Indicator

Figure 1.18 shows a circuit that is used to indicate the presence of an overvoltage by turning on an LED (light-emitting diode). By an adjustment of the gain of the op amp (R_F/R_i) and by proper choice of zener diodes, this circuit can be used to show an over-voltage visually. In this case, if the input voltage goes over +3 V, diode D_1 conducts, turning on D_3. If the input voltage goes lower than −3 V, diode D_4 conducts, turning on D_2.

Op Amp Voltage Regulator

The op amp is also used as a low-current voltage regulator. Figure 1.19 illustrates such an application. Changes in the unregulated input result in a very small change across D_1 (provided it is biased correctly). Therefore, the output voltage remains relatively constant with changing load demands. The output voltage is a function of the zener voltage times the gain ($R_F/R_i + 1$). Since zener voltage, R_F, and R_i remain constant, the output remains constant.

Let us suppose that $R_F = 20$ kΩ, $R_i = 10$ kΩ, $R_L = 1$ kΩ, and the zener is a 5 V device. Since the gain is 3, the output voltage is V_i times the gain, or +15 V. With a 15 V output, the current drawn from the load is 15 V divided by 1000 Ω, or 15 mA.

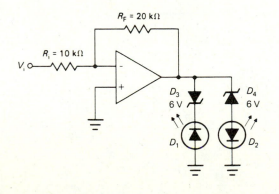

FIGURE 1.18 Op Amp LED Overvoltage Indicator

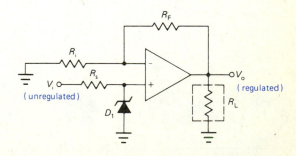

FIGURE 1.19 Op Amp Voltage Regulator

• EXAMPLE 1.9

Given the voltage regulator of Figure 1.19, with $R_F = 10$ kΩ, $R_i = 10$ kΩ, $R_L = 1.5$ kΩ, and the zener a 5 V device, find the regulator output voltage and load current.

Solution

Since the gain is 2, the output voltage is V_i times the gain, or +5 V times 2, which is +10 V. With a 10 V output, the current drawn from the load is 10 V divided by 1500 Ω, or 6.7 mA.

This particular type of voltage regulator works very well, keeping the output voltage constant under changing load conditions. Since the load is drawn from the output terminal of the op amp, the maximum current is limited by the op amp's maximum output current. If the op amp is a 741, for example, the maximum output current is about 25 mA.

This circuit's maximum current can be extended by applying the output of the op amp to a power transistor, as in Figure 1.20. The +5 V potential is applied to the noninverting terminal and is multiplied by 2, the gain factor of the op amp. The +10 V output potential is applied to the emitter of the transistor. If increased output current flows, the voltage across R_4 tends to drop. This drop in voltage is fed back to the inverting terminal, increasing the

error voltage. The increased error voltage at the input to the op amp causes the output voltage to increase, bringing the voltage back up to nearly what it was before loading. The op amp also drives the transistor harder, causing it to conduct more. Note that the unregulated voltage (V_{ur}) is 15 V.

Op Amp Voltage-to-Current Converter

Some applications require a current source rather than a voltage source. Examples include analog meter movements, resistance measurement circuits, and driving inductive loads such as relay coils. Figure 1.21 shows a device that provides a current source, the voltage-to-current converter. In such a circuit, the amount of feedback or load current does not depend on the resistance of the load. Load current depends on the input voltage and current. This effect is important, for example, in metering circuits. Only the input voltage being measured should affect the current through the meter. Any changes in the current through the meter because of meter resistance or voltage changes are undesirable.

We can see this principle at work by assigning values to the circuit in Figure 1.21. What current will be drawn through the load if the input voltage is 10 V and $R = 10$ kΩ? Since the amplifier is basically an inverting amplifier, the entire 10 V potential is dropped across the 10 kΩ resistance, drawing 1 mA through the load.

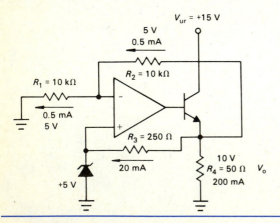

FIGURE 1.20 Voltage Regulator Op Amp Circuit Driving a Power Transistor

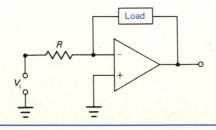

FIGURE 1.21 Op Amp Voltage-to-Current Converter

• EXAMPLE 1.10

Given the amplifier in Figure 1.21, with an input voltage of 15 V and an input resistance of 1 kΩ, find the current drawn through the load.

Solution

The entire 15 V potential is dropped across the 1 kΩ resistance. From Ohm's law, the current through the load is equal to the voltage across the input resistance divided by the input resistance, or

$$I_L = \frac{V_i}{R} \tag{1.18}$$

Thus, we have

$$I_L = \frac{-15\ V}{1000\ \Omega} = 15\ mA$$

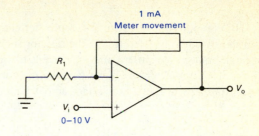

FIGURE 1.22　Op Amp DC Voltmeter

An interesting use of this voltage-to-current principle is illustrated in Figure 1.22. This DC voltmeter uses an inexpensive 1 mA, full-scale, d'Arsonval meter movement as the feedback component. The meter is used to measure 0–+10 V; that is, with a +10 V input, the meter pointer will deflect fully to the right. A +5 V potential at the input will cause the pointer to move to a midscale position. Of course, a 0 V input will leave the pointer stationary at the far left of the meter face. This circuit puts very little load

on the voltage under measurement, a definite advantage. The low loading results from the input voltage seeing only the high op amp input impedance. For example, if we place a +10 V potential at the input of an op amp with a 2 MΩ input impedance, how much loading will occur? The op amp will draw 10 V/2 MΩ, or 5 μA, of current, a very small value. Field effect transistor (FET) input op amps have much higher input impedances and therefore draw even less current from the voltage under measurement.

• EXAMPLE 1.11

Given the circuit shown in Figure 1.23 for measuring two voltage ranges, 0–1 V and 0–10 V, find the resistances for R_1 and R_2 that will give full-scale deflection for 10 V and 1 V.

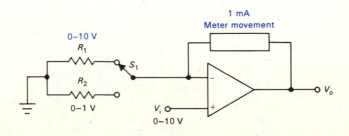

FIGURE 1.23　Two-Range DC Op Amp Voltmeter for Example 1.11

Solution

A +10 V input should cause full-scale deflection with the switch connected to R_1. The resistor should be large enough to cause 1 mA of current to flow with a +10 V input. From Ohm's law, the resistance can be found by dividing the voltage across it by the current through it, or

$$R_1 = \frac{V}{I} = \frac{+10\ \text{V}}{1\ \text{mA}} = 10{,}000\ \Omega$$

The value of R_2 can be found in a similar fashion, substituting 1 V for the 10 V value:

$$R_2 = \frac{V}{I} = \frac{+1\ \text{V}}{1\ \text{mA}} = 100\ \Omega$$

Op Amp Current-to-Voltage Converter

Some transducers, particularly light- and heat-sensitive ones, give an output current proportional to the input variable (light or heat). To work properly with this current output device, we need an amplifier that can process input current rather than input voltage. The op amp shown in Figure 1.24 converts input current to an output voltage.

In this circuit, almost no current flows in the inverting terminal. So any input current from the source will flow through the feedback resistance.

Because of the virtual ground, the voltage drop produced by the feedback resistor is also the output voltage. The output voltage is then determined by the input current, providing that the feedback resistor stays constant.

Op Amp Sample-and-Hold Circuit

A *sample-and-hold* (S/H) *circuit*, as its name implies, is used to sample a voltage at a particular time and hold it constant for another specified time interval. Figure 1.25 shows just such a circuit designed with an op amp.

In the S/H circuit, when a sample control voltage is applied to the switch, the switch turns on. The capacitor charges owing to the input voltage applied across it by the switch. The capacitor continues to charge until the switch is turned off. Since there is very little resistance to current flow in the charging circuit, the capacitor charges instantly to the input voltage.

Depending on the state of the switch, the signal will be transmitted to the capacitor or it will be blocked. The state of the switch will be controlled by the S/H control line, usually a high or low voltage. When the switch is closed, the input signal appearing across the capacitor is buffered by the voltage follower. If the switch is closed for some time while the input signal varies, the S/H circuit is said

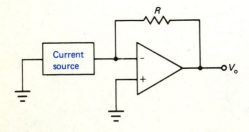

FIGURE 1.24 Op Amp Current-to-Voltage Converter

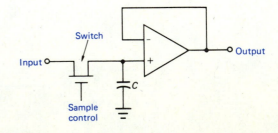

FIGURE 1.25 Op Amp Sample-and-Hold Circuit

to "track" the input. Any changes in the input are transmitted to the output through the voltage follower. When the switch is opened, the capacitor retains the last voltage value of the input as a charge. When the switch opens, the capacitor has no discharge path, and the capacitor "holds" the voltage it has sampled. The follower will read the voltage on the capacitor until the next sample period.

This sample-and-hold behavior can be seen by examining the input, output, and control waveforms shown in Figure 1.26. The waveforms in Figure 1.26A are nearly ideal. They assume perfect switching, tracking, and holding characteristics. In practice, though, we know things are never ideal.

A more practical waveform is shown in Figure 1.26B. A common error, shown here when the S/H circuit switches from hold to sample, is called acquisition time. The *acquisition time* is the time needed for the S/H circuit to acquire and track the input signal after the sample command. It is usually specified for a full-scale-level change, for example, −10 to +10 V. The output must reach the desired level within a stated accuracy range. This accuracy range may be, for example, 0.01% or 0.1%.

Notice also in Figure 1.26B, that the output signal overshoots the input value. Some time later, the output settles down to within a certain error band.

The acquisition time depends to a great extent on the RC charge time of the capacitor. The resistances through which the capacitor must charge are the resistance of the switch and the resistance of the source. The acquisition time is, of course, directly proportional to the size of the capacitance in the sampling capacitor. Acquisition time can be adjusted to produce minimum oscillations by adjusting the RC values.

Another common error in S/H circuits is called droop or tilt. Droop occurs during the hold time. *Droop* is the change in the hold voltage (ΔV) during the hold interval (see Figure 1.26B). This voltage droop is caused by the leakage currents that flow into or out of the capacitor. Droop is normally measured as a rate or slope (in microvolts per microsecond). The droop is therefore

$$\frac{\Delta V}{\Delta t} = \frac{I_{\text{L}}}{C_{\text{H}}}$$ **(1.19)**

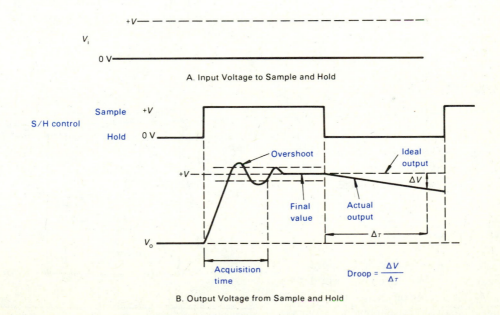

A. Input Voltage to Sample and Hold

B. Output Voltage from Sample and Hold

FIGURE 1.26 Sample-and-Hold Waveforms

where
    ΔV = change in voltage across capacitor
    Δt = time of voltage change
    I_L = leakage current through capacitor
    C_H = capacitance of hold capacitor

Droop can be increased by decreasing either of the two resistances or by reducing the size of the capacitor.

As an example, let us say that we have 1 nA of droop current with a C_H of 100 pF. The droop is then

$$\frac{\Delta V}{\Delta t} = \frac{I_L}{C_H} = \frac{1\ \text{nA}}{100\ \text{pF}} = 10\ \mu V/\mu s$$

Thus, a hold period of 10 μs will cause a total droop of 100 μV.

This calculation may be clearer if we remember that

    1 ampere = 1 coulomb/second (C/s)
and
    1 farad = 1 coulomb/volt (C/V)

Therefore,

$$\text{droop} = \frac{1 \times 10^{-9}\ \text{A}}{100 \times 10^{-12}\ \text{F}}$$

$$= \frac{1 \times 10^{-9}\ \text{C/s}}{100 \times 10^{-12}\ \text{C/V}}$$

$$= 10\ \text{V/s} \quad \text{or} \quad 10\ \mu V/\mu s$$

Droop can be reduced by reducing the hold time, increasing the size of the capacitor, or using capacitances with low amounts of leakage.

LOW-VOLTAGE OP AMP

In 1979, National Semiconductor Corporation introduced a new device into the op amp world, the LM10. The LM10 is a low-voltage op amp. Its two main advantages are a voltage supply range from as low as 1.1 V to as high as 40 V and the ability to function in a floating or conventional mode. A block

diagram of the LM10 is shown in Figure 1.27. Performance characteristics of the LM10 are similar to those of the standard LM108 op amp. The LM10, though, has a high-output drive capability in addition to built-in, thermal-overload circuitry.

An example of the unique capabilities of the low-voltage op amp is the device shown in Figure 1.28. This application is a low-voltage indicator for a +9 V battery power supply. In battery-powered circuitry, there is some advantage to having an indicator to show when the battery voltage is high enough for proper circuit operation. This indicator is especially helpful in circuits that may give inaccurate readings when supplied with a voltage that is too low.

When a +9 V potential supplies the circuit, a 300 mV potential exists at the inverting terminal of the op amp section of the LM10. Connecting pins 1 and 8 produces a voltage follower configuration on the reference amp, placing a 200 mV potential at the noninverting input of the op amp section. The op amp is configured as a comparator, producing an output nearly equal to the 9 V supply potential. This potential forward-biases the LED, turning it on. If the supply voltage falls below +6 V, the voltage at the inverting terminal falls below 200 mV, causing the op amp comparator to go low and turning off the LED.

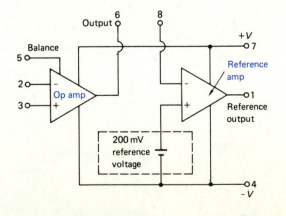

FIGURE 1.27 Block Diagram of LM10 Low-Voltage Op Amp

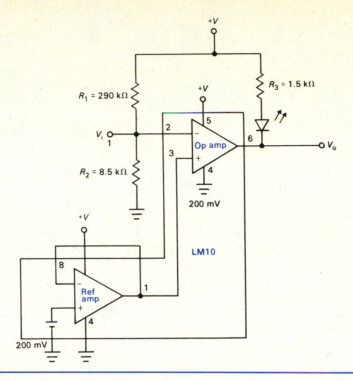

FIGURE 1.28 Battery Charge Indicator

• EXAMPLE 1.12

Given the low-voltage op amp of Figure 1.28, with $R_2 = 8.5$ kΩ, find the voltage at which the LED turns off.

Solution

We know that the input to the noninverting terminal of the op amp is 200 mV; this input comes from the reference amp. The op amp comparator will switch when the voltage across R_2 decreases to less than 200 mV. Using the voltage divider equation, we can solve for the supply voltage that will give a 200 mV potential across R_2. We obtain

$$\left(\frac{R_2}{R_2 + R_1}\right)(V) = \left(\frac{8.5 \text{ k}\Omega}{8.5 \text{ k}\Omega + 290 \text{ k}\Omega}\right)(V)$$

$$= 200 \text{ mV}$$

or

$$V = \frac{(200 \text{ mV})(8.5 \text{ k}\Omega + 290 \text{ k}\Omega)}{8.5 \text{ k}\Omega} = 7 \text{ V}$$

The LED will go out when the threshold voltage of 7 V is reached.

CLIPPER

We may need to limit the output of an op amp stage to a certain maximum voltage. That is, some devices may be damaged with an input voltage higher than a specified maximum voltage. The *clipper*, shown in Figure 1.29, is a circuit that limits its own output voltage. Let us assume that each zener diode has a 5.4 V zener voltage (reverse-biased) and a 0.6 V forward-biased potential. With a positive input, the

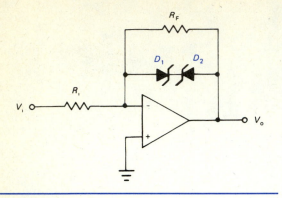

FIGURE 1.29 Op Amp Clipper

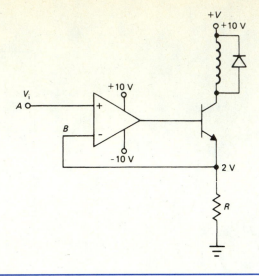

FIGURE 1.30 Op Amp Driving Transistor for Higher Currents

output will go negative, up to a maximum of –6.0 V. The voltage can go no higher than that because D_2 will clamp the output to –5.4 V, and D_1 will clamp at –0.6 V. Together, they clamp the output voltage to –6.0 V. Much the same action is repeated with a negative input. With a negative-going input voltage, the output voltage will go positive, but only as far as +6.0 V. At this point, D_2 will have a forward-biased drop of +0.6 V, and D_1 will have a reverse-biased drop of +5.4 V. Together, the total output voltage can rise no higher than +6.0 V.

OP AMP CURRENT DRIVERS

Low-power op amps, such as the 741, have a limited current output capability. In the 741, the output current is generally restricted to about 25 mA. If more current is desired for a particular application, two choices are available. First, a power op amp may be chosen. Such op amps can handle output powers up to several watts; however, power op amps are sometimes expensive and difficult to procure. A less expensive choice for larger current is to drive a transistor or a power MOSFET (metal-oxide semiconductor field effect transistor) with an op amp. A circuit using the transistor is shown in Figure 1.30. The operational amplifier forces any input voltage at the noninverting terminal to be felt at the inverting

terminal. If a positive 2 V potential is connected to the input at point A, the same potential is felt at point B. The transistor will draw a current equal to

$$I = \frac{V_i}{R}$$

If the resistor is a 10 Ω resistor, the approximate current drawn is

$$I = \frac{+2}{10} = 0.2 \, \text{A} = 200 \, \text{mA}$$

The circuit will then supply 200 mA to the relay coil, energizing it. The op amp, alone and unassisted, may never have been able to generate this much current.

Another power op amp circuit—in this case, a lamp driver—is shown in Figure 1.31. When the input voltage is below the reference voltage, the output is held low by diode D_1. The transistor will not conduct. When the input voltage goes above the reference, the output of the comparator goes high, turning on the transistor. Current will then flow through the lamp L_1, turning it on. The resistor R_2 limits the surge current in the circuit to a safe level

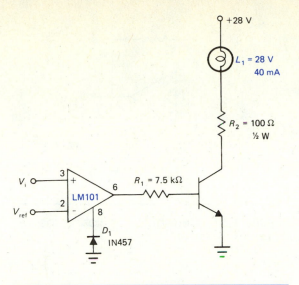

FIGURE 1.31 Op Amp Lamp Driver

CURRENT-DIFFERENCING AMPLIFIER (CDA)

The *current-differencing amplifier* (CDA), or Norton amplifier, developed in the early 1970s, answered a need in the linear device field. Industry required a device that would be linear yet compatible with digital circuitry, and one that would operate from a single power supply. The CDA meets these needs and is relatively inexpensive as well. Four CDAs are housed within a 14-pin, IC, dual in-line package (DIP).

The basic CDA circuit is shown in Figure 1.32. The inverting section of the basic CDA is shown in Figure 1.32A. In Figure 1.32B, the resistors R_C and R_E have been replaced by transistors acting as constant-current sources. Since constant-current sources have very high dynamic resistances, the gain of this circuit reaches 60 dB.

Figure 1.33 illustrates the addition of a circuit that provides a noninverting input. This addition makes a differential input device. Transistor Q_5 and diode D_1 form a *current mirror stage*. Since the base-emitter junction of Q_5 and the cathode-anode junction of D_1 are matched, the application of a bias voltage causes both devices to conduct the same amount of current. Or we say that the current in D_1 is *mirrored* in Q_5. The CDA then tries to keep the

until the filament in the bulb heats up. Resistor R_1 determines the base drive to the transistor. If the base drive needs to be increased, the resistance of R_1 must be decreased.

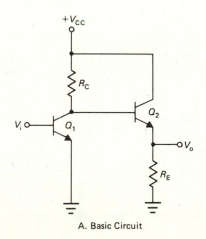

A. Basic Circuit

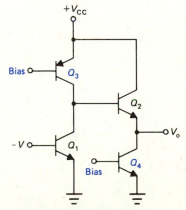

B. With Constant-Current Sources (Transistors) Replacing Resistors

FIGURE 1.32 Inverting Section of CDA

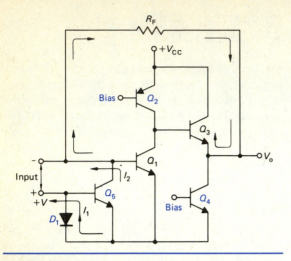

FIGURE 1.33 Simplified CDA Schematic Showing Inverting and Noninverting Terminal Inputs

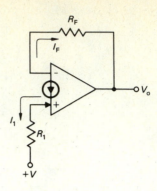

FIGURE 1.34 Current Mirror Biasing the CDA

difference between the two currents equal to zero. The same concept appears in the op amp, which tries to keep the difference in potential between the two inputs equal to zero. The feedback resistor, the only component external to the IC in Figure 1.33, provides a path for current from Q_5 to the output transistor Q_3.

Biasing the CDA

Figure 1.34 shows the biasing arrangement used most for CDAs. When $+V$ is connected to R_1, current flows (I_1). The CDA mirrors the same amount of current through R_F as flows through R_1. The current through R_1 can be found from the following equation:

$$I_1 = \frac{+V - 0.7}{R_1} \approx \frac{+V}{R_1} \qquad (1.20)$$

The output voltage likewise is as follows:

$$V_o = I_F R_F + 0.7 \approx I_F R_F \qquad (1.21)$$

Since $I_1 = I_F$ (from the mirror principle), we have the following equation:

$$V_o = \left(\frac{+V}{R_1}\right)(R_F) \qquad (1.22)$$

The output DC voltage in quiescent conditions is determined by the bias supply ($+V$), R_F, and R_1. For centered operation (where $V_o = +V/2$), $R_1 = 2R_F$. If $2R_F$ is substituted into Equation 1.22, then $V_o = +V/2$ (centered operation).

• EXAMPLE 1.13

Given the CDA biasing circuit in Figure 1.34, with resistance values $R_F = 470$ kΩ and $R_1 = 1$ MΩ, and with a $+15$ V supply, find the bias current and output voltage.

Solution

From Equation 1.20, we have

$$I_b = \frac{+V}{R_1} = \frac{+15 \text{ V}}{1 \text{ M}\Omega} = 15 \text{ μA}$$

And from Equation 1.22, we get

$$V_o = \left(\frac{+V}{R_1}\right)(R_F) = \left(\frac{+15 \text{ V}}{1 \text{ M}\Omega}\right)(470 \text{ k}\Omega)$$

$$= 7.05 \text{ V}$$

CDA Inverting Amplifier

The CDA inverting amplifier is similar in many ways to the comparable op amp circuit. The gain equation is identical. Figure 1.35 shows the basic CDA inverting amplifier. It is generally used in AC applications because the DC bias may be affected by the previous stage if there is no coupling capacitor.

Let us suppose that a positive-going input signal is applied to this circuit. Current i_i (the current due to the AC voltage applied to the input) flows as illustrated. The current mirror, which tries to keep the same current flowing out of the inverting terminal, causes the total current in R_F to decrease by an amount equal to i_i. So the change in R_F current, denoted as i_i, is then equal but opposite to i_F, which is the AC component of the total current in R_F. The gain equation is

$$A_V = -\frac{v_o}{v_i} \qquad (1.23)$$

where

v_i = AC input voltage
v_o = AC output voltage

Thus, we can state v_o and v_i in terms of currents and resistances:

$$\frac{v_o}{v_i} = \frac{-i_F R_F}{i_i R_i}$$

Since $i_i = i_F$, this equation reduces the gain equation to

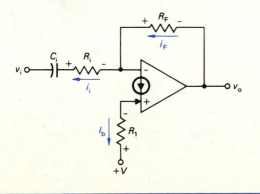

FIGURE 1.35 CDA Inverting Amp

$$A_V = \frac{v_o}{v_i} = \frac{-R_F}{R_i} \qquad (1.23a)$$

The output is now reduced from the $+V/2$ level by the amount A_V times v_i. By rearranging equation 1.23a, we can find the output voltage

$$v_o = A_V v_i \qquad (1.24)$$

• EXAMPLE 1.14

Given the inverting CDA of Figure 1.35, with the values $R_i = 100$ kΩ, $R_F = 1$ MΩ, $R_1 = 2$ MΩ, an input voltage of $+0.2$ V$_{pk}$, and $+V = 12$ V, find the bias current, DC output voltage, AC gain, and AC output voltage.

Solution

From Equation 1.20, we find

$$I_b = \frac{+V}{R_1} = \frac{+12 \text{ V}}{2 \text{ M}\Omega} = 6 \text{ }\mu\text{A}$$

Equation 1.22 yields

$$V_o = \left(\frac{+V}{R_1}\right)(R_F) = \left(\frac{+12 \text{ V}}{2 \text{ M}\Omega}\right)(1 \text{ M}\Omega)$$

$$= 6 \text{ V}$$

Next, Equation 1.23a gives the gain:

$$A_V = \frac{-R_F}{R_i} = \frac{-1 \text{ M}\Omega}{100 \text{ k}\Omega} = -10$$

Finally, Equation 1.24 gives

$$v_o = A_V v_i = -(10)\left(0.2 \text{ V}_{pk}\right) = -2 \text{ V}_{pk}$$

Figure 1.36 illustrates a basic inverting amplifier using a CDA. Current through R_1 (which is designated I_1) is calculated from the current equation (Equation 1.20):

$$I_1 = \frac{+V - 0.7}{R_1} = \frac{+10 \text{ V} - 0.7 \text{ V}}{2 \text{ M}\Omega}$$

$$= 4.6 \text{ }\mu\text{A}$$

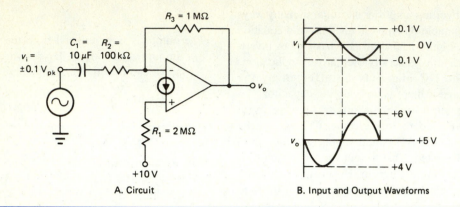

A. Circuit   B. Input and Output Waveforms

FIGURE 1.36   CDA Inverting Amp with AC Input

From our previous discussion, we know that I_3 equals I_1, where I_3 is the current through R_3. The output voltage, then, is (from Equation 1.21)

$$V_o = I_F R_F + 0.7 = 4.6\ V + 0.7\ V$$

$$= 5.3\ V \approx \frac{+V}{2} = 5\ V$$

If we inject a 0.1 V peak signal into this amplifier, the output voltage change, where $A_V = -R_3/R_2 = -10$, is

$$v_o = A_V v_i$$

which, for our example, becomes

$$v_o = 10 \times 0.1\ V = 1\ V_{pk}$$

Thus, a 0.1 V peak input produces a 1 V peak output riding on a +5 V DC level. This relationship is diagramed in Figure 1.36B.

The choice of input capacitance is very important to the circuit designer. If the capacitance is too small, the reactance will be higher at low frequencies. A high series reactance will cause the total circuit gain to be lowered.

An example may be helpful at this point. Figure 1.36A shows a 10 μF input capacitor in series with the 100 kΩ resistor R_2. We wish to find the reactance (X_C) of the capacitor at 1000 Hz. We use the following reactance equation:

$$X_C = \frac{1}{2\pi f C} \qquad (1.25)$$

For our example, we get

$$X_C = \frac{1}{(2\pi)\,(1000\ Hz)\,(10\ \mu F)}$$

$$= 15.9\ \Omega$$

We can clearly see that—at least at 1000 Hz—the reactance of the capacitor is small, compared with the size of resistor R_2.

Now, what would happen if the frequency applied were to decrease to 1 Hz? The reactance would increase by 1000, to almost 16 kΩ. Although this value is still small compared with the 100 kΩ resistance of R_2, it is significant and greater than 10%. So a general rule of thumb designers use is that the reactance should be less than one-tenth of the resistance of R_2 at the lowest frequency to be amplified.

How large would the capacitor need to be if the lowest frequency to be amplified were 20 Hz? Suppose we choose one-tenth of R_2, or 10,000 Ω, as our lower limit. We can solve the reactance equation for C:

$$C = \frac{1}{2\pi f X_C} \qquad (1.26)$$

This equation gives us

$$C = \frac{1}{(2\pi)\,(20\text{ Hz})\,(10{,}000\ \Omega)}$$

$$= 0.796\ \mu F$$

Our input capacitance would have to be larger than 0.8 μF, say 1 μF. This same calculation may be applied to any of the circuits we will study in this book.

• EXAMPLE 1.15

Given the circuit in Figure 1.36A, with the lowest input frequency being 100 Hz, find the input capacitor size needed in this circuit.

Solution

If we chose one-tenth of R_2, or 10,000 Ω as our lower limit, the required capacitance is (from Equation 1.26)

$$C = \frac{1}{2\pi f X_C} = \frac{1}{(2\pi)\,(100\text{ Hz})\,(10{,}000\ \Omega)}$$

$$= 0.159\ \mu F$$

Our input capacitance must be larger than 0.159 μF, say 0.2 μF.

CDA Noninverting Amplifier

The circuit in Figure 1.37 connects a CDA so that it operates as a noninverting amplifier. The gain equation for this amp is the same as that of the inverting amp (except for the sign):

$$A_V = \frac{R_3}{R_2}$$

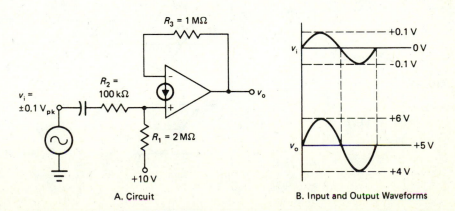

A. Circuit

B. Input and Output Waveforms

FIGURE 1.37 Noninverting CDA Amp with AC Input

As in the inverting amplifier, the output voltage is +5 V, with about 4.6 μA flowing through R_1 and R_3 in the no-input-signal condition. The gain is also identical ($A_V = 10$). A 0.1 V peak input produces a 1 V peak output with no phase difference.

• EXAMPLE 1.16

Given the circuit in Figure 1.38A, find the bias current, DC output voltage, AC voltage gain, AC output voltage, and capacitance needed for a low-frequency input of 20 Hz. Place voltage data in a waveform diagram (called a *synchrogram*), as in Figure 1.37B.

Solution

From Equation 1.20, we find

$$I_b = \frac{+V}{R_1} = \frac{+12\ V}{470\ k\Omega} = 25.5\ \mu A$$

Using Equation 1.22, we get

$$V_o = \left(\frac{+V}{R_1}\right)(R_3)$$

$$= \left(\frac{+12\ V}{470\ k\Omega}\right)(220\ k\Omega) = 5.62\ V$$

Equation 1.23a yields

$$A_V = \frac{-R_F}{R_i} = \frac{-220\ k\Omega}{100\ k\Omega} = -2.2$$

Finally, from Equation 1.24, we find

$$v_o = A_V v_i = -(2.2)\left(\pm 0.5\ V_{pk}\right)$$

$$= \pm 1.1\ V_{pk}$$

CDA Comparator

The CDA comparator is illustrated in Figure 1.39. The output of the device in Figure 1.39 is V_{sat} as long as the input voltage V_i is below the reference voltage V_{ref}. When the input voltage rises above V_{ref}, the output goes to 0 V and stays there as long as the input remains above the reference. Comparators such as these find extensive application in digital systems.

Another type of comparator is shown in Figures 1.40A and 1.40B. In digital electronics, this device is called a *Schmitt trigger*. In analog electronics, it is usually called a *comparator with hysteresis*. A

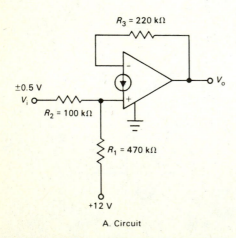

A. Circuit

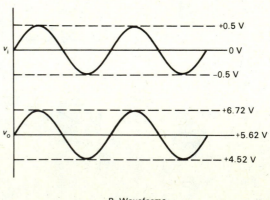

B. Waveforms

FIGURE 1.38 Noninverting Amplifier for Example 1.16

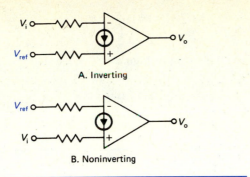

A. Inverting

B. Noninverting

FIGURE 1.39 CDA Comparator

higher than that in the noninverting terminal. The output will trip and go low when the input voltage draws more than 6 μA, which will happen when the input voltage goes above 6 V. When the input voltage goes above 6 V, the output voltage goes low. This 6 V input potential is called the upper trip voltage. Six microamperes is the upper trip current. When the output voltage goes low, the only current drawn from the noninverting terminal is 5 μA. The trip voltage and current have changed. For the comparator to trip and make the output go high again, the current in the inverting terminal will have to go below the 5 μA in the noninverting terminal. This will happen when the input voltage goes below 5 V. This potential is called the lower trip voltage. Five microamperes is the lower trip current.

comparator with hysteresis often is needed to deal with the spurious switching caused by noise. As in an op amp comparator with hysteresis, the CDA circuit changes the trip point every time it is triggered.

The schematic shown in Figure 1.40A is an inverting comparator with hysteresis. Let us assume that the supply voltage is +10 V; and that when the output goes high, it goes to +10 V. Also, we will assume that when the output goes low, it will go to 0 V. The +10 V input at the reference will draw 5 μA of current. If the input is at 0 V, no current will be drawn from the inverting terminal. In this case, the current in the noninverting terminal is higher than the current in the inverting terminal, and the output will be high. A high output will draw 1 μA of current through R_3. The total current drawn from the noninverting terminal is 6 μA. For the output to go low, the current in the inverting terminal must go

We can now write equations to give the predicted upper and lower trip voltages for the inverting comparator. First, we find the reference current by dividing the reference voltage by resistor R_2:

$$I_{ref} = \frac{V_{ref}}{R_2} \qquad (1.27)$$

For the circuit of Figure 1.40A, we obtain

$$I_{ref} = \frac{+10\ V}{2\ M\Omega} = 5\ \mu A$$

Next, the upper trip current is equal to the sum of the current I_{ref} and the feedback current I_F. The feedback current is equal to the output voltage at saturation (+10 V) divided by the feedback resistance R_3, or

$$I_F = \frac{V_o}{R_3} \qquad (1.28)$$

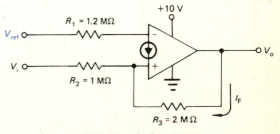

A. Inverting Comparator

B. Noninverting Comparator

FIGURE 1.40 CDA Comparator with Hysteresis

For our circuit, we find

$$I_F = \frac{+10 \text{ V}}{10 \text{ M}\Omega} = 1 \text{ μA}$$

The upper trip current I_{ut} is equal to the reference current plus the feedback current, or

$$I_{ut} = I_{ref} + I_F \tag{1.29}$$

So, for our circuit, we have

$$I_{ut} = 5 \text{ μA} + 1 \text{ μA} = 6 \text{ μA}$$

Finally, the upper trip voltage is equal to the sum of the currents times the resistance of R_1:

$$V_{ut} = (I_{ref} + I_F)(R_1) \tag{1.30}$$

Therefore, we obtain

$$V_{ut} = (5 \text{ μA} + 1 \text{ μA})(1 \text{ m}\Omega) = 6 \text{ V}$$

The upper trip voltage can also be expressed in terms of voltages and resistances as

$$V_{ut} = \left(\frac{V_{ref}}{R_2} + \frac{V_o}{R_3}\right)(R_1) \tag{1.31}$$

The lower trip point, since it does not involve the output voltage, is equal to the reference current times the resistance of R_1, or

$$V_{lt} = (I_{ref})(R_1) = \left(\frac{V_{ref}}{R_2}\right)(R_1) \tag{1.32}$$

For the circuit of Figure 1.40A, we find

$$V_{lt} = \left(\frac{+10 \text{ V}}{2 \text{ M}\Omega}\right)(1 \text{ M}\Omega) = 5 \text{ V}$$

• EXAMPLE 1.17

Given the inverting comparator shown in Figure 1.41A, find the upper and lower trip voltages.

Solution
From Equation 1.31, we find

$$V_{ut} = \left(\frac{V_{ref}}{R_2} + \frac{V_o}{R_3}\right)(R_1)$$

$$= \left(\frac{+15 \text{ V}}{1.5 \text{ M}\Omega} + \frac{+15 \text{ V}}{7.5 \text{ M}\Omega}\right)(1.2 \text{ M}\Omega)$$

$$= 14.4 \text{ V}$$

And using Equation 1.32, we get

$$V_{lt} = (I_{ref})(R_1) = \left(\frac{V_{ref}}{R_2}\right)(R_1)$$

$$= \left(\frac{+15 \text{ V}}{1.5 \text{ M}\Omega}\right)(1.2 \text{ M}\Omega) = 12 \text{ V}$$

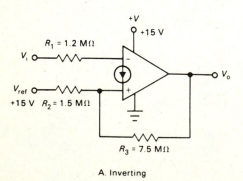

A. Inverting

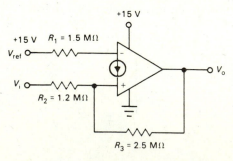

B. Noninverting

FIGURE 1.41 Comparators with Hysteresis for Examples 1.17 and 1.18

Note the difference between the inverting comparator with hysteresis, shown in Figure 1.40A, and the noninverting comparator with hysteresis, shown in Figure 1.40B. We can see that the input and the reference have been switched. There are no other differences. In the noninverting comparator, a high output results from the current in the noninverting terminal exceeding the current in the inverting terminal. As in the inverting comparator, an input voltage causes the current to be drawn from the input terminal, tripping the comparator. The output voltage will go high when the input voltage exceeds a predetermined reference voltage. This reference voltage is called the upper trip point. When the input goes lower than a reference voltage—called the lower trip point—the output voltage goes low.

Upper and lower trip points in the noninverting comparator are calculated in the same manner as for the inverting comparator. When the output is low (0 V), the input current must go above 8.33 µA. This 8.33 µA current will be drawn when the input voltage goes above 8.33 V, driving the output high (+10 V). When the output goes high, 5 µA will be drawn through the 2 MΩ resistor R_3. The input voltage V_i must go below 3.3 V before the output will switch again and go low. In this application, then, the upper trip voltage is 8.33 V and the lower trip voltage is 3.33 V. This variation in trip points is called *hysteresis*.

We can again write equations to give the predicted upper and lower trip voltages for the noninverting comparator in Figure 1.40B. The upper trip voltage is equal to the reference current times the resistance of R_2. The upper trip voltage, expressed in terms of voltages and resistances, is therefore

$$V_{ut} = \left(\frac{V_{ref}}{R_1}\right)(R_2) \tag{1.33}$$

For the circuit of Figure 1.40B, we obtain

$$V_{ut} = \left(\frac{+10 \text{ V}}{1.2 \text{ M}\Omega}\right)(1 \text{ M}\Omega) = 8.33 \text{ V}$$

The lower trip point is equal to the reference current minus the feedback current times the resistance of R_2, or

$$V_{lt} = (I_{ref} - I_F)(R_2)$$
$$= \left(\frac{V_{ref}}{R_1} - \frac{V_o}{R_3}\right)(R_2) \tag{1.34}$$

For our circuit, we get

$$V_{lt} = \left(\frac{+10 \text{ V}}{1.2 \text{ M}\Omega} - \frac{+10 \text{ V}}{2 \text{ M}\Omega}\right)(1 \text{ M}\Omega)$$
$$= 3.33 \text{ V}$$

• EXAMPLE 1.18

Given the noninverting comparator shown in Figure 1.41B, find the upper and lower trip voltages.

Solution

From Equation 1.33, we calculate

$$V_{ut} = \left(\frac{V_{ref}}{R_1}\right)(R_2) = \left(\frac{+15 \text{ V}}{1.5 \text{ M}\Omega}\right)(1.2 \text{ M}\Omega)$$
$$= 12 \text{ V}$$

And from Equation 1.34, we find

$$V_{lt} = (I_{ref} - I_F)(R_2) = \left(\frac{V_{ref}}{R_1} - \frac{V_o}{R_3}\right)(R_2)$$
$$= \left(\frac{+15 \text{ V}}{1.5 \text{ M}\Omega} - \frac{+15 \text{ V}}{2.5 \text{ M}\Omega}\right)(1.2 \text{ M}\Omega)$$
$$= 4.8 \text{ V}$$

CDA Window Comparator

A useful application of the basic CDA comparator is the *window comparator*, shown in Figure 1.42. Before we can analyze this circuit, though, we need to find out what the reference currents are. Note that the reference voltage input is +10 V. This +10 V potential draws a reference current through R_1 and R_4. The current can be approximated by dividing the +10 V potential by the resistance:

$$I_1 = I_4 = \frac{10 \text{ V}}{500 \text{ k}\Omega} = 20 \text{ µA}$$

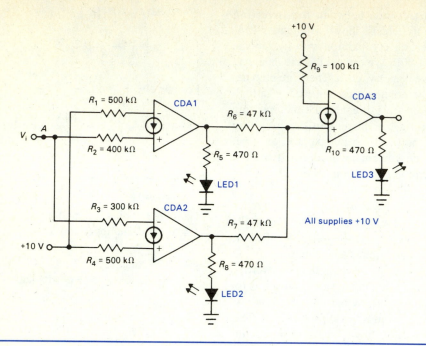

FIGURE 1.42 CDA Window Comparator

We can see from this calculation that the currents through these two resistors are equal to 20 μA.

If we place a 0 V potential at the input to the window comparator, what will the states of the outputs of each comparator be? Let's look at comparator 1 (CDA1) first. With an input of 0 V, no current will flow through R_2. The current through R_1 will be greater than the current in R_2, causing the output to go low. If we assume an ideal CDA, the output will be 0 V. LED1 will be off.

Exactly the opposite condition occurs in comparator 2. The current in the noninverting terminal (flowing through R_4) is greater than the current in the inverting terminal (flowing through R_3). The output of CDA2 will be high, assuming an ideal CDA, at +10 V. The +10 V potential from CDA2 will forward-bias LED2, turning it on.

The high potential (+10 V) at the output of CDA2 draws about 213 μA from the noninverting terminal of CDA3 through R_7. Note that no current is drawn through R_6 since the output of CDA2 is low (0 V). The current in the noninverting terminal of CDA3 is greater than the current in the inverting

terminal (100 μA), so the output of CDA3 will be high. The high output of CDA3 will turn on LED3.

What input voltage will cause the CDA2 output to go low? Since the noninverting terminal current is 20 μA, the current in the inverting terminal must go above 20 μA. What input voltage will provide this current? The input voltage must be greater than 20 μA times the resistance of R_3 (300 kΩ), or 6 V. So if the input voltage goes above 6 V, the output of CDA2 will go from high to low.

What voltage will cause CDA1 to change state? Again, the current through the noninverting terminal must be greater than 20 μA. We can find the voltage that will cause a change of state by multiplying 20 μA by the resistance of R_2, which is 400 kΩ. The trip voltage for CDA1 is thus 8 V.

The result of this analysis is summarized in Table 1.3. When the input voltage is between 6 and 8 V, we say that the input voltage is within the window. Note that when the input voltage is within the window, both the CDA1 and the CDA2 outputs will be low. The two low inputs to CDA3 will keep its output low, turning off LED3.

TABLE 1.3 CDA Window Comparator Outputs

V_i	LED1	LED2	LED3
<6 V	Off	On	On
>6 V and <8 V	Off	Off	Off
>8 V	On	Off	On

• EXAMPLE 1.19

Given the window comparator in Figure 1.42, with resistances $R_1 = 470$ kΩ, $R_2 = 390$ kΩ, $R_3 = 250$ kΩ, and $R_4 = 470$ kΩ, and with supply and reference voltages of +10 V, find the window voltages.

Solution

Since R_1 and R_4 have the same resistance, the reference currents are the same. The reference current can be found by dividing the reference voltage, 10 V, by the resistance of 470 kΩ. Thus, the reference current is approximately 21.3 μA. The input voltage must exceed this current in R_2 to make the CDA1 output go high. The CDA1 output will go high at 8.3 V, the upper edge of the window.

The lower edge of the window is the voltage for which CDA2 goes low. A low output voltage at CDA2 will occur when the current in R_3 exceeds 21.3 μA. So CDA2 will go low when the input voltage exceeds 5.3 V. Therefore, the window lies between 5.3 and 8.3 V.

Window comparators are used in situations where the voltages must lie between two limits. For example, a piece of equipment may operate normally with supply voltages between 4.9 and 5.1 V. In this case, a window comparator may be set up with a window between 4.9 and 5.1 V. The output of CDA3 may then drive an alarm, rather than an LED. If the input voltage is greater than 5.1 V or less than 4.9 V, the output of CDA3 will go high, turning on the alarm. The same window comparator may be built easily by using normal op amps instead of CDAs.

CDA Voltage Follower

One of the simplest applications of the CDA is the buffer, or isolation amplifier, pictured in schematic form in Figure 1.43. A positive DC voltage placed at the input will draw current from the noninverting terminal. That same current will be mirrored in the inverting terminal and converted to a voltage by the 1 MΩ feedback resistor. The output voltage should then follow the input in voltage and phase. This circuit does have some limitations, however. The input must be greater than +0.6 V to turn on the mirror current transistor in the CDA. This disadvantage may be overcome by other biasing techniques.

CDA Oscillator

The CDA is a device that has shown considerable flexibility in application, as does the normal op amp. The diagram in Figure 1.44 shows the CDA used as a square wave oscillator. Capacitor C_1 charges and

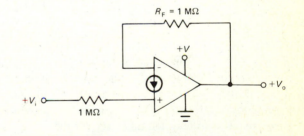

FIGURE 1.43 CDA Buffer Amp

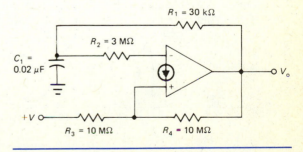

FIGURE 1.44 CDA Square Wave Oscillator

discharges through R_1 between the limits set by R_2, R_3, and R_4. You will recognize this circuit as a modification of the comparator with hysteresis just discussed. When the output is low, the capacitor will discharge through R_1. As it discharges, the voltage across it will be decreasing, as will the current drawn from the inverting terminal. When the current in the inverting terminal goes below the current in the noninverting terminal, the output will go high. This will occur (when $+V = +10$ V) when the capacitor voltage goes below 3.33 V. The output then goes high and the capacitor starts to charge. When the output goes high, the current drawn from the noninverting terminal is doubled, from 1 μA to 2 μA. The capacitor voltage must go to twice the voltage ($+6.67$ V) to trip the CDA and make the output go low. With the component values given, the CDA square wave generator should oscillate at 1 kHz. With the component values given, the output waveform should be relatively symmetrical. If an unsymmetrical output is needed, the ratios of R_2, R_3, and R_4 can be varied.

We can approximate the output frequency (f_o) by the following equation:

$$f_o = \frac{1}{1.4\,R_1 C_1} \tag{1.35}$$

Applied to the circuit in Figure 1.44, the equation yields 1190 Hz. This equation must be used with some caution. The current that charges capacitor C_1 is decreased by the amount of the bias current flowing through resistor R_2. To keep the bias current small, make R_2 at least ten times greater than R_1.

• **EXAMPLE 1.20**
Given the CDA oscillator in Figure 1.44, with a 0.05 μF capacitor, find the frequency of oscillation.

Solution
We use Equation 1.35:

$$f_o = \frac{1}{1.4\,R_1 C_1} = \frac{1}{(1.4)\,(30\text{ k}\Omega)\,(0.05\text{ μF})}$$

$$= 476\text{ Hz}$$

CDA Logic Circuits

Some industrial applications require decision-making circuits, such as those normally provided by digital logic chips like the 7400 TTL series. In many applications, the +5 V power supply is not available to power standard logic chips. The CDA can be used in place of logic chips in low-speed digital and switching applications. The larger voltage swing and the slower speed can be an advantage in some industrial applications.

Recall the digital logic OR function performed by the device schematically represented in Figure 1.45A. High voltages at inputs A, B, or C in this circuit will give a high output at D. Th equivalent CDA circuit is shown in Figure 1.45B. If $+V$ is equal to $+10$ V, the current drawn from the inverting terminal will be approximately 66.7 μA. With no input, this voltage holds the output low since the current in the inverting terminal is higher than that in the noninverting terminal (0 A). A $+10$ V input voltage at either A, B, or C will draw $+10$ V/75 kΩ, or 133 μA. The output voltage will go high since the current in the noninverting terminal is higher than that in the inverting terminal. The Boolean expression for this circuit is $D = A + B + C$.

This device can handle up to 50 inputs, each with its own 75 kΩ resistor, because the input current capability is 10 mA. This capability is called *fan-in*.

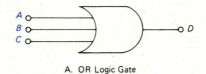

A. OR Logic Gate

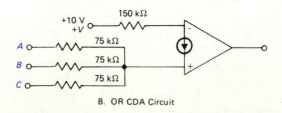

B. OR CDA Circuit

FIGURE 1.45 CDA OR Gate Function

A NOR function can be created by switching the inputs. In this case, the output will be held high until an input pulls it low.

Another logic function, AND, provided by the CDA is symbolized by the schematic diagram in Figure 1.46A. We get a high output at D when highs are present at A, B, and C. The CDA version of this circuit is shown in Figure 1.46B. With a $+V$ of +10 V, the input voltage supply will pull 133 μA from the inverting terminal. This current will hold the output low with no input signal, as in the OR gate just discussed. With an input at A or A and B, there is not enough current to bring the total current in the noninverting terminal higher than in the inverting terminal. The only combination of input voltages that will cause high enough current in the noninverting terminal to trip the comparator is A, B, and C. The output will go high if and only if a high is present at A, B, and C. A NAND gate may be constructed by reversing the inputs, just as a NOR gate was created by switching inputs of an OR gate.

We have seen from this discussion that the CDA can substitute for the op amp in many applications. In fact, most of the op amp configurations can be duplicated by the CDA.

OPERATIONAL TRANSCONDUCTANCE AMPLIFIER (OTA)

A *transconductance amplifier* is a circuit or component that changes an input voltage into an output current. An *operational transconductance amplifier* (OTA) is an IC transconductance amplifier. In many ways, the OTA resembles the op amp device discussed earlier. It possesses a differential voltage input, has a high input impedance, and has a high open-loop gain. The most striking difference in the two devices can be seen in the schematic diagram in Figure 1.47. Note the addition of a terminal labeled I_{ABC} in the OTA. This terminal is a control input that is called the *amplifier bias current control*. Many of the OTA's parameters can be controlled by varying the current in this terminal.

Table 1.4 shows the changes in OTA parameters with an increase in I_{ABC}. Notice that the parameter *transconductance* is given in Table 1.4 instead of voltage gain. Recall that transconductance is the change in output current divided by the change in input voltage, as expressed by the following equation:

$$g_m = \frac{I_o}{V_i} \qquad (1.36)$$

where

g_m = transconductance, in microsiemens, μS (formerly called micromhos)

I_o = output current

V_i = input voltage

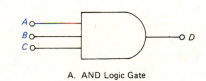

A. AND Logic Gate

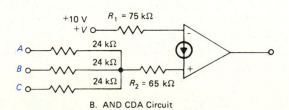

B. AND CDA Circuit

FIGURE 1.46 CDA AND Gate Function

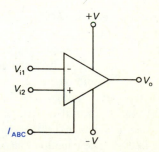

FIGURE 1.47 Schematic Diagram of the OTA

TABLE 1.4 Changes in OTA Parameters with Increased I_{ABC}

Parameters	Changes
Input resistance	Decrease
Output resistance	Decrease
Input capacitance	Decrease
Output capacitance	Decrease
Transconductance	Increase
Input bias current	Increase
Input offset current	Increase
Input offset voltage	No change
Slew rate	Increase

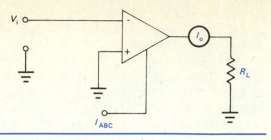

FIGURE 1.48 OTA Gain Control

The OTA, as its name implies, differs from the op amp with regard to its output characteristics. Recall that the op amp's output was defined in terms of voltage. In contrast, the OTA's output is defined in terms of current. In this regard, the OTA resembles an FET (field effect transistor) or a vacuum tube. Therefore, transconductance is the suitable parameter to express this relationship, being the change in output current divided by the change in input voltage.

Like FETs and vacuum tubes, the OTA can be used as a voltage amplifier with the addition of load resistance or impedance. The voltage gain in this case equals the (mutual) transconductance (g_m) times the load resistance (g_mR_L). Or we can say that the output voltage equals the transconductance times the output resistance times the input voltage, as follows:

$$V_o = g_m V_i R_L \qquad (1.37)$$

With an output resistance, the OTA will behave much like the normal op amp.

One of the simplest applications involving the OTA is in a gain control, such as shown in Figure 1.48. The V_{ABC} input voltage controls the transconductance and output current. In OTAs like the CA3080 manufactured by RCA, the transconductance can be calculated if the supply voltage and I_{ABC} are known. The equation that relates these factors is

$$g_m = \left(\frac{300}{V}\right)(I_{ABC}) \qquad (1.38)$$

where 300 is a constant. For example, let us say that we have a supply voltage of ±15 V, with an I_{ABC} of 20 µA. The circuit transconductance is found by substituting these values into Equation 1.38:

$$g_m = \left(\frac{300}{V}\right)(I_{ABC}) = \left(\frac{300}{15\,V}\right)(20\,\mu A)$$
$$= 400\,\mu S$$

By substitution, we can construct an equation that will allow us to predict the output voltage if we know I_{ABC}, V_i, and R_L. The equation that allows us to make this prediction is

$$V_o = \left(\frac{300}{V}\right)I_{ABC}V_iR_L \qquad (1.39)$$

where
V_o = output voltage
V = supply voltage
I_{ABC} = bias current
V_i = input voltage
R_L = load resistance

When using this equation, we will assume that all the output current will flow into R_L, as shown in Figure 1.48. Using this formula, let us calculate the output voltage with an input of 0.5 V, an I_{ABC} of 100 µA, and an R_L of 10,000 Ω. We obtain

$$V_o = \left(\frac{300}{V}\right)I_{ABC}V_iR_L$$
$$= \left(\frac{300}{15\,V}\right)(100\,\mu A)(0.5\,V)(10,000\,\Omega)$$
$$= 10\,V$$

We should see an output voltage of approximately 10 V from this OTA. As expected, the OTA, because of its transconductance properties, has an output impedance much higher than that of the standard op amp.

• EXAMPLE 1.21

Given a CA3080 OTA, with I_{ABC} = 80 μA and a supply voltage of 12 V, find the transconductance of the CA3080.

Solution

We use Equation 1.38, as follows:

$$g_m = \left(\frac{300}{V}\right) I_{ABC} = \left(\frac{300}{12\ V}\right) (80\ \mu A)$$

$$= 2000\ \mu S$$

• EXAMPLE 1.22

Given a CA3080, with I_{ABC} = 150 μA, a load resistor of 10,000 Ω, an input voltage of 0.1 V_{pk}, and a supply voltage of 15 V, find the peak output voltage.

Solution

Again, we use Equation 1.38 to find the transconductance:

$$g_m = \left(\frac{300}{V}\right) I_{ABC} = \left(\frac{300}{15\ V}\right) (150\ \mu A)$$

$$= 3000\ \mu S$$

Then, we use Equation 1.37 to find V_o:

$$V_o = g_m V_i R_L$$

$$= (3000\ \mu S)\ (0.1\ V_{pk})\ (10,000\ \Omega)$$

$$= 3.0\ V_{pk}$$

Figure 1.49 shows the input stage of the OTA. Notice that it is a differential amplifier, just like the op amp. Transistor Q_3 and diode D_1 form the current source for the differential amplifier. This circuit differs from the op amp circuit in that the collector

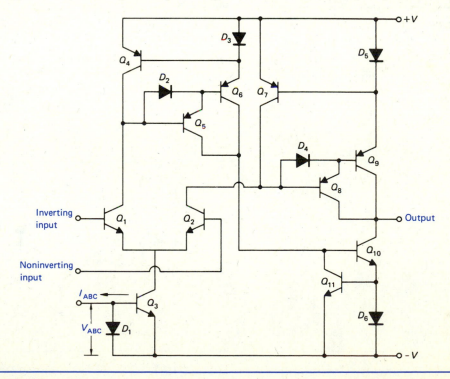

FIGURE 1.49 Internal Construction of the OTA

current of Q_3 can be changed by varying the current in the I_{ABC} terminal. Changing this current varies the transconductance of the differential amplifier.

The OTA bias current is usually controlled in one of several ways. In one method, if the current is to remain constant, the I_{ABC} terminal may be grounded through a resistance, as shown in Figure 1.50. Compare this circuit with the circuit in Figure 1.49. Note the presence of the diode D_1, whose cathode is tied to $-V$. Neglecting the diode drop across D_1, we can approximate the I_{ABC} current by

$$I_{ABC} = \frac{-V}{R_{ABC}} \qquad (1.40)$$

Thus, if we apply a -7.5 V potential to the negative supply terminal through a 100 kΩ resistor, the bias current is

$$I_{ABC} = \frac{-V}{R_{ABC}} = \frac{7.5\ V}{100\ k\Omega} = 75\ \mu A$$

• EXAMPLE 1.23

Given the circuit in Figure 1.50, with a negative supply voltage of 10 V and an R_{ABC} resistance of 200 kΩ, find the I_{ABC} current (in microamperes).

Solution

Using Equation 1.40, we get

$$I_{ABC} = \frac{-V}{R_{ABC}} = \frac{10\ V}{200\ k\Omega} = 50\ \mu A$$

The second way an OTA can be biased is demonstrated in Figure 1.51. We can see from the equivalent circuit that the total voltage across the resistor R_{ABC} is equal to the sum of the negative supply and the control voltage divided by the resistance of R_{ABC}. The amplifier bias current is equal to the sum of the negative supply and the control voltage divided by the resistance of R_{ABC}, or

$$I_{ABC} = \frac{-V + V_c}{R_{ABC}} \qquad (1.41)$$

In the example shown in Figure 1.51, the bias current is

$$I_{ABC} = \frac{-V + V_c}{R_{ABC}} = \frac{7.5\ V + 10\ V}{100\ k\Omega}$$

$$= 175\ \mu A$$

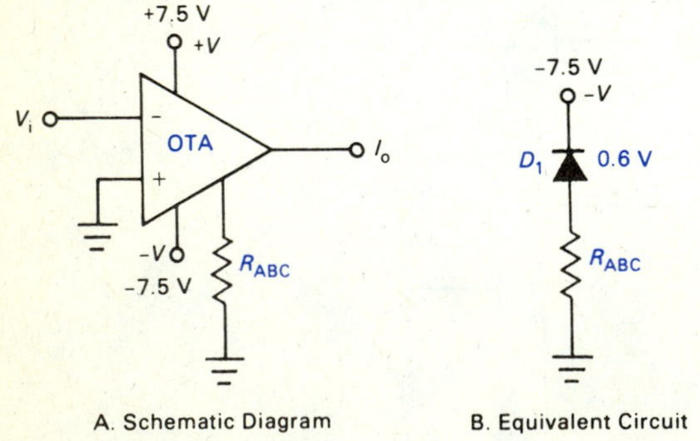

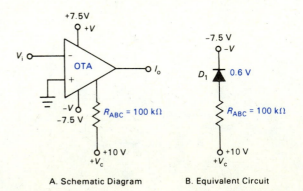

A. Schematic Diagram   B. Equivalent Circuit

FIGURE 1.50 OTA Biasing by Grounding I_{ABC}

A. Schematic Diagram   B. Equivalent Circuit

FIGURE 1.51 OTA Biasing with a Positive Supply

• EXAMPLE 1.24

Given the OTA circuit of Figure 1.51, with a supply of ±10 V, a control voltage of 12 V, and an R_{ABC} of 150 kΩ, find the amplifier bias current.

Solution

We use Equation 1.41:

$$I_{ABC} = \frac{-V + V_c}{R_{ABC}} = \frac{10 \text{ V} + 12 \text{ V}}{150 \text{ k}\Omega}$$

$$= 146.7 \text{ μA}$$

The third way to control the I_{ABC} of an OTA is shown in Figure 1.52. Note that the output current of OTA2 controls the I_{ABC} of OTA1. As long as V_{i2} is constant, the I_{ABC} of OTA1 will be constant. An increase in V_{i2} will cause I_{ABC1} to increase proportionally.

Numerous applications exist for OTAs. It is probably best suited to telemetry and communications use by virtue of the unique control provided by the I_{ABC} terminal. The OTA is also used in measurement circuits in a sample-and-hold configuration. Other applications include nonlinear mixing, gain control (including automatic gain control), waveform generation, and multiplication.

An example of an OTA application is the circuit shown in Figure 1.53, which is an amplitude modulator. Here, the output current is equal to the transconductance times the input voltage. In this amplifier, the level of the unmodulated carrier signal is controlled by the average current into the I_{ABC} terminal. The carrier frequency is amplitude-modulated when the modulating signal V_m causes changes in I_{ABC}. As the input-modulating signal (V_m) goes positive, the current in I_{ABC} increases, thus increasing the transconductance of the OTA. As V_m goes negative, the current in I_{ABC} decreases, decreasing the OTA's g_m. The output, developed across the 5.1 kΩ resistor, is the familiar amplitude-modulated waveform.

PROGRAMMABLE OP AMPS

Most op amps have characteristics that the user has little or no control over. For instance, the input bias current for an op amp is generally fixed for each device according to its internal construction. *Programmable op amps*, however, allow the user to change some of these characteristics. Often, circuit designers are forced to make trade-offs when choosing a particular device for a job. The programmable op amp gives the engineer more flexibility in fitting an op amp to a particular application.

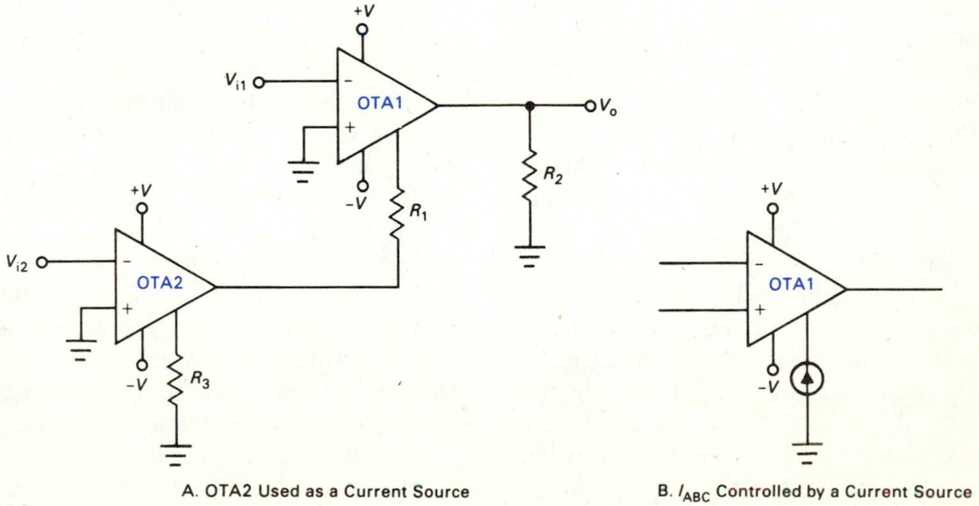

A. OTA2 Used as a Current Source

B. I_{ABC} Controlled by a Current Source

FIGURE 1.52 OTA Biasing with an OTA

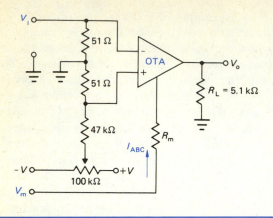

FIGURE 1.53 Amplitude Modulation Using an OTA

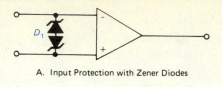

A. Input Protection with Zener Diodes

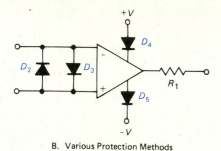

B. Various Protection Methods

FIGURE 1.54 Op Amp Protection Methods

The programmable op amp has a special terminal similar to the I_{ABC} terminal on the transconductance amp. The control terminal on the programmable op amp is called the *set terminal*. Current flow from this terminal can change an op amp's slew rate, gain-bandwidth product, input offset current and voltage, and noise.

PROTECTING OP AMPS

Operational amplifiers, like other ICs, are susceptible to damage from too much voltage in the wrong places. Often, technicians make mistakes when breadboarding (constructing prototypes of) circuits with op amps. In these cases, damage may result to a device that is expensive or difficult to obtain. It is a good idea to try to protect the op amp from damage. Several commonly used methods of op amp protection are shown in Figure 1.54. Remember that although we will be demonstrating these techniques on op amps, they can be used with any linear IC.

Most op amps will be destroyed if too high a voltage is placed across the input terminals. Figure 1.54A shows one input protection method using zener diodes connected back to back. The voltage at the input can go no higher than the zener voltage V_Z plus 0.6 V, regardless of the polarity. An alternate input protection method is shown in Figure 1.54B. This method uses normal semiconductor diodes D_2 and D_3 and operates in a similar fashion to the zener diodes. The difference is that the normal silicon diodes clamp the input voltage to ±0.6 V. The input can go no higher than this value.

Most op amps have built-in, short-circuit protection and therefore need no additional short-circuit protection. For example, the 741 op amp is short-circuit-protected, limiting output current to about 25 mA. For devices that do not have such protection, however, a 200 Ω series output resistor (R_1) will limit any short-circuit current that may be drawn. The device will then be protected at the output.

One of the easiest mistakes to make in breadboarding is to cross the positive and negative supplies. Most ICs, including the op amp, will not tolerate this condition. The device will be destroyed. Protection for this kind of fault is found in diodes D_4

and D_5. Diode D_4 protects against a negative voltage being connected to the positive supply terminal. If a negative voltage is connected to the positive supply terminal, diode D_4 will be reverse-biased and no current will flow. This diode will protect the op amp from damaging reverse voltages. The negative supply is likewise protected from positive voltages by reverse-biasing D_5.

TROUBLESHOOTING OP AMPS

Once a problem has been isolated to an IC op amp, the op amp itself must be checked. Fault analysis of any kind can be broken down into three categories: no output, low output, or distorted output. A simple check of the op amp's output determines in which of these general areas a malfunction is occurring.

No Output

A no-output condition may be caused by any of the following faults:

- Lack of power supply—check the op amp terminals for proper supply voltages.
- Lack of input voltage—check for the proper input signal or voltage.
- Saturated op amp—check for undesired DC inputs that may be causing saturation, check for a shorted input resistor or an open feedback resistor, and check for proper grounding of components and terminals.
- Malfunctioning op amp—take out the op amp and check it in an inverting amp configuration or replace it. (The inverting mode of operation ensures that the op amp is working in order to give an inverted output. A short or open in the noninverting mode may cause it to appear to be working poorly.)

Low Output

A low-output condition may be caused by any of the following faults:

- Low power supply—check the op amp terminals for proper supply voltages.
- Low input voltage—check the op amp input for proper signal or voltage levels.
- Change in component values—check the input and feedback resistances for changes in resistance.
- Malfunctioning op amp—replace it with a good unit.

Distorted Output

A distorted-output condition may be caused by any of the following faults:

- Distortion in the power supply or low power supply—check the op amp terminals for proper supply voltages.
- Distorted input—check the input to the op amp for distortion.
- Malfunctioning op amp—replace the op amp with a good unit.
- Operation outside of published specifications—check the slew rate, gain bandwidth, and so on.

Troubleshooting Breadboarded Circuits

As a technician, you may either breadboard circuits of your own or be required to check breadboarded circuits. In addition to observing the troubleshooting hints just mentioned, you should check closely your circuit's wiring. Both the beginner and the expert often do not wire a circuit correctly. Make sure that the IC pinout from the data sheet is observed.

CONCLUSION

We only have scratched the surface of op amp applications in this chapter. But we have seen that the IC op amp has applications perhaps undreamed of by those who developed it in the 1970s. Since this device is a major analog circuit building block, you should have a firm grasp on its operation.

■ QUESTIONS..

1. The instrumentation amplifier is a _____ amplifier with each input connected to a voltage follower. It amplifies difference-mode signals while _____ common-mode noise.

2. In the log amp, the output is proportional to the _____ of the input. The mathematical operation of _____ can be accomplished with log amps.

3. A filter that passes everything above a specified cutoff frequency is called a _____ filter. The roll-off of a first-order filter is _____ per decade.

4. The CDA is compatible with _____ logic since it can operate on one 5 V supply.

5. In the OTA, the output _____ is proportional to the input voltage.

6. What is the major advantage of a comparator with hysteresis over one without it?

7. Describe the function of a gyrator.

8. Describe the need for and construction of a CDA.

9. What is the phase relationship between input and output in the circuit shown in Figure 1.37A?

10. Compare the characteristics and applications of op amps, CDAs, and OTAs.

■ PROBLEMS..

1. Suppose the instrumentation amp in Figure 1.2 has $R_1 = R_2 = 45$ kΩ, $R_3 = R_4 = 10$ kΩ, and $R_5 = R_6 = 10$ kΩ, with R_A adjusted to 10 kΩ. With a differential input of 1 V, what is the output voltage?

2. The instrumentation amplifier shown in Figure 1.2 has the following component values: $R_1 = R_2 = R_3 = R_4 = 10$ kΩ, $R_5 = R_6 = 50$ kΩ, and $R_A = 1$ kΩ. With the top input (V_1) at +0.1 V and the bottom input (V_2) at +0.025 V, what will the output voltage be?

3. Suppose the instrumentation amplifier in Problem 2 has a variable resistance for R_A, with all the other resistances as in Problem 2. What must the resistance of R_A be for an output voltage of 20 V and V_i ($V_2 - V_1$) = 0.05 V? What is the voltage gain of this circuit?

4. The instrumentation amplifier shown in Figure 1.3 has the following component values: $R_1 = R_2 = R_3 = R_4 = 10$ kΩ, $R_5 = R_6 = 20$ kΩ, and $R_A = 2$ kΩ. Resistor R_9 is adjusted to 121.75 Ω. $A_V = 22$. What will the output voltage be?

5. In the instrumentation amplifier circuit in Figure 1.3, $R_3 = R_4 = R_5 = R_6 = 1$ MΩ. $R_1 = R_2 = 10$ kΩ. $R_A = 1$ kΩ. What is the gain of this stage?

a. Calculate the output voltage with a 55 mV bridge potential at A and B.

b. For $R_7 = R_9 = 1000$ Ω and $R_8 = R_{10} = 500$ Ω, what change in R_9 resistance will give a 55 mV potential difference between A and B?

6. For the low-pass filter of Figure 1.12, with $C_1 = 0.2$ μF, $C_2 = 0.047$ μF, and $R_1 = R_2 = 5$ kΩ, find the filter's cutoff frequency.

7. For the capacitance multiplier in Figure 1.14, with $C_1 = 2$ μF, $R_1 = 10$ MΩ, and $R_3 = 3$ kΩ, find the effective multiplied capacitance.

8. For the gyrator in Figure 1.13A, with $R_1 = R_3 = 2000$ Ω, $R_2 = 150,000$ Ω, and $C = 2$ μF, find the inductance of the gyrator (in henrys).

9. The LED overvoltage indicator in Figure 1.18 has $R_F = 50$ kΩ, $R_i = 10$ kΩ, and zener diodes of 6.2 V. What input voltage will cause the overvoltage indicators to switch?

10. For the voltage regulator in Figure 1.19, with $R_F = 15$ kΩ, $R_i = 2$ kΩ, $R_L = 2$ kΩ, and a zener diode of 6.2 V, find the output voltage and current drawn by the load.

11. The amplifier in Figure 1.21 has an input voltage of 10 V, with an input resistance of 2 kΩ. What is the current drawn through the load?

12. We wish to use the circuit shown in Figure 1.23 to measure two voltage ranges, 0–5 V and 0–15 V. What resistance do we need for R_1 and R_2 to give full-scale deflection for 15 and 5 V?

13. For the low-voltage op amp in Figure 1.28, with $R_2 = 15$ kΩ, find the voltage at which the LED turns off.

14. Assume the following resistance and voltage values in Figure 1.35: $R_i = 100$ kΩ, $R_F = 470$ kΩ, $R_1 = 1$ MΩ, and $+V = +20$ V. Calculate (a) A_V, (b) $V_{o(DC)}$, (c) $V_{o(AC)}$ with a 2 V peak input sine wave, and (d) I_b.

15. With the supply voltage shown in Figure 1.36, a DC voltage of 7.5 V is measured at the output of the CDA.

    a. What is the ratio of R_3 to R_1 that produces this voltage?

    b. With this bias point, what is the maximum peak voltage that can be applied at the input without distorting the output? *Hint*: How far can the output voltage go in a positive direction?

16. Assume the following resistance and voltage values in Figure 1.37: $R_1 = 470$ kΩ, $R_2 = 50$ kΩ, $R_3 = 220$ kΩ, $+V = 5$ V, and the input sine wave is $+0.1$ V peak. Calculate (a) A_V, (b) $V_{o(DC)}$, (c) $V_{o(AC)}$, and (d) I_b.

17. The op amp in Figure 1.30 is driving a transistor so that a 3 V potential is found at the emitter. If the emitter resistance is 100 Ω, how much current is the transistor drawing? Assume $I_C = I_E$. If the transistor beta is 50, how much output current will the op amp need to drive the transistor to get a 3 V potential at the emitter?

18. The lamp driver in Figure 1.31 has an op amp output of +10 V. How much current is the op amp providing to drive the transistor? What is the voltage drop across the resistor R_1? If the

DC beta of the transistor is 50, what is the approximate collector current?

19. For the window comparator in Figure 1.42, with resistances $R_1 = 250$ kΩ, $R_2 = 125$ kΩ, $R_3 = 100$ kΩ, and $R_4 = 250$ kΩ, and with supply and reference voltages of +15 V, find the window voltages.

20. For the CDA biasing circuit in Figure 1.34, with resistance values $R_F = 680$ kΩ and $R_1 = 2$ MΩ, and with a +10 V supply, find the bias current and output voltage.

21. Consider the circuit in Figure 1.36, with the lowest input frequency being 10 Hz. Find the input capacitor size needed in this circuit so that the resistance is $^1/_{10}$ of the reactance at the lowest frequency.

22. Use the circuit in Figure 1.38A. Find the bias current, DC output voltage, AC voltage gain, AC output voltage, and the capacitance needed for a low-frequency input of 30 Hz. Use $R_1 = 1$ MΩ, $R_2 = 150$ kΩ, $R_3 = 470$ kΩ, and $\pm V = +10$ V. Place voltage data in a synchrogram, as in Figure 1.38B.

23. For the inverting comparator shown in Figure 1.41A, with $+V = 10$ V, $R_1 = 1.2$ MΩ, $R_2 = 2$ MΩ, and $R_3 = 5$ MΩ, find the upper and lower trip voltages.

24. For the noninverting comparator shown in Figure 1.41B, with $+V = 10$ V, $R_1 = 2$ MΩ, $R_2 = 1.6$ MΩ, and $R_3 = 3$ MΩ, find the upper and lower trip voltages.

25. For the op amp circuit in Figure 1.17, with $R_{L1} = 250$ Ω, $R_{L2} = 500$ Ω, and $V_{i(DC)} = +10$ V, find the load currents for each load resistance.

26. For the CDA oscillator in Figure 1.44, with $C_1 = 0.01$ μF, find the frequency of oscillation.

27. Consider the inverting CDA in Figure 1.35, with $R_i = 150$ kΩ, $R_F = 2$ MΩ, $R_1 = 5$ MΩ, an input voltage of ± 0.2 V$_{pk}$, and $+V = 12$ V. Find the bias current, DC output voltage, AC gain, and AC output voltage.

28. The comparator in Figure 1.40A has the following resistances: $R_1 = 1.2$ MΩ, $R_2 = 2.5$ MΩ, and $R_3 = 10$ MΩ. Assume $V_o = +10$ V. Find (a) upper and lower trip voltages and (b) upper and lower trip currents.

29. The comparator in Figure 1.40B has the following resistances: $R_1 = 1.0$ MΩ, $R_2 = 1.5$ MΩ, and $R_3 = 1.5$ MΩ. Assume $V_o = +10$ V. Find (a) upper and lower trip voltages and (b) upper and lower trip currents.

30. The OTA in Figure 1.48 has a supply voltage of ± 10 V and an I_{ABC} of 100 μA. Find (a) the OTA transconductance, (b) the output voltage with an input of 10 mV and a load resistor of 1000 Ω, and (c) the I_{ABC} needed to give a 1 V potential at the output with a 120 mV input and a load resistance of 1000 Ω.

31. For a CA3080 OTA, with an I_{ABC} of 55 μA and a supply voltage of 10 V, find the transconductance.

32. For a CA3080, with an I_{ABC} of 100 μA, a load resistor of 15,000 Ω, an input voltage of 0.15 V_{pk}, and a supply voltage of 12 V, find the peak output voltage.

33. For the circuit in Figure 1.50A, with a negative supply voltage of 12 V and an R_{ABC} of 150 kΩ, find the I_{ABC} (in microamperes).

34. For the OTA circuit in Figure 1.51A, with a supply voltage of ± 12 V, a control voltage of 15 V, and an R_{ABC} of 200 kΩ, find the amplifier bias current.

CHAPTER 2

Linear Integrated Circuits for Industrial Applications

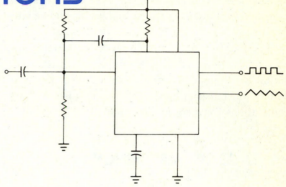

OBJECTIVES

On completion of this chapter, you should be able to:

- Describe the function of the voltage-to-frequency converter (VFC) and calculate its output frequency.
- Describe the function of the phase-locked loop (PLL).
- Define and calculate PLL capture and lock ranges.
- Describe the function of the frequency-to-voltage converter (FVC) and calculate its output voltage.
- Calculate the resolution of a digital system.
- Describe the analog-to-digital (A/D) conversion process, and describe the operation of two analog-to-digital converters (ADCs).
- Describe the digital-to-analog (D/A) conversion process, and describe the operation of two digital-to-analog converters (DACs).
- Describe the operation of the analog switch.
- Define current sourcing and current sinking.

INTRODUCTION

In Chapter 1, we discussed the op amp and some of its applications. As we have seen, few linear ICs can match the op amp for versatility in design. In addition to being versatile, it is inexpensive and reliable. There are some applications, however, that are difficult to achieve with op amps. Consequently, IC designers have developed more specialized ICs, some of which we will discuss here. In this chapter, we will examine voltage-controlled oscillators, phase-locked loops, voltage-to-frequency converters, frequency-to-voltage converters, analog-to-digital converters, and digital-to-analog converters. These devices are used widely in industrial applications. You also will see applications for them in later chapters.

VOLTAGE-TO-FREQUENCY (V/F) CONVERSION

The *voltage-to-frequency converter* (VFC) is a device that changes analog voltages or currents to an AC frequency. When operated within its design parameters, the output signal frequency is proportional to the input voltage or current. The input is a constant or varying DC potential. The output frequency continuously tracks the input signal, responding directly to the input voltage. A block diagram shows the concept of the VFC, with a variable DC input voltage and a variable output frequency.

A common VFC is the NE/SE566 IC made by Signetics. The 566 is a voltage-controlled oscillator (VCO) with two buffered outputs. A block diagram of the 566 is shown in Figure 2.1. Note that the two buffered outputs, pins 3 and 4, provide both a square wave and a triangle wave. An external resistor (R_1) and a capacitor (C_1) determine the oscillator's frequency. The input voltage at pin 5 also will determine, with the resistor and capacitor, the frequency of oscillation. A change in the input voltage will vary the output frequency over a ten-to-one range, with a frequency that is linearly proportional to the voltage.

The block diagram in Figure 2.1 can show us the basic operation of this VCO. The capacitor C_1 charges through the current source and through resistor R_1. The current source provides a constant

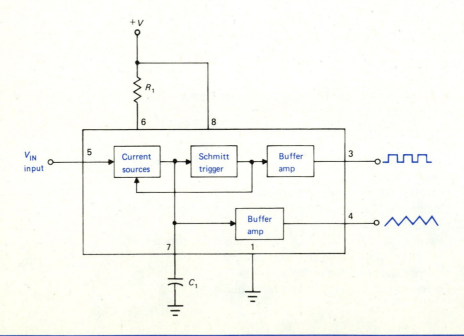

FIGURE 2.1 Voltage-Controlled Oscillator (VCO) Functional Block Diagram

charging current. How fast the capacitor charges depends on the resistance of R_1, the capacitance of C_1, and the input voltage controlling the current source. With a constant capacitance and resistance, an increase in the voltage to the current source will increase the rate at which the capacitor charges. When the capacitor reaches a set point, it starts to discharge at a linear rate through R_1 and controlled by the current source. A capacitor charging and discharging at a constant rate displays a triangle wave voltage signal across itself. A voltage follower–buffer amp will provide a triangle wave output to pin 4 of the 566 VCO. The buffer amp keeps the output from being loaded down and changing the oscillator frequency and waveshape. The triangle wave is also connected to the input of a Schmitt trigger (a comparator with hysteresis). The output of the comparator is a square wave. The square wave is sent to a buffer amp for isolation from loading effects in the output.

A circuit schematic diagram of a 566 VCO is shown in Figure 2.2. Resistors R_2 and R_3 form a voltage divider to provide a bias for pin 5, the control voltage input. The voltage on this pin must be between the values $+V$ and three-quarters of $+V$ for the device to operate properly. For example, if the $+V$ is $+10$ V, the voltage at pin 5 must be between $+10$ V and $+7.5$ V. This feature is a design requirement for the 566 IC. As mentioned before, the resistor R_1 and capacitor C_1 help set the frequency out of the device. Another design requirement for this VCO relates to the size of R_1. Resistor R_1 must be between 2 kΩ and 20 kΩ or the device will not perform as specified.

The output frequency at pins 3 and 4 is approximated by the following formula:

$$f_o = \frac{2[(+V)-(V_i)]}{(R_1)(C_1)(+V)} \tag{2.1}$$

With the resistors as shown in Figure 2.2 and with a $+12$ V supply, the approximate output frequency is

$$f_o = \frac{2[(+12\text{ V})-(+10.43\text{ V})]}{(10\text{ k}\Omega)(0.047\,\mu\text{F})(+12\text{ V})}$$

$$= \frac{3.14}{0.00564} = 557\text{ Hz}$$

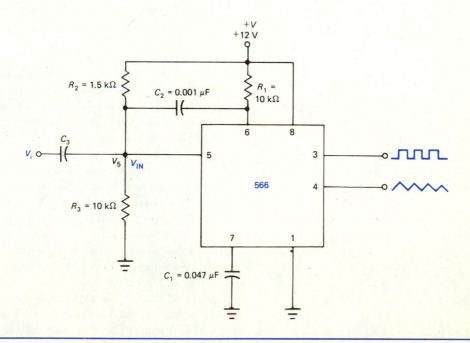

FIGURE 2.2 Basic VCO Circuit

The voltage V_5 is calculated by the voltage divider formula to solve for the drop across R_3:

$$V_5 = \left(\frac{R_3}{R_2 + R_3}\right)(+V) \qquad (2.2)$$

For the circuit of Figure 2.2, we obtain

$$V_5 = \left(\frac{10,000}{1500 + 10,000}\right)(+12 \text{ V})$$

$$= \left(\frac{10,000}{11,500}\right)(+12 \text{ V}) = +10.43 \text{ V}$$

The VCO output frequencies at pins 3 and 4 should be approximately 557 Hz. Let us see what happens to the output frequency if the resistance of R_2 increases to 2 kΩ:

$$f_o = \frac{2\,[(+12 \text{ V}) - (+10 \text{ V})]}{(10 \text{ k}\Omega)\,(0.047\,\mu\text{F})\,(+12 \text{ V})}$$

$$= \frac{4}{0.00564} = 709 \text{ Hz}$$

Note that when the resistor R_2 increases to 2 kΩ, the voltage at V_i decreases from 10.43 to 10.0 V. This decrease in voltage causes an increase in the output frequency from 557 to 709 Hz.

• EXAMPLE 2.1

Given the 566 VCO in Figure 2.2, with $R_1 = 12$ kΩ, $C_1 = 0.027$ μF, $R_2 = 1$ kΩ, $R_3 = 9$ kΩ, and $+V = 12$ V, find the voltage at pin 5 with respect to (a) ground and (b) f_o.

Solution

(a) The voltage at pin 5 can be found by using the voltage divider equation (2.2):

$$V_5 = \left(\frac{R_3}{R_3 + R_2}\right)(+V)$$

$$= \left(\frac{9 \text{ k}\Omega}{9 \text{ k}\Omega + 1 \text{ k}\Omega}\right)(12 \text{ V})$$

$$= 10.8 \text{ V}$$

(b) The frequency of the oscillator can be found by substituting the appropriate values into Equation 2.1:

$$f_o = \frac{2\,[(+V) - (V_i)]}{(R_1)\,(C_1)\,(+V)}$$

$$= \frac{2\,(12 \text{ V} - 10.8 \text{ V})}{(12 \text{ k}\Omega)\,(0.027\,\mu\text{F})\,(12 \text{ V})}$$

$$= \frac{2.4}{0.003888} = 617.3 \text{ Hz}$$

From this discussion, we can see one way to carry information about a sensor that changes resistance. A *sensor* (discussed in a later chapter) is a device that undergoes a change in a characteristic when an environmental parameter changes. For example, some resistors change their resistance as a function of temperature. We can substitute a temperature-sensitive resistance for either R_2 or R_3 in the VCO of Figure 2.2. The output frequency will then be directly related to the temperature changes around the resistor.

In other applications, a voltage is applied to the input through capacitor C_3. The amplitude changes applied to the capacitor C_3 will be converted to frequency changes at the output. For a value of 1.5 kΩ for the resistor R_2, the output frequency will be 557 Hz with no input voltage. This frequency is called the *center* or *rest frequency*.

If we connect a sine wave generator to the input with a frequency of 1 kHz, the output frequency will change at a rate of 1 kHz, as shown in Figure 2.3. Note that as the input signal's amplitude increases, the output square wave frequency decreases. When the input voltage goes less positive, the output frequency increases. We can apply Equation 2.1 to solve for the output frequency at the input voltage peaks. With a ±1 V peak input signal, we know that the voltage at pin 5 will be +10.43 V plus 1 V and +10.43 V minus 1 V. The input voltage at V_i will be +11.43 V and +9.43 V at the input negative and positive peaks, respectively.

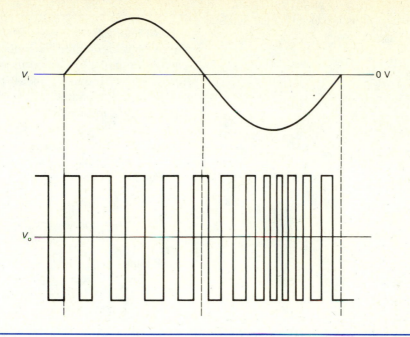

FIGURE 2.3 VCO Basic Input and Output Voltage Waveforms

At a V_i of +9.43 V, the frequency is

$$f_o = \frac{2\left[(+V) - (V_i)\right]}{(R_1)(C_1)(+V)}$$

$$= \frac{2\left[(+12\ V) - (+9.43\ V)\right]}{(10\ k\Omega)\ (0.047\ \mu F)\ (+12\ V)}$$

$$= \frac{5.14}{0.00564} = 911\ Hz$$

At a V_i of +11.43 V, the frequency is

$$f_o = \frac{2\left[(+V) - (V_i)\right]}{(R_1)(C_1)(+V)}$$

$$= \frac{2\left[(+12\ V) - (+11.43\ V)\right]}{(10\ k\Omega)\ (0.047\ \mu F)\ (+12\ V)}$$

$$= \frac{1.14}{0.00564} = 202\ Hz$$

We can see from this illustration that when the input voltage goes from +10.43 V to +11.43 V, the output frequency goes to 202 Hz from the rest frequency (557 Hz). When the input voltage goes down to +9.43 V, the output frequency goes to 911 Hz from the rest frequency. This process is called *frequency modulation* (FM).

We may need to look at our circuit in a different way for some applications. For instance, we may need to know how much input voltage will produce a given range in output frequencies. If we label the high output frequency f_H and the low frequency f_L, we can say that the range (Δf) is the difference between them, or

$$\Delta f = f_H - f_L \qquad\qquad \textbf{(2.3)}$$

Using Equation 2.1, we know that the minimum voltage in ($V_{5(min)}$) will produce f_H, the highest frequency out:

$$f_H = \frac{2\left[(+V) - \left(V_{5(min)}\right)\right]}{(R_1)(C_1)\ (+V)}$$

The lowest frequency, f_L, is produced by the maximum voltage in ($V_{5(max)}$):

$$f_L = \frac{2\left[(+V) - (V_{5(max)})\right]}{(R_1)(C_1)(+V)}$$

By substitution, we can arrive at an equation for Δf, as follows:

$$\Delta f = \left\{ \frac{2\left[(+V) - (V_{5(min)})\right]}{(R_1)(C_1)(+V)} \right\}$$
$$- \left\{ \frac{2\left[(+V) - (V_{5(max)})\right]}{(R_1)(C_1)(+V)} \right\} \qquad \textbf{(2.4)}$$

Simplifying, we get

$$\Delta f = \frac{2\left(V_{5(max)} - V_{5(min)}\right)}{(R_1)(C_1)(+V)} \qquad \textbf{(2.5)}$$

If we think of $V_{5(max)} - V_{5(min)}$ as ΔV, we can solve for ΔV:

$$\Delta V = \frac{(\Delta f)(R_1)(C_1)(+V)}{2} \qquad \textbf{(2.6)}$$

As an example, let us suppose we want to know what change in input voltage (ΔV) will give a 300 Hz change in output frequency (Δf). Using the preceding equation, we get

$$\Delta V = \frac{(300\ \text{Hz})(10\ \text{k}\Omega)(0.047\ \mu\text{F})(+12\ \text{V})}{2}$$

$$= \frac{1.692}{2} = 0.846\ \text{V}$$

According to our calculation, we can get a 300 Hz change in the output frequency with a 0.846 V change in the input voltage.

• EXAMPLE 2.2

Given the values for the 566 in Example 2.1 with C_1 changed to 0.01 µF, find (a) the size of R_1 needed for a center frequency of 10 kHz and (b) the frequency deviation from the center frequency with a 50 mV peak input signal.

Solution

(a) First, we must solve Equation 2.1 for R_1. Equation 2.1 is

$$f_o = \frac{2\left[(+V) - (V_5)\right]}{(R_1)(C_1)(+V)}$$

and thus, we have

$$R_1 = \frac{2\left[(+V) - (V_5)\right]}{(f_o)(C_1)(+V)} \qquad \textbf{(2.1a)}$$

Next, substituting into Equation 2.1a, we get

$$R_1 = \frac{2\ (12\ \text{V} - 10.8\ \text{V})}{(10\ \text{kHz})(0.01\ \mu\text{F})(+12\ \text{V})}$$

$$= 2000\ \Omega$$

(b) The change in frequency can be found by substituting the appropriate values into Equation 2.5:

$$\Delta f = \frac{2\left(V_{5(max)} - V_{5(min)}\right)}{(R_1)(C_1)(+V)}$$

$$= \frac{2\ (0.1\ \text{V})}{(2000\ \Omega)(0.01\ \mu\text{F})(+12\ \text{V})}$$

$$= 833\ \text{Hz}$$

So, f_o is 10,000 Hz ±416.7 Hz, or ±4.14% deviation.

• EXAMPLE 2.3

Given the 566 VCO in Figure 2.2, with $R_1 = 2000\ \Omega$, $R_2 = 1000\ \Omega$, $R_3 = 9000\ \Omega$, and $C_1 = 0.01\ \mu F$, find the input voltage change (ΔV) for a $\pm 15\%$ deviation from the rest frequency.

Solution

With a rest frequency of 10,000 Hz, a deviation of $\pm 15\%$ will be a deviation of ± 1500 Hz, or $\Delta f = 3000$ Hz. Substituting the appropriate values into Equation 2.6, we get

$$\Delta V = \frac{(\Delta f)\,(R_1)\,(C_1)\,(+V)}{2}$$

$$= \frac{(3000\ \text{Hz})\,(2000\ \text{Hz})\,(0.01\ \mu F)\,(12\ \text{V})}{2}$$

$$= 0.36\ \text{V}_{(p-p)} \quad \text{or} \quad 0.18\ \text{V}_{pk}$$

In some cases, the output of the 566 must be compatible with digital transistor-transistor logic (TTL). The circuit shown in Figure 2.4 is preferable for interfacing with TTL. TTL requires a current sink of more than 1 mA. This circuit provides current-sinking capability to 2 mA when the switch is down. When the switch is up, a saturated transistor is inserted between the 566 and the logic circuitry. This circuit is used for TTL circuits that need a fast fall time (<50 ns) and a large current-sinking capability.

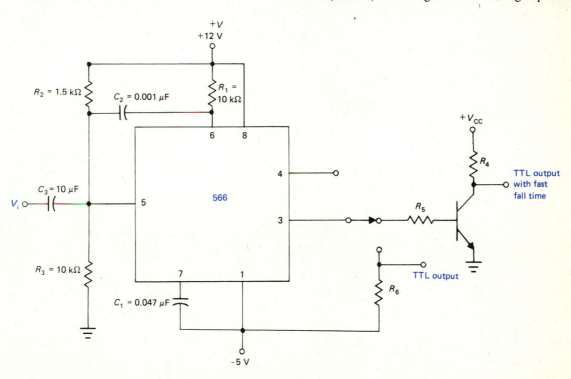

FIGURE 2.4   VCO Circuit with TTL-Compatible Output

VCOs, such as the NE566, are used in the FM modulation process because of the output frequency is a very linear function of the input voltage amplitude. Other applications include tone generators, signal generators, pulse generators, and function generators. An example of a negative ramp pulse generator is shown in Figure 2.5. In the ramp generator portion of this circuit, the transistor is turned on by the output of pin 3. The transistor quickly discharges the capacitor at the end of its charging cycle. The transistor also charges the timing capacitor (C_1) at the end of the discharge period. Because the circuits are reset quickly, the temperature stability of the ramp generator is excellent. The period t is $\frac{1}{2}f_0$, where f_0 is the 566 free-running frequency in normal operation. The period of the output waveform, therefore, is given by the following equation:

$$t = \frac{(R_t)(C_1)(+V)}{5(+V - V_c)} \qquad (2.7)$$

where

t = period of output waveforms

R_t = resistance connected between pin 6 and $+V$

PHASE-LOCKED LOOPS

A *phase-locked loop* (PLL) is a group of electronic circuits arranged in a feedback system, as shown in block diagram form in Figure 2.6. The PLL is composed of three basic circuits: the phase comparator or detector, the low-pass filter, and the VCO. The VCO is the same circuit that we discussed previously, a free-running oscillator that has the capability to lock or synchronize with an incoming frequency. The

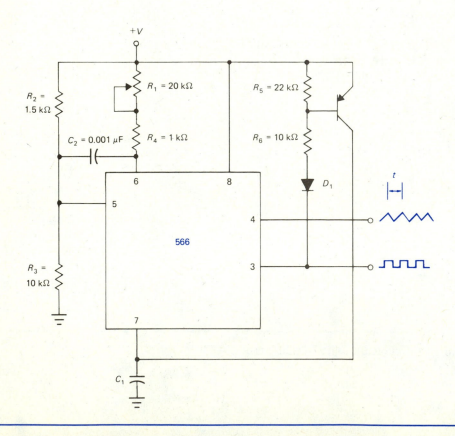

FIGURE 2.5 VCO Negative Ramp Pulse Generator

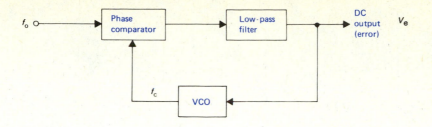

FIGURE 2.6   Phase-Locked Loop (PLL) Functional Block Diagram

VCO center or rest frequency is determined by an external timing resistor and capacitor and by the input voltage. The capacitor and resistor are normally fixed in the PLL, but the input voltage is variable.

The VCO output is connected to the input of the phase comparator. The *phase comparator*, as its name suggests, compares the phase of the VCO signal (f_c) to the phase of the input signal (f_o). The output of the phase comparator is an error correction voltage. The size of this voltage potential is a function of the phase and frequency differences between the two input signals. The error correction voltage is then filtered by the low-pass filter. The filtered output of the low-pass filter is the input to the VCO. A change in the input phase indicates that the input frequency is changing. The phase comparator output voltage increases or decreases just enough to keep the VCO frequency the same as the input frequency. The average voltage applied to the VCO is a function of the input signal frequency.

PLL Capture and Lock Ranges

The PLL has three basic states: the free-running state, the capture state, and the locked or tracking state. In the *free-running state*, the VCO is some distance away from the input frequency f_c.

In this state, the input frequency is too far away from the VCO frequency to have any effect. Changes in the incoming frequency do not affect the VCO frequency; the VCO frequency does not change. If the input frequency is brought closer to the VCO frequency, a point is reached where the low-pass

filter lets through enough voltage to have an effect on the VCO. Remember that the phase comparator puts out a voltage that is proportional to the difference between the two input frequencies, f_o and f_c. The error voltage (V_e) is now large enough to cause the VCO to move toward the input frequency f_c.

If we look at the input to the VCO at this time, we see a waveform such as that shown in Figure 2.7. As the capture process starts, a small sine wave appears. This sine wave represents the *difference frequency* or *beat frequency* between the VCO and f_c. Note that the negative half of the waveform is slightly larger than the positive half. This condition is the DC component of the beat frequency. This DC component is what actually drives the VCO toward the lock state. Every successive cycle causes the VCO to move closer to the locked state, where the VCO and input frequency will be equal. Note that as the VCO moves closer to the input signal, the beat frequency decreases. This decrease allows the low-pass filter to pass a larger DC voltage to the VCO. At the same time, the closer the VCO moves toward the locked state, the longer it lingers near that state. This behavior extends the negative part of the sine wave and reduces the positive alternation. This feedback process continues until the VCO locks on the incoming frequency and the beat or difference frequency is zero.

This positive feedback behavior makes the VCO *snap* into lock when the input frequency is close enough. We can define *close enough* by the term "capture range." The *capture range* of a PLL is the frequency range centered around the free-running

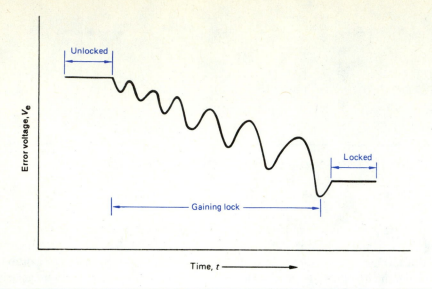

FIGURE 2.7 Error Voltage versus Time during PLL Capture

VCO, over which the device can get a lock on the input signal. The low-pass filter plays an important role in the capturing process. Note that the low-pass filter is made up of an internal resistor and an external capacitor. If the input signal frequency is too far away from the VCO, the beat or difference frequency will be too high to pass through the filter. The low-pass filter will attenuate it. The input frequency is said to be out of the capture range.

Once the VCO has locked on the input frequency, the low-pass filter does not restrict the PLL. The VCO is free to track a signal past the capture range. The low-pass filter does, however, affect the speed at which the VCO can follow the input frequency. If the input frequency changes too fast, the VCO will stop following the input; it will become unlocked. As you might expect, there is a limit to the distance that the VCO can track the input frequency. This limit is called the lock range. The *lock range* is the range of input frequencies over which the PLL will remain in lock. It is always larger than the capture range.

We can get a better picture of the capture and lock ranges of the PLL if we examine the diagram in Figure 2.8. This diagram is a plot of the DC volt-

age input to the VCO versus the input frequency. The input frequency is assumed to be a sine wave, the frequency of which is being swept over a wide frequency range. In Figure 2.8A, the input frequency starts low and is gradually increased. The PLL is not in lock at this time. Note that the PLL does not respond to the input signal until it reaches frequency f_1. This frequency is the lower edge of the capture range. At f_1, the PLL locks on the input signal, causing the error voltage to jump in a positive direction. This positive voltage is the input voltage the VCO needs to bring itself into lock with the input signal. From f_1, the error voltage increases in a linear fashion as the input frequency is increased. If we were to observe the VCO, we would see it track along with the input frequency. The VCO will track until the upper edge of the lock range is reached at f_2. At this point, the PLL loses lock, and the error voltage goes back to zero. When the PLL loses lock, we also should see the VCO go back to its free-running frequency.

We will now sweep the input frequency down, observing the effect on the error voltage. The result is pictured in Figure 2.8B. Note that the same cycle repeats itself, only at different frequencies. The PLL

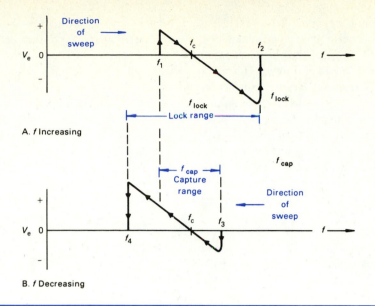

FIGURE 2.8 PLL Capture and Lock Ranges

recaptures the input frequency at f_3, which is the upper edge of the capture range. Note the change in the error voltage as we decrease the input frequency. The error voltage increases, decreasing the VCO and keeping it in lock. When f_4 is reached, the VCO loses lock on the input frequency, and the VCO goes back to its free-running frequency. Note that the capture range is between f_1 and f_3. The lock range is between f_2 and f_4.

Further examination of Figure 2.8 reveals several things. First, the PLL is selective in the frequencies it can lock on. It will capture only frequencies inside the capture range of the device. Second, the VCO has a very linear V/F transfer characteristic. This feature is useful when an application demands a linear correspondence between input frequency (AC) and output voltage (DC). The linear V/F transfer characteristic is an important factor when the PLL is used to demodulate FM signals.

The VCO in the PLL performs the same functions as the VCO discussed earlier in this chapter. When the PLL is locked on to a frequency, the VCO input voltage is a function of the input frequency. When the VCO is used in demodulation applications, the VCO V/F transfer characteristic must be linear.

If the output frequency of the VCO is not linear, distortion will result in the demodulation process. How much the VCO frequency changes for a given input DC voltage change is called *conversion gain*. The conversion gain equation is

$$K = \frac{\Delta f}{\Delta V_e} \tag{2.8}$$

where

 K = conversion gain (in hertz per volt)
 Δf = change in VCO frequency
 ΔV_e = change in VCO DC input voltage

The NE565 IC PLL is one of the more popular IC PLLs. It is a general-purpose PLL that can be used up to frequencies of 1 MHz. Compared with other PLLs, it has an exceptionally wide lock range (up to ±60% of the VCO free-running frequency). Its sensitivity, however, is 1 mV, which is lower than that of the other 560 series PLLs. The sensitivity of the device indicates how large an input signal must be before the VCO can lock on and track. Unlike the other members of the 560 series of PLLs, the 565 does not have a zener at the power supply input to keep the input voltages regulated. When using the 565, you should use a well-regulated power supply.

The diagram in Figure 2.9 shows a circuit that uses the 565 to demodulate an FM input. This input may come from a VCO driven by an input sine wave, such as we discussed earlier in this chapter. In this circuit, the average DC output of the phase comparator is directly proportional to the input frequency (f_i). As mentioned before, the VCO can lock on and track over a range equal to ±60% of its free-running frequency with a linearity of 0.5%.

The capacitor C_2 and resistor R_2 make up the low-pass filter. As you can see from the diagram, C_2 is an external capacitor and R_2 is an internal resistor. This internal resistance (in the 565) is about 3600 Ω ± 20%. This configuration of an internal resistor and an external capacitor is typical of the Signetics 560 series PLLs. The capture range depends on the size of this resistor, as we will see later. Note

that R_1 and C_1 are part of the timing network that defines the VCO free-running frequency (f_o). The free-running VCO frequency is defined by the following formula:

$$f_o = \frac{1}{3.7\,(R_1)\,(C_1)} \tag{2.9}$$

where

f_o = VCO free-running frequency (in hertz)

The VCO free-running frequency should be adjusted so that it is in the center of the input frequency range. As in the VCO, there is no limitation on the size of C_1, but R_1 must be between 2 kΩ and 20 kΩ. A good value is about 4 kΩ.

Note that pins 4 and 5 are shorted. Shorting these pins places the VCO output to one of the phase comparator inputs. Pin 6 is the reference for the demodulated output voltage. This pin provides a potential that

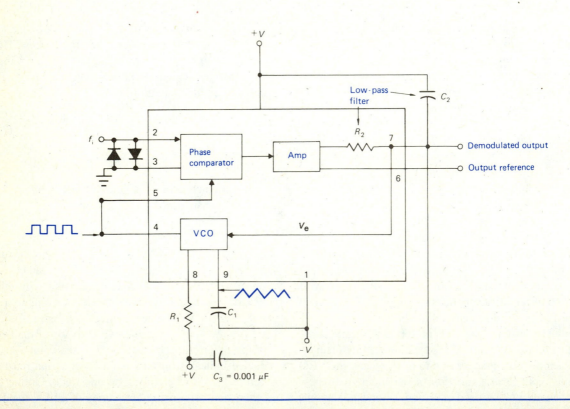

FIGURE 2.9   Basic PLL Circuit

is close to the potential of the demodulated output at pin 7. A resistor connected between pins 6 and 7 will reduce the gain of the output stage without changing the DC voltage at the output significantly. The lock range can then be decreased with little change in the free-running frequency. When a potentiometer is connected across pins 6 and 7, the lock range can be decreased from ±60% of f_o to ±20% of f_o (with a supply voltage of ±6 V). The capacitor C_3, connected between pins 7 and 8, eliminates possible oscillations in the control current source.

The lock range for this circuit is defined as follows:

$$f_L = \pm \frac{8 f_o}{V_{net}} \qquad (2.10)$$

where

f_L = lock range (in hertz)
f_o = free-running VCO frequency (in hertz)
V_{net} = total supply voltage between $+V$ and $-V$ supply voltages

The capture range depends on the lock range and is given by the following equation:

$$f_C = \pm \sqrt{\frac{f_L}{2\pi \, (3600)\,(C_2)}} \qquad (2.11)$$

The schematic diagram in Figure 2.10 shows a PLL with resistor and capacitor values. The free-running frequency is (from Equation 2.9)

$$f_o = \frac{1}{3.7 \, (R_1)\,(C_1)} = \frac{1}{3.7 \, (10{,}000 \,\Omega) \,(0.01 \,\mu F)}$$

$$= 2702.7 \, Hz$$

When not locked on the input frequency, the VCO should be oscillating at about 2703 Hz.

Knowing the free-running frequency, we can now calculate the PLL lock range, using Equation 2.10:

$$f_L = \pm \frac{8 f_o}{V_{net}} = \pm \frac{8 \, (2703 \, Hz)}{12}$$

$$= \pm 1802 \, Hz$$

Finally, we can calculate the capture range of the PLL from Equation 2.11:

$$f_C = \pm \sqrt{\frac{f_L}{2\pi \, (3600)\,(C_2)}}$$

$$= \pm \sqrt{\frac{1802 \, Hz}{2 \, (3.14)\,(3600)\,(10 \,\mu F)}} = 89.25 \, Hz$$

• EXAMPLE 2.4

Given the PLL in Figure 2.10, with $C_1 = 0.02 \,\mu F$, $C_2 = 5 \,\mu F$, $R_1 = 20 \, k\Omega$, and $V_{net} = \pm 12$ V, find the center or rest frequency, lock range, and the capture range.

Solution

Step 1 We need to find the center or rest frequency of the internal VCO first. We can find this frequency by substituting the appropriate values into Equation 2.9:

$$f_o = \frac{1}{3.7 \, (R_1)\,(C_1)} = \frac{1}{3.7 \, (20 \, k\Omega)\,(0.02 \,\mu F)}$$

$$= 676 \, Hz$$

Step 2 After finding the center frequency, we can calculate the lock range, using Equation 2.10:

$$f_L = \pm \frac{8 f_o}{V_{net}} = \pm \frac{8 \, (676 \, Hz)}{24}$$

$$= \pm 225.2 \, Hz$$

Step 3 When the lock range is known, the capture range may be calculated by using Equation 2.11:

$$f_C = \pm \sqrt{\frac{f_L}{2\pi \, (3600)\,(C_2)}}$$

$$= \pm \sqrt{\frac{225.2 \, Hz}{2\pi \, (3600)\,(5 \,\mu F)}} = \pm 44.6 \, Hz$$

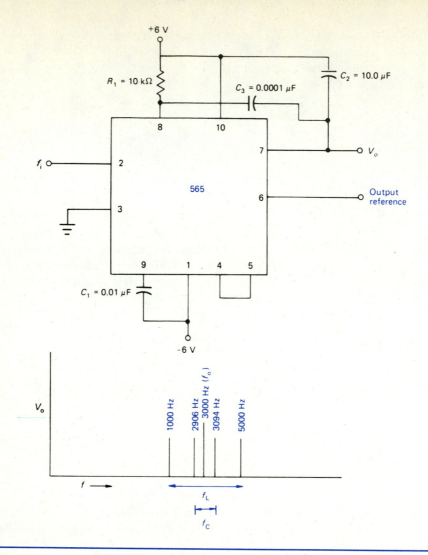

FIGURE 2.10 PLL Circuit Example

The phase comparator in the PLL gives a DC output voltage that is proportional to the difference between the two frequencies. Several different phase comparators are used in the PLL, but the easiest to understand is the Exclusive–OR phase comparator. This phase comparator uses an Exclusive–OR logic gate, in which the output goes high when either of the inputs (but not both) are high (see Figure 2.11A). We see its operation in Figure 2.11B. When the phase difference is 90°, as shown, the output DC or average voltage is $(\frac{1}{2})V_{net}$ (or 2.5 V) since the output duty cycle is 50%. If the output were to shift in phase by 45°, the output would be as shown in Figure 2.11C. Note that the two frequencies are the same but the phase is different, f_o lagging by 90° + 45°, or 135°. Note also that the duty cycle is now 75%, making the output DC or average voltage 3.75 V. The maximum output voltage will occur when f_o lags f_i by 180°, or π radians (rad). The conversion gain is therefore 5 V/180°, or 0.0278 V/°. In radians, the gain is 5 V/π rad, or 1.59 V/rad.

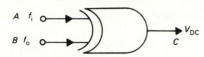

Input A B	Output C
0 0	0
0 1	1
1 0	1
1 1	0

1 = high   0 = low

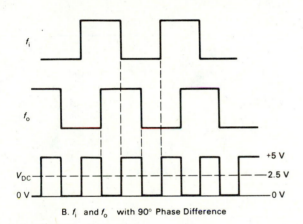

A. Schematic Diagram and Truth Table

B. f_i and f_o with 90° Phase Difference

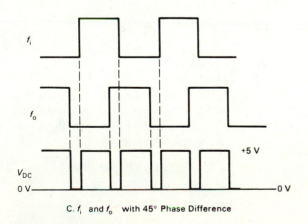

C. f_i and f_o with 45° Phase Difference

FIGURE 2.11 Exclusive–OR Comparator

The output at pin 7 (Figure 2.10) is normally 7/8 of the total supply voltage. This feature is a design criterion for the 566 VCO. The voltage divider input is normally 7/8 of V_{net}. The 566 VCO is identical to the one used in the 565 PLL. With a ±6 V supply, the input to the VCO (and the PLL error voltage output) is

$$V_i = \left(\frac{7}{8}\right)(V_{net}) \qquad (2.12)$$

For our illustration, this voltage is

$$V_i = \left(\frac{7}{8}\right)(12\ V) = 10.5\ V$$

Since the output is referenced to –6 V, the output is 10.5 V above that, or +4.5 V. The PLL DC error voltage output can then be found by using the following equation:

$$V_o = \left(\frac{7}{8}\right)(V_{net}) + (-V) \qquad (2.13)$$

For our illustration, we get

$$V_o = \left(\frac{7}{8}\right)(12\ V) + (-6\ V)$$

$$= 4.5\ V$$

The phase detector sensitivity for the 565 is 0.68 V/rad, or 0.0119 V/°. If $f_i = f_{VCO}$, the output of the phase detector is +4.5 V (with ±6 V supplies). The VCO sensitivity (K_o), however, depends on f_o. The data sheet for the 565 at the back of the text shows a VCO sensitivity of 6600 Hz/V at 10 kHz. The VCO sensitivity may be calculated for any f_o by using the following equation:

$$K_o = \frac{50 f_o}{2\pi (V_{net})} \qquad (2.14)$$

For a VCO free-running frequency of 3000 Hz, the K_o is

$$K_o = \frac{50 f_o}{2\pi (V_{net})} = \frac{50 (3000\ Hz)}{2\pi (12\ V)}$$

$$= 1989\ Hz/V$$

If we increase the f_i from 3000 to 4989 Hz, the change in the error voltage is 1 V. How much will the frequency have to change to produce an increase of 0.5 V? The answer is found by using the following equation:

$$\Delta f = (K_o)(\Delta V) \tag{2.15}$$

For our illustration, we get

$$\Delta f = (1989 \text{ Hz/V})(0.5 \text{ V}) = 995 \text{ Hz}$$

A decrease in frequency of 995 Hz will produce a 0.5 V increase in voltage. For example, a decrease from 3000 to 2005 Hz will produce a change of 0.5 V, from 4.5 to 5.0 V.

• EXAMPLE 2.5
Given a PLL with $f_o = 5000$ Hz and $V_{net} = 10$ V, find (a) the sensitivity of the VCO (in hertz per volt) and (b) the frequency change needed to produce a 1 V change in error voltage.

Solution
(a) For a VCO free-running frequency of 5000 Hz, K_o is

$$K_o = \frac{50 f_o}{2\pi (V_{net})} = \frac{50 (5000 \text{ Hz})}{2\pi (10 \text{ V})}$$

$$= 3979 \text{ Hz/V}$$

(b) The sensitivity is already stated in part (a) in terms of the frequency change that produces a 1 V change in error voltage, that is, 3979 Hz/V. Adding this change in frequency to the 5000 Hz free-running frequency will give us the amount of frequency change needed to produce a 1 V change in error voltage. So if we increase f_i from 5000 to 8979 Hz, the change in the error voltage is 1 V.

Note that the VCO sensitivity is given in hertz per volt. If we take the reciprocal of K_o, we get the number of volts of error voltage change (ΔV_e) per hertz of frequency change. That is,

$$\frac{\Delta V_e}{\Delta f} = \frac{1}{K_o} \tag{2.16}$$

For our illustration, we get

$$\frac{\Delta V_e}{\Delta f} = \frac{1}{1989 \text{ Hz/V}} = 0.503 \text{ mV/Hz}$$

A 500 Hz increase in frequency gives the following decrease in voltage:

$$(500 \text{ Hz})(0.503 \text{ mV/Hz}) = 0.25 \text{ V}$$

If the output at 3000 Hz is 4.5 V, an increase in f_i to 3500 Hz will cause the output to decrease by 0.25 V, to 4.25 V.

• EXAMPLE 2.6
Given the data in Example 2.4, find (a) V_e at $f_o = 750$ Hz, (b) K_o, (c) f_o when $V_e = 9.75$ V (assuming a locked condition), and (d) V_e at the upper and lower limit of the lock range.

Solution
(a) Using Equation 2.13, we find

$$V_e = \left(\frac{7}{8}\right)(V_{net}) + (-V)$$

$$= \left(\frac{7}{8}\right)(24 \text{ V}) + (-12 \text{ V}) = 9 \text{ V}$$

(b) From Equation 2.14, we get

$$K_o = \frac{50 f_o}{2\pi (V_{net})} = \frac{50 (750 \text{ Hz})}{2\pi (24 \text{ V})} = 250 \text{ Hz/V}$$

(c) The change in error voltage will be equal to the current error voltage (V_e) minus the error voltage at the rest frequency (V_{fo}), or

$$\Delta V = V_e - V_{fo} = 9.75 \text{ V} - 9 \text{ V} = 0.75 \text{ V}$$

So we have

$$\Delta f = (K_o)(\Delta V)$$

$$= (250 \text{ Hz/V})(0.75) = 187.5 \text{ Hz}$$

An increase in V_e from 9 V to 9.75 V is caused by a frequency decrease from 750 to 562.5 Hz.

(d) The upper lock range is 250 Hz above f_o. Since K_o is 250 Hz/V, the output will decrease 1 V (from 9 to 8 V) when f_i goes to 1 Hz. The output will increase to 10 V when f_i is 500 Hz.

Our discussion to this point has focused on the output of the low-pass filter (V_e). The filter input comes from the phase comparator output. In the locked condition, when $f_i = f_o$, the output of the VCO and f_i are 90° out of phase, similar to the situation shown in Figure 2.11B. If the phase comparator is an Exclusive–OR gate, the output voltage will look like the waveform shown in Figure 2.12A. A 50% duty cycle gives an average output voltage of 9 V. The phase detector sensitivity, given in the data sheet for the 565, is 0.68 V/rad, or 0.0119 V/°.

If the input frequency (f_i) increases, the PLL responds by driving the phase difference between f_o and f_i so that f_o lags by 45° instead of 90°. The output voltage of the Exclusive–OR gate decreases, as seen in Figure 2.12B. To find the exact amount of voltage change, we use the phase detector sensitivity (K_d) of 0.68 V/rad and the change in angle. A 45° change is 0.785 rad. So to find the output voltage change, we multiply K_d times the change in angle (in radians):

$$(0.68 \text{ V/rad}) (0.785 \text{ rad}) = 0.533 \text{ V}$$

If the error voltage is 9 V when f_o lags behind f_i by 90°, then the error voltage will decrease by 0.533 V when f_o lags behind by only 45° (0.785 rad). The total error voltage output is then 9 V – 0.533 V = 8.4 V.

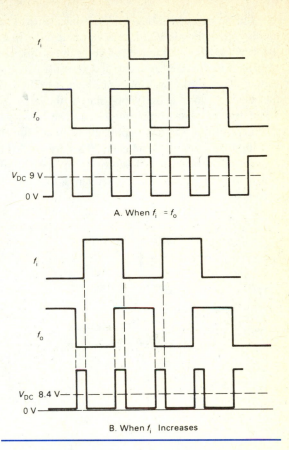

FIGURE 2.12 Phase Comparator Output

• EXAMPLE 2.7
Given a PLL in which f_o lags behind f_i by 60°, find the output voltage V_e.

Solution

Step 1 We convert degrees to radians:

$$(60°) \left(\frac{\pi}{180} \text{ rad} \right) = 1.05 \text{ rad}$$

Step 2 We multiply the phase detector sensitivity by the number of radians:

$$(0.68 \text{ V/rad}) (1.05 \text{ rad}) = 0.714 \text{ V}$$

Step 3 We subtract the change in voltage from 9 V:

$$9 \text{ V} - 0.714 \text{ V} = 8.286 \text{ V}$$

Many areas in modern industrial electronics employ the versatile IC PLL. For instance, the PLL has been used for a number of years in its discrete form in sophisticated military and aerospace applications. The IC PLL was introduced in the 1970s and has become increasingly popular in the area of telemetry and data communications. The low-cost IC PLL has spurred many applications in consumer and industrial areas. In industry, PLLs are used for data communications, frequency dividers and synthesizers, Touch-Tone™ telephones, and motor controls.

Frequency Shift Keying

The PLL in Figure 2.13 demodulates a data communications format called *frequency shift keying* (FSK). In this format, binary information is carried by an oscillator that shifts between two frequencies. A VCO usually is driven by a binary data signal. One frequency represents the *high state* or *binary one* (called a *mark*), and another frequency represents the *low state* or *binary zero* (called a *space*). Before the PLL came on the scene, FSK demodulation was done by bulky audio filters and large, unreliable electromagnetic relays.

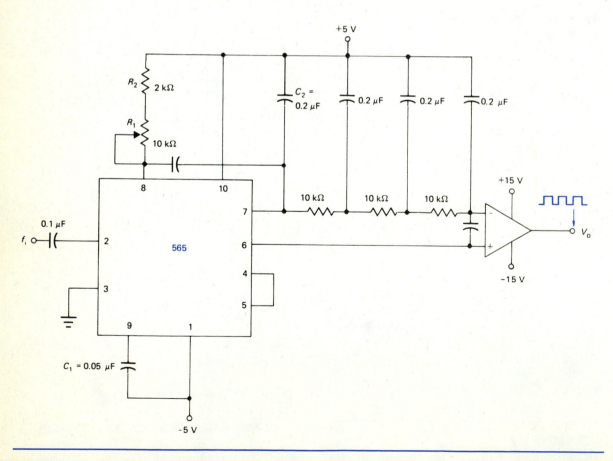

FIGURE 2.13 PLL FSK Decoder

FSK demodulation is achieved when the PLL locks on and tracks the incoming frequency shift. In the circuit in Figure 2.13, the PLL is set to receive FSK signals of 1070 and 1270 Hz. When either frequency appears at the input, the PLL locks on and tracks the carrier as it shifts between the two frequencies. A corresponding two-level DC voltage will appear at the output of the PLL (pin 6). The low-pass filter capacitor C_2 in conjunction with the additional three RC networks filters out all unwanted AC components. All that is left are two clean DC voltage levels at the input to the op amp. The resistor R_1 adjusts the free-running frequency so that the DC voltage level at the output is the same as the voltage at pin 6. The output signal can be made logic-compatible by connecting a voltage comparator between the output and pin 6.

Tone Decoder

The *tone decoder* (TDC) is a specially designed PLL. As its name implies, it is used in applications where it demodulates or decodes different frequency tones. Any frequency entering its detection band will cause the TDC output to go low, supplying current to a load. As in the PLL, the capture range and the center frequency can be adjusted independently by external components. A block diagram of an NE567 is illustrated in Figure 2.14.

Note that several blocks in this device are similar to the PLL we just discussed (compare with Figure 2.6). The VCO, the phase detector (called the I phase detector here), and the low-pass filter all behave the same way they do in a PLL. In fact, this device is a PLL with several components added. Any

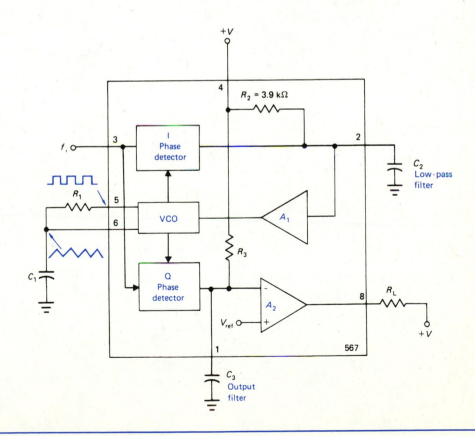

FIGURE 2.14   Tone Decoder (TDC) Functional Block Diagram

input signal within the capture range will cause the VCO to lock on and track. The VCO will then be oscillating at the same frequency as the input signal. The I phase detector gives a zero output voltage only when the input signal frequency is the same as the free-running frequency of the VCO. The Q phase detector (Q stands for quadrature) gives a maximum voltage output when the I phase detector is zero. The high voltage out of the Q phase detector exceeds the reference voltage on the noninverting input to the comparator A_2. When the input frequency is at or near the VCO free-running frequency, the comparator output will go low, causing current to flow in R_L. If the input frequency increases or decreases, moving out of the detection band, the output of the Q phase detector will decrease. When the input frequency moves far enough away in either direction, the input voltage to the comparator will be lower than the reference voltage. The comparator output will go high, interrupting current flow to the load. The TDC can then act as a frequency detector. Whenever a frequency is within the detection band of the TDC,

current will flow through a load, such as a light-emitting diode (LED).

The circuit shown in Figure 2.15 is a basic TDC application. When the input frequency is within the detection band, the output will go low, causing current to flow in R_2. The center frequency (f_0) of the detection band is given by

$$f_0 = \frac{1}{1.1\,(R_1)\,(C_1)} \tag{2.17}$$

For this circuit, the center frequency is

$$f_0 = \frac{1}{1.1(R_1)\,(C_1)} = \frac{1}{1.1\,(11{,}000\,\Omega)\,(0.01\,\mu F)}$$

$$= 8264\ \text{Hz} = 8.26\ \text{kHz}$$

The bandwidth (BW), in hertz, is given by

$$BW = (10.7)\sqrt{\frac{(V_i)\,(f_0)}{C_2}} \tag{2.18}$$

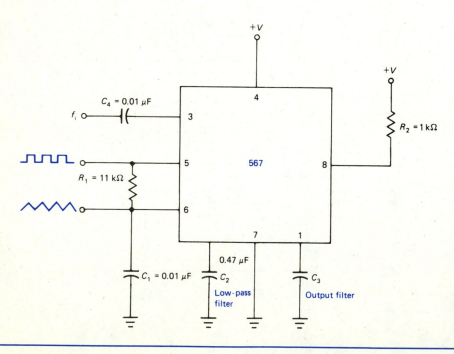

FIGURE 2.15 Basic TDC Circuit

where

BW = detection bandwidth (in hertz)

V_i = input voltage (in volts)

f_o = VCO free-running frequency (in hertz)

C_2 = capacitance of low-pass filter capacitor (in microfarads)

Using this equation for the circuit in Figure 2.15, we can solve for BW (in hertz) with an input signal of 0.1 V:

$$BW = (10.7)\sqrt{\frac{(0.1\ V)\,(8264\ Hz)}{0.47}}$$

$$= 449\ Hz$$

Any frequency that enters into this 449 Hz detection band centered around 8264 Hz will cause the TDC output to go low.

This BW equation may be written in another form:

$$BW = (1070)\sqrt{\frac{V_i}{(C_2)\,(f_o)}} \qquad (2.19)$$

where

BW = detection bandwidth (in percent)

This equation gives BW as a percentage of f_o. In our example, we calculate this percentage as

$$BW = (1070)\sqrt{\frac{0.1\ V}{(0.47)\,(8264\ Hz)}}$$

$$= 5.43\%$$

As a check, we compare our two answers. We found, with Equation 2.18, that the detection bandwidth is 449 Hz with f_o = 8264 Hz. A bandwidth of 5.43% of f_o from Equation 2.19 is 0.0543 times 8264 Hz, or 449 Hz.

• **EXAMPLE 2.8**

Given the TDC in Figure 2.15, with C_1 = 0.05 μF, C_2 = 0.2 μF, R_1 = 10 kΩ, and an input voltage of

0.1 V, find (a) the center frequency (f_o) and (b) the detection bandwidth (BW) of the TDC.

Solution

(a) We use Equation 2.17 to find f_o:

$$f_o = \frac{1}{1.1\,(R_1)\,(C_1)} = \frac{1}{1.1\,(10\ \text{k}\Omega)\,(0.05\ \mu F)}$$

$$= 1818\ Hz$$

(b) We use Equation 2.18 to calculate BW:

$$BW = (10.7)\sqrt{\frac{(V_i)\,(f_o)}{C_2}}$$

$$= (10.7)\sqrt{\frac{(0.1\ V)\,(1818\ Hz)}{0.2}}$$

$$= 323\ Hz$$

• **EXAMPLE 2.9**

Given a need to detect a 5 kHz tone with a bandwidth of 650 Hz and an input signal level of 0.1 V, and given the circuit in Figure 2.15 with C_1 = 0.027 μF, find C_2.

Solution

Step 1 To solve this problem, we must solve Equation 2.17 for R_1, as follows:

$$R_1 = \frac{1}{1.1\,(C_1)\,(f_o)} \qquad (2.20)$$

Substituting into Equation 2.20, we find the value of R_1 for f_o = 5000 Hz:

$$R_1 = \frac{1}{1.1\,(C_1)\,(f_o)}$$

$$= \frac{1}{1.1\,(0.027\ \mu F)\,(5000\ Hz)}$$

$$= 6734\ \Omega$$

We know that R_1 must be around 6734 Ω.

Step 2 To get the correct value for C_2, we need to solve Equation 2.18 for C_2. Equation 2.18 is

$$BW = (10.7)\sqrt{\frac{(V_i)(f_o)}{C_2}}$$

Thus, we find

$$C_2 = \frac{(114.5)(V_i)(f_o)}{BW^2} \tag{2.21}$$

Substituting into Equation 2.21, we get

$$C_2 = \frac{(114.5)(0.1\,\text{V})(5000\,\text{Hz})}{(650\,\text{Hz})^2}$$

$$= 0.135\,\mu\text{F}$$

We need a C_2 of 0.135 μF for this application.

As in the PLL, resistor R_1 should be between 2 kΩ and 20 kΩ. The NE567 TDC f_o limits range from 0.01 Hz to 500 kHz. Detection bandwidths can be adjusted to 14% of f_o. The output can sink up to 100 mA.

An application of the NE567 is shown in Figure 2.16. This circuit is used in a Touch-Tone™ decoding application. The Touch-Tone™ dialing system uses a combination of two frequencies every time a dialing button is depressed. The decoding circuit, as shown in the figure, uses seven TDCs tuned to the following frequencies: 697, 770, 852, 941, 1209, 1336, and 1477 Hz. When the number 1 is pressed, the transmitter will generate a 697 Hz tone and a 1209 Hz tone. When these two tones are received by the TDCs that are tuned to their frequencies, the outputs of the two TDCs will go low. The output to the top NOR gate will go high when each input goes low. The output of the top NOR logic gate will give an output that will enable the switching circuit to dial a 1 digit. Note that if the TDCs are tuned properly, no other NOR gate's output will go high

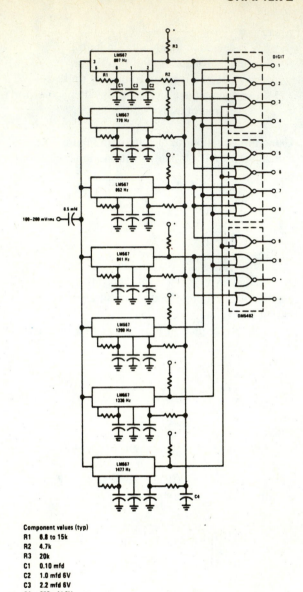

Component values (typ)
R1 6.8 to 15k
R2 4.7k
R3 20k
C1 0.10 mfd
C2 1.0 mfd 6V
C3 2.2 mfd 6V
C4 250 mfd 6V

FIGURE 2.16 Touch-Tone™ Decoder

with 697 and 1209 Hz inputs to the circuit. Note, also, the pull-up resistors between the outputs of the TDCs and the inputs to the NOR gates. These resistors will hold the gate inputs high until the output of the TDC pulls them low.

FREQUENCY-TO-VOLTAGE (F/V) CONVERSION

Previously in this chapter, we covered VFCs, like the VCO. VFCs change analog input voltages and currents to variable-frequency output signals. The output signal is normally proportional to the DC input voltage change. *Frequency-to-voltage converters* (FVCs) perform the inverse function. They change an input frequency to a proportional change in output voltage. (Strictly speaking, the PLL falls into this classification. It gives a proportional DC output voltage change for a given input frequency change.) FVCs offer economical solutions to a wide variety of industrial problems where an analog

frequency needs to be converted to an analog voltage. Some examples of FVC applications are motor speed controls and tachometers, power line frequency controls, and speedometers.

One of the simplest FVCs is the LM2907, made by National Semiconductor Corporation. This device has four main parts, as shown in Figure 2.17: the input Schmitt trigger, a charge pump, an op amp, and an output driver. The input Schmitt trigger converts the input signal into a square wave and applies it to the charge pump. Giving the Schmitt trigger 15 mV of hysteresis prevents false triggering. The charge pump alternately charges and discharges C_1 with a constant 200 µA current. The capacitor C_1 is charged to a preset voltage at a constant rate and then discharged.

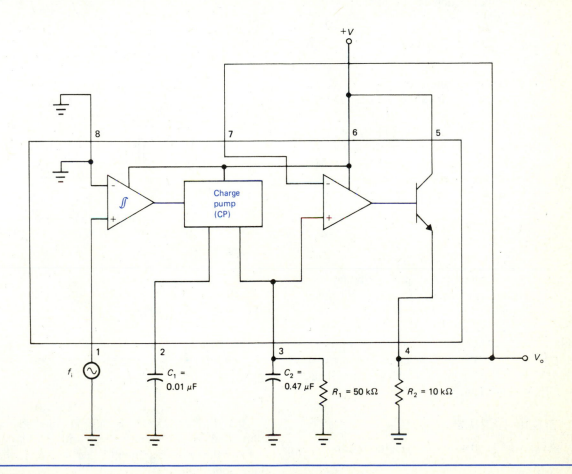

FIGURE 2.17   LM2907 FVC Functional Block Diagram

During the time that C_1 is being discharged, the charge pump charges capacitor C_2 with a 200 µA current. Capacitor C_2 will then be charged by the same current as C_1 for the same amount of time. Capacitor C_2, however, charges to twice the average voltage across C_1. Resistor R_1 (connected to pin 3) provides a discharge path for C_2. The actual average DC voltage found at pin 3 to ground is

$$V_3 = (V_{CC})(R_1)(C_1)(f_i) \qquad (2.22)$$

where

V_3 = voltage to ground at pin 3
V_{CC} = supply voltage
f_i = input frequency (in hertz)

If the op amp–driver section has a gain of 1, the output at pin 4 should be

$$V_o = (V_{CC})(R_1)(C_1)(f_i)(K) \qquad (2.23)$$

where

V_o = F/V output voltage at pin 4
K = gain of the op amp–driver (usually 1)

When working with an LM2907 FVC, we sometimes need to know how high the input frequency can go and still give a proportional output voltage. The maximum input frequency is given by the following equation:

$$f_{i(max)} = \frac{I_2}{(C_1)(V_{CC})} \qquad (2.24)$$

where

$f_{i(max)}$ = maximum input frequency (in hertz)
I_2 = constant current used to charge capacitor C_1 (usually 200 µA)

Using these equations, we can calculate the output voltage from our FVC with an input frequency of 1000 Hz and a supply voltage of +12 V. From Equation 2.23, we calculate

$$V_o = (V_{CC})(R_1)(C_1)(f_i)(K)$$

$$= (+12 \text{ V})(50{,}000 \text{ Ω})(0.01 \text{ µF})(1000 \text{ Hz})(1.0)$$

$$= 6 \text{ V}$$

This circuit should give an output voltage of +6 V with an input frequency of 1000 Hz. Its maximum input frequency, using a minimum (worst-case) estimate of the C_1 charge current of 150 µA from the data sheet, is, from Equation 2.24,

$$f_{i(max)} = \frac{I_2}{(C_1)(V_{CC})} = \frac{150 \text{ µA}}{(0.01 \text{ µF})(12 \text{ V})}$$

$$= 1250 \text{ Hz}$$

The ripple voltage on capacitor C_2 can be calculated from the following equation:

$$V_{r(p-p)} = \left(\frac{V_{CC}}{2}\right)\left(\frac{C_1}{C_2}\right)\left[1 - \frac{(V_{CC})(f_i)(C_1)}{I_2}\right] \qquad (2.25)$$

where

$V_{r(p-p)}$ = peak-to-peak ripple voltage (in volts)

Using this equation, we can calculate the ripple output voltage as follows:

$$V_{r(p-p)} = \left(\frac{V_{CC}}{2}\right)\left(\frac{C_1}{C_2}\right)\left[1 - \frac{(V_{CC})(f_i)(C_1)}{I_2}\right]$$

$$= \left(\frac{12 \text{ V}}{2}\right)\left(\frac{0.01 \text{ µF}}{0.47 \text{ µF}}\right)$$

$$\times \left[1 - \frac{(+12 \text{ V})(1000 \text{ Hz})(0.01 \text{ µF})}{150 \text{ µA}}\right]$$

$$= 0.0255 \text{ V} = 25.5 \text{ mV}_{(p-p)}$$

• EXAMPLE 2.10

Given the FVC in Figure 2.17, with $C_1 = 0.01$ µF, $C_2 = 0.2$ µF, $R_1 = 25$ kΩ, $R_2 = 10$ kΩ, and +12 V supply, find (a) the output voltage with a 1000 Hz input signal, (b) the maximum input frequency, and (c) the ripple voltage.

Solution

(a) Substituting into Equation 2.23, we can solve for V_o as follows:

$$V_o = (V_{CC})(R_1)(C_1)(f_i)(K)$$

$$= (12\text{ V})(25\text{ k}\Omega)(0.01\ \mu\text{F})(1000\text{ Hz})(1)$$

$$= 3\text{ V}$$

(b) We can find the highest input frequency by substituting the appropriate values into Equation 2.24:

$$f_{i(max)} = \frac{I_2}{(C_1)(V_{CC})}$$

$$= \frac{150\ \mu\text{A}}{(0.01\ \mu\text{F})(12\text{ V})}$$

$$= 1250\text{ Hz}$$

(c) The peak-to-ripple voltage can be calculated by using Equation 2.25:

$$V_{r(p-p)} = \left(\frac{V_{CC}}{2}\right)\left(\frac{C_1}{C_2}\right)\left[1 - \frac{(V_{CC})(f_i)(C_1)}{I_2}\right]$$

$$= \left(\frac{12\text{ V}}{2}\right)\left(\frac{0.01\ \mu\text{F}}{0.2\ \mu\text{F}}\right)$$

$$\times\left[1 - \frac{(12\text{ V})(1000\text{ Hz})(0.01\ \mu\text{F})}{150\ \mu\text{A}}\right]$$

$$= 60\text{ mV}_{(p-p)}$$

When we design circuits with the LM2907, certain limitations are placed on the sizes of R_1, C_1, and C_2. First, R_1 should be less than 100 kΩ. Since the output impedance of the charge pump is about 10 MΩ, too large a resistance for R_1 will interfere with the linearity of the charge pump. At a minimum, the resistance of R_1 should be greater than or equal to the maximum voltage at pin 3, divided by the minimum current flowing in pin 3 (about 150 μA), or

$$R_1 > \frac{V_{3(max)}}{I_{3(min)}} \tag{2.26}$$

where

$V_{3(max)}$ = maximum full-scale output voltage required in the application
$I_{3(min)}$ = minimum C_2 charge current

Next, having chosen R_1, we may choose the capacitor C_1 according to the following equation:

$$C_1 = \frac{V_{3(max)}}{(R_1)(V_{CC})(f_{i(max)})} \tag{2.27}$$

where

$V_{3(max)}$ = maximum full-scale output voltage required in the application
$f_{i(max)}$ = maximum or full-scale input frequency

This equation was derived from our original output voltage equation (assuming $K = 1$), which is

$$V_o = (V_{CC})(R_1)(C_1)(f_i)(K)$$

Since the capacitor C_1 also acts as internal frequency compensation for the charge pump, its value should be kept to at least 100 pF. The size of capacitor C_1 may also be found by solving Equation 2.24 for C_1. Equation 2.24 is

$$f_{i(max)} = \frac{I_2}{(C_1)(V_{CC})}$$

So we have

$$C_1 = \frac{I_2}{(f_{i(max)})(V_{CC})} \tag{2.28}$$

We can also solve Equation 2.23 for R_1, if we have calculated or chosen C_1 first. We obtain

$$R_1 = \frac{V_{3(max)}}{(C_1)(V_{CC})(f_{i(max)})} \tag{2.29}$$

We can solve for C_2 in the ripple voltage equation (2.25) and thus calculate a value for the capacitor C_2, provided we know the maximum acceptable ripple voltage. Thus, we have

$$C_2 = \left(\frac{V_{CC}}{2}\right)\left(\frac{C_1}{V_{r(p-p)}}\right)\left[1 - \frac{(V_{CC})(f_i)(C_1)}{I_2}\right] \quad \textbf{(2.30)}$$

Let us use these equations to solve an applications problem. Our application involves a tachometer (Figure 2.18) that will measure a maximum of 3000 revolutions per minute (r/min) on an analog meter movement. The ripple voltage should be no higher than 3% of the full-scale meter reading. With a 10 V full-scale voltage, the ripple should not be greater than 3% of 10 V, or 0.3 $V_{(p-p)}$. We are using a 1 mA meter movement. Thus, 1 mA will cause a full-scale meter deflection. If the maximum shaft rotation is 3000 r/min, then the number of revolutions per second (r/s) of the frequency is 3000/60, or 50 Hz. We can now solve for C_1 by using Equation 2.28:

$$C_1 = \frac{I_2}{(f_{i(max)})(V_{CC})} = \frac{150\,\mu A}{(50\,Hz)(12\,V)}$$

$$= 0.25\,\mu F$$

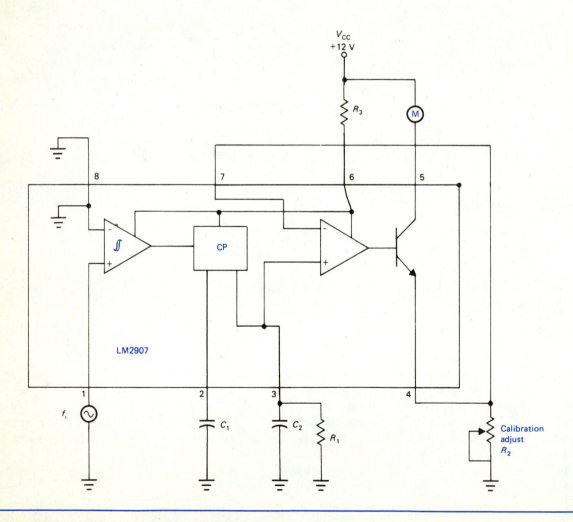

FIGURE 2.18 LM2907 Used as a Tachometer

Assuming a 2 V drop across the output transistor, our maximum output voltage swing will be 10 V. Knowing this value, we can solve for the resistance of R_1, using Equation 2.29:

$$R_1 = \frac{V_{3(max)}}{(C_1)(V_{CC})(f_{i(max)})}$$

$$= \frac{10 \text{ V}}{(0.25 \text{ μF})(12 \text{ V})(50 \text{ Hz})} = 66.7 \text{ kΩ}$$

This value should satisfy the maximum requirement set earlier (Equation 2.26) of less than 100 kΩ. Does it satisfy our minimum requirement? Using a minimum of 150 μA for the charge current of C_2 in pin 3, we make the following calculation:

$$R_1 \geq \frac{V_{3(max)}}{I_{3(min)}} \geq \frac{10 \text{ V}}{150 \text{ μA}} = 66.7 \text{ kΩ}$$

We have indeed satisfied the criterion since the resistance of R_1 is equal to this value.

Finally, C_2 may be calculated. Since the ripple is worst at low frequencies, let us assume that the lowest frequency to be measured is 60 r/min, or 1 Hz. Then, from Equation 2.30, we get

$$C_2 = \left(\frac{V_{CC}}{2}\right)\left(\frac{C_1}{V_{r(p-p)}}\right)\left[1 - \frac{(V_{CC})(f_i)(C_1)}{I_2}\right]$$

$$= \left(\frac{12 \text{ V}}{2}\right)\left(\frac{0.25 \text{ μF}}{0.3 \text{ V}}\right)$$

$$\times \left[1 - \frac{(12 \text{ V})(1 \text{ Hz})(0.25 \text{ μF})}{150 \text{ μA}}\right]$$

$$= 4.9 \text{ μF} \quad \text{or} \quad 5 \text{ μF}$$

The LM2907 FVC has many uses in addition to the tachometer applications just discussed. For example, in Figure 2.19 the FVC is being used to extend the capture and lock ranges on a PLL. Normally, the IC PLL capture and lock ranges are relatively narrow, a considerable device limitation. The PLL capture and lock ranges can be extended by the FVC. The output of the PLL's low-pass filter and the FVC are summed at the summing junction, which may be an op amp summing amplifier. The output of the summing junction is applied to the PLL VCO input. The LM2907 here converts the input frequency to an output voltage that puts the VCO initially at the correct frequency to match the input frequency. The phase detector closes the gap between the VCO and

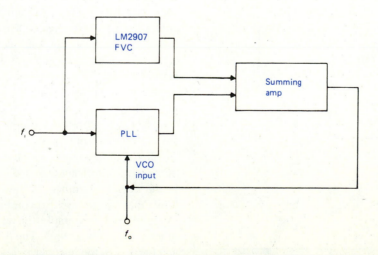

FIGURE 2.19 LM2907 Used to Extend Capture Range of PLL

the input frequency by exerting control through the summing point. If properly adjusted, the PLL can lock over a wide range with the help of the FVC.

A position-sensing application is shown in Figure 2.20. Note that, in this application, timing resistor R_1 has been removed, and $C_2 = 200C_1$. The output current produces a staircase effect instead of a DC voltage. If the input signal comes from a device that senses a passing notch on a rotating shaft, then the output will step up each time the notch is passed. When a specified number of steps is reached, the output can be turned on. In this circuit, the staircase voltage is compared with a reference voltage on the internal comparator, and the output current is turned on when this voltage level is reached.

In the circuit in Figure 2.20, the output current should turn on after 100 input pulses are received. The output may then turn on a relay that can remove power from the rotating device. Applications of this circuit can be found in automated packaging operations and in line printers.

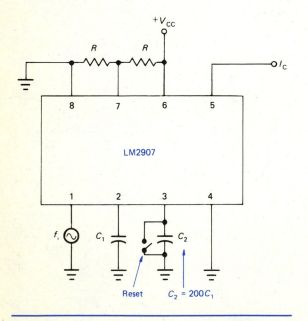

FIGURE 2.20 LM2907 Used as a Position Sensor

ANALOG AND DIGITAL CIRCUITS

Industrial electronics systems are composed of many subsystems. These subsystems have digital as well as linear electronic circuitry in them. As microprocessors become less expensive and more powerful, they will be increasingly popular in industrial circuits and systems. Since analog and digital circuits are fundamentally incompatible, circuits called interfaces must bridge the gap between these two basically different types of circuits. The *interface* provides the necessary function of tying the parts of a system together. Such circuits usually are neither purely digital nor purely analog in their makeup. They usually contain both digital and analog circuitry.

In this section, we will deal with the circuits that convert from digital signals to analog signals and from analog to digital. Recall that the digital signal is made up of bits (the abbreviation for BInary digiT). A *bit* is a voltage level that is either high or low, a 1 or a 0, never a value in between. These high and low levels are grouped together in clumps called words. A *word* is a group of bits. A word may, for example, be made up of four bits. Any number can be represented as a series of bits. For example, the number 3 in the decimal number system is represented in the binary system as 0011, read zero-zero-one-one.

To convert an analog signal to a digital one means converting a voltage or current to a series of high and low voltages. Each high voltage may correspond, for example, to a 1 or a high bit. Each low voltage may correspond to a 0 or a low bit. A series of high and low voltages will then represent a series of bits. The series of bits must faithfully represent the analog value of voltage or current. If, in the conversion process, the binary output is not an accurate representation of the analog input, the system becomes unreliable.

In the binary number system, we must be aware of the most significant digit and least significant digit. In the decimal number system, we have ten digits (0–9) that can be used in combination to represent any number. For example, the number 15 is made up of two digits, a 1 and a 5. The 5 is in the units place and the 1 is in the tens place. Each digit in the units place represents only one unit. As such, the digit in the units place is called the *least significant digit*

(LSD). The number 1 in the tens place is called the *most significant digit* (MSD) since each digit is worth ten units. To extend the concept further, we analyze the number 745. For this number, the MSD is 7 and the LSD is 5.

In digital terminology, the significant digits are called significant bits. Thus, the *most significant bit* is abbreviated MSB. Likewise, the *least significant bit* is LSB.

In the binary number system, the same basic rules apply to the significance of bits. The difference is that, in the binary number system, we have only two digits, 0 and 1. In this system, for example, the binary number 1000 represents the number 8. Bit 1 in the leftmost position is the MSB, and the 0 to the far right is the LSB. Note that the number 8 is represented by four bits. In computer terminology, this group of bits is called a four-bit word. The highest number we can represent with a four-bit word is 1111, corresponding to the decimal number 15. To represent numbers larger than decimal 15, we need words made up of more than four bits. For example, a word with eight bits can represent decimal numbers up to 255.

In the measurement or representation of analog voltages, a voltage may take on any value between two extremes. For example, a voltage with a range of 0–5 V may take any value between 0 and 5 V. This is not true of digital representations of voltage values. For example, if we were to convert the voltage range of 0–5 V, we would find that the digital representation of that range would be spread over that range in discrete steps, or increments. Table 2.1 shows a 5 V range converted to a digital word of four bits. Note from this table that our four-bit word represents 16 individual items of data. We can state this another way by saying that the resolution of our four-bit system is 1 part in 15. *Resolution* is defined as the degree to which values of a quantity can be distinguished or separated. Given as a percentage, a resolution of 1 in 15 is equal to 6.66%.

To find the resolution of a binary system, we can use the following equation:

$$\text{resolution (in percent)} = \frac{1}{2^n - 1} \times 100 \qquad (2.31)$$

where

   n = number of bits in binary system

TABLE 2.1 Conversion of a Range of 0–5 V to Four-Bit Digital Words

Decimal Value	Binary Representation
0.000	0000
0.3125	0001
0.625	0010
0.9375	0011
1.25	0100
1.5625	0101
1.875	0110
2.1875	0111
2.5	1000
2.8125	1001
3.125	1010
3.4375	1011
3.75	1100
4.0625	1101
4.375	1110
4.6875	1111

For example, the resolution of a six-bit system is

$$\text{resolution (in percent)} = \frac{1}{2^n - 1} \times 100$$

$$= \frac{1}{2^6 - 1} \times 100$$

$$= 1.6\%$$

• **EXAMPLE 2.11**

Given a ten-bit binary system, find the percent resolution of this system.

Solution

Substituting 10 for n in Equation 2.31, we get

$$\text{resolution (in percent)} = \frac{1}{2^n - 1} \times 100$$

$$= \frac{1}{2^{10} - 1} \times 100$$

$$= 0.1\%$$

Note also in our 5 V four-bit system that each bit has a value of 0.3125, or one-third of a volt. This value is known as the LSB value since it represents the value of the least significant bit. In general, the value of the LSB may be calculated by using the following equation:

$$\text{LSB} = \frac{V_{\text{fs}}}{2^n} \qquad (2.32)$$

where

V_{fs} = full-scale voltage
n = number of binary digits in the system

Let's use this equation on the system we have just described, a four-bit system with a full-scale voltage equal to 5.0 V. We find

$$\text{LSB} = \frac{5 \text{ V}}{2^4} = 0.3125 \text{ V}$$

• **EXAMPLE 2.12**

Given a six-bit system with $V_{\text{fs}} = 10$ V, find the LSB value (in volts).

Solution

We use Equation 2.32 to find the LSB.

$$\text{LSB} = \frac{V_{\text{fs}}}{2^n}$$

$$= \frac{10 \text{ V}}{2^6} = 156 \text{ mV}$$

The lack of resolution is a limitation on the application of analog-to-digital conversions. If greater accuracy is needed, a word with greater numbers of bits is required. For example, an 8-bit system has a resolution of 1 in 255, or 0.392%. If an analog range of 5 V were broken down in 255 parts, each bit would be equal to about 20 mV instead of 0.3125 V for a system using a 4-bit word. Even greater accuracy is achieved by using a 12-bit word. A 12-bit word system has a resolution of 1 part in 4095, with a 0.024% resolution. In a system with a 12-bit word, a voltage range is broken down into 4095 steps or increments. Even with the increased resolution a 12-bit system offers, however, a digital representation of an analog voltage is never exact.

An important point to notice is that the maximum value of the digital code (all ones) does not correspond with the analog full-scale value. The maximum value of the digital code is 1 LSB less than full-scale. The maximum value of the digital code can be found by taking the full-scale value and multiplying by $1 - 2^{-n}$. A 12-bit converter with a 0–+10 V analog range has a maximum digital code equal to 1111 1111 1111. This digital expression has a maximum analog value of +10 V times $1 - 2^{-12}$ or +9.99756 V. Note that the maximum analog value of the converter never quite reaches the point defined as analog full-scale.

DIGITAL-TO-ANALOG (D/A) CONVERSION

Several different types of circuits can convert digital signals or representations into analog form. A basic *digital-to-analog converter* (DAC) is shown in Figure 2.21. Note that an analog output results with a 4-bit digital signal input. Although we are using a 4-bit word system in this example (and in other examples to follow), you should know that we are using it merely to simplify our illustrations. Actually, 4-bit systems rarely are used in industry due to the lack of resolution. Words with 8, 10, 12, and 16 bits are more popular since they have acceptable resolution for most industrial applications.

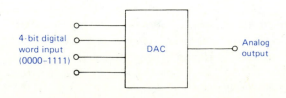

FIGURE 2.21 DAC Functional Block Diagram

Binary-Weighted Ladder DAC

An actual DAC is shown in Figure 2.22. This particular circuit is known as a *binary-weighted ladder*. Note that we have a simple inverting summing amplifier in this circuit. Voltages V_1 to V_3 are either high (+V) or low (0 V). These voltages are applied to each of the three respective inputs to the op amp, and each of the inputs is weighted. For example, the input V_1 is applied across a 40 kΩ resistor. This input gets an attenuation of about one-fifth (8/40) at the output. Now, notice that the input V_2 gets attenuated by a factor of 8/20, less than the amount of attenuation V_1 received. This is appropriate since the digit associated with V_2 has a more significant position than the input V_1. Moving on to the MSB (at V_3), we see that this input has a gain factor of 8/10, four times that of the LSB input.

To see how this circuit works in practice, we will use a 001 digital signal (with +V = +5 V). To obtain this signal, we place S_1 in the +V position, and we place S_2 and S_3 in the 0 V positions. Electrically, this device is a summing amp with only one voltage

input. The output voltage is equal to the input voltage times the gain, or

$$V_o = \left(\frac{8 \text{ k}\Omega}{40 \text{ k}\Omega} \right) (+5 \text{ V}) = 1.0 \text{ V}$$

So a 5 V input will return a 1 V output.

Let us see what happens when we put a 011 in our DAC. Voltages V_1 and V_2 are each +5 V, and V_3 is 0 V. The voltage at V_1 is attenuated by 0.2, to give a 1 V contribution to the output. The 5 V potential at V_2 is attenuated by 0.4, to give a 2 V contribution. Added together, these contributions total 3 V for the output voltage. If we connect a DC voltmeter to the output, we will see an output of 3 V with a 011 input. If we close the final switch, we have a 111 input, which will produce a 7 V output voltage. This potential is also known as the full-scale output voltage (V_{fs}).

Table 2.2 shows the DAC output voltages from 000 to 111. We can see from this table that the output voltage ranges from 0 to 7 V in 1 V steps or increments. The range may be adjusted by

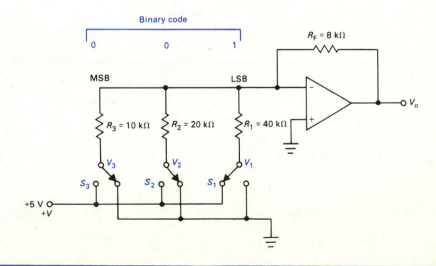

FIGURE 2.22   Three-Bit, Binary-Weighted Ladder DAC

TABLE 2.2 Binary-Weighted Ladder DAC Input and Output Voltages

Binary Input	Analog Output (Volts)
000	0
001	1
010	2
011	3
100	4
101	5
110	6
111	7

changing the size of the feedback resistor. Decreasing the size of the feedback resistor will decrease the range and decrease the size of each individual step or increment.

We can write an equation that will allow us to calculate the output voltage given any set of input voltages. The equation is

$$V_o = -\left[\left(\frac{R_F}{R_3}\right)(V_3) + \left(\frac{R_F}{R_2}\right)(V_2) + \left(\frac{R_F}{R_1}\right)(V_1)\right] \quad (2.33)$$

An input of 101 gives

$$V_o = -\left[\left(\frac{8\ k\Omega}{10\ k\Omega}\right)(+5\ V) + \left(\frac{8\ k\Omega}{20\ k\Omega}\right)(0\ V)\right.$$

$$\left. + \left(\frac{8\ k\Omega}{40\ k\Omega}\right)(+5\ V)\right]$$

$$= -5\ V$$

• EXAMPLE 2.13

Given the DAC in Figure 2.22, find the output voltage with an input of 110.

Solution

Substituting in Equation 2.33, we get

$$V_o = -\left[\left(\frac{8\ k\Omega}{10\ k\Omega}\right)(+5\ V) + \left(\frac{8\ k\Omega}{20\ k\Omega}\right)(+5\ V)\right.$$

$$\left. + \left(\frac{8\ k\Omega}{40\ k\Omega}\right)(0\ V)\right]$$

$$= -6\ V$$

The output voltage is −6.0 V.

If we examine the weights of the input resistors, we can see an interesting relationship. Note that, if $R_3 = R$, then $R_2 = 2R$ and $R_1 = 4R$. Using this relationship and simplifying Equation 2.23, we get an equation that describes the output voltage of any binary-weighted ladder. This equation is

$$V_o = \left[\frac{2\,(R_F)\,(+V)}{R}\right]\left[(B_1)\,(2^{-1})\right.$$

$$+ (B_2)\,(2^{-2}) + (B_3)\,(2^{-3})$$

$$\left. + \dots + (B_n)\,(2^{-n})\right] \quad (2.34)$$

where

R = value of input resistance in MSB input
$+V$ = voltage applied to input
B = value of input bit (either 0 or 1)
n = number of total bits in system

Let us substitute the values for Example 2.13 into Equation 2.34:

$$V_o = \left[\frac{2\,(8\ k\Omega)\,(+5\ V)}{10\ k\Omega}\right]\left[(1)\,(2^{-1})\right.$$

$$\left. + (1)\,(2^{-2}) + (0)\,(2^{-3})\right]$$

$$= (8\ V)\,[(1)\,(0.5) + (1)\,(0.25)$$

$$+ (0)\,(0.125)]$$

$$= (8\ V)[(0.5) + (0.25) + (0)]$$

$$= (8\ V)\,(0.75) = 6\ V$$

• EXAMPLE 2.14

Given a four-bit binary-weighted ladder DAC, with +10 V inputs, $R = 10$ kΩ, and a feedback resistance of 5 kΩ, find the output voltage with a 1110 input.

Solution

Substituting into Equation 2.34, we get

$$V_o = \left[\frac{2\,(R_F)\,(+V)}{R}\right]\left[(B_1)(2^{-1}) + (B_2)(2^{-2})\right.$$

$$+ (B_3)(2^{-3}) + \cdots + (B_n)(2^{-n})\right]$$

$$= \left[\frac{2\,(5\text{ k}\Omega)\,(+10\text{ V})}{10\text{ k}\Omega}\right]\left[(1)(2^{-1})\right.$$

$$+ (1)(2^{-2}) + (1)(2^{-3}) + (0)(2^{-4})\right]$$

$$= (10\text{ V})[(1)(0.5) + (1)(0.25)$$

$$+ (1)(0.125) + (0)(0.0625)]$$

$$= (10\text{ V})\,[(0.5) + (0.25) + (0.125) + (0)]$$

$$= (10\text{ V})(0.875) = 8.75\text{ V}$$

The weighted-ladder approach just described works well for four- to six-bit systems. At higher word lengths, however, the spread in resistive values needed in the inputs becomes much greater. In eight-bit words, the final resistor in the LSB position may be 10 MΩ or more. Resistors spread over so wide a range will change their resistances by different amounts with temperature changes, introducing errors. The R–2R ladder solves this problem by allowing lower-value resistors to be used.

R–2R Ladder DAC

The *R–2R ladder* is so named because resistors are either *R* (in this case 5 kΩ) or 2 times *R* (10 kΩ), as illustrated in Figure 2.23. Relative weighting of the

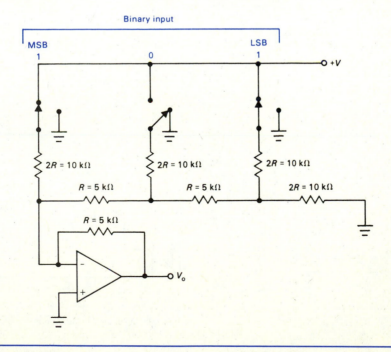

FIGURE 2.23   *R–2R* Ladder DAC

bit inputs is done by using a Thévenin voltage source. The analysis is a somewhat complicated exercise in solving a Thévenin equivalent circuit, and a little will be gained by such an analysis here. The output voltage of this circuit is given by the following equation:

$$V_o = (-R_F)\left(\frac{V_3}{2R} + \frac{V_2}{4R} + \frac{V_1}{8R}\right)$$

MSB $\rightarrow$ LSB

The output voltage with an input of 101 is

$$V_o = (-5\ k\Omega)\left[\frac{+5\ V}{2\ (5\ k\Omega)} + \frac{0\ V}{4\ (5\ k\Omega)} + \frac{+5\ V}{8\ (5\ k\Omega)}\right]$$

$$= 3.125\ V$$

Table 2.3 shows the DAC output voltages from 000 to 111. We can see from this table that the output voltage ranges from 0 to 4.375 V in 0.625 V increments or steps. The $R-2R$ system requires only two values of resistance for its D/A conversion. These resistors are usually precision resistors.

TABLE 2.3 *R–2R* Ladder DAC Input and Output Voltages

Binary Input	Analog Output (Volts)
000	0.0
001	0.625
010	1.250
011	1.875
100	2.500
101	3.125
110	3.750
111	4.375

Although DACs can be constructed from discrete devices, it is usually more economical to use an IC specifically designed to do D/A conversions. Figure 2.24A shows the schematic diagram of an IC DAC. The curve in Figure 2.24B is a graph of output current versus the digital input code for an ideal-current output IC DAC. The three-bit input DAC

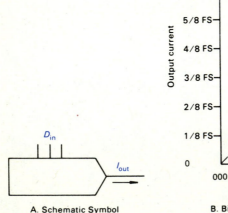

A. Schematic Symbol

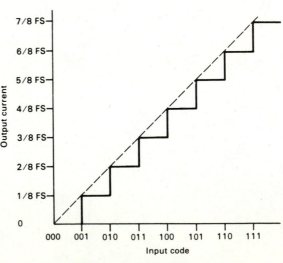

B. Binary Input versus Analog Output

FIGURE 2.24 Three-Bit DAC

shown here has a total of 2^3, or 8, possible binary input codes. In this ideal DAC, each input code step produces an output current change that is exactly ⅛ of the full-scale value. The maximum code of 111 produces an output current that is ⅞ of full-scale.

One of the most common approaches to building IC DACs is to use bipolar transistors as current sources, as in the popular National Semiconductor DAC0800. The output current of the DAC is proportional to the binary input code. The DAC0800 is represented in Figure 2.25A. This device has complementary current outputs that are directly proportional to the binary code input. The actual output current from terminal I_o (pin 4) is

$$I_o = (I_{ref})\left(\frac{D}{256}\right) \tag{2.35}$$

where

I_o = output current from pin 4
D = decimal value of binary input code
I_{ref} = reference current flowing in pin 14

An example may help at this point. Let us suppose that the input code is 11110000. The decimal value of that code (D) is 240. The reference current is approximately

$$I_{ref} = \frac{V_{ref}}{R_{ref}} \tag{2.36}$$

where
R_{ref} = resistance attached to pin 14

For our example, we get

$$I_{ref} = \frac{+10\,V}{5000\,\Omega} = 2\,mA$$

We can now use Equation 2.35 to solve for the actual output current:

$$I_o = (I_{ref})\left(\frac{D}{256}\right)$$
$$= (2\,mA)\left(\frac{240}{256}\right) = 1.875\,mA$$

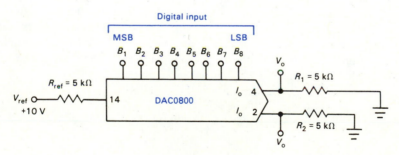

A. Schematic Diagram

	B_1	B_2	B_3	B_4	B_5	B_6	B_7	B_8	I_o (Milliamperes)	I_o (Milliamperes)	V_o (Volts)	V_o (Volts)
Full scale	1	1	1	1	1	1	1	1	1.992	0.000	−9.960	0.000
Full scale − LSB	1	1	1	1	1	1	1	0	1.984	0.008	−9.920	−0.040
Half scale + LSB	1	0	0	0	0	0	0	1	1.008	0.984	−5.040	−4.920
Half scale	1	0	0	0	0	0	0	0	1.000	0.992	−5.000	−4.960
Half scale + LSB	0	1	1	1	1	1	1	1	0.992	1.000	−4.960	−5.000
Zero scale + LSB	0	0	0	0	0	0	0	1	0.008	1.984	−0.040	−9.920
Zero scale	0	0	0	0	0	0	0	0	0.000	1.992	0.000	−9.960

B. Binary Input and Output Currents

FIGURE 2.25 Eight-Bit DAC0800

• EXAMPLE 2.15

Given the DAC0800 represented in Figure 2.25, with an input binary code of 10000001, find the output current from the I_o output terminal (pin 4).

Solution

Step 1 We must find the reference current from Equation 2.36:

$$I_{ref} = \frac{V_{ref}}{R_{ref}} = \frac{+10\ V}{5000\ \Omega} = 2\ mA$$

The reference current is 2 mA.

Step 2 We convert the binary code 10000001 to its decimal form, which is equal to 129.

Step 3 We use Equation 2.35 to solve for the output current:

$$I_o = (I_{ref})\left(\frac{D}{256}\right)$$

$$= (2\ mA)\left(\frac{129}{256}\right) = 1.008\ mA$$

As a check, we can compare the answer for I_o with the chart values in Figure 2.25B. Our input binary code corresponds to half scale plus 1 LSB.

Note two things about the circuit of Figure 2.25A. First, the purpose of the resistors is to convert the output current to an output voltage. For example, if the output current from pin 4 is 1.875 mA, the output voltage will be 1.875 mA times 5000 Ω, or 9.375 V. Second, the sums of the currents in pin 2 ($\bar{I}_o$) and in pin 4 (I_o) are equal to the same value, no matter what the input binary code is. In this circuit (Figure 2.25A), both currents sum to 1.992 mA, a value called I_{fs}. (A circuit where both output currents add to a constant value is said to have complementary current outputs.) The current I_{fs} can be calculated by using the following equation:

$$I_{fs} = \left(\frac{+V_{ref}}{R_{ref}}\right)\left(\frac{255}{256}\right) \tag{2.37}$$

Using Equation 2.37, we find that the reference current is 1.992 mA.

The output currents produced by the DAC may be converted to voltages by resistors, as in Figure 2.25A. Or the output current may go to the summing junction of an op amp, as in Figure 2.26. Note that the output of the circuit in Figure 2.26 is referenced to ground and has a low output impedance. Let's assume that we have a reference current of 2 mA and an input of 11111000 binary (248 decimal). The current output is, from Equation 2.35,

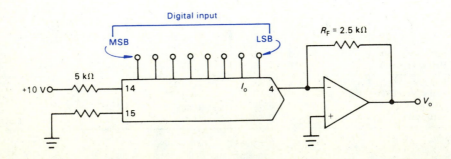

FIGURE 2.26   DAC Driving an Op Amp

$$I_o = (I_{ref})\left(\frac{D}{256}\right)$$

$$= (2\ \text{mA})\left(\frac{248}{256}\right) = 1.94\ \text{mA}$$

The 1.94 mA current flows through the 2.5 kΩ resistor, producing a –4.85 V output from the op amp.

IC DACs like the DAC0800 use internal transistors as current sources. The binary inputs turn on different sources, which all add at the output to produce the required output current. A DAC that uses a different principle is shown in Figure 2.27. This type of DAC is called an *R–2R* CMOS (complementary metal-oxide semiconductor) DAC. It is called an *R–2R* DAC because it uses an *R–2R* ladder internally as a D/A converter. The switching of the different parts of the ladder is done by CMOS switches. The *R–2R* network causes the switch currents to be binary-weighted. Each switch passes one-half the current of the previous one. The current in each leg of the *R–2R* ladder is routed either to ground or to the summing junction (point *A*) of the output op amp.

One of the advantages of the CMOS DAC is that it can be operated like a digitally controlled resistor. The effective resistance of the DAC is given by

$$R_{eff} = \frac{(2^n)(R)}{D} \qquad\qquad (2.38)$$

where

    D = input binary code in decimal form
    R = resistance R in R–$2R$ ladder
    n = number of bits in DAC input

We know that the gain of an op amp is equal to

$$A_V = \frac{R_F}{R_i}$$

Also,

$$A_V = \frac{R_F D}{2^n R}$$

If we substitute R_{eff} (in Equation 2.38) for R_i, we get an equation that describes the output voltage of this circuit (if $R_F = R_i$):

$$V_o = \frac{(V_i)(D)}{2^n} \qquad\qquad (2.39)$$

For example, if we have a three-bit converter with a 101 input to the DAC (decimal 5) and 1.5 V to the circuit, the output voltage of this circuit is

$$V_o = \frac{(V_i)(D)}{2^n} = \frac{(1.5\ \text{V})(5)}{2^3}$$

$$= 0.938\ \text{V}$$

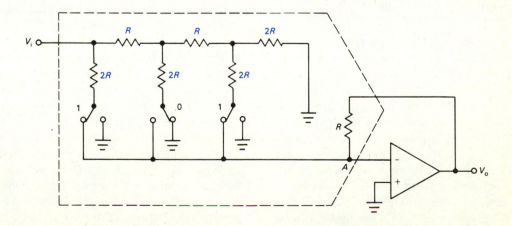

FIGURE 2.27 *R–2R* IC CMOS DAC

• EXAMPLE 2.16

Given an eight-bit R–$2R$ CMOS DAC connected as shown in Figure 2.27, with an input of 2 V to the circuit and an input of 11110000 (decimal 240) to the DAC, find the output voltage from the circuit.

Solution

Substituting the appropriate values into Equation 2.39, we get

$$V_o = \frac{(V_i)(D)}{2^n} = \frac{(2.0\ \text{V})(240)}{2^8}$$

$$= 1.875\ \text{V}$$

The potential applications of this circuit are far reaching. For instance, it allows us to use a digitally or microprocessor-controlled resistance. With a ten-bit DAC, we can change the resistance of a circuit from 1/1024 times the R in the R–$2R$ ladder to 1023/1024 times R, in steps of 1/1024. One potential application is a microprocessor used to control the frequency of an oscillator by changing a resistance in the frequency-determining network.

DACs are found in many industrial applications, such as cathode-ray tube (CRT) character generator and display drivers for computers, programmable power supplies and function generators, and audio tracking, digitizing, and decoding.

ANALOG-TO-DIGITAL (A/D) CONVERSION

An *analog-to-digital converter* (ADC) converts analog voltages or currents to a series of digital bits—just the opposite of what occurs in D/A conversion. A simplified block diagram of the A/D conversion process is shown in Figure 2.28. As in DACs, digital word lengths are commonly 8, 10, 12, and 16 bits. Analog-to-digital converters (ADCs) are somewhat more complicated and difficult to understand than DACs. Also, since most ADCs use DACs as functional blocks, the DACs were presented first in our discussion.

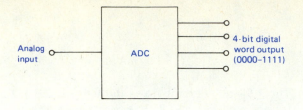

FIGURE 2.28   ADC Functional Block Diagram

Three A/D conversion principles are used in modern ADCs: digital ramp or counting, successive approximation, and parallel or flash. The counting ADC is the easiest to understand and will be presented first.

Counting ADC

A block diagram of the *counting ADC* is shown in Figure 2.29. If we start the up-down counter at zero, the counter will have a digital output that increases every time a clock pulse is received. These digital bits are converted back into an analog form and compared with the input in the comparator. If the analog input voltage is greater than the DAC's output, the counter counts up. When the counter counts upward, the output of the DAC increases in a stepwise fashion. If the analog input voltage is less than the DAC's output, the counter counts down. When the counter counts downward, the output of the DAC decreases in a stepwise fashion, bringing the digital output in line with the analog input. You will recognize this device as a feedback system that keeps the digital output approximately equal to the analog input.

The major disadvantage of this type of ADC is its slow speed. The conversion time for this device is virtually instantaneous for 0 V, but it is equal to 2^n clock periods, where n is the number of bits in the digital word. Thus, a 12-bit DAC will take 2^{12}, or 4096, clock pulses to accomplish the conversion. This conversion time is too long a time interval for many applications. Because of low conversion speed, the counting ADC is not used very widely, especially in 10- and 12-bit ADCs.

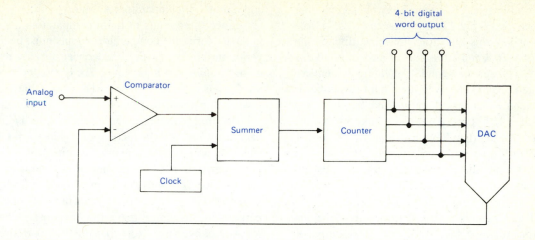

FIGURE 2.29 Counting ADC

Successive-Approximation ADC

More popular than the counting ADC is the *successive-approximation ADC*, shown in simplified block diagram form in Figure 2.30. As in the counting ADC, let us assume that the successive-approximation register (SAR) output is 0000. The operation of this device may be compared with a guessing game. For example, suppose I have a number in mind between 1 and 16. What is the most efficient strategy for you to use to guess the number if you can only ask questions that can be answered with a yes or no? You can start out at 1 and ask, "Is the number greater than 1?" If I answer no, you can conclude that the number is 1. If my answer is yes, you can continue successively from 1 to 2 to 3, and so on, asking the same question. You will have to ask 15 questions, however, if my chosen number is 16. A better way to approach the question is to split the range of possible numbers in half every time you ask a question. For example, you can start out by asking, "Is the number greater than 8?" (Eight is halfway between

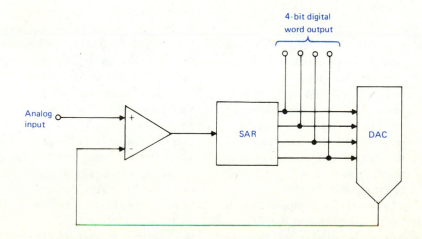

FIGURE 2.30 Successive-Approximation ADC

1 and 16.) I answer yes because my chosen number is 9. Then, you reply, "Is it greater than 12?" (Note that, in asking this question, you have divided the range between 8 and 16 in half.) I answer no to this question. You again respond by dividing the remaining range in half (you know now that the number is greater than 8 but less than 12) and asking, "Is it greater than 10?" I reply no, and you know that the number is greater than 8 but less than 10. Of course, the number 9 fits that description.

This simple illustration demonstrates almost exactly what happens in the successive-approximation ADC. The control logic in the SAR sets the output to 0000 and changes the MSB from 0 to 1, making the output of the SAR 1000. This SAR output is converted to an analog voltage and applied to the comparator. The comparator then asks the question, "Is the input greater than the DAC output?" If it is, the comparator will answer yes by putting out a high voltage. The SAR will move the next MSB from 0 to 1 if the input is high. The same question is asked, and the SAR responds by making trial outputs. If the answer to the first question is no, the comparator output will be low. The SAR will then change the MSB back to a 0 and move on to the next MSB, making it high, 0100. The process continues until the correct binary word is determined. Generally, the successive-approximation ADC is faster than the counting ADC and so is more popular.

Flash ADC

The fastest of the three ADCs we will discuss is the parallel or flash converter, shown in Figure 2.31. The *flash converter* is a series of comparators with a decreasing reference voltage applied to each inverting input. The voltage at the A_1 inverting input is higher than the voltage at A_2. The reference voltages continue to decrease as we move down the diagram, owing to voltage divider action. If the analog input voltage V_i is less than the reference voltage V_{ref}, the comparators with inverting terminal voltages lower than V_i will have high outputs. The comparators with inverting terminal voltages higher than V_i will have low outputs. For example, if V_i is greater than the voltage at point A but less than the voltage at point B, outputs of comparators A_3 through A_7 will be high,

and outputs of comparators A_1 and A_2 will be low.

Since the outputs of these comparators are not in binary form, the comparator outputs must be converted into binary form by a decoder, usually made up of a network of logic gates. The converter in Figure 2.31 is a three-bit word ADC. Note that it has seven comparators to provide this three-bit output. In general, this type of converter needs $(2^n - 1)$ comparators, where n is the number of bits in a word. For example, an eight-bit converter requires $(2^8 - 1)$, or 255, comparators. The number of comparators expands exponentially as the number of bits in the word increases. The extra complexity and numbers of comparators in this circuit make it an expensive choice for A/D conversion. Note that no clock is required in this application.

A six-bit flash ADC made by TRW has a conversion time of 33 nanoseconds (ns) and can provide up to 30 million samples per second. This chip, the TDC1014J, accepts an input up to 1 V peak to peak and is supplied in a 24-pin DIP with an optional evaluation board available. TRW has recently introduced an ADC that has nine-bit output with 511 comparators, decoding logic, and an emitter-coupled logic (ECL)-compatible output on a single chip.

ANALOG SWITCHES

A *switch* is an electronic component that has two stable states: an on state and an off state. The on state typically is a low-resistance condition, where power is controlled by the switch but not consumed by the switch. The ideal switch is one that has 0 Ω resistance when on and infinite ohms resistance when off. The mechanical switch comes closest to this ideal resistance ratio. Mechanical switches are used in applications where current demands are high. In applications where the switch must be operated at a frequency of more than 1 kHz, however, a mechanical switch is not suitable. The bipolar transistor can be used as a switch in these applications. Recall that when no base current flows in a transistor, the device is considered to be an open circuit from emitter to collector. Applying a base voltage will cause the device to conduct current flow from collector to emitter, corresponding to the on state. Many manufacturers make a special switching transistor that has

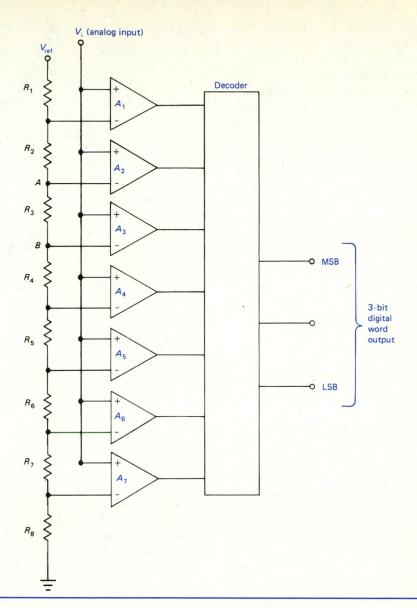

FIGURE 2.31   Flash (Parallel) ADC

low on-state resistance and high switching speed. In general, we may say that most transistors have difficulty with switching speeds above 100 kHz. Higher speeds can be realized with FETs.

Discrete-junction field effect transistor (discrete JFET) and metal-oxide semiconductor field effect transistor (MOSFET) devices can be used as switches in much the same way as the transistor. Applying a high enough voltage to the gate of a depletion-mode MOSFET, for example, will turn off the device. A low voltage will turn it back on. In enhancement MOSFETs, however, the opposite is true. A high voltage on the gate will turn on the MOSFET, and a low gate voltage will turn it off. The IC MOSFET is more popular today than the discrete MOSFET for many switching applications. An example of an IC MOS switch is the 4066 CMOS switch, shown in block diagram form in Figure 2.32.

In the 4066 (and similar switches), when the control line is connected to a high potential (V_{DD}), the resistance between A and B is low (about 100 Ω). When the control line is connected to a low voltage (V_{SS}), the switch is disabled and a high impedance is established between A and B. This type of control is called *active HIGH enable* in digital terminology. A high voltage on the control line turns on the switch, and a low voltage turns it off.

The on-state and off-state resistances normally vary with applied signal voltages and temperature. Table 2.4 summarizes typical parameter values for different types of switches.

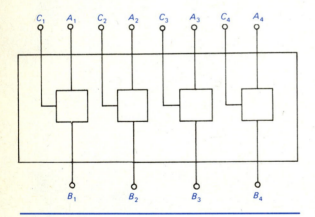

FIGURE 2.32   CMOS Analog Switch

SOURCING AND SINKING ICs

An important topic generally left out of textbooks is sourcing and sinking as related to the IC output. In ICs, two types of output circuits are commonly used: current sourcing and current sinking. The names come from the position of the load in the output circuit. In *current sinking*, shown in Figure 2.33A, the load is connected between the output and the +V supply. In this type of output circuit, the load is isolated from ground when the output is high. Current will flow through the load only when the output is at a lower potential than +V. (Note the use of electron flow notation. In conventional current flow, current moves in the opposite direction.) If the output is high (at or near +V), no current flows through the load. If the output goes low, a difference in potential will exist between the supply and the output of the IC, and current will flow. The IC switches or *sinks* the output current from ground. This output is most often associated with the *open-collector* configuration. Recall that the open collector uses a pull-up resistor. If the IC output is driving a TTL (or similar) input, as shown in Figure 2.33B, the pull-up resistor helps make a solid voltage level at the output. A solid voltage level is necessary since the open-collector output is floating. The pull-up resistor also provides better immunity from noise and faster output rise and fall times.

In *current sourcing*, the load is connected between the output and ground. When the output goes high, a difference in potential exists between the output and ground. Current will then flow through the load, as illustrated in Figure 2.34A. This circuit is called current sourcing because the IC provides a

TABLE 2.4   **Typical Analog Switch Parameters**

Parameter	Mechanical Switch	Transistor	JFET	MOSFET
On-state resistance	10^{-2} Ω	10 Ω	30 Ω	80 Ω
Off-state current	10 pA	100 pA	100 pA	100 pA
Commutation rate	1 kHz	100 kHz	10 MHz	50 MHz

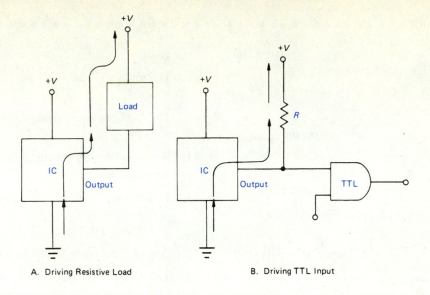

FIGURE 2.33   Current Sinking an IC Output

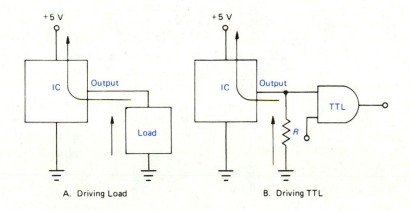

FIGURE 2.34   Current Sourcing an IC Output

source of power to the load. An IC with a current source output will normally be low, and it will go high when power is applied to the load. When the IC output is applied to a TTL device (or one with a similar input), a pull-down resistor is connected as shown in Figure 2.34B. The pull-down resistor is used for reasons similar to the reasons that the pull-up resistor is used in the current-sink output. The current sink, however, is an open-collector output, and the current source is an open emitter.

■ QUESTIONS ·

1. The VCO output frequency _____ when the input voltage (pin 5) increases.

2. Describe the operation of the VCO in a frequency modulation application.

3. The PLL has three basic states: the _____, _____, and _____ states.

4. Define the following terms as they relate to the PLL: (a) capture range and (b) lock range.

5. Describe the PLL in an FM demodulation application.

6. The tone decoder is a specially designed _____.

7. The tone decoder output goes _____ when an input signal is within its capture range.

8. The LM2907 FVC changes an analog input voltage into a _____ output signal.

9. The main application of the LM2907 FVC is in a _____.

10. The DAC converts analog voltages and currents to _____.

11. Explain the disadvantage of the binary-weighted ladder DAC.

12. Describe the operation of the following ADCs: (a) counting ADC, (b) successive-approximation ADC, and (c) flash ADC.

13. Describe the operation of an analog switch.

14. The control pin on the 4066 CMOS analog switch is an active _____ enable.

15. Differentiate between sourcing and sinking in IC outputs.

■ PROBLEMS ·

1. In the VCO shown in Figure 2.2, $R_1 = 15$ kΩ and $C_1 = 0.01$ µF.

   a. Calculate f_o.

   b. Calculate the voltage at pin 5 if R_2 is 1.8 kΩ and R_3 is 10 kΩ.

   c. Calculate the output frequency with the input in part b.

   d. Calculate the output frequencies at the peak voltages with a ±0.1 V peak input.

   e. With the resistances stated in part b, what change in voltage will give a 200 Hz change in output frequency?

2. For the 566 VCO shown in Figure 2.2, with $R_1 = 15$ kΩ, $C_1 = 0.01$ µF, $R_2 = 1$ kΩ, $R_3 = 10$ kΩ, and $+V = 10$ V, find (a) the voltage at pin 5, with respect to ground, and (b) f_o.

3. For the values for the 566 given in Problem 2, with C_1 changed to 0.015 µF, find (a) the size of R_1 needed for a center frequency of 15 kHz and (b) the frequency deviation from the center frequency with a 25 mV peak input signal.

4. For the 566 VCO shown in Figure 2.2, with $R_1 = 2.5$ kΩ, $R_2 = 1.8$ kΩ, $R_3 = 18$ kΩ, and $C_1 = 0.02$ µF, find the input voltage change (ΔV) for a ±15% deviation from the rest frequency.

5. The PLL in Figure 2.10 has the following component values: $R_1 = 15$ kΩ, $C_1 = 0.015$ µF, and $C_2 = 4.7$ µF. Find (a) f_o, (b) the capture range, and (c) the lock range.

6. A PLL is needed with a capture range of 100 Hz, a lock range of 2000 Hz, and a free-running frequency of 5000 Hz. Choose component values that will give these parameters.

7. For the PLL shown in Figure 2.10, with C_1 = 0.027 µF, C_2 = 12 µF, R_1 = 15 kΩ, and V_{CC} = ±10 V, find (a) the capture range, (b) the lock range, and (c) the center or rest frequency.

8. For a 565 PLL, with f_o = 3000 Hz and V_{CC} = 12 V, find (a) the sensitivity of the VCO (in hertz per volt) and (b) how much the frequency must change to produce a 1 V change in error voltage.

9. Use the data of Problem 7 to find (a) V_o at f_o = 740 Hz, (b) K_o, (c) f_o when V_o = 8 V (assuming a locked condition), and (d) V_o at the upper and lower limit of the lock range.

10. For a 565 PLL in which f_o lags behind f_i by 60° and with a phase detector sensitivity of 0.68 V/rad, find the output voltage (V_e) if V_o at f_o is 7.5 V.

11. The tone decoder in Figure 2.15 has the following component values: R_1 = 10 kΩ, C_1 = 0.015 µF, and C_2 = 4.7 µF. Find (a) f_o and (b) the bandwidth (BW) (in hertz).

12. Design a tone decoder with BW = 10% f_o = 1500 Hz.

13. For the TDC in Figure 2.15, with C_1 = 0.047 µF, C_2 = 0.1 µF, R_1 = 15 kΩ, and an input voltage of 0.075 V, find (a) the center frequency and (b) the detection bandwidth of the TDC.

14. You wish to detect a 7 kHz tone with a TDC that has a bandwidth of 600 Hz and an input signal level of 0.1 V. Use the circuit in Figure 2.15, with C_1 = 0.015 µF, to find the value of C_2.

15. The LM2907 FVC in Figure 2.17 has the following component values: C_1 = 0.01 µF, R_1 = 100 kΩ, R_2 = 10 kΩ, and V_{CC} = +12 V. With K = 1, calculate (a) the voltage output with a 200 Hz input frequency, (b) the maximum input frequency if I_2 is 150 µA, and (c) the ripple voltage on C_2.

16. For the FVC in Figure 2.17, with C_1 = 0.015 µF, C_2 = 0.27 µF, R_1 = 20 kΩ, R_2 = 10 kΩ, and a +10 V supply, find (a) the output voltage with a 2000 Hz input signal, (b) the maximum input frequency, and (c) the ripple voltage.

17. Find the percent resolution of a 12-bit binary system.

18. For an eight-bit system with V_{fs} = 12 V, find the LSB value (in volts).

19. Using the DAC in Figure 2.22, calculate the output voltage with an input of 011.

20. For a four-bit, binary-weighted ladder DAC, with +10 V inputs, R = 10 kΩ, and a feedback resistance of 5 kΩ, find the output voltage with a 1010 input.

21. For the DAC0800 of Figure 2.25, with an input binary code of 10000111, find the output current from the I_o output terminal (pin 4).

22. For an eight-bit, R–$2R$ CMOS DAC connected as shown in Figure 2.27, with an input of 1 V to the circuit and 11111001 to the DAC, find (a) the decimal value of the input and (b) the output voltage from the circuit.

23. Design an LM2907 tachometer to give the following parameters: We wish to measure a maximum 2000 Hz frequency with a ripple voltage no higher than 2%.

24. A six-bit system covers a voltage range of 0–5 V. Calculate (a) how much voltage each bit is worth and (b) the resolution of the system, in parts and percentage.

25. For the DAC in Figure 2.23, R is 10 kΩ.

   a. Calculate the output voltage with a 110 input when 1 = +5 V.

   b. Calculate the output voltage with a 101 input when 1 = +5 V.

   c. Find the value of R that will give an output of approximately 0–5 V from 000 to 111.

Wound-field DC Motors and Generators

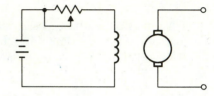

OBJECTIVES

On completion of this chapter, you should be able to:

- Describe the characteristics of ideal, series, shunt, separately excited, and compound DC motors and generators.
- Identify the schematic diagrams of the series, shunt, and separately excited DC motors and generators.
- Predict the performance of DC motors and generators.
- Give examples of each of the configurations of series, shunt, separately excited, and compound DC motors and generators.

INTRODUCTION

In these days of rapid electronic advancement, technicians, especially digital electronics technicians, frequently overlook the importance of the electric motor as a key system element. In fact, in some educational systems, digital electronics students are not required to take industrially related courses, such as courses related to motors and generators. In this chapter on wound-field DC motors and generators and in the next two chapters, we attempt to present a condensed version of the topic of motors and generators. Of course, some topics are omitted here, but the important principles are presented in some detail.

In this chapter, we do not try to derive the equations that are needed. Instead, we use the equations that have long been determined and simplify and adapt them in order to show the concepts of electromechanical action. A second reason for equation simplification is to provide equations whose quantities can be measured easily in the laboratory. For example, magnetic field strength is a quantity that is very important in the understanding of electromechanical devices, but it is very difficult to measure in the laboratory with standard electronic equipment. In some instances, we may not need to measure field strength if it is held constant; and in other instances, field strength may be proportional to another quantity that is easy to measure. This adaptation of the equations, then, allows us to concentrate on the concepts rather than on the memorization of complex equations.

Many applications in industry and in process control (see Chapter 10) require electric motors as critical functional elements. Because such motors are electromechanical rather than purely electrical, their response sets the entire system's performance limits. So technicians who wish to be completely functional in the industrial electronics environment must be familiar with the varieties of motors and generators. In addition, technicians must understand basic operating principles if they are to perform intelligent troubleshooting.

In discussions involving both DC motors and DC generators, it is convenient to refer to a generalized electric machine because much of what is said about electric motors is equally applicable to generators. The generalized electric machine is referred to as a *dynamo*, a machine that converts either mechanical energy to electric energy or electric energy to mechanical energy. When a dynamo is driven mechanically by a power source, such as a gas or diesel engine or a steam or water turbine, and provides electric energy to a load, it is called a *generator*. If electric energy is supplied to the dynamo, and its output is used to provide mechanical motion or *torque* (a force acting through a distance), it is called a *motor*. Generators are rated by the kilowatts they can deliver without overheating at a rated voltage and speed. Motors are generally rated by the horsepower they can deliver without overheating at their rated voltage and speed.

DYNAMO CONSTRUCTION

Figure 3.1A shows a simplified, exploded view of a DC dynamo. Figure 3.1B shows the dynamo in cross section.

The dynamo construction consists of two major subdivisions, the *rotor* (*armature*), or rotating part, and the *stator*, or stationary part. We will discuss these two parts in detail in the following sections.

Rotor

The major parts of the rotor consist of an armature shaft, armature core, armature winding, and a commutator.

The *armature shaft* is a cylinder on which the rotor components are attached. The armature core, armature winding, and commutator (mechanical switch) are attached to the armature shaft. This entire assembly rotates. Sometimes, fins are attached to the shaft to provide cooling of the dynamo.

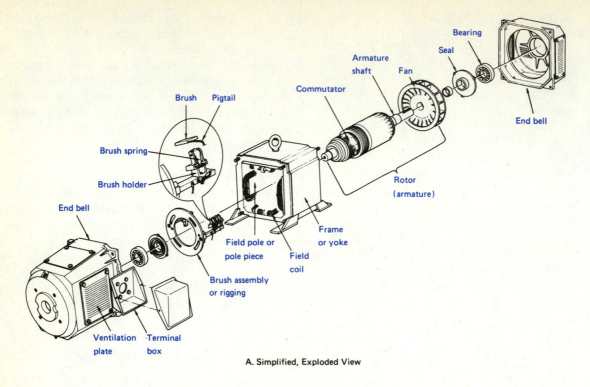

A. Simplified, Exploded View

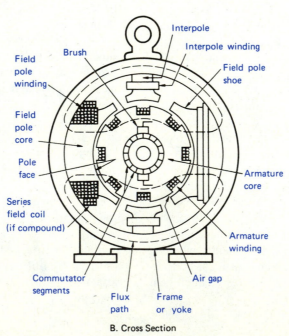

B. Cross Section

FIGURE 3.1 DC Dynamo

The *armature core* is constructed of laminated (thin-sheet) layers of sheet steel. These layers provide a low-reluctance (magnetic resistance) path between the magnetic field poles. The laminations are insulated from each other and attached together securely. The core is laminated to reduce *eddy currents*, which are circular-moving currents. Eddy currents are generally not beneficial, resulting in non-recoverable energy (heat) losses (I^2R losses). The grade of sheet steel used is selected to produce low loss due to hysteresis (lagging magnetization). The outer surface of the core is slotted to provide a means of securing the armature coils (see Figure 3.2B).

Two basic armature, or field pole, forms are available in the construction of the dynamo: salient (standing out from the surrounding material) and nonsalient (minimal projection) poles. Figure 3.2 illustrates these forms for the armature of the dynamo.

The *armature winding* consists of insulated coils, which are insulated from each other and from the armature core. The winding is embedded in the slots of the armature core face, as shown in Figure 3.1B.

The *commutator* consists of a number of wedge-shaped copper segments that are assembled into a cylinder and secured to the armature shaft. The segments are insulated from each other and from the armature shaft, generally with mica. The commutator segments are soldered to the ends of the armature coils. Because of the armature rotation, the commutator provides the necessary switching of armature current to or from the circuit external to the armature.

The armature of the dynamo performs the following four major functions:

1. It permits rotation for mechanical generator or motor action and still maintains electrical continuity.
2. Because of rotation, it provides the switching action necessary for commutation.
3. It provides a solid framework for the armature conductors, into which a voltage is induced or which provide a torque.
4. It provides a low-reluctance flux (magnetic lines) path between the field poles. Recall that *flux* or *flux lines* are invisible magnetic lines of force.

Stator

The stator consists of eight major components: a field yoke, field windings, field poles, an air gap, interpoles, compensating windings, brushes, and end bells. (See Figures 3.1 and 3.4.)

The *field yoke*, or *frame*, is made of cast or rolled steel. The yoke supports all the parts of the stator. It also provides a return path for the magnetic flux produced in the field poles.

The *field windings* are wound around the pole cores and may be connected either in series or in parallel (shunt) with the armature circuit. The windings consist of a few turns of heavy-gauge wire for a series field, or many turns of fine wire for a shunt field, or both for a compound (both series and shunt field) dynamo.

The *field poles* are constructed of laminated sheet steel and are bolted or welded to the yoke after the field windings have been placed on them. The *pole shoe* is the end of the pole core next to the rotor. It is curved and is wider than the pole core. This widening of the pole core spreads the flux more uniformly, reduces the reluctance of the air gap, and provides a means of support for the field coils.

The *air gap* is the space between the armature surface and the pole face. The air gap varies with the size of the machine, but it generally is between 0.6 and 0.16 cm.

The *interpoles*, or *commutating poles* (if used), are small poles placed midway between the main poles. The interpole winding consists of a few turns

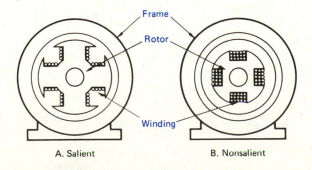

A. Salient   B. Nonsalient

FIGURE 3.2 Pole Construction

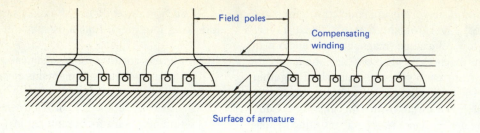

FIGURE 3.3   Location of Compensating Winding in the Face of Field Poles

of heavy wire since interpoles are connected in series with the armature circuit. The interpole magnetic flux, therefore, is proportional to the armature current.

The *compensating windings*, shown in Figure 3.3, are used only in large DC machines. They are connected in the same manner as the interpole windings but are physically located in slots on the pole shoe.

Brushes rest on the commutator segments and are the sliding electrical connection between the armature coils and the external circuit. Brushes are made of carbon of varying degrees of hardness; in some cases, they are made of a mixture of carbon and metallic copper. The brushes are held in place by springs in brush holders, the entire assembly being called the *brush rigging*. Electrical connection between the brush and brush holder is made with a flexible copper braid called a *pigtail*.

End bells support the brush rigging as well as the armature shaft. The end bells generally determine the type of protection the dynamo receives; that is, the construction can be open, semiguarded, dripproof (as in Figure 3.1), or totally enclosed. Semiguarded types have screens or grills to protect rotating or electrically live parts and to provide ventilation. A dripproof housing protects the equipment from dripping liquids. The totally enclosed construction is completely sealed and must dissipate the heat buildup by radiation from the enclosing case.

Figure 3.4 shows a cutaway view of a completely assembled DC machine—in this case, a DC motor. Notice the dual brushes in each brush rigging, provided for high currents. Also, the side access covers are of dripproof construction.

Dynamo Classification

The construction shown in Figures 3.1 and 3.4 is known as a *wound-field* dynamo. There are many other classifications of dynamos. Figure 3.5 charts selected DC motor classifications and shows a rich variety from which to choose. Each class of DC motors has its advantages, depending on the desired applications. Some of the types of DC motors shown in Figure 3.5 are discussed briefly in a table of their characteristics at the end of the chapter (Table 3.6). However, the conventional permanent-magnet (PM) and wound-field DC motors are the subject of most of the discussion in this chapter.

Unlike DC motors, DC generators are primarily of the conventional permanent-magnet and wound-field types. Consequently, they will be the only types discussed.

BASIC PRINCIPLES OF THE DC GENERATOR

The simplest generator that can be built is an AC generator, shown in Figure 3.6. The basic generator principles are most easily explained through the use of the elementary AC generator. For that reason, the AC generator will be discussed first. The DC generator will be developed next.

Elementary AC Generator

The elementary generator of Figure 3.6 consists of a single wire loop (the armature) placed so that it can be rotated in a stationary magnetic field. The

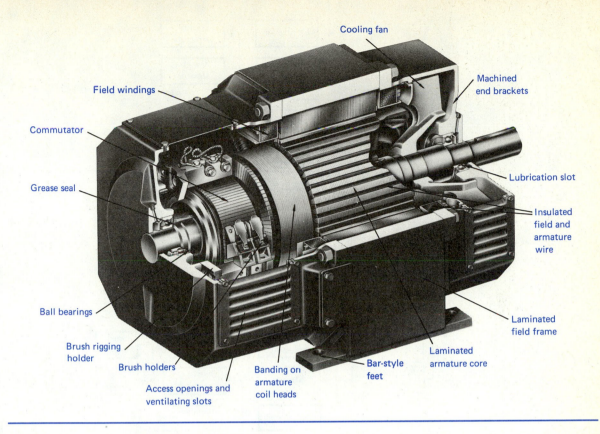

Cooling fan

Machined
end brackets

Field windings

Commutator

Lubrication slot

Grease seal

Insulated
field and
armature
wire

Ball bearings

Laminated
field frame

Brush rigging
holder

Laminated
armature core

Brush holders

Bar-style
feet

Banding on
armature
coil heads

Access openings and
ventilating slots

FIGURE 3.4   Cutaway View of DC Motor

magnetic field is shown in Figure 3.6 with arrows going from the N (north) pole to the S (south) pole. The pole pieces are shaped and positioned as shown to concentrate the magnetic field as close as possible to the wire loop. The rotation of the loop will produce an induced voltage, or emf (electromotive force), in the loop. Brushes sliding on the slip rings connect the loop to an external circuit resistance, allowing current to flow because of the induced emf. *Slip rings* are metal rings around the rotor shaft. Each ring attaches to the separate ends of the rotor coil. If the brushes were not contacting the slip rings but the loop was rotating, an emf would still appear at the slip rings, but no current would flow because the circuit was open.

The elementary generator produces a voltage in the following manner: Assuming the armature loop is rotating in a clockwise direction, its initial or starting position is as shown in Figure 3.7A. This position is called the 0° position, or the *neutral plane*. At the 0° position, the armature loop is perpendicular to the magnetic field, and the black and white portions of the loop are moving parallel to the field. At the instant the conductors are moving parallel to the magnetic field, they do not cut any lines of force, and no emf is induced in the loop. The meter in Figure 3.7A indicates zero current at this time.

As the armature loop rotates from position *A* to position *B* (see Figures 3.7B and 3.7F), the sides of the loop cut through more and more lines of force,

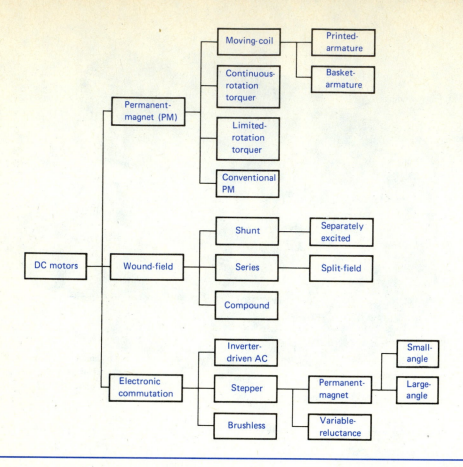

FIGURE 3.5   Selected DC Motor Types

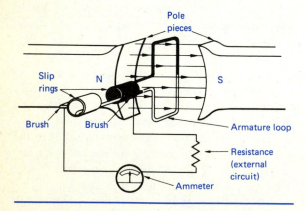

FIGURE 3.6   Elementary AC Generator

at a continually increasing angle. At 90° (position *B*), the sides of the loop are cutting through a maximum number of lines of force and at a maximum angle. The result is that, between 0° and 90°, the induced emf in the conductors builds up from zero to a maximum value. Notice that from 0° to 90° rotation, the black side of the loop cuts down through the field while the white side cuts up through the field. The induced emf's in the two sides of the rotating loop are series adding. Thus, the resultant voltage across the brushes (the armature voltage, V_A) is the sum of the two induced voltages, and the meter in Figure 3.7B indicates a maximum value.

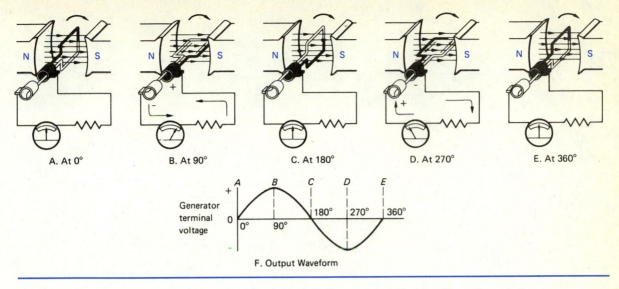

A. At 0°   B. At 90°   C. At 180°   D. At 270°   E. At 360°

F. Output Waveform

FIGURE 3.7   Output Voltage of Elementary AC Generator during One Revolution

As the armature loop continues rotating from position B to position C, the sides of the loop that were cutting through a maximum number of lines of force at position B now cut through fewer lines. At position C (180°), they are again moving parallel to the magnetic field. As the armature rotates from 90° to 180°, the induced voltage will decrease to zero, just as it had increased to a maximum value from 0° to 90°. The meter again indicates zero, as shown in Figure 3.7C.

From 0° to 180°, the sides of the loop were moving in the same direction through the magnetic field. Therefore, the polarity of the induced voltage remained the same. However, as the loop starts rotating beyond 180°, from position C through D to position E, the direction of the cutting action of the sides of the coil through the magnetic field reverses. The black side cuts up through the field as the white side cuts down. As a result, the polarity of the induced voltage reverses, as shown in the graph in Figure 3.7F.

Elementary DC Generator

A single-loop generator with each end of the loop connected to a segment of a two-segment metal ring is shown in Figure 3.8. The two segments of the split metal ring are insulated from each other. The ring is called the *commutator*. The commutator in a DC generator replaces the slip rings of the AC generator and is the main difference in their construction. The commutator mechanically reverses the armature loop connections to the external circuit at the same instant that the polarity of the voltage in the loop reverses. Through this process, the commutator changes the generated AC voltage to a pulsating DC voltage, as shown in the graph of Figure 3.8F. This action is known as *commutation*.

We can follow the rotation of the loop in Figure 3.8 as we did in Figure 3.7. We see that, as the segments of the commutator rotate with the loop, they contact opposite brushes. Also, the direction of current flow through the brushes and the meter is

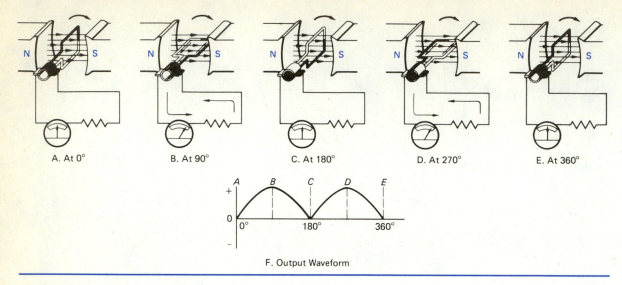

A. At 0°   B. At 90°   C. At 180°   D. At 270°   E. At 360°

F. Output Waveform

FIGURE 3.8   Elementary DC Generator and Effects of Commutation

the same. The voltage developed across the brushes is pulsating and unidirectional, peaking twice during each revolution between zero and maximum. This variation is called *ripple*.

The pulsating voltage of a single-loop DC generator is unsuitable for most applications. Therefore, in practical generators, more armature loops and more commutator segments are used to produce an output voltage with less ripple.

Additional Coils and Poles

The effects of additional coils may be illustrated by the addition of a second coil to the armature. The commutator must now be divided into four parts since there are four coil ends, as illustrated in Figure 3.9A. Since there are four segments in the commutator, a new segment passes each brush every 90° instead of every 180°. This action allows the brush to switch from the white coil to the black coil at the instant the voltages in the two coils are equal.

The graph in Figure 3.9B shows the ripple effect of the voltage when two armature coils are used. As the figure shows, the ripple is limited to the rise and fall between points A and B on the graph.

With the addition of more armature coils, the ripple effect is reduced further. Decreasing ripple in this way increases the effective voltage of the generator output.

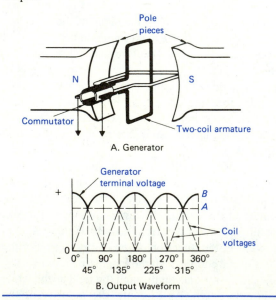

A. Generator

B. Output Waveform

FIGURE 3.9   Effects of Additional Coils in DC Generator

Practical generators use many armature coils. They also use more than one pair of magnetic poles. The additional magnetic poles have the same effect on ripple as the additional armature coils have. In addition, the increased number of poles provides a strong magnetic field. This effect, in turn, allows an increase in output voltage because the coils cut more lines of flux per revolution.

Electromagnetic Poles

Nearly all practical generators use electromagnetic poles instead of the permanent magnets found in the elementary generator. The electromagnetic field poles consist of coils of insulated copper wire wound on soft iron cores, as shown in Figure 3.1. The main advantages of using electromagnetic poles are (1) increased field strength and (2) control over the strength of the field. By variation of the input voltage to the field windings, the field strength is varied. By variation of the field strength, the output voltage of the generator is controlled.

Generator Voltage Equation

The average generated emf (in volts) of a generator may be calculated from the following equation:

$$\text{emf} = \frac{pN_{\text{cond}}\Phi n_A}{10^8 \times 60 b_A} \qquad (3.1)$$

where

p = number of poles

N_{cond} = total number of conductors on armature

Φ = flux per pole

n_A = speed of armature (in revolutions per minute)

b_A = number of parallel paths through armature, depending on type of armature winding

For any given generator, all the factors in Equation 3.1 are fixed values except the flux per pole (Φ) and the speed (n_A). Therefore, Equation 3.1 can be simplified as follows:

$$\text{emf} = C_{\text{emf}}\Phi n_A \qquad (3.2)$$

where

C_{emf} = all fixed values of constants for a given generator

DC MOTOR

Stated very simply, a *DC motor* rotates as a result of two magnetic fields interacting with each other. These two interacting fields are the field due to the current flowing in the field windings and the field produced by the armature as a result of the current flowing through it. As shown in Figure 3.10A, the loop (armature) field is both attracted to and repelled by (depending on relative positions) the field from

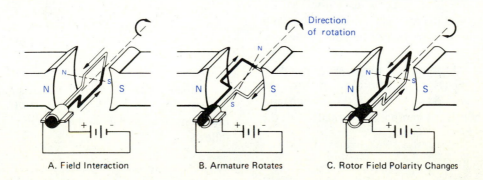

A. Field Interaction   B. Armature Rotates   C. Rotor Field Polarity Changes

FIGURE 3.10   Armature Rotation of Elementary DC Motor

the field poles. This action causes the armature to turn in a clockwise direction (for this illustration), as shown in Figure 3.10B.

After the loop has turned far enough so that its north pole is exactly opposite the south field pole, the brushes advance to the next commutator segment. This action changes the direction of current flow through the armature loop and, therefore, changes the polarity of the armature field, as shown in Figure 3.10C.

Torque

Torque is a twisting action on a body, tending to make it rotate. Torque (T) is the product of the force (F) times the perpendicular distance (d) between the axis of rotation and the point of application of the force:

$$T = F \times d$$

The torque is in newton-meters, the force is in newtons, and the distance is in meters. The unit of measure of torque in the British system of units is the pound-foot.

Recall from physics that work is also defined as a force acting through a distance. However, work requires that the motion be in the direction of the force; torque does not. A distinction between work and torque in the British system of measurement is made by the units: Torque is in pound-feet, and work is in foot-pounds. In the metric system, torque is measured in dyne-centimeters or newton-meters, and work is measured in ergs (which are the same units as dyne-centimeters) or joules (the same as newton-meters).

The metric measurement for torque with units of newton-meters (Nm) will be used in this text. However, conversion to the British system with units of pound-feet (lb-ft) can easily be made with the following conversion factor:

0.73756 lb-ft = 1 Nm

Motor Torque Equation

The torque for the DC motor can be calculated from the following equation:

$$T = \frac{pN_{cond}\Phi I_A}{2\pi b_w} \qquad (3.3)$$

where

I_A = current in external armature circuit
b_w = number of parallel paths through winding

For any given motor, all the factors in Equation 3.3 are fixed values except the flux per pole (Φ) and the armature current (I_A). Therefore, Equation 3.3 can be simplified as follows:

$$T = C_T \Phi I_A \qquad (3.4)$$

where

C_T = all fixed values or constants for a given motor

Ideal DC Machine

In many problems involving the behavior of a DC dynamo as a system component, the machine can be described with satisfactory accuracy in terms of an idealized model having the following properties:

1. The stator has salient poles, and the air gap flux distribution is symmetrical.
2. The armature can be considered as a finely distributed (evenly spread) winding.
3. The brushes are narrow, and the commutation occurs when the coil sides are in the neutral zone.

If these conditions exist, then the air gap flux (Φ) is linearly proportional to field current (I_{field}). Then, Equations 3.2 and 3.4 can be simplified further, as follows:

$$emf = K_{emf}I_{field}n_A \qquad (3.5)$$

$$T = K_T I_{\text{field}} I_A \qquad (3.6)$$

where

K_{emf} = constants C_{emf} and Φ/I_{field}

K_T = constants C_T and Φ/I_{field}

Equations 3.5 and 3.6 are very useful and easily measured. The procedure for evaluating K_T will be shown now. The evaluation of K_{emf} will be given later in this chapter in the section titled "Evaluating Variables."

To determine K_T, we use the following procedure: Attach a lever to the armature shaft, and secure a scale to the end of the lever. Apply and measure the armature current and the field current and note the reading on the force scale. The arrangement is shown in Figure 3.11. In Figure 3.11, the scale should be perpendicular to a line through the motor shaft and the lever on the shaft. The distance d is from the center of the shaft to the point of scale attachment.

The torque is the scale reading times the distance in Figure 3.11. The value of K_T is then determined from the following equation:

$$K_T = \frac{T}{I_{\text{field}} I_A}$$

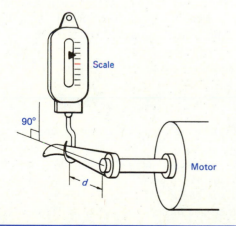

FIGURE 3.11　Arrangement for Measuring Torque

This value of K_T should remain relatively constant for this motor. Factors that could affect the value of K_T are those factors that alter the motor in any way, such as heating of the wire in the motor. When the wires in the armature and field heat because of the current flow, the resistance increases and, therefore, reduces the current (and field strength) for a constant voltage. Also, as the armature turns faster, the air friction increases, causing K_T to change. The value of K_T should not change greatly, however.

After K_T has been determined, any value of torque can be approximated by measuring the values of armature and field currents and solving Equation 3.6 for torque T. Equation 3.6 holds whether the motor armature is rotating or stationary. For purposes of calculation, we will use a value of 0.5 Nm/A^2 in this chapter for K_T of a series motor, unless indicated otherwise. This value of K_T is for a representative 1/8-horsepower (hp) motor.

Counter emf in the Motor

In the development of the torque and voltage equations, recall that the motor had armature current and field current as inputs and torque as an output. Look at Equation 3.5 for emf. Observe that when the armature speed (n_A) and field current (I_{field}) interact, a voltage is generated. This action occurs in the motor as well as the generator. As armature and field currents are applied to the motor to produce torque and armature speed, the speed and the field current produce a voltage in the armature of the motor. This self-generated voltage opposes the external armature voltage that was applied to turn the motor. This self-generated and opposing voltage is termed *counter electromotive force* and is abbreviated cemf. The equation for determining cemf in the motor is exactly the same as the equation for determining emf in the generator:

$$\text{cemf} = K_{\text{emf}} I_{\text{field}} n_A \qquad (3.7)$$

Conveniently, the cemf generally does not equal or exceed the value of armature voltage (V_A) applied to the motor. If the cemf did exceed the applied armature voltage, the motor would be acting like a generator and would drive current back into the line. If a voltmeter were placed across the armature, it would indicate the larger of the two voltages, applied armature voltage or cemf.

Counter Torque in the Generator

Likewise, in the generator, the rotation of the generator causes a current (I_A) to flow in the armature. Then, torque is produced in the generator because armature current (I_A) and field current (I_{field}) produced fields that are interacting. This torque is in opposition to the external torque turning the armature and is termed *counter torque* (cT). The counter torque equation for the generator is the same as the torque equation for the motor:

$$cT = K_T I_{field} I_A$$

Counter torque produced by the generator is generally less than the torque being supplied by the device that is turning the generator. From the previous development, we see that both the torque and voltage equations are in operation in both the motor and the generator.

Let us now apply these equations to the basic motor and generator configurations and observe the results.

DYNAMO CONFIGURATIONS

A cross-sectional view of a dynamo is shown in Figure 3.12A. Schematic diagrams of the various ways to connect the field and armature in the dynamo are shown in Figures 3.12B through 3.12F. The armature is drawn as a circle representing the end view of the armature, with two squares representing the brushes, and the field is represented by the standard inductor symbol.

The general configurations of DC dynamos take their names from the type of field excitation used.

Figure 3.12B shows the *separately excited dynamo*. When the dynamo supplies its own excitation, it is called a *self-excited dynamo* (Figures 3.12C through 3.12F). If the field of a self-excited dynamo is connected in parallel with the armature circuit, it is called a *shunt-connected dynamo* (Figure 3.12C). If the field is in series with the armature, it is called a *series-connected dynamo* (Figure 3.12D). If both series and shunt fields are used, it is called a *compound dynamo*. Compound dynamos may be connected as a *short shunt*, with the shunt field in parallel with the armature only (Figure 3.12E), or as a *long shunt*, with the shunt field in parallel with both the armature and series fields (Figure 3.12F). The series and shunt machines show the electrical placement of the interpole or compensating winding; their physical placement was discussed earlier.

Field potentiometers, as shown in this figure, are adjustable resistors placed in the field circuits to provide a means of varying the field flux, which controls the amount of emf generated or motor torque produced. These resistors are not placed in the armature circuit if that can be avoided because the current is higher in the armature than in the field. In keeping with basic control system principles, it is more efficient and economical to control the low-power circuits which, in turn, control the higher-power circuits. The current control of the lower-power field circuit has a profound effect on the output of the higher-power armature circuit, whether it is generating emf or producing torque.

CONSTANT LINE VOLTAGE

The following analyses are somewhat long and involved, but we feel this part of the chapter on DC motors and generators is the most important. In the following pages, we will produce the DC motor *characteristic curves*, which are graphs of the motor's electrical and mechanical responses. These curves are derived from applications of Equations 3.5 and 3.6, Ohm's law, and Kirchhoff's law. To truly understand DC dynamo principles, you must be very proficient with the characteristic curves and the equations just mentioned.

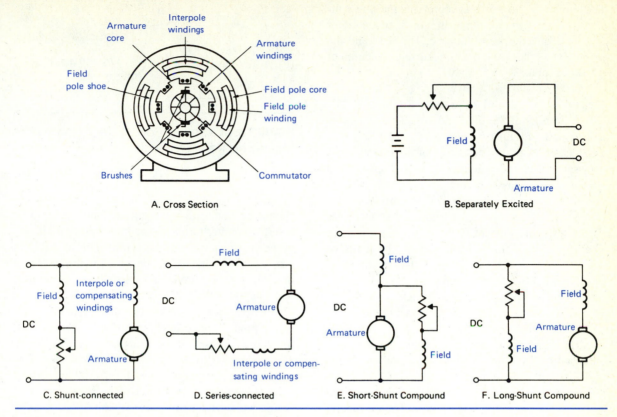

FIGURE 3.12 Cross-Sectional View of Dynamo and Dynamo Connections

The first topic we deal with concerns the series motor with constant line voltage. This condition can occur when the torque load of a motor varies while the line voltage remains fixed. The characteristic curves developed help illustrate what happens in the motor.

Series Motor

The series-connected motor is the first machine considered for analysis. A simplified series motor is shown in Figure 3.13.

Schematic. The field may be split into two windings, in actuality, one on each side of the armature. (Note that this type of motor is listed in Figure 3.5.) But in the schematic, it is generally drawn as one inductor.

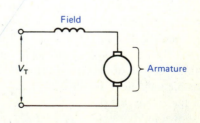

FIGURE 3.13 Basic Series-Connected Motor

The symbol V_T in this text represents the line voltage. It symbolizes the connection made to the motor or generator external to its case. The field and the armature together are the schematic representation of the series motor.

Measuring Field and Armature Resistance. The DC resistances of the field and armature are important factors and must be measured. If the field and armature leads do not come out of the motor, some disassembly of the machine may be necessary to obtain these measurements. In this text, a ⅛ hp machine will be used for illustration purposes.

As shown in Figure 3.13, the series motor has all the armature current flowing through the field. Therefore, the field windings should be made of heavy conductors in order to carry the armature current without heating or dropping an excessive amount of line voltage. A typical value of field resistance in this motor might be 50 Ω, as measured with an ohmmeter.

The resistance of the armature may be measured either with an ohmmeter or by a method called the locked-rotor–voltage-current method. If the ohmmeter method is used, the armature should be rotated *slowly* while measuring resistance. The ohmmeter reading will fluctuate owing to the slight amount of voltage and current being generated in the armature and to the fluctuating resistance caused by the brushes sliding from one commutator segment to another. The armature resistance may be on the order of 10 Ω.

In the *locked-rotor–voltage-current method*, a voltage is applied to the motor while the armature is held stationary (locked rotor). The armature voltage and current are measured, and the armature resistance is determined by applying Ohm's law to these measurements. The test voltage should not be applied for a long time because of the heating of the conductors and possible damage to the insulation.

For the motor just described, we now have the circuit shown in Figure 3.14, where R_{field} is the field resistance and R_A is the armature resistance.

Evaluating Variables. The next procedure is to apply a line or terminal voltage (V_T) and evaluate the value of K_{emf} in the emf equation. Some provision

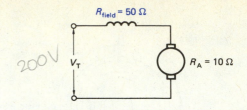

FIGURE 3.14 Series-Connected Motor with Measured Resistances

should be made for applying an armature load or external counter torque to the motor since n_A is one of the variables we want to control. This load can be provided by having the motor drive a hydraulic pump or some other rotational device such as a generator.

The hydraulic pump arrangement is illustrated in Figure 3.15. Here, the pump circulates hydraulic fluid in the pipe. If more load is required on the motor, the valve is closed further to restrict the fluid flow. This restricted flow produces an increased back pressure on the pump, which, in turn, produces an increased load on the drive motor. This loading method would work well for motors of ¼ hp or less.

When considering the torque and emf equations and the schematic of the series motor, we see that there are at least four variables: V_T, I_A or I_{field} (since these currents are the same in a series circuit), torque T, and speed n_A. In most system analyses, an instructive technique is to hold one of the variables constant while allowing a second variable to change and to observe the results on the remaining variables. For example, in Ohm's law, if V_T is held constant and

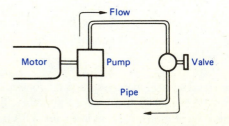

FIGURE 3.15 Hydraulic Pump Arrangement for Providing Load to Motor

resistance R is increased, then current I must decrease in order to keep the equation balanced. This technique will be applied to the motor and generator systems in this text.

For the first case, we will hold V_T constant at 100 V while the torque is varied and observe the effects on I_A and n_A. This result will be called the constant-voltage characteristic. The value of K_T was determined previously to be 0.5 Nm/A^2. We will arbitrarily pick an n_A of 1000 r/min and an I_A of 0.5 A for illustration purposes. These values would, of course, be measured in actual practice.

For the circuit of Figure 3.14, then, where I_A and I_{field} are equal (because it is a series circuit), the voltage drop across the field (V_{field}) is

$$V_{field} = I_{field}R_{field} \qquad (3.8)$$

For our circuit, we have

$$V_{field} = (0.5 \text{ A}) (50 \ \Omega) = 25 \text{ V}$$

The voltage drop across the field winding is due to the resistance of the wire. There is no reactive component of impedance in the field because the current is DC.

Figure 3.16 is a convenient way of representing the armature of the motor as an equivalent circuit. The cemf in the armature in Figure 3.16 is a function of field current (I_{field}) and armature speed (n_A) (see Equation 3.7). (The armature current may produce an inductive reactance, but that factor will be neglected here. See Table 3.6, moving-coil,

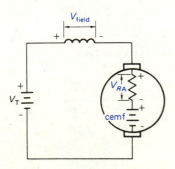

FIGURE 3.16 Voltage Distribution around Series Motor Circuit

printed-armature motor application factors.) In the motor, current flows from the line into the armature against the cemf. Applying Kirchhoff's law to the series motor yields the following equation:

$$V_T = \text{cemf} + I_A R_A + I_{field}R_{field} \qquad (3.9)$$

The line voltage must be balanced by the IR drops in the armature and the field and the cemf at all times. If the cemf were ever to become larger than line voltage, then current would flow back into the line against the line voltage.

The voltages V_{RA} (equal to $I_A R_A$, the drop across R_A, which is 5 V in this case) and V_{field} are voltage drops, and cemf is a voltage rise. Note that a voltmeter placed across the armature would measure V_A (75 V) (V_{RA} and cemf combined). Rearranging Equation 3.9 yields

$$\text{cemf} = V_T - I_A R_A - I_{field}R_{field}$$
$$= 100 \text{ V} - 5 \text{ V} - 25 \text{ V}$$
$$= 70 \text{ V}$$

At this point, the constant K_{emf} in the cemf equation (3.7) can be calculated. Rearranging the cemf equation yields

$$K_{emf} = \frac{\text{cemf}}{I_{field}n_A} \qquad (3.10)$$

For our illustration, we get

$$K_{emf} = \frac{70 \text{ V}}{(0.5 \text{ A}) (1000 \text{ r/min})}$$
$$= 0.14 \text{ V-min/A-r}$$

This value of K_{emf} should remain relatively constant as long as the motor is not taken apart and physically altered.

Torque can also be calculated, as follows:

$$T = K_T I_{field}I_A$$
$$= (0.5 \text{ Nm/A}^2) (0.5 \text{ A}) (0.5 \text{ A})$$
$$= 0.125 \text{ Nm}$$

Now, we have completed the calculations for one set of data. Additional sets of data need to be calculated and graphed so that we can visualize what is occurring when the torque (load demand) of the motor changes but the line voltage remains fixed. This procedure can be repeated for different values of torque. These calculations have been done and are tabulated in Table 3.1. The table also shows that as n_A decreases to zero, torque and armature current increase to some maximum value.

Locked Rotor. The condition when power is applied but n_A is zero is called a *locked-rotor condition*. When n_A is zero, voltage is applied to the motor, and the rotor or armature behaves as if locked. In this condition, maximum armature current and torque are produced because the opposing cemf is zero. This condition also occurs momentarily each time the motor is started from a dead stop.

When power is first applied to the motor, the armature is stationary; maximum current and torque are produced. As the motor starts to turn, the armature speed produces the opposing cemf, which, in turn, reduces current and, therefore, torque. As armature speed continues to increase, the armature current and torque produced are further reduced, until a stable operating condition is reached in which the torque produced by the motor just balances the torque required by the load to turn it.

Another name for the locked-rotor torque is *stall torque*. The motor is stalled until it produces enough torque to overcome the torque demanded and begins to rotate. The maximum rate of change of speed occurs when power is first applied. The rate of change in speed decreases until a stable speed is reached, at which time the rate of change is zero. This curve of armature speed versus time, shown in Figure 3.17, looks very much like the capacitance charge curve. In a typical motor, this action occurs within seconds. (Note the "performance ranges" of the motors listed in Table 3.6 and Example 3.6.)

Zero Torque. The bottom line in Table 3.1 shows the results of the torque demand going to zero. The cemf becomes equal to the line voltage (V_T), and no current flows in the armature or field. Of course,

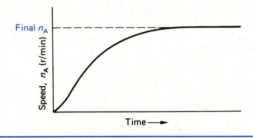

FIGURE 3.17 Time Response of Motor Starting from Standstill

TABLE 3.1   Series Motor, V_T Constant

V_T (V)	Torque (Nm)	$I_A(I_{field})$ (A)	V_{field} (V)	V_A (V)	V_{RA} (V)	cemf (V)	n_A (r/min)
100	1.394	1.67	83	17	17	0	0
100	0.926	1.36	68	32	13.6	18.4	96.6
100	0.500	1.00	50	50	10.0	40.0	286.0
100	0.463	0.962	48	52	9.6	42.0	314.0
100	0.245	0.70	35	65	7.0	58.0	592.0
100	0.125	0.50	25	75	5.0	70.0	1000.0
100	0.080	0.40	20	80	4.0	76.0	1357.0
100	0.045	0.30	15	85	3.0	82.0	1952.0
100	0.020	0.20	10	90	2.0	88.0	3143.0
100	0.005	0.10	5	95	1.0	94.0	6714.3
100	0	0	0	V_T	0	V_T	∞

these conditions can exist only in the ideal machine. The real machine has friction and other losses that limit the maximum armature speed as well as its ability to stay together (see discussion of runaway below).

Characteristic Curve. A graph of torque, current, and armature speed (from Table 3.1) is shown in Figure 3.18. These curves are the characteristic curves, or responses, for the series motor.

As shown in Figure 3.18, armature speed is by no means a linear function of torque. In fact, as T (demand) decreases, n_A increases dramatically. A dangerous condition can occur when the torque demand becomes very low and the speed very high. This condition is generally referred to as *runaway*. Runaway can inadvertently happen when the coupling between the motor and load accidentally breaks.

Two sets of graphs are shown in Figure 3.18: a set for $V_T = 100$ V and a set for $V_T = 50$ V. The resulting curves show that the characteristic curves are similar in shape. These curves are also referred to as *constant–line voltage curves*. A different set of curves results if the line voltage is allowed to vary and the torque is held constant (they are shown later).

Nameplate Specifications. Nameplate specifications are discussed in more detail later in this chapter, but they will be used here to evaluate constants and work examples. The following values were taken directly from the nameplate of a DC motor:

Volts	Hp	Amps	RPM	Duty	°C Rise
115 DC	⅛	0.8	1200	Con.	50

• EXAMPLE 3.1

Given the DC motor described above, find (a) the constant K_{emf} and (b) the constant K_T.

Solution

(a) In addition to the text method (measuring values at a given revolution per minute), another method can be used to evaluate the constants K_{emf} and K_T, using the nameplate data and some resistance measurements. For the motor described above, the field resistance (R_{field}) was measured and found to be 40 Ω and the armature resistance (R_A) was 5 Ω. Assuming the motor to be operating at rated (nameplate) values, we have

$$I_A = I_{field} = 0.8 \text{ A}$$

Using Equation 3.9 but solving for cemf yields

$$cemf = V_T - I_A R_A - I_{field} R_{field}$$
$$= 115 \text{ V} - (0.8 \text{ A})(5 \text{ Ω}) - (0.8 \text{ A})(40 \text{ Ω})$$
$$= 79 \text{ V}$$

Now, solving Equation 3.7 for K_{emf}, we get

$$K_{emf} = \frac{cemf}{I_{field} n_A} = \frac{79 \text{ V}}{(0.8 \text{ A})(1200 \text{ r/min})}$$
$$= 0.0823 \text{ V-min/A-r}$$

(b) To find K_T, we use Equation 3.11 (discussed later) to determine torque (in pound-feet):

$$T = \frac{hp \times 5252}{n_A} \tag{3.11}$$

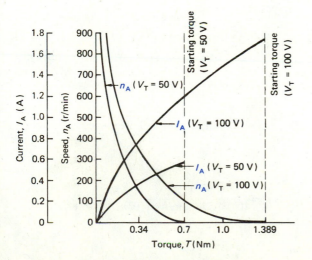

FIGURE 3.18 Characteristic Curves for Series Motor at Constant Voltage

So we have

$$T = \frac{(1/8 \text{ hp}) (5252)}{(1200 \text{ r/min})} = 0.547 \text{ lb-ft}$$

Converting to newton-meters gives

$$T = 0.547 \text{ lb-ft} \times \frac{1 \text{ Nm}}{0.73756 \text{ lb-ft}}$$

$$= 0.742 \text{ Nm}$$

Solving Equation 3.6 for K_T, we get

$$K_T = \frac{T}{I_{\text{field}} I_A} = \frac{0.742 \text{ Nm}}{(0.8 \text{ A}) (0.8 \text{ A})}$$

$$= 1.16 \text{ Nm/A}^2$$

• EXAMPLE 3.2

Given the DC motor and values for K_{emf} and K_T in Example 3.1, find (a) the stall torque and (b) the associated armature current, cemf, and rotor speed.

Solution

Step 1 Sometimes, problems are stated in such a manner that the second part of the solution is needed first. Such is the case here. The torque equation (Equation 3.6) requires that the current be known. In order to find the armature current, the cemf must be known; so we will start with Equation 3.7:

$$\text{cemf} = K_{\text{emf}} I_{\text{field}} n_A$$

$$= K_{\text{emf}} I_{\text{field}} (0 \text{ r/min})$$

$$= 0 \text{ V}$$

Since the armature is not rotating (stalled), armature speed is also zero.

Equation 3.9 can be used to find the current since I_{field} and I_A are equal. We have

$$V_T = \text{cemf} + I_A R_A + I_A R_{\text{field}}$$

So

$$I_A = \frac{V_T}{R_A + R_{\text{field}}} = \frac{115 \text{ V}}{5 \, \Omega + 40 \, \Omega}$$

$$= 2.56 \text{ A}$$

Step 2 And now, from Equation 3.6, we find

$$T = K_T I_{\text{field}} I_A = K_T I_A I_A$$

$$= (1.16 \text{ Nm/A}^2) (2.56 \text{ A})^2$$

$$= 7.60 \text{ Nm}$$

• EXAMPLE 3.3

Given the DC motor and values for K_{emf} and K_T in Example 3.1, find (a) the cemf when armature current is 10% of the rated value and (b) rotor speed at this torque. This question is designed to investigate the torque and rotor speed near the unloaded condition, as if the coupling between the armature and load suddenly broke.

Solution

(a) Since armature current is 10% of the rated value ($I_A = 0.08 \text{ A}$), we can solve Equation 3.9 for cemf:

$$\text{cemf} = V_T - I_A R_A - I_{\text{field}} R_{\text{field}}$$

$$= 115 \text{ V} - (0.08 \text{ A}) (5 \, \Omega) - (0.08 \text{ A}) (40 \, \Omega)$$

$$= 111.4 \text{ V}$$

(b) Now, solving Equation 3.7 for speed, we get

$$n_A = \frac{\text{cemf}}{K_{\text{emf}} I_{\text{field}}}$$

$$= \frac{111.4 \text{ V}}{(0.0823 \text{ V-min/A-r}) (0.08 \text{ A})}$$

$$= 16,900 \text{ r/min}$$

As you can see, this speed is quite high, and the motor may come apart!

Shunt Motor

In the shunt-connected motor, shown in Figure 3.19, the field is connected in parallel, or shunt, with the armature. Field current and armature current are not the same.

Since the field winding no longer carries armature current, it can be made of fine wire and contain many turns. Thus, the shunt motor field has higher resistance than the series motor field. A typical value of shunt field resistance for a ⅛ hp motor may be 1000 Ω.

Evaluating Variables. Again, line voltage is held constant at 100 V, the torque required is varied, and armature current (I_A) and speed (n_A) are observed. The value of K_T is evaluated, as we did previously; it is found to be 0.2 Nm/A². An initial set of values, $n_A = 1000$ r/min and $I_A = 6.25$ A, is measured.

The field voltage drop (V_{field}) is the same as the line voltage (V_T) and armature voltage (V_A) since they are all in parallel. To determine field current (I_{field}), we apply Ohm's law:

$$I_{field} = \frac{V_{field}}{R_{field}} = \frac{100 \text{ V}}{1000 \text{ }\Omega} = 0.1 \text{ A}$$

Then, from Equation 3.6, we have

$$T = K_T I_{field} I_A$$
$$= (0.2 \text{ Nm/A}^2)(0.1 \text{ A})(6.25 \text{ A})$$
$$= 0.125 \text{ Nm}$$

To determine cemf, we first evaluate the internal voltage drop due to R_A. The voltage drop is

$$V_{RA} = I_A R_A \qquad (3.12)$$

For our circuit, we get

$$V_{RA} = (6.25 \text{ A})(10 \text{ }\Omega) = 62.5 \text{ V}$$

The cemf, then, is

$$\text{cemf} = V_T - V_{RA} \qquad (3.13)$$

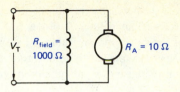

FIGURE 3.19 Basic Shunt-Connected Motor

So

$$\text{cemf} = 100 \text{ V} - 62.5 \text{ V} = 37.5 \text{ V}$$

At this point, K_{emf} for the shunt motor can be evaluated by using Equation 3.10:

$$K_{emf} = \frac{\text{cemf}}{I_{field} n_A} = \frac{37.5 \text{ V}}{(0.1 \text{ A})(1000 \text{ r/min})}$$
$$= 0.375 \text{ V-min/A-r}$$

Knowing the value of K_{emf} now allows us to determine any value of n_A for the shunt motor. First, we select a value of torque required of the motor— for example, 0.1 Nm—and solve for armature current (I_A):

$$I_A = \frac{T}{K_T I_{field}} = \frac{0.1 \text{ Nm}}{(0.2 \text{ Nm/A}^2)(0.1 \text{ A})} = 5 \text{ A}$$

Then,

$$V_{RA} = I_A R_A = (5 \text{ A})(10 \text{ }\Omega) = 50 \text{ V}$$
$$\text{cemf} = V_T - V_{RA} = 100 \text{ V} - 50 \text{ V}$$
$$= 50 \text{ V}$$

We find n_A by using the following equation:

$$n_A = \frac{\text{cemf}}{K_{emf} I_{field}} \qquad (3.14)$$

Therefore,

$$n_A = \frac{50 \text{ V}}{(0.375 \text{ V-min/A-r})(0.1 \text{ A})}$$
$$= 1333 \text{ r/min}$$

Table 3.2 shows the results of this procedure repeated for various values of torque. Similar to the results for the series motor, Table 3.2 shows that as armature speed decreases to zero, torque and armature current increase to some maximum value.

Maximum Armature Speed. How is maximum armature speed determined in the shunt motor? From the cemf equation, we have

$$n_{A(max)} = \frac{cemf_{max}}{K_{emf}I_{field}}$$

$$= \frac{100 \text{ V}}{(0.375 \text{ V-min/A-r}) (0.1 \text{ A})}$$

$$= 2667 \text{ r/min}$$

This value is for an ideal condition, assuming zero armature current.

Maximum speed for the shunt motor can be determined in another way without knowing line voltage. Look at the previous equation:

$$n_{A(max)} = \frac{cemf_{max}}{K_{emf}I_{field}}$$

where

$$cemf_{max} = V_T - I_A R_A$$

But $I_A = 0$ since cemf is maximum. Therefore, cemf $= V_T$.

Solving for I_{field} in terms of voltage and resistance gives

$$I_{field} = \frac{V_T}{R_{field}}$$

Substituting yields

$$n_{A(max)} = \frac{cemf_{max}}{K_{emf}I_{field}} = \frac{V_T}{K_{emf}(V_T/R_{field})}$$

$$= \frac{V_T R_{field}}{K_{emf}V_T}$$

The term V_T cancels. Thus,

$$n_{A(max)} = \frac{R_{field}}{K_{emf}} \tag{3.15}$$

and

$$n_{A(max)} = \frac{1000 \ \Omega}{0.375 \text{ V-min/A-r}}$$

$$= 2667 \text{ r/min}$$

This equation shows that maximum armature speed of the shunt motor is a function not of line voltage but of circuit constants. In other words,

TABLE 3.2 Shunt Motor, V_T Constant

V_T (V)	Torque (Nm)	I_{field} (A)	I_A (A)	V_{field} (V)	V_A (V)	V_{RA} (V)	cemf (V)	n_A (r/min)
100	0.2	0.1	—	100	100	—	—	Decreasing to 0
100	0.15	0.1	7.50	100	100	75.0	25.0	667
100	0.133	0.1	6.65	100	100	66.5	33.5	893
100	0.125	0.1	6.25	100	100	62.5	37.5	1000
100	0.1	0.1	5.00	100	100	50.0	50.0	1333
100	0.075	0.1	3.75	100	100	37.5	62.5	1667
100	0.05	0.1	3.33	100	100	33.3	66.7	1779
100	0.025	0.1	1.25	100	100	12.5	87.5	2333
100	0	0.1	0	100	100	0	V_T	Approaches maximum

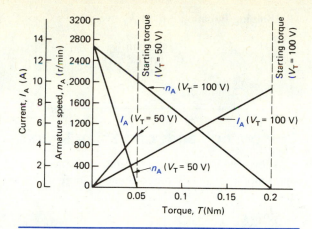

FIGURE 3.20 Characteristic Curves for Shunt Motor at Constant Voltage

theoretically, no matter what line voltage is applied, as torque demand goes to zero, speed will always go to the same value. This result is shown in Figure 3.20 for line voltage of 100 V and 50 V.

Doubling Line Voltage. Observe in Figure 3.20 what occurs when line voltage is doubled. Maximum torque produced is quadrupled because both I_A and I_{field} are doubled. Thus, some high-voltage motors produce disproportionately more available torque.

• EXAMPLE 3.4
Notice that Table 3.2 has some missing values at 0 r/min. Given that the power is applied to the motor but the armature is not rotating, find I_A, V_{RA}, and cemf.

Solution
From Equation 3.7, we have

$$\text{cemf} = K_{emf}I_{field}n_A$$
$$= K_{emf}I_{field}(0 \text{ r/min})$$
$$= 0 \text{ V}$$

Again, since the armature is not rotating, armature speed is also zero.

Equation 3.13 for the shunt motor shows that

$$V_T = \text{cemf} + I_A R_A$$

and solving for I_A gives (since cemf is zero)

$$I_A = \frac{V_T}{R_A} = \frac{100 \text{ V}}{10 \text{ }\Omega}$$
$$= 10.0 \text{ A}$$

Even at first glance, this value appears to be correct since I_A is increasing as armature speed is decreasing.

With armature current determined, we can now find V_{RA} from Ohm's law:

$$V_{RA} = I_A V_{RA} = (10 \text{ A})(10 \text{ }\Omega) = 100 \text{ V}$$

Obviously, $V_{RA} = V_T$ at zero armature speed.

As a check on Table 3.2, we find

$$T = K_T I_{field} I_A = (0.2 \text{ Nm/A}^2)(0.1 \text{ A})(10 \text{ A})$$
$$= 0.2 \text{ Nm}$$

• EXAMPLE 3.5
Given a DC shunt motor with the following name-plate values

Volts	Hp	Amps	RPM	Duty	°C Rise
115 DC	⅛	1.4	1800	Con.	50

find (a) the constant K_{emf} and (b) the constant K_T.

Solution
(a) Measurements were taken to determine the field and armature resistances. They were $R_{field} = 856 \text{ }\Omega$ and $R_A = 15 \text{ }\Omega$.

We now find field current, which remains constant as long as the line voltage does. We obtain

$$I_{field} = \frac{V_{field}}{R_{field}} = \frac{115 \text{ V}}{856 \text{ }\Omega} = 0.134 \text{ A}$$

Armature current can be found by using Kirchhoff's current law for parallel circuits:

$$I_{line} = I_A + I_{field}$$

where

I_{line} = line current as given on nameplate

Solving Kirchhoff's equation for armature current, we get

$$I_A = I_{line} - I_{field} = 1.4\ A - 0.134\ A$$
$$= 1.27\ A$$

The voltage due to armature resistance is

$$V_{RA} = I_A R_A = (1.27\ A)\,(15\ \Omega)$$
$$= 19.0\ V$$

We can now find cemf from Equation 3.13

$$cemf = V_T - V_{RA} = 115\ V - 19\ V$$
$$= 96\ V$$

Solving Equation 3.7 for K_{emf} gives

$$K_{emf} = \frac{cemf}{I_{field}n_A} = \frac{96\ V}{(0.134\ A)\,(1800\ r/min)}$$
$$= 0.398\ V\text{-}min/A\text{-}r$$

(b) To find K_T, we use Equation 3.11 to determine torque (in pound-feet):

$$T = \frac{hp \times 5252}{n_A} = \frac{(1/8\ hp)\,(5252)}{(1800\ r/min)}$$
$$= 0.365\ lb\text{-}ft$$

Converting to newton-meters yields

$$T = 0.365\ lb\text{-}ft \times \frac{1\ Nm}{0.73756\ lb\text{-}ft}$$
$$= 0.495\ Nm$$

Solving Equation 3.6 for K_T, we get

$$K_T = \frac{T}{I_{field}I_A} = \frac{0.495\ Nm}{(0.134\ A)\,(1.27\ A)}$$
$$= 2.91\ Nm/A^2$$

Comparing Series and Shunt Motors

As shown in Figure 3.20, armature current in the shunt motor is directly proportional to torque:

$$I_A = \frac{T}{K_T I_{field}}$$

(The current I_{field} stays constant as long as V_T is constant.)

But in a series motor (Figure 3.18), armature current is a second-order function of torque:

$$I_A = I_{field}$$

Therefore,

$$I_A^2 = \frac{T}{K_T}$$

This equation results in the parabolic-shaped curve of Figure 3.18.

CONSTANT TORQUE

When we examined the series and shunt motors in the previous sections, we observed that there were at least four variables: V_T, I_A or I_{field} (since they are the same in a series motor), torque (T), and n_A. Obviously, there are more variables than can be easily observed on a simple two-dimensional graph. Our approach, then, is to hold one variable constant, vary another, and observe the effect on the remaining variables. In the previous sections, we examined the effects on I_A and n_A as torque was varied and V_T was held constant. In the next investigation, we will maintain torque at a constant value, vary the line voltage (V_T), and again observe the effects on armature current (I_A) and speed n_A.

Series Motor

Since the same series motor will be used in this section as in the previous one, K_T (0.5 Nm/A^2) and K_{emf} (0.14 V-min/A-r) will also be the same because we have not physically altered the motor. Therefore, from Table 3.1, when V_T = 100 V, torque (T) is 0.125 Nm and n_A is 1000 r/min.

The next step is to vary V_T while torque is held at 0.125 Nm. Armature and field currents will be constant at 0.5 A. From Equation 3.6,

$$T = K_T I_{field} I_A$$

Since the field and armature currents are equal, solving for current yields

$$I_A = \sqrt{\frac{T}{K_T}} = \sqrt{\frac{0.125 \text{ Nm}}{0.5 \text{ Nm/A}^2}} = 0.5 \text{ A}$$

Therefore, from Equation 3.8 and from Figure 3.14,

$$V_{field} = I_{field} R_{field} = (0.5 \text{ A})(50 \text{ }\Omega) = 25 \text{ V}$$

Also,

$$V_{RA} = I_A R_A = (0.5 \text{ A})(10 \text{ }\Omega) = 5 \text{ V}$$

Setting V_T to 50 V in Figure 3.14 produces

$$V_A = V_T - V_{field} = 50 \text{ V} - 25 \text{ V} = 25 \text{ V}$$

And from Figure 3.16, we have

$$cemf = V_A - V_{RA} = 25 \text{ V} - 5 \text{ V} = 20 \text{ V}$$

Therefore, rearranging Equation 3.10 yields

$$n_A = \frac{cemf}{K_{emf} I_{field}} = \frac{20 \text{ V}}{(0.14 \text{ V-min/A-r})(0.5 \text{ A})}$$

$$= 286 \text{ r/min}$$

Table 3.3 shows the preceding procedure repeated for different values of V_T. From Table 3.3, we see that as the line voltage continues to increase, so does armature speed; no theoretical limit is reached. In practice, maximum armature speed is determined by how long the motor can hold together.

Minimum line voltage is reached when n_A goes to zero or stalls. At zero armature speed, cemf is also zero. That is,

$$n_A = \frac{cemf}{K_{emf} I_{field}} = \frac{0}{(0.14 \text{ V-min/A-r})(0.5 \text{ A})}$$

$$= 0$$

Also,

$$V_A = V_T - V_{field}$$

Or after rearrangement, we get

TABLE 3.3   Series Motor, Torque Constant

V_T (V)	Torque (Nm)	$I_A(I_{field})$ (A)	V_{field} (V)	V_A (V)	V_{RA} (V)	cemf (V)	n_A (r/min)
Stall voltage	0.125	0.5	25	Decreasing to minimum	5	0	0
50	0.125	0.5	25	25	5	20	286
100	0.125	0.5	25	75	5	70	1000
150	0.125	0.5	25	125	5	120	1714
200	0.125	0.5	25	175	5	170	2429
∞	0.125	0.5	25	∞	5	∞	∞

$$V_T = V_A + V_{\text{field}} = (\text{cemf} + V_{RA}) + V_{\text{field}}$$

$$= (\text{cemf} + I_A R_A) + V_{\text{field}}$$

$$= 0 + (0.5\ \text{A})(10\ \Omega) + 25\ \text{V}$$

$$= 30\ \text{V}$$

This equation shows that when the line voltage decreases to 30 V, the motor will stall if it still is required to produce 0.125 Nm of torque.

The graph in Figure 3.21 is the characteristic curve for the series motor with constant torque. Two armature speed curves have been drawn to show what happens at a higher, fixed torque. Notice that these lines are not parallel. Also, the higher-torque example does not start rotating until a line voltage of 30 V is exceeded.

Shunt Motor

In the shunt motor under varying line voltage but constant torque, there are five possible variables: V_T, I_A, I_{field}, T, and n_A. Field current in the shunt motor varies directly as the line voltage. If the torque is held constant and field current changes, then armature current must vary inversely.

The calculations required to produce the characteristic curves for the shunt motor of Figure 3.19 are

as follows: Using values for K_T of 0.2 Nm/A^2, for T of 0.125 Nm, and for K_{emf} of 0.375 V-min/A-r, increase the line voltage from 100 to 150 V. Then, we have the following calculations:

$$I_{\text{field}} = \frac{V_{\text{field}}}{R_{\text{field}}} = \frac{150\ \text{V}}{1000\ \Omega} = 0.15\ \text{A}$$

$$I_A = \frac{T}{K_T I_{\text{field}}} = \frac{0.125\ \text{Nm}}{(0.2\ \text{Nm/A}^2)(0.15\ \text{A})}$$

$$= 4.17\ \text{A}$$

$$V_{RA} = I_A R_A = (4.17\ \text{A})(10\ \Omega) = 41.7\ \text{V}$$

Recall that in a shunt motor, $V_A = V_T$, from Equation 3.13. Therefore,

$$\text{cemf} = V_A - V_{RA} = 150\ \text{V} - 41.7\ \text{V}$$

$$= 108.3\ \text{V}$$

$$n_A = \frac{\text{cemf}}{K_{\text{emf}} I_{\text{field}}}$$

$$= \frac{108.3\ \text{V}}{(0.375\ \text{V-min/A-r})(0.15\ \text{A})}$$

$$= 1925\ \text{r/min}$$

Table 3.4, which shows the results of repeating the above procedure for various values of line voltage, is another verification that line voltage does not determine maximum possible speed for the shunt motor. Maximum armature speed is determined as follows:

$$n_{A(\text{max})} = \frac{R_{\text{field}}}{K_{\text{emf}}} = \frac{1000\ \Omega}{0.375\ \text{V-min/A-r}}$$

$$= 2667\ \text{r/min}$$

We also need to determine *stall voltage*, the voltage at which the motor stops turning. At stall, we have

$$\text{cemf} = 0$$

$$V_T = V_{\text{field}} = V_{RA} = I_A R_A$$

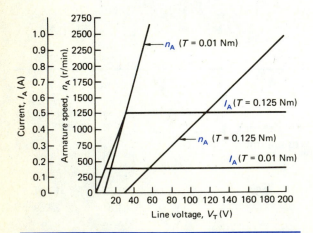

FIGURE 3.21 Characteristic Curves for Series Motor at Constant Torque

TABLE 3.4 Shunt Motor, Torque Constant

V_T (V)	Torque (Nm)	I_{field} (A)	I_A (A)	V_{field} (V)	V_A (V)	V_{RA} (V)	cemf (V)	n_A (r/min)
Stall voltage	0.125	Decreasing to 0	Increasing to maximum	Stall voltage	Stall voltage	Increasing to maximum	Decreasing to 0	Decreasing to 0
90	0.125	0.09	6.94	90	90	69.4	20.6	610
100	0.125	0.1	6.25	100	100	62.5	37.5	1000
120	0.125	0.12	5.21	120	120	52.1	67.9	1509
150	0.125	0.15	4.17	150	150	41.7	108.3	1925
160	0.125	0.16	3.91	160	160	39.1	120.9	2015
200	0.125	0.2	3.125	200	200	31.3	168.8	2250
300	0.125	0.3	2.08	300	300	20.8	279.2	2482
∞	0.125	0	0	—	—	0	—	Maximum

or

$$I_{field}R_{field} = I_A R_A$$

Solving for I_A yields

$$I_A = \frac{I_{field}R_{field}}{R_A}$$

Also,

$$T = K_T I_{field} I_A$$

Substitution gives

$$T = K_T I_{field}\left(\frac{I_{field}R_{field}}{R_A}\right)$$

Solving for I_{field}, we have

$$T = \frac{K_T I_{field}^2 R_{field}}{R_A}$$

$$I_{field} = \sqrt{\frac{TR_A}{K_T R_{field}}}$$

$$= \sqrt{\frac{(0.125\text{ Nm})(10\text{ }\Omega)}{(0.2\text{ Nm/A}^2)(1000\text{ }\Omega)}}$$

$$= 0.07906\text{ A}$$

Also,

$$I_A = \frac{T}{K_T I_{field}} = \frac{0.125\text{ Nm}}{(0.2\text{ Nm/A}^2)(0.07906\text{ A})}$$

$$= 7.905\text{ A}$$

$$V_T = I_A R_A = (7.905\text{ A})(10\text{ }\Omega) = 79.05\text{ V}$$

The data in Table 3.4, along with lower-constant-torque data, are used to obtain the curves in Figure 3.22. Notice that as torque decreases, the curve becomes more angular. Thus, if the torque demand on the shunt motor is low, maximum speed is reached quickly with the application of a relatively small line voltage. In the ideal or theoretical motor with a torque demand of zero, maximum speed will be reached any time the line voltage exceeds 0 V. Figure 3.22 also shows that armature current is linear up to the point where the motor starts rotating. After that point, armature current decreases, owing to increased cemf.

MOTOR CONTROL

Now that we have completed the basic characteristic curves, we must discuss how to control the motor. Methods and circuitry for electronic motor control will be presented in Chapter 7, but a discussion of what parts of a motor lend themselves to control can be undertaken now.

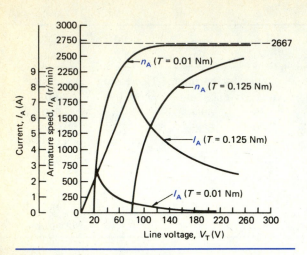

FIGURE 3.22 Characteristic Curves for Shunt Motor at Constant Torque

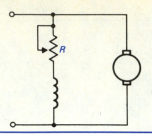

FIGURE 3.23 Shunt Motor with Field Current Control

A general rule for control systems is to place the control mechanism in the low-power part of the system so that the circuit can then control the higher-power part of the system. For example, a signal in a transistor amplifier is generally applied to the base because it is the low-power circuit that controls the higher-power output circuit. However, just as the transistor is not always base-driven, neither are all control systems controlled in the low-power circuitry. Much depends on the application and the component(s) used for the control.

In the shunt motor, to control the speed by controlling field current is relatively easy. Field current is generally much smaller than armature current and, as we will see, has a great effect on armature current. For example, consider the shunt motor in Figure 3.23. A variable resistor R has been placed in the field circuit to control field current. Let us see what effect a changing field current has on armature speed.

For convenience, we consider this motor to be the same as the one in the previous section pertaining to shunt motors with constant line voltage. With $V_T = 100$ V, Table 3.5 shows the measured and calculated values. We will assume that the motor load needs the same amount of torque to turn it, no matter what speed it has.

Now, suppose resistor R is increased suddenly so that field current is cut in half. A synchrogram of this result is shown in Figure 3.24. A *synchrogram* is a vertical alignment of related graphs to show a time relationship.

We will assume that the speed of the motor does not change instantly whenever a change in torque is applied because of inertia of the armature. If at the instant field current changes, n_A is still 2533, then we have the following results: Rearranging Equation 3.14 gives

$$\text{cemf} = K_{\text{emf}} I_{\text{field}} n_A$$
$$= (0.375 \text{ V-min/A-r}) (0.05 \text{ A})$$
$$\times (2533 \text{ r/min})$$
$$= 47.5 \text{ V}$$

TABLE 3.5 Measured and Calculated Values for Shunt Motor

V_T (V)	Torque (Nm)	I_{field} (A)	I_A (A)	V_{field} (V)	V_A (V)	V_{RA} (V)	cemf (V)	n_A (r/min)
100	0.01	0.1	0.5	100	100	5	95	2533

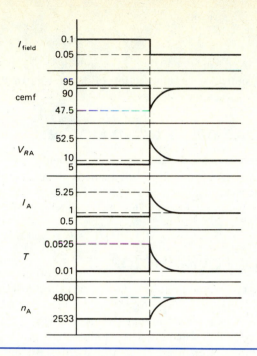

FIGURE 3.24 Synchrogram of Changes that Occur within Shunt Motor

Rearranging Equation 3.13 yields

$$V_{RA} = V_T - \text{cemf} = 100 \text{ V} - 47.5 \text{ V}$$

$$= 52.5 \text{V}$$

Rearranging Equation 3.12 gives

$$I_A = \frac{V_{RA}}{R_A} = \frac{52.5 \text{ V}}{10 \text{ }\Omega} = 5.25 \text{ A}$$

Finally, from Equation 3.6,

$$T = K_T I_{\text{field}} I_A$$

$$= \left(0.2 \text{ Nm/A}^2\right) (0.05 \text{ A}) (5.25 \text{ A})$$

$$= 0.0525 \text{ Nm}$$

With this increased torque produced, the motor will increase in speed. As the motor increases in speed, more cemf is produced, which, in turn, decreases armature current, causing reduced torque.

A stable speed is reached when torque returns to its original value, as the following calculations show:

$$I_{A(\text{final})} = \frac{T}{K_T I_{\text{field}}} = \frac{0.01 \text{ Nm}}{\left(0.2 \text{ Nm/A}^2\right)(0.05 \text{ A})}$$

$$= 1 \text{ A}$$

$$V_{RA(\text{final})} = I_A R_A = (1 \text{ A}) (10 \text{ }\Omega) = 10 \text{ V}$$

$$\text{cemf}_{\text{final}} = V_T - V_{RA} = 100 \text{ V} - 10 \text{ V}$$

$$= 90 \text{ V}$$

$$n_{A(\text{stable})} = \frac{\text{cemf}}{K_{\text{emf}} I_{\text{field}}}$$

$$= \frac{90 \text{ V}}{(0.375 \text{ V-min/A-r}) (0.05 \text{ A})}$$

$$= 4800 \text{ r/min}$$

These calculations show that a 0.05 A change in field current can cause a 0.5 A change in armature current. The surprising aspect is that when field current is reduced, armature speed is *increased*. Thus, if the shunt motor field is accidentally disconnected, the motor could go into runaway.

Observation of the speed curve in Figure 3.24 shows that the previous shunt motor maximum speed of 2667 r/min (Equation 3.15) has been greatly exceeded (4800 r/min). This result occurs because the field resistance must now include the control resistor's resistance:

$$R_{\text{field}} = \frac{V_T}{I_{\text{field}}} = \frac{100 \text{ V}}{0.05 \text{ A}} = 2000 \text{ }\Omega$$

$$n_{A(\text{max})} = \frac{R_{\text{field}}}{K_{\text{emf}}} = \frac{2000 \text{ }\Omega}{0.375 \text{ V-min/A-r}}$$

$$= 5333 \text{ r/min}$$

Note: These last two equations were used to calculate *maximum* possible speed. Figure 3.24 is not representing maximum speeds.

The series motor, shown in Figure 3.25, has a control resistor in parallel with the field. In the series motor, as resistor R is decreased, the actual

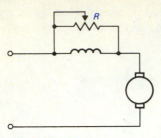

FIGURE 3.25 Series Motor with Control Resistor

field current is reduced, thus reducing field strength. However, armature current is increased because the total field resistance is decreasing. The net effect is that torque is increased in the series motor, just as it was in the shunt motor previously discussed.

CHARACTERISTIC CURVES

The characteristic curves mathematically derived in the previous sections show that the emf and torque equations do represent the motor's physical operation. It is not our intent, though, to have you, as a technician, become proficient in manipulating numbers. Rather, we expect you to perceive basic motor operation through the equations and characteristic curves. For instance, consider the following examples:

1. In the series motor with constant line voltage, what will happen to the armature speed if torque is decreased? From the curve in Figure 3.18, we see that speed will increase. You probably will not need to know how much the speed increases, but you should know whether it increases nonlinearly.

2. If the field current in a shunt motor of constant line voltage is decreased, what happens to the motor armature speed? As discussed in the previous section, speed generally increases because cemf typically is larger than V_{RA}. This action is not easily observable in either Equation 3.6 or 3.7 since cemf is related to speed n_A as follows:

$$n_A = \frac{\text{cemf}}{K_{\text{emf}} I_{\text{field}}}$$

And since cemf and I_{field} both decrease, they both have counteracting effects on n_A or the characteristic curves.

SEPARATELY EXCITED MOTOR

The schematic for the separately excited motor is shown in Figure 3.26. The equations and characteristic curves for this motor are nearly identical to those for the shunt motor. The derivations are left as exercises.

COMPOUND MOTOR

The compound motor has both a shunt and a series field. As a review, Figure 3.27 shows the two possible ways of connecting the fields: long shunt and short shunt. Of these two possibilities, the long shunt is most often used since the shunt field has a relatively constant current—and, therefore, field strength—as long as the line voltage remains constant. The long-shunt arrangement also produces a series field strength that is proportional to armature current.

In most cases, the series winding is connected so that its field aids that of the shunt winding, as shown in Figure 3.28A. Motors of this type are called *cumulative compound motors*. If the series winding is connected so that its field opposes that of the shunt winding, the motor is called a *differential compound*

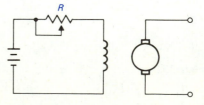

FIGURE 3.26 Separately Excited Motor

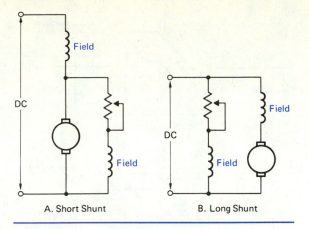

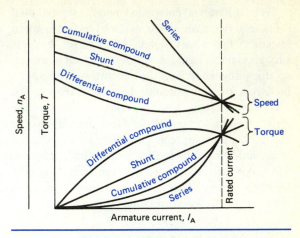

FIGURE 3.27 Compound Motor Configurations

FIGURE 3.29 Characteristic Curves for Compound, Series, and Shunt Motors

motor, as shown in Figure 3.28B. In these figures, Φ represents the flux per pole, and the arrows indicate relative direction of the flux.

The characteristics of a cumulative compound motor are between those of a series motor and those of a shunt motor. If the field strength of the series field is stronger than that of the shunt field, the characteristics are more like those of a series motor, and vice versa. Figure 3.29 illustrates the relationship between the motors.

As the figure indicates, the cumulative compound motor has higher starting torque than the shunt motor and better speed regulation than the series motor. Unlike the series motor, however, it does have a definite no-load speed.

In some operations, use of the cumulative series winding is desirable to obtain good starting torque. When the motor comes up to speed, the series winding is shorted out. The motor then has the improved speed regulation of a shunt motor.

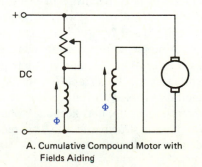

A. Cumulative Compound Motor with Fields Aiding

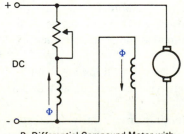

B. Differential Compound Motor with Fields Opposing

FIGURE 3.28 Compound Motors

The differential compound motor is seldom used because of two basic problems. One problem is speed instability under heavy load. The speed regulation is very good under light loads. But as the load increases, so does the speed of the motor. As the speed increases, the armature current also increases. The increased current causes the field strength of the series winding's opposing field to eventually exceed the shunt's field strength, and the motor starts to run backward. As shown in Figure 3.29, the other problem is that starting torque is low for a relatively high armature current. That is, it uses energy in opposing magnetic fields when armature current is high.

GENERAL CONSIDERATIONS

Reversing Direction

Reversing the direction of any DC motor requires that the current through the armature with respect to the field be reversed. If the currents in the armature and field are both reversed, there will be no change in rotational direction. This situation is shown in Figure 3.30, where the direction of *F* (force) remains unchanged when both conductor and field currents are reversed.

To change the direction of a motor, we must reverse the armature rather than the field, for the following four reasons:

1. The field is more inductive than the armature, and frequent reversals produce switch contact arcing and erosion.
2. In a compound motor, both fields must be reversed. Otherwise, the motor will change from a cumulative compound motor to a differential compound motor.
3. The armature circuit connections are usually opened for various types of braking, as will be discussed later.
4. If the field-reversing circuit fails to close, runaway could result.

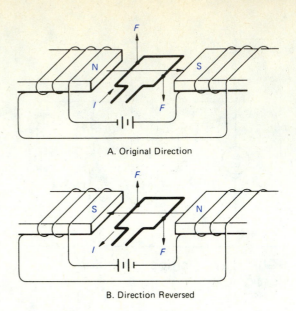

A. Original Direction

B. Direction Reversed

FIGURE 3.30 Magnetic Field and Armature Current Reversed

Motor Starters and Controllers

In large motors, a motor starter is required to prevent excessive starting currents. A *motor starter* is a switching device that is intended to start and accelerate the motor. A *motor controller*, on the other hand, is a device that controls the power flow to a motor, usually resulting in some form of speed control. Controllers will be discussed later in Chapters 6 and 7.

A motor starter generally consists of a resistive bank inserted in the motor circuit to prevent full line voltage from being applied to the motor. Some method is used to reduce this resistance as the motor speeds up, thus keeping motor currents from reaching destructive levels. This reduction may be done manually or automatically.

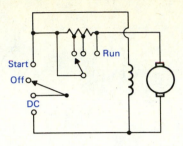

FIGURE 3.31   Manual Motor Starter Circuit

Figure 3.31 shows a typical manual starter circuit. A spring on the starter switch holds it in the off position. But once in the run position, the switch will be held there by an electromagnet. It is up to the operator to provide the necessary time delay between switch positions to allow the speed to stabilize and prevent high currents. An automatic starter typically uses the same resistive configuration, but the switching is done automatically with time-delay relays.

Most starters and controllers include one or more devices known as *overload relays*. These devices protect the motor from overheating caused by loads above the rated value. There are four types of overload protection systems in use: (1) thermal relays, (2) magnetic relays, (3) electronic relays, and (4) thermistors.

Thermal overload relays have a small heating element connected in series with the motor. Some consist of a bimetallic strip with a heating coil around it. As the current heats the strip, it opens and stops the motor. Other devices resemble fuses, which melt to break the circuit. Some method of resetting the circuit, either manually or automatically, is also provided.

Figure 3.32 shows the schematic symbol for the thermal relay. The question-mark shapes represent the heating elements. At the right in Figure 3.32 is the normally closed (NC) contact that the thermal relay would open.

Magnetic overload relays are actuated by an electromagnetic coil, much as the typical circuit breaker is. This relay would be ineffective against heat buildup due to a large number of successive starts and stops.

Electronic overload relays sense both the voltage applied and the current flow through the motor. They electronically simulate the iron loss and copper loss within the motor.

A *thermistor* is a resistor whose resistance changes greatly as the temperature increases. The temperature characteristic of most thermistors is negative; that is, as temperature increases, the thermistor's resistance decreases. However, in heat-sensing applications of motors, the thermistor with a positive temperature coefficient (discussed in Chapter 8) generally is used.

The thermistor is embedded in the windings of the motor and is monitored electronically to detect heat. When the windings get too hot, the power is shut off. Thermistors generally are not placed in the motor rotor because of the connection problem. Thus, a thermistor is not likely to be used in a DC machine.

Stopping a Motor

Large motors with heavy loads may take a very long time to coast to a stop. The motor itself, however, can be used to slow the rotation. Two methods used for electromechanical braking are dynamic braking and plugging.

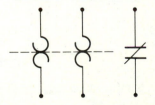

FIGURE 3.32   Schematic Symbol for Thermal Overload Relay

Dynamic braking is illustrated in Figure 3.33. Here, the armature of a shunt motor is disconnected from the line and connected to a resistor. The motor armature is still rotating within a magnetic field and therefore producing an armature voltage due to generator action. If the armature circuit is completed through a resistor, current will flow and a counter torque will be produced. This counter torque will slow the motor. As the motor slows down, the counter torque will decrease because the current is decreasing. Therefore, most of the slowing effect occurs at higher speeds, as indicated in Figure 3.34.

Plugging occurs when the armature is disconnected from the line and then is reconnected to the line in the opposite direction. This technique causes the motor to slow down very rapidly. When the motor comes to rest, it is automatically disconnected from the power line. Figure 3.35 is a simplified diagram illustrating plugging.

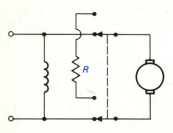

FIGURE 3.33 Dynamic-Braking Circuit

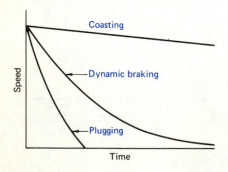

FIGURE 3.34 Time Relationship for Various Braking Methods

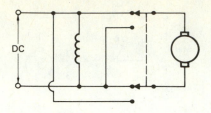

FIGURE 3.35 Plugging Circuit

DC GENERATOR

Classification

DC generators are classified by their method of supplying excitation current to the field coils. The two major classifications are separately excited and self-excited generators. Self-excited generators are further classified by the method in which the field coils are connected. Like the DC motor, the generator can be connected as series, shunt, or compound. The schematic diagrams for these generator configurations are identical to the corresponding motor schematics. These configurations will be discussed in more detail next.

Separately Excited Generator

A DC generator that has its field supplied by another generator, batteries, or some other outside source is referred to as a *separately excited generator*. The circuit for this generator is shown in Figure 3.36.

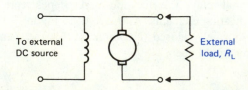

FIGURE 3.36 Separately Excited Generator Connection

The field excitation in this generator is the same as the excitation in the separately excited motor. However, the armature circuit here is different. Whereas the motor has an external voltage applied to the armature so that a torque is produced, the generator has an external torque supplied to rotate the armature, and a voltage (emf) is produced in the armature.

If the armature circuit is completed, current flows through the external load resistor R_L. The voltage distribution is illustrated in Figure 3.37, which shows that emf is the only voltage rise in the armature circuit. Thus, armature voltage must be less than—or, at best, equal to—the emf generated. Therefore, the generator armature voltage (V_{RL}) will also be lower than the emf generated. This result is an obvious departure from the motor characteristics, where the motor armature voltage is always higher than the cemf generated.

Now, let us look at an example that illustrates the development of the separately excited characteristic curve. Assume a constant armature speed drive is turning the generator at 1000 r/min. Armature resistance (R_A) is measured at 10 Ω, load resistor (R_L) is 50 Ω, K_T is 0.5 Nm/A², armature current (I_A) is 0.5 A, and field current (I_{field}) is 0.5 A. With these values determined, the armature terminal voltage, or load resistor voltage, is easily found:

$$V_{RL} = I_A R_L = (0.5\ \text{A})\,(50\ \Omega) = 25\ \text{V}$$

Generated voltage (emf) can also be found, as follows:

$$\text{emf} = I_A R_A + I_A R_L$$

$$= (0.5\ \text{A})\,(10\ \Omega) + 25\ \text{V} = 30\ \text{V}$$

Also,

$$K_{\text{emf}} = \frac{\text{emf}}{I_{\text{field}} n_A} = \frac{30\ \text{V}}{(0.5\ \text{A})\,(1000\ \text{r/min})}$$

$$= 0.06\ \text{V-min/A-r}$$

These calculations show that the emf of 30 V is greater than the armature terminal voltage of 25 V due to the internal drop across R_A. As the load resistor decreases, the armature current increases, and more voltage is dropped across R_A. If the load resistor goes to zero (or shorts), maximum armature current flows:

$$I_A = \frac{\text{emf}}{R_A} = \frac{30\ \text{V}}{10\ \Omega} = 3\ \text{A}$$

This maximum current causes all the generated emf to be dropped internally, and the armature terminal voltage is zero. Figure 3.38 shows this result.

This same situation is encountered when a battery's terminals are shorted with a wire. A high current flows in the shorting wire, causing it to become hot. All the battery's open-circuit terminal voltage is dropped across its own internal resistance.

Maximum counter torque (cT) is also produced in the generator when the output is shorted since the following relation holds:

$$cT = K_T I_{\text{field}} I_A$$

Field current is constant, and armature current is maximum. The characteristic curve is shown in Figure 3.39.

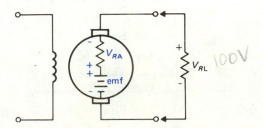

FIGURE 3.37 Voltage Distribution in Armature Circuit of Generator

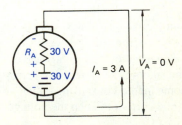

FIGURE 3.38 Voltage Distribution in Armature Circuit of Shorted Generator

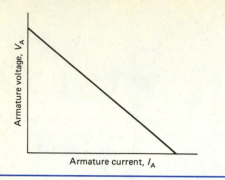

FIGURE 3.39 Characteristic Curve for Separately Excited Generator at Constant Armature Speed

Some applications require the armature speed to vary but the output voltage to remain constant, such as in the automobile charging system. This result can be achieved by changing the field current as an inverse function of speed.

As an example, suppose the output voltage of 25 V of the previous generator is to be maintained but the speed increased to 2000 r/min. Let us find the required field current. As long as the load resistance remains constant, so must armature current and emf. Therefore, the emf equation can be used to solve for I_{field}, as follows:

$$I_{field} = \frac{emf}{K_{emf} n_A}$$

$$= \frac{30\ V}{(0.06\ V\text{-min/A-r})\ (2000\ r/min)}$$

$$= 0.25\ A$$

The automotive charging system is one that has a constantly changing speed drive for the generator but needs a constant output voltage for charging the battery. The automotive system can perform its task by using some method of sensing output voltage and adjusting field current to keep the voltage constant.

Self-Excited Generators

Shunt Generator. When the field windings are connected in parallel with the armature, as in Figure 3.40, the generator is shunt-connected.

Self-excited generators, of which the shunt generator is one, do not produce an output voltage when their armatures are rotated, for several reasons. Consider what happens when the generator in Figure 3.40 starts from 0 r/min. At 0 r/min, armature, field, and load currents are all zero. As the armature starts turning, is an output voltage produced? Ideally, no. Since no field current is flowing, there is no magnetic field across the armature. Remember that voltage is produced only when a conductor moves within a magnetic field.

An output voltage generally is produced, though, because of a small residual magnetism remaining in the soft iron of the field core. Figure 3.41 shows how this residual magnetism appears. Consider the circuit of Figure 3.41A, where field current starts from zero and increases to $I_{field(sat)}$ (field current saturation). As the armature speed is held constant, the generated voltage increases from 0 V to some voltage at point C on the curve in Figure 3.41B. Notice that the curve is approximately linear from point A to point B but then curves to point C. This entire curve is generally called the *magnetization curve*; it shows field strength as a function of field current.

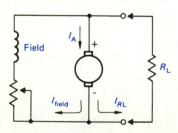

FIGURE 3.40 Shunt-Connected Generator

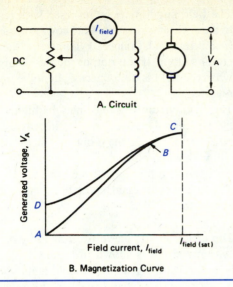

FIGURE 3.41 Determining Generator Magnetization Curve

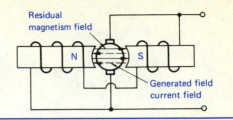

FIGURE 3.42 Field Connected in Incorrect Direction

So field strength increases linearly with field current to point B. After that point, called the *knee* of the curve, it takes proportionately much more field current to produce the same increase in field strength. From point B to point C and beyond is called the *saturation* part of the curve. Realize, though, that there is no point reached where an increase in field current does not produce an increase in field strength.

As field current is decreased to zero, some magnetism is retained because some of the molecular domains (groups of molecules that act together) remain aligned in the direction they were when field current was present. This small magnetism is called *residual magnetism* and is a property of the soft-iron core called hysteresis.

Another cause of no generated output voltage arises in a self-excited generator when the field is connected in the wrong direction. Figure 3.42 shows a shunt-connected generator with residual magnetism.

The solid lines represent the field due to the residual magnetism, and the dashed lines represent the field due to the generated field current. As the generator starts to rotate, current increases in the direction that opposes the residual magnetism. At some point, the field current will completely cancel out the field magnetism, and the output voltage will be zero.

A field resistance that is too large can also prevent voltage buildup in the generator. Figure 3.43 shows a graph of field current and voltage for various field resistances R_{field}. For the same field voltage, as field resistance decreases, field current increases.

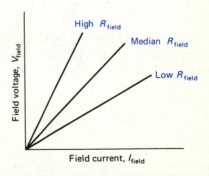

FIGURE 3.43 Voltage-Current Characteristics for Various Resistances

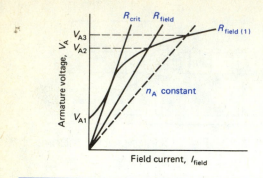

FIGURE 3.44 Magnetization and Field Resistance Curves Superimposed

2. The field circuit may be reversed with respect to the armature. A simple test shows whether the field is reserved. Connect a voltmeter to the generator output. If the output increases slightly when the field is disconnected, then the field is reversed.
3. The field circuit resistance may be higher than the critical value. An open or a high-resistance field circuit may be the problem. Check the circuit with an ohmmeter.
4. An open or a high resistance may exist in the armature circuit. Again, check the circuit with an ohmmeter.

If this graph is superimposed on the magnetization curve (Figure 3.41), the graph in Figure 3.44 results. This graph shows that for a field resistance (R_{field}) and constant speed, a small initial voltage (V_{A1}) will be produced. As current starts to flow in the field, more voltage will be produced, generating more current, and so on. A stable armature voltage (V_{A2}) will be reached when the voltage and current produced are all dropped across the field, assuming no other losses. If the field resistance is decreased to a new value $R_{field(1)}$ (corresponding to the low R_{field} curve in Figure 3.43), armature voltage stabilizes at V_{A3}. If, however, the field resistance is increased above R_{field} to R_{crit} (the critical resistance), the output voltage drops to a value below V_{A2}. If the field resistance is increased above R_{crit}, the generator field will collapse. In other words, the output voltage will drop back down to V_{A1}.

In summary, the following four conditions may cause a self-excited generator not to build an output voltage:

1. There may be a lack of residual magnetism. Residual magnetism may be lost by physical shock (such as being dropped on a floor), heat, vibration, or lack of use for a period of time. This condition can be corrected by *flashing the field*— that is, by applying a DC voltage to the field in the proper direction to reestablish the residual magnetism.

The characteristic curve for a shunt generator with varying speed is shown in Figure 3.45. This graph shows that the generator cannot build voltage until it reaches a speed that overcomes the resistance of the field and load. This point is similar to the critical-resistance point in Figure 3.44.

Figure 3.46 shows the characteristic curve at a constant speed. As the load current increases, the armature voltage drops. At a point called the *breakdown point*, the current drain of the load and the field exceeds what the generator can supply, and so the generator field collapses. Over most of the load current range, however, the armature voltage is relatively constant.

The nonlinearity of the magnetization curve prevents easy use of the emf and torque equations with self-excited generator circuits.

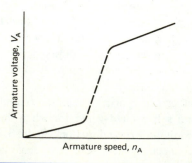

FIGURE 3.45 Characteristic Curve for Shunt Generator with Varying Armature Speed

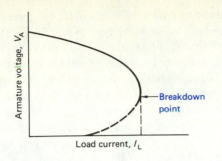

FIGURE 3.46 Characteristic Curve for Shunt Generator at Constant Armature Speed

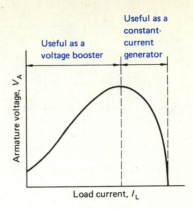

FIGURE 3.48 Characteristic Curve for Series Generator at Constant Armature Speed

Series Generator. When the field is in series with the armature, as in Figure 3.47, the generator is series-connected.

The characteristic curve for the generator at constant armature speed is shown in Figure 3.48. The curve shows that at zero load current (output open-circuited), no armature current can flow to build up field strength. Therefore, the output voltage (V_{A1}) will be a result of residual magnetism only. As the load resistance decreases, more field current can flow and more output voltage is generated. At the peak of the curve in Figure 3.48, an increase in field current does not increase output voltage because the field has become saturated.

Past the peak of the curve, no additional voltage is produced, but more voltage continues to be dropped across R_A and R_{field}. Thus, the output voltage eventually goes to zero. This dropping part of the

curve can be useful in welding generators. In such generators, the constantly changing arc length causes voltage to fluctuate, but the current must be constant in order to produce a consistent heat.

Compound Generator. As we have seen previously, terminal voltages associated with series-connected and shunt-connected generators vary in opposite directions with load current. If both a series and a shunt field were included in the same unit, we would obtain a generator with characteristics somewhere between the characteristics of these two types. The resulting device is a compound generator. The compound configurations and characteristic curves are shown in Figures 3.49 and 3.50.

If the number of turns in the series field is changed, three distinct types of compound generators are obtained: overcompounded, flat-compounded, and undercompounded. See Figure 3.49A.

When the number of turns of the series field is more than necessary to give approximately the same voltage at all loads, the generator is *overcompounded*. Thus, the terminal voltage at full load will be higher than the no-load voltage. This feature is desirable when the power must be transmitted some distance. The rise in generated voltage compensates for the drop in the transmission line.

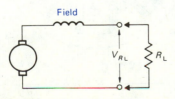

FIGURE 3.47 Series-Connected Generator

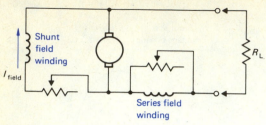

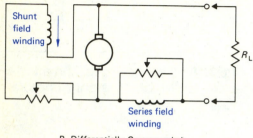

A. Overcompounded, Flat-compounded, and Undercompounded

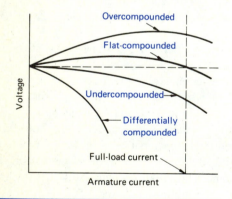

B. Differentially Compounded

FIGURE 3.49 Compound Generators

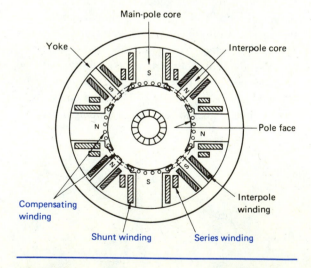

FIGURE 3.50 Characteristic Curves for Compound Generators

When the relationship between the turns of the series and the shunt fields is such that the terminal voltage is approximately the same over the entire load range, the unit is *flat-compounded*.

When the series field is wound with so few turns that it does not compensate entirely for the voltage

drop associated with the shunt field, the generator is *undercompounded*. In this type of generator, the voltage at full load is less than the no-load voltage.

An undercompounded generator in which the series and shunt fields are connected so as to oppose rather than aid one another is referred to as a *differentially compounded generator*. See Figure 3.49B. With this type of generator, the terminal voltage decreases rapidly as the load increases. Undercompounded generators are sometimes used in welding machines.

ARMATURE REACTION

As we noted earlier in the chapter, interpoles, sometimes called commutating poles, are small auxiliary poles placed midway between the main poles, as shown in Figure 3.51. They have a winding in series with the armature. Their function is to improve commutation and to reduce sparking at the brushes to a minimum.

To describe interpole function, we first must discuss *armature reaction*, which is the effect that the armature magnetic field has on the field distribution. Figure 3.52 illustrates how armature reaction is produced. Figure 3.52A shows the flux lines produced by the pair of poles when no current is

FIGURE 3.51 Arrangement of Various Field Windings on DC Machines

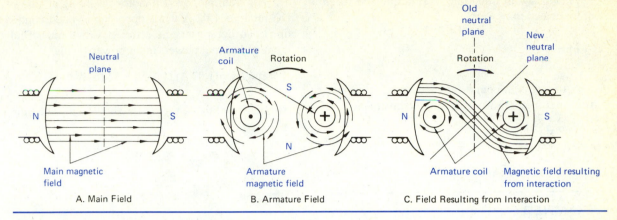

FIGURE 3.52 Armature Reaction

flowing through the armature coils. A vertical line through the field indicates the zero axis (neutral plane) of the field. Figure 3.52B shows the flux lines produced by current flowing through the armature coils alone. Figure 3.52C shows the resultant field of the two fluxes superimposed. Note that the zero axis of the resultant field is displaced, as indicated by the line designated "new neutral plane." This displacement results in a shift in the position of the old neutral plane. Shifting of the neutral plane results in sparking, burning, and pitting of the commutator. As the load current and the resulting armature reaction increase, this effect becomes more pronounced.

The short-circuiting effect can be counteracted by adding interpoles and compensating windings to the generator. Figure 3.53 shows the schematic diagram for a compound generator with interpoles (R_{pole}) and compensating windings (R_{comp}) added. The neutral plane, or load neutral, in the motor is shifted in the opposite direction with the same direction of armature rotation as shown in Figure 3.52. This example shows that motors and generators cannot be interchanged for best results.

POWER AND EFFICIENCY

Power (P) is defined as the rate of doing work. The original unit of power was developed from James Watt's determination of horsepower. Power is such

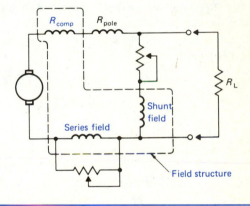

FIGURE 3.53 Compound Generator with Interpoles and Compensating Windings Added

an important and useful concept that the basic electrical unit of power, the watt (W), was named after him, even though he did not work with electricity. Watt defined 1 hp as 550 ft-lb/s. Conversion to metric or SI units uses the following equation:

$$550 \text{ ft-lb/s} = 1 \text{ hp} = 745.7 \text{ Nm/s}$$

$$= 745.7 \text{ W}$$

As we saw in Equation 3.11, torque and speed can be related to power:

$$hp = \frac{T \times n_A}{5252.1}$$

where torque is in pound-feet and speed is in revolutions per minute. In SI units, the conversion is

$$1 \text{ W} = 1 \text{ Nm/s}$$

• EXAMPLE 3.6

Given a conventional PM motor (see Table 3.6) with an output performance range starting at 1 W and a time constant of less than 10 ms, find (a) the horsepower when the motor output is specified in watts and (b) the approximate time it takes for this type of motor to reach full speed if the rated speed is 1800 r/min.

Solution

(a) In the equality stated above in this section, 1 hp = 746 W. Solving for watts, 1 W = 1/746 hp, or 0.00134 hp.

(b) A good assumption is that the time constant for a motor is similar to the capacitor charge curve time constant. Five time constants are needed for a capacitor to charge to over 98% of full voltage. Thus, five time constants are needed for the motor to reach approximately full speed. So we get 5 × 10 ms = 50 ms. Therefore, approximately full speed is reached in less than 50 ms.

The *efficiency*, (η) of any device, such as a motor or generator, is its output power (P_o) divided by its input power (P_i), when they are in the same units:

$$\eta\% = \frac{P_o}{P_i} \times 100$$

(η is the lowercase Greek letter eta.)

For example, a ⅛ hp motor operating at 110 V DC requires 1.02 A when operating at rated conditions. To determine the efficiency, we must find input and output power. Input power (P_i) is

$$P_i = IV = (1.02 \text{ A})(110 \text{ V}) = 112.2 \text{ W}$$

At rated conditions, the motor should be producing ⅛ horsepower, so

$$\frac{1}{8} \text{ hp} = \frac{1}{8}(745.7) = 93.2 \text{ W}$$

Therefore,

$$\eta = \frac{P_o}{P_i} \times 100 = \frac{93.2 \text{ W}}{112.2 \text{ W}} \times 100 = 83.1\%$$

NAMEPLATE SPECIFICATIONS

The nameplate of a dynamo indicates the power, voltage, speed, and so forth, of the machine. These specifications, or nominal characteristics, are the values guaranteed by the manufacturer. The following information punched on the nameplate of a 100 kW generator is an example:

- Power—100 kW
- Speed—1200 r/min
- Voltage—250 V
- Type—compound
- Exciting current—20 A
- Class—B
- Temperature rise—50°C

These specifications show that the machine can deliver, continuously, a power of 100 kW at a voltage of 250 V, without exceeding a temperature rise of 50°C. It can then supply a load current of 100,000/250 = 400 A. It is a compound motor with both series and shunt windings. The shunt field current is 20 A. The class B designation refers to the class of insulation used in the machine. These ratings are designed by the manufacturer and should not be exceeded. Characteristics of selected DC motor types are given in Table 3.6.

TABLE 3.6 **Characteristics of Selected DC Motor Types**

Motor Type and Basic Characteristics	Performance Ranges	Application Areas
CONVENTIONAL **PERMANENT-MAGNET** A simple alternative to wound-field shunt PM field plus wound armature Linear torque-speed relationships in small units Life limited by brushes in high-speed or severe applications Readily controlled by transistors or SCRs	Output from 1 W to a fraction of a horsepower Time constants (to 63.6% of no-load speed) to less than 10 ms Efficiencies from 60% to 70% in 10 W sizes With new magnet materials, can deliver high peak powers (horsepower range)	For full range of inexpensive, good performance drive and control applications With appropriate environmental precautions, suitable for military and aerospace use Preferred as a high-performance general-purpose servomotor
LIMITED-ROTATION DC TORQUER No commutator wear or friction Unlimited life Infinite resolution Smooth, cog-free rotation No electromagnetic interference generation Available as motor elements or fully housed	Travel range typically to 120° Torque from a few ounce-inches to greater than 40 lb-ft Mechanical time constants from 10 to 50 ms	Very high accuracy positioning or velocity control over a limited angle
CONTINUOUS-ROTATION DC **TORQUER** Slow speed, high torque Relatively low power output Available as pancake-shaped components Wide dynamic range Large number of coils give smooth operation	From tens of ounce-inches to hundreds of foot-pounds Moderate mechanical time constants Control to seconds of arc Relatively expensive	For direct coupling to load For very precise control Alternative to geared types
MOVING-COIL, PRINTED-ARMATURE **(IRONLESS ROTOR)** Similar to permanent-magnet DC units Linear torque-speed characteristics Smooth, noncogging rotation Handles very high, short-duration peak loads Fast response (less than 10 ms)	Outputs from less than 1 W to fractional horsepower High efficiencies Very low mechanical and electrical time constants	Computer peripherals where smooth control and fast response are needed Control applications needing high-response bandwidth, fast starting and stopping
VARIABLE-RELUCTANCE STEPPER Brushless and rugged High stepping rate dependent on driver circuitry No locking torque at zero energization Poor inherent damping Low power efficiency Can exhibit resonance Operates open loop Wide dynamic range Easily controlled Very reliable and low in cost in popular frame sizes	Several hundred to thousands of pounds per second Power output up to a few hundred watts	Alternative to synchronous motor Used in control applications where fast response rather than high power is the principal requirement Interfaces well to digital computers
SMALL-ANGLE, **PERMANENT-MAGNET STEPPER** Uses vernier principle to give very small stepping angles High stepping rates High cogging torques with zero input power Efficiency usually very low	Stepping rate from less than 100 lb/s to many thousands Dependent on driver electronics Power up to a few hundred watts Single step takes a few milliseconds	Useful in numerical control and actuator application where control is digital Provides fast slewing and high-resolution tracking
INVERTER-DRIVEN AC Operates from DC line using a switching inverter Somewhat less efficient than AC induction motors; otherwise, similar in performance Single-phase (capacitor) or two-phase versions most common	Outputs from less than 1 W to fractional horsepower Efficiencies from 20% to 80% in larger models Speeds up to 30,000 r/min and higher	Use where DC is only power available For universal applications where AC supplies vary widely, as in foreign applications Use where brushes might not be sufficiently reliable, as in very high speeds or in severe environments Variable-frequency versions used in accelerating high inertial loads
BRUSHLESS DC PM units using electronic commutation of stator armature Exhibits conventional DC motor characteristics, but torque modulation with rotation is higher Lack of brushes gives reliability in difficult applications	From less than 1 W to 1–2 hp Relatively high time constants Speeds to 30,000 r/min Efficiencies to 80% Voltages to hundreds of volts DC	For brushless, long-life applications requiring superior efficiency and control May be operated at very high altitudes or may be totally submerged

TABLE 3.6   (Continued)

Comparisons with Other Motors	Selection and Application Factors
Higher efficiency, damping, lower electrical time constants than comparable AC control motors, except in very low power applications Far more efficient than stepper drives More easily controlled than other motor types	Select for safe operation with acceptable temperature rises Check operating conditions for abrupt starts or reversals, which can demagnetize PM fields Check for altitude or environmental effects on brushes, especially over 10,000 r/min High stall currents are drawn in efficient or high-power motors Low output impedance electronic control required to utilize inherent motor damping
Much simpler than continuous-rotation torquers with or without commutators	Suitable for direct-drive, wide-band, high-accuracy mechanical control Similar wide-angle brushless tachometers available Requires high-power driving amps
Supplies most precise control, smooth and accurate tracking for continuous-rotation applications Requires higher-powered amps compared with geared units	Stiff direct coupling to load preferred Pulse-width modulation amps preferred for high control power
Faster response than iron-rotor motors Excellent brush life Lower starting voltage, limited only by brush friction Much more efficient than stepper motors	Recommended for low-cogging, low starting voltage, fast-response applications Rotor heats up quickly Thermal transients and heat removal can be important factors Larger, high-performance units can be expensive Low armature inductance permits commutation of very high current surges
Power output and efficiency generally very low compared with DC control motors	Care required in application Performance dependent on electronic driver circuitry Heat dissipation a possible problem At certain pulse rates, resonances can occur, which reduce load-handling capability Load inertia reduces performance Friction can improve damping Damping, gearing, and mechanical couplings require special attention
Efficiency of shaft power generation is low More flexible than comparable motors Simple and inexpensive alternative to synchronous or wide-speed-range drives Handles higher load inertia than variable-reluctance stepper and has better damping	Choose where special control characteristics are preferred over efficiency Check for resonances at all pulse rates Gearing can require extra safety margins because of impacts inherent in stepper operation Coupling compliances can help in accelerating load inertia, but additional resonances can be introduced Driver circuit design is critical Standard drivers available
Less efficient than true brushless motors using electronic commutation More complex, expensive, and noisier than brush-type DC motors Less suitable for control than other DC types Very long life with properly designed inverter	Inverter can be separate or packaged with motor High line circuit spikes Electromagnetic interference generation, with bulky filter capacitors required for suppression SCR inverters preferred in higher-power uses, but transistor inverters are easier to switch and more reliable Power supply capacitors can be required and must withstand supply transients
More efficient, easier to control, generates less electromagnetic interference than inverter-type motors Commutating transistors can be used for speed control, reversing current, and torque limiting without a separate controller, unlike other types Delivers highest sustained output in a given package size	Electronics can be packaged externally or within motor housing High peak line currents Bulky line filter required if electromagnetic interference is a problem Power supply capacitors could be required With properly designed electronics, life is limited only by bearings Temperatures can set limits to some commutation sensors

CONCLUSION

The theory of operation of DC motors and generators is quite old and well established, compared with the theory of some electronic components such as the transistor and microprocessor. However, the electric motor is still the major dynamo used in industry.

Because of its widespread use, all electronics technicians need to have a thorough understanding of the basic principles of DC motor and generator operation. With the development of new electronic motor control systems, new motor applications will continue to be created.

■ QUESTIONS ·

In Questions 1–5, fill in the missing words with increase(s) (I), decrease(s) (D), or stay(s) the same (S).

1. The torque on a series motor is held constant. The armature speed _____ and the armature current _____ as the applied voltage is decreased.

2. In a shunt-connected motor, if you hold line voltage constant and increase torque, then armature speed will _____, armature current will _____, and cemf will _____.

3. A separately excited motor has a fixed armature voltage. If the torque demanded of the motor remains constant and the field current is increased, then cemf will _____, armature current will _____, and armature speed will _____.

4. A shunt generator is operating at a constant speed and has unlimited power available from the prime mover. If the load current on the generator is increased, the output voltage of the generator will _____.

5. If the output of any operating DC generator were suddenly shorted, the torque required to keep the speed constant would _____.

6. Define these devices: (a) dynamo, (b) generator, and (c) motor.

7. Name the parts of a dynamo's rotor and stator.

8. Define torque. Name its units (British and SI).

9. When a generator supplies load current, the terminal voltage is not the same as the generated emf. Is it higher or lower? Why?

10. How may the direction of rotation of a DC motor be reversed?

11. Explain why the generator magnetization curve is not a straight line.

12. When a motor is in operation, why is the armature current not equal to the line voltage divided by the armature resistance?

13. What is an interpole? What is its purpose? How is its winding connected?

14. What is the effect of armature reaction in a motor?

15. What four factors may prevent the buildup of voltage for a self-excited shunt generator?

16. Why is it dangerous to open the field of a shunt motor running at no load?

17. When is a compound generator said to be (a) flat-compounded and (b) overcompounded?

18. Why is the speed regulation of a series motor poorer than that of a shunt motor?

19. Define efficiency.

20. Why should a series motor never be operated without load?

21. Describe what is meant by dynamic braking.

22. Draw schematic configurations for series, shunt, separately excited, and long-shunt and short-shunt compound dynamos.

23. Draw the constant-torque and constant–line voltage characteristic curves for the series and shunt motors and generators.

■ PROBLEMS ●

1. Find the armature speed of a series motor with 10 Ω field resistance, 5 Ω armature resistance, K_{emf} = 0.311 V-min/A-r, K_T = 1 Nm/A^2, an applied voltage of 200 V, and T = 0.09 Nm.

2. Given a shunt motor with a field resistance of 500 Ω, an armature resistance of 10 Ω, K_{emf} = 0.311 V-min/A-r, K_T = 0.5 Nm/A^2, a torque demanded of 0.2 Nm, and an applied voltage of 100 V, find the motor speed.

3. In the motor of Problem 2, suppose torque and armature current remain constant. What would the field resistance need to be in order to change the armature speed to 2500 r/min?

4. Given a separately excited motor with R_{field} = 500 Ω, I_{field} = 0.2 A, I_A = 0.45 A, R_A = 20 Ω, V_A = 100 V, K_{emf} = 0.311, and K_T = 0.5 Nm/A^2, a constant torque demand for any n_A, and the field voltage reduced by half (V_A remains fixed), find the new I_A, V_{RA}, cemf, and n_A.

5. In the motor of Problem 4, V_A and V_{field} are returned to what they were originally and what the motor was turning suddenly locks up (stops turning). Find the armature current and torque produced by the motor. What is the power being dropped across R_A?

6. A separately excited generator has a field current of 0.3 A, with R_A = 5 Ω, load resistance (R_L) of 150 Ω, output voltage (V_{RL}) of 100 V, and an armature speed of 1500 r/min. Find the no-load voltage at 1000 r/min (R_L is infinite).

7. With the generator of Problem 6, what armature speed would be needed to give an output voltage of 150 V with a load resistance of 100 Ω?

8. With the generator of Problem 6, if you wished to keep the voltage across the 100 Ω load resistor at 150 V while changing the speed of the armature to 2000 r/min, the field current would have to be changed. What field current would be needed?

9. Use the generator of Problem 6. What is its output voltage when the load resistance goes to 0 Ω? What is the armature current?

10. Derive the characteristic curves for the separately excited motor of Figure 3.26. Use the appropriate motor values, similar to those used in the shunt motor characteristic curves of Figures 3.20 and 3.22.

Brushless and Stepper DC Motors

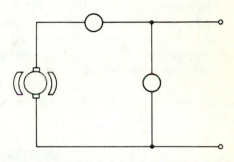

OBJECTIVES

On completion of this chapter, you should be able to:

- Contrast the permanent-magnet (PM) DC motor with the wound-field DC motor.
- Describe the characteristics of the three types of magnets used in PM motors.
- Calculate the factor of merit for a PM motor.
- Identify the parts of a brushless DC motor (BDCM) system.
- Describe the operation of the BDCM.
- Identify the parts of both types of stepper motors, variable-reluctance (VR) steppers and permanent-magnet (PM) steppers.
- Describe and contrast the operation of the VR stepper motor and the PM stepper motor.
- Describe the operation of an optical encoder.
- Contrast the incremental encoder with the absolute encoder.

INTRODUCTION

Chapter 3 examined the popular wound-field DC motor in depth. This chapter discusses three other popular types of DC motors: permanent-magnet (PM) motors, brushless DC motors (BDCMs), and stepper motors. These types of motors have power ratings that range up to about 5 hp (4 kW). Many of the motors are used as servomotors. A *servomotor* is a motor used in remote-control-positioning applications that has identical characteristics in both directions of rotation. Any motor used in a servomechanism application can also be classified as a servomotor. A *servomechanism* is a closed-loop control system whose output is mechanical positioning or rotation. The word *servomotor* came into use around the turn of the century for motors used to turn a ship's rudder.

PERMANENT-MAGNET MOTORS

The *permanent-magnet* (*PM*) *motor* differs from the wound-field DC motor in one respect: The PM motor gets its field from a permanent magnet, as its name implies. The wound-field DC motor gets its name from the fact that the field is wound; that is, the field comes from an electromagnet. In the wound-field DC motor, the field is created when field current flows through the field coils. In the wound-field motor, the flux is constant only as long as the field current is held constant. By contrast, in the PM motor, the flux is always constant. In practice, a constant flux means that the speed-torque curves and the speed-current curves for the PM motor are linear, as shown in Figure 4.1. The diagram in Figure 4.1 is a composite graph showing the relationship between speed, torque, input current, and output power.

The power produced by any motor is given by

$$P_o = \frac{N_{rt}T}{5252} \qquad (4.1)$$

where

P_o = output power (in horsepower, hp)
T = torque (in pound-feet, lb-ft)
N_{rt} = rotor speed (in revolutions per minute, r/min)

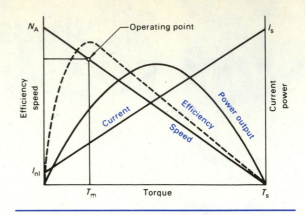

FIGURE 4.1 Graph of PM Motor Parameters

The output power is thus proportional to the product of torque and speed.

Note in Figure 4.1 that, at the no-load armature speed (N_A), the output torque and power are both zero. This result is not surprising since the load is not demanding any torque. Note also that some armature current is flowing, even under a no-load condition. Some armature current must flow to overcome the mechanical and electrical losses of the motor under no-load conditions. Mechanical losses are usually caused by friction. Electrical losses come from I^2R losses and hysteresis and eddy current losses in the rotor. As the motor is loaded down, it produces enough torque to turn the rotor. As the motor is loaded, the output power and armature current increase. The efficiency rises quickly to a peak and then falls off. The maximum output power occurs somewhat later, at a higher torque. As the stall torque (T_s) is approached, both efficiency and output power fall off. When the stall torque value is reached, the motor cannot produce the torque demanded and the rotor stops. The output power and efficiency go to zero at the stall torque point, while armature current is maximum. The armature current is high because the cemf is zero.

Data sheets often do not give enough information to predict how the motor will operate under certain conditions. A simple measurement, as described next, can give us additional information on the motor's behavior.

Connect the motor as shown in Figure 4.2. After applying V_L, and after the motor has warmed up, read the current, voltage, and armature speed. Next, load the motor down and record the same information. The motor should be loaded down to about 80% of its no-load speed.

For purposes of illustration, let's imagine that we have a motor with a no-load speed of 6800 r/min when 9 V is applied and that the motor draws 25 mA of no-load current. After loading to a speed of 5440 r/min, we measure 400 mA of current drawn. We will use these values to calculate a parameter called the *factor of merit (M)*. The factor of merit will also be used to calculate other values. The factor of merit is

$$M = \sqrt{\left[\frac{(N_{nl})(\Delta I)}{(I_{nl})(\Delta N)}\right] + 1} \qquad (4.2)$$

where

N_{nl} = no-load speed of rotor (in revolutions per minute)
I_{nl} = no-load current (in amperes)
ΔI = change in current
ΔN = change in rotor speed

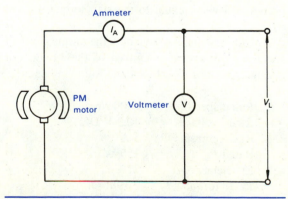

FIGURE 4.2 Basic PM Motor Circuit

Substituting values into Equation 4.2, we get

$$M = \sqrt{\left[\frac{(6800\text{ r/min})(375\text{ mA})}{(25\text{ mA})(1360\text{ r/min})}\right] + 1}$$
$$= 8.7$$

The maximum motor efficiency (E_{max}), shown in Figure 4.1, can be calculated by using the following equation:

$$E_{max} = \left(1 - \frac{1}{M}\right)^2 \qquad (4.3)$$

For our illustration, we get

$$E_{max} = \left(1 - \frac{1}{8.7}\right)^2 = 0.783 \quad \text{or} \quad 78.3\%$$

The power (P_i) at maximum efficiency is

$$P_i = M V_L I_L \qquad (4.4)$$

For our illustration, we have

$$P_i = (8.7)(9\text{ V})(25\text{ mA}) = 1.95\text{ W}$$

The power output (P_o) at maximum efficiency is given by

$$P_o = (M)\left(\frac{M-1}{M+1}\right)(V_L)(I_{nl}) \qquad (4.5)$$

Substituting values into this equation, we get

$$P_o = (8.7)\left(\frac{8.7-1}{8.7+1}\right)(9\text{ V})(25\text{ mA})$$
$$= 1.55\text{ W}$$

The speed (N_m) at maximum efficiency is

$$N_m = (N_{nl})\left(\frac{M}{M+1}\right) \qquad (4.6)$$

For our illustration, we have

$$N_m = (6800\text{ r/min})\left(\frac{8.7}{8.7+1}\right) = 6099\text{ r/min}$$

Finally, the output torque (T_m) at maximum efficiency is (in newton-meters)

$$T_m = \frac{9458(P_o)}{N_m} \qquad (4.7)$$

For our illustration, we get

$$T_m = \frac{(9458)\,(1.55\ \text{W})}{6099\ \text{r/min}}$$

$$= 2.4\ \text{mNm} \quad \text{or} \quad 0.339\ \text{oz-in.}$$

These equations give approximate answers only. They do not, for example, take into account the voltage drop across the brushes or the friction torque demanded by the motor. Since these values are likely to be small, our approximations should be accurate.

• EXAMPLE 4.1

Given a PM DC motor, with $V_L = 24$ V, $N_{nl} = 4500$ r/min, $I_{nl} = 7.9$ mA, loaded speed = 3360 r/min, and loaded current = 205 mA, find M, E_{max}, P_i, P_o, N_m, and T_m.

Solution

The factor of merit is calculated from Equation 4.2:

$$M = \left\{ \left[\frac{(4500\ \text{r/min})\,(197.1\ \text{mA})}{(7.9\ \text{mA})\,(1140\ \text{r/min})} \right] + 1 \right\}^{1/2}$$

$$= 9.97$$

The maximum motor efficiency (E_{max}) is calculated by using Equation 4.3:

$$E_{max} = \left(1 - \frac{1}{M}\right)^2 = \left(1 - \frac{1}{9.97}\right)^2$$

$$= 0.81 \quad \text{or} \quad 81\%$$

The power (P_i) at maximum efficiency is found from Equation 4.4:

$$P_i = M V_L I_L = (9.97)\,(24\ \text{V})\,(7.9\ \text{mA})$$

$$= 1.89\ \text{W}$$

The power output (P_o) at maximum efficiency can be found by using Equation 4.5:

$$P_o = (M) \left(\frac{M-1}{M+1} \right) (V_L)\,(I_{nl})$$

$$= (9.97) \left(\frac{9.97-1}{9.97+1} \right) (24\ \text{V})\,(7.9\ \text{mA})$$

$$= 1.55\ \text{W}$$

The speed (N_m) at maximum efficiency is, from Equation 4.6:

$$N_m = (N_{nl}) \left(\frac{M}{M+1} \right)$$

$$= (4500\ \text{r/min}) \left(\frac{9.97}{9.97+1} \right)$$

$$= 4089\ \text{r/min}$$

Finally, the output torque (T_m) at maximum efficiency is calculated from Equation 4.7:

$$T_m = \frac{9458(P_o)}{N_m} = \frac{(9458)\,(1.55\ \text{W})}{4089\ \text{r/min}}$$

$$= 3.58\ \text{mNm} \quad \text{or} \quad 0.5\ \text{oz-in.}$$

Examination of the power curve in Figure 4.1 shows us that the point of maximum power output occurs at a value equal to the stall torque divided by 2.

Most servomotors give both the torque constant and the cemf constant as part of their data. For instance, a motor might have the following data:

Rated speed	6000 r/min
Rated voltage	40.3 V DC
Rated current	5 A
Rated P_o	155 W
Efficiency	77%
Rated torque	2.19 lb-in.
Maximum torque	21.9 lb-in.
K_{cemf}	5.95 V/1000 r/min
K_T	0.503 lb-in./A

The torque constant (K_T) shows how much torque is developed by the motor per ampere of armature current (I_A). The value of the constant K_T is given by

$$K_T = \frac{T}{I_A} \qquad (4.8)$$

The torque T produced by the motor is then equal to

$$T = I_A K_T \qquad (4.9)$$

In this motor, the rotor produces 0.503 lb-in. of torque when the motor draws 1 A of armature current. How much torque will the motor be producing if we measure a current of 2.3 A? Using Equation 4.9, we get

$$T = I_A K_T = (2.3 \text{ A}) (0.503 \text{ lb-in./A})$$

$$= 1.16 \text{ lb-in.}$$

The current drawn by the motor is equal to

$$I = \frac{V_L - \text{cemf}}{R_A} \qquad (4.10)$$

The motor develops its maximum torque or stall torque when the armature current is at a maximum. The maximum current is determined by the applied voltage (V_L) and the armature resistance (R_A) since there is no cemf in a stalled condition. As speed increases, cemf is developed in the armature, which will reduce the current in the armature. The cemf is equal to the product of the armature speed (N_A) and the cemf constant (K_{cemf}):

$$\text{cemf} = N_A K_{\text{cemf}} \qquad (4.11)$$

When the motor reaches no-load speed, the cemf is almost equal to the applied voltage. The no-load current will then depend only on the friction torque of the motor.

Sometimes, the armature resistance is given in the data sheet. If it is not, it may be calculated by using Ohm's law, as in the following equation:

$$R_A = \frac{V_L - \text{cemf}}{I} \qquad (4.12)$$

To use Equation 4.12, we must first find the cemf from Equation 4.11. Using the data from the table, we get

$$\text{cemf} = N_A K_{\text{cemf}}$$

$$= (6000 \text{ r/min}) (5.95 \text{ V}/1000 \text{ r/min})$$

$$= 35.7 \text{ V}$$

Substituting into Equation 4.12, we get

$$R_A = \frac{V_L - \text{cemf}}{I} = \frac{(40.3 \text{ V}) - (35.7 \text{ V})}{5 \text{ A}}$$

$$= 0.92 \ \Omega$$

• EXAMPLE 4.2

Given a PM DC motor with the following information

Rated speed	3000 r/min
Rated voltage	40 V DC
Rated current	7 A
Rated P_o	200 W
Rated torque	5.66 lb-in.
Maximum torque	56.6 lb-in.
K_{cemf}	10.15 V/1000 r/min
K_T	0.858 lb-in./A
R_A	1.36 Ω

find (a) the rotor speed and (b) the torque produced by the motor if we apply rated voltage to the armature and draw 3 A of current.

Solution

(a) With a current flow of 3 A, the voltage drop across the armature resistance (V_{RA}) is equal to the current times the armature resistance:

$$V_{RA} = I_A R_A \qquad (4.13)$$

Thus, we get

$$V_{RA} = (3 \text{ A}) (1.36 \ \Omega) = 4.08 \text{ V}$$

The cemf is equal to the line voltage (V_L) minus the drop across the armature resistance:

$$\text{cemf} = V_L - V_{RA} \qquad (4.14)$$

So we have

$$\text{cemf} = 40 \text{ V} - 4.08 \text{ V} = 35.92 \text{ V}$$

To get the speed, we solve Equation 4.11 for N_A:

$$N_A = \frac{\text{cemf}}{K_{\text{cemf}}} = \frac{35.92 \text{ V}}{10.15 \text{ V}/1000 \text{ r/min}}$$

$$= 3539 \text{ r/min}$$

(b) The torque produced is found from Equation 4.9:

$$T = I_A K_T = (3 \text{ A})(0.858 \text{ lb-in.}/\text{A})$$

$$= 2.574 \text{ lb-in.}$$

PM Motor Classification and Characteristics

PM motors can be classified into three basic types: the conventional PM motor, the moving-coil PM motor, and the torque motor. These motors all use the same operating principle. Torque is developed by the interaction of two fields, one in the rotor (armature) and one in the stator.

PM motors are available in sizes up to the low-integral-horsepower range. They are efficient, are easy to control, and have linear performance characteristics. Only one power supply is needed since there is no field to be excited. The efficiency of the PM motor is better than the efficiency of the wound-field DC motor since there are no I^2R losses in the field. Behavior in changing temperature conditions depends on the type of magnet used.

Generally, three different types of magnets are used: the alnico type, the ferrite or ceramic type, and the rare-earth type. An alnico magnet has a very high flux density but is easily demagnetized. (Resistance to demagnetization is called *coercivity*. The alnico magnet has low coercivity.) The flux density of an alnico magnet is less affected by temperature than the

flux density of a ceramic magnet. Also, the flux density of a ceramic magnet is low compared with the flux density of an alnico magnet, but the ceramic magnet has higher coercivity. Ceramic magnets are inexpensive, both in materials and production costs.

The rare-earth magnet has become more cost-effective recently. More PM motors are using this type of magnet since it has high coercivity and high flux density. This new magnet, usually made out of samarium or neodymium combined with cobalt, cuts the weight of the motor by 30% to 50%. It also reduces the diameter greatly without increasing cost or decreasing performance. Especially in the alnico magnet, overloads produce high armature currents, which can demagnetize the magnet. Demagnetization causes a loss of field strength or flux density.

Conventional PM Motors

The armature in the *conventional PM motor* is similar to the rotor in the wound-rotor DC motor. It has an iron core with wound coils placed in slots in the rotor. The type of magnet in the field determines the construction of the stator.

When alnico is used, the field has a structure similar to the field shown in Figure 4.3. Because they have low coercivity, alnico magnets must be magnetized lengthwise, as shown in Figure 4.3. Magnets can be placed in either a two-pole structure or a four-pole structure. Although two- and four-pole alnico structures are common, six- (or more) pole motors are not uncommon.

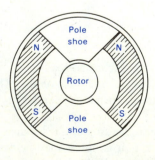

FIGURE 4.3 PM Field Structure with Alnico Magnet

When ceramic or ferrite magnets are used, a structure similar to the one in Figure 4.4 is used. Since the ceramic magnet has a lower flux density than the alnico magnet, it is made longer than the rotor. This construction places a higher flux density in the armature. This motor can have a very small air gap. The smaller the air gap, the more efficient the motor is.

PM motors with rare-earth magnets often have the construction shown in Figure 4.5. These magnets can be magnetized along their width, like ferrite magnets. This feature helps keep the cost of the motor down since the magnets can be thin. A samarium-cobalt magnet of the same length and width as a ferrite magnet has about twice the flux density.

The conventional PM motor is the motor of choice for electric vehicle applications. It is chosen for its low-weight and minimum-space requirements, key features for efficient vehicle operation. PM motors are used to power wheelchairs, forklifts, and

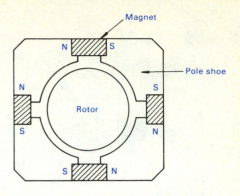

FIGURE 4.5 PM Field Structure with Rare-Earth Magnet

other electric vehicles. In general, they are preferred for applications demanding high efficiency, high peak power, and fast response.

Moving-Coil Motors

The *moving-coil motor* (MCM), although still a PM motor, differs from the conventional PM motor primarily in the armature. The MCM is a result of an engineering requirement that motors have a high torque output, low rotor inertia, and low electrical time constant. These requirements are met in the MCM.

The illustrations in Figures 4.6A and 4.6B show the two different types of MCMs. The MCM is classified by the type of armature construction used. The two different types of armatures are the disc armature and the shell (or cup) armature. Both armatures use the same basic concept to generate torque—that is, a current-carrying conductor interacting with a permanent magnetic field. In contrast, the conventional PM motor uses the same type of armature as the wound-rotor motor—a series of coils wound on an iron base. The iron base makes the rotor of the conventional PM heavy and slow to respond to changes in applied voltage. The MCM armature, in contrast, contains no iron. Because of this unique construction, this motor is sometimes called the *ironless-rotor* motor. The lack of iron does two

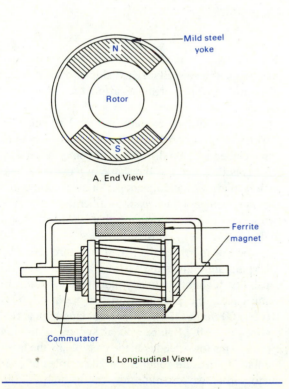

A. End View

B. Longitudinal View

FIGURE 4.4 Ferrite Magnet Field Structure

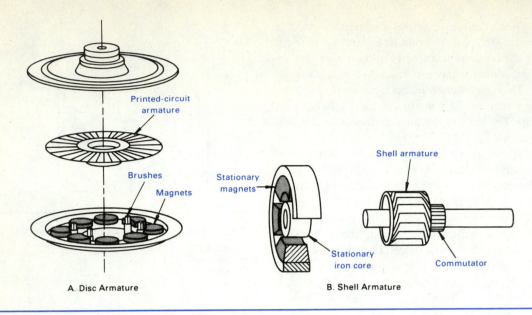

FIGURE 4.6   Two Types of MCM Armatures

things. First, it makes the armature much lighter, giving it low inertia. The lower inertia makes the MCM respond more quickly to changes in speed. Second, the lack of iron decreases the electrical time constant by lowering the armature inductance. Lowering the electrical time constant helps the motor respond more quickly to changes in the applied voltage.

The disc armature, shown in Figure 4.6A, has flat conductors arranged as if they were placed on a plate. Today, only one type of disc armature is used in this construction—the printed-circuit armature. Its conductors are stamped from a sheet of copper, welded together, and placed on a disc. The conductor segments are then joined with a commutator at the center of the disc. In Figure 4.6A, the conductors radiate out from the center of the disc. The conductors serve the same function as the armature coils in a conventional PM DC motor. When current from a supply is passed through them, the conductors generate a field. This field interacts with the PM field (from alnico, ferrite, or rare-earth magnets) to create the torque needed to turn the armature. The field, in

this case, is generated by two sets of eight permanent magnets arranged on each side of the armature. This motor is sometimes called the "pancake" motor because it is very thin.

The second type of rotor, the shell armature, is shown in Figure 4.6B. The conductors are copper wires embedded in a disc made of epoxy resin. The rotor is hollow and has the shape of a cylinder. The shell is made by bonding copper or aluminum wires together, using polymer resins or fiberglass. Note the two sets of permanent magnets that provide the field in this MCM.

The low inertia of the MCM armature gives the MCM armature gives the MCM a high acceleration capability, higher than in any other type of motor. In some cases, the acceleration can be as high as 10,000,000 r/min/s. In addition to having high acceleration, the MCM runs well at speeds as low as 1 r/min. This low speed is made possible by the large number of armature conductors. The number of conductors in a disc armature may run as high as 200. The larger the number of conductors, the smoother is the torque produced by the rotor.

MCMs with no iron in the rotor do not have favored positions, as conventional motors do. This favored-position effect may be demonstrated by rotating a conventional motor shaft between your fingers without power applied. The rotor lobes are attracted to the field magnets, causing a "cogging" effect. The lack of iron in a disc or shell armature results in no favored positions and, therefore, no cogging.

The MCM has significantly less loss due to hysteresis and eddy currents. These losses are called iron losses because they are caused by the presence of iron in the rotor. Since the MCM minimizes these losses, the MCM is potentially more efficient than the conventional PM motor.

Finally, the MCM has a lower armature inductance than a comparable conventional PM motor. The lower inductance does two things. First, it lowers the L/R time constant, allowing the motor to respond more quickly to changes in electric signals. Second, it eliminates inductive kick. High inductances generate large inductive kicks when the commutator disconnects a coil. These large inductive kicks cause arcing, which destroys both the commutator and the brushes.

MCMs have gained popularity because of their advantages over conventional DC motors. Because of their smooth torque output and fast acceleration, they are used in computer peripheral devices and tape transport systems.

Torque Motors

A case may be made that all motors produce torque. All motors could, therefore, be called torque motors. However, a *torque motor* is different from most other DC motors in that it is required to run for long periods in a stalled or low-speed condition. Not all DC motors are designed for this operation. Recall from Chapter 3 that cemf is directly proportional to armature speed. A low cemf means that a large amount of armature current will flow. Most conventional DC motors are not designed to dissipate the heat that this large current will create. But torque motors are designed to be run under a low-speed or a stalled condition for long periods of time and are used in such applications as spooling or tape drives.

In spooling applications, the tension is often controlled by a torque motor.

Torque motors are found in three major applications in industry. First, they are used in applications like spooling, where the motor is stalled when no rotation is required. In this case, the torque motor operates much like a spring, exerting tension or pressure. The tension can easily be controlled by adjusting the amount of current in the motor. Second, the torque motor, or torquer, is found in applications that require only a few revolutions, usually at a low speed. Examples of this application include opening and closing switches, valves, and clamping devices. Third, torquers are found in applications where they must run constantly at a low speed. Reel drives in a tape transport system are a good example of this application.

BRUSHLESS DC MOTORS

Recall that in the conventional wound-field DC motor, the commutator is divided into segments. Generally, the effect of winding inductance when the current starts and stops in the winding segments is reduced when the number of segments increases. Increasing the number of segments will improve performance. Another reason for increasing the number of winding segments is to control the torque ripple. The more windings there are, the lower is the amount of torque ripple. Thus, we find that a fractional-horsepower DC motor may have anywhere from 7 to 32 commutator bars per armature, and an integral-horsepower motor of 5–10 hp may have less than 100 bars. Larger motors may have more than 100 bars per armature. Although we can see the effect of commutation by making a connection across the armature windings, we cannot see what is going on inside the armature. Reaching proper internal test points in a rotating structure like the armature is not easy.

Much of the designer's effort for wound-field motors goes into choosing the proper brush and commutator. A great variety of brush compositions have enabled designers to cope with the widely varying operational and environmental conditions that exist in modern technology. Even so, brush and commutator wear is the primary maintenance requirement for wound-field DC motors.

The problems associated with a commutator may be avoided if the switching is done by semiconductor devices instead of by mechanical means. Using semiconductor devices to commutate a PM DC motor generates a new set of problems, however. A simple translation of the brush-type motor design to a brushless type is not practical. For example, if we were to take a 16-bar, two-pole DC motor and substitute semiconductor devices for the brush and commutator assembly, we would end up with 32 power transistors, two of which would be conducting at any one given time. This design would result in an inefficient use of semiconductors—an efficiency factor of about 6%. Obviously, the problem has to be solved in a new way to get cost-effective performance.

A DC motor in which the brushes and commutator have been replaced by functionally equivalent semiconductor switches is called a *brushless DC motor* (BDCM). A brushless DC motor system should have the torque-speed characteristics of the conventional PM DC motor.

BDCM Construction

The diagram in Figure 4.7 illustrates the speed-torque characteristics of a conventional PM DC motor. The brushless DC motor should have the same basic characteristics.

When BDCMs are designed, one of the requirements to be met is the reduction of the number of semiconductor switches to as low a number as possible. The second design requirement is that, wherever possible, permanent-magnet materials should be used in the rotor or armature. This requirement is necessary to eliminate the need for slip rings in the rotor assembly.

In a BDCM, the most practical design uses a stator structure, as shown in Figure 4.8, where the windings are placed in an external, slotted stator. The rotor consists of the shaft and a hub assembly with a PM structure. The figure shows a two-pole magnet in the rotor.

Contrast Figure 4.8 with Figure 4.9, which shows the equivalent cross-sectional view of a conventional PM DC motor. Figure 4.9 shows permanent magnets situated in the stator structure; the rotor carries the various winding coils. We see that there are significant differences in winding and magnet locations. The conventional DC motor has the active conductors in the slots of the rotor structure; in contrast, the BDCM has the active conductors in slots in the outside stator. The removal of the heat produced in the active windings is easier in the BDCM since the thermal path to the environment is shorter. Since the PM rotor does not produce heat, the result is that the BDCM is a more stable mechanical device from a thermal point of view.

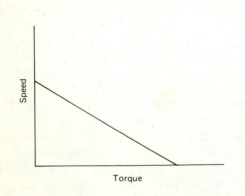

FIGURE 4.7 Speed-Torque Curve for PM DC Motor

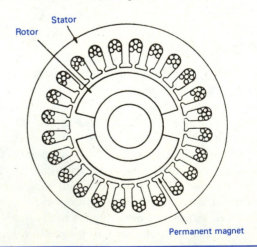

FIGURE 4.8 Cutaway View of BDCM Assembly

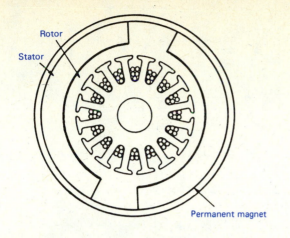

FIGURE 4.9 Cutaway View of Conventional PM DC Motor Assembly

In spite of the advantages just discussed, there are cases where a BDCM uses the configuration shown in Figure 4.9. In these cases, the roles of the two parts are reversed so that the PM outside structure rotates, and the wound lamination part is the stator. This design produces a high rotor moment of inertia, which is useful in applications where a high mechanical time constant is needed to smooth out any existing torque ripple. In such cases, the thermal advantage of the design in Figure 4.8 is not used.

To illustrate the similarities and differences between the conventional and brushless DC motor systems, we compare the sketches of the two systems in Figures 4.10 and 4.11. Figure 4.10 shows the elements of a conventional DC motor and control. The bidirectional controller and driver stage is shown together with a power supply and control logic. The

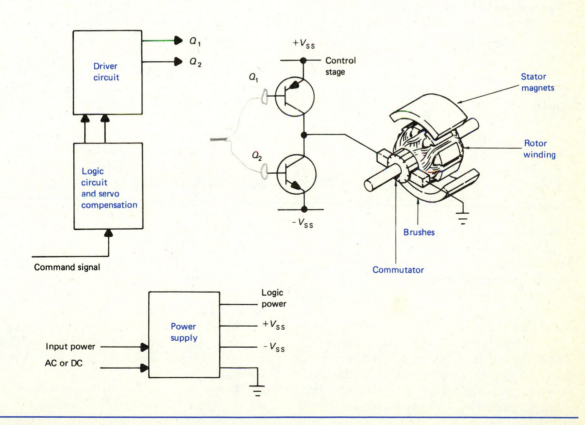

FIGURE 4.10 Main Parts of DC Motor Servo Control

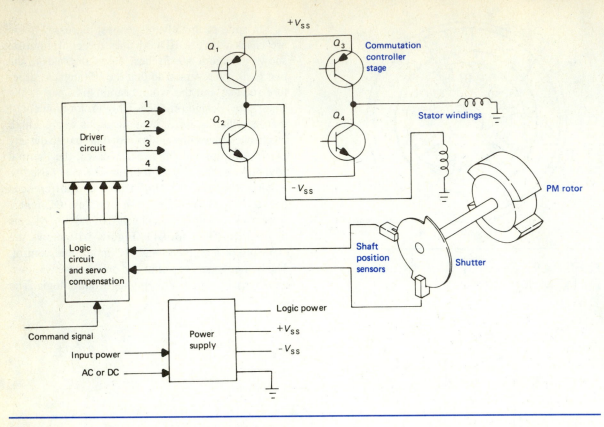

FIGURE 4.11   Main Parts of BDCM Control

bidirectional controller enables the motor to turn in both directions. The control logic signals the driver circuit to turn on either Q_1 or Q_2, depending on which direction the rotor needs to turn. The equivalent BDCM system is shown in Figure 4.11, where the main differences are seen to be the stator windings, the PM rotor, four transistors instead of the two found in the conventional motor, and a shaft position encoder. The shaft encoder tells the control logic circuitry about the rotor's position. Remember that the logic signals control the commutation of the windings. It is vitally important, therefore, that the control logic knows where the rotor is so that proper commutation can take place. This is not a problem in the conventional mechanical commutator since the commutation process is fixed by the position of the brushes and the commutator.

BDCM Operation

BDCMs have four basic parts: the rotor, the stator, the commutator, and the rotor position sensor. In the conventional wound-field DC machine, the field is stationary, generated by a permanent magnet or an electromagnet. This field is called the *stator* since it is stationary. In the normal DC motor, the DC power is fed to the armature, which is free to rotate. In the BDCM, the field rotates, and we call the field the *rotor*. The rotor is usually a permanent magnet. DC power is applied to the armature, or stator, which is stationary.

Two types of magnetic rotors are shown in Figure 4.12. In the foreground (left, center) is a rare-earth samarium-cobalt magnet rotor. This particular rotor is a patented skewed-arc design made by Pacific

FIGURE 4.12 Two Types of BDCM Magnetic Rotors

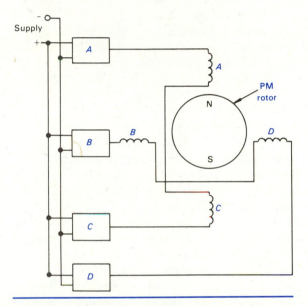

FIGURE 4.13 Basic BDCM with Driving Circuits

Scientific. Other items on the rotor include bearings at either end, a commutation magnet rotor, and a tachometer rotor. In the background is a ferrite magnet rotor. The ferrite magnet shows a notching that is part of an anticogging technique.

The basic operation of the BDCM is shown in Figure 4.13. Blocks *A* through *D* in Figure 4.13 contain switches that are usually transistor, thyristor, or power MOSFET pairs. Current from the power supply flows in either direction through the coils, depending on which switch is on.

Let us suppose that current flows up through coils *A* and *C*, placing a north pole at the bottom of *C* and a south pole at the bottom of *A*. The rotor will then position itself as in Figure 4.14A. Next, keeping the current flow in *A* and *C*, we switch the current on in coils *B* and *D*, creating the magnetic fields as shown in Figure 4.14B. The rotor will move to align itself with the total resultant field, shown by the dotted arrow. Next, the current in *A* and *C* is shut off, and the rotor proceeds to the position shown in Figure 4.14C. Notice that the rotor has moved 90°. No other parts in the motor have moved. Also, all switching has been done with semiconductor devices instead of a mechanical commutator. The stepper motor, discussed later in this chapter, works in a similar way.

The major difference between a stepper motor and the BDCM is the presence of rotor position feedback in the BDCM. The BDCM uses feedback to switch the current to the correct coils at just the right time. The switches must be turned on relative to the

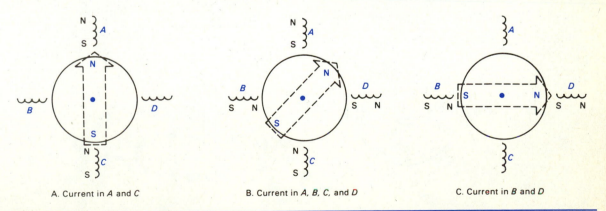

A. Current in *A* and *C*   B. Current in *A*, *B*, *C*, and *D*   C. Current in *B* and *D*

FIGURE 4.14 BDCM Turning through 90°

rotor position. Two types of rotor position sensors are used in BDCMs today. Most popular is the *Hall effect device*, which senses the presence of a magnetic field. The Hall effect device (discussed in more detail in Chapter 8) gives an output voltage when a magnetic field passes close by. The Hall effect device is relatively inexpensive and is used in low-voltage, low-power applications. An example of a Hall effect sensor is shown in Figure 4.15. In this figure, three Hall effect sensors are located 60° apart, with a four-pole commutation magnetic rotor in the center. The electronic input-output cable may be seen at the top of the photo.

The other sensor used in rotor position detectors is *optoelectronic*. This sensor uses a phototransistor and a light source, as shown in Figure 4.16. A light source illuminates the phototransistor at the proper time. The phototransistor turns on the device that

allows current to flow through the proper coil. These optical sensors are small and lightweight and can react to changes in rotor position quickly.

A phototransistor-switched, three-phase BDCM is shown in Figure 4.17. The three phototransistor sensors are placed on the end plate of the motor at 120° intervals. A shutter attached to the rotor always blocks light from two of the sensors. As shown in Figure 4.17, the south pole of the rotor magnet is aligned with pole S_2 on the stator. With the shutter in the position shown, light strikes the base of the phototransistor Q_1, turning it on. Transistor Q_1 then turns on transistor Q_4, causing current to flow in stator coil S_1. The north pole created at the edge of the pole face will attract the south pole of the PM rotor, pulling it clockwise. The rotor will then align itself with stator pole S_1. When the shaft rotates, the shutter will revolve, shading Q_1 and Q_2. Phototransistor Q_3 will thus be exposed to light and will turn on. Transistor Q_3 will then turn on transistor Q_6, which will force current through stator pole S_3, creating a north pole on the pole face. The south pole of the rotor will turn clockwise to align with stator pole S_3. When the rotor turns to this position, the shutter shades Q_3 and exposes Q_2. The switching continues, turning the rotor clockwise continuously.

A four-phase BDCM controller is shown in Figure 4.18. In this controller, two Hall effect devices are used to detect the rotor position. The Hall effect device produces a voltage potential output in the presence of a magnetic field. With the rotor in the position shown, HE_2 will detect the rotor's presence,

FIGURE 4.15 Hall Effect Rotor Position Sensors for a BDCM

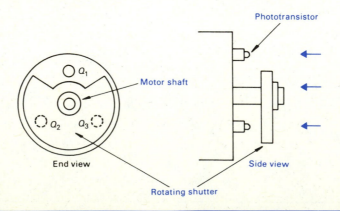

FIGURE 4.16 Phototransistor Rotor Position Sensor

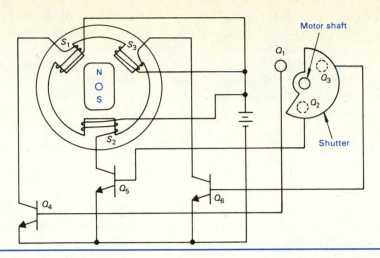

FIGURE 4.17   Three-Phase BDCM Operation with Photoelectric Switching

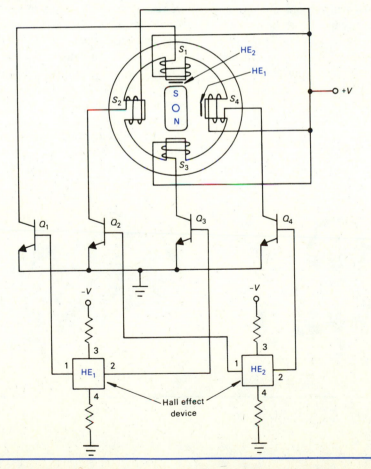

FIGURE 4.18   Four-Phase BDCM Controller Using Hall Effect Sensors

turning on Q_2. The north pole created at S_2 will pull the rotor through 90°. As the pole leaves S_1, the flux decreases in HE_2, decreasing its output voltage and turning off Q_1. At the same time that HE_2 turns off Q_2, HE_1 turns on Q_3, pulling the rotor south pole to the bottom of the motor. In this way, the rotor is pulled around the motor, alternately being attracted to the next pole.

Since the BDCM produces very little starting torque, it is most useful on low-torque loads, such as blowers and fans. Torque may be increased by using a gear reduction system. A lower gear ratio will increase torque as well as decrease speed. When the BDCM is used in a low-voltage system, it is not very efficient. The switching circuit, usually a transistor or a thyristor, can drop up to 1.5 V across its output terminals, which reduces efficiency. Since the motor has no brushes or mechanical commutator, the life of the motor is limited by the life of the bearing lubricant.

A cutaway view of a brushless DC motor showing the stator windings and the ceramic PM rotor is illustrated in Figure 4.19A. An assembled BDCM is shown in Figure 4.19B.

Commutation-Sensing Systems

As mentioned previously, the brushless DC motor system must have some method for letting the control circuit know the position of the rotor. Commutation cannot be accomplished without this information. Several methods are available for sensing the rotor's position. The most commonly used methods are Hall effect sensors, electro-optical sensors, and radio frequency (RF) sensors. The theory behind these sensors is explained in Chapter 8.

The *Hall effect sensing system* uses a sensor that detects the size and polarity of a magnetic field. The signals are amplified and processed to form logic-compatible signal levels. The sensors usually are mounted in the stator structure, where they sense the polarity and magnitude of the PM field in the air gap. The outputs of these sensors control the logic functions of the controller to provide current to the proper coil in the stator at the right time. The system can provide some compensation for the effect of armature reaction, which occurs in some motor designs. One drawback with such a location of the angular position sensor is that it is subject to stator temperature conditions. At times, temperature variations can be rather severe in high-performance applications. It is not unusual, for instance, for a winding temperature to reach 160°–180°C under peak load conditions. Such a temperature change may affect the Hall effect switching performance and therefore affect the system performance.

The Hall effect device can, of course, be located away from the immediate stator structure. It may use

A. Cutaway View

B. Assembled BDCM

FIGURE 4.19 BDCM with Ceramic Magnet

a separate magnet for sensing angular position. In such a case, the sensor is not necessarily subject to the severe operating conditions just mentioned. On the other hand, the sensor does not compensate for armature reaction problems in this position.

The second method for angle sensing is the *electro-optical switch*. This system most commonly uses a combination of an LED (light-emitting diode) and a phototransistor (discussed in Chapter 9). A shutter mechanism controls light transmission between the transmitter and the sensor. The sensor voltages can be processed to supply logic signals to the controller. The electro-optical system lends itself well to generation of precise angular encoding signals.

The third angle-sensing method, the *RF sensor*, is based on inductive coupling between RF coils. Several varieties of such devices have been used with varying degrees of circuit complexity. The angle-sensing accuracy of such devices depends on several design factors which may limit their use in high-performance systems. In addition, their switching time depends on oscillator frequency and may cause switching time delays that can be undesirable at higher shaft speeds.

The three systems discussed have their own advantages and drawbacks, and the requirements of each application usually dictate which system is best.

Power Control Methods

So far, we have discussed commutation control without reference to power control of the BDCM. One method of power control is to *vary the supply voltage to the commutation system*. An example of this method is shown in Figure 4.20. The six switching transistors control commutation at the proper time. The series-connected power transistor handles velocity and current control of the brushless motor. This control can be accomplished either by linear (class A) control or by pulse-width or pulse-frequency modulation. If directional control is needed, the commutation sequence must be adjusted from 0° to 180°. In effect, then, we have a series regulator controlling the power supply voltage for a switching-stage commutation controller.

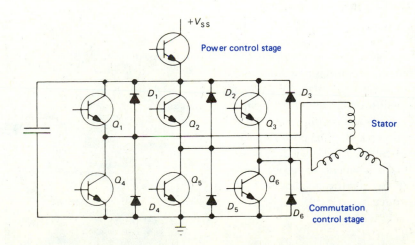

FIGURE 4.20   Series Regulator Power Control for Three-Phase BDCM

Another way to control voltage and current in the BDCM is to *let the commutation transistors control the motor current* either by linear control means or by pulse-width or pulse-frequency modulation. (Again, modulation methods and circuits are covered in later chapters.) Such a control method results in better use of available semiconductor devices. Proper attention, however, must be paid to power dissipation in the controller stage.

In the case of *linear transistor control*, the control stage must be operated in a constant-current mode rather than constant-voltage mode. Otherwise, the transistor stage held at a zero output voltage will tend to conduct cemf-induced currents during the inactive part of the cycle. Induced currents, in turn, cause a viscous damping effect that is detrimental to motor operation.

The *pulse-width or pulse-frequency control scheme* is well suited for control of voltage and current to a brushless motor. Since logic circuitry already in place is capable of switching the appropriate transistors on and off, the use of such control is easily accomplished. The switching rate must be within the switching capability of the power transistors so that undue dissipation losses do not occur during the transistor turnoff and turnon times.

Any high-performance BDCM will require some form of current limit control, either to protect the controller stage or to protect the magnetic circuit. With either of the controller schemes discussed, applying such current limit control is easy. However, the switching-type controller can sustain current limit conditions without significant circuit power dissipation—as compared with a linear control system where the excess power is dissipated in the power transistors. For the reasons mentioned, pulse-width or pulse-frequency modulation is a superior control scheme for BDCMs in which any significant amount of DC power is being controlled.

Advantages and Disadvantages of BDCMs

BDCMs can be used for a variety of applications. The advantages of state-of-the-art magnet materials have made possible BDCM designs that have very high torque-to-inertia ratios. Since commutation is performed in circuit elements external to the rotating parts, there are no items that will suffer from mechanical wear, except the motor bearings. The BDCM, therefore, will have a life expectancy limited only by mechanical bearing wear and the reliability of the electronic controller.

The BDCM can be controlled with very efficient amplifier configurations. In cases of severe environmental conditions, the controller can be located away from the motor. The control system can interface easily with digital and analog inputs. Therefore, BDCMs are well suited for motion control requiring rotor movement in discrete steps. This kind of motor control is called *incremental motion control*. The motor and control also have a lower level of RF emission than the conventional DC motors and controls.

The BDCM controller has, in general, a more complex configuration than the controller for an equivalent conventional DC motor, but it may be similar in size and complexity to a closed-loop step motor controller. A small BDCM with a controller is shown in Figure 4.21. The commutation sensor

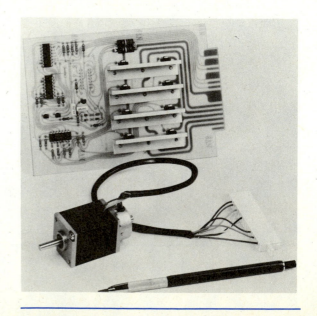

FIGURE 4.21   BDCM with Controller

system, in some cases, can provide incremental motion applications, but, usually, an encoder system is added to the commutation system to suit the application.

Table 4.1 gives a brief comparison of the conventional DC motor and the brushless DC motor. Because of the many available variations of the two kinds of motors, the comparison is very general. However, it may give more insight into the basic characteristics of the two types of motors.

Characteristics and Applications of BDCMs

Some of the advantages of BDCMs have already been mentioned. BDCMs have a long life and a high reliability, due to the lack of brushes and mechanical commutator. No maintenance is required. Since the motor lacks a mechanical commutator, it produces little RF noise interference. BDCMs also respond very quickly to applied voltages. The fast response is due to the low mass of the rotor.

Efficiencies of BDCMs are high, often exceeding 75%. Although some wound-field DC motors have as high an efficiency, wound-field DC motors more often have efficiencies as low as 30%. Most of the heat generated in a BDCM is produced in the stator windings, located on the outside of the motor. This construction makes it easy to heat-sink—or conduct heat away from—the stator coils. Heat dissipation is much more difficult in the rotor of the wound-field DC motor since the armature must be free to turn inside the motor.

As in every electronic device, the advantages of the BDCMs are offset by some disadvantages. We have already mentioned the low starting torque of these motors. BDCMs are also expensive because of the semiconductor switching and sensing circuitry. The size of the BDCM is approximately the same as the size of a conventional DC motor.

Generally, BDCMs have top speeds up to 30,000 r/min, and some manufacturers claim speeds as high as 100,000 r/min. These speeds are coupled with high reliability and low maintenance. Two areas that need

TABLE 4.1   Brief Comparison of Conventional DC Motors and Brushless DC Motors

Type of Motor	Good Points	Problem Areas
Conventional PM DC Motor	Provides control over a wide range of speeds Capable of rapid acceleration and deceleration Convenient control of shaft speed and position by servo amplifiers	Commutation (brushes) causes wear and electrical noise; this problem can be kept under control by selecting the best materials for each application
Brushless DC Motor	Provides control over a wide range of speeds Capable of rapid acceleration and deceleration Convenient control of shaft speed and position No mechanical wear problem due to commutation Better heat dissipation arrangement	More semiconductor devices are required than for the brush-type DC motor for equal power rating and control range

these characteristics are the aerospace and biomedical fields. Biomedical applications include cryogenic coolers and artificial heart pump motors. An aerospace example is a gyroscope motor. Gyroscopes need high speeds, and AC power is usually not available. Aerospace applications also demand low maintenance and high reliability. Remember that several thousand hours of operation is all one can reasonably expect from a wound-rotor DC motor. BDCMs are also used as tape drives for video recorders and high-reliability tape transport systems. Some stereo turntables also use this motor. As the popularity of these motors increases, the price should make them more competitive with conventional drives.

STEPPER MOTORS

The *stepper motor* is the only motor that is truly digital in its output. Most motor rotors, including those in AC motors, turn at a rate proportional to the voltage (or frequency) applied to them. The stepper motor, as its name implies, turns in discrete movements called *steps*. After the rotor makes a step, it stops turning until it receives the next command. Stepper motor operation can be likened to a series of electromagnets or solenoids arranged in a circle, as shown in Figure 4.22. When energized in sequence, the electromagnets react with the rotor, causing it to turn either clockwise or counterclockwise, depending on the input commands. The stepping angle (ψ) is determined by the design of the motor, but it should not be greater than 180°.

There are two large classes of steppers: mechanical, which rely on ratchet-and-pawl arrangements or other linkages, and magnetic (true motors). Since magnetic stepper motors are more common, we will be dealing with them only. There are two basic types of magnetic stepper motors: permanent-magnet (PM) and variable-reluctance (VR).

PM Stepper Motors

The *PM stepper motor* operates on the reaction between an electromagnetic field and PM rotor. In its simplest form, the PM unit consists of a two-pole PM rotor revolving within a four-pole slotted stator (Figure 4.23). Although the rotor is shown in the diagram as round and smooth, the actual PM rotor has teeth. The stator likewise is a toothed construction. Current is applied to each of the stator windings successively, creating a series of magnetic fields. The PM rotor, interacting with the magnetic reaction of the field, is pulled into alignment as each winding is energized. Each successive realignment produces a stepping motion of the rotor, in this case steps of 90°, since the four stator poles are 90° apart. By a change in the excitation sequence to the windings, the motor can be made to operate either clockwise or counterclockwise.

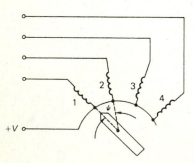

FIGURE 4.22 Series of Electromagnets Driving Stepper Motor

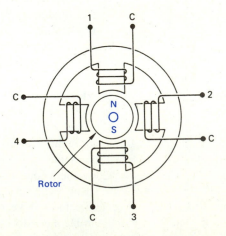

FIGURE 4.23 Construction of Four-Phase (90°) PM Stepper Motor

Because of the rotor reaction and the relatively large step angle of the PM stepper, the rotor has little tendency to overshoot. *Overshoot* is the condition when the rotor goes past the pole that is attracting it. The pole, still attracting the rotor, pulls it back. The rotor then overshoots again (but not as much), this time in the opposite direction. This oscillating behavior continues until all the rotor energy is absorbed. The rotor then stops. The oscillating waveform shown in Figure 4.24 illustrates the oscillations *damping*, or dying out. Sometimes, when the number of oscillations cannot be tolerated, external means of damping must be used. In the case of the PM stepper, however, damping is rarely necessary.

A new type of PM stepper motor has overcome the rotor size and weight problems that limit the maximum speed the motor can achieve. The rotor of this new type of stepper motor is a disc rather than the more typical cylinder. The rotor is a thin disc made from rare-earth magnetic material (Figure 4.25). Because the disc is thin, it can be magnetized with up to a hundred individual tiny magnets, evenly spaced around the edge of the disc. Conventional PM steppers are generally limited to a minimum step angle of 30°, for a maximum of 12 steps per revolution. The new thin-disc motors are generally half the size of the hybrid motors and weigh 60% less.

The disc of the stepper motor is supported on a nonmagnetic hub, the disc and hub together making up the rotor. The disc magnets are polarized with alternating north and south poles, as shown in Figure 4.25. A simple C-shaped electromagnet forms the field poles. When one of the phases is energized, the rotor will align itself with the electromagnetic field generated. Then, when the first phase is turned off and the second is turned on, the rotor will turn by one-half of a half (or one-quarter) of a rotor pole to align itself with the field from the second phase. So that the rotor keeps turning in the same direction, the second phase is turned off and the first phase is turned on again. As in other steppers, disc motors are half-stepped by turning on both coils at the same time every other half step.

PM Stepper Motor Parameters. The proper amount of excitation voltages and currents must be applied to the stator windings. The rotor may become demagnetized by excessive excitation beyond normal ratings. The best sources of information about proper excitation voltage and current are manufacturers' data sheets. For example, we may learn from a data sheet for an Airpax L82101–P2 stepper that it is designed to be operated with a stator voltage of 12 V. Some manufacturers give the rotor resistance and the maximum power that the rotor can dissipate instead of the current and voltage ratings. The same Airpax stepper has a stator resistance of 118 Ω.

Rotor inertia generally is higher in PM motors than in VR steppers. Recall from physics that *inertia* is the resistance a body possesses to starting motion or changing the direction of motion. Inertia is directly proportional to the mass of an object; the more mass a body has, the more inertia it possesses. *Rotor inertia* (also called *moment of inertia*) is

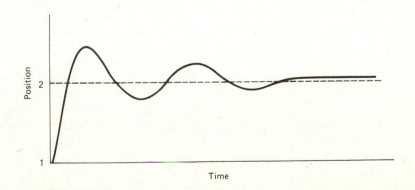

FIGURE 4.24 Stepper Motor Oscillations

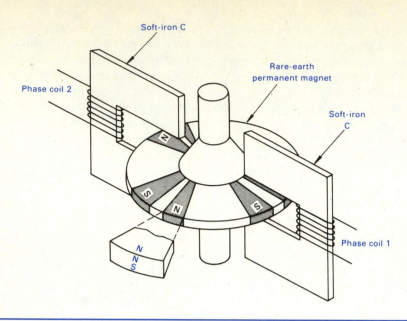

FIGURE 4.25 Disc PM Stepper Motor

generally given in gram-meters squared (g-m²). The following units also are used for rotor inertia: kilogram-centimeters squared (kg-cm²) and, in the English system, ounce-inches squared (oz-in.²) and pound-inches squared (lb-in.²). PM motors permit higher rotor speeds than other types of motors and, conversely, lower stepping rates.

Another parameter usually seen on data sheets is the step angle. The *step angle* is defined as the specific amount of shaft rotation in degrees caused by a change in winding polarity. Step angles generally are larger in the PM stepper. Step angles range from 0.72° to 90°, with 7.5°–18° most common. PM steppers normally have 12 or 24 poles, permitting stepping angles of 3.75°, 7.5°, or 15°. Normally, only 45° and 90° steps are possible with a two-pole rotor. Increasing the number of poles results in smaller stepping angles and higher maximum stepping speeds.

Along with step angle, the maximum number of steps per second (steps/s) is given in manufacturers' data sheets. This parameter replaces the maximum revolutions-per-minute rating of a DC or AC motor. The *maximum stepping speed* of a stepper motor is inversely proportional to the mass (and inertia) of the rotor; as mass and inertia increase, the maximum stepping speed decreases.

PM stepper motors exhibit an interesting and useful parameter called holding torque. Holding torque is produced when the stator windings are energized. When the stator is energized and the motor has stepped to its new position, the rotor will be held in place by the attraction between the two magnetic fields. *Holding torque* is defined as the torque needed to move the rotor a full step with the stator energized but at a standstill. The holding torque generally is larger than the running torque. The holding torque then acts as a powerful braking mechanism in holding a load.

Even when the stator field is not energized, it takes some torque to move the rotor since the rotor produces a cemf when cutting the stator windings. This torque is called *detent torque* and is usually about one-tenth of the holding torque. This useful feature of PM steppers holds the load in the proper position, even when the motor is off. The position will not, however, be held as accurately as when the motor is energized.

The accuracy of a stepper motor is expressed either in a number of degrees or as a percentage per step taken. The *stepping accuracy* is the total error made by the stepper in a single step movement. For example, the accuracy of a motor may be ±6.5%. A

7.5° stepper motor will position a load to within ±0.5°. This error is said to be noncumulative, which means that the error does not build up with each step taken. The accuracy will be within ±6.5% whether 1 step or 1000 steps are taken.

VR Stepper Motors

Variable-reluctance (VR) steppers use a ferromagnetic multitoothed rotor with an electromagnetic stator similar to the PM stepper. A typical three-phase design (Figure 4.26) has 12 stator poles spaced 30° apart; the rotor has eight poles spaced at 45° intervals. The stator poles are energized sequentially by the three-phase winding. When current is applied to phase 1, the rotor teeth closest to the four energized (magnetized) stator poles are pulled into alignment. The four remaining rotor teeth align midway between the nonenergized stator poles. This position is the position of least magnetic reluctance between rotor and stator field.

Energizing phase 2 produces an identical response. The second set of four stator poles magnetically attracts the four nearest rotor teeth, causing the rotor to advance along the path of minimum reluctance into a position of alignment. This action is repeated as the stator's electromagnetic field is sequentially shifted around the rotor. Energizing the poles in a definite sequence produces either clockwise or counterclockwise stepping motion.

The exact increment of motion (step angle) is the difference in angular pitch between stator and rotor teeth, in this case 30° and 45°, respectively, for a net difference of 15°. The VR stepper's step angles are small, making possible finer resolution than can be obtained with the PM type. Maximum stepping rates generally are higher than in the PM stepper. Also, because of the nonretentive rotor, VR steppers do not have detent torque when unenergized.

A typical VR stepper uses a stator with 12 fields. Poles are set about 30° apart and grouped for three-phase operation where each phase has four coils set 90° apart. The rotor has eight teeth spaced 45° apart. VR steppers have a maximum stepping speed of about 18,000 steps/s, much higher than the PM stepper can produce. At high speeds, however, the VR stepper tends to overshoot and must be damped.

Stepper Operation Modes

From our discussion of the theory of stepper motor operation, we can divide the motor operation into four modes: rest, stall, bidirectional, and slew. These modes of operation are illustrated by the speed-torque curves in Figures 4.27A and 4.27B for both VR and PM stepper motors.

The first mode of operation is the *rest mode*. Unenergized PM steppers, due to the interaction between the PM rotor and the stator, exhibit a resistance to movement called *detent*, or *residual*, torque. Sometimes called position memory, this characteristic is valuable when final position must be known in the event of electrical failure in a system. VR steppers do not exhibit this characteristic.

The second mode of operation is the *stall mode*. When a stator winding is energized, both VR and PM steppers resist movement. In data sheets, this mode is sometimes called *static stall torque*.

The third mode of stepper operation is called the *bidirectional mode*. In the bidirectional mode, the shaft continually advances and then stops momentarily (start-stop). The direction of rotation can be reversed instantaneously. Performance curves given in data sheets indicate the maximum speed at which a given load may be run bidirectionally without losing a step. The maximum load that the motor can drive occurs at a rate of about 5 steps/s and is listed on some data sheets as *maximum running torque*. The maximum speed (in steps per second) that a stepper

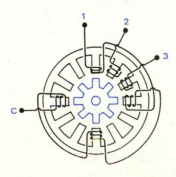

FIGURE 4.26 Three-Phase VR Stepper Motor

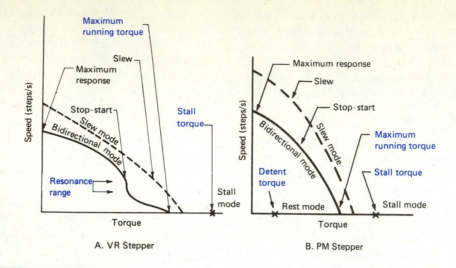

FIGURE 4.27   Stepper Motor Speed-Torque Curves

will run in bidirectional mode occurs with zero load, and it is shown as *maximum stepping rate* or *maximum response* on the data sheets.

The fourth and last mode of stepper operation is called *slewing*. The stepper motor can be accelerated beyond its start-stop bidirectional range in a unidirectional slew mode. The rotor pulls into synchronism with the rotating stator field, much like a standard AC synchronous motor, to be discussed in Chapter 5. In the slew mode, the motor is beyond its bidirectional start-stop speed range; it cannot be instantaneously reversed and still maintain pulse and step integrity. Nor can the motor be started in this range. For the motor to attain slewing speed, whether from rest or from bidirectional mode, it must be carefully accelerated (or ramped). In a similar fashion, for the motor to stop or to reverse in the slew mode, it must first be decelerated carefully to some speed within its bidirectional capability. When ramping is supplied for acceleration and deceleration, there is no loss of pulse-to-step integrity in slew mode.

Driving Stepper Motors

Early stepper systems used mechanical commutation switches to energize a stepper's stator windings in

sequence. Applications typically required transmission of bidirectional shaft rotation for remote indicators such as repeater compasses.

The modern stepper motor usually is driven by a high-speed, solid-state circuit (the controller) that sequences commands to the motor for either clockwise (CW) or counterclockwise (CCW) rotation, as shown in Figure 4.28. The subject of stepper motor-driving circuitry will be discussed in more detail in later chapters.

Stepper Excitation Modes

Depending on the stator winding and performance desired, a stepper motor can be excited in several different modes: two-phase and two-phase modified; three-phase and three-phase modified; and four-phase and four-phase modified. *Phase* refers to a stator winding, and *modified* means that two windings are driven simultaneously. All of these modes of excitation are compatible with most stepper controllers except the two-phase modes.

PM steppers normally are made with two stator windings that are center-tapped, as shown in Figure 4.29. When the center taps are connected to the controller, this stepper is considered to be a four-

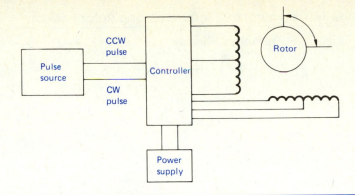

FIGURE 4.28 Block Diagram of Stepper Motor Controller

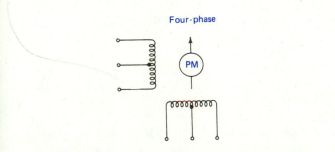

FIGURE 4.29 Four-Phase Stepper Motor with Two Center-Tapped Stator Windings

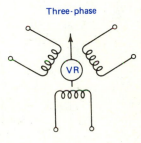

FIGURE 4.30 VR Three-Phase Stepper Motor with Y-Shaped Winding Arrangement

phase motor. Two-phase motors have the center taps either eliminated or open-circuited, and only the end taps are connected.

VR steppers, due to geometry and construction, lend themselves to a convenient Y-shaped (wye) winding arrangement, illustrated in Figure 4.30. They are excited in the three-phase or three-phase modified mode.

Two-Phase Mode. One entire phase of the motor, end tap to end tap, is energized at any given moment. Compared with standard four-phase excitation (the usual way of rating four-phase motors on data

sheets), two-phase resistance is doubled; thus, input current and power are halved. Heat dissipation is increased since more copper is used in the motor winding. Because of the reduced input and greater heat dissipation, the output of the motor can be improved by as much as 10% over the standard four-phase mode of excitation. See Table 4.2 and Figure 4.31 for input sequence and rotor position.

Two-Phase Modified Mode. Both windings (that is, considering one winding to be end tap to end tap and ignoring the center taps shown in Figure 4.29) are energized simultaneously. Energy input in this mode

TABLE 4.2   Two-Phase Motor Energizing Sequence and Rotor Position

Excitation Mode	Energized Winding (Figure 4.31)	Rotor Position (Figure 4.31)	Motion Sequence
Two-phase	3–1	f	Index
commutation of	6–4	h	CCW
$B+$ and $B-$	1–3	b	CCW
	4–6	d	CCW
Two-phase modified	3–1 and 6–4	g	Index
commutation of	1–3 and 6–4	a	CCW
$B+$ and $B-$	1–3 and 4–6	c	CCW
	3–1 and 4–6	e	CCW

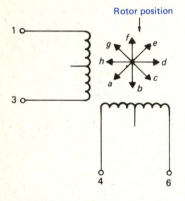

FIGURE 4.31   Rotor Position and Winding Orientation for Two-Phase Excitation Mode

is the same as in the standard four-phase mode, but output performance is increased by about 40%. The controller is more complex and costly than a four-phase system for this mode. For critical applications, maximum performance is obtained from minimal energy inputs.

Three-Phase Mode. Most VR steppers use three-phase windings. Windings are excited one phase at a time to obtain the ratings shown in data sheets. The energizing sequence and consequent motion are shown in Table 4.3. Figure 4.32 illustrates the rotor position and winding orientation.

TABLE 4.3   Three-Phase Motor Energizing Sequence and Rotor Position

Excitation Mode	Energized Winding (Figure 4.32)	Rotor Position (Figure 4.32)	Motion Sequence
Three-phase	2–1	a	CCW
commutation of	3–4	c	CCW
$B+$ only	5–6	e	CCW
Three-phase modified	2–1 and 3–4	b	CCW
commutation of	3–4 and 5–6	d	CCW
$B+$ only	5–6 and 2–1	f	CCW

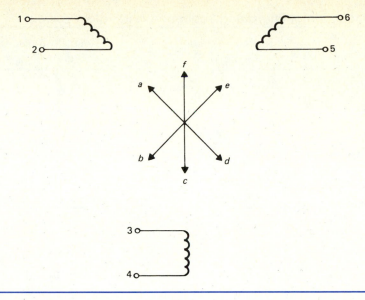

FIGURE 4.32   Rotor Position and Winding Orientation for Three-Phase Excitation Mode

Three-Phase Modified Mode. In this mode, two adjoining phases of a three-phase motor are excited simultaneously. The rotor steps to a minimum-reluctance position corresponding to the resultant of the two magnetic fields. Since two windings are excited, twice as much power as the standard mode (one phase at a time) is required, as shown in Table 4.3. The output is not significantly increased, but damping characteristics are noticeably improved.

Four-Phase Mode. In this mode of excitation, each half winding (end tap to center tap, as in Figure 4.29) is regarded as a separate phase, and phases are energized one at a time. The sequence of excitation is shown in Table 4.4. Rotor position and winding orientation are shown in Figure 4.33. Performance curves for four-phase motors shown on most data sheets use this mode of excitation. Although this mode is less efficient than others, the controller is uncomplicated, being a simple four-stage ring counter.

Four-Phase Modified Mode. Phases (halves) of different windings are simultaneously energized in this mode, as shown in Table 4.4. Since two phases are simultaneously excited, twice the input energy of single-phase excitation is required. Torque output is increased by about 40%, and the maximum response rate is also increased, compared with the single-phase excitation.

Load Torque and Inertia

The maximum response rate of a stepper motor is measured in a no-load condition. As the torque load on the motor is increased, the maximum stepping or response rate will decline proportionally. The increased torque ratings are plotted on a continuous curve. The resulting curve describes the bidirectional operating areas of a stepper motor over its entire torque capacity range. The curve in Figure 4.34 is such a curve. Read from left to right, torque is shown to be greatest at 0 steps (pulse)/s, and response is

TABLE 4.4   Four-Phase Motor Energizing Sequence and Rotor Position

Excitation Mode	Energized Winding (Figure 4.33)	Rotor Position (Figure 4.33)	Motion Sequence
Four-phase	2–1	*f*	Index
commutation of	5–4	*h*	CCW
B+ only	2–3	*b*	CCW
	5–6	*d*	CCW
Four-phase modified	2–1 and 5–4	*g*	Index
commutation of	2–3 and 5–4	*a*	CCW
B+ only	2–3 and 5–6	*c*	CCW
	2–1 and 5–6	*e*	CCW

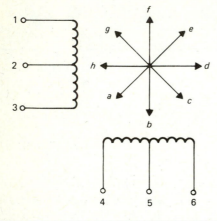

FIGURE 4.33   Rotor Position and Winding Orientation for Four-Phase Excitation Mode

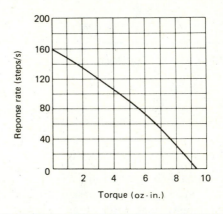

FIGURE 4.34   Speed-Torque Curve for Four-Phase Stepper Motor

highest at the no-load point. Operation at rates beyond those indicated on the curve in Figure 4.34 is in the unidirectional slow range discussed earlier.

Methods of Damping

A factor called resonance can be a problem with VR steppers and, occasionally, with PM motors, especially when load inertia is high. *Resonance* is the inability of the rotor to follow the step input command. Every object has a natural resonant fre-

quency. When a motor's natural resonant frequency is reached, the motor will lose steps and oscillate about a point.

A brief description of the dynamic movement of a stepper motor will be helpful in further defining resonance. As shown in Figure 4.35, when the stepper goes from position 1 to position 2, the kinetic energy of the rotor must be dissipated, producing an oscillating motion. In the PM types, these oscillations are damped by the interaction of the rotor with the energized magnetic field, eddy currents, and

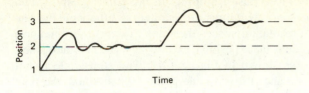

FIGURE 4.35 Stepper Motor Oscillations after Stepping

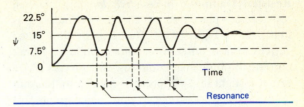

FIGURE 4.37 Graphic Representation of Stepper Motor Resonance

hysteresis loss. However, in VR motors, which have no permanent magnet, eddy currents and hysteresis loss alone provide little damping effect. Under resonant conditions, the rotor will oscillate at random and lose pulse step integrity. Figures 4.36 and 4.37 graphically illustrate this situation in a 15° VR stepper. Rotor pole 2 is in a position 7.5° away from its final position. It is in line with energized stator winding 2. In this condition, rotor poles 2 and 3 are the same distance away from stator winding 3, 22.5° apart on either side. If, at this precise moment, current is switched from stator winding 2 to winding 3, the rotor can step in either direction regardless of programmed command.

In a 15° VR stepper, resonance typically may occur at several points, as shown in Figure 4.37. If resonance is to be eliminated in VR steppers, the amplitude of oscillation must be kept within 7.5° deflection from each 15° step position. Several

methods of damping may be employed to eliminate resonance.

Slip-Clutch Damping. A *slip-clutch damper* is a mechanical device that uses a heavy inertia wheel sliding between two collars. As the rotor moves, the inertia wheel resists movement and adds friction load to the system. This friction has the effect of reducing rotor speed and consequently decreasing overshoot and undershoot. The amount of friction is controlled by spring pressure, and wear is reduced by using Teflon discs to separate the steel members.

Most VR stepper motors are supplied with a rear-shaft extension to accommodate this type of damper assembly. Figure 4.38 shows a typical slip-clutch damper installed on the rear shaft of a VR stepper motor. The disadvantage in using system friction to damp resonance is that resonance will vary as the system wears, slowing down response.

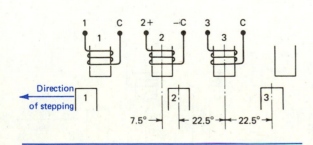

FIGURE 4.36 Resonance in a 15° Stepper Motor

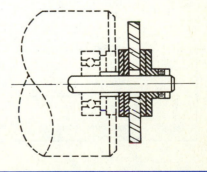

FIGURE 4.38 Slip-Clutch Damper

Resistive Damping. In *resistive damping*, external resistors are placed across the stator windings, as illustrated in Figure 4.39. The resistors allow rated current to pass through one phase while they limit current through the remaining two phases. As a result, a slight reverse torque is applied to the rotor by the two windings not carrying rated current. This reverse torque prevents the rotor from accelerating quickly and limits overshoot. Resistive damping increases power consumption an average of 20% for most inertia loads.

Capacitive Damping. In place of resistors, capacitors can be used to apply reverse torque to the rotor in *capacitive damping*, as shown in Figure 4.40. At the moment phase one is deenergized and field two is energized, the capacitor on phase one slowly discharges. The discharging current applies retrotorque on the rotor. This action is repeated as the remaining phases are shifted. This damping method offers the advantage of lower power consumption than in the resistive-damping technique.

Two-Phase Damping. In *two-phase damping*, two stator windings are excited simultaneously. The excitation causes the rotor to step to a minimum-reluctance position, midway between the stator poles, as shown in Figure 4.41. The step angle is not changed from single-phase excitation, but the final rotor position will be different. While advancing to this final step position, the two adjacent rotor teeth exert equal and opposing torque. In combination with the stator's magnetic field, this system produces twice the damping effect of single-phase excitation. The disadvantage of two-phase damping is that power consumption is approximately double that of single-phase excitation. The speed and torque delivery, however, remain essentially unchanged.

Retrotorque Damping. The most satisfactory method of damping the stepper motor is through the use of reverse-torque controller electronics. The *retrotorque controller* drives the stepper motor in a single-phase mode and effects damping by supplying a pulse of power to the stator winding previously (last) energized. Both overshoot and undershoot tendencies are eliminated. Damping in this manner does not increase power consumption or noticeably affect performance. However, the greater complexity and increased expense of the retrotorque controller may be a factor for consideration.

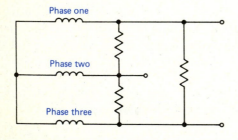

FIGURE 4.39   Resistive Damping

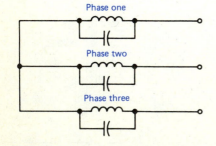

FIGURE 4.40   Capacitive Damping

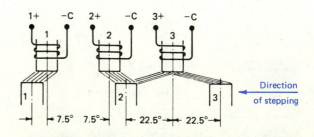

FIGURE 4.41   Two-Phase Damping

Advantages and Disadvantages of Steppers

Compared with the usual closed-loop analog servomotors, stepper motors offer significant advantages:

1. Feedback is not ordinarily required when the stepper motor is applied properly. However, the stepper motor is perfectly compatible with feedback, whether analog or digital, whether for velocity or position, or both.
2. Error is noncumulative as long as pulse-to-step integrity is maintained. A stream of pulses can be counted into a stepper, and its final shaft position is known within a very small percent of one step. DC servo accuracy, in contrast, is subject to the sensitivity and phase shift of the loop.
3. Stepper motors do not have a null position. Bidirectional rotation is continuous.
4. Maximum torque occurs at low pulse rates. The stepper can readily accelerate its load. When the desired position is reached and command pulses cease, the shaft stops. There is no need for clutches and brakes. Once the stepper is stopped, there is little tendency to drift. Indeed, PM steppers are magnetically detented in the last position. In sum, a load can be started in either direction, moved to a position, and it will remain there until commanded again.
5. A wide range of step angles is available off-the-shelf in the stepper motor line—1.8°, 7.5°, 15°, 45°, and 90° angles—without special drivers and logic manipulation.
6. Inherent low velocity is available without gear reduction. For example, a Tormax model 200 stepper motor driven at 500 pulses/s turns at 150 r/min. Many AC and wound-rotor DC motors have difficulty turning at such a slow speed.
7. Steppers are true digital devices. They do not need a D/A conversion at the input as do conventional servos. Steppers may then be used with computers without any additional D/A conversions.
8. They offer close speed control and reversibility over a wide range.
9. Starting current is low.
10. The rotor inertia usually is low.
11. Multiple steppers driven from the same source maintain perfect synchronization, a problem in other types of DC motors.

These advantages are offset by certain disadvantages:

1. Efficiency is low. Much of the input energy must be dissipated as heat.
2. Loads must be analyzed carefully for optimum stepper performance. Inputs (pulse sources and controllers) must also be matched to the motor and load.
3. When load inertia is exceptionally high, damping may be required.

Table 4.5 summarizes typical parameter values for stepper motors. The category "Special PM Type" has been included. Special hybrid steppers have been produced that have improved performance in some areas of operation. Table 4.5 reflects only typical values and not maximum values.

OPTICAL ENCODERS

In servo control systems (discussed in Chapter 10), where mechanical position is required to be controlled, some form of position-sensing device is needed. There are a number of types in use—magnetic, contact, resistive, and optical—but for accurate position control, the most commonly used is the *optical encoder*. There are two forms of this encoder: absolute and incremental.

Optical encoders operate by means of a *grating* that moves between a light source and a detector. When light passes through the transparent areas of the grating, an output is seen from the detector. For increased resolution, the light source is collimated and a *mask* is placed between the grating and the detector. The grating and the mask produce a shuttering effect so that only when their transparent sections are in alignment is light allowed to pass to the detector (Figure 4.42).

TABLE 4.5 Parameters for Stepper Motors

Parameter	PM	VR	Special PM Type
Sizes	5–32	8–25	23–34
Mode	2, 4	3, 4	4
Step angle (degrees)	45, 90	7.5, 15	1.8
Steps per revolution	8, 4	48, 24	200
Maximum steps per second (pps)	90–500	350–1025	250–600
Detent torque	Yes	No	Minimal
Maximum running torque (oz-in.)	0.17–24	0.15–12	25–100
Stall torque (oz-in.)	0.3–50	0.25–35	45–200
Resonance	Minimal	Some	Minimal

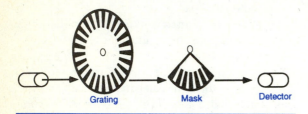

FIGURE 4.42 Optical Encoder System

An *incremental encoder* generates a pulse for a given increment of shaft rotation (rotary encoder) or a pulse for a given linear distance traveled (linear encoder). Total distance traveled or shaft angular rotation is determined by counting the encoder output pulses. An *absolute encoder* has a number of output channels so that every shaft position may be described by its own unique code. The higher the resolution, the more output channels are required.

Basics of Incremental Encoders

In most industrial applications, cost is an important factor. Resetting to a known zero point following power failure is seldom a problem. The rotary incremental encoder is the type that fits these two requirements. Industrial system designers favor the rotary incremental encoder. Its main element is a shaft-mounted disc carrying a grating. The shaft and grating rotate between a light source and a masked detector. The light source may be a light-emitting diode or an incandescent lamp. The detector is usually a phototransistor or, more commonly, a

photovoltaic cell. Such a simple system, providing a single low-level output, is unlikely to be frequently encountered. It has a DC offset, which is temperature dependent, making the signal difficult to use (Figure 4.43).

In practice, two photovoltaic cells are used with two masks, arranged to produce signals with 180° phase difference for each channel. The two diode outputs are subtracted and cancel the DC offset (Figure 4.44). This sinusoidal output may be used unprocessed. More often, it is either amplified or used to produce a square wave output. Rotary incremental encoders, then, may have sine wave or square wave outputs, and they usually have up to three output channels.

A two-channel encoder, as well as giving position of the encoder shaft, can also provide information on the direction of rotation. It does this by examining the output signals to identify the leading channel. This is possible when the channels are arranged in quadrature (90° phase shifted), as shown in Figure 4.45.

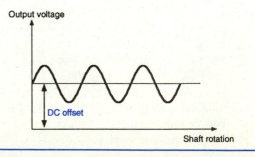

FIGURE 4.43 Encoder Output Voltage

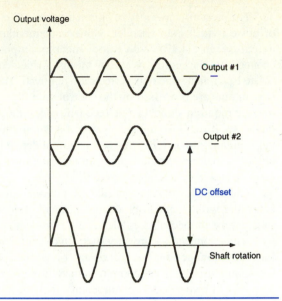

FIGURE 4.44 Output Voltage for Dual-Photocell System

occur. This allows the resolution of the encoder to be quadrupled by processing the A and B outputs to produce a separate pulse for each square wave edge. For this process to be effective, however, it is important that quadrature is maintained within the necessary tolerance so that the pulses do not run into one another.

Square wave output encoders are generally available in a wide range of resolutions (up to about 5000 lines/revolution) and with a variety of different output configurations. Some of the different outputs are as follows:

- TTL—This is a commonly available output, is compatible with TTL logic levels, and normally requires a 5 V supply. TTL outputs are also available in an open-collector configuration, allowing the system designer a choice of pull-up resistor value.
- CMOS—This output is available for compatibility with the higher logic levels normally used with CMOS devices.
- Line driver—These outputs use low output impedance devices designed for driving signals over a long distance and are usually used with a matched receiver.
- Complementary outputs—These outputs are derived from each channel. Each channel gives a pair of signals, 180° out of phase. These outputs are useful where maximum immunity to interference is required.

For most machine tool or positioning applications, a third channel, known as the index channel or Z channel, is also included. This gives a single output pulse per revolution and is used when establishing the zero position. Reference to Figure 4.45 shows that for each complete square wave from channel A, if channel B output is also considered during the same period, then four pulse edges may be seen to

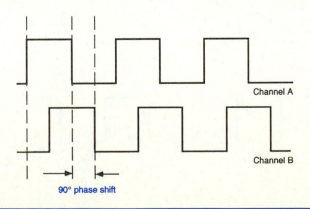

FIGURE 4.45 Quadrature Output Signals

Noise Problems

The control system for a machine is normally screened and protected within a metal cabinet. An encoder may be similarly housed, but unless suitable precautions are taken, the cable connecting the two can be a source of trouble. The cable can pick up electrical noise. This noise may result in the loss or gain of signal counts, giving rise to incorrect data input and loss of position.

Figure 4.46 shows how the introduction of two noise pulses has converted a three-pulse train into one of five pulses. A number of techniques are available to overcome problems due to noise, the most obvious being to use shielded interconnecting cables. Since the signals may be at low level (5 V) and may be generated by a high impedance source, it may be necessary to go to further lengths to eliminate the problem.

The most effective way of eliminating the effects of noise is to use an encoder with complementary outputs (Figure 4.47). With complementary outputs connected to the control system by means of shielded, twisted-pair cable, the noise effects are removed. The two outputs are processed by the control circuitry so that the required signal can be reconstituted without the noise. If the A signal is inverted and is fed with the A signal into an OR gate (whose output depends on one signal or the other being present), the resultant output will be a square wave.

The simple interconnection of encoder and controller with channel outputs at low level may be satisfactory in electrically "clean" environments or where interconnections are very short. In cases where long interconnections are necessary or where the environment is "noisy," complementary line-driver outputs will be needed. Any interconnections should be made with shielded, twisted-pair cable.

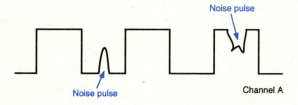

FIGURE 4.46 Noise Degrading an Encoder Signal

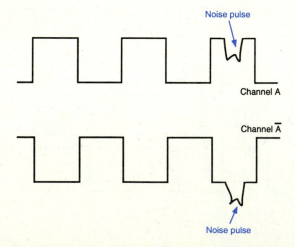

FIGURE 4.47 Complementary Output Signals

Accuracy of Encoders

Several factors will determine the accuracy of an encoder. A rotary incremental encoder will have a maximum frequency at which it will operate. For most encoders, the maximum frequency is typically 100 kHz. The maximum rotational speed, or *slew rate*, will be determined by this frequency. Beyond this frequency of operation, the output will become unreliable and the accuracy affected.

Let us consider an example of a 600-line encoder rotated at 1 r/min. This encoder will give an output of 10 Hz.

$$f = \left(600 \, \frac{\text{lines}}{\text{r}}\right)\left(1 \, \frac{\text{r}}{\text{min}}\right)\left(\frac{1 \, \text{min}}{60 \, \text{s}}\right)\left(1 \, \frac{\text{cycle}}{\text{line}}\right) \qquad (4.15)$$

$$= 10 \, \frac{\text{cycles}}{\text{s}} \, (\text{Hz})$$

where

f = output frequency at 1 r/min

If the maximum operating frequency of the encoder is 50 kHz, then its speed will be limited to 5000 times 1 r/min, or 5000 r/min.

$$n_{\text{max}} = \frac{f_{\text{max}}}{f} \qquad (4.16)$$

$$= \frac{50 \, \text{kHz}}{10 \, \text{Hz}} = 5000 \, \text{r/min}$$

• EXAMPLE 4.3

Given a 360-line encoder with a maximum frequency of 25 kHz, find the maximum shaft speed.

Solution

$$f = \left(360 \, \frac{\text{lines}}{\text{r}}\right)\left(1 \, \frac{\text{r}}{\text{min}}\right)\left(\frac{1 \, \text{min}}{60 \, \text{s}}\right)\left(1 \, \frac{\text{cycle}}{\text{line}}\right)$$

$$= 6 \, \frac{\text{cycles}}{\text{s}} \, (\text{Hz})$$

$$n_{\text{max}} = \frac{f_{\text{max}}}{f} = \frac{25 \, \text{kHz}}{6 \, \text{Hz}} = 4167 \, \text{r/min}$$

If an encoder is rotated at speeds higher than its design maximum, there may be conditions set up that will be detrimental to the mechanical components of the assembly. This could cause damage, affecting encoder accuracy.

In common with all digital systems, it is not easy to interpolate between output pulses so that knowledge of position will be accurate only to the grating width. This error is called *quantization error*. The quantization error is equal to one-half of the angular rotation between two successive bits. For example, an encoder that has 360 lines and that produces 360 pulses per revolution will have a quantization error of ±0.5°, or ±½ of an angular degree. A disc with 600 lines would have

$$e_Q = \pm\frac{1}{2}\left(\frac{360}{N_{\text{lines}}}\right) = \pm\frac{1}{2}\left(\frac{360}{600}\right) = \pm0.3° \qquad (4.17)$$

• EXAMPLE 4.4

Given a 720-line encoder, find the quantization error.

Solution

$$e_Q = \pm\frac{1}{2}\left(\frac{360}{N_{\text{lines}}}\right) = \pm\frac{1}{2}\left(\frac{360}{720}\right) = \pm0.25°$$

Basics of Absolute Encoders

An absolute encoder is a position-verification device that provides unique position information for each shaft location. The location is independent of all other locations, unlike the incremental encoder. In the incremental encoder, a count from a reference is required to determine position.

In an absolute encoder, there are several concentric tracks. The incremental encoder has a single track. Each track in the absolute encoder has an independent light source. As the light passes through a slot, a high state (true "1") is created. If light does not pass through the disc, a low state (false "0") is created. The position of the shaft can be identified through the pattern of ones and zeros.

An example of an absolute encoder system is shown in Figure 4.48. When light from an LED passes through the clear area on the disc, a detector receives the light, converting it to an electrical signal. For reliability, it is desirable to have the discs constructed of metal rather than glass. A metal disc is not as fragile and has lower inertia.

The tracks of an absolute encoder vary in slot size, moving from smaller at the outside edge to larger toward the center. The pattern of slots is also staggered with respect to preceding and succeeding tracks. The number of tracks determines the amount of position information that can be derived from the encoder disc—its resolution. For example, if the disc has 10 tracks, the resolution of the encoder would usually be 1024 positions per revolution, or 2^{10}. A four-bit system, similar to the one shown in Figure 4.49, has 2^4, or 16 positions per revolution. Note the four tracks, labeled 0 through 3; the letter B indicates the presence of a sensor behind the track. If the disc rotates one position clockwise, none of the sensors will receive light. Thus, this position would be 0000. When the disc is moved to position 2, the B0 sensor is uncovered, producing a binary 0001. Position 3 gives a binary 0010, corresponding to decimal 2.

A problem comes up with this system when, for example, we go from decimal 11 (1011) one count up to decimal 12 (1100). Note that three of the four binary bits change state in this transition. If these bits do not all change at the same time, a false count will occur. To solve this problem, the Grey code was invented. Table 4.6 shows both the binary and Grey

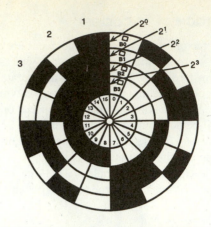

FIGURE 4.49 Optical Encoder Disc

TABLE 4.6 Four-Bit Binary and Grey Codes

Decimal Number	Binary Code	Grey Code
0	0000	0000
1	0001	0001
2	0010	0011
3	0011	0010
4	0100	0110
5	0101	0111
6	0110	0101
7	0111	0100
8	1000	1100
9	1001	1101
10	1010	1111
11	1011	1110
12	1100	1010
13	1101	1011
14	1110	1001
15	1111	1000

codes. No more than one bit changes in the Grey code when counting adjacent tracks.

Gearing an additional absolute disc to the primary high-resolution disc provides for turns counting so that unique position information is available over multiple revolutions. An example of how an encoder with 1024 counts per revolution becomes an absolute device for 524,288 discrete positions follows.

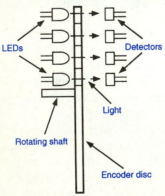

FIGURE 4.48 Cross Section of Absolute Encoder System

The primary high-resolution disc has 1024 discrete positions per revolution. A second disc with 3 tracks of information will be attached to the high-resolution disc geared 8:1. The absolute encoder now has 8 complete turns of the shaft or 8192 discrete positions. Adding a third disc geared 8:1 will provide for 64 turns of absolute positions. In theory, additional discs could continue to be incorporated. But, in practice, most encoders stop below 512 turns. Encoders using this technique are called *multiturn absolute encoders*. This same technique can be used in a rack-and-pinion style of linear encoder, resulting in long lengths of discrete absolute locations.

Advantages of Absolute Encoders

Rotary and linear absolute encoders offer a number of significant advantages in industrial motion control and process control applications:

1. There is no position loss on power down or loss of power. An absolute encoder is not a counting device like an incremental encoder because an absolute system reads actual shaft position. The lack of power does not cause the encoder to lose position information. Whenever power is supplied to an absolute system, it is capable of reading the current position immediately. In a facility where frequent power failures are common, an absolute encoder is a necessity.

2. An absolute encoder operates in electrically noisy environments. Equipment such as welders and motor starters often generate electrical noise that can look like encoder pulses to an incremental counter. Electrical noise does not alter the discrete position that an absolute system reads.

3. It can use high-speed, long-distance data transfer. Use of a serial interface such as the RS-422 gives the user the option of transmitting absolute position information over as much as 4000 feet.

4. It can eliminate a go-home or referenced starting point. The need to find a home position or a reference point is not required with an absolute encoding system since an absolute system always knows its location. In many motion control applications, it is difficult or impossible to find a home reference point. This situation occurs in multiaxis machines and on machines that cannot reverse direction. This feature is particularly important in a "lights-out" manufacturing facility. Significant cost saving is realized in reduced scrap and setup time resulting from a power loss.

5. It can provide reliable position information in high-speed applications. The counting device is often the factor limiting the use of incremental encoders in high-speed applications. Often, the counter is limited to a maximum pulse input of 100 kHz. An absolute encoder does not require a counting device or continuous observation of the shaft or load location. This attribute allows the absolute encoder to be applied in high-speed and high-resolution applications.

CONCLUSION

In this chapter, we have seen three DC motors that present alternatives to the conventional wound-field DC motor: the permanent-magnet motor, the stepper motor, and the brushless DC motor. The PM DC motor is similar to the shunt-connected field DC motor with a constant field current. The speed is then directly proportional to the applied armature voltage. The stepper motor is a truly digital device. The rotor steps in discrete increments each time voltage is applied to the stator. The stepper is, therefore, the motor of choice in digital applications, such as may be required in a microcomputer. The BDCM is not a digital device. It is similar in function to the PM DC motor. The BDCM is commutated by semiconductors, unlike the mechanical commutator of the wound-field and PM DC motors. Both the stepper motor and the BDCM have specific applications for which their advantages give them a valuable place in the DC motor family. We have also seen how encoders can be used to sense the position of a motor shaft. Of the four types of encoders, the optical encoder is the most popular. Optical encoders can be either incremental or absolute encoders. For more accurate position information, most absolute optical encoders use the Grey code.

■ QUESTIONS ●

1. List the parts of a BDCM.

2. How does the rotor on a BDCM get its electromagnetic field?

3. List several of the differences between the BDCM and the PM DC motor.

4. Describe the commutation process in a BDCM.

5. How are the semiconductors protected from damage in a BDCM circuit?

6. List the sensors used in a BDCM controller to detect the rotor position. What are the advantages and disadvantages?

7. Compare the advantages and disadvantages of BDCMs with those of PM DC motors.

8. Define the following terms as they relate to the stepper motor: (a) stepping angle, (b) maximum stepping rate, (c) rotor inertia, (d) holding torque, and (e) residual or detent torque.

9. Describe the differences in construction between VR and PM stepper motors.

10. Describe the operation of the PM and VR stepper motors.

11. List and describe the four modes of stepper motor operation.

12. Draw the schematic diagram for a two-phase stepper motor, and describe how it is operated.

13. Describe the relationship between torque, speed, and rotor inertia in a stepper motor.

14. Describe the condition of resonance and how it is reduced in stepper motor controllers.

15. Explain the differences between resistive and capacitive damping in a stepper motor controller, and draw diagrams to illustrate both damping methods.

16. Explain the difference between incremental and absolute optical encoders.

17. Describe the Grey code and contrast it with a straight binary coding scheme.

■ PROBLEMS ●

1. Given a PM DC motor, with $V_L = 12$ V, $N_{nl} = 5500$ r/min, $I_{nl} = 2$A, loaded speed = 4520 r/min, and loaded current = 10 A, find M, E_{max}, I_{max}, P_o, N_m, and T_m.

2. Given a PM DC motor with the data in the accompanying table, find (a) the torque produced by the motor and (b) the rotor speed, if we apply rated voltage to the armature and draw 0.5 A of current.

Rated speed	1500 r/min
Rated voltage	79 V DC
Rated current	1.5 A

Rated P_o	72 W
Rated torque	4.07 lb-in.
Maximum torque	28.1 lb-in.
K_{cemf}	33.5 V/1000 r/min
K_T	2.83 lb-in./A
R_A	52.6 Ω

3. Given a 720-line encoder with a maximum frequency of 20 kHz, find the maximum shaft speed.

4. Given a 360-line encoder, find the quantization error.

AC Motors

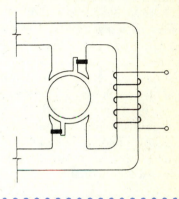

OBJECTIVES

On completion of this chapter, you should be able to:

- Draw the schematic symbols for wye and delta connections, and calculate phase and line voltage and currents.
- Define power factor, and explain how power factor relates to the operation of AC motors.
- Contrast apparent power and true power, and state the measurement units of each.
- Classify AC motors by horsepower and internal construction.
- Explain the concept of the rotating field and calculate its speed.
- Explain how torque is produced in an induction motor.
- Calculate the slip of an induction motor.
- List and describe the different methods of starting single-phase motors.
- Justify the need for special starting methods for synchronous motors.

INTRODUCTION

Alternating current (AC) has one property that allows it to be transported long distances by wire more efficiently than direct current. Alternating current circuits can be stepped up (transformed) in voltage and at the same time stepped down in current. AC transmission lines thus can carry high voltages at low currents and keep I^2R (power) losses much lower than the equivalent DC power lines can. Because of this property, most modern power-generating systems produce AC power. Consequently, the majority of motors used throughout industry are designed to use AC power.

There are other advantages to AC motors besides the wide availability of AC power. In general, AC motors of the same horsepower rating are smaller and, therefore, less expensive than DC motors. AC induction motors do not use brushes and commutators. Thus, they are less prone to mechanical wear and sparking. This feature decreases maintenance requirements and the possibility of igniting explosive gases. AC motors are well suited to constant-speed applications; several AC motors can be made to run in synchronization at the same speed.

Industry uses AC motors in a wide variety of shapes, sizes, and power ratings. One type of AC motor in use is shown in Figure 5.1.

In this chapter, we will examine the characteristics of AC motors and what makes them different from DC motors. We will consider briefly three-phase AC and then the universal motor, which is basically a series DC motor. As its name implies, the universal motor can be operated with either DC or AC voltage. The remainder of the chapter will deal with the two other classifications of AC motors, the induction motor and the synchronous motor.

FIGURE 5.1   AC Motor

THREE-PHASE AC

Before examining AC motors, let us take a moment to review the basic principles of three-phase AC.

Three-phase AC power is produced by a three-phase generator, sometimes called an *alternator*. The three-phase alternator has three single-phase windings. Each winding is spaced so that the voltage induced in each winding is 120° out of phase with the voltages in the other two windings. A schematic diagram of the three-phase alternator is complex, and, therefore, seeing how the alternator works is difficult. A simplified schematic diagram, showing all the windings of a single-phase lumped together as one winding, is shown in Figure 5.2A. A magnetic rotor (not shown) turns inside the stator illustrated in the diagram. The voltage waveforms produced by this generator are given in Figure 5.2B. Note that the three waveforms are spaced 120° apart.

Wye-Connected Systems

Rather than six leads coming out the generator, one lead from each phase may be connected together to form a common junction. Alternators connected like this are sometimes called *wye-* or *star-connected generators*. The wye connection is shown in the schematic diagram of Figure 5.3A. The common lead may or may not be brought out of the machine. The common lead is sometimes called the *neutral*. Each of the three voltages are called *phase voltages*, symbolized V_A, V_B, and V_C. The terminals at which these voltages are present are labeled A, B, and C, with the neutral labeled N. The neutral serves as a common return circuit for each of the three phases. This system is called a *three-phase*, *four-wire* system and is very popular in industry.

The three-phase wye system can be better understood by examining the vector analysis shown in Figure 5.3B. Note that phase voltage B lags behind phase voltage A by 120°. Furthermore, phase voltage C lags behind B by 120° and behind A by 240°. The order of the phases, A followed by B followed by C, is called the *phase sequence*. The only other possibility for the phase sequence is A–C–B, which can be achieved by interchanging the wiring at any two terminals (but not the neutral). The phase sequence is also determined by the direction in which the AC generator rotates.

In the wye connection, the line voltage (V_L) is equal to the sum of the two phase voltages (V_{AN} and V_{BN}). So

$$V_L = V_{AB} = V_{AN} + V_{BN} \qquad (5.1)$$

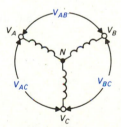

A. Schematic Diagram

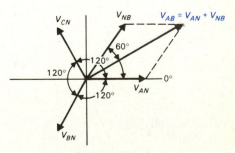

B. Vector Diagram

FIGURE 5.3 Wye-Connected System

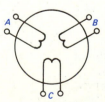

A. Simplified Schematic

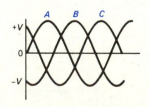

B. Waveforms

FIGURE 5.2 AC Generator Schematic and Output

We must be careful here. Let us suppose that we have phase voltages equal to 120 V AC. The line voltage is *not* equal to 240 V AC. That is, when we add these two voltages, we must consider that they are out of phase by 120°. So we must perform the vector sum of the voltages to get the correct answer. The line voltage, then, is equal to the square root of 3 times any one of the phase voltages, or

$$V_L = 1.73 V_P \qquad (5.2)$$

where

V_L = line voltage
$1.73 = \sqrt{3}$
V_P = any phase voltage

Let us use the 120 V AC system as an example. If the phase voltage is 120 V AC, what is the line voltage (the voltage between any of the two phase voltages)? Using Equation 5.2, we get

$$V_L = 1.73 V_P$$

$$= (1.73)(120 \text{ V AC}) = 207.6 \text{ V AC}$$

In the wye-connected generator, the current in any line is in phase with the current in the winding that it feeds. So we can say that

$$I_P = I_L \qquad (5.3)$$

where

I_P = current flowing in any phase
I_L = line current

• EXAMPLE 5.1

Given a 240 V AC three-phase, wye-connected generator, with a phase current of 10 A, find the (a) line voltage and (b) line current.

Solution

(a) If the phase voltage is 240 V AC, the line voltage is, from Equation 5.2,

$$V_L = 1.73 V_P$$

$$= (1.73)(240 \text{ V AC}) = 415.2 \text{ V AC}$$

(b) In the wye-connected generator, the current in any line is equal to the phase current, according to Equation 5.3:

$$I_P = I_L = 10 \text{ A}$$

Delta-Connected Systems

The *delta-connected generator* is shown in Figure 5.4. Note that there is no neutral, and the windings are connected in the shape of the Greek letter delta (Δ). In a circuit that has equal phase voltages (called a *balanced circuit*), the voltage between any two phases (the line voltage) is equal to that of a single phase. So the line voltage and the voltage across any winding are in phase. The line current, however, is 120° out of phase with the current in any of the phases. This situation is the opposite of what happens in the wye-connected generator.

The relationships between line current and line voltage in the delta-connected generator are

$$I_L = 1.73 I_P \qquad (5.4)$$

and

$$V_P = V_L \qquad (5.5)$$

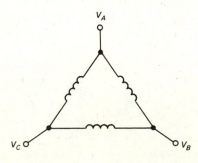

FIGURE 5.4   Schematic Diagram of Delta Connection

As an example, suppose we have a 120 V AC system with a 5 A phase current. The line voltage is, according to Equation 5.5,

$$V_P = V_L = 120 \text{ V AC}$$

The line current is found from Equation 5.4:

$$I_L = 1.73 I_P$$

$$= (1.73)(5 \text{ A}) = 8.65 \text{ A}$$

• EXAMPLE 5.2

Given a 240 V AC, three-phase, delta-connected generator, with a 10 A phase current, find the (a) line voltage and (b) line current.

Solution

(a) If the phase voltage is 240 V AC, the line voltage, from Equation 5.5, is

$$V_P = V_L = 240 \text{ V AC}$$

(b) The line current is calculated from Equation 5.4:

$$I_L = 1.73 I_P$$

$$= (1.73)(10 \text{ A}) = 17.3 \text{ A}$$

In single-phase power systems, the power produced by a phase is equal to

$$P_P = (V_P I_P)(\cos \theta) \tag{5.6}$$

where

P_P = power produced by phase
$\cos \theta$ = power factor

The *power factor* ($\cos \theta$) is the cosine of the phase angle between the phase current and voltage. Power factor will be discussed in more detail later in the chapter.

Equation 5.6 also holds for the power produced by each of the three phases of a three-phase system, in both wye and delta connections. The total power in a three-phase system is, then,

$$P_{total} = 3(V_P I_P)(\cos \theta)$$

If we substitute Equations 5.2 and 5.3, or 5.4 and 5.5, into Equation 5.6, we get the total power in terms of line voltage and current:

$$P_{total} = 1.73(V_L I_L)(\cos \theta) \tag{5.7}$$

The term $(V_P I_P)(\cos \theta)$ from Equation 5.6 is the actual, *real power* the generator produces (in watts). Since $(\cos \theta)$ is the power factor of the generator, then $(V_P I_P)$ is the *apparent power per phase*. (Real power versus apparent power will be discussed in more detail later in the chapter.) The total generator apparent power is $3(V_P I_P)$, or, in terms of line voltage and current, $1.73(V_L I_L)$. The total generator apparent power (in kilovolt-amperes, kVA) is called its *kVA rating* and is given by

$$kVA = \frac{1.73(V_L I_L)}{1000} \text{ (in kilovolt-amperes)}$$

$$\tag{5.8}$$

An example may be helpful at this point. Let us suppose that we are working with a three-phase, wye-connected alternator that has a line output voltage of 240 V AC and a full-load current of 100 A, with a lagging power factor of 0.75. Let us find the phase voltage, the phase current, the full-load power output (in watts), and the kVA rating of the generator. If the phase voltage is 240 V AC, the line voltage is, from Equation 5.2,

$$V_L = 1.73 V_P$$

$$= (1.73)(240 \text{ V AC}) = 415.2 \text{ V AC}$$

In the wye-connected generator, the current in any line is equal to the phase current, from Equation 5.3. So

$$I_P = I_L = 100 \text{ A}$$

The total generator apparent power, its kVA rating, is calculated from Equation 5.8:

$$kVA = \frac{1.73(V_L I_L)}{1000}$$

$$= \frac{(1.73)(415.2 \text{ V AC})(100 \text{ A})}{1000}$$

$$= 71.8 \text{ kVA}$$

The real power (in watts) is given by Equation 5.7:

$$P_{total} = 1.73(V_L I_L)(\cos \theta)$$

$$= (1.73)(415.2 \text{ V AC})(100 \text{ A})(0.75)$$

$$= 53,872 \text{ W} \quad \text{or} \quad 53.9 \text{ kW}$$

Note that the term $(P_{VA})(\cos \theta)$ can be substituted into Equation 5.7 to give the power in watts,

$$P_{total} \text{ (in watts)} = (P_{VA})(\cos \theta) \tag{5.9}$$

or the power in kilowatts,

$$P_{total} \text{ (in kilowatts)} = (P_{kVA})(\cos \theta) \tag{5.10}$$

• EXAMPLE 5.3

Given a three-phase, delta-connected generator that supplies a full-load current of 200 A at a phase voltage of 480 V AC and a power factor of 0.75 lagging, find (a) the line voltage, (b) the line current, (c) the apparent power (in kilovolt-amperes), and (d) the true power output (in kilowatts).

Solution

(a) If the phase voltage is 480 V AC, the line voltage is also 480 V AC.
(b) In the delta-connected generator, the current in any line is calculated from Equation 5.4:

$$I_L = 1.73 I_P$$

$$= (1.73)(200 \text{ A}) = 346 \text{ A}$$

(c) The total generator apparent power (in kilovolt-amperes) is, from Equation 5.8,

$$kVA = \frac{1.73(V_L I_L)}{1000}$$

$$= \frac{(1.73)(480 \text{ V AC})(346 \text{ A})}{1000}$$

$$= 287.3 \text{ kVA}$$

(d) The real power (in kilowatts) is given by Equation 5.10:

$$P_{total} \text{ (in kilowatts)} = (P_{kVA})(\cos \theta)$$

$$= (287.3 \text{ kVA})(0.75)$$

$$= 215 \text{ kW}$$

Up to this point, we have been discussing the output voltage, power, and current of three-phase generators. We may connect loads—either resistive, capacitive, or inductive (or any combination)—across the output terminals of our generator. The equations for the current, voltage, and power are identical to the equations we used for the delta- and wye-connected generators.

AC MOTOR CONSTRUCTION AND CHARACTERISTICS

Classification of AC Motors

AC motors may be classified in several different ways. For instance, they may be classified by power rating as fractional- or integral-horsepower machines. As the name suggests, a *fractional-horsepower motor* has a power rating under 1 hp. Fractional-horsepower motors use three-phase and single-phase AC power sources. Single-phase voltages are typically either 120 or 240 V AC. The three-phase voltages found in industry are 208, 240, 480, and 600 V AC. The *integral-horsepower motor* has a rating above 1 hp.

Large-apparatus AC motors are usually found in sizes from 200 to 100,000 hp. Operating voltages are typically high, starting at 480 V AC and continuing to 2300, 4000, 6900, and 13,200 V AC.

The number of phases (usually one or three) applied to the stator windings is another basis for classification. Single-phase motors are found in domestic, business, farm, and small-industry applications. Three-phase power generally is used in heavy industry.

In this chapter, we divide motors into three areas: (1) universal, (2) induction, and (3) synchronous. This classification is based on the internal construction of the motor. We cannot cover all aspects of AC motors in one chapter. Consequently, we will deal with the most important operating principles of the most common types of AC motors. For more information on AC motors, consult the bibliography for this chapter at the end of the book.

Stator Construction

The diagram in Figure 5.5 shows a typical *squirrel-cage induction motor* (SCIM) stator. The name *squirrel cage* comes from the unique construction of the machine. The SCIM stator usually has a laminated iron core. Slots are cut into this cylindrical iron core, and the stator coil windings are placed in the slots. The coils are held in place in the core by means of

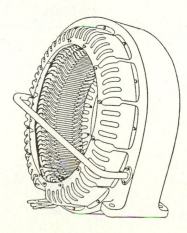

FIGURE 5.5 Large-Apparatus SCIM Stator

sticks or wedges. Coils of large AC motors consist of rectangular wire made to fit within the rectangular slots cut in the stator core. Smaller SCIMs use normal circular wire wound into a coil. After the coils are set into the slots, the whole stator assembly is coated with varnish or epoxy resin. The epoxy resin has two functions: First, it helps hold the stator coils to the core. Second, it prevents moisture from attacking the insulation.

As in the DC motor, the proper stator coil insulation must be used. The classes of insulation are the same for both AC and DC motors. Classes A and B are most commonly found, with B predominant. Class A insulation is made of cotton, silk, paper, or other organic compounds and is filled with an insulator such as varnish or resin. Class A insulation is classified as 105°C insulation. Class B insulation is more common today, especially on large motors. Class B, rated at 130°C, is usually made of fiberglass, mica, or some other inorganic insulator. Classes F and H are reserved for special applications. Class F insulation (155°C) and class H insulation (180°C) are made of special materials that do not break down at high temperatures.

Motor Characteristics

The *temperature rise* is an important part of the motor's temperature characteristics. The temperature rise is based on a 40°C ambient temperature. Manufacturers determine the temperature rise of a motor from the difference between a deenergized motor and one that has been running for several hours at full load. Class B insulation allows a 90°C rise in temperature from the base temperature of 40°C, which is why the class B is called 130°C insulation.

Long motor life and safe operation can be ensured by the proper type of motor enclosure. The National Electrical Manufacturers Association (NEMA) has classified motor enclosures according to the method of cooling and the type of environmental protection offered. In general, there are two types of motor enclosures—open and totally enclosed. *Open motor enclosures* have ventilation passages that allow air to circulate freely. The free air circulation removes the heat, keeping the motor cool. The *totally enclosed* motor does not allow air to flow freely from outside the motor enclosure.

According to NEMA standards, each of the following types of enclosures are considered to be open:

- Dripproof
- Splashproof
- Semiguarded
- Guarded
- Dripproof, fully guarded
- Weather-protected

Open motors are the most common type of enclosure in industry. Guarded or protected enclosures limit the size and shape of the openings in the motors. Guarded motors are found in industrial environments where pieces of material may occasionally be launched into the air. Guarded motors also protect personnel from coming into contact with the motor's rotating parts.

NEMA defines the following enclosures as totally enclosed:

- Totally enclosed, nonventilated
- Totally enclosed, fan-cooled
- Explosionproof
- Dust-ignition-proof
- Waterproof
- Totally enclosed, fan-cooled, guarded

Each of these types of enclosures meets a particular standard associated with protection from environmental factors and cooling. For example, a guarded machine is an open machine in which all openings giving direct access to rotating parts are limited in size by screens, baffles, or grills. The screens, baffles, and grills prevent accidental contact with hazardous parts. Openings that give access to the motor's rotating parts should not permit the passage of a cylindrical rod ¾ in. in diameter. The splashproof machine is an open machine constructed so that drops of liquid or solid particles striking the enclosure at an angle not greater than 100° from the vertical do not prevent the motor from working properly. The explosionproof motor enclosures are designed so that if explosive gas or vapor gets into

the motor and ignites, the enclosure will be able to withstand the explosion. Also, internal explosions in this type of motor will not cause gases or vapors outside the motor to explode. These are some examples of the definitions NEMA gives to motor enclosures. You may study these different types of motor enclosures further by referring to the appropriate NEMA standards publications.

UNIVERSAL MOTOR

The *universal motor* operates equally well on AC or DC power. The construction of the universal motor is shown in Figure 5.6. Note that it is electrically the same as the series DC motor. Current flows from the supply through the field, through the armature windings, and back to the supply.

If you use the left-hand rule for coils, you can determine the directions of the fields shown in the figure. The polarities of the armature and field flux oppose each other. If current is reversed, both fields reverse. The magnetic flux of the armature and the field still oppose one another. The motor, therefore, tends to turn in one direction no matter which way current flows. When 60 Hz AC is applied to the input terminals, the current (and the magnetic fields) changes direction 60 times per second.

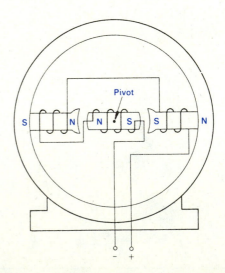

FIGURE 5.6 Universal Motor Construction

Universal motors differ somewhat from DC motors in their physical construction. Recall that transformers have eddy currents and hysteresis losses. AC motors tend to have these same losses. So that these losses are reduced, AC motors are constructed with special metals, laminations, and windings. We can say, therefore, that a universal motor works equally well on AC or DC. A series DC motor, however, does not work well on an AC supply because DC motors are not constructed to reduce the losses mentioned.

The operating characteristics of the universal motor are similar to those of the DC motor. For example, starting torque is high in the universal motor. As in the DC motor, speed can become dangerously high if the load is taken off. Universal motors are normally fractional-horsepower devices powered by single-phase AC.

FIGURE 5.7 Single-Phase Stator Winding

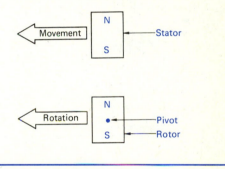

FIGURE 5.8 Rotating Rotor by Physically Moving Stator

PRINCIPLE OF THE ROTATING FIELD

Synchronous and induction motor construction can be divided into two basic parts: the rotor and the stator. Stators for both types of motors are similar in construction. As discussed previously, the stator is made of many laminated steel discs in order to reduce eddy currents and hysteresis losses. Coil slots are punched around the inside bore. The stator windings are wound around these slots. A typical single-phase stator winding is shown in Figure 5.7.

Since stators differ very little in induction and synchronous motors, the operating principle of rotating fields is basically similar for both. The *principle of the rotating field* is the key to the operation of synchronous and induction AC motors. The idea is relatively simple. Let us, for the moment, assume that the rotor is a permanent magnet. (Rotor construction will be discussed later in the chapter.) One way we can get the rotor to rotate is to place a magnet near it and move the magnet (see Figure 5.8).

Obviously, this technique is unsatisfactory. One of these two parts, the magnet or rotor, must be stationary. Normally, the magnet that pulls the rotor around is a series of fixed windings called the stator windings. If these windings are fixed and cannot move physically, how can we get the rotor to rotate?

The answer is to use for the stator a magnetic field that can be made to rotate electrically—the *rotating field*.

Let us see how this result can be obtained. Refer to Figure 5.9A. The principle of the rotating field is most easily seen in a two-phase system, as shown in the figure. But note that this principle is most frequently used in three-phase power systems.

Let us begin with a motor with two stator windings, each displaced from the other by 90°. Stator 1 pole pairs are vertical, and stator 2 pole pairs are horizontal. If the voltages applied to stator windings

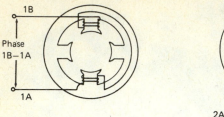
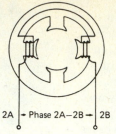
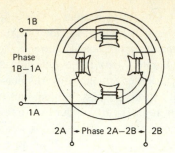

A. Two Phases Applied to Opposite Poles

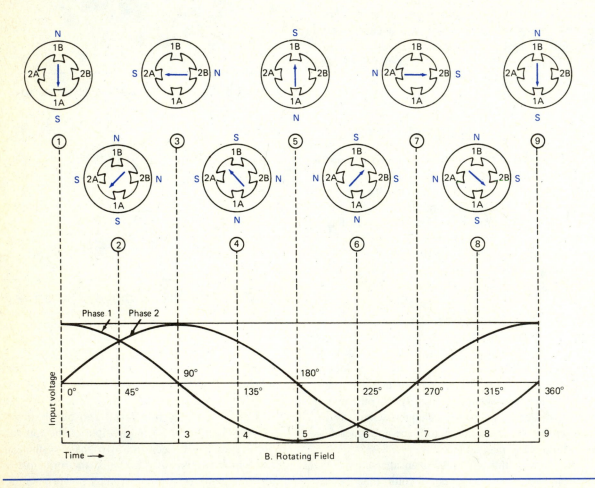

B. Rotating Field

FIGURE 5.9 Rotating Field in a Two-Phase System

are 90° out of phase, then the currents that flow in those windings are 90° out of phase also. If the currents are out of phase by 90°, the magnetic fields produced by those currents are 90° out of phase.

To see how the fields rotate, refer to Figure 5.9B. Remember that phase 1 is applied to stator winding 1 and phase 2 is applied to stator winding 2. At time 1, note that the voltage at stator 1 is maximum while the voltage across stator 2 is zero. The current in stator 1 is then maximum, producing the vertical magnetic field shown. Note that it is north on top and south on the bottom. Also note that there is no horizontal field. The total resultant field is shown by the arrow.

At time 2, the current has decreased in stator 1 and increased in stator 2. Thus, there are two fields present, one in stator 1 and one in stator 2. The total resultant field is again shown by the arrow.

At time 3, the current in stator 1 has decreased to zero, while stator 2 current has reached a maximum. There is a strong horizontal field but no vertical field. The resultant field is again indicated by the arrow.

Looking at the total resultant field direction, we see a clockwise movement of the field. It has moved a total of 90°. The field starts out in the six o'clock position and ends in the nine o'clock position. In the figure, this rotation continues to one full cycle. When the two-phase voltages have finished one complete cycle, the resultant field has moved one complete 360° rotation.

We have, then, created a rotating field by doing two things. First, we placed two field windings (called the stator) at right angles to each other. Second, we excited those windings with voltages 90° out of phase.

Two-phase motors are rarely used except in specialized equipment. We have discussed them here only as an aid to understanding the rotating field. This same principle of the rotating field is used, however, in three-phase and single-phase systems. In the three-phase system, the phases are 120° out of phase, not 90°. In the single-phase system, the required phase difference most commonly is produced by adding inductance or capacitance to one of the pairs of windings.

Note in Figure 5.9B that if the supplied current completes 60 cycles each second, the resultant field rotates at 60 r/s, or 3600 r/min. If we double the number of stator coils, the stator field rotates only half as fast. In a motor with a rotating stator field, the field travels one rotation per pole pair, per winding, for each cycle of applied voltage. If we double the number of poles for each winding, the magnetic field has more pole pairs to travel. Therefore, it takes twice as long to complete one rotation of all the pole pairs in one winding. This relationship can be stated as shown in the following equation:

$$n_{st} = \frac{120f}{p} \qquad\qquad \textbf{(5.11)}$$

where

n_{st} = stator speed (in revolutions per minute)
f = frequency of voltage applied to stator windings
p = number of magnetic poles per phase

For example, let us say we have a three-phase motor with four magnetic poles. If the frequency of the voltage applied to the stator is 60 Hz, what is the stator speed? We can calculate the stator speed from Equation 5.11:

$$n_{st} = \frac{(120)(60\ \text{Hz})}{4} = 1800\ \text{r/min}$$

The speed at which the stator field moves is called the *synchronous speed*. This name is used because the field is synchronized to the frequency of the supply voltage at all times. Note that as we increase the number of poles, the stator speed decreases. Thus, we see that the speed of this type of motor can be changed by switching in different numbers of poles. However, this method of speed control is limited to a small number of discrete speed changes.

• EXAMPLE 5.4

Given a three-phase induction motor, with eight poles and a line frequency of 50 Hz, find the synchronous speed (in revolutions per minute).

Solution

Substituting the appropriate values into Equation 5.11, we get

$$n_{st} = \frac{120f}{p} = \frac{(120)\,(50\ \text{Hz})}{8} = 750\ \text{r/min}$$

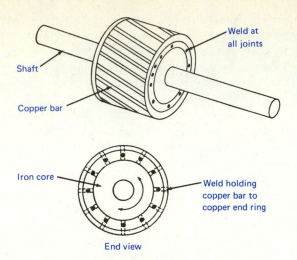

FIGURE 5.10 Squirrel-Cage Rotor Construction

FIGURE 5.11 Squirrel-Cage Rotor

INDUCTION MOTOR

The driving torque of both DC and AC motors comes from the interaction of current-carrying conductors in a magnetic field. Recall that, in the DC motor, the armature moves and the field is stationary. Currents in both field and armature windings create the magnetic fields that produce the torque. The rotor current comes from the supply through the brushes and commutator. In the *induction motor*, electromagnetic induction produces the rotor currents.

Rotor Construction

Before we turn to the operation of the induction motor, we will examine the rotor construction, as shown in Figure 5.10. The rotor conductors are usually made of heavy-gauge copper. All these bars are connected at either end by copper or brass end rings. The end rings short-circuit the copper bar at both ends. No insulation is needed between the bars because the voltages generated are low and the bars represent the lowest-resistance path for the current to flow through. The core of the rotor (not shown in the diagram) is made of iron. This construction technique reduces air gap reluctance. It also concentrates the flux lines between the rotor conductors. Because of its unique appearance, this rotor is often called the *squirrel-cage rotor*. An actual squirrel-cage rotor is shown in Figure 5.11.

Motor Operation

Let us proceed now to the operation of the induction motor. Assume that we have a motor with a two-phase rotating field. If this field rotates over the stationary rotor windings, current is induced in the copper bars. One of these bars is shown in Figure 5.12.

The current generation follows the left-hand rule for generators. Remember that although the stator field (shown as a magnet) is moving right, the relative motion of the conductor is to the left. From the

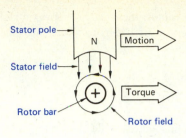

FIGURE 5.12 Interaction of Stator Field and Rotor Bars

dot convention, current is generated, flowing into the page. This current flow generates a magnetic field around the conductor.

The interaction of the stator field and the rotor field produces a torque. The torque generated tends to move the conductor toward the right. We see, then, that the rotor tends to move in the direction of the stator rotation. Note that there is no physical contact between the rotor and stator. The only thing that links them is the electromagnetic field.

Since this motor has no physical connection between rotor and stator, it is relatively maintenance-free. There are no brushes to replace; since no brushes exist, there is no sparking to ignite explosive gases and vapors.

Slip. Note that the rotor of an induction motor cannot turn at the same speed as the stator. If it could turn at the same speed, no current would be produced in the rotor. No rotor current means no torque, so the motor will not run on its own. There must be relative motion between the rotor and the stator field for the motor to operate. The rotor, therefore, runs at a slower speed than the stator.

The difference between the rotor speed and the stator field speed is called *slip*. Slip (s) is normally expressed in equation form as

$$s = \frac{n_{st} - n_{rt}}{n_{st}} \qquad (5.12)$$

where

n_{st} = stator speed (in revolutions per minute)
n_{rt} = rotor speed (in revolutions per minute)

In this equation, slip will range between 0 and 1. Slip sometimes is expressed as a percentage. In that case, we multiply the calculated value by 100. In this equation, slip ranges between 0% and 100%. In actual loaded AC motors, slip ranges between 1% and 10%.

When the motor starts, the stator immediately goes to synchronous speed. At the first instant of starting, the rotor is stationary. The slip at starting is, therefore, 1, or 100%. Gradually, the rotor picks up speed, rotor currents decrease, and the rotor reaches a point where the torque produced equals the torque demanded by the load. The rotor speed ceases to increase but remains constant as long as this condition exists.

• EXAMPLE 5.5

Given a three-phase induction motor, with a stator synchronous speed (n_{st}) of 900 r/min and a rotor speed (n_{rt}) of 850 r/min, find the slip of this motor (in percent).

Solution

The slip (s) can be calculated from Equation 5.12:

$$s = \frac{n_{st} - n_{rt}}{n_{st}}$$

$$= \frac{900 \text{ r/min} - 850 \text{ r/min}}{900 \text{ r/min}}$$

$$= 0.0556 \quad \text{or} \quad 5.56\%$$

Rotor Reactance. As the motor gains speed, the rate at which the rotating field cuts the rotor conductors decreases. The amount of rotor voltage and current then decreases. We observe that the frequency of the rotor current varies directly with slip. As slip increases, so does rotor current frequency.

Recall the equation for inductive reactance (X_L) from basic electronics:

$$X_L = 2\pi f L \qquad (5.13)$$

where
  L = inductance
  f = frequency of current flow through that inductance

In our motor, if L is the inductance of the rotor and f is the stator frequency, we can add slip (s) to the equation because the reactance decreases as rotor frequency decreases. Our new equation for *rotor reactance* is, then,

$$X_{rt} = 2\pi f L s \qquad (5.14)$$

where
  X_{rt} = rotor reactance at slip s

Torque. Note that when the two speeds, rotor and stator, are close to each other, slip is nearly zero. As slip approaches zero, so does rotor reactance. When slip is near zero, the only impedance to rotor current flow is resistive. Thus, rotor current and voltage are in phase. At starting, however, slip is 1, and the impedance is mostly reactive. In this case, rotor current lags rotor voltage by about 90°.

With this value of slip (100%), the high rotor reactance produces a low power factor. The *power factor* of an AC circuit is the power delivered to or absorbed by the circuit, divided by the apparent power of the circuit. (These terms will be discussed in detail late in the chapter.) A resistor, because it does not have any reactance, has a power factor of 1. A perfect coil or capacitor has a power factor of 0 because it consumes no DC power.

In the near-zero-slip motor application, a low power factor means low torque due to the small rotor current. At the other end of the scale, when slip approaches one, torque is low because the power factor angle is large. The torque produced by an induction motor, then, varies approximately as the power factor of the rotor current. The torque equation is

$$T = K I_{st} I_{rt} \cos \theta \qquad (5.15)$$

where
  K = constant (as in DC motor)
  I_{st} = stator current
  I_{rt} = rotor current
  $\cos \theta$ = power factor of rotor current

Equation 5.15 implies that the product of the rotor current and the rotor power factor for a given strength of magnetic field is maximum when the phase angle between rotor current and voltage is 45° lagging. When rotor current lags rotor voltage by 45°, the reactance of the rotor equals the resistance of the rotor. The rotor power factor is, therefore, 0.707, or 70.7%. Beyond this point, sometimes called the *pull-out* or *breakdown point*, if the load on the rotor is increased, the motor will stall. The maximum torque the motor can produce is therefore called the *breakdown* or *stall torque*.

Another implication from Equation 5.15 relates to the torque that the induction motor rotor produces. The flux in the stator is created by the stator current. Stator currents are, in turn, caused by the applied stator voltage. Since the torque produced by the rotor is the result of the interaction of the rotor and stator fields, the torque produced is directly proportional to the applied stator voltage. (Actually, the rotor torque is proportional to the square of the stator voltage.) As stator voltage increases, so does stator current and the stator flux. Increasing flux strength produces more rotor torque.

The full-load torque of a motor is defined as the turning force generated by the rotor when the motor develops a full-rated load at rated speed. The maximum torque of an induction motor is usually two to three times the full-load torque. Induction motors can, for a short time, carry loads greater than the full-load torque. A load in excess of the full-load is called an *overload*. Momentary overloads can be tolerated for a short time. Overloads that exist for a long time can overheat the motor and damage it. The maximum torque (T_{max}) an SCIM can produce can be calculated if we know the speed (n_{BR}) at which breakdown occurs and if we know the rated speed (n_{rated}) and torque (T_{rated}) at full load. This relationship is

$$T_{max} = (T_{rated})\left[\frac{(n_{BR}/n_{rated}) + (n_{rated}/n_{BR})}{2}\right] \quad \textbf{(5.16)}$$

where

T_{max} = maximum torque SCIM produces
T_{rated} = torque at full load
n_{rated} = rated speed
n_{BR} = breakdown speed

An example will show how to use this equation. A 5 hp, three-phase SCIM has a synchronous speed of 1800 r/min and a rotor whose speed is 1725 r/min at full load and that stalls at 1260 r/min. What is the maximum torque the motor will produce with rated voltage applied? We must find the rated torque produced by the motor when it is operating at rated speed. But first, we need the following equation, which describes the mechanical power output of a motor at rated speed and torque:

$$P_{rated} = \frac{(n_{rt})(T_{rated})}{5252} \quad \textbf{(5.17)}$$

where

P_{rated} = mechanical output power produced by rotor (in watts)
n_{rt} = rotor speed (in revolutions per minute)
T_{rated} = rated torque of motor (in pound-feet)

Now, we solve Equation 5.17 for T_{rated}:

$$T_{rated} = \frac{5252(P_{rated})}{n_{rt}} \quad \textbf{(5.18)}$$

Substituting the appropriate values into Equation 5.18, we get

$$T_{rated} = \frac{(5252)(5 \text{ hp})}{1725} = 15.22 \text{ lb-ft}$$

Before we can use Equation 5.16 to calculate the maximum torque produced, we need to calculate the slip at breakdown and the slip when the motor is operating under rated conditions. We obtain, from Equation 5.12,

$$n_{BR} = \frac{n_{st} - n_{rt}}{n_{st}}$$

$$= \frac{1800 \text{ r/min} - 1260 \text{ r/min}}{1800 \text{ r/min}}$$

$$= 0.3 \quad \text{or} \quad 30\%$$

$$n_{rated} = \frac{n_{st} - n_{rt}}{n_{st}}$$

$$= \frac{1800 \text{ r/min} - 1725 \text{ r/min}}{1800 \text{ r/min}}$$

$$= 0.0417 \quad \text{or} \quad 4.17\%$$

We now have enough information to solve for the maximum torque, using Equation 5.16:

$$T_{max} = (T_{rated})\left[\frac{(n_{BR}/n_{rated}) + (n_{rated}/n_{BR})}{2}\right]$$

$$= (15.22 \text{ lb-ft})$$

$$\times \left[\frac{(0.3/0.0417) + (0.0417/0.3)}{2}\right]$$

$$= 55.9 \text{ lb-ft}$$

• EXAMPLE 5.6

Given a 15 hp, three-phase SCIM with a synchronous speed of 1800 r/min and a rotor whose speed is 1745 r/min at full load and that stalls at 1360 r/min, find (a) the torque the motor produces at rated load and (b) the maximum torque the motor will produce with rated voltage applied.

Solution

(a) We must find the rated torque produced by the motor when it is operating at rated speed. We use Equation 5.18:

$$T_{rated} = \frac{5252(P_{rated})}{n_{rt}}$$

$$= \frac{(5252)(15 \text{ hp})}{1745 \text{ r/min}} = 45.15 \text{ lb-ft}$$

(b) Before we can use Equation 5.16 to calculate the maximum torque produced, we need to calculate the slip at breakdown and the slip when the motor is operating under rated conditions. We use Equation 5.12 to obtain

$$n_{BR} = \frac{n_{st} - n_{rt}}{n_{st}}$$

$$= \frac{1800 \text{ r/min} - 1360 \text{ r/min}}{1800 \text{ r/min}}$$

$$= 0.244 \quad \text{or} \quad 24.4\%$$

$$n_{rated} = \frac{n_{st} - n_{rt}}{n_{st}}$$

$$= \frac{1800 \text{ r/min} - 1745 \text{ r/min}}{1800 \text{ r/min}}$$

$$= 0.0306 \quad \text{or} \quad 3.06\%$$

We now have enough information to solve for the maximum torque, using Equation 5.16:

$$T_{max} = (T_{rated}) \left[\frac{n_{BR}/n_{rated} + (n_{rated}/n_{BR})}{2} \right]$$

$$= (45.15 \text{ lb-ft})$$

$$\times \left[\frac{(0.244/0.0306) + (0.0306/0.244)}{2} \right]$$

$$= 182.8 \text{ lb-ft}$$

The starting torque of the average SCIM is from 1.25 to 1.5 times the normal rated torque when the motor is started at full voltage. The average SCIM takes about 5 to 6 times the full-load current at starting. In large SCIMs, starting at full voltage causes an excessive starting current. Usually, the large SCIM starts at a reduced voltage. You should be aware that the starting torque is reduced when the motor is started with a reduced stator voltage.

As we said previously, the starting torque varies as the square of the voltage. For example, let's say that we start a motor at 50% of the full-rated voltage. The motor will start with $(0.5)(0.5) = 0.25$, or 25%, of the starting torque developed at full stator voltage. We can express this relationship as

$$T_A = \left(\frac{V_A^2}{V_{st}^2} \right)(T_S) \qquad \text{(5.19)}$$

where

T_A = actual starting torque
V_A = actual applied stator voltage
V_{st} = full stator voltage
T_S = starting torque developed at rated voltage

For instance, let's say that a 240 V AC SCIM produces 100 lb-ft of torque at start-up. If we reduce the voltage at start-up to 132 V AC, what will the starting torque be? Using Equation 5.19, we can solve for the new starting torque:

$$T_A = \left(\frac{V_A^2}{V_{st}^2} \right)(T_S) = \left(\frac{132^2}{240^2} \right)(100 \text{ lb-ft})$$

$$= (0.3025)(100 \text{ lb-ft}) = 30.25 \text{ lb-ft}$$

• EXAMPLE 5.7

Given a three-phase, 480 V AC, 2 kW, 60 Hz SCIM, with a starting torque of 50 Nm when the rated voltage (480 V AC) is applied, find (a) the starting torque when 500 V AC is applied and (b) the stator voltage needed to produce a starting torque of 60 Nm.

Solution

(a) Using Equation 5.19, we can solve for the new starting torque:

$$T_A = \left(\frac{V_A^2}{V_{st}^2}\right)(T_S) = \left(\frac{500^2}{480^2}\right)(50 \text{ Nm})$$

$$= (1.0851)(50 \text{ Nm}) = 54.26 \text{ Nm}$$

(b) To find the voltage needed to produce a starting torque of 60 Nm, we must solve Equation 5.19 for V_A. We get

$$V_A = (V_{st})\sqrt{\frac{T_A}{T_S}} \qquad (5.20)$$

Substituting the appropriate values into Equation 5.20, we get

$$V_A = (480 \text{ V AC})\sqrt{\frac{60 \text{ Nm}}{50 \text{ Nm}}}$$

$$= 525.8 \text{ V AC}$$

Operating Characteristics

The SCIM compares closely with a transformer with a rotating secondary. In a no-load condition, the rotor cuts the turns of the stator winding. This action generates a cemf in the stator winding, which limits line current to a small value. This no-load value is called the *exciting current*. The exciting current maintains the rotating field. Because the circuit is highly inductive, the power factor of the motor with no load is poor. The power factor for an unloaded SCIM may be up to 30% lagging. When unloaded, the rotor runs close to synchronous speed, and rotor current is very small. The interaction between the rotor flux on the rotating field is also small.

When a load is placed on the rotor shaft, the rotor slows down slightly. The rotating field, however, continues to rotate at synchronous speed. When the rotor is loaded, the SCIM slip increases, which causes rotor current to increase. The motor torque increases when the speed decreases and the power output increases. The increased rotor flux, caused by the increased rotor current, opposes the stator flux and lowers it slightly. The primary cemf, therefore, decreases slightly and stator current increases. You will recognize this behavior as similar to what happens in a transformer. Because of the low impedance of the stator windings, a small reduction in speed and cemf may produce large increases in motor current, torque, and power output. We can, therefore, conclude that the SCIM has variable-torque, constant-speed characteristics.

If the motor is stalled, the resulting increase in rotor current lowers the stator cemf and causes excessive stator current. This excessive current can damage the stator windings. When a motor is deliberately stalled or operated in a locked-rotor condition, the voltage applied to the stator should not exceed 50% of the rated voltage. For example, if we are operating a 230 V AC motor under a locked-rotor condition, the maximum voltage we should apply to the stator is 115 V AC.

When an SCIM is operated at full load, the stator voltage and current are more in phase than they are when the SCIM is operated in a no-load condition. The power factor of a motor operated at full load is, therefore, better than one that is operated at no load.

The operational characteristics of the SCIM are shown graphically in Figure 5.13, which gives typical torque and current curves for a three-phase SCIM. Rotor reactance increases with slip and increases its effects on rotor current and power factor as the motor load is increased. The pull-out point on the torque curve occurs at a slip of about 0.25, or 25%. The maximum torque at the pull-out point is about 3.5 times the full-load torque. At this point, as previously mentioned, the rotor resistance equals the rotor reactance. The rotor power factor is 0.707, or 70.7%. Any additional load added to the motor beyond this torque value causes the motor to stall. In the stall condition, the stator current is about five times the full-load value.

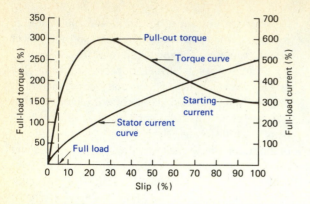

FIGURE 5.13 Torque and Current Curves for Three-Phase SCIM

Performance curves of a four-pole, three-phase, 440 V AC, 15 hp SCIM are shown in Figure 5.14. Note that the full-load slip is only about 3.5%. In a stalled condition, the rotor reactance of this type of motor is nearly five times the rotor resistance. At full load, however, the rotor reactance is much less than the rotor resistance. When the motor is running, the rotor current is determined mainly by rotor resistance. The torque and slip both increase up to the pull-out point. Beyond the pull-out point, the torque decreases and the motor stalls. Because the change in speed from no load to full load is small, the motor torque and the power output are considered to be directly proportional.

The full-load current of a three-phase SCIM can be approximated by

$$I_{FL} = \frac{600P_o}{V_{st}} \qquad (5.21)$$

where

I_{FL} = full-load current (in amperes)
P_o = power output of motor (in horsepower)
V_{st} = voltage applied to stator of SCIM

As an example, let us consider a 5 hp, 240 V AC SCIM. The approximate full-load current is

$$I_{FL} = \frac{600P_o}{V_{st}} = \frac{(600)\,(5\ hp)}{240\ V\ AC} = 12.5\ A$$

Two other rules of thumb may help in approximating the behavior of an SCIM. First, the no-load current is about 30% of the full-load current. Second, the starting current is about six times the full-load current. In the 5 hp motor just mentioned, the no-load current is 30% of the full-load current, or $(0.3)(12.5\ A) = 3.75\ A$. The starting current is $(6)(12.5\ A) = 75\ A$ with full voltage applied.

• EXAMPLE 5.8
Given a 100 hp, 480 V AC SCIM, find the approximate full-load current, no-load current, and starting current.

Solution
The full-load current is approximated by Equation 5.21:

$$I_{FL} = \frac{600P_o}{V_{st}} = \frac{(600)\,(100\ hp)}{480\ V\ AC} = 125\ A$$

The no-load current is 30% of the full-load current, or $(0.3)(125\ A) = 37.5\ A$. The starting current is $(6)(125\ A) = 750\ A$ with full voltage applied.

Speed Regulation

The SCIM speed regulation is excellent, changing no more than a few percent over the range from full load to no load. How does this good speed regulation happen? As an example, let us suppose that we have an

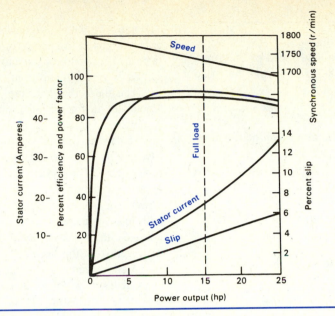

FIGURE 5.14   Performance Curves for Four-Pole, Three-Phase SCIM

SCIM running at no load. When we load down the motor, slip increases and the frequency of the current induced in the rotor increases. The current in the rotor increases as well, producing a stronger field in the rotor. The rotor torque is, therefore, increased by an amount necessary to handle the extra load. The speed stays approximately the same.

At starting, the slip in an SCIM is maximum. Large currents are induced in the rotor, which produces the torque needed to start the motor. As the motor picks up speed, the frequency induced in the rotor decreases, and so does the current in the rotor bars. The torque continues to increase to its maximum value. As the motor continues to accelerate, the rotor current continues to decrease until a point is reached where enough current flows to produce the torque needed to handle the motor load at constant speed. This same type of equilibrium system is seen in the shunt-connected, wound-field DC motor discussed in Chapter 3. Increases in load cause increases in the torque needed to handle the extra load at a relatively constant speed.

The speed regulation of a motor is given in percentage and is calculated from the following equation:

$$\% \text{ speed regulation} = \frac{n_{\text{NL}} - n_{\text{FL}}}{n_{\text{FL}}} \times 100 \qquad \textbf{(5.22)}$$

where

n_{NL} = no-load speed
n_{FL} = full-load speed

An example will clarify the use of this equation. Suppose we are working with a six-pole, 60 Hz, three-phase SCIM, with a no-load speed of 1190 r/min and a full-load speed of 990 r/min. The speed regulation is

$$\% \text{ speed regulation}$$

$$= \frac{1190 \text{ r/min} - 990 \text{ r/min}}{990 \text{ r/min}} \times 100$$

$$= 20.2\%$$

• **EXAMPLE 5.9**

Given an SCIM with a no-load speed of 880 r/min and a full-load speed of 800 r/min, find the speed regulation of this motor.

Solution

The speed regulation of the motor can be found by using Equation 5.22:

$$\% \text{ speed regulation}$$

$$= \frac{n_{NL} - n_{FL}}{n_{FL}} \times 100$$

$$= \frac{880 \text{ r/min} - 800 \text{ r/min}}{800 \text{ r/min}} \times 100$$

$$= 10\%$$

Power Factor and Efficiency

Power Factor. In our discussion so far, we have assumed that the current and voltage produced by the AC generator will be in phase in the load. If the load is a pure resistance, this assumption is true. In a system with a resistive load, all the power produced by the generator can be consumed by the load. To determine the power consumed by a resistive load, we simply measure voltage and current in the load and multiply the two together.

However, no actual load is purely resistive (even a straight piece of wire has some inductance, for example). All loads have a mixture of resistance and reactance. Consequently, we can say that a load made up of resistance and reactance will usually have a phase difference between current and voltage. If we were to measure current and voltage in this type of load and multiply them together, we would not have the real power consumed by the load. We would have only the power that appears to be consumed by the load. (Remember that a reactive load consumes no power.) This power that appears to be consumed by a load is called *apparent power*. It is measured in volt-amperes (VA) to distinguish it from *real power*, which is measured in watts.

In a purely resistive load, the real power and the apparent power are equal. In a purely reactive load (a perfect capacitor or inductor), the real power consumed is zero. Actually, all loads fall somewhere between these two extremes. Current through the reactances in the circuit produces reactive power, sometimes symbolized by the letter Q. Reactive power can be useful and necessary in a system. *Reactive power* is power that is stored in a capacitor or inductor. The inductor or capacitor ultimately returns it to the circuit, as in the motor armature coils. Power (I^2R) losses increase because current increases with reactive increases. Since we normally regulate voltages, the voltage stays constant and the power delivered stays the same.

A convenient way to represent these two parameters, real and apparent power, is through the concept of power factor. The *power factor* is the ratio of real power to apparent power:

$$\text{power factor} = \frac{\text{real power}}{\text{apparent power}} \quad (5.23)$$

The real power—or true power, as it is sometimes called—is the power used by the load, as measured by a wattmeter. The apparent power, as discussed earlier, is the product of the current and voltage produced by the generator.

In electrical terms, it so happens that the power factor is the cosine of the angle between the current and voltage. In a purely resistive circuit, the current and voltage are exactly in phase. The phase angle difference is then 0°. The cosine of 0 is 1. This circuit then has a power factor of 1. This result agrees with what we said earlier about resistive circuits; that is, their power factor is 1. In a purely reactive circuit, the difference between current and voltage is 90°. The cosine of 90° is 0, so the power factor is 0. This result again agrees with what we said earlier about reactive loads not consuming power.

There are two kinds of reactive components: capacitors and inductors. When a load is capacitive, the current leads the voltage. In this situation, the circuit has what is called a *leading power factor*. When a load is inductive, the current lags the voltage. This situation is called a *lagging power factor*.

Most loads in industry are inductive because they are made up of coils of wire. Coils are found in motors, transformers, solenoids, relays, and the like. Since the loads in industry are primarily inductive, industrial power consumption has a low, lagging power factor. A low, lagging power factor is not desirable for many reasons. One important reason is power cost. Power companies supplying energy to industry have established penalties in their rates for low, lagging power factors. So power users pay more for consumption with a low, lagging power factor. When power is purchased under a power factor clause, the user benefits when the power factor is kept high. A low power factor is also undesirable because most loads require a certain amount of true power. The apparent power will be higher than the true power, causing current to be higher. This extra current adds to the heating of conductors, which, in turn, limits their current-carrying ability. Power factors usually range from 0.7 to 1.0. We will discuss this topic in more detail in a later chapter.

To summarize, power factor is the real power divided by the apparent power (Equation 5.23):

$$\text{power factor} = \frac{\text{real power}}{\text{apparent power}}$$
$$= \text{cosine of phase angle}$$
$$\text{between voltage and}$$
$$\text{current}$$

In electric circuits, the power factor is equal to the cosine of the angle between current and voltage. Apparent power (measured in volt-amperes) is due to the applied voltage and current; true power (measured in watts) is the useful power extracted by devices. (Motors, for instance, have little resistance but draw and convert real power, as well as provide a reactive load.) Since true power is usually less than apparent power, the power factor will be a percentage or a decimal number less than 1. A low, lagging power factor is generally unacceptable in industry since it makes utility bills higher and can limit circuit performance.

The power factor of an induction motor equals the power factor of the current the motor draws from the line. For the SCIM, data sheets usually give the motor power factor. The power factor rating is the power factor that exists when the motor meets all rated conditions, such as rated voltage, torque, and speed. Power factors for SCIMs are usually lagging. Thus, the motor appears inductive to the line voltage, with stator current lagging applied voltage. The power factor (pf) of a motor is usually given as a percentage. It is the ratio between true power and apparent power:

$$pf = \frac{P_{i(kW)}}{P_{i(kVA)}} \times 100 \qquad (5.24)$$

where
pf = power factor (in percent)
$P_{i(kW)}$ = true input power drawn by motor (in kilowatts)
$P_{i(kVA)}$ = apparent power input to motor (in kilovolt-amperes)

As an example, let us say we are working with a motor that has an apparent input power of 75 kVA and that draws 56 kW from the line. What is the power factor of the motor? Using Equation 5.24, we get

$$pf = \frac{P_{i(kW)}}{P_{i(kVA)}} \times 100$$
$$= \frac{56 \text{ kW}}{75 \text{ kVA}} \times 100 = 75\%$$

Calculating the kVA input to a three-phase induction motor is not done simply by multiplying the stator voltage by the stator current. Recall that the kVA input currents and voltage will be out of phase, with current lagging voltage in the three-phase induction motor. The motor draws rated current from the line when rated voltage is applied to the stator windings. The motor then receives a certain amount of apparent power from the line supply. Since apparent power is measured in kilovolt-amperes, you may find the motor input called the *kVA input*. The apparent kVA input depends on the number of phases in the line, the stator voltage, and the stator current.

The kVA input to a three-phase induction motor can be calculated by using Equation 5.8:

$$kVA = \frac{1.73(V_L I_L)}{1000}$$

For example, if an SCIM has an applied voltage of 240 V AC and a current of 10 A, the kVA input is

$$kVA = \frac{(1.73)\,(240\text{ V AC})\,(10\text{ A})}{1000}$$

$$= 4.15\text{ kVA}$$

The real power output of the motor depends on the power factor. If our motor has a power factor of 0.8, the power output (in watts) is given by Equation 5.7:

$$P_{total} = 1.73(V_L I_L)\,(\cos\theta)$$

$$= (1.73)\,(240\text{ V AC})\,(10\text{ A})\,(0.8)$$

$$= 3322\text{ W} \quad\text{or}\quad 3.322\text{ kW}$$

• EXAMPLE 5.10

Given a 100 hp, 480 V AC SCIM that draws 100 A from the line under rated load conditions, find (a) the kVA input to the motor and (b) the power factor of the motor.

Solution

(a) The kVA input to a three-phase induction motor can be calculated by using Equation 5.8:

$$kVA = \frac{1.73(V_L I_L)}{1000}$$

$$= \frac{(1.73)\,(480\text{ V AC})\,(100\text{ A})}{1000}$$

$$= 83\text{ kVA}$$

(b) The power factor of the motor can be calculated by using Equation 5.24. But if we are to use this equation, both power values must be in the same units. The output power is in horsepower and can be converted into watts by using the following equation:

$$P_{o(W)} = 746 P_{o(hp)} \tag{5.25}$$

With an input power of 100 hp, we get

$$P_{o(W)} = (746)\,(100\text{ hp})$$

$$= 74{,}600\text{ W} \quad\text{or}\quad 74.6\text{ kW}$$

We can now substitute the appropriate values of power into Equation 5.24:

$$pf = \frac{P_{o(W)}}{P_{VA}} \times 100$$

$$= \frac{74.6\text{ kW}}{83.0\text{ kVA}} \times 100 = 89.88\%$$

• EXAMPLE 5.11

Given an SCIM with an input kVA of 200 VA, or 0.2 kVA, find the real power the motor draws (in watts) if the power factor is 0.9 (90%).

Solution

To solve this problem, we must first solve Equation 5.24 for $P_{i(kW)}$. Equation 5.24 is

$$pf = \frac{P_{i(kW)}}{P_{i(kVA)}} \times 100$$

So we get

$$P_{i(kW)} = \frac{(pf)\,(P_{i(kVA)})}{100}$$

And substituting the appropriate values gives

$$P_{i(kW)} = \frac{(90\%)\,(0.2\text{ kVA})}{100}$$

$$= 0.18\text{ kW} \quad\text{or}\quad 180\text{ W}$$

Efficiency. The *efficiency* of a motor is the ratio of the mechanical power output to the electric power input:

$$\% \text{ efficiency} = \frac{P_{o(W)}}{P_{i(W)}} \times 100 \tag{5.26}$$

The output power is normally rated in horsepower, and input power is normally given in watts (or can be calculated easily by measuring voltage and current). So we must convert the horsepower rating of the motor to power in watts. Since 1 hp = 746 watts, we can use Equation 5.25 to solve for the output power in watts. Equation 5.25 is

$$P_{o(W)} = 746 P_{o(hp)}$$

By substituting Equation 5.25 into Equation 5.26, we get

$$\% \text{ efficiency} = \frac{746 P_{o(hp)}}{P_{i(W)}} \times 100 \qquad (5.27)$$

As an example, let's calculate the efficiency of a 10 hp induction motor that requires an input of 8 kW:

$$\% \text{ efficiency} = \frac{746 P_{o(hp)}}{P_{i(W)}} \times 100$$

$$= \frac{(746)(10 \text{ hp})}{8 \text{ kW}} \times 100$$

$$= 93.3\% \quad \text{or} \quad 0.0933$$

When the rated output, efficiency, and power factor are known, these factors can be used to calculate the kVA input to the motor, regardless of how many phases the motor has. The rated kVA can be calculated thus:

$$\text{kVA input} = \frac{0.746 P_{o(hp)}}{(pf)(\eta)} \qquad (5.28)$$

where
$P_{o(hp)}$ = output power (in horsepower)
pf = rated power factor as decimal
η = efficiency as decimal

The symbol for efficiency, η, is the Greek letter eta.

As an example, suppose we are working with a 10 hp induction motor that has a rated power factor of 87% and a rated efficiency of 85%. What is the rated kVA input for this system? Using Equation 5.28, we get

$$\text{kVA input} = \frac{0.746 P_{o(hp)}}{(pf)(\eta)}$$

$$= \frac{(0.746)(10 \text{ hp})}{(0.87)(0.85)} = 10.1 \text{ kVA}$$

• EXAMPLE 5.12

Given a 60 hp SCIM with a power factor of 0.9 and an efficiency of 0.85 (85%), find (a) the kVA input and (b) the real power drawn from the load.

Solution
(a) The kVA input may be found from Equation 5.28:

$$\text{kVA input} = \frac{0.746 P_{o(hp)}}{(pf)(\eta)}$$

$$= \frac{(0.746)(60 \text{ hp})}{(0.9)(0.85)} = 58.5 \text{ kVA}$$

(b) The input power drawn (in watts) must be found by solving Equation 5.27 for $P_{i(W)}$. Equation 5.27 is

$$\% \text{ efficiency} = \frac{746 P_{o(hp)}}{P_{i(W)}} \times 100$$

So we obtain

$$P_{i(W)} = \frac{74,600 P_{o(hp)}}{\% \text{ efficiency}}$$

For this example, the input power drawn is then

$$P_{i(W)} = \frac{(74,600)(60 \text{ hp})}{85\%}$$

$$= 52,658 \text{ W} \quad \text{or} \quad 52.7 \text{ kW}$$

• EXAMPLE 5.13

Given a 100 hp, 480 V AC, eight-pole, three-phase, 60 Hz, 880 r/min SCIM with a full-load efficiency of 90% and a power factor of 80%, find (a) the stator synchronous speed, (b) the full-load slip, (c) the full-load torque, (d) the kVA input, (e) the real power drawn by the motor from the line, and (f) the full-load current.

Solution

(a) Equation 5.11 gives us the synchronous speed:

$$n_{st} = \frac{120f}{p} = \frac{(120)(60 \text{ Hz})}{8} = 900 \text{ r/min}$$

(b) Equation 5.12 gives us the full-load stator slip:

$$s = \frac{n_{st} - n_{rt}}{n_{st}} = \frac{900 \text{ r/min} - 880 \text{ r/min}}{900 \text{ r/min}}$$

$$= 0.022 \quad \text{or} \quad 2.2\%$$

(c) The full-load torque can be found from Equation 5.18:

$$T_{rated} = \frac{5252(P_{rated})}{n_{rt}}$$

$$= \frac{(5252)(100 \text{ hp})}{880 \text{ r/min}} = 597 \text{ lb-ft}$$

(d) The kVA input may be found from Equation 5.28:

$$\text{kVA input} = \frac{0.746P_{o(hp)}}{(pf)(\eta)}$$

$$= \frac{(0.746)(100 \text{ hp})}{(0.8)(0.9)} = 103.6 \text{ kVA}$$

(e) The input power drawn (in watts) may be found by using Equation 5.29:

$$P_{i(W)} = \frac{74,600P_{o(hp)}}{\% \text{ efficiency}}$$

$$= \frac{(74,600)(100 \text{ hp})}{90\%}$$

$$= 82,889 \text{ W} \quad \text{or} \quad 82.9 \text{ kW}$$

(f) The full-load current may be calculated by solving Equation 5.8 for I_L. Equation 5.8 is

$$\text{kVA} = \frac{1.73(V_L I_L)}{1000}$$

So we obtain

$$I_L = \frac{1000(\text{kVA})}{1.73V_L} = \frac{(1000)(103.6 \text{ kVA})}{(1.73)(480 \text{ V AC})}$$

$$= 124.8 \text{ A}$$

As a check on our answer, the full-load current may be approximated by Equation 5.21:

$$I_{FL} = \frac{600P_o}{V_{st}} = \frac{(600)(100 \text{ hp})}{480 \text{ V AC}} = 125 \text{ A}$$

The answers are nearly the same.

Classifications of SCIMs

Squirrel-cage induction motors have several classifications, all based on differing types of rotor construction. Changes in the rotor construction produce differences in torque, speed, slip, and current characteristics. These motor types have been standardized by the National Electrical Manufacturers Association (NEMA). Examples of this classification scheme are NEMA motor types B, C, and D.

The NEMA type B motor is one with normal starting torque and current. A low rotor resistance produces low slip and high running efficiency. Generally, applications include constant-speed and constant-load requirements, where the motor does not need to be turned on and off frequently. Centrifugal pumps, blowers, and fans are examples of such applications. The type B is the most commonly used NEMA motor.

The NEMA type C motor has the same slip and starting currents but a higher starting torque and poorer speed regulation than the type B. This motor is used with loads that are hard to start, such as compressors, conveyors, and crushers.

The NEMA type D motor has a high-resistance rotor. Slip increases and efficiency decreases in this motor. It is used in applications that require frequent starting, stopping, and reversing. Since the type D motor has the highest starting torque, it is used to drive loads with very high torque demand, such as cranes, hoists, and punch presses. Because of its specialized rotor construction, the type D motor is the most expensive of the three.

Double Squirrel-Cage Construction

Another special type of rotor construction is illustrated in Figure 5.15. This rotor construction is known as the *double squirrel-cage rotor*. In this rotor, the set of bars closest to the surface has a small cross-sectional area and a resistance of a few tenths of an ohm. The bottom bars have a much lower resistance, on the order of several thousandths of an ohm.

When the motor starts, the frequency induced in the bars is high. Since the lower bars have more inductance, the high frequency causes them to have a higher reactance than the top bars have. The current flow at starting, therefore, is higher in the top bars. Since these bars have a high resistance, they cause a reduction in the phase angle between the stator flux field and the rotor current. This reduction tends to increase starting torque. As speed picks up, the reactance of the bottom bars decreases, causing more current to flow in them. The effect of adding these extra bars is more torque to the motor under varying load conditions. The NEMA type C induction motor is an example of the double squirrel-cage construction.

Single-Phase Motors

As their name implies, *single-phase motors* run when supplied with only one phase of input AC power. They are used in fractional-horsepower sizes in small commercial, domestic, and farm applications. Single-phase AC motors are generally less expensive to manufacture in fractional-horsepower sizes than three-phase motors. Of course, they eliminate the need to run three-phase power lines, which can be expensive and may not be necessary. Single-phase motors are used in blowers, fans, compressors in refrigeration equipment, portable drills, grinders, and so on.

The single-phase induction motor does not produce a rotating field and, therefore, has no starting torque. But if the rotor is accelerated, it behaves as if it were driven by a rotating field. With a single phase supplied to the stator, the stator field reverses at each alternation of the supply voltage. Thus, we can visualize the stator field as rotating in half-turn steps in either direction.

Because the single-phase induction motor does not produce any starting torque, it must depend on another system to start it. Consequently, single-phase induction motors are classified by the method used to start them. These classifications are discussed next.

Shaded-Pole Motor. The *shaded-pole motor* uses one of the lowest-cost starting methods. Shaded-pole motor construction is illustrated in Figure 5.16.

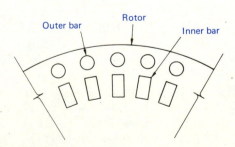

FIGURE 5.15 Double Squirrel-Cage Rotor Construction

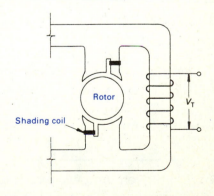

FIGURE 5.16 Shaded-Pole Motor Construction

Notice that a portion of each pole is surrounded by a copper strap, or shading coil. When current is increasing through the windings, the flux increases also. But as the flux travels through the portion of the pole piece shaded by the copper bar, the current induced into the copper strap builds up a field of its own. This field opposes the field buildup caused by the stator winding. The total flux is concentrated on the left side of the pole, as shown in Figure 5.17A.

As the applied voltage reaches a peak, the rate of change of flux is zero. At this time, the flux is evenly spread over the pole piece, as shown in Figure 5.17B.

When the applied voltage decreases, the flux decreases also. At this time, however, the flux around the copper bar reverses direction, opposing the decrease in flux around the bar. This reversal produces a total flux concentration on the right side of the pole piece, as shown in Figure 5.17C.

Note that the total flux density shifts from left to right with respect to time. In other words, the flux moves toward the shaded pole. The rotor then follows this shift as if it were a rotating field.

The shaded-pole motor produces starting torques in the range of 50% of full-load torque. It is manufactured in fractional-horsepower sizes to about $\frac{1}{20}$ hp, and it is not very efficiency. A typical application of this motor is in a small fan or inexpensive phonograph.

Split-Phase Motor. The *split-phase motor* has more starting torque than the shaded-pole motor. From Figure 5.18A, we see that the split-phase motor has a stator with two sets of windings. One set, the start winding, has a few turns of thin wire. It has a high resistance and a low reactance compared with the other set, the run winding. The axes of these windings are displaced by 90°.

The current in the start winding (I_{start}) lags the line voltage by about 30°, as shown in Figure 5.18B. It is also smaller than the current in the run winding because the impedance of the start winding is larger. The current in the run winding (I_{run}) lags the applied voltage by about 45°. This phase shift produces a rotating field effect such as we discussed earlier in the chapter. The interaction of this field and the one produced by induction in the rotor causes a starting torque to be generated. This motor is sometimes called the *resistance-start motor*.

When the motor comes up to about 75% of its running speed, a centrifugal switch opens. A *centrifugal switch* is a device that disconnects the start winding in an AC motor at about three-fourths of the full-load rotor speed. The centrifugal switch is mounted on the rotor and, as its name implies, is operated by centrifugal force. This switch disconnects the start winding. The motor continues to turn as long as a single phase is applied to the stator. Thus, the split-phase motor has good speed regulation. Starting torques range from 150% to 200% of full-load torque.

More starting torque is produced by adding an inductor in series with the run winding, as shown in Figure 5.18C. When the centrifugal switch opens, it not only disconnects the start winding but also shorts out the inductor. This motor is sometimes called the *reactor-start motor*. Direction of rotation is reversed by reversing either the start or run windings.

Applications of split-phase motors include washers, ventilating fans, and oil burners.

Another type of split-phase motor is the *capacitor-start motor*. This motor has more starting torque than the reactor-start motor because a capacitor is added in series with the start winding. The capacitor-start motor is illustrated in Figure 5.19A.

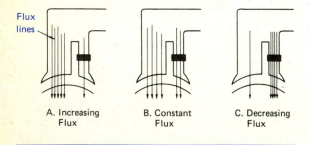

| A. Increasing Flux | B. Constant Flux | C. Decreasing Flux |

FIGURE 5.17 Movement of Flux in Shaded-Pole Motor

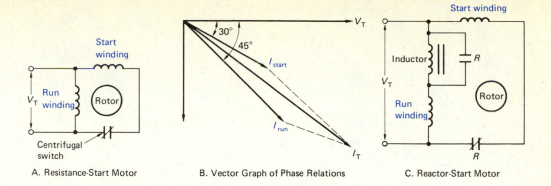

A. Resistance-Start Motor   B. Vector Graph of Phase Relations   C. Reactor-Start Motor

FIGURE 5.18 Split-Phase Motor

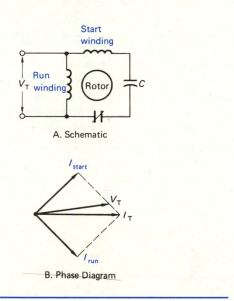

A. Schematic

B. Phase Diagram

FIGURE 5.19 Capacitor-Start Motor

The capacitor produces a larger phase difference between currents in the start and run windings than the resistance-start motor does. The current in the start winding is about 90° out of phase with the current in the run winding, as shown in Figure 5.19B. This phase difference produces a larger starting torque than is available in the resistance-start motor. Starting torque can reach 350% of full-load torque.

If the capacitor is disconnected by a centrifugal switch, the motor is called a *capacitor-start motor*. If the capacitor is left in the circuit, the motor is called a *capacitor-run motor*.

The capacitor in the capacitor-start motor must be a nonpolarized one. Sizes range from 80 μF for a ⅛ hp motor to 400 μF in a 1 hp motor. The capacitor-start motor normally is used in applications that demand high starting torque where DC motors cannot be used.

Wound-Rotor Motor

Another type of induction motor, called the *wound-rotor motor*, is sometimes seen in industry, although not as often as the squirrel-cage. Its rotor construction is more like the DC motor armature than the squirrel-cage rotor construction. The windings are also similar to the stator windings. They are usually wye-connected, with the free ends of the windings connected to three slip rings mounted on the rotor shaft, as indicated in Figure 5.20. An externally variable wye-connected resistance is connected to the rotor circuit through the slip rings. In Figure 5.20, ϕ_1, ϕ_2, and ϕ_3 indicate the various phases in a three-phase system.

Rotor resistance is high at starting, producing high starting torque. The higher rotor resistance increases the rotor power factor, which accompanies the higher starting torque. As the motor accelerates,

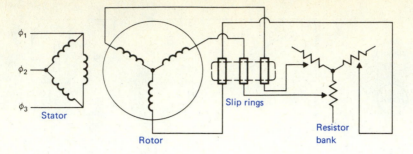

FIGURE 5.20 Wound-Rotor Induction Motor

the variable resistance is reduced. When the motor reaches full-load speed, the resistance is shorted out. The motor then acts like a squirrel-cage motor.

The advantages of the wound-rotor motor over the squirrel-cage motor are (1) smoother acceleration under heavy loads, (2) no excessive heating when starting, (3) good running characteristics, (4) adjustable speed, and (5) high starting torque. There are two major disadvantages of the wound-rotor motor: Its initial cost is high, and it has higher maintenance costs.

The wound-rotor motor is not used as much today as it was in the past. Its decline is due to improvements in speed control in other motors by varying stator frequency and to the development of the double squirrel-cage rotor.

SYNCHRONOUS MOTOR

The *synchronous motor* (SM) is one of the most efficient electric motors in industry. The unique features of the SM allow efficient, economical conversion of electric power to mechanical power. Synchronous motors are effectively applied over a wide range of speeds and loads. Because SMs can operate at a leading power factor, they are used in industry to reduce the cost of electric power. The savings in power cost commonly will pay for the motor in several years.

The SM functionally is different from the induction motor in that the SM rotor rotates at synchronous speed. Recall that the induction motor rotates at a speed slightly lower than synchronous speed. In the induction motor, the difference between rotor and

stator is called slip. In the SM, there is no slip since the rotor rotates at synchronous speed.

Motor Operation and Construction

You will recall that the induction motor used the principle of the rotating field in motor operation. The increasing and decreasing currents in the stator windings cause the magnetic field to rotate, although the physical poles remain stationary. The rotor of the SM uses this same rotating field principle. In fact, the SM can be converted to an induction motor by replacing the synchronous rotor with an induction motor rotor. The primary difference between SMs and induction motors lies in the rotor.

Integral-horsepower SM rotors are constructed as shown in Figure 5.21. This type of motor is said to have salient poles. The word *salient* means "projecting out." *Salient poles* are basically electromagnets, with the pole pieces projecting outward, as shown in the Figure 5.21. The pole pieces are laminated, as in the wound-field DC motors. Coils of wire are wound around the pole pieces and are connected to slip rings on the rotor shaft. Most SMs have salient poles. The salient poles on an SM are similar in construction to the poles on a wound-field DC motor. The coils are wound around the center of the pole core.

In rotor construction, the SM is similar to the synchronous generator or alternator. In addition to the coil windings on the rotor, many SMs have squirrel-cage bars. As we will see later, one of the disadvantages of the SM is that it does not produce starting torque. The squirrel-cage windings allow the SM to

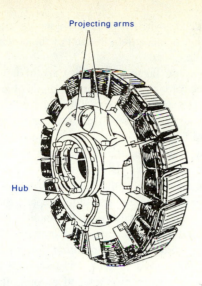

Projecting arms

Hub

FIGURE 5.21 SM Rotor

start. The squirrel-cage bars are embedded in the rotor as in the induction motor and are shorted by end rings. These squirrel-cage bars are sometimes called *damper* or *amortisseur windings*.

There are two differences between rotor construction in a squirrel-cage rotor and an SM with damper windings. First, in the induction motor, the rotor bars are evenly spaced and located all around the rotor. In the SM, the damper windings are located only on the pole faces. Second, no current flows in the damper windings. Current only flows when there is a difference between rotor and stator speeds. In the SM, a difference in rotor and stator speeds occurs during starting. As soon as the rotor turns at synchronous speed, all current in the damper windings stops. Furthermore, since no current flows in the damper windings when the rotor turns at synchronous speed, no power losses occur in the damper windings. Thus, the SM runs more efficiently than a comparable induction motor.

The motor pole pieces are typically laminations punched out of metal, as shown in Figure 5.22A. In large motors, the laminations are assembled, riveted together, and wound with wire, as shown in Figure 5.22B. Each pole piece is fitted into a steel or iron form called a *spider*. A dovetailed end on each pole piece attaches to the spider (Figure 5.23). A bar of metal called a *key* holds the pole piece to the spider. This construction technique is used in high-speed SMs. In low-speed SMs, the rotor is bolted to the spider.

At the other end of the pole piece, the rotor is fitted with squirrel-cage bars. These bars are either soldered or brazed to end pieces. The end pieces short

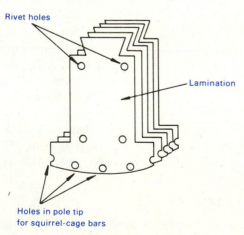

Rivet holes

Lamination

Holes in pole tip
for squirrel-cage bars

A. Laminated Pole Pieces

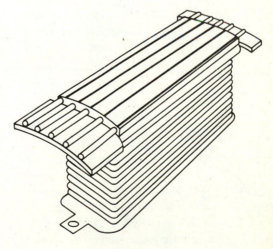

B. Assembled Rotor Arm

FIGURE 5.22 SM Rotor Construction

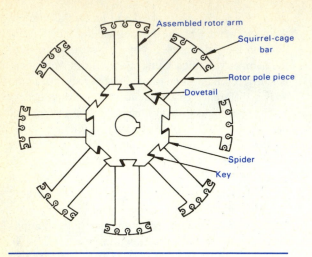

Assembled rotor arm
Squirrel-cage bar
Rotor pole piece
Dovetail
Spider
Key

FIGURE 5.23 Spider and Key of Rotor

each of the squirrel-cage bars. Squirrel-cage bars are made of a nonmagnetic material, such as brass or copper. As in induction motors, the starting characteristic of the SM with damper bars is influenced by the resistance of bars. The high-resistance brass bars produce starting characteristics different from those of low-resistance copper bars.

Another construction feature distinguishing the synchronous motor from the induction motor is the exciter. Recall that the induction motor has a field caused by current induced in bars from the rotating stator field. Large SMs have a rotor field created by an electromagnet. Current flows from an external source into the rotor windings, creating a magnetic field. The current usually comes from a source called the *DC exciter.*

Rotor Excitation

The rotor in the SM gets its magnetic field from one of several different methods. One method is to use a wound-field *DC generator exciter,* usually mounted on the rotor shaft. A schematic diagram of this type of device is shown in Figure 5.24A. As the rotor shaft rotates, the generator also rotates and produces a DC voltage. The DC voltage is then fed into the rotor via brushes and slip rings. The current flowing in the rotor windings produces the field in the rotor. The end of the motor attached to the load is called the

drive end. On synchronous motors with exciters, the DC exciter is located on the end of the motor opposite the drive end.

On some large SMs, the current needed to create the rotor field can be in hundreds of amperes. In this case, another exciter, called the *pilot exciter,* controls the main exciter. This arrangement is shown in Figure 5.24B.

The exciter using the DC generator has fallen out of favor in industry. Downtime and maintenance costs for replacing brushes and repairing commutators make DC generator excitation expensive. As the brushes wear, conductive dust gets into the machine, causing further maintenance problems.

Today, *brushless DC excitation systems* are more popular. They avoid the problems associated with the DC generator with brushes and a commutator. A schematic diagram of one type of brushless excitation is shown in Figure 5.25A. The exciter is a three-phase alternator instead of a DC generator. A rectifier changes AC from the alternator to DC. The DC voltage is then sent to power the SM rotor, creating magnetic flux. Again, since the excitation currents are high, a pilot exciter (Figure 5.25B) may be needed to provide control at lower values of current and voltage.

Some modern excitation methods get rid of the DC and AC generator entirely. This type of excitation is called *static excitation.* The AC from the main power bus is rectified by one or more high-power semiconductors called *silicon-controlled rectifiers* (SCRs). The SCR and SCR control circuits are discussed in detail in Chapters 6 and 7. The SCRs do two jobs at the same time. They rectify by changing AC to DC, and they control the amount of voltage and current from the exciter. SCRs are simple and fast, and they can carry currents in thousands of amperes. Also, static DC excitation using SCRs has great cost advantages over other types of excitation.

Starting the Motor

The stator windings of the three-phase SM and the three-phase induction motor are, for all practical purposes, the same. Application of three-phase AC power to the stator of a three-phase induction motor produces a rotating field. At starting, the rotor is

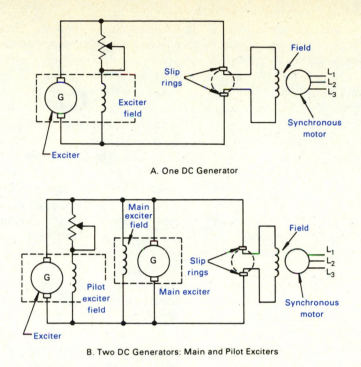

A. One DC Generator

B. Two DC Generators: Main and Pilot Exciters

FIGURE 5.24 DC Generator Excitation for SMs

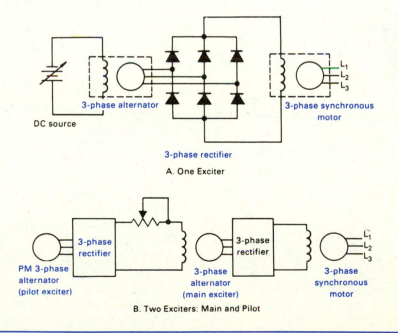

A. One Exciter

B. Two Exciters: Main and Pilot

FIGURE 5.25 Brushless Excitation for Three-Phase SMs

stationary and the DC excitation voltage is zero. If the poles were excited with DC, a magnetic field would be created. However, no starting torque is produced when the rotor is excited with DC. We can see why by examining Figure 5.26 (only one set of stator poles is shown). As the rotating field approaches the pole, both being north poles, they repel each other. Thus, rotor torque is created in the clockwise direction, as shown in Figure 5.26A. With the stator field rotating clockwise, the same north pole moves into the position shown in Figure 5.26B. Note that, in this position, the two north poles repel again, forcing the rotor counterclockwise. The total net rotor torque is zero in one 360° stator field rotation. Similar results occur in other positions, as shown in Figures 5.26C and 5.26D.

The SM, therefore, needs a special system to start the motor. Starting systems can be classified into two basic types: external and internal. *Internal starting systems* generate torque from windings embedded in the rotor—the damper or amortisseur

windings. *External starting systems* depend on some type of motor to drive the SM rotor to near synchronous speed. These starting motors are called *pony motors*. The pony motor can be an AC induction motor or a DC wound-field motor.

If the SM has damper windings, the motor starts as an induction motor. The rotor is, therefore, driven to almost synchronous speed by the squirrel-cage bars in the rotor. The DC excitation voltage is not applied until the motor has accelerated to almost synchronous speed. When the rotor's DC excitation voltage is applied, the rotor becomes magnetized. The rotor is constructed so that one of the salient poles is a north pole and the adjacent pole is south. North and south poles alternate around the salient pole rotor. When, for example, a north pole on the stator passes over a particular pole on the rotor, DC power is applied. The south pole of the rotor is attracted to the north pole of the stator. The rotor then locks into step with the stator field. The salient pole rotor produces torque only after locking into step with the rotating stator field. The speed of the rotor is then identical to the stator synchronous speed—which is why the motor is called a *synchronous* motor. The speed of the SM depends only on the frequency of the applied stator voltage.

The SM is classified as a constant-speed machine. As long as the line frequency does not change, the rotor rotates at a constant synchronous speed. As you might guess, the DC excitation voltage must be applied at the correct time and with the correct polarity. For example, Figure 5.27 shows an undesirable situation. What if power were applied when two north poles are created as shown? Instead of the rotor pulling into step, the rotor will be slowed down suddenly. At this point, the air gap flux is reduced, and a large amount of current rushes into the stator from the line. If DC rotor excitation continues to be applied, large fluctuations in stator currents occur as north and south rotor poles pass under the stator. Thus, the best time to apply DC excitation voltage is when a rotor pole of opposite polarity passes under a stator pole. This discussion brings up an important point about SM construction: An SM should have the same number of stator and rotor poles.

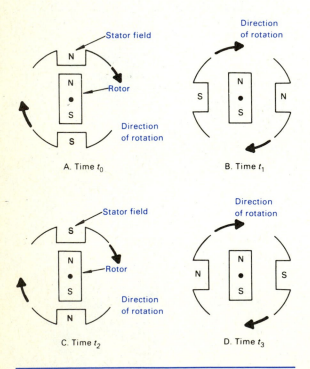

FIGURE 5.26   Starting of SMs

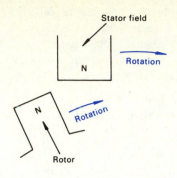

FIGURE 5.27   Field Application to SM Rotor

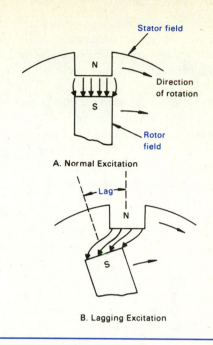

A. Normal Excitation

B. Lagging Excitation

FIGURE 5.28   Relationship between Rotor and Stator in SM

Another point to bear in mind concerns the torque produced by the motor when it pulls into synchronization. Obviously, there is a certain attractive force that pulls the rotor into synchronization with the stator field. If the force produced by this attraction is not large enough, the rotor will never be pulled into lock. The amount of torque needed to pull the rotor into synchronization is called the *pull-in torque*. Once the rotor is synchronized with the stator, the torque load on the rotor can be increased. If the load torque is too great, the magnetic tie between the rotor and stator is broken. The motor quickly comes to a stop when this tie is broken. The torque at which the rotor pulls out of synchronization with the stator is called the *pull-out torque*.

This torque action is illustrated in Figure 5.28A. Note that the attraction between the rotor and stator fields is strong enough to hold the rotor directly beneath the stator pole. If the load torque demand on the rotor is increased, the rotor will fall slightly behind the stator, as shown in Figure 5.28B. At this point, we can pull the rotor back under the stator, but we will need to increase the strength of one of the fields. The rotor field is the easiest to change. We can pull the rotor field back under the stator field by increasing the DC excitation current. Increasing DC excitation current increases the attractive force between the rotor and stator fields. Further increases in the load torque without an excitation increase will reach a point where the rotor pulls out of synchronization with the stator.

Load Effects

The power factor of an induction motor depends on the load and varies with it. The power factor of an SM may be 1 or less than 1, either leading or lagging. In the induction motor, the power factor can only be lagging. As in the induction motor, the power factor in the SM depends on the load. Unlike the induction motor power factor, the SM power factor depends on the amount of DC excitation.

Recall that, in the wound-field DC motor, the armature current depends on the cemf. The cemf comes from the rotating armature coils cutting flux in the stationary field. The SM also develops cemf. The cemf is developed in the armature, but the armature in the SM is actually the stationary stator coil. The stationary stator coils are cut by the flux from the rotor. The cemf in the stator is maximum when the rotor poles are directly under the stator poles. When the SM is unloaded, the cemf is 180° out of phase with the applied stator voltage. If the field is adjusted so that the cemf almost equals the applied voltage, the stator current is small. This small stator current corresponds to the exciting current in an unloaded transformer secondary. The stator current lags behind the resultant stator voltage by an angle θ.

When a load is applied to the rotor, it causes the poles to be pulled a certain number of mechanical degrees behind their no-load position. The cemf induced into the stator is, then, the same amount of electrical degrees behind the applied voltage. The resultant voltage in the stator causes the stator current to lag behind the applied voltage. In this condition, with the stator current lagging behind the stator voltage, the power factor is lagging. The SM then appears inductive to the line. If the load is increased further, the rotor poles are pulled further behind the stator poles. Consequently, the stator current lags further behind the stator voltage. So the power factor decreases, and the stator current increases. We should point out here that the load cannot continually be increased. A point will be reached where the flux link between the rotor and stator will be broken. The link is broken when the rotor pole is about half-way between the stator pole it is attached to and the stator pole behind it. As mentioned previously, the point where the motor loses synchronization with the load is called the pull-out torque.

Because the speed is constant, if we decrease the rotor excitation, the cemf decreases. The resultant stator voltage, therefore, increases. The stator current increases and lags the applied voltage by a greater amount. In contrast, if we increase the rotor field excitation until the current I_{st} is in phase with the applied voltage, the power factor is 1. The power factor will be 1, however, only for this particular value of load. For one particular load value, the stator current and resultant voltage are minimum when the power factor is unity. If excitation is increased past this point (when power factor is unity), stator current increases and leads the applied voltage. For a particular load, therefore, the power factor is controlled by the amount of excitation in the rotor. A weak rotor field produces a lagging motor power factor, and a strong rotor field produces a leading power factor. When the stator current is in phase with the applied voltage, the power factor is 1.

Synchronous motor curves, called *V*-curves, are shown in Figure 5.29. These curves indicate the changes of current for a constant load and varied rotor excitation. The corresponding changes in power factor are also shown.

Power Factor Correction

Synchronous motors are frequently used in certain systems to change the power factor of the system. Industrial power users have many reasons for wishing to change the power factor of the system. For instance, large numbers of induction motors and transformers make the power factor of an industrial plant low and lagging. Plant managers may wish to change the power factor because three problems arise when a power system has a low, lagging power factor. First, low, lagging power factors tend to lower the supply voltage. A lowered supply voltage can cause motors to overload and other electronic equipment to work improperly.

Second, a low power factor reduces the ability of the system to carry power. The kVA rating of a piece of equipment determines the amount of kVA

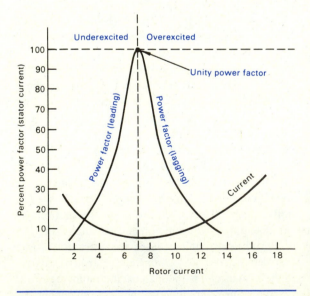

FIGURE 5.29 SM *V*-Curves

power it can carry—that is, its capacity. When the power factor is low, larger generators, transformers, and switches must be used for each kilowatt of load. Also, the power factor of the load determines the power-handling capability of transformers. Transformers are rated in kilovolt-amperes. Their kilowatt-carrying capacity depends directly on the power factor of the load. At low power factors, the effective capacity of the transformer is reduced. A reduced power factor means larger power distribution lines and increases in the voltage drop in those lines.

A third problem of a low power factor relates to the cost of electricity. Power company kilowatt-hour meters register the energy used by industrial loads. With a low, lagging power factor, the kilowatt-hour meter registers the use of energy even though no useful work is done by the reactive current. Most power companies, in addition, assess penalties for a low, lagging power factor and give incentives for a high power factor.

Probably the best way to deal with a low, lagging power factor is to prevent it from happening. For induction motors, a lightly loaded motor has a high proportion of lagging, reactive kVA power. So a first step in preventing a low, lagging power factor is to make sure that all induction motors are loaded as fully as possible. Excessively high voltages can also decrease power factors. The second step, then, is to make sure line voltages are correct. Capacitors may also be used to improve the power factor. Finally, the SM can be used to adjust the power factor.

Power factor is adjusted by adjusting the excitation current of the SMs attached to the line. If the motor is operated without a load, the SM power factor may be adjusted to a value of, say, 10% leading. When operated in this condition, the motor is called a *synchronous condenser*. Its operation is similar to a capacitor's (or condenser's) because it makes current lead voltage. The synchronous condenser takes only enough true power from the line to supply its losses. At the same time, it supplies a high, leading power factor. The high, leading power factor cancels out the lagging power factor from inductive loads like induction motors. The entire power factor of the line is therefore improved. With enough SM excitation, the line power factor may be adjusted to 1.

Synchronous condensers are built to operate without a mechanical load. It makes sense, however, to use SMs for all appropriate applications, such as compressors. The SM in this type of application can be adjusted so that it has unity power factor. It will not contribute to the lagging power factor problem, as an induction motor will. It adds to a plant load without requiring reactive kVA power. Also, when it is partially loaded, it operates at a leading power factor, helping boost the power factor of a site. Synchronous motors, then, do double duty. They can drive a load economically, efficiently, and precisely. They also improve the power factor of the system.

Fractional-Horsepower Motors

Up to this time, we have been discussing integral-horsepower synchronous motors. But there are many fractional-horsepower (fractional-hp) synchronous motors in use today. Most do not use a separately excited rotor, however. It is too expensive and usually not necessary. Motors without a separately excited rotor are called *nonexcited synchronous motors*. There are two major types in use: reluctance motors and hysteresis motors.

Reluctance Motor. The *reluctance motor* uses differences in reluctance between areas of the rotor to produce torque. The reluctance principle utilizes a force that acts on magnetically permeable material. When such a material is placed in a magnetic field, it is forced in the direction of the greatest flux density. A diagram of the special rotor construction is shown in Figure 5.30.

The rotor has slots cut into it at regular intervals, as shown in the figure. These slots form salient poles. Proper motor operation demands that the number of salient poles be equal to the number of stator poles. The rotor also has bars, which cause it to accelerate as an induction motor. When the motor nears synchronous speed, the field magnetizes to the rotor and pulls it into step. The torque to drive the motor is generated because the reluctance is greater in the areas that do not have any bars. The areas of the rotor that have bars have a lower reluctance.

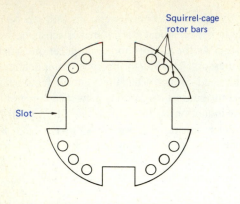

FIGURE 5.30 Rotor Construction of Reluctance Motor

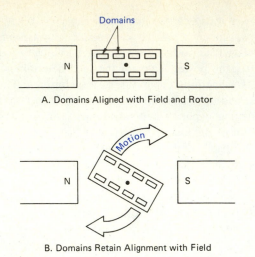

A. Domains Aligned with Field and Rotor

B. Domains Retain Alignment with Field

FIGURE 5.31 Rotor Behavior of Hysteresis Motor

This motor has very low starting torque as well as low pull-out torque. Although the reluctance motor is normally found in fractional-hp sizes, it is also encountered in integral sizes up to 100 hp.

Hysteresis Motor. Another interesting synchronous motor uses the hysteresis effect as a driving principle. The rotor of the *hysteresis motor* is made of a cobalt steel alloy. This alloy retains a magnetic field well and has high permeability.

Recall from basic electricity that when a magnetically permeable material is placed in a magnetic field, the domains in the metal align with the field. If the fields rotate, the domains still align themselves with the field, as shown in Figure 5.31. Power is consumed when the domains maintain their alignment against the rotation of the rotor. This power loss is called *hysteresis loss*. During the time these domains are changing, starting torque is produced by another method, such as squirrel-cage bars. As the

rotor approaches synchronous speed, current in the squirrel-cage bars decreases, producing less torque. At this time, the rotor domains lock into alignment, resulting in a fixed magnetic polarity in the rotor. The retained magnetic field is now pulled into synchronization with the rotating field.

One interesting feature of the hysteresis motor is that the torque produced remains constant, regardless of load or speed changes. Consequently, it is used in clock and timer motors as well as audio turntables and tape drives.

CONCLUSION

The AC motor has special characteristics that differ from those of the DC motor. The induction motor, for example, is one of the most common motors in industry because it needs little maintenance. Since many AC motors are used in industry, technicians must have a general understanding of how they operate.

■ QUESTIONS ·

1. The universal motor is similar in construction to the _____-connected DC motor. The universal motor, then, can run on _____ as well as AC power.

2. Both induction and synchronous motors use the principle of the _____ stator field to generate torque. This principle requires that separate pairs of windings have applied voltages that differ in _____.

3. In the induction motor, energy is coupled from the stator to the rotor by _____. Therefore, there is no _____ connection between rotor and stator.

4. Because only one phase is present, single-phase motors must have special circuitry to _____ them.

5. The rotors of synchronous motors run at _____ speed. Synchronous motors develop very little torque at _____. This problem is sometimes solved by placing squirrel-cage bars in the synchronous motor's _____.

6. Explain how the universal motor can be used in AC and DC circuits. Explain the difference in construction of a series DC motor and a universal motor.

7. Describe how a rotating magnetic field is created.

8. Is it possible to change the direction of an AC motor? If so, explain how.

9. What factors influence the rotor speed of an AC motor?

10. Describe how an induction motor produces torque.

11. Define slip in an induction motor.

12. Explain the NEMA classification system for induction motors.

13. Describe the construction of a double squirrel-cage motor. What effect does this construction have on motor operation?

14. Rank the types of single-phase induction motors in terms of the torque produced as a percentage of full-load torque.

15. Explain how starting torque is produced in the following engines: (a) split-phase, (b) capacitor-start, and (c) shaded-pole.

16. Describe the operation of the centrifugal switch.

17. Describe the method of rotor excitation in the wound-rotor motor.

18. List the advantages and disadvantages of the wound-rotor motor compared with the squirrel-cage motor.

19. At what speed does the synchronous motor run?

20. Why doesn't the synchronous motor develop enough torque to start on its own? How is the synchronous motor started?

21. What purpose does the synchronous motor serve other than the obvious one of turning mechanical loads?

22. Explain the method by which torque is produced in the hysteresis and reluctance motors.

■ PROBLEMS ·

1. For a 480 V AC, three-phase, wye-connected generator with a phase current of 20 A, find (a) the line voltage and (b) the line current.

2. For a 240 V AC, three-phase, delta-connected generator with a phase current of 25 A, find (a) the line voltage and (b) the line current.

3. Given a three-phase, delta-connected generator that supplies a full-load current of 150 A at a phase voltage of 240 V AC and has a power factor of 0.85 lagging, find (a) the line voltage and current, (b) the true power output (in kilowatts), and (c) the apparent power (in kilovolt-amperes).

4. Given a three-phase induction motor, with 12 poles and a line frequency of 400 Hz, find the synchronous speed (in revolutions per minute).

5. For a three-phase induction motor, with a stator synchronous speed of 450 Hz and a rotor speed of 400 Hz, find the slip of this motor (in percent).

6. Given a 15 hp, three-phase SCIM that has a synchronous speed of 1800 r/min and a rotor speed of 1720 r/min at full-load and that stalls at 1300 r/min, find (a) the torque the motor produces at rated load and (b) the maximum torque the motor will produce with rated voltage applied.

7. Given a three-phase, 240 V AC, 2.5 kW, 60 Hz SCIM, with a starting torque of 100 Nm when rated voltage (240 V AC) is applied, find (a) the starting torque when 250 V AC is applied and (b) the stator voltage needed to produce a starting torque of 110 Nm.

8. Suppose you are working with a 45 hp, 240 V AC SCIM. Find (a) the approximate full-load current, (b) the no-load current, and (c) the starting current.

9. For an SCIM with a no-load speed of 1750 r/min and a full-load speed of 1700 r/min, find the speed regulation of the motor.

10. A 50 hp, 240 V AC SCIM draws 100 A from the line under rated load conditions. Find (a) the kVA input to the motor and (b) the power factor of the motor.

11. An SCIM has an input kVA of 500 VA, or 0.5 kVA. If the power factor is 0.85 (85%), how much real power does the motor draw (in watts)?

12. Suppose you are working with a 100 hp SCIM with a power factor of 0.95 and an efficiency of 0.92 (92%). Find (a) the kVA input and (b) the real power drawn from the load.

13. Given a 75 hp, 240 V AC, four-pole, three-phase, 60 Hz, 1750 r/min SCIM, with a full-load efficiency of 91% and a power factor of 87%, find (a) the stator synchronous speed, (b) the full-load slip, (c) the full-load torque, (d) the kVA input, (e) the real power drawn by the motor from the line, and (f) the full-load current.

14. Calculate the synchronous speed of a 12-pole motor with three-phase, 60 Hz power applied to its stator.

15. How many poles would a three-phase motor need to turn at a synchronous speed of 900 r/min?

16. If an induction motor's rotor is turning at 1750 r/min, what is the most likely stator synchronous speed? *Hint*: Examine Equation 5.11. On the basis of this speed, what is the slip?

CHAPTER 6

Industrial Control Devices

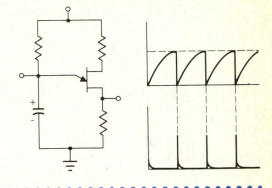

OBJECTIVES

On completion of this chapter, you should be able to:

- Categorize individual control devices into one of five areas.
- Summarize the advantages and disadvantages of electromagnetic relays and reed relays.
- Summarize the advantages and disadvantages of solid-state relays.
- Identify the parts of the electromagnetic relay.
- Identify the schematic symbols of thyristors in common industrial use.
- Calculate the firing angle of an AC thyristor circuit.
- Calculate the junction temperature and power dissipation of a thyristor.
- Give examples of manually operated switches and their applications.
- Give examples of mechanically operated switches and their applications.
- List four methods by which thyristors are triggered.
- Describe the principles and applications of the solenoid.
- List two types of motor protection circuits and describe how they work.
- Contrast thermal and magnetic overload relays in terms of operation and applications.
- Contrast the electromagnetic relay and the solid-state relay in terms of operations and applications.
- List two methods of thyristor transient protection.

INTRODUCTION

Control devices lie at the heart of modern industrial systems. Very simply, a *control device* is a component that governs the power delivered to an electric load. The load may be a motor or generator load, an electronic circuit, or even another control device. Indeed, every industrial system uses a control device. Since control devices are so common, a good background in the different types of control devices—how to recognize them and how they work—is essential.

Basically, control devices can be broken down into five categories: manually operated switches, mechanically operated switches, solenoids, electromagnetic switches (relays), and electronic switches (thyristors). A *switch* may be described as a device used in an electric circuit for making, breaking, or changing electrical connections. In this chapter, we discuss each of these basic categories in terms of the control devices in each category and the operating principles of these devices.

MANUALLY OPERATED SWITCHES

A *manually operated switch* is one that is controlled by hand. Many types and classifications of switches have been developed. Frequently, switches are classified by the number of poles and throws they have. The *pole* is the movable part of the switch. The *throw* of a switch signifies the number of different positions in which the switch can be set. The number of poles can be determined by counting the number of movable contacts. The number of throws can be found by counting how many circuits can be connected to each pole. Figure 6.1 shows several schematic diagrams of different combinations of poles and throws.

Toggle Switches

A *toggle switch* is an example of a manually operated switch. One type of toggle switch is shown in Figure 6.2. Toggle switches have many uses, most of them centered around energizing low-power lighting and other low-power electronic equipment. Electrical ratings are in terms of volts and amperes. These ratings should not be exceeded; otherwise, damage to the switch may result. If a switch needs

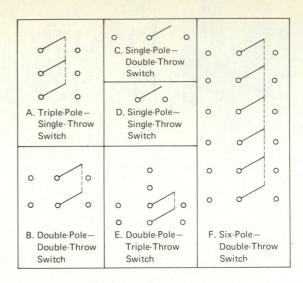

FIGURE 6.1 Switch Schematic Diagrams

A. Triple-Pole—Single-Throw Switch

B. Double-Pole—Double-Throw Switch

C. Single-Pole—Double-Throw Switch

D. Single-Pole—Single-Throw Switch

E. Double-Pole—Triple-Throw Switch

F. Six-Pole—Double-Throw Switch

to be replaced but an exact replacement cannot be found, substitute a switch with higher voltage and current ratings.

Push-Button Switches

Another common manually operated switch in industrial use is the *push-button switch*. This switch and its schematic diagram are shown in Figure 6.3.

Push-button switches can be broken down into two categories: momentary-contact and maintained-contact. In the *momentary-contact switch*, actuation occurs only when the switch is pressed down. This push-button switch resets itself when released. (Resetting is the process of returning the switch back

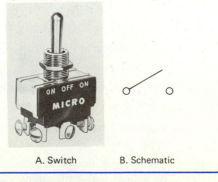

A. Switch   B. Schematic

FIGURE 6.2 Toggle Switch

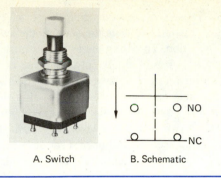

A. Switch   B. Schematic

FIGURE 6.3 Push-Button Switch

A. Switch   B. Schematic

FIGURE 6.4 Knife Switch

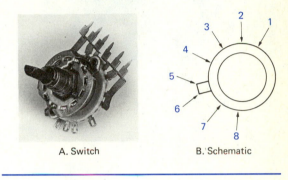

A. Switch   B. Schematic

FIGURE 6.5 Rotary Switch

to its original state.) In the *maintained-contact switch*, as its name suggests, the switch is actuated even when the button is released. It must be pressed again for it to change states. This switch is most commonly used in turning electric equipment on and off. A good example of its application is the dimmer switch in a car.

Note the abbreviations NO and NC in Figure 6.3B. These letters represent the electrical state of the switch contacts when the switch is not actuated. The abbreviation NC means that the switch contacts are normally closed; NO means that they are normally open.

Knife Switches

The *knife switch* is a manually operated switch used extensively for controlling main power circuits. The knife switch is shown in Figure 6.4. The knife switch typically is a bar of copper with an attached handle. The switch is actuated by pushing the copper bar down into a set of spring-loaded contacts. Normally, the knife switch is enclosed in a box to prevent accidental contact with high voltage.

Rotary-Selector Switches

The *rotary-selector switch* is another common manually operated switch. This switch is shown in Figure 6.5. As the knob of the switch is rotated, circuits can be opened and closed according to the construction of the switch. Some rotary switches have several layers, or wafers, as shown in Figure 6.5B. The

addition of wafers makes it possible to control more circuits with a single change of position. The rotary switch is often used on test equipment and other multifunction low-power equipment. The range-selector switch of a voltmeter is an example of the use of a rotary switch.

Manual Motor Starters

The *manual motor starter* is a special manually operated switch that is used to start or stop AC and DC motors. It is illustrated in Figure 6.6. A mechanical linkage usually transmits the movement of the toggle or handle to the actual switch contacts. The wiring diagrams, shown in Figure 6.7, do not show the operating mechanism since it is not electrically controlled.

Figure 6.7A is a simplified diagram of a manual starter wired to control an AC motor. Note the three normally open (NO) switch contacts labeled S_1, S_2,

FIGURE 6.6 Manual Motor Starter

and S_3. These switches are said to be normally open because this term reflects the state of the switch in its off position. Throwing the switch or pushing the start button mechanically closes these contacts, connecting the three-phase AC motor to the line. The contacts can be opened by pressing the stop button. The contacts can also be opened by tripping one of the two overload relays. An *overload relay* (OL) is a device that will open a set of NC contacts when a certain preset temperature is reached. It will be discussed a little later in this chapter. When an overload relay trips (caused by a motor overheating), the starter mechanism unlatches and the S contacts open. The open switch removes power from the motor. Usually, the contacts cannot be reclosed until the starter mechanism has been reset. Resetting can be accomplished by pushing the stop or reset button or by moving the handle to the reset position. The overload relay should be allowed to cool before resetting. The relay should not be reset, however, until the problem that caused the relay to trip has been corrected.

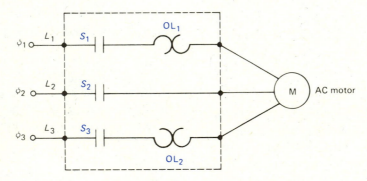

A. Starter Wired to Control an AC Motor

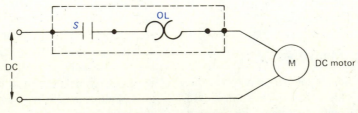

B. Starter Wired to Control a DC Motor

FIGURE 6.7 Simplified Diagrams of Manual Motor Starters

A simplified diagram of a manual starter wired to control a DC motor is shown in Figure 6.7B. Note the difference in wiring between this starter and the one used to control an AC motor. Only one overload relay is necessary since only one path for current exists.

Manual starters generally are used in conveyors, compressors, pumps, blowers, fans, and small machine tools. They are low in cost, are simple to use and install, and do not give off the annoying AC hum often found in devices with a coil. The manual starter is used in situations where equipment can run continuously and automatic control is not required. As a general rule of thumb, if a motor has to be started and stopped more than 12 times an hour, week after week, manual control should not be used. They also are used as power disconnects and overload protectors where these two functions are not provided elsewhere. Manual starters normally are used on small AC and DC machines. If the motor is large, the designer will have no choice; another type of controller must be selected. In single-phase AC applications, manual starters are limited to 5 hp (3.7 kW) at voltages up to 230 V. In three-phase applications, 15 hp motors may run on manual switches to 600 V AC.

MECHANICALLY OPERATED SWITCHES

Mechanically operated switches are used extensively in industry. A *mechanically operated switch* (sometimes called an *automatic switch*) actuates automatically in the presence of some specific environmental factor, such as pressure, position, or temperature.

Limit Switches

The *limit switch*, shown in Figure 6.8, is a very common industrial device. It usually is actuated by horizontal or vertical contact with an object, such as a cam. Note in the figure that the actuation is accomplished by moving a lever. Often, the lever is the roller type, as shown in Figure 6.8A. Or as in the case of the limit switch in Figure 6.8B, a wand type of arm is used. Note the difference between the wand

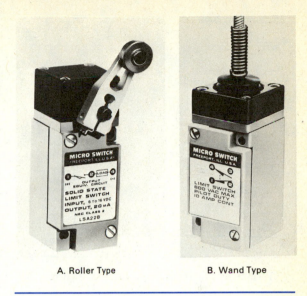

A. Roller Type   B. Wand Type

FIGURE 6.8   Limit Switches

and lever types. The roller lever may be actuated only by movement in one direction, but the wand may be actuated by movement in any direction.

Limit switches have two basic uses. First, they serve as safety devices to keep objects between certain physical boundaries. For instance, in a machine that uses a moving table, a limit switch can keep the table from moving past its safe physical limits by shutting off the table's drive motor. Second, limit switches are used to control industrial processes that use conveyors or elevators.

Mercury Switches

The *mercury switch* is nothing more than a glass enclosure containing mercury and several switch contacts placed so that the mercury will flow over and around them. Several examples of mercury switches are shown in Figure 6.9.

Because of the simplicity of the mercury switch, it is very versatile. It can be used as a position sensor; it can sense the amount of centrifugal or centripetal force and changes in speed, inertia, and momentum. In addition, it has little contact bounce (a major problem in relays and mechanical switches), is hermetically sealed against environmental impact, and has a relatively long life.

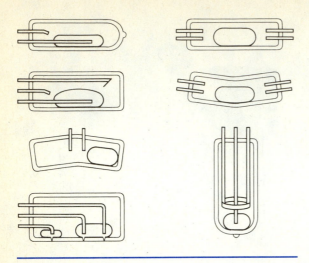

FIGURE 6.9 Mercury Switch Configurations

Most mercury switches are used with currents up to 4 A and voltages up to 115 V. When the switch is actuated, the contact resistance can be less than a tenth of an ohm. As shown in Figure 6.9, the mercury switch comes in many different configurations. The glass envelope is usually filled with hydrogen gas to prevent arcing.

An application of a mercury switch is shown in Figure 6.10. When the mercury switch closes, trigger voltage is provided to the silicon-controlled rectifier (SCR), which turns on. The SCR is an electronic switch; its operation is discussed later in this chapter.

Snap-Acting Switches

A *snap-acting switch* is a switch in which the movement of the switch mechanism is relatively independent of the activating mechanism movement. In a snap-acting switch, as in a toggle switch, no matter how quickly or slowly the toggle is moved, the actual switching of the circuit takes place at a fixed speed. In other switches, like the knife switch, the speed with which the switch closes is determined by the mechanism driving it. In the snap-acting switch, the switching mechanism is a leaf spring that snaps between positions.

A *precision snap-acting switch* is one in which the operating point is preset and very accurately known. The operating point is the point at which the plunger or lever causes the switch to change position. The precision snap-acting switch is commonly called a *microswitch*, which is actually a trade name of the Micro Switch Division of Honeywell Corporation. An example of a microswitch is shown in Figure 6.11. Frequently, limit switches, like those in Figure 6.8, contain snap-acting switches as their contact assembly. Much less arcing results from the rapid movement of the contacts.

The basic precision snap-acting switch is used in many applications as an automatic switch. It is most often used as a safety interlock to remove power from a machine when there is danger to the operator or technician.

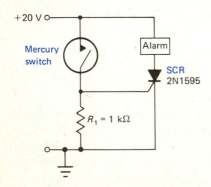

FIGURE 6.10 Mercury Switch Controlling SCR

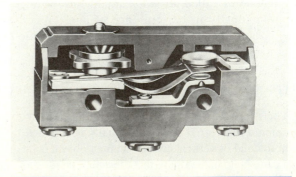

FIGURE 6.11 Cutaway View Showing Construction of Precision Snap-Acting Switch

Several different methods are used to actuate this type of switch. Some of the more common actuators and their uses are shown in Figure 6.12.

Motor Protection Switches

The motor is the most important source of rotating power in industry. We learned in earlier chapters that the motor tries to meet any torque demand placed on it. Unfortunately, the motor cannot detect unaided when a demand is too high; it is incapable of recognizing its own limitations. Motors must be protected from potentially damaging conditions. Industrial motor protection is based on two conditions: excessive line current and high motor temperatures. Too much line current is called a *short-circuit condition*, usually ranging from 10 to 100 times normal current flow. It is the more serious of the two conditions. If not interrupted, short circuits can damage motors, lines, and branch circuits, can cause fires, and can possibly injure personnel. Short circuits normally

occur when physical contact is made between two or more line conductors or a line conductor and ground. The resulting damage caused by the short circuit may be thermal, physical (due to magnetic forces), or electrical-arcing damage. Short circuits normally are prevented by fuses or circuit breakers.

Historically, *fuses* have been recognized as the best short-circuit protection, due to their fast response time. The fuse, however, in some situations is not the best protection for a motor or a motor circuit. System designers must protect motor control circuits and feeder lines, as well as the motors themselves, according to National Electrical Code (NEC) requirements. Ordinary fuses may protect the motor from damage but may let through currents high enough to damage starters or other associated equipment. A good fuse should respond to an increase in current within a few milliseconds.

A device that is an improvement over the fuse is called a *motor short-circuit protector* (MSCP). This device responds more quickly than a fuse, thus letting through a fraction of the energy a fuse lets through. When the MSCP opens, a trigger protrudes from the end of the device. The trigger indicates whether the device has been opened by an over-current. The MSCP also is used in motor starters to remove power from the starter in the event of a short circuit. Power is removed when the trigger presses on a bar, which opens the power switch. MSCPs are available in ratings up to 480 V and 100,000 A.

A *circuit breaker* is a device that protects against short circuits by unlatching a breaker switch. The response times of circuit breakers range from 8 to 10 ms. For motor protection circuits, this response may be too slow. If enough current gets through the circuit breaker before the breaker responds, some equipment may be damaged. To combat this slow response time, several manufacturers have developed a new design. Called the *blow-apart design*, this new type of breaker will respond to overcurrents in about 2 ms. The breaker arms, shown in Figure 6.13, are held together by a special calibrating spring. Current flows through each breaker arm, creating like magnetic fields. When the repulsive force of the two electromagnetic breaker arms overcomes the force of the spring, the breaker arms blow apart. The faster the current increases, the faster the arms blow apart.

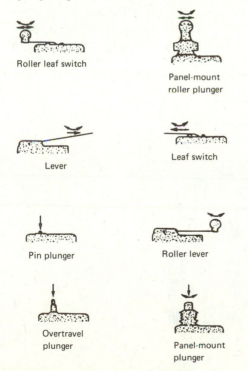

Roller leaf switch

Panel-mount roller plunger

Lever

Leaf switch

Pin plunger

Roller lever

Overtravel plunger

Panel-mount plunger

FIGURE 6.12 Common Actuators Used in Precision Snap-Acting Switches

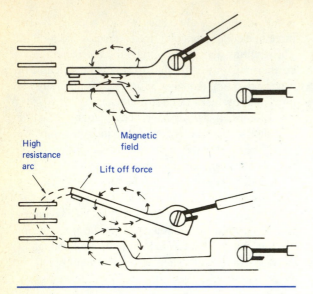

FIGURE 6.13 Blow-Apart Breaker

SOLENOIDS

The principle of the solenoid is an important one in industrial electronics. The *solenoid* is a device used to convert an electric signal or an electric current to linear mechanical motion. As shown in Figure 6.14,

the solenoid is made up of a coil with a movable iron core. When the coil is energized, the core—or armature, as it is sometimes called—is pulled inside the coil. The amount of pulling or pushing force produced by the solenoid is determined by the number of turns of copper wire and the amount of current flowing through the coil.

One factor often overlooked in solenoid design is the heating factor. Heat makes the coil less efficient at producing flux, lowering the flux density and the solenoid pulling force. Heat may be controlled by properly venting the enclosure, by using a heat sink, by forcing air past the solenoid, or by using a larger solenoid. The amount of heat produced by a solenoid is related to the duty cycle. The higher the duty cycle, the higher is the temperature of the device. Solenoids in continuous-duty applications may need a special circuit to reduce current and, therefore, power dissipation. Such a circuit is shown in Figure 6.15. This method of current reduction depends on using a solenoid that has a set of normally closed (NC) switch contacts. These contacts are placed across resistor R, as shown in the diagram. When the solenoid coil K is energized, current flows through the normally closed switch contacts. With power applied, the armature will start to close. When the

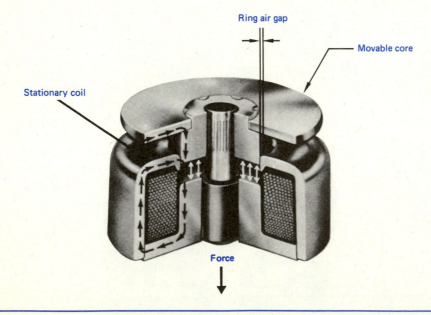

FIGURE 6.14 Linear Solenoid Construction

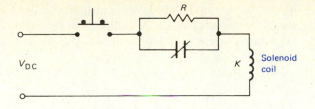

FIGURE 6.15 Circuit for Continuous-Duty Solenoid Action

armature reaches the end of its travel, it opens the NC switch. Current through the solenoid will then be reduced since the circuit has a higher total resistance.

Solenoids are used in a wide variety of applications, such as braking motors, closing and opening valves, shifting the position of objects, and providing electrical locking. The mechanical movement caused by the solenoid is usually between ½ and 3 in. Solenoids are available for use in either AC or DC circuits. In the selection of a solenoid, the important considerations are the stroke length and force required, the length of time the solenoid is to be energized, how frequently the device will be turned off and on, and the ambient temperature conditions. You should bear in mind when working with solenoids that too large a force in solenoid operation is almost as bad as not enough force. Too great a pull in the solenoid will cause it to pound itself into premature failure.

ELECTROMAGNETIC RELAYS (EMRS)

Relays are undoubtedly one of the most widely used control devices in industry. A *relay* is an electrically operated switch. Relays may be divided into two basic categories: control relays and power relays. There also are additional, more specialized relays.

Control relays, as their name implies, are most often used to control low-power circuits or other relays. Control relays find frequent use in automatic relay circuits, where a small electric signal sets off a chain reaction of successively acting relays performing various functions.

Power relays are sometimes called *contactors*. The power relay is the workhorse of large electrical systems. The *power relay* controls large amounts of power but is actuated by a small, safe power level. In addition to increasing safety, power relays reduce cost since only lightweight control wires are connected from the control switch to the coil of the power contactor.

The components of both power and control relays are the same; see Figure 6.16A. Each has a coil wound around an iron core, a set of movable and stationary relay contacts (the switch), and the mounting. A switch usually is used to start or stop current through the coil. When current flows through the coil, a strong magnetic field is created. This electromagnet pulls the armature, which, in turn, moves the relay contact down to make electrical connection

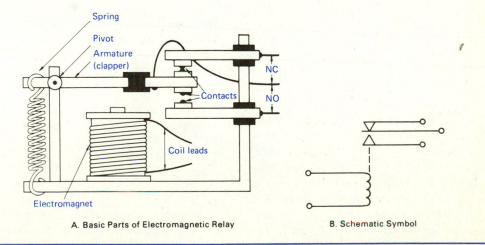

A. Basic Parts of Electromagnetic Relay

B. Schematic Symbol

FIGURE 6.16 Electromagnetic Relay

with the stationary contact. The physical movement of the armature occurs only when current flows through the coil. Any numbers of sets of contacts may be built onto a relay. Thus, a relay can control many different circuits at the same time. Note the schematic symbol for the relay shown in Figure 6.16B.

Like switches, the relay contacts have maximum voltage and current ratings. If these ratings are exceeded, the life of the contact may be decreased greatly. Relays also have pull-in voltage and current ratings. *Pull-in current* is the amount of current through the coil needed to operate the relay. *Pull-in voltage* is the voltage needed to produce that current. Usually, the actual steady-state (continuous) voltage applied to the relay coil is somewhat higher than the pull-in voltage. This condition guarantees enough current so that the relay actuates when needed and holds under vibration.

The pull-in current is sometimes called the *inrush current* in large relays. When the coil is de-energized and the armature is not pulled in, there is a large air gap in the magnetic circuit. The air gap causes the coil impedance to be low. When voltage is applied to the coil, the low impedance allows a large current to flow. As the armature moves closer to the magnetic assembly, the air gap gets smaller. The reduction in air gap causes the coil impedance to increase. The coil impedance reaches a maximum when the armature is seated firmly to the magnetic assembly. The final current that flows when the armature is seated is called the *sealed current*. The higher current that flows when the coil is first energized is the inrush current. The inrush current is usually six to ten times the sealed current. The actual ratio of sealed current to inrush current depends on the design of the relay.

A graph of the current in the relay plotted against time is shown in Figure 6.17A for an intermittent-duty relay. The intermittent-duty relay is only meant to be energized for a specified period of time. During the time the relay is shut off, the coil dissipates the heat energy built up during the time the current flowed through the coil. In contrast, the continuous-duty relay is meant to be energized continuously. Its current graph is shown in Figure 6.17B.

In large relays, the coil specifications are often given in volt-amperes. Suppose we are working with a coil that is rated at 500 VA inrush and 50 VA sealed. If the coil is a 120 V coil, we can calculate the inrush current by using the inrush volt-ampere rating:

$$\text{inrush current} = \frac{\text{volt-ampere inrush rating}}{\text{coil voltage rating}}$$

$$(6.1)$$

For our example, we obtain

$$\text{inrush current} = \frac{500\,\text{VA}}{120\,\text{V}} = 4.17\,\text{A}$$

The sealed current can be calculated in a similar fashion:

$$\text{sealed current} = \frac{\text{volt-ampere sealed rating}}{\text{coil voltage rating}}$$

$$(6.2)$$

The sealed current for our example is

$$\text{sealed current} = \frac{50\,\text{VA}}{120\,\text{V}} = 417\,\text{mA}$$

• **EXAMPLE 6.1**

Given a 240 V coil with a sealed value of 200 VA and an inrush value of 1200 VA, find the sealed and inrush currents.

Solution

The sealed current is found by using Equation 6.2:

$$\text{sealed current} = \frac{200\,\text{VA}}{240\,\text{V}} = 833\,\text{mA}$$

The inrush current can be calculated by substituting the appropriate values into Equation 6.1:

$$\text{inrush current} = \frac{1200\,\text{VA}}{240\,\text{V}} = 5\,\text{A}$$

Another rather specialized type of relay is the magnetic motor starter. Many industrial applications require the starter to be located some distance from the operator. But remote location is not possible with

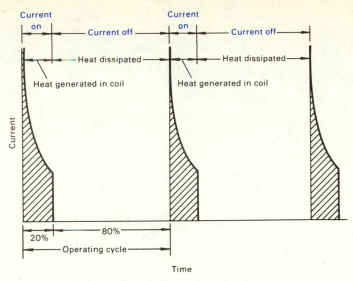

A. Cycling Chart of AC Intermittent-Duty Solenoid

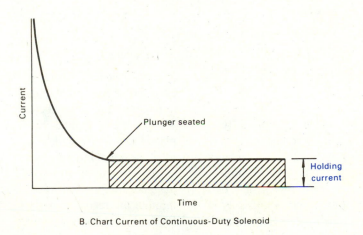

B. Chart Current of Continuous-Duty Solenoid

FIGURE 6.17 Solenoid Currents versus Time

the manual starter. Also, many processes require automatic control of motors that the manual starter cannot provide. Magnetic motor starters are thus used in these applications. *Magnetic starters* usually are relays that work on a variation of the solenoid principle, as shown in Figure 6.18. Current through the coil of an electromagnet generates a magnetic field. The field attracts the moveable armature, pulling it up. The relay contacts are attached to the armature,

closing when the armature is pulled in. Like the relays just mentioned, the magnetic motor starter has a pull-in voltage and current rating, which is larger than needed to keep the starter energized.

Magnetic motor starters usually have built-in motor overload protection circuits. (These protective switches are discussed later in this chapter.) The overload circuits are the main difference between a motor starter and a contactor. A contactor, being a

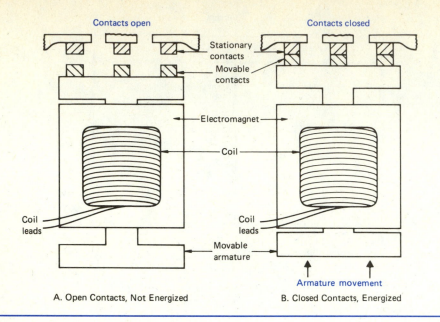

Contacts open   Contacts closed

Stationary contacts
Movable contacts
Electromagnet
Coil
Coil leads
Coil leads
Movable armature
Armature movement

A. Open Contacts, Not Energized   B. Closed Contacts, Energized

FIGURE 6.18   Magnetic Motor Starter Operation

general, high-power relay, does not have motor overload protection built in. Magnetic starters often are controlled by many of the switches we have discussed in this chapter—for example, limit switches, push buttons, and timing relays.

You may recall from Chapter 5 on AC motors that to reverse the direction of a three-phase, squirrel-cage motor, we need only reverse any two of the power leads. Such a reversing task usually is done by a special starter called a *reversing starter*. A simplified diagram of the reversing starter is shown in Figure 6.19. You will notice that the starter actually contains two contactors, one for forward operation and the other for reverse operation. A mechanical interlock (not shown) keeps only one contactor on at a time. The L_1, L_2, and L_3 lines are for three-phase AC. The three F (forward) and three R (reverse) contacts are part of a motor starter, with a relay similar to the one shown in Figure 6.18. Both the F and the R relay contacts are normally open. Thus, current must flow through the F and R coils to make the relay contacts close.

Not all relays have the same construction as the armature relay of Figure 6.16. Another type of relay, called the *reed relay*, is also widely used in industry. The reed relay, like the armature relay, uses an electromagnetic coil, but its contacts consist of thin reeds made of magnetically permeable material. These contacts, called the *reed switch*, usually are placed inside the coil. A cutaway view of a reed relay is shown in Figure 6.20A.

When current flows through the relay coils, a magnetic field is produced. The field is conducted through the reeds, magnetizing them. As shown in Figure 6.20B, the magnetized reeds are attracted to one another, and the relay contacts are closed.

Reed relays are superior to armature relays in many ways. Reed relays are faster, more reliable, and produce less arcing than the armature relay. However, the current-handling capabilities of the reed relay are limited. In addition, because of the glass case that houses the reed relay contacts, the reed relay is more susceptible to damage by shock than the armature relay is.

Arcing and pitting can be further reduced in the reed relay by wetting the relay contacts with mercury. Since mercury is a liquid metal, it is drawn up the bottom relay contact by capillary action. The mercury completely covers the relay contact, thus protecting it from arcing and extending contact life.

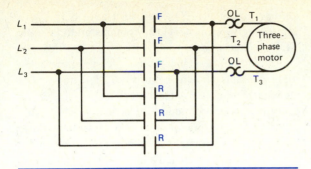

FIGURE 6.19 Reversing Starter

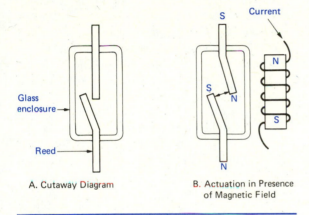

A. Cutaway Diagram

B. Actuation in Presence of Magnetic Field

FIGURE 6.20 Reed Relay

Timers

The proper timing of events in a process represents an important part of industrial electronics. In industrial processes, certain actions must occur at a certain time interval after another operation occurs. For example, in a DC motor system, there is often a time delay before full power is applied to the armature. This time delay gives the motor a chance to accelerate to a high speed. The actual time delay varies widely according to the application and device doing the timing. Delays may range from as little as a microsecond to as long as a day.

Timing Methods. Timing functions are performed by devices called *timers*. Timers fall into three basic groups: electromechanical (EM) timers, solid-state timers, and pneumatic timers. *Electromechanical timers* have a motor and mechanical timing elements.

Solid-state timers use resistor-capacitor (*RC*) networks, line frequency counters, and internal oscillators for timing. Some solid-state timers use the versatile 555 IC timer discussed in Chapter 11. *Pneumatic timers* measure time by admitting air into an evacuated chamber through an adjustable orifice. When the chamber reaches a certain pressure, it triggers the timer's output.

Outputs for EM and pneumatic timers are mechanical contact devices, such as relays or miniature switches. Solid-state timers have mechanical contacts as well as solid-state outputs, such as triac or transistor output signals.

Pneumatic timers typically are used in short time range applications when precise accuracy is not needed. In addition, they do not require electric power for the timing cycle, a useful feature in many applications.

Solid-state timers offer extreme accuracy and long time ranges. They reset about ten times faster than EM devices, and they are available in small sizes if space is critical. Their cost has been reduced recently because of the introduction of integrated circuits. Today, some solid-state timers do not cost any more than EM timers. Solid-state timers are often used in applications that demand short or accurate time delays.

EM timers are still popular for applications that do not need extremely fast duty cycles or reset actions. In many instances, users prefer to use EM timers for very long timing ranges, such as from 30–50 hours.

Reset Functions. Timers can also be grouped by function as well as by the timing method. Two common functional classifications are repeat cycle and reset. The reset classification can be further divided into interval and delay timers.

Repeat-cycle devices time two or more periods and repeat the pattern endlessly. These timers are used for many nonindustrial applications, such as traffic signals, airport runway lights, and warning flashers.

Reset timers are started by an external signal, such as a manual-start push button or a control input, and are timed for a single period. A reset timer can be in one of three states: reset, timing, or timed-out. When in the *reset state*, the timer does not perform any timing function. The *timing state* starts when an external signal is detected. The *timed-out state* is the period between the end of the timing state and the point when the timer returns to the reset state. A timer can return to reset when an external signal, such as a manual-reset push button or a control input, is detected. Some timers automatically reset after the timing state. They are in the timed-out state for only a fraction of a second.

A reset timer can be further grouped into two or more divisions, depending on when the timer's output occurs. In *interval reset timers*, the control output changes during the timing state. The interval timer applies power to or removes power from a component or circuit for a certain specified amount of time. In *delay reset timers*, the control output changes at the end of the timing state. As its name implies, the delay timer removes power from or applies power to a component or circuit after a preset time interval has elapsed. Table 6.1 shows how the timer's output changes as the timer moves through reset, timing, and timed-out states.

The delay timer is the more common device. Delay timers can be classified into two basic groups: on-delay and off-delay. The *on-delay timer* is shown in Figure 6.21A. When the coil is energized, the timing mechanism causes the time-delay relay TDR_1 contact to close after a preset period of time. The red (r) lamp, P_1, will turn on when the TDR_1 closes. The timing function of TDR_1 contacts is called on-delay, timed-closed. Also, when the coil is energized, the TDR_2 contact opens after a set time delay. The green (g) lamp, P_2, will stay on after the coil is energized for an amount of time equal to the time delay. This set of contacts is called on-delay, timed-open.

De-energizing the TDR coil in Figure 6.21A will immediately open the TDR_1 contacts and close the TDR_2 contacts.

Note the schematic symbols for the time-delay relay contacts shown in Figure 6.21. The symbol shown in Figure 6.21A for the TDR_1 contact is called a normally open, timed-closed contact. The arrow indicates the direction of the time delay—in this case, an on-delay. The symbol shown in Figure 6.21A for the TDR_2 contact is called a normally closed, timed-open contact.

The diagram in Figure 6.21B illustrates an *off-delay timer*. In this timer, when the coil is energized, the TDR_1 contact is closed immediately. It remains on as long as the coil remains energized. The red lamp P_1 will also turn on when the coil is energized since power is applied through the TDR_1 contacts. The TDR_2 contacts, on the other hand, will open and remain open as long as the coil is energized. This action turns off green lamp P_2. When the coil is de-energized, the delay mechanism will cause the TDR_1 contacts to open after a time delay. The TDR_1 contacts in this relay are called off-delay, timed-open. Following the same time interval after the coil loses power, the TDR_2 contacts, which were open, will close. These contacts are called off-delay, timed-closed.

Again, note the schematic symbols for the TDR contacts in Figure 6.21B. The symbol shown in Figure 6.21B for the TDR_1 contact is called a normally open, timed-open contact. The symbol shown in Figure 6.21B for the TDR_2 contact is called a normally closed, timed-closed contact. The direction of the arrow indicates that the contacts are off-delay.

TABLE 6.1   Timing States of Interval and Delay Reset Timers

Type	Mode	Timer Output		
		Reset	Timing	Timed-out
Interval reset timers	On while timing	Off	On	Off
	Off while timing	On	Off	On
Delay reset timers	On in timed-out state	Off	Off	On
	Off in timed-out state	On	Off	Off

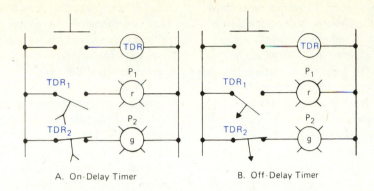

A. On-Delay Timer   B. Off-Delay Timer

FIGURE 6.21 Delay Timers

Setting the Timer. Several methods are available for setting timers, and each has characteristics that are useful for specific applications.

Analog setting methods use a knob and reference dial. Setting accuracy depends upon the design of the knob and the accuracy of the reference marks. Accuracy also may depend upon the operator's ability to set a time between the reference marks. Parallax may reduce setting accuracy if the operator is not directly in front of the unit. (Recall that *parallax* is the apparent change in position of an object, such as a pointer, resulting from the position from which it is viewed.)

Thumbwheel or *push-button switches* are the most precise method for setting the timer. Parallax problems do not exist with thumbwheel switches, and estimating a setting is not necessary. However, setting accuracy is limited to the least significant digit of the thumbwheel.

Accuracy and Reset Time. In most cases, the *accuracy* of a timer far exceeds its ability to be set. In other words, the accuracy of a timer is more dependent upon the operator's ability to set the time period than upon the timer itself.

For timers, repeat accuracy is more important than overall accuracy. *Repeat accuracy* refers to the timer's ability to produce repeated results within tolerance limits. A bottling machine, for example, requires repeated accuracy sufficient to hold the fill quantity between shortage and overfill limits.

Reset time refers to the time required for a timer to move from the timing or timed-out state to the reset state. Electromechanical parts react very differently in EM and solid-state timers. Typical reset times are 0.1 to 0.7 s for an EM timer, and 0.02 to 0.08 s for a solid-state timer.

Multiple-Circuit Timers. Many applications require that loads be turned on and off at various points during a timing cycle. One method for solving this problem is to use individual timers, each connected to a load, and start all of them at the same time. Another way is to use a multiple-circuit timer. *Multiple-circuit timers* allow up to 12 more loads to be turned on and off in a timing period. These timers come in both reset and repeat-cycle models, with a variety of setting methods.

In most cases, the timers allow the user to select different output sequences. Each channel begins timing at time zero, but it can have different on and off sequences than the other channels. Multiple-circuit timers are very expensive and are difficult to adjust. As a result, most of these timers have been replaced by the programmable controller, discussed in Chapter 14.

Overload Relays

A mechanical motor load imposes a certain torque demand on a motor. The motor tries to meet that demand by converting electric power to mechanical energy. If the load demands more torque than the motor is designed to supply, the motor will draw above its rated current. A motor is considered to be *overloaded* when it draws too much current. You will recall from an earlier chapter that a large amount of current may be drawn by a motor at starting, due to the low cemf. Although this large starting current is permissible for a short time at starting, if it continues for any length of time, the motor will overheat. Overheating damages the insulation on windings and interferes with the normal lubrication process. Generally, if a motor draws more than its full-load rated current, it is being overloaded. Actually, the term *overload* is almost a misnomer. The motor is not protected from an overload; rather it is protected from the high motor temperatures that result from an overload.

A means must be designed to monitor the motor and protect from excessive loads. The NEC specifically directs designers "to protect the motors, the motor control apparatus, and the branch-circuit conductors against excessive heating due to motor overloads or failure to start." Section 430 of the NEC details the requirements for motor circuit protection. One device used to give a motor protection from intermittent or sustained overloads is called an *overload relay*. Overload relays are divided into two classifications: thermal and magnetic.

Magnetic Overload Relays. The magnetic overload relay was one of the original methods of providing overload protection. It is still used, although it is not as popular as some of the thermal overload relays. It is used presently as a protection method for large DC motors where commutation, rather than heating, is the problem. The *magnetic overload relay* has a movable magnetic core located inside a coil. Through the coil passes all or a part of the motor current. As current flows in the coil, the core is pulled upward. When the core moves up far enough, it trips a set of contacts at the top of the relay. The contacts open, removing power to the motor. The movement of the core is controlled by a piston inside an oil-filled dash-pot. The piston-dashpot arrangement is similar to a shock absorber in an automobile. The piston is slowed down by the resistance of the oil in the dash-pot, producing a time delay. The time delay is

adjusted by increasing the size of the holes in the piston. The tripping current is adjusted by moving the core on a threaded rod. Since both time-delay and tripping currents can be adjusted, this relay is sometimes used to protect motors with long acceleration times or unusual duty cycles.

The magnetic overload relay, as we have seen, responds to the amount and duration of motor current. It does not measure the temperature of the motor directly. The magnetic relay does not respond to overloads lasting for very short time intervals. If too many overloads occurred, the motor windings could get dangerously hot; the magnetic relay would never trip. We would then need a method to detect motor temperatures directly. Sometimes, temperature is detected with thermistors embedded into the motor case. Thermistors with a positive temperature coefficient are used because of their extreme nonlinearity. However, these circuits are generally expensive and restricted to certain appropriate applications.

Thermal Overload Relays. A more economical device for general use in detecting motor temperatures is the externally mounted *thermal overload relay*. This device does not need to be in contact with the motor. It simulates the heating conditions that are occurring in the motor. It is adjusted so that the relay trips before the motor reaches a dangerous temperature. This thermal relay has two basic parts. First, it has a resistive heating element that is heated when motor current passes through it. Second, it has a sensor switch element that detects the temperature and opens the switch when the temperature is too high.

Many types of sensor switch elements have been used in motor overload relays. The two that are used today are the bimetal strip and the eutectic alloy. The *bimetal strip*, explained in Chapter 8, deflects when the heater increases its temperature. A certain deflection corresponding to a preset temperature opens a set of contacts, removing power from the motor. The *eutectic alloy* melts when it reaches a precise temperature. The alloy (similar to solder) releases a wheel when it melts. The turning wheel actuates a switch, which removes power from the motor.

Bimetal strips sometimes detect temperature changes directly on the motor case. An example of this kind of device is shown in Figure 6.22. Note that

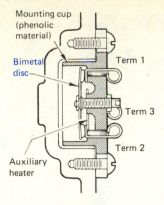

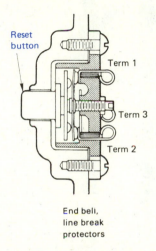

FIGURE 6.22 Bimetal Strip Overload Relay

the protector senses the temperature at the end bell. The bimetal strip, when heated to a certain temperature, will snap open, removing power from terminals 1 and 2.

For most motors, especially small motors, the simply types of protectors are economically justified. Thermal and magnetic overload devices protect the motor from damage due to excessive temperatures. In larger motors, a current transformer senses the excessive current and turns on an electromechanical relay to remove power from the motor.

Large motors and some small specialty motors require more protection than overload and short-circuit protection can provide. For example, ground faults normally are not detected by fuses. A *ground fault* is a current imbalance between the hot and neutral leads of a line supplying power. It usually occurs when a current-carrying conductor accidentally touches a grounded conducting material. Depending on the nature of the materials, the current can be small or large. In motors, the ground fault is usually caused by the breakdown of insulation on the motor windings. Among other things, ground faults can damage a motor's stator laminations and can also be a safety hazard. Ground faults are detected and dealt with by devices called *ground fault interrupters* (GFIs). The GFI detects the current imbalance and removes power from the circuit. You should be aware that the GFI does not prevent electrical shocks; it provides protection only for an adjustable amount of current. The GFI also does not offer any protection from overloads because this protection is provided by the overload relay. Furthermore, it does not offer any protection to the person who comes in contact with two power lines in a polyphase system.

Another form of protection needed in three-phase circuits is phase-loss protection. If a three-phase motor loses a phase, the motor may still run (depending on the load), but the motor will definitely overheat, causing damage. *Phase-loss detection circuits* are available to sense the loss of a phase and to de-energize the circuit if that loss occurs. The same problem of overheating occurs when one of the three-phase voltages increases or decreases. An increased or decreased phase voltage causes a phase imbalance, which also can be detected by appropriate circuitry.

Most of these functions have been accomplished in the past by EMRs with appropriate sensing circuits attached to detect the problems. Providing most of these forms of protection using electromechanical hardware is not as feasible today as it once was. Providing all of these functions we have mentioned (and some we have not) would be expensive and would require a large enclosure. It is more cost- and space-effective today to use a single *solid-state motor protector*, such as the one shown in Figure 6.23 manufactured by Sprecher & Schuh, Inc. This single solid-state unit provides protection against phase loss, thermal overload, ground fault, motor stall, phase reversal, and underload. The cost of this unit, when compared with the cost of its electromechanical counterparts, is considerably less. In addition to being less expensive, the solid-state motor protector is easier

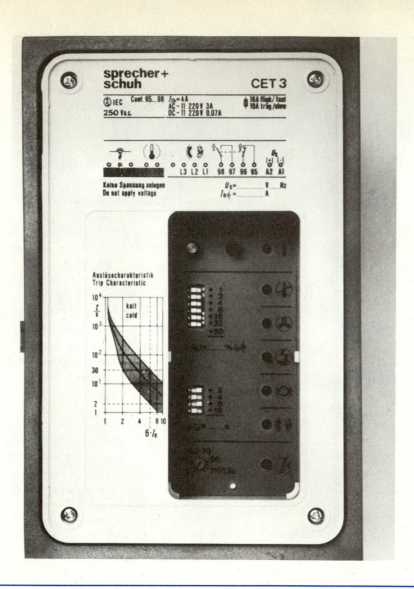

FIGURE 6.23 Solid-State Motor Protector

and less time-consuming to install. Most of these units are extremely flexible in allowing the operator to select the points at which power will be removed from the system. They are also very rugged and allow installation in environments hostile to EMRs.

EMR Load Types. One of the most common mistakes technicians make in using relays is to assume that a relay contact can switch its rated current regardless of what type of load it sees. High inrush currents and high induced counter emfs can erode or even weld contacts to the point where the life of the contacts is reduced greatly. So that you will better understand the types of loads a relay can have, a brief discussion follows on each type of load.

Incandescent Lamps: The cold resistance of a tungsten filament lamp is extremely low, resulting in inrush currents as much as 15 times the steady-state current. Such high inrush currents can cause contacts to erode rapidly or even weld.

Capacitive Load: The charging current to a capacitive circuit can be very high. The capacitor at first acts as a short circuit, and the current is limited only by the circuit resistance. Sometimes, the technician may not be aware that the load is capacitive. Technicians should remember that long transmission lines, filters, power supplies, and the like, are highly capacitive.

Motors: A motor load draws high inrush current because, at standstill, the motor input impedance is very low. As current is drawn, a torque is developed owing to the interaction of the current and the magnetic field. When the motor starts to rotate, it develops an internal counter emf that tends to reduce the current. Depending on the mechanical load, the starting time may be very short or very long, and the inrush current will continue for the duration of the starting time. When the motor is de-energized, a high-voltage inductive kick will be seen by the contacts. This inductive kick tends to produce an arc that can wear away the contact.

Inductors, Solenoids, and Contactor Coils: These loads are highly inductive and do not usually produce high inrush currents. In fact, they may actually limit the rise of inrush current because of their inductance. When highly inductive loads are turned off, however, the magnetic field in the coil collapses. The stored energy in this magnetic field must be dissipated across the open contacts, which will cause arcing.

DC Loads: DC loads are more difficult to turn off than comparable AC loads because the voltage in a DC load never passes through zero. As the contacts open, an arc is struck and may be sustained by the applied voltage.

Contact Materials. When relay contacts are made or are broken during switching, many chemical and metallurgical effects occur. These effects may result in contact wear, poor contact conductivity, excess heating, and welding. Different contact materials have different characteristics and are thus used with different types of loads. We will examine some of the more common contact materials in the following paragraphs.

Silver is one of the most popular contact materials. It has high electrical and thermal conductivity and is relatively low in cost for a precious metal. It is easily formed and resists oxidation and tarnishing. Silver contacts are used for currents in the light to medium range (1–10 A). It is not suitable for switching voltages below 6 V.

Silver cadmium oxide is a very common contact material for currents in the medium to high range (5–25 A). Its thermal and electrical conductivities are almost as good as those for silver, but it is more resistant to erosion and welding. It is not recommended for switching voltages below 12 V.

Gold contacts have been used extensively in communications systems because they work well in low-energy, dry switching. This characteristic is due to gold's inert nature and resistance to the formation of a surface film. Because gold is soft, it sometimes tends to stick or cold-weld. It is not recommended for currents above 1 A.

The metal paladium, in combination with gold, has been used effectively in low-energy circuits because it does not form surface films and does not cold-weld.

Rhodium has excellent resistance to corrosion and oxidation. It will not cold-weld, and its contact resistance stays the same over long periods of time. It is used most often in reed relays and in contacts with a current capability between 10 mA and 1.5 A.

Mercury is probably one of the most interesting of the contact materials. It has a stable contact resistance, does not erode, and does not have contact bounce. All these characteristics are due to the fact that mercury is a liquid, not a solid. Every time the relay contacts, which are covered by mercury, open, the contacts are renewed by fresh mercury. Mercury-wetted contacts are available in ratings between 60 mA and 100 A.

Tungsten has high resistance to mechanical wear, electrical erosion, and welding. Its electrical conductivity is low, and contact resistance is high because it has a tendency to form thick oxide-film coatings. Because of this high contact resistance, the highest current practical for tungsten is less than 10 A. Since tungsten can handle high inrush currents of short duration, it is used in contacts of motor starters and incandescent lamp switches. It should not be used for switching voltages below 24 V.

SEMICONDUCTOR ELECTRONIC SWITCHES: THYRISTORS

While the EMR has advantages in the numbers of circuits controlled, the *electronic switch* is faster, cheaper, and more energy-efficient, and it lasts longer. The electronic switch does not exhibit contact bounce, a major problem in digital circuits. Also, the electronic switch is not acceleration-sensitive or vibration-sensitive, works better in hostile environments, and does not arc or spark. It is, therefore, explosionproof. Because of these advantages, electronic switches, or thyristors, have replaced many EMRs and thyratron tubes in industry. In addition, thyristors are used in rectifiers, alarms, motor controls, heating controls, and lighting controls. Thyristors are used wherever large amounts of power are being controlled.

A *thyristor* is a four-layer, PNPN device. This PN sandwich, then, has a total of three junctions. As we will see later, this device has two stable switching states: the on state (conducting) and the off state (not conducting). There is no linear area in between these two states, as there is in the transistor. That is, a switch is either on or off, never (we hope) in between.

Silicon-Controlled Rectifier (SCR)

The most common thyristor you will encounter as a technician is the *silicon-controlled rectifier* (SCR). It is widely used because it can handle higher values of current and voltage than any other type of thyris-

tor. Presently, SCRs can control currents greater than 1500 A and voltages greater than 2000 V.

The schematic symbol for the SCR is shown in Figure 6.24A. Notice that the schematic symbol is very much like that of the diode. In fact, the SCR resembles the diode electrically since it conducts only in one direction. In other words, the SCR must be forward-biased from anode to cathode for current conduction. It is unlike the diode because of the presence of a gate (G) lead, which is used to turn the device on.

The SCR switching operation is best understood by visualizing its PNPN construction, as shown in Figure 6.24B. If we slice the middle PN junction diagonally, as shown in Figure 6.25, we end up with an NPN transistor connected back to back with a PNP transistor. A forward-bias voltage between the gate-to-cathode lead turns on transistor Q_1. Transistor Q_1 conducts, turning on Q_2. Transistor Q_2 drives Q_1 further into conduction. This regenerative process,

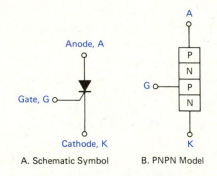

FIGURE 6.24 Silicon-Controlled Rectifier (SCR)

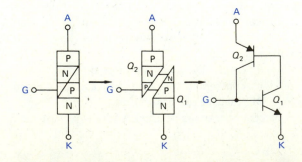

FIGURE 6.25 SCR Two-Transistor Analogy

shown in Figure 6.26, continues until both transistors are driven into saturation. Only microseconds are needed to complete the process. This regenerative action is called *latching* because the transistors continue to conduct even if the gate-to-cathode, forward-bias voltage is removed.

Why does the SCR handle power efficiently? When it is in the off state, it does not draw much current. When it is on, the voltage across it is around 1 V, regardless of the current through the device. Recall the power equation:

$$P = IV$$

where

    P = power dissipated by device
    I = current through device
    V = voltage across device

If the voltage remains at about 1 V, the power is approximately equal to the size of the current flow. For instance, if the voltage drop across the device stays at 1 V with 1000 A flowing through it, the power dissipated is only 1000 W. All the rest of the power is transferred to the load.

Turning on the SCR. The gate-to-cathode voltage required to turn on the SCR is referred to as the *gate turnon voltage* (V_{GT}). The amount of gate-to-cathode current needed to turn the device on is called *gate turnon current* (I_{GT}). Voltage V_{GT} ranges from 1 to 3 V, and I_{GT} ranges from 1 to 150 mA. The SCR can be turned on by methods other than the gate voltage. However, with the exception of the radiation method of triggering, little use is made of other triggering methods.

Refer to Figure 6.27, which shows the current-voltage behavior of an SCR with no gate current flowing and the gate grounded. When the SCR is forward-biased (positive voltage on the anode with respect to the cathode), very little current flows. This region is called the *forward blocking region*. However, if the *forward voltage* (V_{AKF}) is high enough, minority carriers will be accelerated to high velocities and create carriers in the gate-cathode junction. When enough carriers are generated, the device breaks into conduction, just as if a gate current were drawn. The voltage at which the device breaks over is represented by $V_{(BO)}$ and is called the *breakover voltage*.

Regardless of how the current flow starts, once it begins, it keeps on flowing. However, the diagram in Figure 6.27 also gives a clue as to how to shut off the SCR. Note the part of the curve labeled I_H. If anode current falls below this current, called the *holding current*, the device goes back into the off state.

Note that, in the reverse direction, the SCR curve looks like the curve for a normal PN diode. Very little reverse current flows when the SCR is reverse-biased. This region is called the *reverse blocking region*. When the reverse breakdown voltage $V_{(BR)R}$ is reached, the device will quickly go into avalanche conduction, which usually destroys the device.

Although the most common method of turning on an SCR is gate voltage control, there are other ways to turn it on. SCRs are often turned on by

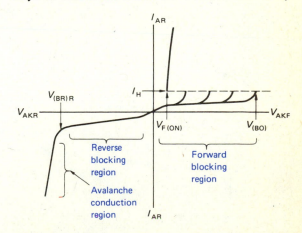

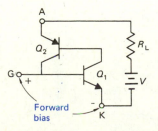

FIGURE 6.26 SCR Two-Transistor Analogy with Regenerative Feedback

FIGURE 6.27 SCR Current-Voltage Characteristic Curve

radiation, particularly light. When radiation falls on the middle junction, it creates electron-hole pairs. If enough pairs are created, the device breaks into conduction. Such a device is called a *light-activated SCR* (LASCR).

The LASCR is similar in construction to the SCR, except that it has a window to let in light. The construction of the LASCR is shown in Figure 6.28A. The schematic symbol for the LASCR is shown in Figure 6.28B. Light is directed through the lens to fall on the silicon pellet. When light of great enough intensity falls on the middle PN junction, the device conducts. The LASCR is often used in industry to detect the presence of opaque objects. It is also used in cases where electrical isolation between circuits is needed.

There are two remaining methods by which SCRs are turned on. The first is called the *rate effect turnon method*, which operates in the following manner. Recall that every transistor has PN junctions, which act as capacitors. If a quickly rising voltage is applied between the anode and gate of an SCR, charge current flows through the device. If this charge current is high enough, it can trigger the SCR into conduction. Most specification sheets specify this parameter in terms of volts per microsecond. It is called the *critical rate of rise*, or *dV/dt*. For example, the 2N1595 SCR has a critical rate of rise of 20 V/μs. In other words, if the anode voltage rises any faster than 20 V in 1 μs, the device turns on. The critical rate of rise also limits the maximum line frequency that can be connected to the device.

Engineers usually try to prevent rate effect turnon because the device turns on when they do not want it to. Firing by exceeding the critical rate of voltage rise is prevented by the addition of an *RC snubber circuit*, shown in Figure 6.29. The *RC* network delays the rate of the anode voltage rise.

Finally, the SCR can be turned on by temperature. As junction temperature increases, minority carrier current flow (leakage) increases. In fact, it doubles every 14°F (8°C). If the temperature rises high enough, this increased current is enough to turn on the SCR. This thermal effect is usually prevented by using a heat sink.

Another problem with SCRs and temperature is the temperature dependence of some of the SCR parameters. Specifically, V_{GT}, I_{GT}, and $V_{(BO)}$ all decrease with increasing temperature. This decrease will cause a change in firing angle with temperature fluctuations in circuits where the firing angle depends on these parameters.

Motorola, the semiconductor manufacturer, has made available a new type of SCR, the MOS SCR. The schematic is shown in Figure 6.30. This device has the advantage of the high input impedance and fast turnon of the MOS family of devices with the switchlike action of an SCR. Because of the high input impedance and low gate current drain, the MOS SCR can be driven by standard CMOS and TTL logic. Figure 6.30 shows the MOS SCR directly controlled by a logic gate. Note the complete isolation between the 120 V AC line and the logic gate.

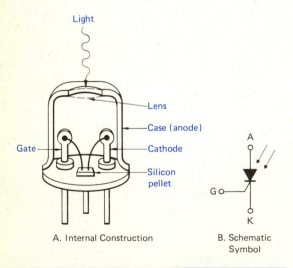

A. Internal Construction

B. Schematic Symbol

FIGURE 6.28 Light-Activated SCR (LASCR)

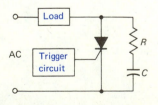

FIGURE 6.29 SCR Circuit with Snubber to Decrease *dV/dt*

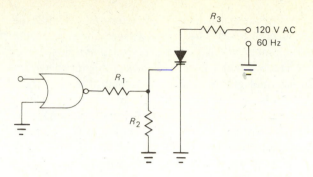

FIGURE 6.30 TTL Driving MOS SCR

Another MOS device, called the *MOS-controlled thyristor* (MCT), has been developed by General Electric. Not only does this device have the high input impedance of the MOS devices, but it can be turned off easily with a gate pulse in a DC circuit. Typical devices require a 15 V gate signal to turn off an MCT with a current density of 1000 A/cm^2.

Heat-Sinking the SCR. The reliability and behavior of electronic equipment often depend on how well the semiconductor devices within are cooled. One manufacturer estimates that the life of a piece of electronic equipment is cut in half for every 10°C rise in temperature. Although this estimate is possibly overstated, it is true, nevertheless, that a semiconductor's life and reliability are affected by its operating temperature. Also, the efficiency of a device is inversely proportional to the efficiency of the circuit. With the same input power, an increase in temperature means that less power is being delivered to the load and more wasted on generating heat. This wasted power will lower efficiency. For these two reasons, excessive heat should be removed from semiconductor devices.

Most SCRs are devices that handle large amounts of power. Consequently, the case of the SCR is often not large enough to dissipate the heat generated by the thyristor. When the case of the semiconductor is not effective in carrying away the heat from the junction, a heat sink is used. A *heat sink* is a device that, when attached to the semiconductor, will aid the device in getting rid of the heat it generates.

The maximum power that the thyristor can dissipate safely is a function of (1) the temperature at its junction, (2) the ambient (air) temperature, and (3) the size of its thermal resistance (R_θ). *Thermal resistance* is defined as the quotient of the temperature drop between two points and the power passing between them. The temperature drop usually is measured between the point where heat is being generated and some other point of reference. In the semiconductor, the heat usually is generated at the junction. The reference points are the case, the heat sink, and the ambient temperature. Thermal resistance is, then, a measurement of how well the device will conduct heat away from the junction. Its units of measure are degrees Celsius per watt (°C/W).

Power semiconductors and heat sinks are rated thermally by a statement of their thermal resistance. We will need to find out the correct size of heat sink to cool a semiconductor adequately for a particular application. Although we will be discussing the application of the heat sink with reference to the SCR, these principles apply to the cooling of any semiconductor device—for example, transistors, power MOSFETs, and voltage regulators.

A diagram of a semiconductor with a heat sink and three thermal resistances is shown in Figure 6.31A. These resistances are considered to be in series, like electric resistors. In fact, the behavior of the thermal resistances may be understood best by drawing the thermal circuit as an electric one, as shown in Figure 6.31B.

1. *Junction-to-Case Thermal Resistance*: The first thermal resistance is between the junction and the case, symbolized by $R_{\theta JC}$. The heat created by the device starts its journey to the outside world at the junction. This thermal resistance is a function of the manufacture and design of the semiconductor itself and usually is reported in the manufacturer's data sheet. It cannot be changed or influenced by any factor, including heat sinks. Since some semiconductors come in different packages, obtaining a package with a higher or lower thermal resistance may be possible. Table 6.2 shows some typical thermal resistances for selected semiconductor packages.

2. *Case-to-Sink Thermal Resistance*: The next thermal resistance the heat generated at the junction encounters is between the case and the heat sink. This

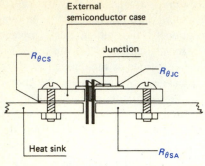

A. Semiconductor with Heat Sink

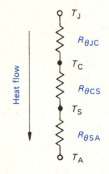

B. Thermal Resistance Equivalent Circuit

FIGURE 6.31 Semiconductor with Heat Sink

TABLE 6.2 Thermal Resistances for Selected Semiconductor Packages

Package	$R_{\theta JC}$ (°C/W)
TO-3	0.5–6.0
TO-3 (typical 2N3055)	1.5
TO-66	1.5–15
TO-66 (typical 2N3054)	7.0
TO-220	1.7–5
TO-220 (typical D44C8)	4.2
TO-92	175–200
TO-92 (typical 2N2222)	175
TO-5 (metal header)	20–40
TO-5 (typical 2N1595)	25
TO-5 (glass header)	30–50
TO-5 (typical 2N2322)	32

thermal resistance is symbolized by $R_{\theta CS}$. Unlike the junction-to-case thermal resistance, the case-to-sink thermal resistance can be controlled to some degree. The two major factors that affect this resistance are thermal grease and the presence of an electrical insulator. The thermal resistance across any two metals is a function of (1) the cross-sectional area between the two, (2) the condition of the surfaces, (3) how flat the surfaces are, (4) how much force is pressing the surfaces together, and (5) the thermal conductivity of the medium between the two metals.

For a reduction of the thermal resistance between the case and sink (thus allowing heat to pass better), four factors must be kept in mind. First, keep the surfaces—in this instance, the sink and semiconductor—as smooth and flat as possible. Second, keep the surface areas between the two as high as possible. Third, use thermal joint compound to keep the thermal conductivity of the medium between the surfaces as great as possible. *Thermal grease*, usually made of a silicon compound, fills the small voids that occur between surfaces that are not perfectly smooth. The thermal grease has a higher thermal conductivity (lower thermal resistance) than the air that would normally be filling the spaces between the metals. Fourth, where bolts, screws, or studs are used for mounting the sink, apply tightening torques according to the manufacturer's recommendations. Increased tightness beyond the recommended torque will increase thermal conductivity. Do not overtighten screws or bolts; stress limits may be exceeded and cause damage.

Sometimes, electrical insulators are needed between the semiconductors and heat sinks. These insulators are necessary since collector or anode voltages often are present on the case. Bringing the case into electrical contact with the heat sink may cause a short circuit since the sink is usually at ground potential. An appropriate insulator should be chosen—that is, one with the correct insulating properties and the lowest thermal resistance. Table 6.3 shows the thermal resistances of certain selected insulators used with three semiconductor packages. Note that the thermal resistances given are based on the use of thermal grease and the proper tightening of screws and bolts.

TABLE 6.3 Thermal Resistances of Insulators

Insulator	Package	R_θ (°C/W)
Beryllium oxide	TO-3	0.22
	TO-220	1.4
Mica	TO-3	0.8
	TO-220	5.2
Plastic	TO-3	0.8
	TO-220	5.2
Silicone rubber	TO-3	1.2
	TO-220	7.9
Bare package	TO-3	0.15–0.25
	TO-220	0.55–0.65

Note: All measurements are taken with thermal grease in use, except for silicone rubber.

The method most overlooked by technicians in reducing case-to-sink thermal resistance is the proper torquing of mechanical hardware. Exerting proper torque is the least expensive and most efficient method to reduce this thermal resistance. The thermal resistance of a TO-3 package may range from a low of 0.1°C/W with no insulator or thermal grease to 2°C/W with an insulator mounted dry. A semiconductor dissipating 15 W with a TO-3 case is 29°C hotter in the latter case (insulator mounted dry), a considerable temperature difference.

3. *Sink-to-Ambient Thermal Resistance*: The last thermal resistance in our heat transfer example is the resistance from the heat sink surface to the ambient (usually air). This resistance is symbolized by the designation $R_{\theta SA}$. It is probably the most important thermal resistance of the three in keeping the semiconductor junction cool. The smaller this thermal resistance, the cooler the device will be under steady-state conditions. Put another way, the lower $R_{\theta SA}$ is, the more power the device can dissipate without exceeding maximum junction temperatures. This thermal resistance may be expressed as a function of several parameters. A discussion of each follows.

First, the sink-to-ambient thermal resistance is a function of the convection heat transfer coefficient. This parameter is affected by the amount of air moving past the surface of the sink. If the air is still

at the surface, this condition is called *natural convection*. More heat can be carried away if air is forced to move past the surface. The amount and type of convection change the heat transfer coefficient. Generally, the larger the amount of airflow, the larger the heat transfer coefficient is and the lower the thermal resistance is.

Another factor that affects the sink-to-air thermal resistance is emissivity. You know from your studies in physics that a blackbody absorbs and gives off heat better than any other substance. For this reason, heat sinks are usually made with substances that have good emissivity. Some substances, such as lacquer and oil paint, approximate the blackbody standard and improve the heat transfer coefficient. The factor called *emissivity*, usually symbolized by the Greek letter epsilon (ε), is the ratio of the emissive power of a body compared with the emissive power of a standard blackbody. The emissivity of a standard blackbody is equal to 1. Materials that do not emit as much heat have an emissivity less than 1. So emissivity is a ratio, usually given in decimal form, between 0 and 1. For example, lacquer of any color has an emissivity of 0.8 to 0.95, and oil paint of any color has an emissivity of 0.92 to 0.96. You should be aware that the term *blackbody* has little to do with the optical color of a material. Bodies of any color can have high emissivities and be referred to as blackbodies. For example, anodized aluminum has an emissivity of 0.8, regardless of its optical color. Emissivity is affected more by the type of surface. A matte or dull surface will have a higher emissivity than a shiny, specular surface.

Another factor that determines the sink-to-air thermal resistance is the area of the sink. The heat given off by the sink is directly proportional to the amount of heat sink area. The larger the amount of sink area, the more effective it is in dissipating heat. Greater surface area will lower the sink-to-ambient thermal resistance. In most applications involving natural convection, the only way to decrease the sink-to-ambient thermal resistance is to increase the surface area of the heat sink.

Looking at the thermal system shown in Figure 6.31B, we can see that as the heat flows from the junction, it encounters the three resistances just discussed: the junction-to-case thermal resistance, $R_{\theta JC}$; the case-to-sink thermal resistance, $R_{\theta CS}$; and the sink-to-ambient thermal resistance, $R_{\theta SA}$. As shown in the diagram, these resistances are considered to be in series. We can find the total thermal resistance between the junction area and the ambient condition by adding these resistances:

$$R_{\theta JA} = R_{\theta JC} + R_{\theta CS} + R_{\theta SA} \qquad (6.3)$$

The following basic equation describes the heat transfer in this system:

$$P_D = \frac{T_J - T_A}{R_{\theta JC} + R_{\theta CS} + R_{\theta SA}} \qquad (6.4)$$

where

P_D = power dissipated by semiconductor (in watts)
T_J = junction temperature (in degrees Celsius)
T_A = ambient temperature (in degrees Celsius)
$R_{\theta JC}$ = junction-to-case thermal resistance
$R_{\theta CS}$ = case-to-sink thermal resistance
$R_{\theta SA}$ = sink-to-ambient thermal resistance

We can use this equation to solve for the maximum power a junction can be called upon to dissipate safely without exceeding a specified junction temperature. Let us consider an example at this point. Let us assume that we are using an SCR that is in a TO-3 package. In the data sheet for this device, we find it has an $R_{\theta JC}$ of 1.5°C/W. From the heat sink manufacturer's data, we find an $R_{\theta SA}$ of 2.5°C/W and an $R_{\theta CS}$ of 0.3°C/W. How much power will the SCR be able to dissipate and not exceed a junction temperature of 200°C if the ambient temperature is 50°C? Using the power dissipation equation (6.4), we can calculate the resultant maximum power dissipation as follows:

$$P_D = \frac{T_J - T_A}{R_{\theta JC} + R_{\theta CS} + R_{\theta SA}}$$

$$= \frac{200°C - 50°C}{1.5°C/W + 0.3°C/W + 2.5°C/W}$$

$$= 34.9 \text{ W}$$

In this system, the semiconductor will safely dissipate 34.9 W with a junction temperature of 200°C.

• EXAMPLE 6.2
Given a 2N1595 SCR in a TO-5 package, with $R_{\theta JC}$ = 25°C/W and a heat sink with $R_{\theta SA}$ = 2°C/W and $R_{\theta CS}$ = 0.5°C/W, find the power the SCR will dissipate if it does not exceed 125°C at an ambient temperature of 40°C.

Solution
Substituting into Equation 6.4, we get

$$P_D = \frac{T_J - T_A}{R_{\theta JC} + R_{\theta CS} + R_{\theta SA}}$$

$$= \frac{125°C - 40°C}{25°C/W + 0.5°C/W + 2°C/W}$$

$$= 3.09 \text{ W}$$

The system will dissipate 3.09 W of power.

Rearranging Equation 6.4, we can solve for the junction temperature, given the thermal resistances, ambient temperature, and the power dissipated:

$$T_J = (P_D)(R_{\theta JA}) + T_A \qquad (6.5)$$

where

$R_{\theta JA}$ = junction-to-ambient thermal resistance

Using the thermal resistances from the previous discussion and a semiconductor dissipating 20 W at an ambient temperature of 50°C, we calculate the junction temperature as

$$T_J = (20 \text{ W})(4.3°C/W) + 50°C$$

$$= 136°C$$

• EXAMPLE 6.3
Given a semiconductor dissipating 2 W at 50°C and the thermal resistances from Example 6.2, find the junction temperature of the semiconductor.

Solution

Substituting into Equation 6.5, we get

$$T_J = (P_D)(R_{\theta JA}) + T_A$$

$$= [(2\ W)(25°C/W + 0.5°C/W + 2°C/W)]$$

$$+ 50°C$$

$$= 105°C$$

We can also solve for the particular size of heat sink necessary for a specific application. Assume we have an application that requires a TO-220 package, which we are calling upon to dissipate 5 W. We wish to keep the junction temperature to a maximum of 150°C. In the data sheet for the TO-220, we find that it has an $R_{\theta JC}$ of 3.0°C/W. With the semiconductor mounted on an insulator and using thermal grease, we have an $R_{\theta CS}$ of 1.0°C/W. Solving the original equation for $R_{\theta SA}$, we get

$$R_{\theta SA} = \frac{T_J - T_A}{P_D} - (R_{\theta JC} + R_{\theta CS}) \qquad (6.6)$$

Substitution of the appropriate values yields

$$R_{\theta SA} = \frac{150°C - 50°C}{5\ W}$$

$$- (3.0°C/W + 1.0°C/W)$$

$$= 16°C/W$$

We need a heat sink with a thermal resistance of 16°C/W to meet this application.

Referring again to Figure 6.31A, we see that we cannot make an actual measurement of the junction temperature to check on whether we are exceeding the maximum junction temperature specifications. However, we can measure the case temperature with a temperature probe. Since case temperature is a more practical measurement, some manufacturers give maximum allowable case temperature as a function of the current drawn by the device. For the SCR, the information usually is in the form of a graph of maximum allowable case temperature plotted against the on-state anode current. The case temperature may be found from the following equation:

$$T_C = (P_D)(R_{\theta CA}) + T_A \qquad (6.7)$$

Let us calculate the case temperature, using the data from our example:

$$T_C = (P_D)(R_{\theta CA}) + T_A$$

$$= [(P_D)(R_{\theta CS} + R_{\theta SA})] + T_A$$

$$= [(5\ W)(1.0°C/W + 13.2°C/W)]$$

$$+ 50°C$$

$$= 121°C$$

The case temperature in this application is 121°C. This temperature can then be compared with the maximum allowable case temperature graph to see whether it is excessive.

• EXAMPLE 6.4

Given a semiconductor that dissipates 10 W and uses a TO-3 package with a thermal resistance of 2°C/W, find the size of heat sink needed if we wish to keep the maximum junction temperature below 125°C in an ambient temperature of 40°C, using a mica insulator with thermal grease.

Solution

Substituting the appropriate information into Equation 6.6, we get

$$R_{\theta SA} = \frac{T_J - T_A}{P_D} - (R_{\theta JC} + R_{\theta CS})$$

$$= \frac{125°C - 40°C}{10\ W}$$

$$- (2°C/W + 0.8°C/W)$$

$$= 5.7°C/W$$

Note: The 0.8°C/W value is taken from Table 6.2.

• **EXAMPLE 6.5**

Given a semiconductor dissipating 2 W at 50°C, and using the thermal resistances from Example 6.2, find the case temperature of the semiconductor.

Solution

Substituting into Equation 6.7, we get

$$T_C = (P_D)(R_{\theta CA}) + T_A$$

$$= [(2 \text{ W}) (0.8°C/W + 2°C/W)] + 50°C$$

$$= 56°C$$

In some instances, bolting a semiconductor to the chassis of a piece of equipment may provide an acceptable heat sink. Or if a commercial heat sink is not available, a sink may be manufactured from a flat piece of copper or aluminum. In either case, the thermal resistance must be known. The nomogram in Figure 6.32 gives an estimate of the thermal resistance of a flat piece of vertically or horizontally mounted copper or aluminum. Let us say we have a 25 in.² vertically mounted piece of copper ³⁄₁₆ in. thick. Using a straightedge, we draw a vertical straight line down from the 25 on the surface area

scale to find a thermal resistance of about 2°C/W on the scale for vertically mounted copper ³⁄₁₆ in. thick.

In the foregoing calculations, we have assumed a steady-state condition, with the SCR (or semiconductor) conducting all the time. As we will see in the next chapter, this assumption will not hold in certain kinds of power control circuits in AC applications. In most cases, however, the manufacturer will provide information in data sheets that will allow the estimation of the maximum case temperature under different dynamic conditions. Thermal-resistance calculations in those situations where the SCR is on for only a portion of the AC cycle are difficult and complex and are outside the scope of this text.

Installation, Mounting, and Cooling Considerations. Before installing power semiconductors, technicians should understand why so much emphasis is placed on using proper mounting techniques. For maximum efficiency and reliability in the use of power semiconductors, the heat developed by power dissipation at the junction must be removed as fast as possible. As stated earlier, there are usually three distinct obstacles in the path of this heat transfer: (1) interface and material between the semiconductor element and the semiconductor device mounting

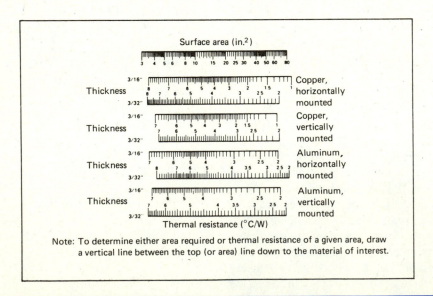

FIGURE 6.32 Heat Sink Nomogram

surface; (2) interface between the device mounting surface and the heat sink; and (3) transfer of heat through the heat sink to the ambient, which might be natural convection air, forced air, oil, or water. These obstacles are the thermal resistances previously discussed: (1) $R_{\theta JC}$, (2) $R_{\theta CS}$, and (3) $R_{\theta SA}$. Thermal transfer is depicted by using an electrical analog in Figure 6.31B.

The $R_{\theta JC}$ generally is a function of device construction and design and is determined by the semiconductor manufacturer. Except for disc-mount devices that require the user to apply the correct mounting force to ensure the proper $R_{\theta JC}$, the user has no control over $R_{\theta JC}$ once a device has been selected. The user's only option then is to select semiconductors that offer low junction-to-case thermal resistances. The other two thermal resistances, however, vary quite a lot. The user must carefully consider the various methods and techniques available to ensure long life and efficiency.

Primary considerations in obtaining a low $R_{\theta CS}$ are the degree of flatness and surface finish of the device and heat sink mating surfaces, the use of a thermal joint compound, and the proper application of force (torque). Here are two suggestions: (1) Use a torque wrench to ensure proper mounting force, and (2) use thermal joint compounds sparingly. Only a thin film of thermal joint compound is needed between the device and heat sink mounting interface. Most technicians apply too much compound. Remember when using insulating hardware that the $R_{\theta CS}$ will increase about tenfold.

The final obstacle in the heat removal path is the $R_{\theta SA}$. The semiconductor user can minimize this value by choosing the proper heat sink and most effective cooling method. A semiconductor can be no better than its heat sink. A wide range of heat sink materials, configurations, sizes, and finishes are available that will meet almost any space, cost, and heat dissipation requirement.

The most cost-effective and popular air-cooled heat sinks are made from aluminum extrusion. For higher output current at less cost, forced convection is the answer. Forced air can frequently allow the user to more than double the output current of a given device–heat sink assembly over natural convection rating capability. Why buy more semiconductors than

are really needed to do the job? A few fans and some baffling may provide a substantial savings in system design costs. Use forced air in place of natural convection air cooling when possible.

Water cooling is the most efficient type of cooling in general use today. So that the thermal efficiency of a water-cooled heat sink is optimized, pure copper heat sinks should be used. Copper alloys (bronze, aluminum, etc.) generally have much lower thermal conductivities and thus are not as efficient. The water-cooled assembly offers from 1.5 to 4 times more output current than the air-cooled assembly. It weighs only one-third as much, occupies about 40% of the space, and costs about the same. When heat sinks are compared, therefore, water cooling (if available) is usually the most economical choice; forced convection air is next; and natural convection air is last.

A final consideration when installing power semiconductors is to place them where the surrounding ambient temperature is as low as possible. Do not mount semiconductors near or above transformers or other heat-generating components in a cabinet, and avoid "dead air" pockets. Mount semiconductor heat sinks vertically to take advantage of the natural chimney effect or updraft of airflow. When using forced air or water to cool components in a cabinet, make sure the cooling medium reaches the power semiconductors first. Good installation practices when mounting and cooling power semiconductors will ensure good operating performance and long life.

Critical Rate of Current Rise. Some data sheets specify a parameter called *critical rate of current rise*. Critical rate of current rise is expressed in amperes per microsecond. For example, the critical rate of current rise of the 2N1595 is about 25 A/µs. If this parameter is exceeded, hot spots develop within the device, which may destroy it.

One method used to prevent current damage is shown in Figure 6.33. The inductor L in the anode lead opposes any change in current, thus slowing down the rise of anode current.

Gate Triggering for the SCR. Figure 6.34 shows a simple SCR circuit with a DC supply. Note that the SCR is forward-biased from anode to cathode.

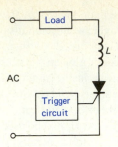

FIGURE 6.33 SCR Circuit with Inductor to Limit Current Rise

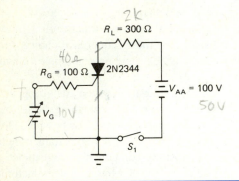

FIGURE 6.34 SCR in DC Circuit

Assume that gate voltage V_G is 0 V. If we close S_1, no current flows because of the absence of gate current I_G. What value of V_G will fire the SCR? Assume that gate turnon voltage V_{GT} is 0.6 V, and I_{GT} is 20 mA. Notice that the current I_G flows through R_G. If I_G flows through the 100 Ω resistor, it should drop about 2 V. Therefore, by Kirchhoff's law, V_G is

$$V_G = V_{GK} + V_{RG} = 0.6\ V + 2\ V$$
$$= 2.6\ V$$

where
V_{GK} = DC voltage from gate to cathode
V_{RG} = DC voltage drop across R_G

By this calculation, we find the gate voltage necessary to fire this device.

• EXAMPLE 6.6

Given a gate voltage clamped to 0.6 V, find the gate voltage necessary to draw a gate current of 50 mA. Assume a gate resistor of 100 Ω.

Solution

A gate current of 50 mA drawn through a gate resistance of 100 Ω will drop a voltage equal to

$$V_{RG} = I_G R_G \qquad (6.8)$$

which, after substitution, is

$$V_{RG} = (50\ mA)\ (100\ \Omega) = 5\ V$$

If we add the 0.6 V drop across the gate, we obtain a gate voltage of 5.6 V.

Also important is that the maximum gate power of the SCR not be exceeded. The maximum gate power that a device can withstand is normally found in the data sheet. The actual gate power dissipated can be calculated by multiplying the gate current drawn times the gate voltage. For Example 6.6, the gate power (P_G) can be found by using the following equation:

$$P_G = I_G V_G \qquad (6.9)$$

After substitution, we get

$$P_G = (50\ mA)\ (0.6\ V) = 30\ mW$$

Since the SCR has an anode-to-cathode voltage drop across it, it will dissipate some power. The amount of power the device will dissipate equals the product of current through it and the voltage drop across it. For example, an SCR that drops 1.8 V and draws 2 A of current will dissipate 3.6 W. As in the gate circuit, the maximum power dissipation capabilities of the SCR (and its heat sink, if one is used) must not be exceeded. In Example 6.2, we calculated the maximum power that an SCR with a given heat sink could dissipate; that power was 3.09 W. By knowing the anode-to-cathode voltage drop across the SCR and the maximum power the SCR can dissipate, we can calculate the maximum current that the SCR can safely handle. If we assume that the SCR drops 1.5 V, with a maximum power dissipation of 3.09 V, we can find the maximum current through the device by dividing the maximum power by the voltage V_T:

$$I_{max} = \frac{P_{max}}{V_T} \qquad (6.10)$$

or

$$I_{max} = \frac{3.09 \text{ W}}{1.5 \text{ V}} = 2.06 \text{ A}$$

• EXAMPLE 6.7

Given an SCR with a heat sink that can safely dissipate 30 W of power and $V_T = 2$ V, find the maximum average current the device can draw.

Solution

If we substitute the appropriate values into Equation 6.10, we get

$$I_{max} = \frac{P_{max}}{V_T} = \frac{30 \text{ W}}{2 \text{ V}} = 15 \text{ A}$$

The next question we might ask is, "Will the SCR stay on after the gate voltage is removed?" Anode current must rise to a value called the latching current for the SCR to remain on. The *latching current* is defined as the amount of anode current necessary to keep the device on after switching from the off state and after the trigger has been removed. The latching value is normally two to three times the holding current value. If we assume that the latching current is 300 mA, then the anode current must be more than this value if the SCR is to remain on.

Anode current (I_A) is calculated by assuming that the voltage drop across the SCR is zero (actually above 1 V). Thus, we assume that the entire supply voltage (V_{AA}) is dropped across the 300 Ω load (R_L). By Ohm's law, then, we have

$$I_A = \frac{V_{AA}}{R_L} = \frac{100 \text{ V}}{300 \text{ Ω}} = 333 \text{ mA}$$

So when the SCR fires, about 333 mA of current flows. Since this value exceeds the latching current value (300 mA), the device stays latched when the gate voltage is reduced to zero.

The gate trigger current and voltage in this example are assumed to be DC values. They may, however, be pulses with a certain pulse width. This pulse width must be long enough to allow the anode current to build up to the latching current value. The time it takes the anode current to build up to the latching current level is called *turnon time*.

In this example, we have also assumed that the load is resistive. Many times, the load is inductive, such as in a relay or a motor. This inductance may keep the SCR from firing. As we know, an inductor opposes any change in current. With pulse triggering, the anode current may be kept below the latching current for the duration of the pulse. One way to overcome this problem is to make the pulse width longer, which is not always possible. Another solution is to bypass the inductive load with a resistor. This solution allows the anode current to rise quickly to the latching value.

Suppose the load in Figure 6.34 is inductive instead of resistive. How large does the resistor need to be? We can assume that 100 V is applied across the load when the SCR fires. If we divide this voltage by the latching current value, we have the maximum value of resistance needed to keep the SCR on. In this case, the resistance bypassing the inductive loads needs to be at least 333 Ω.

When we are triggering SCRs with a gate signal, care must be taken that the maximum gate power dissipation is not exceeded. The actual gate power dissipated in a steady-state condition can be determined by multiplying the gate current drawn times the gate-to-cathode voltage. When the gate is triggered by repetitive pulses, the average gate power dissipation must be calculated. It can be found by first calculating or measuring the peak power of the triggering pulse. Once this power is known, the average gate power can be determined by multiplying the peak power times the duty cycle of the pulse. You will recall that the duty cycle of a pulse is found by dividing the amount of time the pulse is on by the period of the repeating pulses.

Let us try an example. Suppose we are working with a triggering circuit for an SCR where repetitive pulses are used to trigger the SCR. The pulse repetition rate (PRR) is 200 Hz, with a pulse width of 1 ms. If the peak power in the pulse is 1 W, what is the average power dissipated by the gate? First, the period of the repeating pulses must be calculated. The pulse period (T_P) is equal to the inverse of the PRR:

$$T_P = \frac{1}{\text{PRR}} \qquad \qquad (6.11)$$

or

$$T_P = \frac{1}{200 \text{ Hz}} = 5 \text{ ms}$$

Next, the duty is calculated. The duty cycle is equal to the time the pulse is on divided by the pulse period:

$$\text{duty cycle} = \frac{T_{\text{ON}}}{T_P} \qquad \qquad (6.12)$$

or

$$\text{duty cycle} = \frac{1 \text{ ms}}{5 \text{ ms}} = 0.2$$

Finally, the average gate power dissipation ($P_{G(av)}$) is the duty cycle times the peak pulse power ($P_{P(pk)}$):

$$P_{G(av)} = (\text{duty cycle})\left(P_{P(pk)}\right) \qquad (6.13)$$

or

$$P_{G(av)} = (0.2)(1 \text{ W}) = 200 \text{ mW}$$

Under these conditions, then, our gate is dissipating 200 mW of power. This value can then be compared with the maximum average gate power dissipation of the SCR to see whether it is excessive.

• EXAMPLE 6.8

Given an SCR whose gate is pulsed by a signal with a peak of 500 mW, a PRR of 300 Hz, and a pulse width of 0.75 ms, find the average power dissipated by the gate.

Solution

The pulse period is given by Equation 6.11:

$$T_P = \frac{1}{\text{PRR}} = \frac{1}{300 \text{ Hz}} = 3.33 \text{ ms}$$

The duty cycle is calculated from Equation 6.12:

$$\text{duty cycle} = \frac{T_{\text{ON}}}{T_P} = \frac{0.75 \text{ ms}}{3.33 \text{ ms}} = 0.225$$

Now that we know the duty cycle, we use Equation 6.13 to find the average power dissipated by the gate:

$$P_{G(av)} = (\text{duty cycle})\left(P_{P(pk)}\right)$$

$$= (0.225)(500 \text{ mW}) = 112.5 \text{ mW}$$

Turning Off the SCR. In most DC circuits, the problem is not how to turn an SCR on; the problem is how to turn it off. The only way to turn an SCR off is to reduce anode current below the holding current value. In DC circuits, this reduction usually is accomplished by adding a manual reset switch (S_1 in Figure 6.34). When the reset switch is opened, the blocking junction reestablishes itself.

Another common turnoff method for DC circuits is the commutation capacitor illustrated in Figure 6.35A. To simplify the schematic diagrams in this text, we represent the thyristor trigger circuits with a T at the gate. When SCR_2 is turned on by a gate pulse, anode current flows through the load and also charges C through R (SCR_1 is off). Because SCR_2 drops so little voltage in its on state, the capacitor C charges to about 100 V. SCR_2 may be turned off by triggering SCR_1 into conduction. When SCR_1 turns on, it reverse-biases SCR_2 by placing −100 V across it from anode to cathode.

Alternatively, the SCR may be turned off by turning transistor Q on, as shown in Figure 6.35B. When Q is forward-biased to saturation, it conducts current around the SCR. If the circuit is properly designed, the anode current falls below the holding current, and the thyristor turns off. This method depends on the use of a power transistor with a low collector-to-emitter saturation voltage.

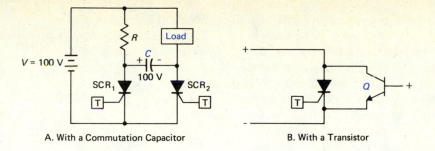

A. With a Commutation Capacitor   B. With a Transistor

FIGURE 6.35 Turning Off an SCR

The problem of SCR turnoff does not occur in AC circuits. The SCR is automatically shut off during each cycle when the AC voltage across the SCR approaches zero. As zero voltage is approached, anode current falls below the holding current value. The SCR stays off throughout the entire negative AC cycle since the SCR is reverse-biased.

A simple AC application of the SCR is shown in the circuit in Figure 6.36A. In Figure 6.36B, the positive AC half cycle is shown. As the supply goes positive, an increasing amount of gate current is drawn. The SCR blocks current flow (between points A and B) until V_T rises to V_{GT} (point B). At point B, the SCR quickly turns on and load current flows. The current through the SCR rapidly rises to a peak and

decreases sinusoidally as the AC supply voltage decreases toward zero. Finally, at point C, the anode current falls below the holding current value. The SCR turns off.

Note that, as shown in Figure 6.36B, the SCR blocks current flow during the negative half cycle. No current flows through the load at this time. Thus, this circuit is sometimes called a *half-wave controlled rectifier*. Reverse-biased diode D_1 prevents potentially damaging current from flowing from the gate during the negative half cycle. Resistor R_1 is placed in this circuit so that the total resistance cannot be reduced to a point where damaging gate current flows.

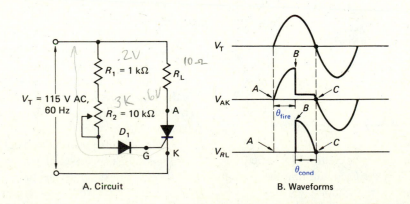

A. Circuit   B. Waveforms

FIGURE 6.36 SCR in AC Circuit

The distance from point *A* to *B* in Figure 6.36B represents the time in the positive half cycle when the SCR is off. The distance between *A* and *B*, in degrees, is called the *firing angle* (θ_{fire}). Note that the SCR is conducting between points *B* and *C*. This angle is called the *conduction angle* (θ_{cond}). As this circuit is drawn, the firing angle can range between 10° and 90°.

Decreasing the resistance of R_2 in the circuit allows more gate current to flow, thus firing the SCR earlier in the cycle. This technique applies a higher anode current to the load by increasing the average DC current and voltage to the load. In many cases, an ammeter is used to monitor the current through the load. Recall that most DC ammeters are calibrated to read average current. The following equation allows calculation of the average load voltage:

$$V_{\text{L(av)}} = \frac{V_{\text{pk}}}{2\pi}\left(1 + \cos\theta_{\text{fire}}\right) \qquad \textbf{(6.14)}$$

where
$V_{\text{L(av)}}$ = average load voltage (in volts)
V_{pk} = peak supply voltage (in volts)
θ_{fire} = firing angle (in degrees)

An example may help at this point. Suppose we are working with an SCR circuit with 120 V AC, 60 Hz applied, and the SCR is triggered at 45°. From Equation 6.14, the load voltage is

$$V_{\text{L(av)}} = \frac{V_{\text{pk}}}{2\pi}\left(1 + \cos\theta_{\text{fire}}\right)$$

$$= \frac{170}{(2)\,(3.14)}\left(1 + \cos 45°\right)$$

$$= 46.2 \text{ V}$$

• EXAMPLE 6.9

Given an SCR in a half-wave controlled rectifier, triggered at 68° in a 240 V AC circuit, find the average voltage applied across the load.

Solution

The average DC voltage across the load may be calculated by using Equation 6.14:

$$V_{\text{L(av)}} = \frac{V_{\text{pk}}}{2\pi}\left(1 + \cos\theta_{\text{fire}}\right)$$

Before we can use Equation 6.14, we must calculate the peak voltage for a root-mean-square (rms) voltage of 240 V AC:

$$V_{\text{pk}} = (1.414)\,(V_{\text{rms}})$$

$$= (1.414)\,(240 \text{ V AC})$$

$$= 340 \text{ V}$$

We can now substitute the V_{pk} value of 340 V into Equation 6.14:

$$V_{\text{L(av)}} = \frac{V_{\text{pk}}}{2\pi}\left(1 + \cos\theta_{\text{fire}}\right)$$

$$= \frac{340 \text{ V AC}}{2\pi}\left(1 + \cos 68°\right)$$

$$= 74.4 \text{ V}$$

Adjusting the AC circuit to a specified firing angle is difficult without using an oscilloscope. How can we adjust the SCR for a specified firing angle by using only a voltmeter? We can perform the calculation in Example 6.9. A voltmeter placed across the load will read 74.4 V when a firing angle of 68° is reached.

We may also want to find out what firing angle will give us a specified amount of power delivered to a load. For example, what firing angle will we need to deliver 100 W of average power to the load in

Figure 6.36A, with $R_L = 15\ \Omega$? The average power can be calculated from Ohm's law:

$$P_{av} = \frac{V_{av}^2}{R_L} \qquad (6.15)$$

Therefore,

$$V_{av} = \sqrt{(P_{av})(R_L)} \qquad (6.16)$$

or after substituting,

$$V_{av} = \sqrt{(100\ \text{W})(15\ \Omega)}$$

$$= 38.7\ \text{V}$$

We may now solve Equation 6.14 for the firing angle:

$$\theta = \cos^{-1}\left[\frac{(2\pi)(V_{L(av)})}{V_{pk}} - 1\right] \qquad (6.17)$$

$$= \cos^{-1}\left[\frac{(2\pi)(38.7\ \text{V})}{170\ \text{V}} - 1\right]$$

$$= 64.5°$$

• **EXAMPLE 6.10**

Given the half-wave phase control shown in Figure 6.36A, find the firing angle necessary to deliver 50 W into a 10 Ω load.

Solution

From Equation 6.16, we calculate the average voltage:

$$V_{av} = \sqrt{(P_{av})(R_L)}$$

$$= \sqrt{(50\ \text{W})(10\ \Omega)}$$

$$= 22.4\ \text{V}$$

Now from Equation 6.17, we can find the firing angle:

$$\theta = \cos^{-1}\left[\frac{(2\pi)(V_{L(av)})}{V_{pk}} - 1\right]$$

$$= \cos^{-1}\left[\frac{(2\pi)(22.4\ \text{V})}{170\ \text{V}} - 1\right]$$

$$= 100°$$

Note that—for the application in Example 6.10, at least—we need to fire the SCR beyond the 90° point on the input AC waveform. This firing angle is impossible for this circuit but not for the circuit shown a bit later in Figure 6.38.

Alternatively, we can calculate the firing angle if we know R_1, R_2, I_{GT}, and V_{GT}. For $I_{GT} = 10$ mA, $V_{GT} = 1$ V, the forward-biased diode drop of D_1 at 0.6 V, $R_1 = 1$ kΩ, and $R_2 = 5$ kΩ, we can calculate the supply voltage at which the SCR fires. At the firing point, I_{GT} is flowing through the gate circuit. This current drops 10 V across R_1, 50 V across R_2, 1 V across V_{GT}, and 0.6 V across D_1. These voltages total 61.6 V. In other words, the SCR will trigger when the supply reaches 61.6 V. Recall that any instantaneous voltage on a sine wave can be calculated from the following equation:

$$v = V_{pk}\sin(2\pi ft) \qquad (6.18)$$

where
   v = instantaneous voltage
   V_{pk} = peak sine wave voltage
   $2\pi ft$ = firing angle (in degrees)

Solving for the firing and using 162.6 V for V_{pk}, we get a firing angle of 22.3°.

The SCR phase-control circuit shown in Figure 6.36A is called a *half-wave phase control* since it can only control half of the input AC. The circuit shown in Figure 6.37 is a *full-wave phase control*. It can control both halves of the input AC, doubling the controlled power to the load. The average voltage across the load in this circuit is

$$V_{L(av)} = \frac{V_{pk}}{\pi} \left(1 + \cos \theta_{fire}\right) \qquad \textbf{(6.19)}$$

Note that the 2π term in Equation 6.14 is only π here since Equation 6.19 returns values twice as large as Equation 6.14. The next example shows that the use of this equation (6.19) is identical to the use of the half-wave equation (6.14).

• EXAMPLE 6.11
Given an SCR in a full-wave controlled rectifier, triggered at 68° in a 240 V AC circuit, find the average voltage applied across the load.

Solution
The average DC voltage across the load may be calculated by using Equation 6.19:

$$V_{L(av)} = \frac{V_{pk}}{\pi} \left(1 + \cos \theta_{fire}\right)$$

Before we can use Equation 6.19, we must calculate the peak voltage for an rms voltage of 240 V AC:

$$V_{pk} = (1.414)\left(V_{rms}\right)$$
$$= (1.414)(240 \text{ V AC})$$
$$= 340 \text{ V}$$

We can now substitute the V_{pk} value of 340 V into Equation 6.19:

$$V_{L(av)} = \frac{V_{pk}}{\pi} \left(1 + \cos \theta_{fire}\right)$$
$$= \frac{340 \text{ V AC}}{\pi} \left(1 + \cos 68°\right)$$
$$= 148.8 \text{ V}$$

Note that the value of 148.8 V is exactly twice the value of the voltage for the identical problem in the half-wave circuit in Example 6.9.

An improvement to the circuit in Figure 6.36A is shown in Figure 6.38. Here, a capacitor has been added between R_2 and the cathode of the SCR. The delayed gate voltage rise across the capacitor allows the SCR to fire past the 90° point. This circuit can control the SCR firing angle between about 15° and 170°. This type of AC control is called *phase control*.

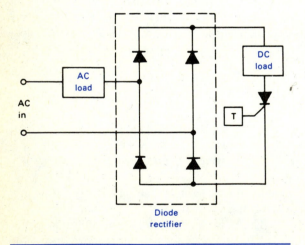

FIGURE 6.37 Full-Wave SCR Circuit Showing Possible Placement of AC or DC Loads

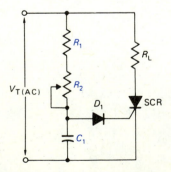

FIGURE 6.38 SCR Phase Control in AC Circuit

Triac

After the SCR, the most commonly used thyristor is the triac. Like the SCR, the *triac* acts as a switch, with a gate that controls the switching state. Unlike the SCR, the triac can conduct in both directions. Thus, this device is a *bilateral*, or *bidirectional*, device. Recall that the SCR triggers only on a positive gate voltage (with respect to the cathode). The triac, on the other hand, fires on either positive or negative polarity gate voltages. The triac is superior to the SCR in that it can be used in full-wave AC control applications, such as control of AC motors and AC heating systems. (The SCR can also be used in full-wave control applications, but only in specially designed circuits, such as the one shown in Figure 6.37.) Triac current and voltage ratings do not yet approach those of the SCR, though.

The triac schematic symbol is shown in Figure 6.39A. Note that none of the terminals are labeled anode or cathode. The bidirectional nature of this device makes that labeling impossible. Instead, the two comparable terminals are called *main terminals* 1 (MT$_1$) and 2 (MT$_2$). The gate lead retains the same name and function as in the SCR.

The electrical behavior of the triac is depicted in Figure 6.39B. Quite simply, the triac can be thought of as two SCRs in inverse, parallel connection. When both high-current and full-wave operation are required, two inverse, parallel SCRs are used.

Notice that the gate leads of both SCRs are connected together. When SCR$_1$ fires, current flows from MT$_1$ to MT$_2$. When SCR$_2$ fires, current flow is in the opposite direction. Thus, we have a bidirectional device. Its bidirectional nature is further shown in the current-voltage graph in Figure 6.40. Notice that the curve in quadrant III of the graph has the same shape as the one in quadrant I.

The triac is ideal for use in AC control circuits. Since the SCR is a unidirectional device, it has no control over half of each input cycle. The triac, however, can control current flow through a load driving both halves of the input cycle. A simple triac-controlled AC circuit is shown in Figure 6.41. The capacitor C_1 charges in either direction until $V_{(BO)}$ is reached. The triac then fires and stays on throughout the remainder of that half cycle. The triac turns off when the AC input voltage nears zero. The main-terminal current falls below the holding current.

Phase control is achieved by varying the resistance of R_2. Increasing R_2 resistance causes the capacitor to charge up more slowly, thus firing the triac later in the cycle. This increase thus delivers less energy to the load. Decreasing R_2 resistance causes the capacitor to charge up quickly, firing the thyristor earlier in the cycle. This decrease thus delivers more energy to the load.

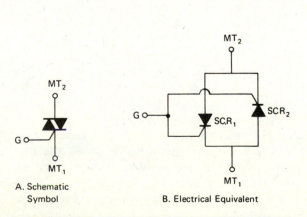

A. Schematic Symbol

B. Electrical Equivalent

FIGURE 6.39 Triac

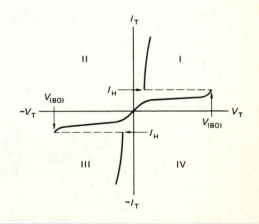

FIGURE 6.40 Triac Voltage-Current Characteristic Curve

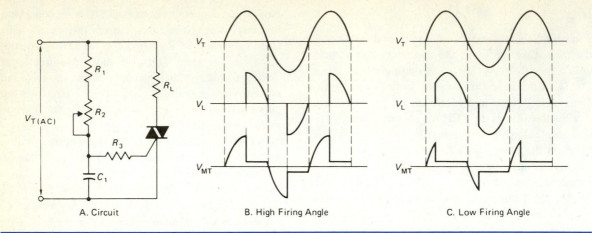

FIGURE 6.41   Triac in AC Circuit

Gate-Controlled Switch

As we have mentioned previously, the SCR and the
triac can be turned off only by low-current dropout.
In other words, anode (or main-terminal) current must
fall below the holding current. The *gate-controlled
switch* (GCS), or *gate-turnoff SCR* (GTO), as it is
sometimes called, turns on like a normal SCR—that
is, with a gate signal. But unlike the SCR, it may be
turned off by a negative trigger voltage. Positive and
negative trigger spikes can be provided by differen-
tiating a square wave and applying the signal to the
gate. (Recall from Chapter 1 that a differentiated
square wave is a series of alternating positive and
negative voltage spikes.) The GCS schematic
symbols are shown in Figure 6.42.

One problem with the GCS is that 10 to 20 times
more gate current is needed to turn off the GCS than
to turn it on. However, this disadvantage is more than
outweighed when the GCS is used in DC circuits. At
best, an additional SCR and commutation capacitor
are required when SCRs are used in DC circuits (see
the discussion on commutation earlier in this chapter). But the GCS may be used in DC circuits without these extra components.

The GCS finds applications in automobile igni-
tion systems, hammer drivers in computer printers,
and television horizontal deflection circuits.

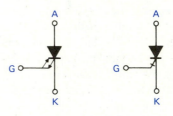

FIGURE 6.42   Schematic Diagrams for
Gate-Controlled Switch

Silicon-Controlled Switch

The final gate-controlled thyristor we will discuss is
the *silicon-controlled switch* (SCS). Figure 6.43A
gives a schematic diagram of the SCS. The SCS has
two gates: an anode gate (AG) and a cathode gate
(KG).

Figure 6.43B reveals an interesting fact. The
electrical equivalent for Figure 6.43A is just the
basic SCR circuit with both transistor bases acces-
sible to trigger pulses. The device may be turned on
with a positive gate pulse (with respect to the
cathode), just as in the SCR. The anode gate exercises control over the thyristor when a negative gate
voltage (with respect to the anode) is received. The
SCS is used in logic applications and in counters and
lamp drivers.

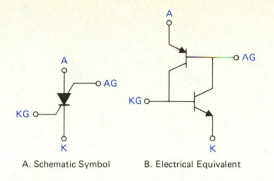

A. Schematic Symbol   B. Electrical Equivalent

FIGURE 6.43   Silicon-Controlled Switch

As a technician, you do not need to know how to design thyristors. However, you should know in a general sense how much current a thyristor is designed to carry. As an aid to understanding thyristors, we have included drawings, approximate case sizes, and approximate current-handling capability of the most popular case configurations in Figure 6.44.

THYRISTOR TRIGGERING

The thyristors discussed so far are used primarily as power control devices—high power in the case of the SCR and the triac and somewhat lower power in the others. We also have seen that these devices are triggered by gate signals. The next thyristor devices to be discussed are breakover devices. In *breakover devices*, there are usually no gate structures, in contrast with the SCR. Breakover devices are triggered by placing a voltage across the device high enough to break over the middle junction. Once the breakover occurs, the device quickly turns on by regeneration, just as in the SCR.

Shockley Diode

The *Shockley diode* (or *four-layer diode*) is a unidirectional thyristor used as a breakover device. Its structure is similar to that of the SCR without the gate, as shown in Figure 6.45A. The commonly used schematic symbol is shown in Figure 6.45B.

A circuit using the Shockley diode is illustrated in Figure 6.46. This circuit represents a relaxation oscillator. When the +15 V supply is connected, capacitor C_1 charges exponentially through the resistor R_1. When the breakover voltage is reached (5 V in this case), the capacitor discharges very quickly through the thyristor. At this time, the voltage falls quickly toward zero. Near 0 V, the current through the device will fall below the holding current value. The output waveform approaches that of the sawtooth oscillator (see Figure 6.46B).

The oscillator period is calculated by using the following equation:

$$T = R_1 C_1 \ln \frac{1}{1 - (V_{(BO)}/V_{AA})} \qquad (6.20)$$

where

T = time period (in seconds)
V_{AA} = voltage applied to circuit
$V_{(BO)}$ = breakover voltage of device
R_1 = resistance (in megohms)
C_1 = capacitance (in microfarads)

If we use the circuit in Figure 6.46A as an example, we have the following calculation:

$$T = R_1 C_1 \ln \frac{1}{1 - (V_{(BO)}/V_{AA})}$$

$$= (1 \text{ M}\Omega)(1 \text{ }\mu\text{F}) \ln \frac{1}{1 - (5 \text{ V}/15 \text{ V})}$$

$$= 0.405 \text{ s}$$

Since the frequency (f) of oscillation is the reciprocal of the period, we also have the following equation:

$$f = \frac{1}{T} \qquad (6.21)$$

or

$$f = \frac{1}{0.405 \text{ s}} = 2.47 \text{ Hz}$$

This relaxation oscillator, then, is oscillating at 2.47 Hz.

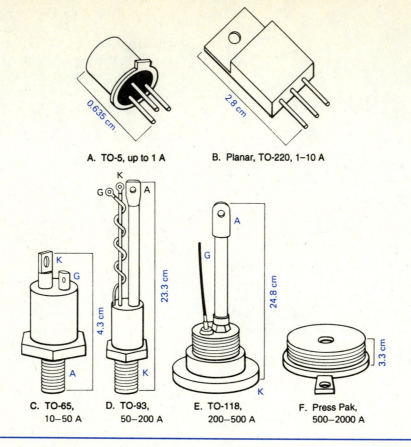

A. TO-5, up to 1 A   B. Planar, TO-220, 1–10 A

C. TO-65, 10–50 A   D. TO-93, 50–200 A   E. TO-118, 200–500 A   F. Press Pak, 500–2000 A

FIGURE 6.44   Thyristor Case Configurations

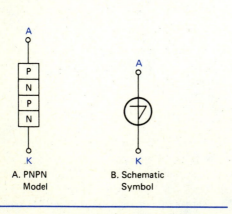

A. PNPN Model   B. Schematic Symbol

FIGURE 6.45   Shockley Diode

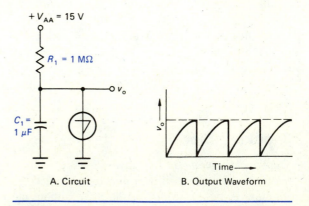

A. Circuit   B. Output Waveform

FIGURE 6.46   Relaxation Oscillator Using a Shockley Diode

• EXAMPLE 6.12

Given the relaxation oscillator circuit in Figure 6.46A, with $R_1 = 500$ kΩ, $C_1 = 0.2$ μF, an applied voltage of 50 V, and a breakover voltage of 30 V, find the frequency at which the oscillator will operate.

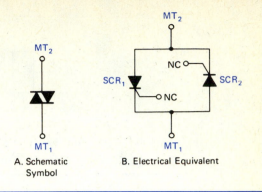

A. Schematic Symbol

B. Electrical Equivalent

FIGURE 6.47   Diac

Solution

We must first find the period of the oscillator, using Equation 6.20:

$$T = R_1 C_1 \ln \frac{1}{1 - (V_{(BO)}/V_{AA})}$$

$$= (500 \text{ k}\Omega)(0.2 \text{ μF}) \ln \frac{1}{1 - (30 \text{ V}/50 \text{ V})}$$

$$= 0.0916 \text{ s} \quad \text{or} \quad 91.6 \text{ ms}$$

Now the frequency of the oscillator can be calculated by taking the reciprocal of the period (Equation 6.21):

$$f = \frac{1}{T} = \frac{1}{91.6 \text{ ms}} = 10.9 \text{ Hz}$$

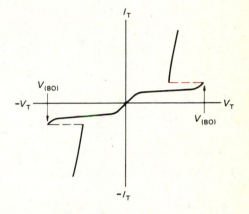

FIGURE 6.48   Diac Voltage-Current Characteristic Curve

Diac

We have seen that the thyristor family contains a unilateral breakover device called the Shockley diode. The *diac* is the bidirectional part of the family. Like the Shockley diode, the diac does not have a gate. The only way to get it to conduct is by exceeding the breakover voltage. The diac schematic symbol is shown in Figure 6.47A. Notice in Figure 6.47B that, electrically, the diac is similar to the triac, with no gate leads available. The diac characteristic curve is shown in Figure 6.48.

The diac is used most often to trigger a triac into conduction, an application discussed in Chapter 7. However, it may also be used in the same relaxation oscillator circuit as the Shockley diode.

Unijunction Transistor

Although the *unijunction transistor* (UJT) is not strictly a thyristor, we consider it here because it is often used to trigger an SCR or a triac. The UJT schematic symbol and construction are shown in Figure 6.49. The UJT's physical construction consists of an evenly doped block of N material with a portion of P material grown into its side.

Electrically, the UJT is illustrated by the circuit in Figure 6.50. The resistance between the PN junction and base B_1 is designated by r_{b1}; the resistance between the PN junction and base B_2 is called r_{b2}. The PN junction is represented by a diode symbol.

Since the UJT is used often as a relaxation oscillator, we will discuss that circuit, as shown in

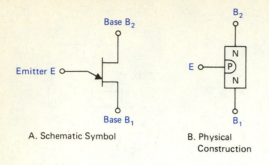

Base B$_2$

Emitter E

Base B$_1$

A. Schematic Symbol

B$_2$

E

B$_1$

B. Physical
Construction

FIGURE 6.49 Unijunction Transistor (UJT)

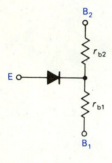

B$_2$

r_{b2}

E

r_{b1}

B$_1$

FIGURE 6.50 UJT Electrical Equivalent

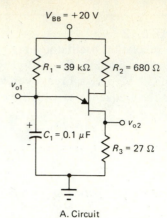

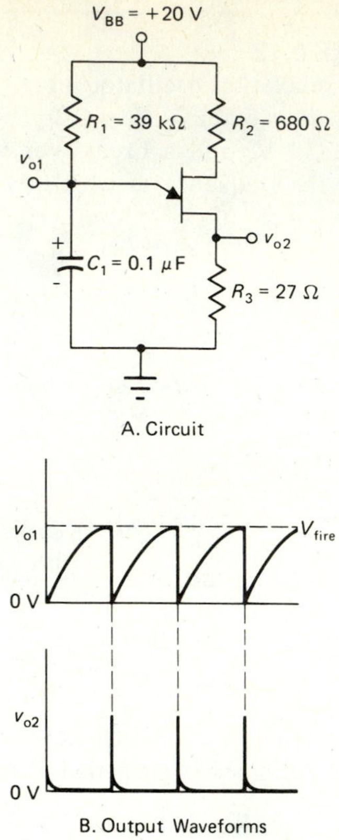

$V_{BB} = +20$ V

$R_1 = 39$ kΩ $R_2 = 680$ Ω

v_{o1}

$C_1 = 0.1$ μF v_{o2}

$R_3 = 27$ Ω

A. Circuit

v_{o1} V_{fire}

0 V

v_{o2}

0 V

B. Output Waveforms

FIGURE 6.51 UJT Relaxation Oscillator

Figure 6.51A. When supply voltage V_{BB} is applied to this circuit, the capacitor C_1 starts to charge at a rate determined by the time constant R_1C_1. Let us say that the UJT's P section is midway between base B$_1$ and base B$_2$. Since the block of N material is evenly doped, a voltage of about +10 V (with respect to ground) exists at the PN junction. The PN junction is reverse-biased until the capacitor voltage rises to about 0.6 V above the +10.6 V potential, or about +10.6 V. When this +10.6 V potential is reached across the capacitor, the PN junction becomes forward-biased. The capacitor discharges quickly through the forward-biased junction and R_3. When the capacitor voltage goes down near zero, the current flow goes below the holding current value, and the diode becomes reverse-biased again. The waveforms at the two outputs are shown in Figure 6.51B.

Notice that the output taken across R_3 is a spiked waveform created by the rapid discharge of C_1 through R_3.

The period T of the oscillations (in seconds) is calculated from the following equation:

$$T = R_1 C_1 \ln \frac{1}{1 - \eta} \qquad (6.22)$$

where

η = intrinsic standoff ratio

The *intrinsic standoff ratio*, electrically, is

$$\eta = \frac{r_{b1}}{r_{b1} + r_{b2}} \qquad (6.23)$$

This parameter tells us basically how far the PN junction is from B_1. For instance, if the length of the N material is 10 mils, then an intrinsic standoff ratio of 0.6 means that the PN junction is 6 mils from B_1. UJTs have intrinsic standoff ratios that vary from 0.5 to 0.8.

The waveform's period (using $\eta = 0.5$), then, is as follows:

$$T = R_1 C_1 \ln \frac{1}{1 - \eta}$$

$$= (39 \text{ k}\Omega)(0.1 \text{ μF}) \ln \frac{1}{1 - 0.5}$$

$$= 2.7 \text{ ms}$$

Taking the reciprocal of the period gives a frequency of 370 Hz.

The firing voltage may also be calculated. The intrinsic standoff ratio η times the supply voltage V_{BB} gives the voltage potential at the PN junction. Adding 0.6 V to this value yields the firing voltage—that is, the voltage at which the PN junction is forward-biased. The equation for the firing voltage (V_{fire}) is

$$V_{fire} = \eta V_{BB} + 0.6 \text{ V} \tag{6.24}$$

In the example shown in Figure 6.51A, where η is 0.5 and V_{BB} is 20 V, the PN junction is forward-biased when the capacitor charges to 10.6 V. Note also that the oscillation frequency of this circuit can be changed by varying the resistance of R_1.

• EXAMPLE 6.13

Given the relaxation oscillator shown in Figure 6.51A, with a UJT whose intrinsic standoff ratio is 0.7 and with $R_1 = 100 \text{ k}\Omega$, and $C_1 = 0.5 \text{ μF}$, find the waveform's period and the frequency at which the oscillator will operate.

Solution
From Equation 6.22, we calculate

$$T = R_1 C_1 \ln \frac{1}{1 - \eta}$$

$$= (100 \text{ k}\Omega)(0.5 \text{ μF}) \ln \frac{1}{1 - 0.7}$$

$$= 0.06 \text{ s} \quad \text{or} \quad 60 \text{ ms}$$

The frequency of the oscillator can be calculated by taking the reciprocal of the period:

$$f = \frac{1}{T} = \frac{1}{60 \text{ ms}} = 16.6 \text{ Hz}$$

Programmable Unijunction Transistor

The *programmable unijunction transistor* (PUT) is a thyristor that has a function similar to that of the UJT. It is commonly used in relaxation oscillators. Unlike the UJT, the PUT is a thyristor with a gate connected to the anode PN junction. The schematic symbol and electrical equivalent are shown in Figure 6.52. If you conclude from looking at the figure that the PUT is more like an SCR than a UJT, you are right. The PUT has an anode gate instead of a cathode gate. Thus, to trigger the PUT, we need a negative voltage on the gate with respect to the anode.

A relaxation oscillator using a PUT is shown in Figure 6.53A. When supply voltage V_{BB} is applied to the circuit, capacitor C_1 charges through R_1 at a rate determined by the time constant. The R_2–R_3 voltage divider applies 10 V to the gate. When the voltage at the anode reaches 10.6 V, the PN junction is forward-biased and the PUT fires. The capacitor discharges quickly toward zero, eventually turning off the thyristor when anode current falls below the holding current.

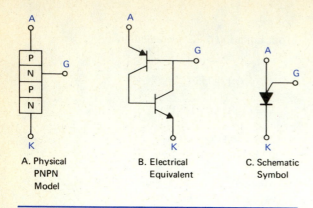

A. Physical
PNPN
Model

B. Electrical
Equivalent

C. Schematic
Symbol

FIGURE 6.52 Programmable Unijunction Transistor (PUT)

The functional difference between the PUT and UJT is that the PUT has an intrinsic standoff ratio that is variable or programmable, whereas the UJT does not. For the circuit of Figure 6.53, the intrinsic standoff ratio is

$$\eta = \frac{R_3}{R_2 + R_3} \qquad (6.25)$$

Calculations of peak firing voltage and frequency are identical to those for the UJT relaxation oscillator circuit discussed previously.

• **EXAMPLE 6.14**

Given the PUT circuit in Figure 6.53A, find (a) the intrinsic standoff ratio, (b) the waveform's period, and (c) the frequency of oscillation.

Solution

(a) The intrinsic standoff ratio is equal to the voltage divider ratio of R_2 and R_3 (Equation 6.25):

$$\eta = \frac{R_3}{R_2 + R_3} = \frac{15 \text{ k}\Omega}{30 \text{ k}\Omega + 15 \text{ k}\Omega} = 0.33$$

(b) We substitute into Equation 6.22 to find the period:

$$T = R_1 C_1 \ln \frac{1}{1 - \eta}$$

$$= (1 \text{ M}\Omega)(0.2 \text{ μF}) \ln \frac{1}{1 - 0.33}$$

$$= 0.081 \text{ s} \quad \text{or} \quad 81 \text{ ms}$$

(c) The frequency of the oscillator can be calculated by taking the reciprocal of the period:

$$f = \frac{1}{T} = \frac{1}{81 \text{ ms}} = 12.3 \text{ Hz}$$

The PUT is often used to trigger an SCR into conduction, as shown in Figure 6.54. When the switch is closed, the voltage at the gate of the PUT is equal to the voltage drop across R_3:

$$V_{R3} = \left(\frac{R_3}{R_2 + R_3}\right)(V_{AA}) \qquad (6.26)$$

For the circuit in Figure 6.54, the voltage drop is

$$V_{R3} = \left(\frac{22 \text{ k}\Omega}{39 \text{ k}\Omega + 22 \text{ k}\Omega}\right)(+40 \text{ V})$$

$$= 14.4 \text{ V}$$

with a standoff ratio of 0.36. The firing voltage is 0.6 V higher and is equal to 15 V.

We may also wish to know the amount of delay between the closing of the switch and the firing of the SCR. We can calculate the amount of delay by using Equation 6.22:

$$T = R_1 C_1 \ln \frac{1}{1 - \eta}$$

$$= (220 \text{ k}\Omega)(0.033 \text{ μF}) \ln \frac{1}{1 - 0.36}$$

$$= 3.24 \text{ ms}$$

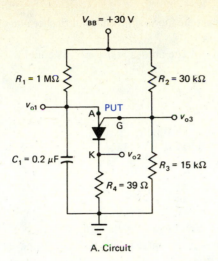

A. Circuit

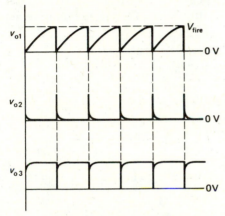

B. Output Waveforms

FIGURE 6.53 PUT Relaxation Oscillator

The SCR will fire 3.24 ms after the switch is closed.

The frequency of the oscillator can be calculated by taking the reciprocal of the period:

$$f = \frac{1}{T} = \frac{1}{3.24 \text{ ms}} = 309 \text{ Hz}$$

• **EXAMPLE 6.15**

Given the PUT circuit in Figure 6.54, with $R_3 = 20 \text{ k}\Omega$, find (a) the firing voltage across the

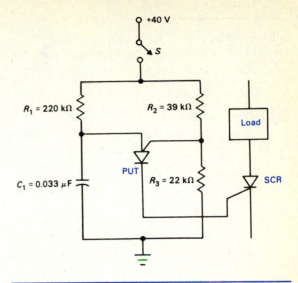

FIGURE 6.54 PUT Driving SCR

capacitor C_1 and (b) the delay in firing the SCR after the switch is closed.

Solution

(a) When the switch is closed, the voltage drop across R_3 is given by Equation 6.26:

$$V_{R3} = \left(\frac{R_3}{R_2 + R_3} \right) (V_{AA})$$

$$= \left(\frac{20 \text{ k}\Omega}{39 \text{ k}\Omega + 20 \text{ k}\Omega} \right) (+40 \text{ V})$$

$$= 13.56 \text{ V}$$

with a standoff ratio of 0.339. The firing voltage is 0.6 V higher and is equal to 14.16 V.

(b) We can calculate the amount of delay by using Equation 6.22:

$$T = R_1 C_1 \ln \frac{1}{1 - \eta}$$

$$= (220 \text{ k}\Omega) (0.033 \text{ }\mu\text{F}) \ln \frac{1}{1 - 0.339}$$

$$= 3.00 \text{ ms}$$

• EXAMPLE 6.16

Given the PUT circuit in Figure 6.54, find the resistance of R_1 that will give a 5 ms time delay after the switch is closed.

Solution

First, we need to solve Equation 6.22 for R_1. Equation 6.22 is

$$T = R_1 C_1 \ln \frac{1}{1 - \eta}$$

Therefore, we obtain

$$R_1 = \frac{T}{C_1 \ln\left(\frac{1}{1 - \eta}\right)} \qquad (6.27)$$

Now we can substitute appropriate values into Equation 6.27:

$$R_1 = \frac{5 \text{ ms}}{(0.033 \ \mu\text{F}) \ln\left(\frac{1}{1 - 0.339}\right)}$$

$$= 366 \text{ k}\Omega$$

The circuit shown in Figure 6.55 is another example of a circuit using PUTs and SCRs. This circuit functions as a rear turn signal flasher for an automobile. When S_1 is thrown, power is immediately applied to lamp 1, turning it on. At the same time power is applied to this lamp, power is applied to R_1 and C_1. Capacitor C_1 will continue to charge until the anode of Q_1 is more positive than the cathode by about 0.6 V. When this potential is reached, Q_1 will fire, applying a positive voltage at the gate of SCR_1 and turning it on. When SCR_1 fires, lamp 2 will turn on, and a potential is applied to the Q_2 circuit. Capacitor C_2 charges until the anode of Q_2 is 0.6 V more positive than its anode. When the anode reaches this potential, Q_2 will fire, turning on SCR_2, and turning on the final lamp 3. After lamp 3 has been on for some time, the thermal mechanical flasher (usually a bimetal strip) will disconnect all power from the circuit. The cycle will start all over again when the thermal mechanical flasher unit closes the circuit.

The voltage across resistor R_3 is equal to

$$V_{R3} = \left(\frac{R_3}{R_2 + R_3}\right)(V_{AA})$$

$$= \left(\frac{100 \text{ k}\Omega}{10 \text{ k}\Omega + 100 \text{ k}\Omega}\right)(+12 \text{ V})$$

$$= 10.9 \text{ V}$$

with a standoff ratio of 0.9. The firing voltage across capacitor C_1 is equal to the voltage across R_3 plus 0.6 V, which is equal to 11.5 V. As stated earlier, lamp 1 goes on immediately after S_1 closes. The delay in firing lamp 2 can be found by applying Equation 6.22:

$$T = R_1 C_1 \ln \frac{1}{1 - \eta}$$

$$= (1 \text{ M}\Omega)(0.5 \ \mu\text{F}) \ln \frac{1}{1 - 0.9}$$

$$= 1.15 \text{ s}$$

A delay of about 1 s will separate the lighting of lamp 1 and lamp 2.

• EXAMPLE 6.17

Given the PUT turn signal flasher in Figure 6.55, find the size of R_1 in timer 1 that will give a 0.75 s delay between lamp 1 and lamp 2.

Solution

We can find the correct size of R_1 by substituting the appropriate values into Equation 6.27:

$$R_1 = \frac{0.75 \text{ s}}{(0.5 \ \mu\text{F}) \ln\left(\frac{1}{1 - 0.9}\right)}$$

$$= 651 \text{ k}\Omega$$

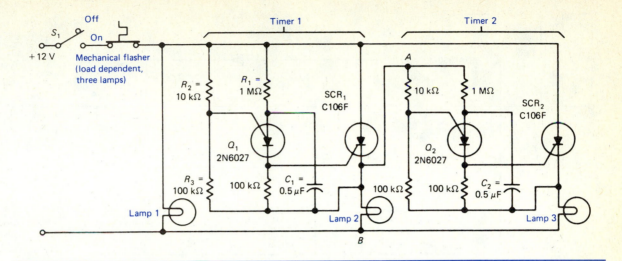

FIGURE 6.55 Turn Signal Flasher

Another circuit using SCRs and a PUT is shown in Figure 6.56. Here, the thyristors form a short-circuit-proof battery charger. This 12 V lead-acid battery charger provides an average charging current of 8 A. In addition to providing short-circuit protection, the charger will shut down if the battery is connected incorrectly. With the battery correctly connected as shown, a small current is drawn through R_2, R_3, and R_4. A small current also charges C_1 through R_1. When the voltage on the anode of the PUT reaches the PUT firing potential, the PUT will conduct. The PUT conduction through T_2 will fire the SCR and start applying charge current to the battery. As the battery charges, its terminal voltage will increase, raising the firing potential of the PUT. The capacitor, therefore, must charge to a slightly higher voltage to fire the PUT. The firing potential on the capacitor will increase until the voltage on the capacitor starts to go above the zener diode voltage. The zener diode will keep the voltage across the capacitor at 10 V, preventing the relaxation oscillator from oscillating, and charging stops. In this circuit, the charging voltage can be set between 10 and 14 V, 14 V being the voltage of a fully charged, lead-acid battery. The lower charging voltage limit is set by the zener diode. The 14 V upper limit is set by T_2 and is adjusted downward by setting R_2. This

circuit is easily modified to charge lead-acid batteries with other voltages. The charging voltage may be lowered by replacing the zener diode with one that has a lower zener voltage. Charging current may be limited by adding a resistor in series with the SCR or by replacing transformer T_1 with one having a lower secondary voltage. Resistor R_4 prevents the PUT from being destroyed in the event that the wiper of R_2 is adjusted to the top.

Solid-State Relays

After performing switching tasks for several decades, the EMR has been replaced in some applications by a new type of relay, the solid-state relay (SSR). A *solid-state relay* differs from an EMR in that the SSR uses a thyristor to do the switching. An SSR has no moving parts.

A block diagram of an SSR is illustrated in Figure 6.57. A DC voltage turns on the LED. The diode turns on the phototransistor, which triggers the triac. The triggering circuit usually contains a zero-voltage switch that activates when the output voltage AC crosses zero. Zero-voltage switching is advantageous since it cuts down on electromagnetic interference (EMI), a common problem in EMRs. Well-designed SSRs have snubbers included across

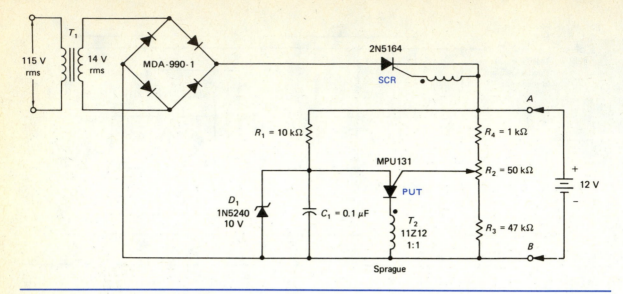

FIGURE 6.56 Battery Charger

the output terminals. You will recall from an earlier discussion that the snubber prevents false thyristor triggering caused by exceeding the critical rate of voltage rise.

SSRs are available in a great variety of shapes and sizes. Maximum currents of up to 50 A are possible at voltage ratings of 120, 240, and 480 V AC. The input voltages usually range from 3 to 32 V in the DC versions and 80 to 280 V in the AC versions.

The SSR has many advantages over the EMR. The SSR is more reliable and has a longer life because it has no moving parts. It is compatible with IC circuitry and does not generate as much EMI. The SSR is more resistant to shock and vibration, has a much faster response time, and does not exhibit contact bounce. Like the EMR, the SSR finds application in isolating a low-voltage control circuit from a high-power circuit. The degree of isolation between the control circuit and the power circuit depends on two factors: the dielectric characteristics and the distance between the source and sensor.

As shown in Figure 6.58, the SSR can be triggered by TTL or CMOS digital circuitry, including the output ports of microcomputers. In the application shown, the logic gate is used in its sinking mode.

The SSR triggers when the logic gate output goes low, drawing current through the LED in the SSR. The SSR parameters are slightly different from the EMR parameters. The *must-operate voltage* is a parameter that specifies the minimum level of input control voltage required to change the output from the off state to the on state. In EMR terms, this would be equivalent to the pull-in voltage. The *must-release voltage* is the maximum input control voltage required to change the output from the on state to the off state.

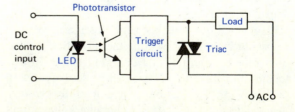

FIGURE 6.57 Functional Diagram of Optically Coupled SSR

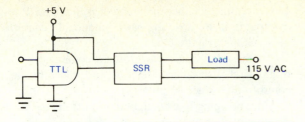

FIGURE 6.58 TTL Controlling SSR

Many of the same circuits currently used in EMRs can be duplicated with SSRs. One example is the latching relay configuration shown in Figure 6.59. Current will flow from the supply through contact 4, contact 3, and the push button to the other side of the supply. This flow closes the circuit between contacts 1 and 2. When the push button is released, current will flow through contacts 3 and 4, through R_1, and then through contacts 1 and 2 to the other side of the supply.

Start and stop push buttons have been added to the latching circuit in Figure 6.60. Note that the start circuit is essentially identical to the circuit shown in Figure 6.59. Note further that the stop push button is in series with the control current that flows when the start push button is pressed. When the stop push button is pressed, the current is interrupted in the control circuit, opening the output circuit. Current will not flow in the input or output circuit until the start push button is pressed.

SSRs may also be used in three-phase as well as single-phase circuits, as shown in Figure 6.61. The three-phase power is applied to the load when the control voltage is applied to the input circuit. Although the load is shown wye-connected, it may be delta-connected.

As with every device, SSRs do have some disadvantages. The SSR contains semiconductors that are susceptible to damage from transient-voltage and -current spikes. Such spikes are prevalent in the industrial AC line (especially on 480 V AC service), being generated by motors, solenoids, EMRs, and lightning. Some manufacturers add metal-oxide varistors (MOVs) that provide some protection for the delicate semiconductors within the SSR. (MOVs are discussed under the heading "Thyristor Protection" later in this chapter.) Unlike the EMR contacts, the thyristor has a significant on-state resistance, enough to cause a 1 V potential to be felt across the output thyristor when it is on. This 1 V potential may not seem like a significant voltage. Consider, however, that the power this device generates when drawing 10 A is 10 W. The heat generated by this 1 V potential must be dissipated or the output thyristor may be damaged. Other disadvantages include a significant off-state current, possibly as high as 1 mA, and damage to the device by nuclear radiation.

In most applications, SSRs are used to interface between a low-voltage control circuit and a higher AC line voltage. These devices are used as switches

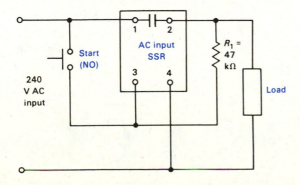

FIGURE 6.59 Latching Relay Circuit Using SSR

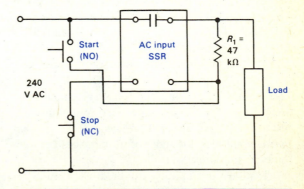

FIGURE 6.60 Latching Relay Circuit Using SSR with Stop Function Added

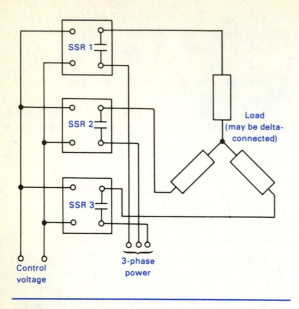

FIGURE 6.61 SSRs Controlling Three-Phase Load

in all kinds of industrial equipment and systems. Specifically, SSRs are used with incandescent lamps, motors, solenoids, transformers, motor starters, and industrial heaters. Several manufacturers offer I/O (input-output function) systems for computers made up of banks of SSRs. These systems usually consist of a mounting rack that contains from 4 to 32 separate plug-in SSR modules. These modules can be switched by input logic voltages of 5, 15, or 24 V DC. The outputs can be either AC or DC. A very important advantage of this type of system is that it allows remote location of the SSR modules, keeping them away from the central processing unit (CPU). The newest generation of SSR I/O racks have an on-board microprocessor and a serial data link between the processor in the I/O rack and the processor in the computer. This system reduces wiring to a minimum since only four wires are needed to control up to 512 relays. All relay actuation is controlled through coded messages between the two microprocessors.

Malfunctions in SSRs can be divided into three problem areas: The SSR will not turn on, it will not turn off, or it will not turn on and off reliably. Let us address each one of these situations to see how we

can determine whether the problem is in the SSR or external to it. Refer to the basic functional diagram of the SSR in Figure 6.62.

If the relay fails to turn off on command, you might first disconnect the SSR and measure the control voltage across points A and B. The voltage may not be going low enough to turn off the device. Another way of saying this, using SSR terminology, is that the control voltage may be exceeding the must-release voltage mentioned previously. In this case, the malfunction is probably a control system malfunction, not a problem in the SSR. If the measured control voltage is lower than the SSR must-release voltage, the relay is probably damaged and should be replaced.

If the relay fails to turn on, a good starting place is a check of all circuit connections. Connections should be checked for tightness and correct wiring, including proper polarity. If no problem is discovered, measurements must be taken. Again, the input voltage should be measured across points A and B on the diagram in Figure 6.62. If the input voltage is too low (under the must-operate voltage), the SSR will not turn on. If the input voltage is found to be too low, the malfunction is probably in the control system. This problem should be rectified before you take any further action. The problem can be confirmed, if necessary, by simulating a control voltage with a special external power supply. If the SSR behaves properly, applying power to the load, the problem is definitely isolated to the input control circuit. A proper control circuit voltage, however, does not necessarily mean that the input circuit is operating properly. Since the internal LED at the input (see Figure 6.59) is basically a current device, current must be allowed to flow through the LED.

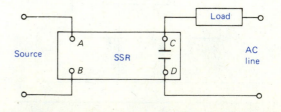

FIGURE 6.62 Basic SSR Functional Diagram

Something in the input circuit may be keeping the proper current from flowing through the LED, even though proper voltage may be applied to the input. An input current measurement will confirm the proper amount of current through the LED. A lack of current indicates an open circuit in the relay input circuit and requires SSR replacement. A low current is more likely to be a control circuit problem, which should be diagnosed before continuing to use the SSR. If the current and the input voltage are normal, you should proceed to the output circuit.

With no input signal, measure across the relay terminals at points C and D (Figure 6.62). Full line voltage should be seen across the SSR output terminals. If no voltage is present, the load is probably open, preventing voltage from being felt at the output terminals of the SSR. The load should be thoroughly checked; the load may be shorted. In this case, the SSR will probably be defective, due to an overcurrent situation caused by the defective load. If the load is defective, it may be checked by first removing power from the circuit. Then, disconnect the output wires from the SSR and connect them together. Apply power to the circuit. The load, having the proper voltage applied to it, should behave normally. The SSR may be checked by replacing the output connections and applying the proper input signal and output power. The SSR should now control the load in a normal fashion.

TROUBLESHOOTING THYRISTORS

The most appropriate in-circuit test for thyristors is one performed with the oscilloscope. You should become familiar with the waveforms shown with each of the thyristors in this chapter. Bear in mind that most thyristors are open circuits when off and have an anode-to-cathode drop of about 1 V when on.

A simple static test for SCRs and triacs may be made with an ohmmeter connected as in Figure 6.63. If the anode-to-cathode is forward-biased by the ohmmeter's internal battery, the middle junction should block current flow. The ohmmeter should read infinite ohms. When lead L is connected to the gate, the middle junction will be forward-biased and the

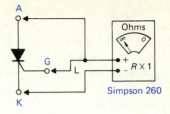

FIGURE 6.63 Checking SCR with Ohmmeter

thyristor should turn on. The resistance should decrease and should stay low if the gate lead is removed.

Care is required with this method of checking SCRs and triacs. If the voltage applied by the ohmmeter to the gate-cathode junction is too high, the thyristor could be destroyed. Also, the resistance in the ohmmeter may be too high. The thyristor may turn on but turn off again immediately if the anode current is lower than the holding current value. The ohmmeter scale should be changed to lower the resistance.

Troubleshooting relaxation oscillators is made more difficult by the fact that circuit resistances can determine whether the oscillator will work. For example, in Figure 6.51, the value of resistor R_3 is critical to circuit operation. If the UJT appears to be good, try checking the value of each of the resistors around the circuit.

Thyristor Protection

Thyristors are semiconductor devices and, as such, they are susceptible to damage by transient-voltage and -current spikes. Years ago, the technology, being dominated by sturdy relay and vacuum tubes, was relatively immune to such damage. With the increased popularity of more delicate semiconductor controls, transient-voltage and -current protection techniques are now necessary.

Transient-voltage and -current surges come from many sources but can be classified into two broad categories: internal and external. Internal sources of transients come from de-energizing a transformer primary. Opening a transformer secondary can generate extremely high transient-voltage spikes as the magnetic field built up in the core collapses. The collapsing core field induces a high-voltage spike in the secondary winding. In many cases, spikes can be greater than ten times the normal secondary voltage. Switches opening and closing are also known to produce high-voltage transients, especially when switches close and contact bounce is present. Spikes resulting from contact bounce in relay contacts and switches can reach levels exceeding 3 kV in 120 V AC circuits.

Transients can enter a circuit externally from the commutation of motors and generators, from welding equipment, and from natural sources such as lightning. These sources of transient voltages and currents usually enter circuits through the power supply.

Most semiconductors will be damaged or suffer a shortening of their useful life when operated with voltage and current above normal values. Relay contacts also may be damaged by the arcing across them during contact bounce. The insulation found on motors, generators, and transformers may be damaged as well by the application of transient-voltage spikes. This damage may be prevented—or at least reduced—by the application of a special transient-suppressing device called a *metal-oxide varistor* (MOV).

Transient suppression is accomplished in two ways. First, transients can be suppressed by decreasing the size of the transient voltage or current. Low-pass filters placed in series with a load will attenuate high frequencies and pass the lower-signal frequencies. The second method of dealing with transient voltage or current diverts the transient from delicate devices. This diversion is normally done with a voltage-clamping device. The MOV belongs under this classification. The MOV behaves in a nonlinear fashion; its impedance depends on the voltage across it. Simply put, the MOV has a high impedance to low voltage and a low impedance to high voltage. In normal circuit operation, the MOV has a high impedance that does not affect circuit operation. In the presence of a high-voltage spike, however, the MOV's high impedance changes to a low impedance, and the MOV clamps the voltage to a safe level.

The schematic symbols of the MOV shown in Figure 6.64 are similar in function to back-to-back zener diodes. The MOV, however, is more rugged than back-to-back zeners because it can absorb more energy. The MOV is made from zinc oxide with small amounts of cobalt, bismuth, manganese, and other metallic oxides added. The zinc oxide grains in the MOV form an interlocking matrix. The boundaries between the grains give the MOV a characteristic similar to that of a PN junction. Since many boundaries exist in the material, any energy absorbed by the device is dissipated across the whole device. You will recall that, in the zener diode, the energy absorbed is concentrated near the junction. The characteristics of the MOV are varied by the manufacturer by altering the size of the device. Generally, the higher the zinc oxide content, the more energy the MOV can absorb.

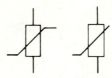

FIGURE 6.64 Metal-Oxide Varistor (MOV) Schematic Symbols

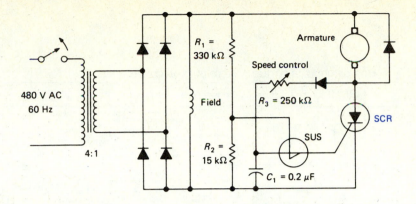

FIGURE 6.65   Circuit in Need of MOV Protection

A good example of an application for an MOV is illustrated in Figure 6.65. During design and testing, the SCR and diodes in this motor control were destroyed when the power was removed from the primary. The manufacturer used components with voltage ratings up to 600 V without success. Placing an MOV across the secondary will clamp any transient to a safe level.

Another example of the use for an MOV is in the power supply shown in Figure 6.66. In this case, the MOV is used to protect the rectifier diode from damage transmitted through the power line. The 100 mH (millihenry) choke and the 0.1 μF capacitor are used here as a filter to reduce RF interference. The filter will also cut in half the amplitude of any transient introduced. The MOV will reduce any remaining transients to a safe level.

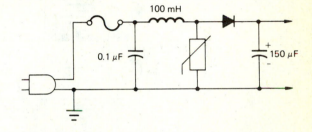

FIGURE 6.66   MOV Protection in Power Supply

CONCLUSION

In recent years, more has been heard about the new IC technology than about power semiconductor devices, such as the thyristor. However, the thyristor is still the workhorse of power control applications. Recent developments in power semiconductors have increased the voltage and current ratings to levels where vacuum tubes are virtually unnecessary. Even though the thyristor is far from an ideal switch, its usefulness can only increase in industrial applications. The entire thyristor family is summarized in tabular form in Figure 6.67.

MAJOR GENERAL ELECTRIC SEMICONDUCTOR COMPONENTS

NAME OF DEVICE	CIRCUIT SYMBOL	COMMONLY USED JUNCTION SCHEMATIC	ELECTRICAL CHARACTERISTICS	MAJOR APPLICATIONS
GE-MOV® Varistor V series		WIRE LEAD, ELECTRODE, INTER-GRANULAR PHASE, ZINC OXIDE GRAINS, ELECTRODE, EPOXY ENCAPSULANT, WIRE LEAD	When exposed to high energy transients, the varistor impedance changes from a high standby value to a very low conducting value, thus clamping the transient voltage to a safe level.	Voltage transient protection High-voltage sensing Regulation
Diode or Rectifier 1N and A series	ANODE CATHODE	ANODE, p, n, CATHODE	ANODE I; V_{ANODE} (-); V_{ANODE} (+) Conducts easily in one direction, blocks in the other.	Rectification Blocking Detecting Steering
Tunnel Diode 1N3712-21	POSITIVE ELECTRODE NEGATIVE ELECTRODE	POSITIVE ELECTRODE, p, n, NEGATIVE ELECTRODE	I_p; V_{ANODE} (+); I Displays negative resistance when current exceeds peak point current I_p	UHF converter Logic circuits Microwave circuits level sensing
Back Diode BD2-7 400 Series	ANODE CATHODE	ANODE, n, p, CATHODE	I; V_{ANODE} (-); V_{ANODE} (+) Similar characteristics to conventional diode except very low forward voltage drop.	Microwave mixers and low-power oscillators
n-p-n Transistor	COLLECTOR, I_C, BASE, I_B, EMITTER	COLLECTOR, n, p, n, BASE, EMITTER	I_C; I_{B5}, I_{B4}, I_{B3}, I_{B2}, I_{B1}; 0; $V_{COLLECTOR}$ (+) Constant collector current for given base drive	Amplification Switching Oscillation
p-n-p Transistor	COLLECTOR, I_C, BASE, I_B, EMITTER	COLLECTOR, p, n, p, BASE, EMITTER	$V_{COLLECTOR}$ (-) 0; I_{B1}, I_{B2}, I_{B3}, I_{B4}, I_{B5}; $I_{COLLECTOR}$ (-) Complement to n-p-n transistor	Amplification Switching Oscillation
Unijunction Transistor (UJT) 2N2646,7 2N4871	EMITTER, BASE 2, BASE 1	BASE 2, EMITTER, p, n, I_E, BASE 1	VOLTAGE BETWEEN EMITTER & BASE 1; V_P; 0; EMITTER I_E Unijunction emitter blocks until its voltage reaches V_p; then conducts	Interval timing Oscillation Level Detection SCR Trigger
Complementary Unijunction Transistor (CUJT) D5K1, 2	EMITTER, BASE 1, BASE 2	BASE 1, p, n, p, n, EMITTER, BASE 2	V_E; PEAK POINT; VALLEY POINT; I_E Functional complement to UJT	High-stability timers Oscillators and level detectors
Programmable Unijunction Transistor (PUT) 2N6027, 8	ANODE, GATE, CATHODE	ANODE, p, n, p, n, GATE, CATHODE	I_A; A; C; G; VALLEY POINT; PEAK POINT; V_{AC} Programmed by two resistors for V_p, I_p, I_v, Function equivalent to normal UJT.	Low-cost timers and Oscillators Long-period timers SCR Trigger Level Detector
Photo Transistor L14 series	COLLECTOR, BASE, I_B, EMITTER	COLLECTOR, n, p, n, BASE, EMITTER	I COLLECTOR; H4, H3, H2, H1; V_{CE} Incident light acts as base current of the photo transistor	Tape readers Card readers Position sensor Tachometers

FIGURE 6.67 Thyristor Family Characteristics

MAJOR GENERAL ELECTRIC SEMICONDUCTOR COMPONENTS

NAME OF DEVICE	CIRCUIT SYMBOL	COMMONLY USED JUNCTION SCHEMATIC	ELECTRICAL CHARACTERISTICS	MAJOR APPLICATIONS
Opto Coupler (1) Transistor (H11A, H15A) (2) Darlington (H11B, H15B) Outputs			Output characteristics are identical to a normal transistor/Darlington except that the LED current (I_F) replaces base drive (I_B)	Isolated interfacing of logic systems with other logic systems. Power semiconductors and electro-mechanical devices Solid-state relays.
Opto Coupler SCR Output (H11C)			With Anode voltage (+), the SCR can be triggered with a forward LED current. (Characteristics identical to a normal SCR except that LED current (I_F) replaces gate trigger current I_{GT}	Isolated interfacing of logic systems with AC power switching functions. Replacement of relays, microswitches
AC Input Opto Coupler (H11AA)			Identical to a "standard" transistor coupler except that LED current can be of either polarity	Telecommunications — ring singnal detection, monitoring line usage. Polarity insensitive solid-state relay. Zero voltage detector.
Silicon Controlled Rectifier (SCR)	ANODE GATE CATHODE	ANODE CATHODE GATE	With anode voltage (+), SCR can be triggered by I_g, remaining in conduction until anode I is reduced to zero.	Power switching Phase control Inverters Choppers
Complementary Silicon Controlled Rectifier (CSCR)	ANODE GATE CATHODE	ANODE GATE CATHODE	Polarity complement to SCR	Ring counters Low-speed logic Lamp driver
Light-Activated SCR* L8, L9	ANODE GATE CATHODE	ANODE CATHODE GATE	Operates similar to SCR, except can also be triggered into conduction by light falling on junctions	Relay Replacement Position controls Photoelectric applications Slave flashes
Silicon Controlled Switch* (SCS) 3N83-6	CATHODE GATE ANODE GATE CATHODE	CATHODE GATE ANODE ANODE GATE CATHODE	Operates similar to SCR except can also be triggered on by a negative signal on anode-gate. Also, several other specialized modes of operation	Logic applications Counters Nixie drivers Lamp drivers
Silicon Unilateral Switch (SUS) 2N4983-90	ANODE GATE CATHODE	ANODE GATE CATHODE	Similar to SCS but zener added to anode gate to trigger device into conduction at ~8 volts. Can also be triggered by negative pulse at gate lead.	Switching Circuits Counters SCR Trigger Oscillator
Silicon Bilateral Switch (SBS) 2N4991,2	ANODE 2 GATE ANODE 1	GATE ANODE 2 ANODE 1	Symmetrical bilateral version of the SUS. Breaks down in both directions as SUS does in forward	Switching Circuits Counters TRIAC Phase Control
Triac SC92-265	ANODE 2 GATE ANODE 1	ANODE 2 GATE ANODE 1	Operates similar to SCR except can be triggered into conduction in either direction by (+) or (−) gate signal	AC switching Phase control Relay replacement
Diac Trigger ST2			When voltage reaches trigger level (about 35 volts), abruptly switches down about 10 volts.	Triac and SCR trigger Oscillator

FIGURE 6.67 (Continued)

■ QUESTIONS •

1. The toggle switch, normally operated by hand, is an example of a _____ switch. Switches that are actuated automatically are called _____ _____ switches. A switch often used to start and stop motors by the switch's being contacted by a moving part of the system is called a _____ switch.

2. The relay, quite commonly used in industry, is an example of an _____ switch. In many applications, the use of relays has given way to the electronic switch, sometimes referred to as a _____.

3. When the thyristor is triggered, it turns on quickly, or latches, by means of _____ feedback. The SCR, the most popular thyristor, is usually triggered by a voltage on its _____.

4. The SCR is a thyristor that conducts in only one direction. Such a device is called a _____ device. The triac can conduct in both directions, so it is a _____ device.

5. The diac and Shockley diode are examples of _____ devices; they are used to turn on other thyristors.

6. The UJT has a _____ intrinsic standoff ratio that determines relaxation oscillator behavior. The PUT has an intrinsic standoff ratio that is variable or _____.

7. List the four classifications of control devices.

8. Give examples of each of the four switch classifications and their applications.

9. Describe each of the main parts of the armature type of EMR.

10. List the advantages of thyristors over EMRs.

11. List the advantages of reed relays over EMRs.

12. Describe the regenerative switching action that makes a thyristor latch.

13. Draw the schematic symbols for the thyristors listed, and describe the usual method of triggering each: (a) SCR, (b) triac, (c) diac, (d) UJT, (e) PUT.

14. Describe how an LASCR is triggered.

15. Why is a snubber necessary across a thyristor? How does a snubber work?

16. Distinguish between the PUT and UJT in terms of function and construction.

17. Define the intrinsic standoff ratio for a UJT.

18. List four advantages of a solid-state relay over the electromagnetic version.

19. Define critical rate of voltage and current rise. How do these parameters affect the operation of thyristors?

20. Describe how temperature affects the following parameters: (a) I_{GT}, (b) V_{GT}, and (c) $V_{(BO)}$.

21. Draw the schematic symbol of the following devices, and describe how each one is triggered: (a) GCS, (b) SCS, and (c) Shockley diode.

■ PROBLEMS •

1. A 240 V coil has a sealed value of 250 VA and an inrush value of 1000 VA. Find the inrush and sealed current.

2. Assume you have the following voltages and resistances for the circuit shown in Figure 6.34: $R_L = 2$ kΩ, $R_G = 40$ Ω, $V_{AA} = 50$ V, and $V_{F(ON)}$ = 1 V. The SCR used is a 2N2344 with I_{GT} = 200 mA, $V_{GT} = 0.8$ V, and $I_H = 5$ mA.

   a. Will a V_G of 10 V fire the SCR?

   b. Calculate the power dissipated by the SCR, assuming that it turns on.

3. Assume you are given the following values: $R_{\theta JA} = 345°C/W$, $R_{\theta JC} = 124°C/W$, an ambient temperature of 40°C, and $T_{J(max)} = 200°C$.

   a. Calculate the junction temperature when the thyristor is dissipating 300 mW without a heat sink.

   b. Calculate $R_{\theta CA}$ of the heat sink needed to dissipate 1.2 W at the junction. Assume $R_{\theta CS} = 1°C/W$.

   c. Calculate the power dissipated at the junction with a sink having an $R_{\theta CA}$ of 30°C/W and a junction temperature of 100°C.

   d. Calculate the case temperature when the device is dissipating 1.2 W at the junction.

4. An SCR with a TO-3 package has an $R_{\theta JC}$ of 2.0°C/W, a heat sink with an $R_{\theta SA}$ of 3.5°C/W, and an $R_{\theta CS}$ of 0.75°C/W. How much power can the system dissipate and not exceed 150°C in an ambient temperature of 50°C?

5. An SCR is dissipating 10 W at an ambient temperature of 45°C. Given $R_{\theta JC} = 2.5°C/W$, $R_{\theta CS} = 1°C/W$, and $R_{\theta SA} = 4°C/W$, find the actual junction temperature of the SCR.

6. An SCR dissipates 15 W in a 220 case structure. If we want to keep the junction temperature below 150°C in an ambient temperature of 60°C, and the system uses a BeO_2 insulator with thermal grease, what size heat sink is necessary for this system?

7. An SCR dissipates 5 W at 55°C. The thermal resistances are $R_{\theta JC} = 4.0°C/W$, $R_{\theta CS} = 1.4°C/W$, and $R_{\theta SA} = 2.5°C/W$. Find the case temperature of the SCR.

8. Suppose the load in Figure 6.34 is resistive. What size bypass resistor is necessary to ensure that the SCR will turn on? Assume $I_L = 15$ mA, $V_{F(ON)} = 1$ V, and $V_{AA} = 50$ V.

9. If R_L in Figure 6.36A is 10 Ω, what is the firing angle with R_2 set at 3 kΩ? Assume $I_{GT} = 200$ μA and $I_H = 10$ mA.

10. At the firing angle calculated in Problem 9, what is the average current flowing through the load?

11. For the following values for the circuit shown in Figure 6.46A, calculate the frequency and the period of the relaxation oscillator: $R_1 = 25$ kΩ, $C_1 = 0.1$ μF, $V_{(BO)} = 30$ V, and $+V_{AA} = 50$ V.

12. If you assume that the gate voltage on an SCR is clamped to 0.6 V, what gate voltage will be necessary to draw a gate current of 75 mA? Assume you have a gate resistor of 1000 Ω.

13. An SCR has a heat sink that can safely dissipate 25 W of power. If V_T of the SCR is 1.5 V, what is the maximum average current the device can draw?

14. The gate of an SCR is pulsed by a signal with a peak of 250 mW, a PRR of 350 Hz, and a pulse width of 0.6 ms. Find the average power dissipated by the gate.

15. An SCR in a half-wave circuit is triggered at 80° in a 120 V AC circuit. Find the average voltage applied across the load.

16. Find the firing angle necessary to deliver 75 W into a 5 Ω load in the half-wave phase control shown in Figure 6.36A.

17. Calculate the peak firing voltage and the oscillator frequency for the circuit shown in Figure 6.53A. Use the following values: $R_1 = 220$ kΩ, $C_1 = 0.01$ μF, $R_2 = 16$ kΩ, $R_3 = 27$ kΩ, $R_4 = 22$ Ω, and $V_{BB} = 25$ V.

18. A triggering circuit for an SCR has a pulse repetition rate of 100 Hz with a pulse width of 1 ms, and the peak power is 2 W.

   a. Calculate the average power dissipated by the device.

   b. Calculate the duty cycle.

19. An SCR circuit has 230 V AC and 60 Hz applied to the SCR. The SCR circuit has a firing angle of 25°.

a. Calculate the average voltage at the load.

b. What is the conduction angle?

c. Find the average current in the load if the load resistance is 100 Ω.

20. In Figure 6.51A, R_1 = 50 kΩ and η = 0.6.

    a. Calculate the frequency and period of the oscillator.

    b. Calculate the potential across the capacitor when the UJT fires.

21. In Figure 6.51A, what will the value of R_1 need to be for the oscillator to oscillate at 100 Hz?

22. What will the value of R_1 need to be in Figure 6.53A to give an output frequency of 200 Hz?

23. If R_3 in Figure 6.53A is increased to 200 kΩ, what will the new output frequency be?

24. In Figure 6.55, how long a delay will occur between the time power is applied to Q_1 and the time Q_2 is triggered?

25. For the relaxation oscillator circuit in Figure 6.46A, with R_1 = 600 kΩ, C_1 = 2.5 μF, an applied voltage of 35 V, and a breakover voltage of 20 V, find the frequency at which the oscillator will operate.

26. Given the relaxation oscillator shown in Figure 6.51A, with a UJT that has an intrinsic standoff ratio of 0.5, R_1 = 150 kΩ, and C_1 = 0.47 μF, find the firing voltage and the frequency at which the oscillator will operate.

27. For the PUT circuit in Figure 6.53A, with R_2 = 25 kΩ, R_3 = 10 kΩ, R_1 = 750 kΩ, and C_1 = 0.27 μF, find (a) the intrinsic standoff ratio, (b) the firing potential, and (c) the frequency of oscillation.

28. Given the PUT circuit in Figure 6.54 with R_2 = 25 kΩ, calculate the firing voltage across the capacitor C_1 and the delay in firing the SCR after the switch is closed.

29. For the PUT circuit in Figure 6.54, find the resistance of R_1 that will give a 25 ms time delay after the switch is closed.

30. Given the PUT turn signal flasher in Figure 6.55, find the size of R_1 in timer 1 that will give a 1.25 s delay between lamp 1 and lamp 2.

Power Control Circuits

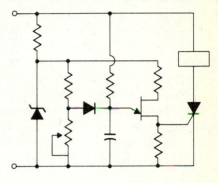

OBJECTIVES

On completion of this chapter, you should be able to:

- Describe how hysteresis produces the snap-on effect in phase control.
- Describe the operation and advantages of the ramp-and-pedestal circuit, zero-voltage switching, and chopper motor control.
- Describe open-loop and closed-loop motor control.
- Explain how pulse-width modulation can accomplish motor speed control.
- List and describe the methods by which an AC motor's speed may be varied.
- Describe how different levels of power are applied to a load with phase-control circuitry.
- Define a converter, and describe the conversion that takes place.
- Describe how the PLL is used to control the speed of a DC motor.
- List the advantages of transistors and power MOSFETs as power control devices.

INTRODUCTION

In the preceding chapter, we discussed the basic building blocks of semiconductor power control circuits, the thyristors. Now that we have provided a background in the operation of these devices, we are ready to discuss their applications. Applications fall into two basic categories: (1) general power control circuits, such as phase control and zero-voltage switching, and (2) motor control. Motor control will be broken down further into AC and DC motor control. Since motor control is one of the major applications of thyristors in industry, we will use most of this chapter to discuss this important subject.

As a technician, you will need to be able to analyze power control circuits. Consequently, in our presentation, we not only show simplified schematic diagrams but also give an idea of what the voltage waveforms should look like around the circuit.

PHASE CONTROL

In the previous chapter, we discussed AC phase-control circuits using SCRs and triacs as power control devices. Examples of these circuits are shown in Figures 6.38 and 6.41. These circuits work well but do have some problems. In this section, we will show circuits that overcome some of the problems inherent in the circuits presented in Chapter 6.

The resistor R and the capacitor C in Figure 7.1A are called a *phase-shift network*. The capacitor delays the rise of voltage across the capacitor past the 90° point, giving greater control. The waveforms across the load, triac, and capacitor are shown in Figure 7.2. Note that the capacitor, at the beginning of every new cycle, starts out at the opposite polarity. For example, during the early part of the positive alternation of the supply, the capacitor has a negative voltage across it. The capacitor must first discharge before it can start to charge in a positive direction. We must take this feature of the capacitor into account in any attempt to estimate the firing angle of the circuit in Figure 7.1A.

Let us assume that the trigger device is a diac with a breakover voltage of 32 V and that the

voltage across the diac goes to a minimum of 10 V when the capacitor discharges. If the diac triggers at 32 V, the additional capacitor voltage needed to switch on the diode is 32 V + 10 V, or 42 V. We use the charts in Figure 7.1B and 7.1C to calculate the firing (or delay) angle. The chart in Figure 7.1C is just an expanded version of the one in Figure 7.1B, where the normalized-voltage axis (the y axis) is expanded to cover the range from 0 to 0.35. To use the chart in Figure 7.1B, we need to calculate the *normalized voltage*, which is approximately $V_C/V_{T(AC)}$. For our example, we have 42 V/120 V AC, or 0.35. Now we use the equation

$$\tau = 2RCf$$

where

τ = ratio of RC time constant to period of a half cycle
R = resistance of resistor in RC network (in ohms)
C = capacitance of capacitor in RC network (in microfarads)
f = frequency of applied voltage (in hertz)

For R = 120 kΩ, C = 0.1 μF, and f = 60 Hz applied, we get

$$\tau = 2RCf = (2) (120 \text{ k}\Omega) (0.1 \text{ μF}) (60 \text{ Hz})$$

$$= 1.44$$

Examination of the graph in Figure 7.1C shows a firing angle of about 108° for this value (1.44) of τ.

• EXAMPLE 7.1

Given the circuit in Figure 7.1A for a diac trigger device, with a breakover voltage of 32 V, a minimum diac voltage of 10 V when the capacitor discharges, R = 90 kΩ, C = 0.1 μF, and f = 60 Hz applied, find the firing angle of the triac.

Solution

If the diac triggers at 32 V, the additional capacitor voltage needed to switch on the diode is 32 V + 10 V,

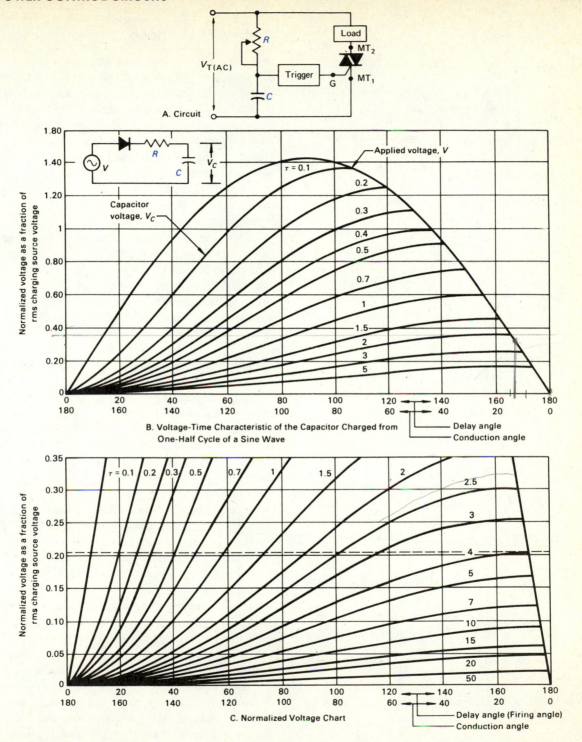

A. Circuit

B. Voltage-Time Characteristic of the Capacitor Charged from
One-Half Cycle of a Sine Wave

C. Normalized Voltage Chart

FIGURE 7.1   Triac Phase Control

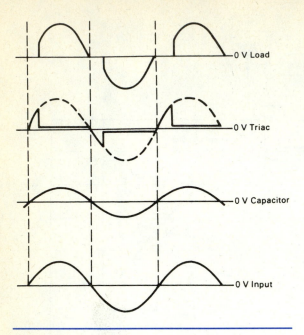

FIGURE 7.2 Waveforms in Triac Phase-Control Circuit

or 42 V. To use the chart in Figure 7.1C, we need to calculate the normalized voltage $V_C/V_{T(AC)}$, which is 42 V/120 V AC, or 0.35. For

$$\tau = 2RCf$$

with $R = 90$ kΩ, $C = 0.1$ µF, and $f = 60$ Hz applied, we get

$$\tau = 2RCf = (2)\,(90\,\text{k}\Omega)\,(0.1\,\text{µF})\,(60\,\text{Hz})$$

$$= 1.08$$

Examination of the graph in Figure 7.1C shows a firing angle of about 82.5° for $\tau = 1.08$.

Hysteresis in Phase Control

One major problem with phase-control circuits is one that designers call the *hysteresis effect*. The hysteresis effect can be explained through the use of the triac phase-control circuit in Figure 7.1A. From examining this circuit, we would expect that the load (say, a lamp) would have AC power applied to it gradually; but it does not. In reality, the lamp will turn on suddenly with moderate brightness. After the lamp comes on, its intensity can be varied from low to high levels. This *snap-on effect* is due to the hysteresis inherent in this design.

During the time when the triac is off, the capacitor charges through the variable resistor R. Since the capacitor is a reactive component, the capacitor voltage lags the line voltage by about 90°. When the triac fires, the capacitor discharges through the gate and main terminal 1 junction of the triac and the trigger device. During the next half cycle, the capacitor voltage will exceed the trigger device's breakover voltage sooner because it will start to charge from a lower voltage.

This undesirable hysteresis effect can be reduced or eliminated in several ways. Two ways of accomplishing the reduction are shown in Figure 7.3. The addition of the extra RC network in Figure 7.3A allows C_1 to partially recharge C_2 after the diac has fired, thus reducing hysteresis. This addition also adds a greater range of phase shift across C_2.

The second method of reducing hysteresis, illustrated in Figure 7.3B, uses a device called an *asymmetrical trigger diode*. This diode has a greater breakover voltage in one direction than in the other. It is so made that when it triggers for the first time, the capacitor discharges into the triac gate. During the next half cycle, however, the triggering voltage is equal to the original breakover voltage plus the capacitor's voltage decrease. This result allows the capacitor voltage to have the same time relationship to the applied voltage. Thus, hysteresis is reduced.

The circuits shown in Figure 7.3 can be used in virtually any AC low-power control circuits, such as universal and induction motors, lamps, and heaters. In fact, one designer calls this circuit the "universal power controller."

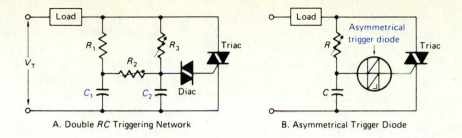

A. Double *RC* Triggering Network   B. Asymmetrical Trigger Diode

FIGURE 7.3   Reducing Hysteresis

UJT Phase Control

As discussed in Chapter 6, when thyristors are used in power control circuits, the thyristor is fired at some variable phase angle. But if we want maximum power delivered to a load, the thyristor must fire as soon as the AC voltage across it goes positive (and/or negative in the triac). The thyristor must conduct for 180° after firing for maximum power. If less than full power is needed, the thyristor firing must be delayed past 0° or 180°. A control circuit is needed, therefore, to control the firing angle between 0° and 180° for the SCR and also between 180° and 360° for the triac.

We have seen that the *RC* lag network can accomplish this control. These circuits, however, are affected by loading when connected to thyristor gate leads. They also are affected by supply voltage variations. The UJT oscillator, on the other hand, has none of the disadvantages of the *RC* networks. Thus, the UJT can be used as a phase-control device in control circuits using either triacs or SCRs.

Figures 7.4A and 7.4B show a UJT oscillator controlling both an SCR circuit and a triac circuit. Both circuits function in essentially the same way. The thyristor is in the off state until the UJT fires. In Figure 7.4A, when the UJT fires, the capacitor will

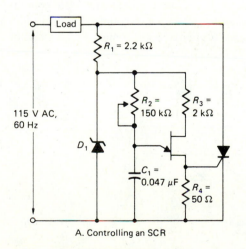

A. Controlling an SCR

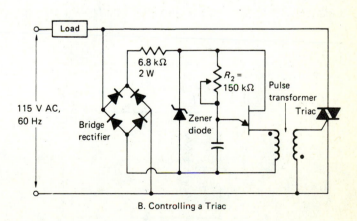

B. Controlling a Triac

FIGURE 7.4   UJT Oscillator Control Circuits

discharge quickly through the resistor R_4. This positive pulse will turn on the SCR. The SCR will remain on until the line voltage approaches zero. At this point, the SCR will turn off and remain off throughout the entire negative half cycle.

The triac circuit in Figure 7.4B works in the same way. The positive spike is coupled to the triac's gate by the pulse transformer. When the triac turns on, the voltage to the control section of the circuit is taken away. No more pulses can occur during that half cycle of the input voltage. Since the line voltage is rectified by the bridge, the operation is the same for both positive and negative half cycles. This circuit can control up to 97% of the power available to the load.

Ramp-and-Pedestal Phase Control

Most of the preceding phase-control circuits work well in manually controlled applications. But signals in automatic-control situations are generally not large enough to produce adequate control. We might say that these circuits have relatively low gain. That is, in the RC networks used in phase control, a relatively large change in resistance is needed.

An alternative to the RC network is the *ramp-and-pedestal control circuit*, which provides much higher gain. Such a control circuit is illustrated in Figure 7.5A. When the line voltage goes positive, the zener diode keeps point A at a positive 20 V. The capacitor C_1 charges quickly through diode D_2, R_2,

and R_1. Let us say that resistor R_3 is adjusted to 3.3 kΩ. With this resistance, R_3 will drop about 10 V. The capacitor will continue to charge until the voltage at point C reverse-biases the diode D_2, which will occur at about 10 V. The capacitor, which was charging through about 8.3 kΩ, will now charge much more slowly through the 5 MΩ resistance of R_4. Eventually, the voltage at point C will reach the firing potential of Q_1. The UJT will then turn on and fire the SCR.

The voltage rise at point C is diagrammed in Figure 7.5B. Note that the voltage rises very quickly between points 1 and 2 (the pedestal). At point 2, diode D_2 becomes reverse-biased and charges more slowly from point 2 to 3 (the ramp). At point 3, the UJT will fire, turning on the SCR. If we increase the resistance of R_3, the capacitor will charge quickly to a higher voltage before D_2 becomes reverse-biased. This higher value is represented by point 4. Note that since the capacitor started charging slowly from a higher voltage, it will reach the firing potential sooner in the input cycle. Thus, more power will be delivered to the load.

IC Ramp-and-Pedestal Phase Control

The simple ramp-and-pedestal circuit we have just discussed can be improved by adding more components. Of course, these extra components add more cost as well as improved performance. Since one of

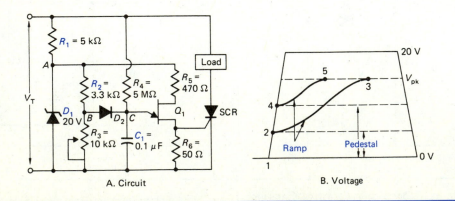

A. Circuit B. Voltage

FIGURE 7.5 Discrete-Component, Ramp-and-Pedestal Control

the advantages of ICs is their ability to replace large and sometimes costly collections of discrete devices, ICs are well suited to this application.

Several manufacturers make an IC that does the phase-control task we discussed. One such IC is the PA436 phase controller manufactured by General Electric. Figure 7.6A shows the ramp-and-pedestal waveform used by the PA436. Note the difference between this waveform and the UJT waveform (Figure 7.5B). The UJT ramp-and-pedestal has a positive-going ramp, but the PA436 has a negative-going cosine ramp and a positive pedestal and reference. (A *cosine ramp* is a waveform whose average DC level is changing at an inverse sine rate.) In the UJT circuit, the reference (or firing) voltage remains constant and the pedestal changes. In the PA436 system, the pedestal remains constant and the reference changes.

The block diagram of the PA436 is illustrated in Figure 7.6B. The input signal is a DC voltage that establishes the pedestal level. The cosine ramp is developed from the supply voltage and is adjustable externally by the gain adjust. The ramp-and-pedestal waveform is then compared with a reference waveform by the differential comparator. The differential comparator produces an output waveform only when the ramp is below the reference level. The lockout gate keeps the differential comparator signal from reaching the trigger circuit until the AC supply voltage passes through zero.

Heater Control

The circuit in Figure 7.7 is a heater controlled by ambient temperature. By using phase control, this circuit can reduce the power to the heating element when the ambient temperature reaches the correct value. The bridge rectifies the line voltage and applies DC through R_1 to the 20 V zener. The thermistor and the variable resistor R_1 control the base current of the transistor Q_1. Resistor R_2 is adjusted so that Q_1 is off at the desired temperature. Since

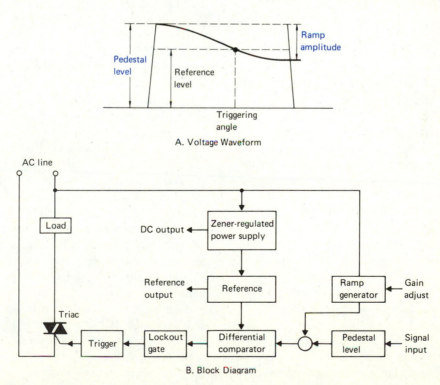

FIGURE 7.6   IC Ramp-and-Pedestal Control

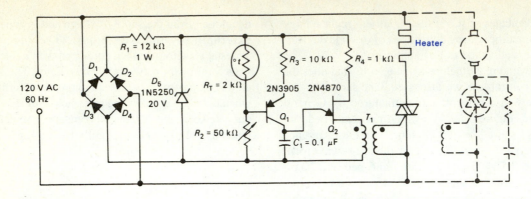

FIGURE 7.7 Heater Control

transistor Q_1 controls the charging current of C_1, no charging current can flow with Q_1 off, and the UJT cannot fire. The triac remains off. If the temperature decreases, the resistance of the thermistor increases. At some point, Q_1 is turned on and C_1 charges to the firing voltage of Q_2. Transistor Q_2 fires and turns on the triac through the pulse transformer T_1. If the temperature continues to decrease, the Q_1 base-emitter bias increases, increasing the charging current. Capacitor C_1 charges more quickly, firing the triac earlier in the cycle and applying more power to the heater. When the temperature increases, transistor Q_1 will conduct less, turning the triac on later in the cycle. Finally, when the temperature reaches the desired value, transistor Q_1 will be off, keeping the triac off.

The circuit shown in Figure 7.7 works well for a heater load. However, it can also be used for a blower motor control. Interchanging the positions of R_2 and R_T makes this circuit suitable for a cooling application.

ZERO-VOLTAGE SWITCHING

Several problems arise in the design of thyristor power control circuits such as the ones we have described previously. Opening and closing switches with applied voltages cause large variations in line current during short periods of time. These rapid current variations produce unwanted RF interference and potentially damaging inductive-kick effects. (Recall that an inductor with current flowing through it will produce a large magnetic field around it. The field will collapse when current stops, generating a very high voltage.)

One solution to this problem is to make sure that the thyristor fires only when the supply is at or near 0 V. With no voltage across the switch, no current should flow. Several manufacturers provide ICs that accomplish this type of switching, called *zero-voltage switching*.

One such circuit is the CA3058 manufactured by RCA. This device is a monolithic IC designed to act as a triggering circuit for a thyristor. Internally, the CA3058 contains a threshold detector, a diode limiter, and a Darlington output driver. Together, these circuits provide the basic switching action. The DC operating voltages for the device are provided by an internal power supply. This power supply is large enough to supply external components, such as transistors and other ICs.

An example of a circuit using this IC zero-voltage switch (ZVS) is shown in Figure 7.8. Note that the ZVS output is connected directly to the gate of the thyristor. Whenever the DC logic level is high, the output to the gate of the triac is disabled. A low-level input enables the gate pulses, thus turning on the triac and providing current to the load.

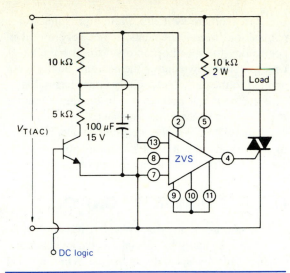

FIGURE 7.8 IC Zero-Voltage Switch (ZVS) Controlling a Triac

When the voltage at pin 13 is equal to or less positive (more negative) than the voltage at pin 9, the output of pin 4 is inhibited and the triac gets no firing voltage. When the voltage at pin 13 is more positive than the voltage at pin 9 or equal to it, the output of pin 4 is enabled and the triac gets firing voltage pulses. When pins 10 and 11 are connected to pin 9, as in Figure 7.8, a precision voltage divider is formed, placing a potential equal to about half of the DC voltage at pin 2, which is about 6.5 V. In the diagram in Figure 7.8, the voltage at pin 9 is about 3.25 V. If the voltage at pin 13 goes above the 3.25 V potential on pin 9 or is equal to it, the triac will be turned on, applying power to the load. A low voltage on the DC logic input will turn the transistor off, placing 6.5 V on pin 13 and turning on the triac. A high voltage on the base of the transistor will pull the voltage at pin 13 lower than the voltage at pin 9, turning the triac off.

An on-off temperature controller using a ZVS is shown in Figure 7.9. Let's assume that R_1 is adjusted to 1000 Ω, R_2 is a GB32J2 thermistor, and the voltage at pin 9 is 3.25 V. At what temperature will the triac go off? Let's say we turn the device on at 25°C. The resistance of the thermistor at 25°C is 2000 Ω. From the voltage divider equation, the voltage at pin 13 is

$$V_{13} = \left(\frac{R_2}{R_2 + R_1}\right)(V)$$

$$= \left(\frac{2000 \ \Omega}{2000 \ \Omega + 1000 \ \Omega}\right)(6.5 \text{ V})$$

$$= 4.33 \text{ V}$$

The voltage at pin 13 is greater than the voltage at pin 9. The triac will be turned on, and power will be applied to the heater. As the load produces heat, the temperature will rise, lowering the resistance of the thermistor and the voltage at pin 13.

When will the ZVS turn the triac off? It will turn the triac off when the voltage at pin 13 goes below the voltage at pin 9, at 3.25 V. The voltage at pin 13 will equal the voltage at pin 9 when $R_1 = R_2 = 1000$ Ω. From Figure 8.15 (in Chapter 8), we can see that the thermistor will equal 1000 Ω at about 40°C.

• EXAMPLE 7.2
Given the ZVS in Figure 7.9, with $R_1 = 100$ Ω and R_2 a GB32J2 thermistor, find the temperature at which the triac will turn off if we energize the heater at room temperature.

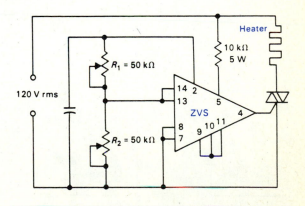

FIGURE 7.9 On-Off Temperature Controller Using a ZVS

Solution

The resistance of the thermistor at 25°C is 2000 Ω. From the voltage divider equation, the voltage at pin 13 is

$$V_{13} = \left(\frac{R_2}{R_2 + R_1} \right)(V)$$

$$= \left(\frac{2000\,\Omega}{2000\,\Omega + 100\,\Omega} \right)(6.5\,\text{V})$$

$$= 6.2\,\text{V}$$

The voltage at pin 13 is greater than the voltage at pin 9. The triac will be turned on, and power will be applied to the heater. As the load produces heat, the temperature will rise, lowering the resistance of the thermistor and the voltage at pin 13. The voltage at pin 13 will equal the voltage at pin 9 when $R_1 = R_2 = 100\,\Omega$. From Figure 8.15, we can see that the thermistor will equal 100 Ω at about 125°C.

Since the ZVS turns on only at or near the zero-voltage point of the applied voltage, only complete half or full cycles are applied to the load. This point is illustrated in Figure 7.10. The amount of power delivered to the load depends on how many full or half cycles are applied during a particular time. The average power delivered to the load, then, depends on the ratio of how long the thyristor was on to how long the thyristor was off.

Zero-voltage switching, although suitable for controlling heating elements, is not very useful in controlling the speed of motors. Motors have a tendency to slow down during the time that power is not applied to the circuit, especially under heavy loads. This factor is not bothersome in heating elements because of their long time constant.

The IC ZVS is a very versatile device. It is used in industry in relay and valve control applications as well as in heating and lighting controls.

ELECTRICAL MOTOR CONTROLS

As we saw in the previous chapter, motors can be controlled manually by special manual switches and magnetically by special contactors. When manual switches are used to start a motor, the device is called a *manual motor starter*. Manual motor starters usually have overload protection for the motor as well as the switch to apply power to the motor. When motors cannot be operated by hand or when they must be operated by other switches, a special contactor called a *magnetic motor starter* is used. Circuits using manual motor starters generally are not complicated or difficult to understand. Magnetic motor starters, on the other hand, can be quite complex. In this section, we will discuss several common circuits that use magnetic motor starters.

Before we examine these circuits, we will explain the special types of diagrams we will be using in the text. These same types of diagrams are

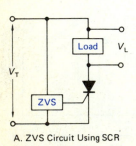

A. ZVS Circuit Using SCR

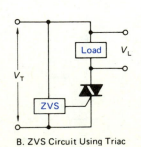

B. ZVS Circuit Using Triac

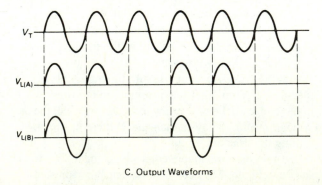

C. Output Waveforms

FIGURE 7.10 ZVS Applying Voltage to Load

widely used in industry. Two of the diagrams are illustrated in Figure 7.11: the wiring diagram and the simplified diagram.

Wiring diagrams, like the one shown in Figure 7.11A, include all of the devices in the system and show their physical relation to each other. All poles, terminals, coils, and so on, are shown in their proper place on each device. If you are involved in wiring a motor starter, you will find this type of diagram helpful since connections can be made exactly as they

are shown on the diagram. The wiring diagram, however, is not very useful in explaining how the system works. The connections are not easy to follow.

If the wiring diagram is rearranged and simplified, the result is a line diagram (Figure 7.11B). The *line diagram*—also referred to as a *simplified*, or *ladder*, *diagram*—is a representation of the system showing everything in the simplest way. No attempt is made to show the various devices in their actual positions. All the control devices are shown between vertical lines. The lines represent the source of power, either AC or DC. Circuits are shown connected as directly as possible from one of these lines to the other. All connections are made in such a way that the functions of the various devices can be easily traced. Note that the same terminal identification letters and numbers are used in both the wiring and line diagrams. They designate the control and power connections. For troubleshooting, a line diagram is much easier to work with than a wiring diagram.

The line diagram in Figure 7.11B illustrates a three-wire circuit with low-voltage protection. When the start button is pushed, current flows through the start switch, through the motor starter coil labeled M, through the overload relay contacts, and back to the other side of the line. All relay contacts associated with the M coil change state. Thus, all NO switches close and all NC switches open. Note that the three NO contacts that supply power to the motor close. Power is then applied to the motor. When the start button is pushed and current flows through the M coil, the contacts around the start switch will close. Current will then flow through the circuit even when the start switch is released. This configuration is called a *latching relay circuit*. This circuit is protected against a low-voltage or a voltage-loss condition. If the supply voltage goes low or is lost, the contactor will open. When proper line voltage is restored, the operator must press the start button to energize the contactor. The contactor can be de-energized by pressing the stop button. The contacts will then go back to their normal state, removing power from the motor.

As mentioned in our discussion of AC motors, the three-phase, squirrel-cage AC motor is well suited to reversing rotation. This reversal is accomplished by switching any two of the three line conductors.

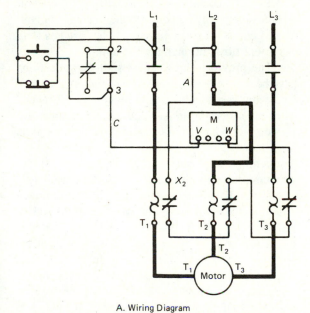

A. Wiring Diagram

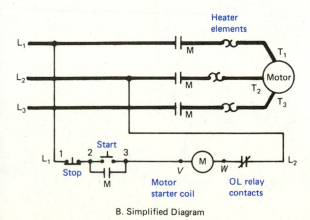

B. Simplified Diagram

FIGURE 7.11 Motor Starter Diagrams

Reversing starters usually use two separate contactors, one for forward rotations and the other for reverse. The reversing starter is electrically and mechanically interlocked so that both contactors cannot be energized at the same time and cause a short circuit.

A *reversing circuit* is shown in Figure 7.12. In this circuit, the motor is brought to a stop before the direction of rotation is changed. Pressing the forward push button will energize the forward contactor coil (F), causing the motor to rotate in the forward direction. At the same time, the F coil opens the normally closed F contact in the reverse contactor coil circuit and closes the normally open F contact around the forward push button. As long as the F contactor is energized, pressing the reverse push button will have no effect because the F contact is open in the reverse coil circuit. The operator must press the stop button before the direction of rotation can be changed. When the forward contactor drops out, the NC contact F in the reverse coil circuit is reclosed. The rotation of the motor now can be started in the reverse direction. The limit switches (*LS*) are shown in this circuit since they are sometimes used for equipment such as overhead doors, which are stopped with a limit switch at the end of the door travel.

Some motors must have the capability to inch or jog. A motor is said to be *jogged* when power is applied briefly in controlled bursts. Jogging motors are necessary when motor loads need to be precisely positioned prior to starting the motor. A *jogging circuit* applies power to the motor only when a push button is held down. The motor should stop when the

push button is released. The circuit shown in Figure 7.13 is such a circuit. When the jog button is pressed, current flows through the M coil, energizing it. Although not shown, the M coil will have NO contacts that, when closed, will apply power to the motor. Note that when the jog power button is released, power is removed from the M coil and any NO contacts will open. The M coil is not in a latching relay configuration. When the start button is pushed, current flows through the control relay (CR) coil. Both CR contacts close, one latching the CR coil and the other applying power to the M contactor coil. The M contactor coil will remain energized as long as current flows through the CR coil.

The previous three motor starter applications have been based on AC motors. DC motors use similar circuits, such as the one in Figure 7.14. This circuit uses a resistor as its basic control mechanism. As you will recall from Chapter 3 on DC motors, a large inrush of current occurs at starting, due to the low cemf. If a resistor is placed in series with the armature, the voltage and current will be decreased at starting. The resistor, however, must be removed after the motor has picked up speed. Both M and A_1 in Figure 7.14 are TDRs. When the start push button is pressed, the M contacts close immediately, applying supply voltage across the full acceleration resistance and the field. After a certain delay, the A_1 contacts close, shorting out part of the acceleration resistance. Of course, by the time that M times out, the motor has accelerated to a speed short of its full speed. After another time delay, the motor will have reached its final speed. At this time, A_1 times out, and

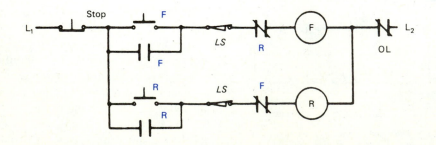

FIGURE 7.12 Reversing Starter Circuit

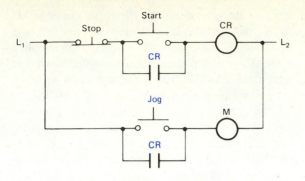

FIGURE 7.13   Jogging Circuit

DC MOTOR CONTROL

In the process of converting factories to automatic controls, engineers are faced with a bewildering variety of new, high-tech motor controls. In some cases, the choice is between old, reliable, well-understood techniques and newer, less understood, and possibly less reliable techniques. For many years, the choice was simple. If you wanted a variable-speed motor drive, you chose a DC motor and a Ward Leonard drive control system. Invented in the 1890s and patented by H. Ward Leonard, the Ward Leonard system held sway in motor control until the late 1950s, when other types of control became prominent.

Even after the introduction of thyristor-controlled motors, however, motor controls were still constructed component by component to control field current and armature voltage. Today, though, everything needed to control the speed of DC motors is in one easy-to-install package. All rectification and control functions are self-contained in the DC drive. There is no need to assemble a drive component by

the A_1 contacts close, applying power to A_2. Since A_2 is a normal control relay, the A_2 contacts close immediately, shorting out the remainder of the acceleration resistance. The motor has accelerated to full speed with an adjustable resistance, which lowers the armature current and torque at starting.

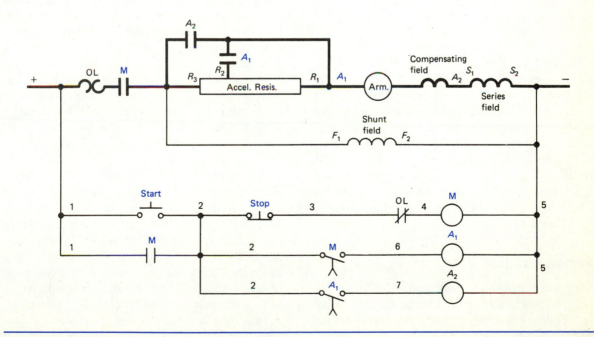

FIGURE 7.14   DC Motor Starter Circuit

component. The only connections necessary are an input source of AC power, either single- or three-phase, and an output to field and armature leads.

DC motors are widely used in modern industry for many reasons. Speed is relatively easily controlled with precision in DC motors. Furthermore, speed can be controlled continuously over a wide range, both above and below the motor's rated speed. Many industrial applications call for a machine that has very high starting torque, such as those found in traction systems. The DC motor satisfies most of these requirements adequately. Compared with AC drives, DC drives are generally less expensive, an important advantage in keeping initial equipment costs down.

As we have seen in this chapter, the thyristor is the major power control element in power controllers. And the SCR is the thyristor most often used in DC motor control applications. SCRs are used in two types of DC motor control systems: phase-control circuits and chopper circuits. The phase-control circuits, as we have discussed, control power by changing the firing angle (α) (also called θ_{fire} in an earlier chapter). When the SCR fires early in the cycle, corresponding to a small α, more average power is delivered to the load. As the firing angle increases, the SCR fires later in the cycle, delivering less power to the load.

Some DC motors have power supplies that are also DC. A good example of this application is the traction motor found in industrial delivery vehicles and forklifts. In the past, variable-resistance control was used in these situations. But such control is very inefficient. Today, the electronic chopper is used to control machines with DC power sources. A *chopper* is a thyristor switching system in which an SCR is turned on and off at variable intervals, producing a pulsating DC. The longer the SCR is on, or the higher the switching frequency, the more power is delivered to the load.

Phase Control

Many small industrial plants, farms, and homes are supplied with only single-phase power. DC motor speed control systems in these cases depend on the rectifying capabilities of the SCR to change the AC into the DC needed to develop torque in the motor.

Because the reverse-blocking thyristor converts AC to DC, this circuit is often referred to as a *converter*.

Since in these cases the thyristors are supplied with AC voltage, commutation of the thyristor is not a problem. The SCR naturally commutates when the line voltage passes near 0 V. In addition to supplying natural commutation with AC line voltages, reverse-blocking, thyristor converter circuits are very efficient. Because of the low voltage drop across the device, as much as 95% of available power can be delivered to the load.

Basic Converter. The simplest of the converters, the *half-wave converter*, is illustrated in Figure 7.15A. During the time that the supply goes negative, SCR$_1$ blocks current flow. As a result, no power is applied to the motor (M). When the input goes positive, SCR$_1$ is forward-biased. It will conduct when a positive trigger pulse is applied to the gate (T represents the trigger circuit in the figure). The SCR will remain in conduction until the supply goes negative.

When SCR$_1$ conducts, it applies power to the motor. Diode D_1 is a flywheel, or freewheeling, diode. After current flows through the armature and the SCR turns off, the magnetic field built up around the armature will collapse. The field reverses polarity and conducts through the diode. The collapsing field supplies energy to the load during the time the SCR is off.

As shown in the armature voltage (V_A) waveform (Figure 7.15C), the half-wave converter has a large amount of ripple. As a result, this form of control is not used very much in DC motor speed control.

Half-Controlled Bridge Converter. Figure 7.15B shows a more useful converter, the *half-controlled bridge converter*. When the supply goes positive, SCR$_1$ and D_2 are forward-biased. Current will flow (when SCR$_1$ is triggered) through D_2, the armature, and SCR$_1$. Near 0 V, SCR$_1$ commutates (at time t_1 in Figure 7.15C) and the armature field collapses, causing current to flow in the flywheel diode. At α_2, SCR$_2$ is triggered, causing current to flow through D_1 and the motor armature. Current will continue to flow until time t_2, when SCR$_2$ commutates. From that point, the cycle starts over again.

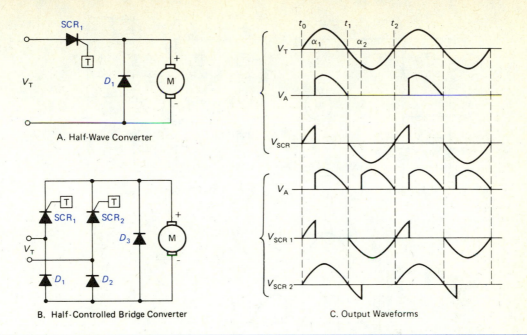

A. Half-Wave Converter

B. Half-Controlled Bridge Converter

C. Output Waveforms

FIGURE 7.15 Half-Wave and Half-Controlled Bridge Converters Driving DC Motors

If you compare the armature waveforms of these two systems, you will note that the half-controlled bridge has far less ripple than the half-wave system. As in any phase-control system, decreasing the firing angle will cause more power to be delivered to the load, increasing motor speed.

Full-Bridge Converter. A *full-bridge converter* is illustrated in Figure 7.16A. When the line voltage goes positive, SCR_1 and SCR_4 are forward-biased. They will conduct when triggered. Note in Figure 7.16C that the armature voltage goes negative between time t_1 and α_2. At time t_1, the SCRs have commutated. Beyond this point, the motor armature field collapses and reverses polarity, supplying power back to the supply.

If the firing angle goes past 90°, the average armature voltage will be negative. That is, the motor will be returning more power than it is using. Any

rotating system that returns more power than it uses will decrease rotational speed. In fact, some motors use this system as a brake. Such a system is called a *regenerative braking system*.

The average DC voltage applied across the armature can be easily calculated if a flywheel diode is applied across the motor. The flywheel diode will change the armature voltage (V_A) waveform by clamping the negative voltage excursions to zero. When the line goes through 0 V, the flywheel diode will conduct, rather than force current to flow through the SCRs. The average armature voltage can be found using the following equation:

$$V_A = 0.637 V_{pk} \cos \alpha$$

where

V_A = average armature voltage
V_{pk} = maximum value of the line voltage
α = firing angle

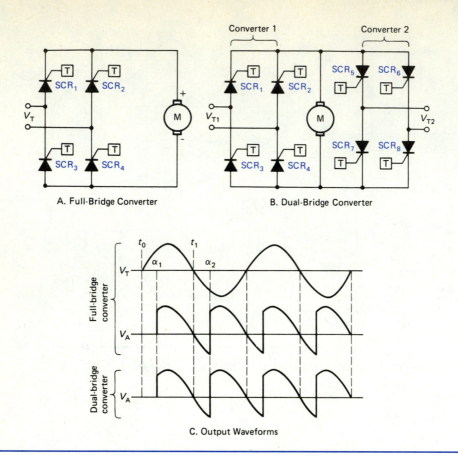

A. Full-Bridge Converter

B. Dual-Bridge Converter

C. Output Waveforms

FIGURE 7.16 Full-Bridge and Dual-Bridge Converters Driving DC Motors

As an example, what is the average armature voltage if a 120 V AC line is applied to this circuit with a firing angle of 30°? We have

$$V_{pk} = 1.414\, V_{rms}$$

$$= (1.414)\,(120\text{ V AC})$$

$$= 170\text{ V}$$

$$V_A = 0.637 V_{pk} \cos \alpha$$

$$= (0.637)\,(170\text{ V})\,(\cos 30°) = 93.8\text{ V}$$

A change to a firing angle of 45° would give a new average voltage:

$$V_A = 0.637 V_{pk} \cos \alpha$$

$$= (0.637)\,(170\text{ V})\,(\cos 45°) = 76.6\text{ V}$$

Note that the average voltage decreases as the firing angle increases, as we would expect.

Recall from Chapter 3 that a DC motor can reverse direction by reversing the direction of current in the armature. One way to accomplish this reversal is to have a switch that, when energized, will reverse the armature connections. In a shunt or a separately excited motor, we could reverse the direction of the current in the field winding by the same method. But mechanical or electromechanical switching is often too slow for some applications. However, almost instantaneous reversals of current flow can be achieved by using the circuit shown in Figure 7.16B, the *dual-bridge converter*. In this circuit, changing from converter 1 to converter 2 will reverse the direction of current flow in the motor.

An example of a full-bridge converter (or *full-wave bridge*, as it is sometimes called) is shown in Figure 7.17. This illustration was taken from an instruction manual on a MiniPak Plus DC motor drive, manufactured by Reliance Electric Company. The drive controls a wound-field, shunt-connected DC motor. It allows adjustable speed and gives constant torque output from 50% of base speed to base speed, provided the current to the shunt field is held constant. The controller changes the speed by adjusting the average armature voltage. Armature speed is closely proportional to the armature voltage. If this converter is used with PM DC motors, no field excitation is required. The amount of armature current drawn is proportional to the load torque on

the motor shaft. The MiniPak Plus system contains an adjustable-voltage armature supply, a fixed-voltage field supply, and a regulator to control and adjust armature voltage under varying torque and speed demands.

The four SCRs (labeled THY in the figure) are contained in a package called a *power cube*. The armature voltage is regulated by adjusting the firing angle of these thyristors. Each of the SCRs receives a gate pulse at a point on the incoming line voltage where the anode is positive with respect to its cathode. Each SCR turns on when a low-power gate pulse is applied for a duration of 350 μs. This pulse will turn the device on and will keep it on as long as the anode is positive with respect to the cathode. When the anode goes negative with respect to the cathode, this device will turn off, or commutate.

As we have noted earlier, the gate pulse timing controls the firing angle. If the SCR fires early in the cycle, the SCR will conduct for a relatively long time. The longer the SCR remains on, the more average power is delivered to the load. Firing the thyristor late in the positive line alternation allows only short periods of conduction through the load. The less time the SCR conducts, the less average power is delivered to the load.

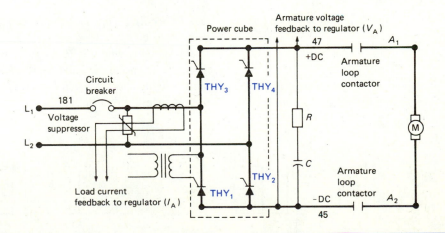

FIGURE 7.17 Reliance Electric Company DC Motor Drive

When L_1 is positive with respect to L_2, current flows through thyristor 3, through the motor armature, through thyristor 2, and back to the supply. When L_2 is positive with respect to L_1, current flows through thyristor 4, through the motor armature, through thyristor 1, and back to the supply. Reliance has also supplied special circuitry to prevent thyristors 3 and 2 and thyristors 4 and 1 from turning on at the same time. Energizing these pairs would cause a short circuit, possibly damaging the motor and controller.

Note the transient protection in the MOV connected across the line. The MOV is part of a module for easy replacement when it fails.

Three-Phase Systems. Large DC motors in industry use a three-phase power source. One reason three-phase power is used to supply DC drive systems is ease of filtering. The three-phase system has significantly less ripple voltage than a single-phase system and therefore takes less filtering to achieve the same result.

We will not discuss three-phase DC motor drive systems because they do not differ significantly in theory of operation from single-phase systems. Of course, there will be two additional phases to contend with, each displaced 120° from the other. Because of the additional number of phases, circuit complexity increases. Also, waveform analysis of such a system becomes very complex, but little added understanding is gained through such analysis.

Chopper Control

In those applications where the power source is DC and the load is a motor, the DC chopper motor speed control is one of the most efficient methods of control. Chopper controls are used to control the speed of electric vehicles such as forklifts, delivery cars, and electric trains and trolleys. Chopper control is used not only because it is the most efficient but also because it provides smooth acceleration characteristics.

Basic Chopper Control. The basic *chopper control circuit* is illustrated in Figure 7.18A. Note that the SCR provides DC power to the motor by switching on and off. The average DC voltage presented to the motor is controlled by keeping the frequency constant and increasing and decreasing the amount of time that the SCR is on. When the SCR is on 50% of the time and off by the same amount (Figure 7.18B), an average of 50% of the voltage is delivered to the load. With a duty cycle of 25% (the SCR on 25% of the time), only a quarter of the applied voltage is delivered to the load (Figure 7.18C). Changing from a duty cycle of 50% to one of 25% will be accompanied by a decrease in the motor's speed. You may recognize this technique as pulse-width modulation.

Another method of chopper control, which we will not discuss, keeps the pulses the same width and increases or decreases the frequency. This method is not often used in controlling DC motors because it requires too great a frequency change to be practical.

Jones Circuit. Figure 7.19 shows a circuit often used in chopper control of DC motors, the *Jones circuit*. In this circuit, when SCR_1 is fired, current flows through the motor, L_1, and SCR_1. Because of the mutual inductance between L_1 and L_2, current flow in L_1 induces current flow in L_2. Current flows down

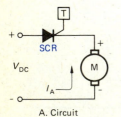

A. Circuit

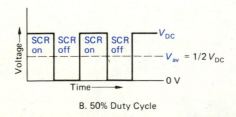

B. 50% Duty Cycle

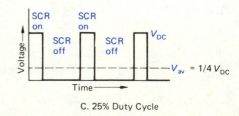

C. 25% Duty Cycle

FIGURE 7.18 Basic Chopper Control Circuit

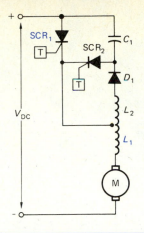

FIGURE 7.19 Jones Circuit

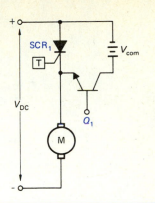

FIGURE 7.20 External Commutation Circuit

through D_1 and C_1, resulting in the charge shown. As long as SCR_1 is conducting, power is applied to the motor.

SCR_1 is commutated as follows: SCR_2 is triggered into conduction, causing C_1 to discharge through SCR_1. This discharge path is opposite the path of the load current flowing through SCR_1. Total anode current is reduced in SCR_1 to the point where it falls below the holding current level. SCR_1 then turns off.

External Commutation. Chopper circuits may be commutated externally. An *external commutation circuit* is shown in Figure 7.20. When SCR_1 is fired, current flows through the motor from the supply, developing torque to turn the motor. The thyristor is commutated by turning on the transistor Q_1. The transistor applies a reverse-biased voltage V_{com} across the thyristor, turning it off. Power to the motor is regulated by proper choice of the transistor turnon point.

Full-Bridge, Pulse-Width–Modulated Speed Control

The chopper control circuit just discussed is a basic pulse-width–modulated motor speed control. The power control device may be a thyristor, as shown in Figure 7.18, a transistor, a power MOSFET, or an

IC, as shown in Figure 7.21B. The UDN2954 shown in Figure 7.21B, manufactured by Sprague Electric Company, was designed for bidirectional control of PM DC or stepper motors. It can provide 2 A of output current at 50 V. The truth table in Figure 7.21A gives a brief summary of the operation of this device. When the output enable (pin 9) is low and the phase (pin 8) is high, the output A (pin 10) goes high and output B (pin 2) low, forcing current through the motor from left to right. Bringing the phase input low causes output B high and output A low, forcing current (and turning the motor) in the opposite direction. Bringing the output enable high opens both output currents to zero, thus removing power from the motor. Eventually, the motor will coast to a stop. For quick braking of the motor, a low is placed on pin 5.

Closed-Loop Speed Control

Up to this point, we have assumed that the control of the motor's speed was adjusted in an open-loop situation. *Open-loop control*, as we will see in a later chapter, is control without feedback. *Feedback control* of DC motors is an excellent method to regulate the motor's speed precisely. There are many methods used to control a motor by using feedback. The one we will discuss uses a device called a *phase-locked loop*.

Output Enable	Phase	V_{ref}/Brake	Output A	Output B
Low	High	>2.4 V	High	Low
Low	Low	>2.4 V	Low	High
High	X	>2.4 V	Open	Open
X	X	<0.8 V	High	High

X = irrelevant.

A. Truth Table

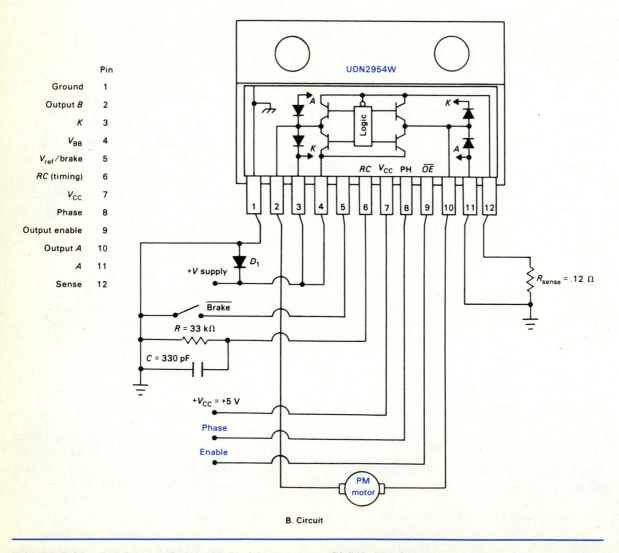

Pin	
Ground	1
Output B	2
K	3
V_{BB}	4
V_{ref}/brake	5
RC (timing)	6
V_{CC}	7
Phase	8
Output enable	9
Output A	10
A	11
Sense	12

B. Circuit

FIGURE 7.21   Full-Bridge Pulse-Width Modulation (PMW) PM Speed Control

Phase-Locked Loop (PLL). The PLL, discussed in detail in Chapter 2, is an electronic feedback control system having a phase detector, a low-pass filter, and a voltage-controlled oscillator (VCO). A PLL can also be considered a frequency-to-voltage converter (FVC).

Every system using negative feedback has the same characteristics. Negative (180° out of phase) feedback is fed from the output back to the input. The feedback signal or voltage is then compared with a reference. If the feedback is not equal to the reference, an error voltage or signal is generated. This error component is proportional to how different the two values are. This error component is then used to control the system and reduce the error to zero.

In the PLL system shown in Figure 7.22, the reference is a VCO. The VCO, as discussed in Chapter 2, is nothing more than a free-running multivibrator whose frequency of oscillation is controlled by an *RC* network. This frequency is compared with a frequency fed back from the motor. These two frequencies are compared in the phase comparator. If they are different in frequency, the phase comparator will produce an output signal proportional to the difference. This signal is then rectified by the low-pass filter and used to control a DC amplifier. The DC amplifier adjusts the motor's speed until there is little or no difference between the reference signal and the feedback signal.

This type of control can regulate motor speed to within 0.001% of the desired speed. Other methods of analog feedback are used to control DC motor speed, but none is as efficient as the PLL.

Pulse-Width Modulation. Figure 7.23 illustrates *pulse-width modulation* (PWM) for DC motor control. Here, pulse-width modulation is generated by the LM3524 IC. The pulse-modulated output of the LM3524 drives a transistor, which, in turn, controls the speed of the DC motor. A variable-reluctance sensor acts as a tachometer, converting the motor's mechanical motion to a variable frequency. This frequency is converted to a voltage by the LM2907 FVC. Both the voltage produced by the LM2907 and the speed-adjust potentiometer are connected to the input of the LM3524, where they are compared. Any error voltage generated by this comparison is used to adjust the amount of pulse-width modulation applied to the DC motor.

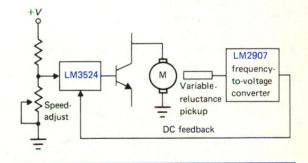

FIGURE 7.23 Pulse-Width Modulation Motor Control

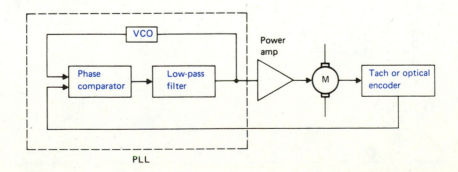

FIGURE 7.22 PLL Motor Control

Stepper Motor Control

You will recall from Chapter 4 that the stepper motor is the only motor with a truly digital output. The stepper motor converts electric pulses applied to the stator into discrete rotor movements called *steps*. Some low-power stepper motors are directly driven by TTL circuits; usually, however, TTL circuits do not have enough power to drive stepper motors. In this section, we present a brief discussion of stepper motor drive circuits.

A basic stepper motor drive circuit is illustrated in Figure 7.24. In this circuit, the switching transistor energizes the stator coil. Each coil in the stator would have a circuit identical to this one. Although not shown, the transistor is usually driven by the output of a TTL gate or through an optoisolator, if isolation is required. The schematic in Figure 7.25A shows a TTL circuit driving the transistor. The optoisolator in Figure 7.25B provides the isolation needed by some circuitry.

The stepper motor stator is an inductive load. As such, when the transistor interrupts current flow, the stator coil collapses, producing a transient-voltage spike. This spike can destroy the switching transistor if it is unprotected. A transistor protection circuit is shown in Figure 7.26. The diode D is reverse-biased when Q is conducting and current is flowing

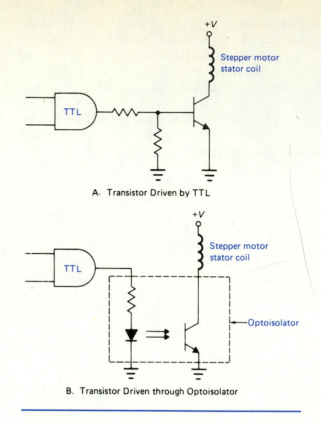

FIGURE 7.25 Methods of Driving Transistor in Stepper Control Circuit

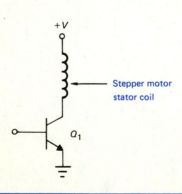

FIGURE 7.24 Transistor Driving Stepper Motor

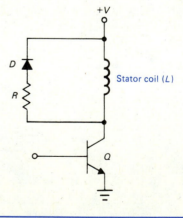

FIGURE 7.26 Transistor Protection Circuits in Stepper Motor Control

through the stepper motor stator. When the transistor turns off, the field built up around the coil collapses and forward-biases the diode. The pulse is then shorted out. The resistor R, connected in series with the diode, shortens the time constant. You will recall that, unlike the RC circuit, an LR circuit has a time constant inversely proportional to resistance. The time constant is L divided by R (L/R). If R were not included, the time constant would be long. The diode would be less effective as it shorts the transient pulse. Adding R decreases the time constant. By the same token, many stepper motor drive circuits include a resistance in series with the stepper motor stator coil. The resistor in series with the stator coil reduces the time constant and increases the rotor torque. When resistance is added to the stator, the supply voltage must be increased to make up for the voltage dropped across the resistor.

The power MOSFET (metal-oxide semiconductor field effect transistor) has been used recently in stepper drive circuits instead of the transistor. The power MOSFET, described later in this chapter, is a special high-current FET (field effect transistor). The current through the device can be controlled by nanoamperes of current. Power MOSFETs can be driven directly from TTL or CMOS circuitry. The use of power MOSFETs and CMOS circuitry in stepper motor drives simplifies control circuitry and allows considerable flexibility in control.

A circuit using power MOSFETs and CMOS control is illustrated in Figure 7.27. A full-step, center-tapped stepper motor is controlled by four power MOSFETs. The power MOSFETs are controlled by an MC14194 CMOS four-bit presettable shift register. Tapping the appropriate shift register outputs gives the required phasings to turn the rotor.

The motor rotates clockwise by right-shifting the MC14194. Left-shifting causes the rotor to turn counterclockwise. Control signals on the S_0 and S_1 inputs plus the clock (CLK) input control the stepping.

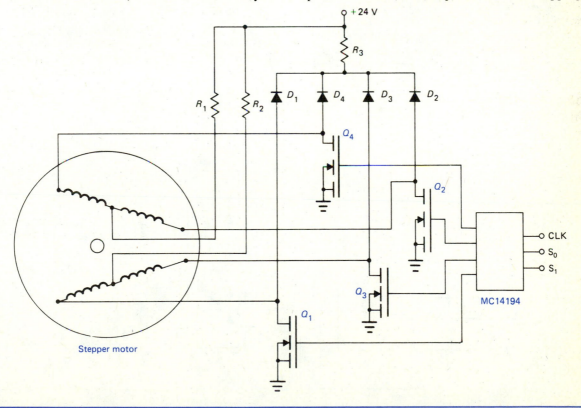

FIGURE 7.27 Power MOSFET Stepper Motor Driver

When power is applied to the device, the MC14194 needs to be preset by putting highs at S_0 and S_1 and by receiving a leading clock pulse. These three requirements, if met, will put the logic in a known state. The other control functions are shown in Table 7.1. Stepping will occur when a leading-edge clock pulse is received. The diodes serve the same purpose as in the transistor circuit previously described. They protect the power MOSFET from damage due to spikes. Power MOSFETs switch very quickly. As a result, the diodes may not be able to turn on fast enough to prevent damage to the power MOSFETs. To prevent this damage, a small capacitor (0.01–0.01 μF) is sometimes placed across the motor windings.

Several manufacturers make drives that use custom ICs. These chips develop the required pulse trains and require a minimum of external components. A stepper driver made by Sprague Electric Company is shown in Figure 7.28. This IC, the UCN5804B, controls a four-coil (bifilar-wound), two-phase stepper. The bipolar outputs are capable (with proper heat sinking) of sinking 1.5 A per phase. The voltage at pins 9 and 10 determine the type of excitation. As shown in Figure 7.28, pin 9 has a high (+5 V) applied to it, and pin 10 has a low or ground potential applied to it. If we examine the truth table in Figure 7.29A, we see that this arrangement corresponds to a one-phase or wave drive (see Chapter 4 to review this type of drive).

The *wave drive sequence* is shown in Figure 7.29B. Starting with step 1, we see from this table that output A (pin 8) is on and the other three outputs are off. Current will flow through the right coil of phase 1, creating a field as shown. The south pole at the top of the rotor will be pulled to the 9 o'clock position, as shown by the dotted S. The north pole is also pulled to the 3 o'clock position at the same time. In the next step, output A goes off and output B goes

TABLE 7.1 Control Functions for MC14194

S_0, S_1	Result
0, 0	Hold
0, 1	Shift right
1, 0	Shift left
1, 1	Preset

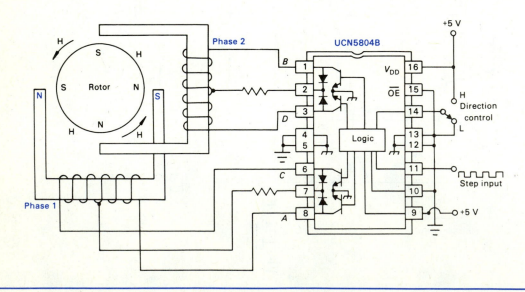

FIGURE 7.28 IC Stepper Motor Control Using UCN5804B

	Pin 9	Pin 10
Two phase	L	L
One phase	H	L
Half step	L	H
Step inhibit	H	H

A. Truth Table

← Direction = L →   ← Direction = H →

Step	A	B	C	D
POR	On	Off	Off	Off
1	On	Off	Off	Off
2	Off	On	Off	Off
3	Off	Off	On	Off
4	Off	Off	Off	On

B. Wave Drive Sequence (Half Step = L, One Phase = H)

← Direction = L →   ← Direction = H →

Step	A	B	C	D
POR	On	Off	Off	Off
1	On	Off	Off	Off
2	On	On	Off	Off
3	Off	On	Off	Off
4	Off	On	On	Off
5	Off	Off	On	Off
6	Off	Off	On	On
7	Off	Off	Off	On
8	On	Off	Off	On

C. Half-Step Drive Sequence (Half Step = H, One Phase = L)

← Direction = L →   ← Direction = H →

Step	A	B	C	D
POR	On	Off	Off	On
1	On	Off	Off	On
2	On	On	Off	Off
3	Off	On	On	Off
4	Off	Off	On	On

D. Two-Phase Drive Sequence (Half Step = L, One Phase = L)

FIGURE 7.29 UCN5804B Stepper Motor Driver Modes

on. Current flows through the phase 2 coil, producing a north pole in the stator at the 6 o'clock position and a south pole at the 12 o'clock position. This stator polarity pulls the rotor another 90° counterclockwise. The south pole of the rotor is now at the 6 o'clock position. In the third step, the current in phase 2 goes off and the C output causes current to flow in the left winding of phase 1, producing a north pole at the 3 o'clock position and a south pole at the

9 o'clock position. The rotor rotates another 90°. After output C goes off and D comes on, the rotor will be in its original position, with a south pole on top and a north pole on the bottom. This sequence may be reversed by switching pin 14 to ground. In this arrangement, the coil will rotate clockwise instead of counterclockwise.

A *half-step drive sequence* (Figure 7.29C) can be selected by pulling pin 9 low and pin 10 high. Sometimes called the *one-two excitation mode*, the half-step mode energizes coil A first, as in the wave drive. In step 2, however, output A is left on while output B is turned on. This causes the resultant stator field to be aligned at the H marking in Figure 7.28. In the third step, output A is turned off and output B remains on, as in step 2 of the wave drive. After step 3, the half-step excitation mode will have moved the rotor 90°. The half-step excitation mode has the disadvantage of producing uneven torque. The second step, where both phases are energized, produces more torque than the steps when only one phase is on. The result is a strong step followed by a weak step.

The third mode of excitation is called the *two-phase mode* (see Figure 7.29D). In this mode, two phases are always energized at the same time. This mode produces a full step and also produces twice the torque of the wave drive since two stators are always on at the same time.

Stepper Motor Control Example. As we have discussed previously, a stepper motor easily lends itself to microcomputer control. Figure 7.30 shows an example of an interface circuit that can be used to drive a small stepper motor. For convenience the microcomputer used, but not shown in the diagram, was an Intel 8085–based system. Any single-board microcomputer could be used, but the control lines may be different from the ones shown in the diagram.

The three inputs to the system are (1) the address line, or decoded port, to select this particular port, (2) the control line $IO/\overline{M}$, and (3) the control line $\overline{WR}$, used in conjunction with the OUT port instruction to enable the 74116 four-bit latch. [The $IO/\overline{M}$ is a status line input, used to differentiate between I/O (input/output) and memory operations. When the voltage on this pin is high, it indicates an I/O opera-

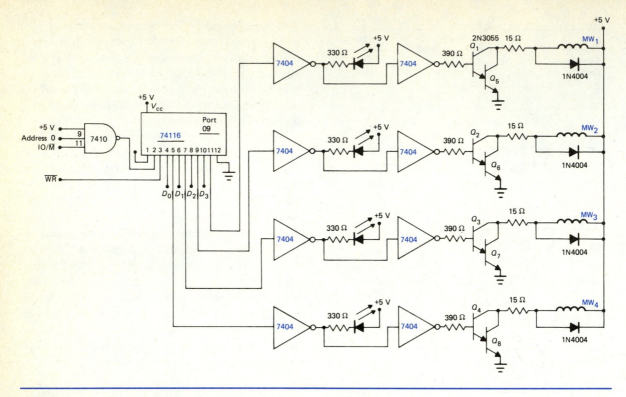

FIGURE 7.30   Microprocessor Controlling Stepper Motor

tion; when the voltage is low, it indicates a memory operation. The $\overline{WR}$ is a write control signal line with an active low enable.] The 74116 latches the data that were presented to it during the OUT port instruction and uses these data to drive the 7404 inverters. The inverters, in turn, drive other inverters and LEDs that display the data code that was presented to the 74116 latch. The second set of 7404 inverters drive the

2N3055 transistors that drive the stepper motor coils. The motor coils have flywheel diodes that prevent inductively generated voltages from destroying the transistors. The data codes and the time between changes in these codes determine the direction and speed of the stepper rotation.

Figure 7.31 shows various combinations of codes that could be used to rotate the stepper motor.

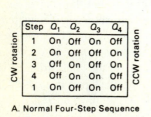

Step	Q_1	Q_2	Q_3	Q_4
1	On	Off	On	Off
2	On	Off	Off	On
3	Off	On	Off	On
4	Off	On	On	Off
1	On	Off	On	Off

CW rotation / CCW rotation

A. Normal Four-Step Sequence

Step	Q_1	Q_2	Q_3	Q_4
1	On	Off	On	Off
2	On	Off	Off	Off
3	On	Off	Off	On
4	Off	Off	Off	On
5	Off	On	Off	On
6	Off	On	Off	Off
7	Off	On	On	Off
8	Off	Off	On	Off
1	On	Off	On	Off

CW rotation / CCW rotation

B. Half-Step, Eight-Step Sequence

Step	Q_1	Q_2	Q_3	Q_4
1	On	Off	Off	Off
2	Off	Off	Off	On
3	Off	On	Off	Off
4	Off	Off	On	Off
1	On	Off	Off	Off

CW rotation / CCW rotation

C. Wave Drive Four-Step Sequence

FIGURE 7.31   Changing Stepper Motor Direction

AC MOTOR SPEED CONTROL

In recent years, the AC motor has become increasingly popular in certain areas of industry. Most of the interest in AC motors comes, at least in part, from the advantages it has over DC motors. The AC motor is smaller, and therefore less expensive, than the DC motor of equivalent horsepower rating. Generally, the AC motor is less costly to maintain than the DC motor. In the induction motor, reduced maintenance is due to the absence of a mechanical connection between rotor and stator. Not only does this feature mean lower maintenance costs, but it also means increased safety. Machines with commutators and brushes generate sparks, which could ignite in an explosive atmosphere. It is true that large synchronous motors have slip rings and, therefore, mechanical contacts. However, motors with slip rings produce less sparking and are easier to maintain than mechanically commutated motors. AC motors also can be run at generally higher speeds. Most general-purpose, wound-field DC motors have a top speed of about 2500 r/min. A comparable AC induction motor typically has a top speed twice that.

Just a few years ago, the SCR was the leading component in power control in AC and DC drives. Today, large high-voltage (>1000 V) transistors are used in some AC drive applications. The use of transistors also eliminates the need for bulky and expensive commutation circuits in DC drives. The GTO (gate-turnoff SCR), a thyristor discussed in Chapter 6, has been used as a replacement for SCRs, owing to recent advances in GTO power-handling capability. The Westinghouse Accutrol 300 uses GTOs exclusively as the power driver.

Several years ago, the AC motor was used predominantly in fixed-speed applications. Recall from Chapter 5 on AC motors that the synchronous stator speed is directly proportional to the line frequency and inversely proportional to the number of magnetic poles. It is relatively easy to change the stator speed by varying the number of poles per phase. This type of speed control, however, results in speed control by steps only. There is no way to adjust the speed proportionally by altering the number of poles.

Speed may also be changed in an AC motor by varying the frequency of the voltage applied to the stator field. Although this type of speed control was known for a number of years, it was not economically sound until the advent of the thyristor in the 1950s. Research and development was soon undertaken to apply the thyristor to adjustable-frequency AC drives.

Universal Motor Speed Control

One circuit that will control the power applied to a universal motor is illustrated in Figure 7.32A. In this circuit, an open-loop control circuit, the capacitor charges up to the firing voltage of the diac in either direction. Once fired, the diac will apply a voltage

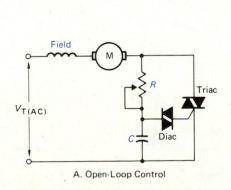

A. Open-Loop Control

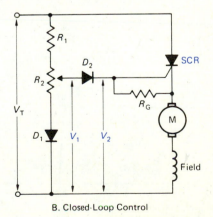

B. Closed-Loop Control

FIGURE 7.32 Universal Motor Control

to the gate of the triac. The triac will conduct and apply power to the motor. Note that the triac will conduct in either direction. Since this device is basically a series DC motor, current flowing in either direction will tend to cause rotation in only one direction. Speed may be changed by varying the resistance of the potentiometer, as discussed previously under phase control.

Closed-loop control for the universal motor is shown in Figure 7.32B. When line voltage is applied, the voltage at R_2 will rise proportionally to the voltage divider network. When the motor is turning at the desired speed, it develops a voltage V_2, which is mostly the cemf of the motor. If V_1 and V_2 are equal, no voltage is applied to the gate of the SCR. In this condition, the reverse-biased diode D_2 blocks the voltage. If the motor's speed decreases, however, the cemf will decrease, causing voltage V_2 to decrease. When V_1 is greater than V_2, a gate voltage is applied to the SCR. This voltage generates torque in the motor. The motor will then pick up speed, returning to near-normal speed.

The firing angle of the SCR is inversely proportional to the load on the motor. As the load on the motor increases, the speed of the motor will decrease. The more the motor speed decreases, the earlier the thyristor fires in the cycle. The earlier the thyristor fires, the more power is applied to the motor. Thus, the universal motor speed is kept relatively constant under varying load conditions.

Adjustable-Frequency AC Drives

With the exception of the universal motor, the remainder of the AC motor drives covered in this chapter operate on the principle of adjustable-frequency control (AFC). As we have seen, in this type of control the rotor speed is directly proportional to the stator frequency. The AC motor can, therefore, be run at any speed, from a few revolutions per minute to the top speed of the motor.

We can see the effect of a change in stator frequency on the induction motor by looking at the speed-torque curve in Figure 7.33. Curve A is a typical curve for a four-pole machine with an applied stator frequency of 60 Hz. The motor will operate where the curve crosses the line labeled 100% load torque. If the stator frequency is decreased to 40 Hz, the motor will shift its operating point to curve B. Further reduction in frequency to 30 Hz will take the motor to curve C. This reduction in frequency could continue until the rotor is at a standstill. Although we moved the stator frequency in steps, the actual control is infinitely variable up to the maximum motor speed.

The induction motor is named for the way the rotor gets its power—that is, electromagnetic induction. Like the transformer, the AC induction motor needs to have a constant flux in the rotor and stator. Unless this requirement is met, the motor will not be able to generate full torque. If the frequency applied

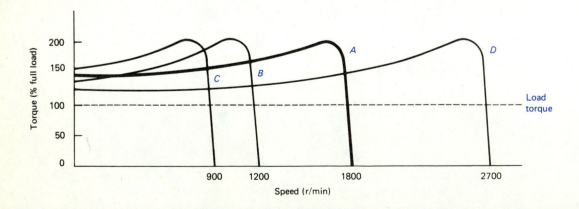

FIGURE 7.33 Adjusting AC Motor Speed

to the stator is decreased, the voltage applied to the stator must be decreased by the same amount. Another way to state this requirement is that the ratio between the voltage applied to the stator and the frequency of the voltage applied to the stator must be constant. This ratio is called the *volts/hertz ratio*, or the *constant volts/hertz characteristic*, of an AC induction motor. To find the volts/hertz (V/Hz) ratio for an induction motor, simply divide the rated voltage on the motor nameplate by the rated frequency. For example, a motor nameplate rates an AC induction motor for 230 V AC, 60 Hz operation. The V/Hz ratio is

$$\frac{\text{volts}}{\text{hertz}} = \frac{230 \text{ V AC}}{60 \text{ Hz}} = 3.83 \text{ V/Hz}$$

Above a frequency of about 15 Hz, the amount of voltage needed to keep the flux constant is a constant value, as shown in Figure 7.34. To get the voltage needed above 15 Hz, multiply the V/Hz constant by the applied frequency. Below 15 Hz, the voltage applied to the stator must be boosted above the value predicted by the constant V/Hz curve. The extra voltage is needed to make up for the I^2R losses in the motor at low speeds. The amount of voltage

boosting will depend on the motor. As an example, we can find the stator voltage needed at 30 Hz by multiplying 30 Hz by the 3.83 V/Hz ratio as follows:

$$V_{st} = f(\text{V/Hz}) = (30 \text{ Hz}) (3.83 \text{ V/Hz})$$

$$= 115 \text{ V AC}$$

where
V_{st} = stator voltage required
f = frequency applied to stator
V/Hz = volts/hertz ratio

Another consequence of operating in the constant V/Hz ratio relates to the power available from the machine. Since the torque remains constant, the power from the machine will be directly proportional to the speed of the machine. As we learned in Chapter 5, the power from a rotating machine is related to speed and torque as follows:

$$P_o = \frac{Tn_{rt}}{5252}$$

As an example, let us consider a 1 hp machine with a base speed of 1750 r/min at 230 V AC and 60 Hz putting out 3 ft-lb of torque. At 875 r/min (one-half of rated speed), the power out is

$$P_o = \frac{(3 \text{ ft-lb}) (875 \text{ r/min})}{5252} = 0.5 \text{ hp}$$

At one-third the rated speed, the maximum power available would be 0.33 hp. You should also bear in mind that the motor may need extra cooling when operated below rated speed with full-load torque.

We can now calculate the voltage and frequency needed to operate a motor at any given rotor speed. First, we determine the speed at which the rotor would be turning the load. Next, we add the value of the slip at that load to get the synchronous speed. Recall that the induction motor must rotate at a slower speed than the stator for induction to take place. We next solve for the frequency applied to the stator. Finally, we multiply the stator frequency by the V/Hz ratio.

Let's consider an example. We have a four-pole 5 hp motor with a rated voltage of 230 V AC at

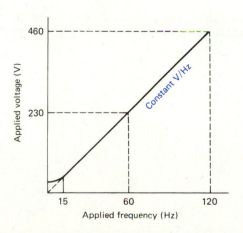

FIGURE 7.34 AC Induction Motor Applied Frequency-Voltage Curve

60 Hz. We wish to operate the motor at 850 r/min at full-load torque. If slip is 5.55% at full-load torque, then the stator speed is 900 r/min. Rearranging the formula for finding stator speed (see Chapter 5) and solving for frequency, we get

$$f = \frac{n_{st}p}{120} = \frac{(900 \text{ r/min}) (4)}{120} = 30 \text{ Hz}$$

The V/Hz ratio is found by dividing the rated voltage by the rated stator frequency:

$$\frac{\text{volts}}{\text{hertz}} = \frac{230 \text{ V AC}}{60 \text{ Hz}} = 3.83 \text{ V/Hz}$$

Finally, we find the stator voltage needed with a V/Hz ratio of 3.83 V/Hz:

$$V_{st} = f (\text{V/Hz}) = (30 \text{ Hz}) (3.83 \text{ V/Hz})$$

$$= 115 \text{ V AC}$$

We can see an interesting relationship in these calculations. The synchronous speed at 30 Hz is exactly half of the speed at 60 Hz. The voltage needed with a 30 Hz stator frequency is also exactly half of the voltage needed for operation at 60 Hz. We should expect these relationships to hold true because of the proportionality between applied stator frequency and stator voltage.

An induction motor's rated speed may be exceeded in one of two ways. First, the frequency and voltage may be increased according to the V/Hz ratio to about 125% of the full-load speed. Higher speeds are not recommended without specially constructed motors. Induction motors may also be operated above rated speed in the constant-voltage mode, as shown in Figure 7.35. As the graph shows, the line voltage is held at the rated level and the applied stator frequency is increased past 60 Hz. In this area of operation, the motor does not have enough voltage to produce full-load torque as speed increases. As speed increases, therefore, the torque decreases. In this constant-voltage range, both the rated torque and the breakdown torque will decrease. Breakdown and rated torque decrease as the square of the increase

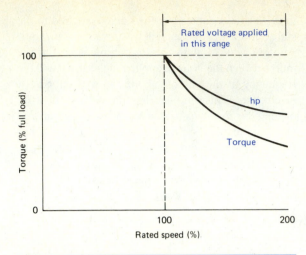

FIGURE 7.35 Operating an AC Induction Motor above Rated Speed

in speed past 60 Hz. The new torque can be found by using the following equation:

$$T = \left(\frac{60 \text{ Hz}}{f}\right)^2$$

where

 T = torque available as percentage of rated torque
 f = stator frequency

As an example, what percentage of rated torque would be available at 75 Hz? The torque is calculated as

$$T = \left(\frac{60 \text{ Hz}}{75 \text{ Hz}}\right)^2 = 0.64 \quad \text{or} \quad 64\%$$

The motor power output decreases proportionately as speed increases:

$$P = \frac{60 \text{ Hz}}{f}$$

where

 P = percentage of output power at rated speed and frequency
 f = new frequency (in hertz)

In the example just given, the percentage of output power is

$$P = \frac{60\ \text{Hz}}{75\ \text{Hz}} = 0.80 \quad \text{or} \quad 80\%$$

Changing the frequency of the voltage applied to the stator field will give smooth linear control of motor speed. Frequency control of synchronous and induction motors can be broken down into two methods: rectifier-inverter systems and cyclo-converter systems. Figure 7.36A is a block diagram of a *rectifier-inverter system*. Note that the frequency conversion takes place in two steps. First, the fixed 60 Hz line frequency is rectified to DC. Second, the DC is converted to variable-frequency AC by the inverter. Thyristor inverters can supply an output frequency of up to 1 kHz over a range of about 15:1. The *cycloconverter system* is shown in block diagram form in Figure 7.36B. Cycloconverters give output frequencies of up to 25 Hz.

Rectifier-Inverter System. Three types of adjustable-frequency controls are in common use today, all using the rectifier-inverter as a basic building block. These three types are the variable-voltage input (VVI), the current-source inverter (CSI), and the pulse-width modulation (PWM).

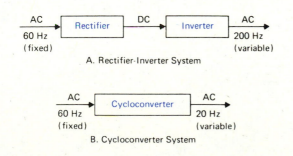

A. Rectifier-Inverter System

B. Cycloconverter System

FIGURE 7.36 Frequency Control Methods for AC Motors

The VVI type of inverter, shown in Figure 7.37A, receives DC input power from the phase-controlled bridge rectifier. The phase-controlled bridge varies the amount of DC input voltage to the inverter. The inverter switches the output circuit between the plus and minus bus. Each of the three-phase outputs is displaced from one another by 120 electrical degrees (Figure 7.37B). These waveforms (A to B, B to C, and C to A) are created by the algebraic summing of each output with respect to the DC bus. During time periods T_1 and T_2, the voltages from outputs A and B are equal to V_{adj} since B is at the $-V_{adj}$ bus potential. The entire V_{adj} voltage would be seen across whatever motor winding is connected across A to B. During the time period T_3, the A-to-B voltage is zero since both A and B are at the same potential. Note that the AC voltages applied to A, B, and C are displaced by 120 electrical degrees and, therefore, will give the rotating field effect when applied to the stator windings. Direction of rotation may be changed by firing the inverter thyristor diodes in different sequences. The stator frequency is changed by firing and commutating each thyristor for different time periods. For example, if we wanted to increase the frequency, we would shorten time periods T_1 to T_7 by decreasing the time the appropriate thyristors are gated on. The voltage applied to the stator is increased by increasing V_{adj}. Voltage V_{adj} is increased, in turn, by gating on the rectifier thyristors earlier in the cycle. The VVI drives are the most popular general-purpose AFC drive today. They are used in applications up to 400 hp in pumps, fans, grinders, and saws.

The CSI drive uses the same basic components as the VVI drive. The CSI drive, shown in Figure 7.38, has a variable DC bus controlled by a thyristor regulator, a large DC filter inductor, and an inverter. The rectifier and filter provide a controlled, regulated DC current to the inverter. The source of power to the inverter is a controlled current source, rather than a controlled voltage source, as in the VVI. Current is usually detected by a sensor after the inductor. The amount of current sensed is used to provide feedback for current adjustment. The CSI drive system is most often found in applications above 10 hp.

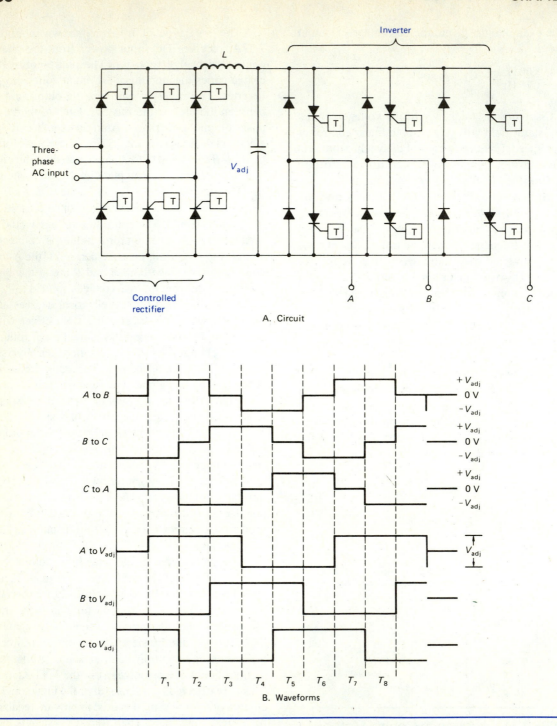

FIGURE 7.37 Variable-Voltage Inverter (VVI), Three-Phase AC Motor Drive

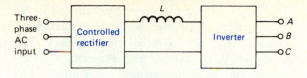

FIGURE 7.38 Current-Source Inverter (CSI), Three-Phase AC Motor Drive

Instead of turning on only once in a cycle, the SCRs trigger and commutate many times. Pulse-width modulation also is used to reduce the amount of undesired harmonic frequencies generated by other methods of inversion. You may have noticed that we have used the PWM system in DC motor control earlier in this chapter. PWM is used frequently in both AC and DC applications. As in DC applications, PWM AFC drives are very responsive to speed changes. They find applications in conveyors, fans, and pumps.

The final AFC drive method is achieved by pulse-width–modulating the input to the inverter. As you can see from Figure 7.39A, the PWM system converts AC to DC by using a diode rectifier system, not a thyristor rectifier, as in the CSI and VVI drives. The thyristors are not necessary because the PWM system uses a fixed DC voltage input to the inverter. A pulsed output is constructed by varying the gating and commutation times, as shown in Figure 7.39B.

Cycloconverter System. The cycloconverter method is the second way to provide variable-stator-frequency control. The cycloconverter changes the fixed-frequency input voltage to a variable-frequency output voltage. The cycloconverter commutates naturally from the input line voltage. Thus, the cycloconverter is less complicated than systems requiring forced commutation, such as the PWM inverter just discussed. The cycloconverter has

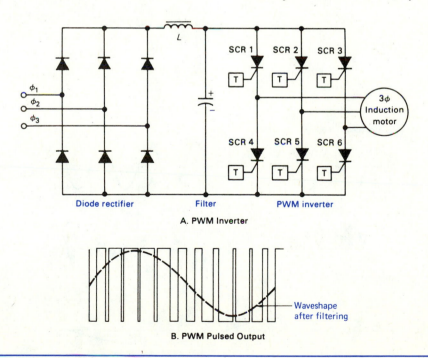

A. PWM Inverter

B. PWM Pulsed Output

FIGURE 7.39 Pulse-Width Modulation (PWM)

another advantage over the inverter. The inverter requires a DC input, which necessitates rectification before it can be used. The cycloconverter, in contrast, operates from the AC input directly.

A simplified circuit diagram of a cycloconverter is shown in Figure 7.40A. During the first 3½ input cycles, SCR₁ and SCR₂ are gated on, producing the positive pulses shown in the output of Figure 7.40B. Thyristors SCR₃ and SCR₄ are gated on during the next 3½ input cycles, producing the set of negative pulses shown in Figure 7.40B. The process then repeats. The output voltage, after it is filtered, is sinusoidal in shape. The cycloconverter shown is actually a 7:1 frequency divider.

The cycloconverter finds applications in low-speed synchronous motors where high torques are demanded. Previous to the advent of thyristors, low speeds were achieved by gearing, especially when the application demanded an induction motor. The cycloconverter allows the induction motor to run at a low speed while developing maximum torque.

Before moving on to another topic, we should mention another method of controlling induction motor speed. Varying the voltage applied to the stator winding will give a limited amount of speed control while keeping the frequency constant. This type of speed control can be achieved by phase-control techniques such as those discussed earlier in

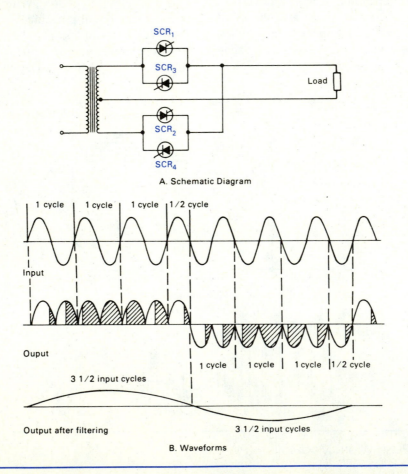

A. Schematic Diagram

B. Waveforms

FIGURE 7.40 Cycloconverter

the chapter. For example, the circuit illustrated in Figure 7.32A can be used for this type of control. Control is achieved in this case by changing the torque produced by the motor. Recall that the torque produced by an induction motor is directly proportional to the square of the stator voltage. Thus, changing this voltage produces a change in torque. This method of control is not used often in industry, for several reasons. First, it gives only limited speed control. Second, it changes the breakdown torque of the motor. Third, decreasing the speed by this method can result in motor overheating.

THYRISTORS IN MOTOR CONTROLS

Two semiconductors were used extensively prior to 1978 in power control applications: the SCR and the transistor. The SCR was used in AC drive applications greater than 20 kVA, and transistors were used for the lower ratings. In the years since 1978, we have been seeing a greater variety of semiconductors in power control applications, including power FETs, GTOs, and higher-powered transistors and Darlingtons. The power GTOs and transistors are used extensively in PWM AC drives. The power-handling capabilities of GTOs, transistors, and Darlingtons are a result of improvements in the methods of doping, device geometry, and packaging. Gains in power Darlingtons have increased to 500 with an increase in reliability. Better and more efficient packaging of semiconductors improves not only the reliability of the devices but also the cost.

Semiconductors in Power Control

Special switching transistors can turn on very quickly and have low collector-to-emitter resistance. They are, therefore, very efficient. Also, transistors do not have the disadvantage that SCRs have in the ability to turn off in DC circuits. In DC circuits, transistors turn off easily by a removal of base current or voltage.

One of the major problems in using transistors in switching circuits involves the base current requirements. Transistors require that base current flow to provide collector current. Normally, power transistors have low current gain (ß), some as low as 10. Such a low ß means that considerable base current may be needed to keep the device on. This current is not doing useful work. Thus, because of this situation, the transistor is inefficient in power applications.

The power MOSFET (metal-oxide semiconductor FET) has taken a firm hold on the power control market for power levels up to 30 hp (about 22 kW). Recall that, unlike the transistor, the MOSFET is basically a voltage-controlled device, like the vacuum tube. The power MOSFET, which can presently control up to 40 A, draws only nanoamperes of gate current. Power MOSFETs are presently being used in motor controls, especially choppers and inverters. Chopper and inverter applications are illustrated in Figure 7.41. The power MOSFET has advantages over transistors in the areas of ruggedness and much lower drive requirements. Circuit requirements are

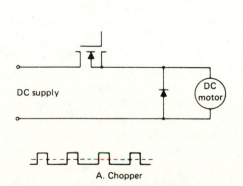

A. Chopper

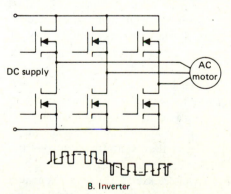

B. Inverter

FIGURE 7.41 Power MOSFET Motor Control Applications

simplified, thereby reducing costs and improving reliability and response time. Another advantage of the power MOSFET lies in its ability to increase power-handling capability by connecting power MOSFETs in parallel. Connecting transistors in parallel to get higher current ratings provides designers with an adventure in fusing and current-sharing techniques. Power MOSFET choppers with currents in the hundreds of amperes are not beyond the capabilities of the power MOSFET. Power MOSFETs also offer significant advantages in gain and switching times over standard power transistors and Darlingtons.

Applications for power MOSFETs are increasing every day. These hardy devices are used in switching power supplies, audio amplifiers, AM transmitters, induction heaters, high-frequency welding machines, and fluorescent-lighting controls. The hammer driver application shown in Figure 7.42 is an interesting application demonstrating the strength of the power MOSFET. The power MOSFET in this application drives the coils that move the hammers in a computer dot matrix printer. You have no doubt seen the output of this type of printer. Each letter is made up of usually five to seven dots, providing the typical printed computer output. In this application, the drive coil is energized by a power MOSFET when it gets a triggering voltage on its gate. Removing the triggering voltage stops the current through the MOSFET, turning it off. Before the advent of the power MOSFET, this switching was done by Darlingtons. The Darlington solution to this application has some problems. Darlingtons are slow and have high losses, especially when pulse-width–modulated. The collector-to-emitter voltage drop of the Darlington may be as high as 2.5 V, increasing the heat dissipated by the TO-220 package usually used. The TO-220 package can dissipate the heat but must be hand-inserted, making the manufacturing of printed-circuit boards containing TO-220 packages more expensive. The power MOSFET has a voltage drop ten times less than that found across the Darlington, improving efficiency and lessening heat buildup. The power MOSFET switches faster and can be automatically inserted in printed-circuit boards in a dual in-line package (DIP), thus reducing assembly costs.

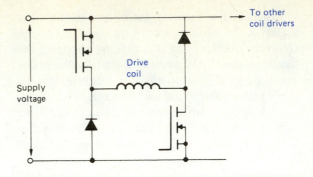

FIGURE 7.42 Printer Hammer Driver Application for Power MOSFET

Operating Power Semiconductor Equipment

Thyristor Identification. Technicians are often called on to replace semiconductors that have failed. Semiconductor failure may be caused by external overloads or by an internal semiconductor fault. In either case, the correct replacement part may not be available. The competent technician should be able to suggest a replacement part that will perform as well as the part being replaced. Suggesting a substitute part is not always an easy task. Every technician should be aware of the proper techniques for substituting semiconductors. This section will give some suggestions on how to make substitutions effectively.

Many semiconductors are identified by a JEDEC (Joint Electronic Device Engineering Council) number. JEDEC numbers for rectifiers begin with a "1N" prefix (for example, 1N4001, 1N1206A), whereas SCRs and transistors begin with a "2N" prefix (for example, 2N681, 2N3055). A JEDEC number is an attempt by the industry to standardize electrical and mechanical parameters so that products of one manufacturer will be interchangeable with those of another manufacturer. The technician should be careful, however, when making a direct JEDEC substitution of another manufacturer's part because the JEDEC registered parameters do not always cover all of the critical parameters. As a result, manufacturers using completely different manufacturing processes often sell devices to meet the same JEDEC number. Although these differences may not pose a problem in most general-purpose and phase-control applications, the user must recognize that various manufacturers' devices marked with the same JEDEC number can exhibit completely different secondary characteristics and safety margins.

An engineer may unknowingly design a circuit around a particular JEDEC-type number that was manufactured using process A, which built in a large safety factor. For example, our engineer chooses a 6 A rated device that really has a 12 A actual capability. Everything works fine until the user chooses a replacement JEDEC from another manufacturer. Then the new devices keep failing in the circuit. The circuit works properly with one manufacturer's device but not with another's. No one knows why because the replacement part was not carefully researched. To be safe, all JEDEC substitutions should be evaluated thoroughly before use.

Most power semiconductors (especially those rated over the 40–100 A range) are marketed under their respective manufacturer's part number. Even though each manufacturer uses its own device nomenclature, the mechanical packages and electrical ratings are reasonably well standardized throughout the industry. Device substitution is not difficult, therefore, providing the user has a good cross-reference guide and the technical data sheets for the devices being evaluated. As with any cross-reference, the technician must determine an acceptable substitution by reviewing the detailed electrical and mechanical characteristics of the devices being considered. This comparison will ensure that the manufacturer's suggested replacement will perform properly in the given application.

Thyristor Replacement. Be sure to order the exact part number of the device you are replacing. If in doubt, include the entire device marking on your order. The most likely place to get off-the-shelf delivery of an exact replacement is directly from the equipment manufacturer. The original equipment manufacturer (OEM) normally carries a good supply of spare parts to service its equipment market. If you cannot get a replacement from the OEM, contact a local industrial or electronics distributor that handles power semiconductors. A user should buy a device having a particular semiconductor-type number only from an OEM authorized distributor outlet or directly from the OEM factory. The user could receive inferior or counterfeit devices if purchased from an unknown source. Unless otherwise directed, deal only with a manufacturer or an authorized distributor.

The technician must be thoroughly familiar with equipment service manuals, with the power semiconductors and their technical data sheets, and with the various tools and instruments available for testing and replacing semiconductors. It is wise to carry spare semiconductor components in stock in case of equipment breakdown, especially for critical pieces of equipment.

Many phase-control applications—for example, welders, battery chargers, DC motor controls, and general-purpose power supplies—can use rectifiers, SCRs, and transistors with fairly broad parameters. As long as the device selected has an adequate current and voltage rating and fits mechanically, it should work. The technician must be extremely careful in selecting a replacement semiconductor when general-purpose or phase-control devices are used in series and/or parallel combination, when fast-recovery rectifiers and fast-switching SCRs are used in inverters, choppers, and the like, or when semiconductor devices are marked with special (noncatalog) part numbers. Specially selected, tested, and matched units may be needed for the semiconductor to operate properly in the equipment. Failure to use the correct semiconductor device could result in device failures, equipment damage, and plant downtime.

When selecting a device for replacement, the technician can always use one with a higher voltage rating, provided all other device ratings are equal or better. Likewise, a higher-current-rated unit can be selected as long as all of the other ratings are equal or better and the mechanical package is the same.

By following these guidelines, the user often can locate a suitable replacement faster. Also, spare-parts inventories may be reduced by standardizing a smaller number of replacement semiconductors to gain increased reliability with greater voltage and/or current safety factors. Of course, the additional cost for a higher-rated semiconductor must be weighed against the savings in inventory reduction, fewer failures, and reduced downtime.

In the absence of any other information, a good rule of thumb for specifying the proper device voltage rating (based upon the supply voltage to the semiconductor equipment) is as follows: 110 V line—use a 300 V device; 220 V line—use a 600 V device; and 440 V line—use a 1200 V device. If in doubt about what device to use, call the semiconductor manufacturer and ask for a recommendation.

The user must realize that power semiconductors, like any other components, fail for a reason. It is acceptable merely to replace a suspect semiconductor if the semiconductor is the problem. Frequently, however, a semiconductor fails because of a current or voltage overload elsewhere in the circuit. Simply replacing the suspected semiconductor in these instances will result only in destruction of more semiconductors. Therefore, always look for the cause of failure before replacing any devices.

There is still a tendency to consider the state of the art in the design, manufacture, and application of power semiconductors as something new, mysterious, and glamorous. Technicians should avoid yielding to this tendency. Industry has long been using silicon power semiconductors, and a vast knowledge in the use of these devices has evolved. Along with this knowledge has come a new state of the art for using and maintaining semiconductor equipment. The emphasis placed on selecting, purchasing, and installing devices is part of this new knowledge. Of equal importance are the proper use and maintenance of equipment using power semiconductors, such as we have discussed in these last two chapters.

Power Semiconductor Heat Protection. Semiconductors are made from materials that have different rates of thermal expansion. Therefore, both short- and long-term temperature changes not only impair the electrical characteristics of these devices but also set up internal mechanical stresses at each of the material interfaces. These combined effects of high-temperature changes can lead ultimately to device malfunction and destruction. The maximum temperature limits of various materials used in semiconductors vary from 150°C for soft solder to 1300°C for silicon. Once these materials are combined into a

fabricated device, however, electrical deterioration begins at lower temperatures.

For example, the forward- and reverse-blocking capability of a junction begins to decrease and leakage currents increase when allowable maximum junction temperatures are exceeded. These temperatures are generally between 125° and 200°C, far below the maximum temperature limit of silicon itself.

Periodic checks of ambient temperatures, of cooling fan operation, of cooling water temperatures, and of case temperatures of the semiconductors themselves will pinpoint many malfunctions before they cause catastrophic failures. Because of the importance of preventing high temperatures from developing, this monitoring must be emphasized in the design, testing, and operation of electronic equipment.

Equipment Overload Protection. Avoid overloading the semiconductor—which, in effect, means you must protect against overloading the equipment that the semiconductors are feeding and/or controlling. Overloading a motor operated with semiconductor controls puts a direct overload on the semiconductors themselves. Even if an overload occurs only for a short time, a semiconductor's useful life can be shortened. Controls are designed to include motor overloads, and the maximum values of overload should never be exceeded. Even short-term overload should be avoided. The damage caused by overload accumulates and is not detectable. The result is random, often unexplainable semiconductor failures.

Operators may cause overload conditions that could result in excessive semiconductor currents. Operator overloads are caused by a high on/off duty cycle or jogging for high production needs, rapid reversing, motor jam, and long acceleration time. Some mechanical problems that can cause motor overloads are high equipment temperature due to lack of cooling, wiring or bearing failure due to improper installation or maintenance, phase failure due to blown fuses or loose connections, phase imbalance, overvoltage, transient overvoltage, contaminants, and dirty environment.

Table 7.2 shows various motor stress conditions and the common protection method employed. This

TABLE 7.2   Motor Overloads and Their Effect on Semiconductors

Problem	Cause	Motor Current Variance	Semiconductor Current or Voltage Variance	Common Semiconductor Protection	Comments
Overload	Operator's choice	Increases three-phase current, approaching 200%	Increases semiconductor current; eventual motor burnout can cause surge on semiconductors	Thermal overload on semiconductors and motors; design for semiconductor current overload	Operation at thermal overload conditions can reduce semiconductor life
Motor jam	Load blockage	High operating time at locked rotor current to 600%	Increases semiconductor current; eventual motor burnout can cause surge on semiconductors	Thermal overload on semiconductors and motors; design for semiconductor current overload	Depending on the design, up to 600% increases in semiconductor currents are possible, resulting in reduced life or shorted semiconductors
High on/off duty cycle	Jogging for high production needs	High operating time at locked rotor current to 600%	Increases semiconductor current; eventual motor burnout can cause surge on semiconductors	Thermal overload on semiconductors and motors; design for semiconductor current overload	Surge currents can also reduce semiconductor lifetime and cause semiconductor failures
Rapid reversing	Production needs	High operating time at locked rotor current to 600%	Increases semiconductor current; eventual motor burnout can cause surge on semiconductors	Thermal overload on semiconductors and motors; design for semiconductor current overload	High ambient temperatures produce the same results as high currents – i.e., reduced life and failed semiconductors
Long acceleration time	High-inertia, slow-starting loads	High operating time at locked rotor current to 600%	Increases semiconductor current; eventual motor burnout can cause surge on semiconductors	Thermal overload on semiconductors and motors; design for semiconductor current overload	Never operate semiconductors without proper cooling
High equipment temperature	High ambient temperatures; lack of cooling	No increase but can cause wiring and insulation failure	Increases semiconductor current; eventual motor burnout can cause surge on semiconductors	Thermal overload on semiconductors and motors; design for semiconductor current overload	
Motor wiring or bearing failure	Overcurrent; improper installation or maintenance	Locked rotor current to 600%	Increases semiconductor current; eventual motor burnout can cause surge on semiconductors	Thermal overload on semiconductors and motors; design for semiconductor current overload	
Phase failure	Blown fuse; loose connection	Decrease in current until motor core reaches saturation and current increases	Increases semiconductor current; eventual motor burnout can cause surge on semiconductors	Thermal overload; phase failure relay	
Phase unbalance	Unbalanced single-phase loads on same line; poorly regulated service	Decrease in one phase and increase in the other two phases of current	Increases semiconductor current; eventual motor burnout can cause surge on semiconductors	Thermal overload; phase failure relay	
Overvoltage	High source	Slight voltage increase decreases current; large increase may saturate core and increase current	Decrease or increase semiconductor current	Overvoltage relay	Increase of currents can cause semiconductor failure and reduced life
Transient overvoltage	Lightning; opening inductive switches; etc.	Slight average current increases	High transient voltages across semiconductor when semiconductor is in the off position	Capacitor-resistor networks, voltraps, MOVs, etc.	Short-duration high transients can cause semiconductor failure
Underload	Operator's choice	Decrease in motor current	Decrease in semiconductor current	None	Light load increases semiconductor life
Contaminants	Dirty environment	Causes corona	Causes corona	Clean room filters	Clean equipment periodically even with filters; corona carbonizes dirt and causes eventual semiconductor shorts

table can be used to determine adverse load conditions and the type of protection that should be used. In a properly designed system, overload conditions are anticipated and ways are found to deal with the overload. The user or operator has an obligation to operate the equipment within the design ratings. The table shows the equipment user and designer methods to prevent overloading a motor. Following these guidelines will help avoid situations that cause semiconductor failure or reduce semiconductor life. A good maintenance program should include a check of all the protective devices mentioned in Table 7.2.

Protection against many of the adverse conditions shown in Table 7.2 are often used in the design of semiconductor controls. They include water flow thermal relays, thermal overloads on the semiconductors and motors, phase failure relays, overvoltage relays, transient suppressors such as metal-oxide varistors (MOVs), capacitor-resistor networks, air filters, and fuses with blown-fuse indicators.

In a properly designed system, overload conditions are considered and provisions are made to accommodate the overloads. However, the user and the operator have an obligation to operate the equipment within the design ratings. Table 7.2 gives the equipment user and the designer methods by which to prevent the overloading of a motor and to avoid the situation that would cause semiconductor failure or reduce semiconductor life.

Power semiconductors must be protected from transient overloads. Silicon semiconductors will have a long, useful life if they are adequately protected against surge currents and transient-voltage spikes. Protection against failure due to current and voltage transients is relatively simple. Normally, this protection is designed into the circuit. In industrial plant environments, most voltage transients in semiconductor circuits arise from three major causes: switching, commutation transients, and regenerative surges.

Switching is the most common source of voltage transients. Whenever current is switched on or off in an inductive circuit, a transient voltage is generated at the switch terminals. Transformers and motor windings are highly inductive components, and the circuit wiring itself is inductive. Surges caused

by lightning also add to switching transients. These transients can originate in remote circuits and still feed back to the semiconductor circuit through the power supply line. Protecting semiconductors from such transients is primarily a matter of reducing the surge to a level the semiconductor can tolerate.

Commutation transients are associated with the reverse recovery characteristic of a rectifier junction. In normal use, a semiconductor is continually switching from a conducting state to a nonconducting state. This switching process causes rapid changes of current (high di/dt). Especially where fast-switching rectifiers are used, keep circuit inductance to a minimum. Here again, suitable suppression networks should be designed into the circuit.

Regenerative surges in inductive or dynamic loads is the third major source of voltage transients. Such loads include motors, lifting magnets, solenoids, relays, and many other devices involving stored inductive energy in the form of high induced voltage. Reducing or eliminating these transient voltages usually requires protective devices with high-energy-storage capacity.

Some of the common devices used for transient protection are capacitors, zener diodes, freewheeling diodes, and varistors (MOVs). A discussion of the MOV is included near the end of Chapter 6.

Power semiconductor circuits must be protected from current overloads caused by short circuits or other component breakdowns. The following methods are available for this type of protection.

1. Semiconductor fuses protect against overloads and, when properly applied, remove the semiconductor from the power source when an overload occurs. Fuses, however, cause more downtime and higher operating costs. Therefore, fusing is generally limited to applications where the power source can damage components before the slower circuit breakers remove the power.

2. Magnetically operated breakers provide an inexpensive means of limiting current. This type of mechanical breaker operates best under short-circuit conditions and offers speedy, low-cost restarting and relatively fast circuit interruption.

3. Thermal breakers employ heating elements to operate bimetallic contact actuators. This type of breaker offers reasonable protection for wires and good protection for components with high thermal capacity. Their ability to protect semiconductors is limited, however.

4. Overrated semiconductors—semiconductors with current ratings high enough to accommodate anticipated current overloads—offer another method of protection. This approach allows the cost of fuses, wiring, and mounting to be put into the semiconductor. It also gives only minimal protection. Adequate safety measures require a conventional circuit breaker to disable the circuit in extreme malfunctions.

5. Combinations of circuit breakers with fuses, oversized semiconductors, or current feedback circuits are frequently applied protection techniques. Another approach is using special branch protection. For example, each leg of a single-phase bridge might be separately protected instead of a single fuse or breaker to fuse the mains. This arrangement prevents overstressing a motor while the semiconductor devices are operating at high current peaks that are within fuse limits.

6. Semiconductors and semiconductor equipment must be kept clean. Why is cleanliness so important? It is important because the glass or ceramic seal on a high-power semiconductor package can be a source of trouble when dirt accumulates. Good engineering practice calls for the positioning of components and support material to reduce or eliminate excess voltage stress and voltage gradients. These gradients can change with moisture, dirt, pressure, temperature, and aging. In a clean system, high transients can initiate corona that will then be extinguished upon return to normal voltage. However, with any buildup of moisture or dirt on the insulating glass or ceramic surface, the corona will remain when the voltage returns to normal. This residue can cause extensive damage to the circuit. Even semiconductors without a visible layer of contaminants may be covered with conductive particles that can cause corona. Therefore, periodic cleaning is advisable. The time between cleanings can be lengthened by the proper use of filters, but filters will not eliminate the need for cleaning. A good maintenance program is essential to ensure proper operation of semiconductors.

Preventive Maintenance

The key to successful and efficient operation of high-power semiconductors is a good, well-planned maintenance program. Routine maintenance of equipment using semiconductors must include good operating procedures. Emphasis should be placed on detection and prevention of (1) temperature buildup, (2) dirt accumulation, and (3) loose mountings and connections. A regular maintenance schedule should be followed for checking and eliminating these conditions.

Temperature Buildup. Temperature buildup may be caused by excess ambient temperature, poor or blocked air circulation, failure of cooling devices such as fans or water-circulating equipment, dirty heat sinks and air filters, and equipment overloading. Since temperature buildup usually is gradual, it should be constantly monitored with strategically placed thermometers. Thermocouples often are used to measure semiconductor temperatures. One method of attaching a thermocouple to a semiconductor base or heat sink is first to drill a small shallow hole in the device. Then fit the thermocouple into it and peen the surface around the hole to secure it. Equipment designed for forced-air cooling or water cooling should never be energized without proper air or water flow.

Dirt Accumulation. Dirt accumulation on the glass or ceramic surfaces of semiconductor packages must be regularly and thoroughly removed. The safest method of cleaning semiconductors is the one that gives the best results with the fewest disadvantages. Solvents can be hazardous and should be used with care. Even a solvent that is considered nontoxic can kill or injure when used with improper ventilation. One of the safest methods of cleaning is simply by wiping with clean cloths. Often, the areas that need to be cleaned cannot be reached properly, however, and a liquid cleaner must be used. Table 7.3 lists solvents that can be used as cleaners.

The safest method of cleaning semiconductors is to use a detergent wash and then flush with distilled water. The system must be completely dry before the reapplication of power. Regular tap water may contain many conductive impurities and should not be used as a final rinse.

Methyl, ethyl, and isopropyl alcohol can be used for cleaning, but they can attack sleevings and insulations. Before using alcohol or any solvent, test the cleaner on a small sample of sleeving and insulation to determine whether they are attacked. The amount of time a material is exposed to a solvent determines adverse solvent effects. The time of exposure during the test, therefore, should be the same as the time of exposure expected during actual cleaning. If you observe any stretching, softening, or deterioration of the material being cleaned, do not use the solvent, although a temporary softening may not necessarily be detrimental. Remember that alcohol is toxic and should be used with care.

Follow all safety precautions when using any solvents. Read the labels on all the chemicals. Do not mix solvents (or any chemicals) unless recom-

mended. The information in Table 7.3 regarding the fire, explosion, and toxicity hazards of solvents is believed to be accurate, but it cannot be guaranteed. This information provides very general guidelines only; always carefully follow any precautions and directions for use printed on a solvent label.

Loose Mountings and Connections. Along with scheduled periodic cleaning, a good maintenance program should include regular checks for mounting and terminal connection tightness. If a loosely mounted device is discovered, it should be removed, with both mounting surfaces thoroughly cleaned and thermal compound reapplied, and then remounted to the specified torque.

Safety

Safety in operating and servicing solid-state equipment is especially important. Unlike mechanical equipment that has rotating parts, moving contactors, and so on, semiconductor equipment has nothing to warn the unwary technician that it is energized. There also is a tendency to believe that voltage and current levels associated with semiconductor electronics are

TABLE 7.3   **Solvents for Cleaning Semiconductors**

Solvent	Fire Hazard	Explosion Hazard	Toxicity Hazard	Electrically Conductive	Attack on Rubber Insulation
Water, tap	None	None	None	Yes	None
Precautions: Rinse with distilled water and dry thoroughly.					
Water, distilled	None	None	None	No	None
Precautions: Dry thoroughly.					
Water and detergent	None	None	None	Yes	None
Precautions: Rinse with distilled water and dry thoroughly.					
Methyl alcohol	High	High	High	No	Slight
Precautions: Avoid skin contact and use ventilation.					
Ethyl alcohol	High	High	High	No	Slight
Precautions: Can cause internal damage; ingestion can cause blindness.					
Isopropyl alcohol	High	High	High	No	Slight
Precautions: Mix with distilled water to reduce hazards.					
Paint thinner (mineral spirits)	Low	Low	Low	No	Slight
Precautions: Use proper ventilation and limit exposure to rubber.					
Acetone	High	Moderate	Low	No	Slight
Precautions: Limit exposure to prevent rubber degeneration.					
Perchloroethylene (dry-cleaning solvent)	Low	Low	Moderate	No	Slight
Precautions: Use short exposure to prevent rubber damage; irritates eyes and causes headaches; heating causes fumes.					
Trichloroethylene	Low	Low	Moderate	No	Slight
Precautions: Use adequate ventilation and limit exposure time to prevent rubber damage.					
Freon	Low	Low	Low	No	Slight
Precautions: Use adequate ventilation and limit exposure time.					
Maltier XL-100	Low	Low	Low	No	Slight
Precautions: Use adequate ventilation and limit exposure time.					
Miller Stephenson MS-180	Low	Low	Low	No	Slight
Precautions: Use adequate ventilation and limit exposure time.					

too low to be hazardous. This is a dangerous assumption and not true. Thyristor voltage can be in the thousands of volts. Obvious safety measures must be carefully observed. Most high-power equipment includes built-in safety interlocks. These interlocks ensure that the equipment is turned off before anyone can gain access to the high-voltage areas.

Maintenance personnel, however, often defeat these interlocks to simplify their service work. This action is both careless and foolish. In cases when circuits must be checked while energized, there are a few simple rules that could prevent injury and equipment damage. First, if possible, always work with another person present—one who knows how to shut down the equipment quickly. Next, make sure that the floor is covered with a rubber mat, and follow the good practice of working with one hand in your pocket. This guideline sounds simple, but it could prevent you from completing an electrical circuit. Always use insulated tools to avoid short circuits that could further damage electronic components and circuits. The following common shop practices for safety are too often overlooked or ignored:

- Always wear safety glasses. Hot metal from a short or a soldering iron can ruin an eye as quickly as a metal chip.
- Keep long hair contained with a cap or net.
- Lock out the disconnect or breaker for the circuit you are about to service.

- Check with a voltmeter to be sure the circuit is completely dead.
- Maintain off-limit areas for high-voltage equipment. Access should be allowed only to authorized personnel.

When measuring high voltage, use the following procedure:

- De-energize equipment and tag breakers so that power will not be turned on accidentally.
- Discharge any capacitors that may have a dangerous voltage present.
- Attach meter probes.
- Energize equipment and take a reading.
- De-energize equipment, discharge capacitors, and remove probes.

Too many accidents are caused by carelessness or laziness. Bear in mind that accidents can happen to anyone, so be careful.

CONCLUSION

We have seen that the thyristor is a versatile and powerful control device. Its most prominent disadvantage derives from its use in DC circuits. Despite the problems associated with its use, it is very popular in industry. Because of its widespread applications, a thorough understanding of the thyristor is a necessity for technicians.

■ QUESTIONS •••

1. The ramp-and-pedestal control circuit is classified as a _____ control. The ramp-and-pedestal has a greater control over the _____ _____ than other types of phase control.

2. The snap-on effect in some phase controls is due to _____. This effect can be reduced by adding a special diode or by adding another _____ network.

3. The type of switching that occurs only when the supply is at or near 0 V is called _____ switching. This type of switching is appropriate for heating controls but not for _____ controls.

4. The chopper motor control converts pure DC to _____. When the thyristor's on time increases compared with its off time, the _____ voltage at the load increases.

5. Closed-loop, or feedback, control is used to control a motor's speed by using an IC called a _____. This type of control can regulate a motor's speed to within _____% of desired speed.

6. The speed of AC motors is controlled by varying the number of poles and by varying the _____ _____. In the rectifier-inverter speed control, the supply voltage is converted to DC,

and the DC is converted to variable-frequency _____.

7. Explain how phase control varies the power delivered to a load.

8. Describe hysteresis, and explain how it affects some phase-control circuits.

9. Why is the UJT used as an oscillator to trigger thyristor circuits?

10. Explain how the ramp-and-pedestal circuit works.

11. Explain the advantages and disadvantages of using zero-voltage switching in control circuits. Define zero-voltage switching.

12. Describe basic converter action, and explain how this action changes the power delivered to a load.

13. Define chopper. Describe its operation. How are choppers commutated?

14. Describe how pulse-width modulation can be used to control a DC motor's speed.

15. List the methods by which an AC motor's speed may be controlled for (a) the universal motor and (b) the synchronous and the induction motors.

16. Explain how a closed-loop speed control system works.

17. Explain the operation of the following circuits: (a) rectifier-inverter and (b) cycloconverter.

18. Match the types of conversion with the circuits that perform them.

   a. AC to AC   **1.** Inverter
   b. DC to AC   **2.** Rectifier
   c. AC to DC   **3.** Chopper
   d. DC to DC   **4.** Converter
                          5. Cycloconverter

19. List the applications of the cycloconverter. What are its main advantages?

20. What are the advantages of using transistors as power switches?

21. What problems occur if transistors are used as power switches?

22. Describe how the power MOSFET overcomes one of the transistor problems you listed in your answer to Question 21.

23. Explain how a PLL can be used as a motor speed controller.

■ PROBLEMS ●

1. In the full-bridge converter with a flywheel diode, the applied voltage is 230 V AC. With a firing angle of 60°, what is the average armature voltage?

2. A four-pole, 100 hp, 440 V AC motor is designed to operate at 60 Hz and has a base speed of 2000 r/min.

   a. Calculate the V/Hz ratio.

   b. Determine the amount of voltage needed to keep the flux constant when operating the motor at 20 Hz.

   c. Calculate the rotor speed with a slip of 5%.

   d. Find the power output in watts when the motor produces a torque of 100 Nm.

3. A six-pole, 1000 W motor has a rated voltage of 230 V AC at 60 Hz. We would like to operate the motor at 1000 r/min at full-load torque. Slip is 3% at full-load torque.

   a. Calculate the V/Hz ratio.

   b. Calculate the stator voltage needed to operate the motor at 1000 r/min.

   c. Determine the stator frequency needed to operate the motor at 1000 r/min.

   d. Find the power output of this machine when producing 8 Nm of torque.

   e. Calculate the percentage that full-load torque and power will decrease at the operating frequency.

4. Given the circuit in Figure 7.1A for a diac trigger device, with a breakover voltage of 35 V, a minimum diac voltage of 8 V when the capacitor discharges, $R = 100$ kΩ, $C = 0.2$ μF, and $f = 60$ Hz applied, find the firing angle of the triac.

5. For the ZVS in Figure 7.9, with $R_1 = 200$ Ω and R_2 a GB32J2 thermistor, find the temperature at which the triac will turn off it we energize the heater at room temperature.

8

Transducers

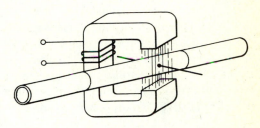

OBJECTIVES

On completion of this chapter, you should be able to:

- Define a transducer in terms of input, conversion, and output.
- Explain the transduction principle behind individual transducers.
- Classify individual transducers into one of seven basic areas.
- Calculate the unknown resistance in and the sensitivity of a bridge circuit.

INTRODUCTION

Control and regulation of industrial systems and processes depend on accurate measurement. It would not be going too far to state that a variable must be measured accurately to be controlled. In industrial systems, the device that does this measurement is the transducer, or sensor. *Transduction* is the process of converting energy from one form to another. We can define a *transducer* as a device that converts a *measurand* (that which is to be measured) into an output that facilitates measurement. In other words, a transducer converts a variable (fluid flow, temperature, humidity) into an analog of the variable.

Let's use an example: the mercury thermometer. A thermometer is a transducer. The measurand is temperature. The transducer converts temperature into an analog of temperature: fluid level in a glass tube. The level of mercury is directly proportional to (an analog of) ambient temperature.

In this chapter, we describe several transducers and the principles by which they operate. Our treatment of this topic is in no way exhaustive. Rather, we intend to describe the major sensors in use in industry today. The sensors covered are primarily electrical sensors—that is, sensors that give an electrical output. Sensors giving other outputs are discussed when appropriate. In our treatments, we have classified sensors according to the variable measured. In this chapter, we also briefly discuss bridges, which are common devices for detecting sensor changes.

TEMPERATURE

Temperature is undoubtedly the most measured dynamic variable in industry today. Many industrial processes require accurate temperature measurement because temperature cannot be precisely controlled unless it can be accurately measured.

Before we discuss temperature sensors, we must define temperature. Very simply, *temperature* is the ability of a body to communicate or transfer heat energy. Alternatively, we can define temperature as the potential for heat to flow. Recall that heat will flow from a hotter body to a cooler one.

Temperature generally is measured on three arbitrary scales: Fahrenheit, Celsius, and Kelvin. The *Fahrenheit scale* is referenced to the boiling point of water at 212°F and the freezing point of water at 32°F. The *Celsius scale* references the boiling point at 100°C and the freezing point at 0°C. The *Kelvin scale* is based on the divisions of the Celsius scale, with absolute zero at 0°. To convert from Celsius to Kelvin, you simply add 273°.

Temperature can be measured in many different ways. For purposes of simplicity, we divide temperature sensing into two areas: mechanical and electrical.

Mechanical Temperature Sensing

Mechanical temperature sensing depends on the physical principle that gases, liquids, and solids change their volume when heated. Furthermore, different substances change volume in differing amounts. For example, liquid mercury expands about 0.01% per degree Fahrenheit. Methyl alcohol, on the other hand, expands at about 0.07% per degree Fahrenheit (0.1% per degree Celsius). In this section, we discuss mechanical temperature sensors—devices that convert temperature to position or motion.

Glass-Stem Thermometers. The *glass-stem thermometer* is one of the oldest types of thermometers. Its invention is credited to Galileo in about 1590. As shown in Figure 8.1, it consists of a bulb filled with liquid, a capillary tube, and a supporting glass stem. The capillary tube is quite thin, which helps increase the sensitivity of the device. Some type of magnifying lens is usually included in the glass stem to aid in reading temperature measurement. Linearity of this device is improved by evacuating the remaining air from the tube.

The glass stem has several advantages. It is inexpensive to manufacture and has excellent linearity and accuracy characteristics. It also has several disadvantages. It is fragile and difficult to read, it is not easily used for remote measurement or control, and it allows considerable time lag in temperature measurement. This last disadvantage arises because of the poor thermal conductivity of glass. In spite of

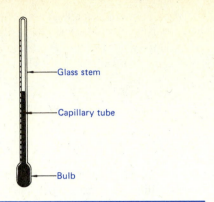

FIGURE 8.1   Glass-Stem Thermometer

these disadvantages, the glass-stem thermometer is still popular in industry today.

Filled-System Thermometers. The *filled-system thermometer* works on the same basic principle as the glass-stem thermometer. The bulb, shown in Figure 8.2, is filled with a gas or a liquid. As the bulb is heated, the gas or liquid expands, exerting pressure on the Bourdon tube (to be described later in the chapter) through the capillary tubing. The Bourdon tube uncoils, moving the pointer. Since this system exerts a pressure, it can be used for chart-recording displays of temperature.

Although this system is mechanical, it can be converted into an electric transducer by using a potentiometer, as shown in Figure 8.3, or a linear variable differential transformer (LVDT). Potentiometers and LVDTs convert position or displacement into an electrical parameter. These devices are discussed later in the chapter.

Filled systems react very quickly to temperature changes, can be as accurate as 0.5%, and can be used for remote measurement up to 300 ft (100 m). The major disadvantage with filled systems is the need for temperature compensation. Although the sensing element in the filled system is the bulb, the Bourdon tube and capillary tube are also temperature-sensitive. Thus, ambient temperature changes around the Bourdon and capillary tubes can give a false temperature reading. Filled systems can be temperature-compensated by use of bimetal strips or systems using dual elements.

Bimetallic Thermometers. The *bimetallic thermometer* operates on the principle of differential expansion of metals; that is, metals increase their volume when heated and different metals expand by different amounts. How much a metal expands when heated is indicated by a parameter called the *linear expansion coefficient*. This parameter shows volume expansion in millionths per degree Celsius. A list of the parameter values for a few metals is given in Table 8.1.

When two metals are joined together and heated, physical displacement occurs. In designing these devices, engineers try to ensure maximum movement

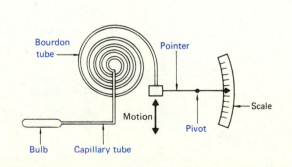

FIGURE 8.2   Filled-System Thermometer

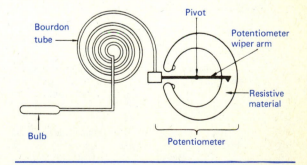

FIGURE 8.3   Filled System Connected to Potentiometer

TABLE 8.1   Linear Expansion Coefficient of Various Metals

Substance	Linear Expansion Coefficient (Millionths/°C)
Aluminum	23.5
Brass	20.3
Copper	16.5
Invar (copper-nickel alloy)	1.2
Kovar (copper-nickel-cobalt alloy)	5.9

with a given temperature change. They do so by joining a metal with a low expansion coefficient to one with a high expansion coefficient. A popular combination is brass and invar (a copper-nickel alloy). From Table 8.1, we see that this combination meets the design criteria.

A bimetallic thermometer is shown in Figure 8.4. The bimetal strips usually are wound in a spiral or helix (coil). The helical type is more sensitive.

The bimetallic thermometer is one of the most popular thermometers in industry. It is relatively inexpensive, has a wide range (800°F, or 400°C), is rugged, and is easily installed and read. Its major disadvantages are that it cannot be used in remote measurement or in analog process control. Bimetallic sensors are used in simple on-off control systems. For this use, a mercury switch is placed on the bimetal strip, as in home thermostats.

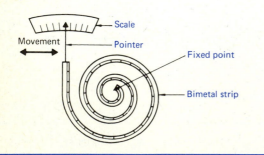

FIGURE 8.4   Bimetallic Thermometer

An important calculation when using mercury switches concerns how much the strip moves with a given temperature change. The amount of deflection of a straight bimetallic strip is given by

$$y = \frac{3(c_A - c_B)(T_2 - T_1)\,l^2}{4d} \tag{8.1}$$

where

y = amount of deflection
c_A = linear expansion coefficient of metal A
c_B = linear expansion coefficient of metal B
T_1 = lower temperature
T_2 = higher temperature
l = length of strip
d = thickness of strip

This formula tells us that the deflection of a straight bimetal strip is directly proportional to the change in temperature and inversely proportional to the thickness of the strip. The amount of deflection is directly proportional to the square of the length of the strip.

For example, suppose we have a 20 cm long, 0.05 cm thick, brass-invar strip that is straight at room temperature (25°C). How much will the strip deflect if the temperature changes 5°C? From Table 8.1, we see that invar and brass have coefficients of 1.2 and 20.3, respectively. Using Equation 8.1, we find that the deflection is

$$y = \frac{3\left[(2.03 \times 10^{-5}) - (1.2 \times 10^{-6})\right](5°C)(20\ \text{cm})^2}{4\,(0.05\ \text{cm})}$$

$$= 0.573\ \text{cm}$$

Note that the amount of deflection is directly proportional to the differences between the temperatures and the coefficients of expansion and proportional to the square of the length. Also, the deflection is inversely proportional to the strip's thickness.

• EXAMPLE 8.1

Given a copper-invar bimetal strip 15 cm long and 0.25 cm thick that is straight at 30°C, find the deflection of the strip if the temperature increases to 80°C.

Solution

The deflection can be found by substituting the appropriate values into Equation 8.1:

$$y = \frac{3\left[\left(16.5 \times 10^{-6}\right) - \left(1.2 \times 10^{-6}\right)\right](50°C)(15 \text{ cm})^2}{4(0.25 \text{ cm})}$$

$$= 0.516 \text{ cm}$$

The equation for finding the deflection of a spiral strip is similar to the equation for a straight strip:

$$y = \frac{9\left(c_A - c_B\right)\left(T_2 - T_1\right)rl}{4d} \tag{8.2}$$

where

r = spiral's radius
l = length of strip if it were extended

We can see the value of the spiral strip if we use the data from the brass-invar example but convert the straight strip to a spiral. Let us say we wind the 20 cm strip into a spiral with a radius of 4 cm. Using Equation 8.2 and substituting values, we get

$$y = \frac{9\left(c_A - c_B\right)\left(T_2 - T_1\right)rl}{4d}$$

$$= \frac{9\left[\left(2.03 \times 10^{-5}\right) - \left(1.2 \times 10^{-6}\right)\right](5°C)(4)(20)}{4(0.05 \text{ cm})}$$

$$= 0.340 \text{ cm}$$

You will notice that, all things being equal, the spiral bimetallic strip is more sensitive than the straight strip. Spiral winding makes the response of this device linear over a 200°C range.

• **EXAMPLE 8.2**

Given a spiral copper-invar bimetal strip 40 cm long and 0.25 cm thick whose radius is 5 cm at 30°C, find the deflection of the strip if the temperature increases to 80°C.

Solution

The deflection can be found by substituting the appropriate values into Equation 8.2:

$$y = \frac{9\left[\left(16.5 \times 10^{-6}\right) - \left(1.2 \times 10^{-6}\right)\right](50°C)(5 \text{ cm})(40 \text{ cm})}{4(0.25 \text{ cm})}$$

$$= 1.38 \text{ cm}$$

The bimetallic strip is a good example of a practical application of the principle of the thermal expansion of materials. These sensors are used in single–temperature control circuits to make or break an electrical contact that produces corresponding changes in temperature.

Electrical Temperature Sensing

We have considered several methods of mechanical sensing. Because mechanical rather than electrical principles are used, these sensors are not suitable for use in analog process control systems. In this section, therefore, we discuss four commonly used electrical sensors: the thermocouple, the thermistor, the resistance temperature detector, and the semiconductor temperature sensor.

Thermocouple. Of the electrical temperature-sensing devices, the *thermocouple* enjoys the widest use in industry. Its discovery dates back to 1821, when Thomas Seebeck, a German physicist, joined two wires made of different metals. He found that when he heated one end, electric current flowed in the loop formed by the wires. This effect is called the *Seebeck effect* and is illustrated in Figure 8.5.

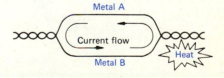

FIGURE 8.5 Seebeck Effect

When the circuit was broken, as shown in Figure 8.6, Seebeck observed that there was a voltage between the two terminals. He discovered that the size of the voltage varied with heat. An increase in heat caused an increase in voltage. He also found that different combinations of metals produced different voltages.

Thermocouples come in many different forms, as shown in Figure 8.7. The thermocouple junction may be exposed, grounded, fashioned into washers, or placed in a well for protection. Each of these methods has its own advantages.

Generally speaking, the thermocouple is simple, rugged, inexpensive, and capable of the widest temperature measurement range (about 4500°F, or 2500°C) of any electrical temperature transducer. On the other hand, it is the least sensitive and stable of the electrical temperature transducers. Perhaps its biggest disadvantage is the need for a reference junction.

Consider the schematic diagram in Figure 8.8. As shown in the figure, we must connect a readout device to a thermocouple to read the voltage produced. But in doing so, we create two more thermoelectric junctions, X_1 and X_2. These extra junctions interfere with the voltage produced by the iron-constantan thermocouple. (Constantan is an alloy of copper and nickel.)

This problem can be overcome by adding an iron lead to the constantan side of the thermocouple and connecting the new iron lead to the readout device, as shown in Figure 8.9. The two iron-copper junction potentials will cancel out, leaving the difference in potential between the measuring junction and the

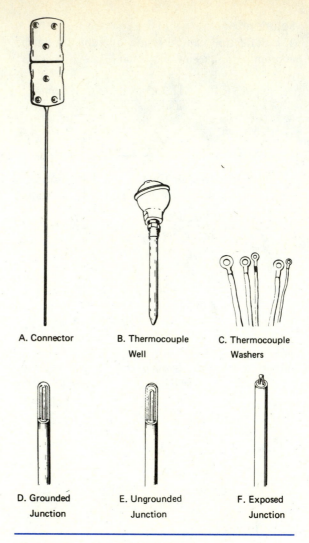

A. Connector   B. Thermocouple Well   C. Thermocouple Washers

D. Grounded Junction   E. Ungrounded Junction   F. Exposed Junction

FIGURE 8.7 Thermocouples

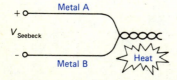

FIGURE 8.6 Potential Difference across Heated Junction

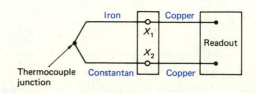

FIGURE 8.8 Indicating Device Connected to Thermocouple

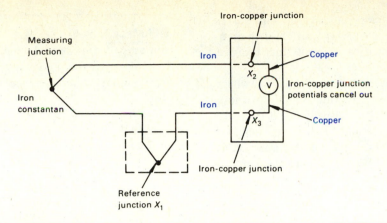

FIGURE 8.9 Additional Thermocouple Junctions Created

reference junction at the output. If the reference junction is held at a constant temperature, it is a simple matter of converting the total voltage produced at the readout device to a temperature. As a matter of fact, most conversion tables are based on a reference junction temperature of 32°F (0°C). This reference temperature has been chosen because for many years the reference junction was immersed in an ice bath to keep its temperature constant.

Because ice baths are often inconvenient to use and therefore impractical, several other methods are used to achieve the same result. One such method is shown in Figure 8.10. Both reference junctions and the bridge resistor R_2 are thermally integrated on a substrate. The resistor R_2 is a temperature-sensitive resistor. This resistor will change its resistance with ambient temperature changes, producing a voltage that is opposite to the reference voltage change. The remaining resistors in the bridge are not sensitive to temperature changes. As the temperature of the cold junction varies, the bridge produces a voltage to cancel out the cold junction potential produced. The only voltage that varies is the one that is produced by the thermocouple.

Another method for compensating thermocouples, popular in recent years, uses a computer. The reference junction temperature is measured accurately with an electrical transducer. An ADC changes the analog temperature to a digital representation of the temperature and sends it into the input port of a computer. A computer program uses the reference junction measurement to calculate the required correction voltage and makes an estimate of the temperature of the measuring junction.

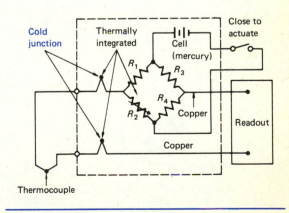

FIGURE 8.10 Thermocouple System with Thermally Integrated Cold Junctions and Resistors

Thermocouple temperature versus emf curves are shown in Figure 8.11. The different thermocouples are represented by different letter designations.

Tables are available that give the output voltage of a particular type of thermocouple at various temperatures. These charts usually use a reference junction of 0°C, as mentioned previously. Table 8.2 is an example of a table for a type J thermocouple. Notice that the type J device puts out 5.431 mV at 103°C.

• EXAMPLE 8.3
Given a type J thermocouple, with a reference junction at 0°C and an output voltage of 5.431 mV, find the temperature of the reference junction.

Solution
Using Table 8.2, we find that the corresponding temperature is 103°C.

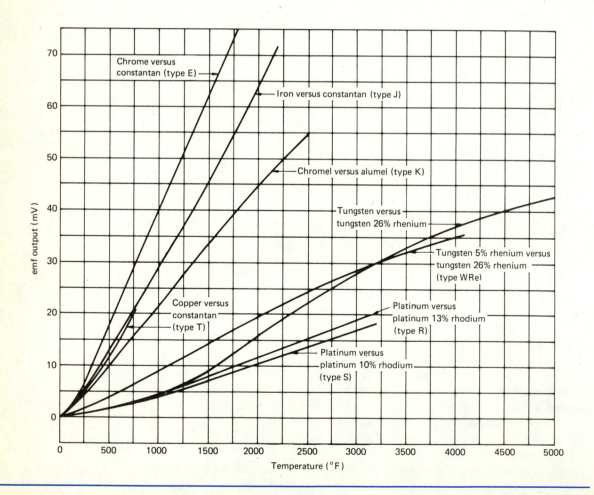

FIGURE 8.11 Temperature versus emf (Reference Junction at 32°F)

TABLE 8.2 Output Voltage for Type J Thermocouple

THERMOELECTRIC VOLTAGE IN ABSOLUTE MILLIVOLTS

DEG C	0	1	2	3	4	5	6	7	8	9	10	DEG C
−210	−8.096											−210
−200	−7.890	−7.912	−7.934	−7.955	−7.976	−7.996	−8.017	−8.037	−8.057	−8.076	−8.096	−200
−190	−7.659	−7.683	−7.707	−7.731	−7.755	−7.778	−7.801	−7.824	−7.846	−7.868	−7.890	−190
−180	−7.402	−7.429	−7.455	−7.482	−7.508	−7.533	−7.559	−7.584	−7.609	−7.634	−7.659	−180
−170	−7.122	−7.151	−7.180	−7.209	−7.237	−7.265	−7.293	−7.321	−7.348	−7.375	−7.402	−170
−160	−6.821	−6.852	−6.883	−6.914	−6.944	−6.974	−7.004	−7.034	−7.064	−7.093	−7.122	−160
−150	−6.499	−6.532	−6.565	−6.598	−6.630	−6.663	−6.695	−6.727	−6.758	−6.790	−6.821	−150
−140	−6.159	−6.194	−6.228	−6.263	−6.297	−6.331	−6.365	−6.399	−6.433	−6.466	−6.499	−140
−130	−5.801	−5.837	−5.874	−5.910	−5.946	−5.982	−6.018	−6.053	−6.089	−6.124	−6.159	−130
−120	−5.426	−5.464	−5.502	−5.540	−5.578	−5.615	−5.653	−5.690	−5.727	−5.764	−5.801	−120
−110	−5.036	−5.076	−5.115	−5.155	−5.194	−5.233	−5.272	−5.311	−5.349	−5.388	−5.426	−110
−100	−4.632	−4.673	−4.714	−4.755	−4.795	−4.836	−4.876	−4.916	−4.956	−4.996	−5.036	−100
−90	−4.215	−4.257	−4.299	−4.341	−4.383	−4.425	−4.467	−4.508	−4.550	−4.591	−4.632	−90
−80	−3.785	−3.829	−3.872	−3.915	−3.958	−4.001	−4.044	−4.087	−4.130	−4.172	−4.215	−80
−70	−3.344	−3.389	−3.433	−3.478	−3.522	−3.566	−3.610	−3.654	−3.698	−3.742	−3.785	−70
−60	−2.892	−2.938	−2.984	−3.029	−3.074	−3.120	−3.165	−3.210	−3.255	−3.299	−3.344	−60
−50	−2.431	−2.478	−2.524	−2.570	−2.617	−2.663	−2.709	−2.755	−2.801	−2.847	−2.892	−50
−40	−1.960	−2.008	−2.055	−2.102	−2.150	−2.197	−2.244	−2.291	−2.338	−2.384	−2.431	−40
−30	−1.481	−1.530	−1.578	−1.626	−1.674	−1.722	−1.770	−1.818	−1.865	−1.913	−1.960	−30
−20	−0.995	−1.044	−1.093	−1.141	−1.190	−1.239	−1.288	−1.336	−1.385	−1.433	−1.481	−20
−10	−0.501	−0.550	−0.600	−0.650	−0.699	−0.748	−0.798	−0.847	−0.896	−0.945	−0.995	−10
0	0.000	−0.050	−0.101	−0.151	−0.201	−0.251	−0.301	−0.351	−0.401	−0.451	−0.501	0

DEG C	0	1	2	3	4	5	6	7	8	9	10	DEG C
0	0.000	0.050	0.101	0.151	0.202	0.253	0.303	0.354	0.405	0.456	0.507	0
10	0.507	0.558	0.609	0.660	0.711	0.762	0.813	0.865	0.916	0.967	1.019	10
20	1.019	1.070	1.122	1.174	1.225	1.277	1.329	1.381	1.432	1.484	1.536	20
30	1.536	1.588	1.640	1.693	1.745	1.797	1.849	1.901	1.954	2.006	2.058	30
40	2.058	2.111	2.163	2.216	2.268	2.321	2.374	2.426	2.479	2.532	2.585	40
50	2.585	2.638	2.691	2.743	2.796	2.849	2.902	2.956	3.009	3.062	3.115	50
60	3.115	3.168	3.221	3.275	3.328	3.381	3.435	3.488	3.542	3.595	3.649	60
70	3.649	3.702	3.756	3.809	3.863	3.917	3.971	4.024	4.078	4.132	4.186	70
80	4.186	4.239	4.293	4.347	4.401	4.455	4.509	4.563	4.617	4.671	4.725	80
90	4.725	4.780	4.834	4.888	4.942	4.996	5.050	5.105	5.159	5.213	5.268	90
100	5.268	5.322	5.376	5.431	5.485	5.540	5.594	5.649	5.703	5.758	5.812	100
110	5.812	5.867	5.921	5.976	6.031	6.085	6.140	6.195	6.249	6.304	6.359	110
120	6.359	6.414	6.468	6.523	6.578	6.633	6.688	6.742	6.797	6.852	6.907	120
130	6.907	6.962	7.017	7.072	7.127	7.182	7.237	7.292	7.347	7.402	7.457	130
140	7.457	7.512	7.567	7.622	7.677	7.732	7.787	7.843	7.898	7.953	8.008	140
150	8.008	8.063	8.118	8.174	8.229	8.284	8.339	8.394	8.450	8.505	8.560	150
160	8.560	8.616	8.671	8.726	8.781	8.837	8.892	8.947	9.003	9.058	9.113	160
170	9.113	9.169	9.224	9.279	9.335	9.390	9.446	9.501	9.556	9.612	9.667	170
180	9.667	9.723	9.778	9.834	9.889	9.944	10.000	10.055	10.111	10.166	10.222	180
190	10.222	10.277	10.333	10.388	10.444	10.499	10.555	10.610	10.666	10.721	10.777	190
200	10.777	10.832	10.888	10.943	10.999	11.054	11.110	11.165	11.221	11.276	11.332	200
210	11.332	11.387	11.443	11.498	11.554	11.609	11.665	11.720	11.776	11.831	11.887	210
220	11.887	11.943	11.998	12.054	12.109	12.165	12.220	12.276	12.331	12.387	12.442	220
230	12.442	12.498	12.553	12.609	12.664	12.720	12.776	12.831	12.887	12.942	12.998	230
240	12.998	13.053	13.109	13.164	13.220	13.275	13.331	13.386	13.442	13.497	13.553	240
250	13.553	13.608	13.664	13.719	13.775	13.830	13.886	13.941	13.997	14.052	14.108	250
260	14.108	14.163	14.219	14.274	14.330	14.385	14.441	14.496	14.552	14.607	14.663	260
270	14.663	14.718	14.774	14.829	14.885	14.940	14.995	15.051	15.106	15.162	15.217	270
280	15.217	15.273	15.328	15.383	15.439	15.494	15.550	15.605	15.661	15.716	15.771	280
290	15.771	15.827	15.882	15.938	15.993	16.048	16.104	16.159	16.214	16.270	16.325	290

TABLE 8.2   (Continued)

DEG C	0	1	2	3	4	5	6	7	8	9	10	DEG C
300	16.325	16.380	16.436	16.491	16.547	16.602	16.657	16.713	16.768	16.823	16.879	300
310	16.879	16.934	16.989	17.044	17.100	17.155	17.210	17.266	17.321	17.376	17.432	310
320	17.432	17.487	17.542	17.597	17.653	17.708	17.763	17.818	17.874	17.929	17.984	320
330	17.984	18.039	18.095	18.150	18.205	18.260	18.316	18.371	18.426	18.481	18.537	330
340	18.537	18.592	18.647	18.702	18.757	18.813	18.868	18.923	18.978	19.033	19.089	340
350	19.089	19.144	19.199	19.254	19.309	19.364	19.420	19.475	19.530	19.585	19.640	350
360	19.640	19.695	19.751	19.806	19.861	19.916	19.971	20.026	20.081	20.137	20.192	360
370	20.192	20.247	20.302	20.357	20.412	20.467	20.523	20.578	20.633	20.688	20.743	370
380	20.743	20.798	20.853	20.909	20.964	21.019	21.074	21.129	21.184	21.239	21.295	380
390	21.295	21.350	21.405	21.460	21.515	21.570	21.625	21.680	21.736	21.791	21.846	390
400	21.846	21.901	21.956	22.011	22.066	22.122	22.177	22.232	22.287	22.342	22.397	400
410	22.397	22.453	22.508	22.563	22.618	22.673	22.728	22.784	22.839	22.894	22.949	410
420	22.949	23.004	23.060	23.115	23.170	23.225	23.280	23.336	23.391	23.446	23.501	420
430	23.501	23.556	23.612	23.667	23.722	23.777	23.833	23.888	23.943	23.999	24.054	430
440	24.054	24.109	24.164	24.220	24.275	24.330	24.386	24.441	24.496	24.552	24.607	440
450	24.607	24.662	24.718	24.773	24.829	24.884	24.939	24.995	25.050	25.106	25.161	450
460	25.161	25.217	25.272	25.327	25.383	25.438	25.494	25.549	25.605	25.661	25.716	460
470	25.716	25.772	25.827	25.883	25.938	25.994	26.050	26.105	26.161	26.216	26.272	470
480	26.272	26.328	26.383	26.439	26.495	26.551	26.606	26.662	26.718	26.774	26.829	480
490	26.829	26.885	26.941	26.997	27.053	27.109	27.165	27.220	27.276	27.332	27.388	490
500	27.388	27.444	27.500	27.556	27.612	27.668	27.724	27.780	27.836	27.893	27.949	500
510	27.949	28.005	28.061	28.117	28.173	28.230	28.286	28.342	28.398	28.455	28.511	510
520	28.511	28.567	28.624	28.680	28.736	28.793	28.849	28.906	28.962	29.019	29.075	520
530	29.075	29.132	29.188	29.245	29.301	29.358	29.415	29.471	29.528	29.585	29.642	530
540	29.642	29.698	29.755	29.812	29.869	29.926	29.983	30.039	30.096	30.153	30.210	540
550	30.210	30.267	30.324	30.381	30.439	30.496	30.553	30.610	30.667	30.724	30.782	550
560	30.782	30.839	30.896	30.954	31.011	31.068	31.126	31.183	31.241	31.298	31.356	560
570	31.356	31.413	31.471	31.528	31.586	31.644	31.702	31.759	31.817	31.875	31.933	570
580	31.933	31.991	32.048	32.106	32.164	32.222	32.280	32.338	32.396	32.455	32.513	580
590	32.513	32.571	32.629	32.687	32.746	32.804	32.862	32.921	32.979	33.038	33.096	590
600	33.096	33.155	33.213	33.272	33.330	33.389	33.448	33.506	33.565	33.624	33.683	600
610	33.683	33.742	33.800	33.859	33.918	33.977	34.036	34.095	34.155	34.214	34.273	610
620	34.273	34.332	34.391	34.451	34.510	34.569	34.629	34.688	34.748	34.807	34.867	620
630	34.867	34.926	34.986	35.046	35.105	35.165	35.225	35.285	35.344	35.404	35.464	630
640	35.464	35.524	35.584	35.644	35.704	35.764	35.825	35.885	35.945	36.005	36.066	640
650	36.066	36.126	36.186	36.247	36.307	36.368	36.428	36.489	36.549	36.610	36.671	650
660	36.671	36.732	36.792	36.853	36.914	36.975	37.036	37.097	37.158	37.219	37.280	660
670	37.280	37.341	37.402	37.463	37.525	37.586	37.647	37.709	37.770	37.831	37.893	670
680	37.893	37.954	38.016	38.078	38.139	38.201	38.262	38.324	38.386	38.448	38.510	680
690	38.510	38.572	38.633	38.695	38.757	38.819	38.882	38.944	39.006	39.068	39.130	690
700	39.130	39.192	39.255	39.317	39.379	39.442	39.504	39.567	39.629	39.692	39.754	700
710	39.754	39.817	39.880	39.942	40.005	40.068	40.131	40.193	40.256	40.319	40.382	710
720	40.382	40.445	40.508	40.571	40.634	40.697	40.760	40.823	40.886	40.950	41.013	720
730	41.013	41.076	41.139	41.203	41.266	41.329	41.393	41.456	41.520	41.583	41.647	730
740	41.647	41.710	41.774	41.837	41.901	41.965	42.028	42.092	42.156	42.219	42.283	740
750	42.283	42.347	42.411	42.475	42.538	42.602	42.666	42.730	42.794	42.858	42.922	750
760	42.922											760
DEG C	0	1	2	3	4	5	6	7	8	9	10	DEG C

• EXAMPLE 8.4

Given a type J thermocouple with a reference junction at 0°C, find the voltage output you would expect from the thermocouple at 200°C.

Solution

Again, using the information in Table 8.2, we see that the measuring junction output is equal to 10.777 mV.

Keep in mind that the voltage shown in Table 8.2 is a function of the temperature difference between the two junctions. Knowing this fact, we can use this table for references other than 0°C. Using type J as an example, we can find the voltage when our measuring junction is 250°C with the reference at 25°C. At 250°C, the output from the measuring junction is 13.553 mV, while the reference junction's output is 1.277 mV. The thermocouple's output voltage is the difference between the two voltages; that is, 13.553 mV – 1.277 mV = 12.276 mV. Reversing this process allows us to find the temperature given the voltage output of the thermocouple.

As an example, we observe an output of 2.69 mV from a type J thermocouple (iron-constantan) with a reference junction at 0°C. What is the temperature at the measuring junction? Looking down through the table of voltages in Table 8.2, we see that 2.69 mV corresponds to a temperature of 52°C. A similar procedure is used when a measurement is taken with a reference junction at another temperature. If we observe an output voltage of 16.05 mV from a type J thermocouple with a reference junction at a room temperature of 25°C, what is the measuring junction temperature? We cannot use the 16.05 mV voltage in Table 8.2 since the table assumes a reference junction of 0°C. We must first get a corrected voltage by adding to the output voltage the voltage produced by a type J thermocouple at 25°C. The corrected voltage is 16.05 mV + 1.277 mV, which equals 17.327 mV. Looking in our type J table, we see that this value corresponds to a measuring junction temperature of 308°C.

• EXAMPLE 8.5

Given a type J thermocouple with a reference junction at 100°C and a measuring junction that is producing 22.122 mV, find the temperature of the measuring junction.

Solution

Since Table 8.2 assumes a reference of 0°C, we must subtract the potential of the reference junction from the measuring junction. A reference junction at 100°C will produce a potential of 5.268 mV. If we add that value to 22.122, we get 27.390 mV. From Table 8.2, a measuring junction at a temperature of 500°C will have a potential of 27.390 mV with a reference junction at 0°C.

• EXAMPLE 8.6

Given an iron-constantan thermocouple that produces an output voltage of 9.279 mV with the reference junction at 24°C, find (a) the output voltage we would expect if the thermocouple reference junction was at 0°C and (b) the temperature of the measuring junction.

Solution

(a) Looking up the potential the reference junction produces at 24°C, we find a value equal to 1.225 mV. To find the potential produced at a reference junction temperature of 0°C, we add the 9.279 mV potential to the 1.225 mV potential, which leads us to a final value of 10.504 mV.

(b) The temperature of the measuring junction is then approximately 195°C from Table 8.2.

Sometimes, we may wish to estimate a thermocouple output voltage without reference to the more accurate but less convenient tables. Note from Figure 8.11 that the output voltage versus temperature curve is fairly linear for most thermocouples. In the case of the iron-constantan (type J) device, the slope of the output curve is about 50 microvolts (μV) per degree Celsius. We can get a quick estimate of the output voltage by multiplying the sensitivity of the device (in microvolts per degree) by the number of degrees we are above 0°C. For the data from Example 8.4, what potential would we expect from a type J thermocouple at 200°C? Multiplying 200°C by 50 μV/°C, we get a value of 10 mV. Looking at Table 8.2, we see that this value corresponds to an actual potential of 10.777 mV, close enough for an approximation.

• **EXAMPLE 8.7**

Given a type J thermocouple with a reference junction at 0°C and a measuring junction at 345°C, find the estimated output voltage from the thermocouple.

Solution

The sensitivity of the type J thermocouple is 50 μV/°C. To find the estimated output voltage, we multiply the sensitivity of the device by the number of degrees above 0°C. In this case, we multiply 50 μV/°C by 345°C, to get 17.25 mV. As a check, we note from Table 8.2 that a potential of 18.813 mV is actually produced. The difference between the estimated value and the actual value is about 10%.

At this point, we should note that the temperature of a thermocouple (or any sensor) takes a certain amount of time to react to a temperature change. The thermocouple's response time depends on the mass of the thermocouple, the specific heat of the thermocouple, the coefficient of heat transfer from one boundary to another, and the area of contact between the thermocouple and the material to be measured. These factors are related in the following equation:

$$t_c = \frac{mc}{kA} \tag{8.3}$$

where

t_c = thermal time constant
m = mass of sensor (in grams)
c = specific heat (in calories per gram-degree Celsius, cal/g°C)

k = heat transfer coefficient (in calories per centimeter-second-degree Celsius, cal/cm-s°C)
A = area of contact between sensor and sample (in square centimeters)

Just as in a capacitor, the rate of change of a thermocouple's temperature follows the time constant curve. Using Equation 8.3, we can calculate the time constant for any thermocouple. For example, let us suppose a thermocouple has a mass of 0.002 g, a specific heat of 0.08 cal/g°C, a heat transfer coefficient of 0.01 cal/cm-s°C, and an area of 0.02 cm². The time constant, based on this information, is

$$t_c = \frac{(0.002 \text{ g}) (0.08 \text{ cal/g°C})}{(0.01 \text{ cal/cm-s°C}) (0.02 \text{ cm}^2)} = 0.8 \text{ s}$$

• **EXAMPLE 8.8**

Given a thermocouple with a mass of 0.01 g, a specific heat of 0.05 cal/g°C, a heat transfer coefficient of 0.02 cal/cm-s°C, and an area of 0.1 cm², find the thermal time constant.

Solution

We can find the thermal time constant by using Equation 8.3:

$$t_c = \frac{mc}{kA} = \frac{(0.01 \text{ g}) (0.05 \text{ cal/g°C})}{(0.02 \text{ cal/cm-s°C}) (0.1 \text{ cm}^2)}$$

$$= 0.25 \text{ s}$$

When temperature changes quickly, the temperature of a thermocouple at any time is calculated by using

$$T - T_2 = (T_1 - T_2) e^{-t/t_c} \tag{8.4}$$

where

T = thermocouple temperature
T_1 = starting temperature
T_2 = final temperature
t_c = thermocouple time constant
t = time from start

For example, suppose a thermocouple at 20°C is placed into a 100°C water bath. With the time constant calculated previously (0.8), the temperature after 1 s is

$$T - 100° = (20° - 100°) \, e^{-1/0.8} = 77°C$$

We can see from these calculations that the thermocouple does not react instantly to a temperature change. All sensors take some time to respond to a change in the measurand.

• EXAMPLE 8.9

Given a thermocouple with a time constant of 1 s at a temperature of 25°C that is immersed in boiling water ($T = 100°C$) for 1 s, find the temperature of the sensor (a) after 1 s and (b) after 4 s.

Solution

(a) To find the temperature of the sensor, we need to solve Equation 8.4 for T, as follows:

$$T = \left[(T_1 - T_2) \, e^{-t/t_c} \right] + T_2$$

Substituting into this equation, we get

$$T = \left[(25°C - 100°C) \, e^{-1/1} \right] + 100°C$$

$$= [(-75°C) \, (0.368)] + 100°C$$

$$= -27.6°C + 100°C$$

$$= 72.4°C$$

(b) After 4 s, the temperature is

$$T = \left[(25°C - 100°C) \, e^{-4/1} \right] + 100°C$$

$$= [(-75°C) \, (0.0183)] + 100°C$$

$$= -1.37°C + 100°C$$

$$= 98.6°C$$

• EXAMPLE 8.10

Given a thermocouple with a time constant of 1.25 s at 25°C that is immersed in boiling water ($T = 100°C$), find the time it will take the thermocouple to heat to 50°C.

Solution

We must first solve Equation 8.4 for t, as follows:

$$t = (t_c) \left[-\ln \left(\frac{T - T_2}{T_1 - T_2} \right) \right] \tag{8.5}$$

Substituting the appropriate values into Equation 8.5, we get

$$t = (1.25 \text{ s}) \left[-\ln \left(\frac{50°C - 100°C}{25°C - 100°C} \right) \right]$$

$$= (1.25 \text{ s}) \, [-(-0.405)]$$

$$= 0.5 \text{ s}$$

A short time constant is a desirable feature for a thermocouple since the thermocouple can monitor changing temperatures without a large time delay. Equation 8.3 indicates four areas where the design of the thermocouple and the system can keep the time constant small. First, the mass of the thermocouple should be kept small. Second, materials with a low specific heat should be used, if possible. This choice often conflicts with the need for a large thermoelectric voltage, a design trade-off. Third, the area of contact between the sensor and the object being measured should be as large as possible. Fourth and last, the thermocouple and the measurand should have a good thermal bond between them.

In some applications, a compromise must be made between the thermal time constant and the ruggedness of the sensor. For example, in an application where there is a large amount of vibration, a larger and more rugged device would be chosen. A device with a larger mass to withstand shock and vibration will necessarily have a larger mass and, therefore, a larger time constant.

A recent development in thermocouple technology is the *thin-film thermocouple*. The advantage of the thin-film thermocouple is its fast speed of response. A look at Equation 8.3 will tell us why this device responds more quickly to temperature changes. As we have just discussed, in the conventional two-wire thermocouple, two metal wires are joined together, usually by welding or soldering. The two-wire thermocouple is then attached to the sample, usually with a thermal grease to improve heat transfer. In the thin-film thermocouple, two metals are vapor-deposited directly on the material in layers about 10^{-5} cm thick. The thermal mass of this kind of thermocouple is much lower than that of the conventional two-wire thermocouple. Also, since the metals are directly deposited on the material, there is no need to use thermal grease. Both of these factors improve the heat transfer from the sample to the thermocouple. Thin films also may be spread over a larger area than conventional thermocouples.

Thermal time constants on thin-film thermocouples have been measured to be as low as 1 µs. These low-mass devices offer significant improvements in response time over conventional thermocouples. Conventional thermocouples with small wires have response times between 0.1 and 0.4 s, considerably longer than the times of thin-film thermocouples. The cost of thin-film thermocouples is greater than that of conventional thermocouples, however, and thus, these devices may not be cost-effective for all applications.

Many of the inaccuracies that result from using thermocouple sensors come from improper applications and inadequate system design criteria. One system consideration not often taken into account is lead length. Thermocouples must be connected to measuring instrumentation by lead wires. As a general rule of thumb, lead lengths between measuring junction and measuring instrument should be kept as short as possible. If the total resistance of the thermocouple and leads exceed 100 Ω, you should check to see what resistance your measuring instrument requires. Some instruments will not work properly with resistances over 100 Ω.

Thermocouples are used in industry to measure oven and furnace temperatures. Sometimes, they are used as part of a process control system where temperature is not merely measured but also carefully controlled. Thermocouples also are used to measure the temperature of flowing fluids, especially when their temperature fluctuates widely. A more critical application involves measuring the temperatures in nuclear reactor cores. Sensors need to respond quickly and accurately to allow the system to keep temperature within acceptable limits. Thermocouples, because of their ruggedness and wide temperature range, are also used to measure the temperatures of exhaust gases of nose cones in missile and rocket applications.

Thermistor. A *thermistor* is a thermally sensitive resistor, usually having a negative temperature coefficient. As temperature increases, the thermistor's resistance decreases, and vice versa.

We have talked about the fact that the thermistor has a negative temperature coefficient. The temperature coefficient of resistance of a material, signified by α (the Greek letter alpha), is represented by

$$\alpha = \left(\frac{\Delta R}{R_s}\right)\left(\frac{1}{\Delta T}\right) \qquad \text{(8.6)}$$

where

ΔR = change in resistance caused by given temperature change

R_s = resistance of material at reference temperature

ΔT = change in temperature above or below reference temperature

Let us compare the temperature coefficient of iron, which is 50×10^{-4}, with that of nichrome (a nickel-chromium alloy), which is 2×10^{-4}. We can see from these coefficients that iron increases its resistance about 25 times more than nichrome per degree Celsius of temperature rise. Contrast these coefficients with the coefficient of the thermistor, which can be as high as 600×10^{-4} per degree Celsius, corresponding to a resistance change of 6%.

Thermistors are made out of oxides of nickel, manganese, cobalt, copper, and other metals. Figure 8.12 shows the almost bewildering variety of packages in which thermistors come. Modern thermistors have many of the requirements system designers want

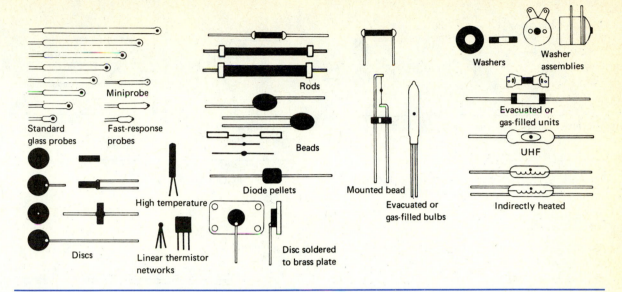

FIGURE 8.12 Thermistors

for temperature sensors. They have excellent sensitivity, a wide range of impedance levels, small and varied sizes, high precision, and, recently, much better long-term stability.

A typical thermistor's temperature-resistance curve is shown in Figure 8.13. As you can see, there are several differences between this curve and that for the thermocouple, which is also shown in Figure 8.13. First, the thermocouple curve is relatively linear. The thermistor's resistance varies exponentially. As a result, the thermistor's thermal sensitivity is very high, as much as 5% resistance change per degree Celsius. Thus, the thermistor is the most sensitive temperature sensor in common use. The thermistor is therefore reasonably linear over a relatively narrow range of temperatures.

Thermistors often are used in conjunction with a parallel resistor to improve linearity, as shown in Figure 8.14A. Resistor R_s (a shunt resistance) is chosen to equal the resistance of the thermistor at the median temperature expected to be measured. Figure 8.14D shows the change in the response curve when a resistor is added.

There are many ways to use thermistors. Figure 8.14 shows several circuits using thermistors. Note the schematic representation of the thermistor as a resistor with a T inside a circle. The circuit shown in Figure 8.14A represents a potentiometric circuit. As temperature increases, the resistance of the thermistor decreases, increasing the current through the ammeter A. Figure 8.14B depicts a more sensitive

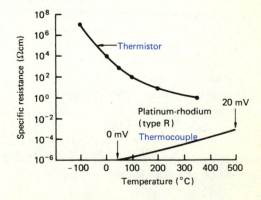

FIGURE 8.13 Temperature-Resistance Curves for Thermistor and Thermocouple

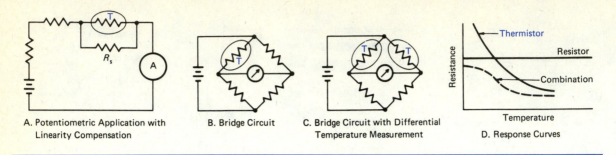

A. Potentiometric Application with
   Linearity Compensation

B. Bridge Circuit

C. Bridge Circuit with Differential
   Temperature Measurement

D. Response Curves

FIGURE 8.14 Thermistors in Potentiometric and Bridge Circuits

bridge circuit, where changes in temperature unbalance the bridge, causing a meter indication. Differential temperature is measured with the circuit shown in Figure 8.14C. Bridge unbalance comes only when there is a difference in temperature between the two sensors. Notice that none of these circuits use an amplifier. Because of the large voltage output produced by a typical thermistor bridge, amplification is normally unnecessary. A 4000 Ω thermistor in a bridge network at 25°C produces approximately 18 mV/°C.

The thermistor is the most sensitive of the electrical temperature sensors. Some thermistors actually double their resistance with a temperature change of 1°C. Thermistors are relatively inexpensive, react quickly to temperature changes, and require only very simple circuitry. Resistances range from 0.5 Ω to 80 MΩ. Thermistors are not without disadvantages, however. They are extremely nonlinear and fragile, and they have a limited temperature range.

The most accurate way to use thermistors is to refer to the resistance-temperature curves that are included with each device. The graph in Figure 8.15 shows a manufacturer's thermistor temperature-resistance curve for a specific device. Note that the resistance decreases as temperature increases.

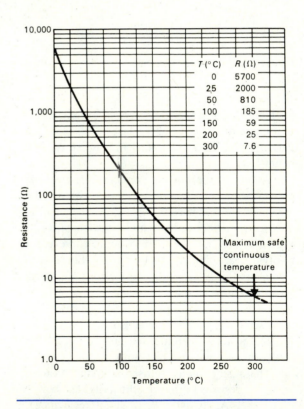

T (°C)	R (Ω)
0	5700
25	2000
50	810
100	185
150	59
200	25
300	7.6

FIGURE 8.15 Thermistor Resistance-Temperature Curve

As an example, consider the circuit shown in Figure 8.16. What will the output voltage be when the thermistor is at a temperature of 100°C? From Figure 8.15, the resistance at that temperature is 185 Ω. The voltage divider equation will help us find the correct output voltage:

$$V_o = \left(\frac{R_2}{R_1 + R_2}\right)(+V) = \left(\frac{100\ \Omega}{185\ \Omega + 100\ \Omega}\right)(5\ V)$$

$$= 1.75\ V$$

The same thermistor can be used in an op amp circuit, with the thermistor replacing the feedback resistor, as shown in Figure 8.17. The output voltage will equal the gain times the input voltage. Let us calculate the output voltage for the circuit in Figure 8.17 at 25°C. From Figure 8.15, the resistance of the thermistor at 25°C is 2000 Ω. Knowing this resistance, we can then calculate the gain of the inverting op amp:

$$A_V = \left(\frac{R_F}{R_i}\right) = -\left(\frac{R_1}{R_2}\right) = -\left(\frac{2000\ \Omega}{1000\ \Omega}\right) = -2$$

The output voltage is the input voltage multiplied by the gain:

$$V_o = A_V V_i = (-2)\,(+5\ V) = -10\ V$$

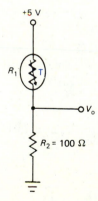

FIGURE 8.16 Thermistor Voltage Divider

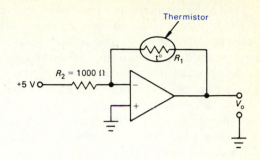

FIGURE 8.17 Op Amp Thermistor Circuit

• EXAMPLE 8.11

Given the op amp thermistor thermometer in Figure 8.17, find the output voltage with the thermistor at 50°C.

Solution

The resistance of the thermistor at 50°C is 810 Ω. Knowing this resistance, we can then calculate the gain of the inverting op amp:

$$A_V = -\left(\frac{R_F}{R_i}\right) = -\left(\frac{R_1}{R_2}\right) = -\left(\frac{810\ \Omega}{1000\ \Omega}\right)$$

$$= -0.81$$

The output voltage is the input voltage multiplied by the gain:

$$V_o = A_V V_i = (-0.81)\,(+5\ V) = -4.05\ V$$

Not all thermistors have a negative temperature coefficient. Some specialized thermistors have a positive temperature coefficient (PTC) of resistance. These PTC thermistors have a very nonlinear resistance-temperature curve. Their application is usually in motors to indicate an overload condition.

Most of the applications we have been discussing up to this point have used the thermistor in what is called the *externally heated mode*. In this mode of operation, the thermistor changes its resistance due to heating from an external source. Another mode of thermistor operation, frequently used in time-delay circuitry, is called the *self-heating mode*. In this mode, the thermistor uses the heating effect of its own current flow. If a thermistor is connected in series with a variable resistor, a source, and a relay, a variable time-delay circuit can be constructed. When current is allowed to flow in the circuit, the amount allowed to flow is small, limited by the high resistance of the cold thermistor. As current flows in the circuit, the thermistor gradually heats up, allowing more current to flow, until the relay is triggered into conduction. Increasing the variable resistance will increase the time lag. With careful circuit design, a wide range of time delays are possible, owing to the lag in the response of the self-heated thermistor.

Over the past several years, thermistors have become popular as sensors in industry. They are used as primary sensors; that is, they are used to measure temperature as a variable. Many applications in the chemical industry require temperature control to 1°C. For example, the chemical activity of some enzymes may change as much as 4% with each degree Celsius change in temperature. Thermistors are ideal for this kind of application because of their extremely high sensitivity. Thermistors are also used in compensation circuits for thermocouples, making the thermocouples more accurate and linear.

Thermistors find applications in more than temperature measurement and control. They also are used as secondary sensors to measure liquid level or fluid flow. If a thermistor is placed in a tank, its resistance changes, depending on whether or not the device is immersed in a liquid.

Resistance Temperature Detector (RTD). All metals change resistance when subjected to a temperature change. Pure metals, such as platinum, nickel, tungsten, and copper, have positive temperature coefficients. Thus, for pure metals, temperature and resistance are directly proportional. As temperature increases, a pure metal's resistance will increase. This result is the idea behind the *resistance temperature detector* (RTD).

The resistance of the RTD can be calculated at temperatures above 0°C by using the following formula:

$$R_T = R_0 (1 + \alpha T) \qquad (8.7)$$

where
R_T = resistance at temperature T
R_0 = resistance at 0°C
α = temperature coefficient of resistance

Let us take, as an example, an RTD with a resistance of 100 Ω at 0°C and a temperature coefficient of resistance of 0.00392. We can calculate the resistance of the RTD at 50°C:

$$R_T = R_0 (1 + \alpha T)$$
$$= (100\ \Omega)\ [1 + (0.00392\ \Omega/\Omega/°C)\ (50°C)]$$
$$= 119.6\ \Omega$$

At 50°C, this RTD would be expected to have a resistance of 119.6 Ω.

• EXAMPLE 8.12
Given a platinum RTD with a temperature coefficient of 0.00392 and an original resistance of 150 Ω at 0°C, find the resistance at 75°C.

Solution
The resistance of the RTD at 75°C can be found by using Equation 8.7:

$$R_T = R_0 (1 + \alpha T)$$
$$= (150\ \Omega)\ [1 + (0.00392\ \Omega/\Omega/°C)\ (75°C)]$$
$$= 194.1\ \Omega$$

RTDs come in many packages; the most popular are shown in Figure 8.18. The helical RTD (Figure 8.18A) consists of a platinum wire wound in a tight spiral (helix) threaded through a ceramic cylinder. A newer construction technique is found in the metal film RTD (Figure 8.18B). Platinum is usually deposited, or screened, on a small, flat ceramic substrate. It is then etched with a laser trimming system and sealed. The film RTD offers a substantial reduction in cost over other forms of RTDs and reacts more quickly to temperature changes.

Common resistance values for RTDs range from 10 Ω (for platinum) to several thousand ohms. The most common value of resistance is 100 Ω. Because of their low resistance, long lead wires can produce inaccurate temperature measurements. This problem can be solved by using a three-wire bridge configuration, as shown in Figure 8.19. If wires A and B are identical in resistance, their resistive effects cancel because each is in an opposite leg of the bridge. The third wire (C) carries no current; it is only a sense lead.

Of the temperature sensors we have presented, the RTD is the most accurate and stable. Furthermore, as can be seen from the curve in Figure 8.20, it is very linear and covers a high temperature range. Disadvantages include slowness of response, small resistance changes (usually requiring amplification), and high cost.

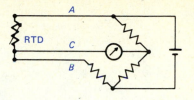

FIGURE 8.19 Three-Wire Bridge Configuration

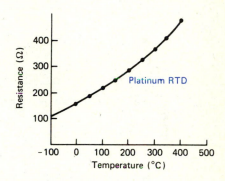

FIGURE 8.20 Temperature-Resistance Curve for Platinum RTD

A circuit using an RTD is shown in Figure 8.21. It measures temperature between 0° and 266°C, the output ranging between 0 and 1.8 V. The 2.5 V reference is amplified to 6.25 V by the first op amp. The span-adjust sets the output to 1.8 V at 266°C, while the offset is adjusted to 0 V at 0°C.

We can calculate the output voltage from the circuit shown in Figure 8.21. Let us assume that the temperature of the platinum RTD is 70°C and the R_6 span-adjust is set to 26 Ω. First, we must calculate the resistance of the RTD, assuming that the RTD has a resistance of 100 Ω at 0°C. The resistance is, from Equation 8.7,

$$R_T = R_0 (1 + \alpha T)$$

$$= (100\ \Omega)\ [1 + (0.00392\ \Omega/\Omega/°C)\ (70°C)]$$

$$= 127.4\ \Omega$$

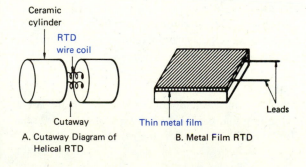

A. Cutaway Diagram of Helical RTD

B. Metal Film RTD

FIGURE 8.18 Two Popular RTDs

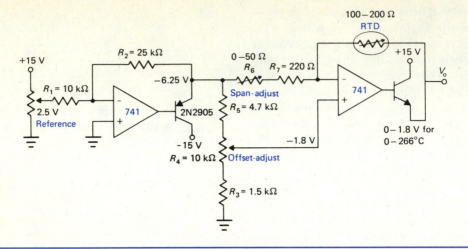

FIGURE 8.21 Low-Cost Temperature Sensor Using RTD with Fixed Reference and Op Amp

Next, we calculate the gain of the op amp circuit. The gain of the inverting input is

$$A_V = -\left(\frac{R_F}{R_i}\right) = -\left(\frac{R_{RTD}}{R_6 + R_7}\right)$$

$$= -\left(\frac{127.4\ \Omega}{26\ \Omega + 220\ \Omega}\right)$$

$$= -0.518$$

The output voltage contribution of the inverting input is the input voltage multiplied by the gain:

$$V_{o(inv)} = A_V V_i = (-0.518)\ (-6.25\ V) = 3.24\ V$$

Next, we must find the contribution to the output from the noninverting terminal. The noninverting gain is

$$A_V = \frac{R_F}{R_i} + 1 = \frac{R_{RTD}}{R_6 + R_7} + 1$$

$$= \frac{127.4\ \Omega}{26\ \Omega + 220\ \Omega} + 1$$

$$= 1.52$$

The output voltage contribution of the noninverting input is the input voltage multiplied by the gain:

$$V_{o(ninv)} = A_V V_i = (1.52)\ (-1.8\ V)$$

$$= -2.74\ V$$

The total output voltage is the sum of both contributions, or $3.24\ V + (-2.74\ V) = 0.5\ V$.

• EXAMPLE 8.13

Given the RTD thermometer circuit in Figure 8.21 with the sensor at a temperature of 32°C, find the circuit output voltage.

Solution

First, we must calculate the resistance of the RTD, assuming that the RTD has a resistance of 100 Ω at 0°C. From Equation 8.7, the resistance is

$$R_T = R_0\ (1 + \alpha T)$$

$$= (100\ \Omega)\ [1 + (0.00392\ \Omega/\Omega/°C)\ (32°C)]$$

$$= 112.5\ \Omega$$

Next, we calculate the gain of the op amp circuit. The gain of the inverting input is

$$A_V = -\left(\frac{R_F}{R_i}\right) = -\left(\frac{R_{RTD}}{R_6 + R_7}\right)$$

$$= -\left(\frac{112.5\ \Omega}{26\ \Omega + 220\ \Omega}\right)$$

$$= -0.457$$

The output voltage contribution of the inverting input is the input voltage multiplied by the gain:

$$V_{o(inv)} = A_V V_i = (-0.457)\ (-6.25\ V) = 2.86\ V$$

Next, we must find the contribution to the output from the noninverting terminal. The noninverting gain is

$$A_V = \frac{R_F}{R_i} + 1 = \frac{R_{RTD}}{R_6 + R_7} + 1$$

$$= \frac{112.5\ \Omega}{26\ \Omega + 220\ \Omega} + 1$$

$$= 1.46$$

The output voltage contribution of the noninverting input is the input voltage multiplied by the gain:

$$V_{o(ninv)} = A_V V_i = (1.46)\ (-1.8\ V)$$

$$= -2.63\ V$$

The total output voltage is the sum of both contributions:

$$2.86\ V + (-2.63\ V) = 0.23\ V$$

The three most common metals that make up the sensing element of the RTD are platinum, copper, and nickel. Platinum has the greatest temperature range and stability with moderately good linearity, at about ±0.4°C over the 0°–100°C range. Copper provides nearly perfect linearity, and nickel offers low cost, high resistance, and sensitivity.

Semiconductor Temperature Sensors. In the semiconductor area, evenly doped crystals of germanium can be used to sense temperature near absolute zero. The temperature-resistance response curve of the device resembles that of the thermistor. It has a negative temperature coefficient and is very nonlinear.

Silicon crystals are also employed as temperature sensors. They usually are shaped in the form of discs or wafers for measuring surface temperature. Their useful range extends from −67° to 275°F. Unlike germanium crystals and thermistors, silicon crystals have a positive linear temperature coefficient in this range. Below this range, the temperature-resistance response curve becomes negative and very nonlinear.

Recall that a reverse-biased PN junction conducts a small amount of leakage, or minority carrier, current flow. The amount of current flow depends on temperature. As temperature increases, leakage current increases exponentially. This effect is used to measure temperature with transistors and reverse-biased diodes. Normally, germanium semiconductors are used because their leakage is much greater than that of silicon. In general, a germanium diode is used in this application since germanium diodes produce around 10 μA at room temperature. Silicon diodes produce nanoamperes of current at this temperature, a value too low to be useful.

The diode may be used in a CDA, as illustrated in Figure 8.22. Note that the germanium diode is reverse-biased by the positive supply. As temperature increases, the leakage current will increase. The increase in current will be mirrored in the inverting terminal, increasing the output voltage of the CDA. The CDA output will then be directly proportional to temperature; as temperature increases, the output voltage will increase, and vice versa.

Let us do a calculation. Suppose we have a feedback resistance R_F of 1 MΩ and a diode current I_D at room temperature of 10 μA. For the circuit shown in Figure 8.22, the output voltage is

$$V_o = I_D R_F = (10\ \mu A)\ (1\ M\Omega) = 10\ V$$

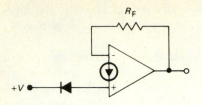

FIGURE 8.22 CDA with Diode as Temperature Sensor

The temperature sensitivity (s) of the germanium diode is about 0.0333 µA/°C. Let us suppose that the temperature goes from 25° to 75°C. Then the change in current (ΔI_D) equals the change in temperature (ΔT) times the sensitivity:

$$\Delta I_D = (\Delta T)(s) = (50°C)(0.033 \text{ µA/°C})$$

$$= 1.67 \text{ µA}$$

The total diode current at 75°C equals the change in current plus the current at 25°C:

$$I_{D(75°C)} = \Delta I_D + I_{D(25°C)} = 1.67 \text{ µA} + 10 \text{ µA}$$

$$= 11.67 \text{ µA}$$

The output voltage at 75°C is then

$$V_o = I_D R_F = (11.67 \text{ µA})(1 \text{ M}\Omega) = 11.67 \text{ V}$$

The bijunction transistor, shown in Figure 8.23, is sometimes used in thermometers. The transistor base-emitter voltage V_{BE} depends on temperature, varying −2.25 mV per degree Celsius. For every degree Celsius increase in temperature, the base-emitter voltage will decrease by 2.25 mV.

One of the most recent innovations in thermometry is the IC temperature sensor. The output of the IC (either voltage or current) is directly proportional to temperature. Although limited in temperature range (below +300°F), it produces a very linear output over the operating range. Linearity exceeds that of the RTD. A typical sensitivity value is 1 µA/°C or 10 mV/°C. Figures 8.24A and 8.24B show the current and voltage modes of the AD590 manufactured by Analog Devices.

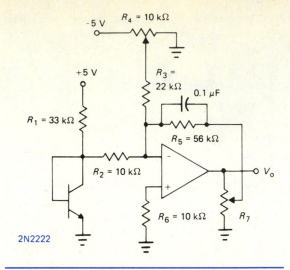

FIGURE 8.23 Op Amp Thermometer with Transistor as Temperature Sensor

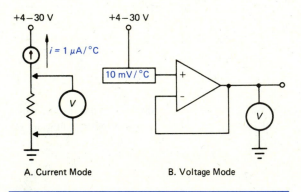

A. Current Mode   B. Voltage Mode

FIGURE 8.24 IC Temperature Sensor

An example may help at this point. Suppose we are using an LM335, which has an output equal to 10 mV/K. What voltage will be produced at 27°C? First, we must convert 27°C into degrees kelvin (K) by adding 273. A temperature of 27°C corresponds to 300 K. We then multiply the number of degrees kelvin by the sensitivity (s) (in millivolts per degree kelvin) of the device to get the output of the sensor at 300 K:

$$V_o = s \times \text{current temperature}$$
$$\text{(in degrees kelvin)} \qquad \textbf{(8.8)}$$

Thus, after substitution, we obtain

$$V_o = (10 \text{ mV/K}) (300 \text{ K}) = 3 \text{ V}$$

• EXAMPLE 8.14
Given an LM335 at a temperature of 100°C, find the output voltage you would expect to see at that temperature.

Solution
First, we must convert 100°C into degrees kelvin by adding 273. A temperature of 100°C corresponds to 373 K. We then multiply the number of degrees kelvin by the sensitivity of the device to get the output of the sensor at 373 K, using Equation 8.8:

$$V_o = s \times \text{current temperature}$$

$$= (10 \text{ mV/K}) (373 \text{ K}) = 3.73 \text{ V}$$

Suppose an LM335 produces an output voltage of 2.84 V. What is the temperature of the sensor? To find the actual temperature, we divide the output voltage by the sensitivity and then subtract 273 to convert the temperature into degrees Celsius:

$$T_{°C} = \frac{V_o}{s} - 273°C \qquad (8.9)$$

Therefore, we get

$$T_{°C} = \frac{2.84 \text{ V}}{10 \text{ mV/K}} - 273°C = 11°C$$

• EXAMPLE 8.15
Given an LM335 with an output voltage of 3.35 V, find the temperature of the sensor (in degrees Celsius).

Solution
To find the actual temperature, we divide the output voltage by the sensitivity and then subtract 273 to

convert the temperature into degrees Celsius, using Equation 8.9:

$$T_{°C} = \frac{V_o}{s} - 273°C = \frac{3.35 \text{ V}}{10 \text{ mV/K}} - 273°C$$

$$= 62°C$$

National Semiconductor Corporation manufactures two new sensors, the LM34 and LM35. The LM34 is a device having an output voltage that is linearly proportional to the Fahrenheit temperature. Its output voltage sensitivity is +10 mV/°F. No trimming or calibration circuits are required. A simple thermometer using the LM34 is shown in Figure 8.25. The supply voltage necessary is between +5 and +30 V. The output voltage is calculated in the same way as it is for the LM335.

As an example, suppose an LM34 was heated to 55°F. The output voltage is given by

$$V_o = s \times \text{current temperature}$$
$$\text{(in degrees Fahrenheit)} \qquad (8.10)$$

Thus, we obtain

$$V_o = (10 \text{ mV/°F}) (55°F) = 0.55 \text{ V}$$

Given an output voltage, say 0.88 V, the temperature (in degrees Fahrenheit) may be calculated:

$$T_{°F} = \frac{V_o}{s} \qquad (8.11)$$

So we obtain

$$T_{°F} = \frac{0.88 \text{ V}}{10 \text{ mV/°F}} = 88°F$$

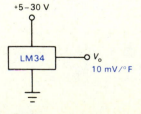

FIGURE 8.25   LM34 IC Temperature Sensor

Note that the value in millivolts (880) is equal to the temperature when divided by 10.

The LM35 is similar to the LM34, but it has a sensitivity of 10 mV/°C. All other characteristics are the same. Both devices have a low output impedance and a linear output voltage, and both can be used with single or dual power supply voltages.

Generally speaking, semiconductor temperature sensors produce good linearity, are small and low in cost, and produce a high impedance output.

Tables 8.3 and 8.4 list the temperature ranges of the thermometers we have discussed in this section and give a summary of information about them.

HUMIDITY

Many industrial processes depend on accurate assessment and control of humidity. *Humidity* is defined as the amount of water vapor in the air. How much water vapor air can hold depends on several factors, the most important of which is the air temperature. Warm air can hold more water vapor than cold air can. The most common measure of water vapor in air is called relative humidity. *Relative humidity* is a ratio of the amount of moisture air holds at a certain temperature compared with how much it could hold at that temperature.

TABLE 8.3   **Temperature Ranges of Mechanical and Electrical Temperature Transducers**

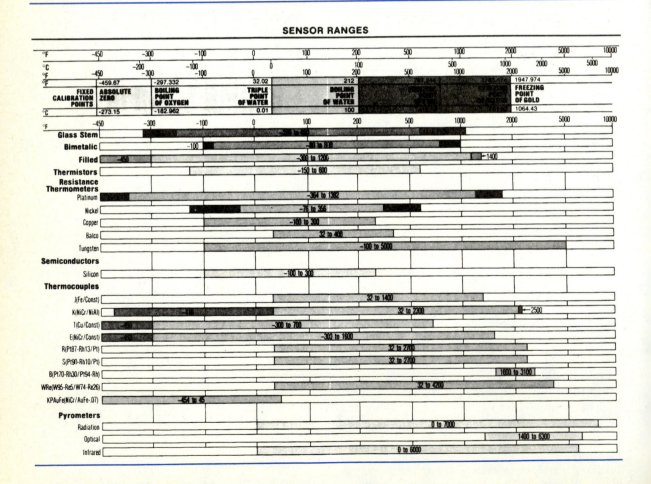

TABLE 8.4 Summary of Temperature Sensors

Type/Name	Temperature Range	Linearity	Advantages	Disadvantages
Mechanical				
Glass stem	−100°–500°C	Good	Inexpensive Linear Accurate (0.05°C)	Local measurement only Fragile Hard to read Time lag
Bimetallic	−100°–600°C	Good	Inexpensive Wide range Rugged Easy to read and install	Local measurement only
Filled	−200°–650°C	Fair	Reacts quickly Accurate (0.5%) Remote measurement (300 ft) Can be used with chart recorders	Often needs temperature compensation
Electrical				
Thermocouple	−273°–2000°C	Good	Simple, low cost Rugged Wide range Self-powered	Low sensitivity Reference needed Poor stability
RTD	−200°–800°C	Good	Very stable Very accurate Linear Wide range	Slow response Low sensitivity Expensive Self-heating Limited range Subject to thermal runaway
Semiconductor	−50°–150°C	Poor	Inexpensive Small Output impedance high	
IC	−50°–150°C	Excellent	Excellent linearity Inexpensive Very sensitive	Slow response
Thermistor	−100°–300°C	Very poor	Small size Low cost High sensitivity Fast response	Very nonlinear Poor stability at high temperatures Limited range

In this section, we discuss humidity sensors in two classifications: direct methods through hygrometers and indirect methods by comparing changes in temperature.

Generally, the measurement of the amount of moisture in a gas is accomplished by one of three techniques. First, the water can be extracted from the sample and either weighed or detected in some other manner. Second, some property of the water may be measured to estimate how much water is in a sample. The properties of water sometimes used are conductivity, dielectric constant, and infrared, ultraviolet, or microwave absorption. Third, water content may be analyzed by a change it makes in another material. This principle is used in the hair hygrometer, discussed a little later in this chapter. When you are reading through this section on humidity sensors, try to identify each sensor with one of these three techniques.

Some areas of industry use other methods to measure the amount of water in a material. The mass or volume ratio is often used. This ratio is specified in a percentage or in parts per million (ppm) by weight or volume of a sample. This measure is also called the *specific humidity*. Volumetric specific humidity is usually expressed as parts per million by volume (ppm_V). Specific humidity also may be expressed on a weight basis. Like the volumetric measure, the weight, or gravimetric, measure is expressed in parts per million by weight (ppm_W). Another measure involves the dew, or frost, point. The *dew point* is the temperature at which water vapor condenses and deposits on a solid surface. It is measured in degrees Celsius or Fahrenheit.

Psychrometers

The *psychrometer*, shown in Figure 8.26, is a common means of measuring relative humidity. Note that its construction is based around two bulbs, one wet and one dry. Air must be passed over the wet bulb, causing the water to evaporate.

Both bulbs contain temperature sensors. If the air temperature remains constant, the dry-bulb sensor remains at a constant temperature. However, the wet bulb's temperature varies with the relative humidity. More evaporation takes place as relative humidity

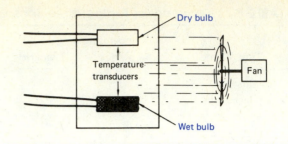

FIGURE 8.26 Psychrometer Humidity Sensor

decreases. Thus, as humidity drops, the wet bulb becomes cooler. The difference between the temperature of the two bulbs, then, reflects the relative humidity.

Conversion tables are usually necessary to estimate relative humidity with the psychrometer. However, relative humidity conversion may be accomplished automatically with appropriate hardware or software.

Hygrometers

A *hygrometer* is a device used to measure humidity. These devices contain a material whose properties change when moisture is adsorbed. As we have seen, the psychrometer indirectly measures humidity by comparing the temperatures of two bulbs. The hygrometer, on the other hand, measures humidity directly. In this section, we will describe several common hygrometers in use today.

Hair Hygrometer. The simplest and oldest hygrometer is made from human hair or an animal membrane. A simple *hair hygrometer* is shown in Figure 8.27. Human hair lengthens about 3% over a range from 0% to 100% humidity. This change in length is detected by the pointer, giving a readout in relative humidity.

The hair hygrometer is accurate to within 3%. It is used for measuring relative humidity only between 15% and 90% over a temperature range from 1° to 40°C. Although the hair hygrometer is strictly mechanical, it can be converted to an electrical

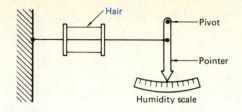

FIGURE 8.27   Hair Hygrometer

sensor by attaching the hair to the core of, say, an LVDT or any appropriate position-to-voltage transducer.

Impedance Hygrometer. Several hygrometers use a change in impedance to detect levels of humidity. One, the *resistance hygrometer*, varies its resistance (and, hence, impedance) as humidity changes. The resistance hygrometer, as shown in Figure 8.28, is composed of two electrodes separated by a thin layer of lithium chloride. The lithium chloride film is *hygroscopic*, which means that it adsorbs moisture from the air. As relative humidity increases, the device's resistance decreases. The change is then sensed by potentiometric or bridge methods.

A similar principle is used in the *electrolytic hygrometer*. In this device, two wires, usually made of platinum or rhodium, are wound in a spiral and encased in a tube. The gas under measurement is continually circulated through the tube. The metal wires are coated with phosphorous pentoxide, a hygroscopic material with properties similar to those of lithium chloride. A high fixed voltage is applied across the two wires, which electrolyzes the water adsorbed by the coating. The amount of current flow is directly proportional to the concentration of water vapor in the gas.

Another hygrometer that uses an impedance change is made out of aluminum oxide. This device, sometimes called a *resistance-capacitance hygrometer*, is shown in Figure 8.29. The surface of the aluminum base is anodized to form a layer of aluminum oxide. Gold is then vapor-deposited on top of the oxide coating. Note that one electrode is attached to the gold film and another is attached to the aluminum base. Since the gold is very thin, water vapor diffuses through to the oxide layer. An increase in relative humidity causes the impedance (resistance and capacitance) of the oxide coating to decrease. The change in impedance is detected by an impedance bridge calibrated to read percent relative humidity.

The aluminum oxide element operates over the range from 0% to 100% relative humidity. The element is very sensitive, small in size, and very linear. Condensation of moisture on the surface of the element does not affect it.

Sorption Hygrometer. The *sorption hygrometer* uses the principle of an oscillating crystal to measure humidity. The moisture increases the mass of the crystal and decreases the frequency of oscillation.

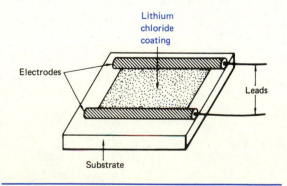

FIGURE 8.28   Resistance Hygrometer

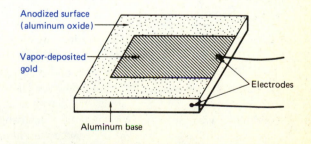

FIGURE 8.29   Impedance Hygrometer

Frequencies used are normally around 9 MHz. Commercial units use two crystals: One is the sensor, and the other is exposed to a dry gas and acts as a reference. This device is also called a *piezoelectric hygrometer*, and it can measure moisture in gases from 1 to 25,000 ppm.

Capacitive Hygrometer. The *capacitive hygrometer* works on the principle that the dielectric constant of a capacitor changes in the presence of water. Recall that changes in the dielectric constant vary the capacitance of a capacitor. Measurements of capacitance are made through impedance bridges or oscillator frequency changes.

Microwave Absorption. The absorption of microwave radiation by water vapor is also used in humidity measurements. Microwave radiation is part of the electromagnetic spectrum and has values between 1 and 100 GHz. Water absorbs thousands of times more microwave radiation than a dry gas does. More water vapor in the air increases microwave energy absorption and decreases energy transmitted to a sensor. The level of radiation transmitted is inversely proportional to the relative humidity.

A summary of the humidity sensors discussed in this section is presented in Table 8.5.

DISPLACEMENT, STRESS, AND STRAIN

In industrial process control, accurate information about an object's position is often needed. *Displacement transducers* provide this kind of information. In measurement systems, displacement and the force that produces the displacement are inseparably linked. We can, then, measure either the displacement or the force (stress) producing the displacement. In some applications, we are interested in measuring the deformation of an object caused by force or stress. This deformation is called *strain*. In this section, we deal with transducers that can be used to measure displacement, stress, or strain.

Displacement Transducers

Displacement is defined as an object's physical position with respect to a reference point. Displacement breaks down into two categories: linear displacement and angular displacement. *Linear displacement* is defined as the position of a body in a straight line with respect to a reference point. *Angular displacement* is the angular position of an object with reference to a fixed point. Industry uses transducers for both types of displacement.

TABLE 8.5   Summary of Humidity Sensors

Name	Operating Principle	Relative Humidity Range	Temperature Range	Accuracy
Psychrometer	Evaporation rate of water	2%–98%	32°–212°F	±5%
Hair hygrometer	Hair changes length with humidity	15%–90%	0°–160°F	±5%
Resistance hygrometer	Lithium chloride changes resistance with humidity	1.5%–99%	−40°–160°F	±1.5%
Resistance-capacitance hygrometer	Aluminum oxide changes impedance with humidity	0%–100%	0°–300°F	±2%
Capacitive hygrometer	Dielectric constant changes with humidity	0%–100%	0°–200°F	±0.5%

Angular-Displacement Transducers. One of the most common angular-displacement transducers is the potentiometer. You are probably familiar with the basic structure of a potentiometer. It is composed of a resistor shaped in a circle with a wiper sliding on it, as shown in Figure 8.30. If the shaft is rotated clockwise, the resistance between contacts 1 and 2 increases. This increasing resistance can be used to indicate the position of a rotating shaft, for example, on a motor. The resistive element may be carbon, conductive plastic, or thin wire wrapped around a nonconductive form.

Linear-Displacement Transducers. Linear displacement can be measured in many ways. For example, if the resistive element in the angular-displacement transducer is straightened out, we have a linear-displacement transducer. The position of the tap or wiper is directly controlled by linear displacement.

Variations of capacitance and inductance are also used to indicate linear displacement. Inductance normally is changed by withdrawing the core from the windings, as shown in Figure 8.31A. Capacitance is varied by moving the plates or by withdrawing or inserting the dielectric, as shown in Figure 8.31B.

By far, the most common displacement transducer is the *linear variable differential transformer (LVDT)*. The LVDT is basically a transformer with two secondaries, as illustrated in Figure 8.32A. A movable core is connected to the shaft (Figure 8.32B). An object attached to the shaft moves the core. The primary winding is excited by an AC frequency between 50 Hz and 15 kHz with an amplitude of up to 10 V. The secondary windings usually are connected series opposing so that when the core is centered, there is no output from the secondary. If the core material moves in either direction, an output voltage is produced since the mutual inductance changes.

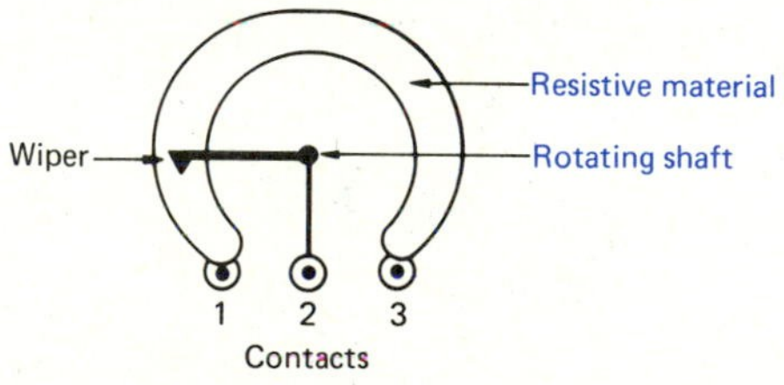

![Potentiometer diagram with Wiper, Resistive material, Rotating shaft, Contacts 1 2 3]

FIGURE 8.30 Potentiometer as Indicator of Angular Displacement

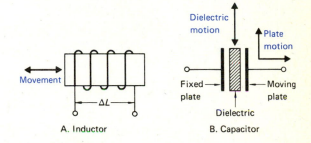

A. Inductor   **B. Capacitor**

FIGURE 8.31 Capacitor and Inductor as Indicators of Linear Displacement

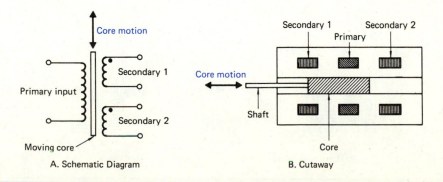

A. Schematic Diagram   B. Cutaway

FIGURE 8.32 LVDT as Indicator of Linear Displacement

The LVDT gives an output voltage that is linear with changes in the core position. The diagram in Figure 8.33A shows a plot of output voltage and the core position. Note how straight the line is, within the ±0.06 core range. Outside this range, the LVDT output is less linear. The diagram in Figure 8.33B shows the LVDT output voltage as a function of core position with reference to the phase of the output voltage. When the core is in position A, the phase relationship is shown below the horizontal axis. As the core is moved to position B, the phase changes 180°.

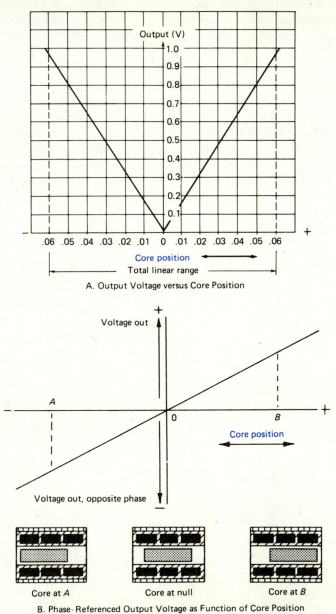

A. Output Voltage versus Core Position

B. Phase-Referenced Output Voltage as Function of Core Position

FIGURE 8.33   Relationship of LVDT Output Voltage and Core Position

The LVDT is a passive transducer; that is, it does not produce an output voltage without some form of excitation. The excitation usually is provided by an oscillator that is called the *carrier generator*. This AC voltage used by the LVDT is at a frequency and amplitude not normally generated by other electronic circuits and, therefore, must be generated by special signal-conditioning circuits. Since most readout and control devices operate on DC, a circuit called a *demodulator* converts the AC output of the LVDT to DC. This small DC potential may need to be amplified or otherwise conditioned to make an appropriate input for other circuitry.

A block diagram of a typical LVDT circuit is shown in Figure 8.34. The LVDT is supplied with an AC voltage from the carrier generator. When the core is moved within the form, an AC potential is produced by the LVDT and rectified and filtered by the demodulator. The DC voltage produced by the demodulator is amplified by the DC amplifier. An example of a passive demodulator is illustrated in Figure 8.35. In this circuit, the difference between the two DC voltages is sensed by the differential amplifier. Since both the secondary winding outputs, S_1

and S_2, are positive when rectified, the differential amplifier reverses the polarity as the core passes through null. Therefore, if more of the core is between primary winding P and S_1, the DC produced by the rectifier associated with it will produce more DC voltage than the bottom circuit. Since the S_1 DC input is connected to the noninverting terminal of a differential amplifier, the output will be a positive voltage. If, however, the core position is such that the signal from S_2 is larger than that from S_1, then the output of the differential amplifier will be negative.

LVDTs are used extensively in industry as displacement transducers. Displacements of as little as 50 μin. are detected by LVDTs. In addition to measuring displacement, LVDTs can measure weight and pressure in combination with mechanical transducers.

Figure 8.36 shows an application where an LVDT is used in a Bourdon tube and a diaphragm to sense pressure. The LVDT is mounted on a block to which the fixed end of the Bourdon tube (covered later in this chapter) is attached also. A spring attached to the LVDT core keeps the core centered in the form. When pressure increases, the Bourdon

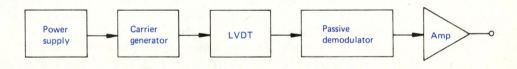

FIGURE 8.34 LVDT Driver Functional Block Diagram

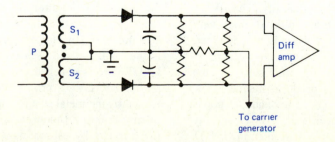

FIGURE 8.35 LVDT Driver Circuit

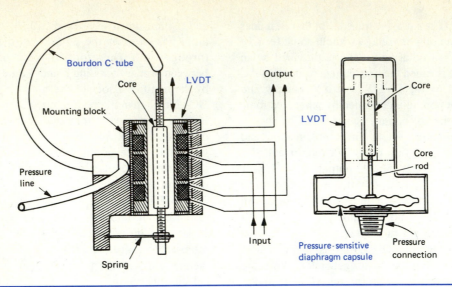

FIGURE 8.36 LVDT Being Driven by a Bourdon Tube and a Diaphragm

tube tries to uncoil, pulling the core of the LVDT upward. With small changes in pressure, the output of the LVDT output voltage is linear with respect to changes in pressure. The diaphragm (also covered later in this chapter) also is used in conjunction with the LVDT to change pressure to a voltage. An increase in pressure felt on the diaphragm will cause the diaphragm to expand, moving the LVDT core.

The LVDT has many features that are helpful when a transducer is used in automatic-control applications. The LVDT exhibits little frictional drag since the core is not mechanically connected to the coil. Because the coil and core do not touch, friction and mechanical wear are reduced, thus reducing maintenance and improving reliability. The LVDT output has infinite resolution, unlike displacement transducers such as the wire-wound potentiometer. The wire-wound potentiometer gives an output resistance curve made up of many small steps. As the wiper slides over a wire, it produces a small resistive change. The LVDT, on the other hand, responds to the smallest change in motion with essentially stepless resolution. Like any transformer, the LVDT has good isolation between the excitation circuitry and the output.

No device is without some disadvantages and limitations. The LVDT is susceptible to interference from strong electromagnetic fields. Such fields are created by induction welding and brazing machinery. LVDTs are not cost-effective for linear displacements of greater than 1 ft and require a high-frequency excitation source.

We can see the versatility of the LVDT by examining its industrial applications. One use of the LVDT in a rolling mill is shown in Figure 8.37. The rolling mill is an essential part of any factory that makes metal in sheets. The metal is melted, then pressed and rolled into sheets of uniform thickness. Part of the quality control of such a factory is the measurement of the thickness of the sheets to ensure that the thickness is within allowable limits. A key element in rolling mill production is the LVDT, used as a position transducer. The LVDT core is attached to a chromium-steel-tipped probe. The LVDT probe senses the displacement caused by the metal as it fills up the roll. A computer usually makes a running computation of how thick the metal is by estimating what the displacement should be at that point with metal of a uniform, known thickness.

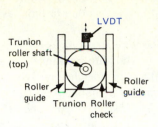 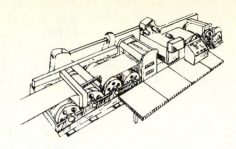

FIGURE 8.37   LVDT in Rolling Mill Application

Figure 8.38 illustrates another LVDT application where ceramic tiles are sorted according to thickness. The LVDT probe is mounted above a moving conveyor belt. Thickness of the tiles is measured when the LVDT core is pressed down on the tile. The core is spring-loaded, moving upward as it touches the tile. The greater the upward movement of the core, the greater the tile thickness is. Once the tiles are accurately measured, they can be sorted appropriately according to thickness. LVDTs allow sorting to thousandths of an inch in such applications. This same type of system presently is being used to sort semiconductor chips as small as 2.5 mm^2.

If you have ever adjusted the valves on an automobile, you know the great care required to make the difficult adjustments and readjustments with feeler gauges. Some automobile manufacturers have automated this process on assembly lines with LVDTs, as shown in Figure 8.39. A valve clearance may need to be set at 0.33 mm with a tolerance of 0.025 mm.

In this application, the engine camshaft is rotated until the lifter is riding on the lowest part of the camshaft lobe. The adjusting nut is loosened under computer control until the valve is fully seated, a position indicated by the LVDT. The locking nut is then tightened until the LVDT indicates that the valve is starting to move downward. This, on some engines, is the correct valve setting.

Stress and Strain Transducers

Stress, strain, and force are included in this section on displacement because displacement is the primary means of measuring these quantities. *Force* is defined as the quantity that changes the motion in a body. *Stress*, a similar concept, is the force acting on a

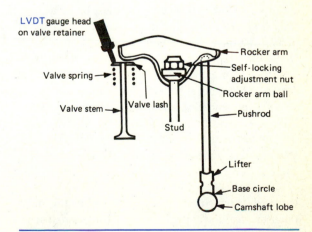

FIGURE 8.38   LVDT in Conveyor Application

FIGURE 8.39   LVDT in Automatic Valve Adjustment Application

solid's unit area. *Strain*, on the other hand, is the change in shape or form resulting from stress. The change in shape may be a change in length or width. Thus, the force applied to a solid material is the stress, and the deformation of that material is the strain.

Strain is defined as the change in length of an object per unit of length of the object. The unit of strain in the English system is microinches per inch (μin./in.). In the SI system, strain can be given in microcentimeters per centimeter (μcm/cm) or micromillimeters per millimeter (μmm/mm). The unit of strain in either case is called *microstrain*. You will notice that strain is actually a dimensionless number, like voltage or current gain. Like gain, strain is a ratio.

When you are working with semiconductor materials, temperature and power dissipation are critical, due to the temperature dependencies of semiconductor parameters. Bridge excitation voltages, therefore, must be chosen with care. The maximum excitation voltage of the piezoresistive semiconductor strain gauge is limited by its maximum power dissipation. Factors that affect heat dissipation are gauge size, amount of glue that fixes the gauge to the object, size of the object being measured, ambient temperature, and amount of air circulation. For gauges with lengths of 0.15 to 0.25 in., power dissipations of from 20 to 50 mW are acceptable.

In many scientific and industrial applications, the strain capabilities of new alloys must be measured. Measurements may need to be made of mechanical parts under different operating conditions. How much stress and strain a material can endure before permanent distortion occurs is valuable knowledge in the design of mechanical components. Companies manufacturing automobiles, for example, test the stress and strain on axles of new models.

In this section, we discuss two popular types of strain gauges: the wire gauge and the semiconductor gauge.

Bonded-Wire Strain Gauge. Probably the most popular strain-sensing device is the *bonded-wire strain gauge*. Strain gauges are made of metal, in the form of either wire (about 0.001 in. in diameter) or foil (about 3 μm thick). Strain gauges usually are shaped in a serpentine fashion, as shown in Figure 8.40.

The wire strain gauge is cemented firmly to a paper or Bakelite backing. The metal foil gauge normally is photoetched on an epoxy resin backing, as a printed-circuit board is. The backing serves two purposes. First, it supports the delicate strain gauge and protects it from damage. Second, it insulates the gauge from the object being measured.

Both wire and foil gauges react to force with a change in resistance. Recall that the resistance of a conductor depends on length and cross-sectional area, among other factors. As length increases and diameter decreases, resistance increases. This principle is the one on which the wire and foil strain gauges work.

The strain gauge is bonded firmly to the surface of the device under test, usually with a cement developed specifically for this purpose. As force is applied to the object under test, the surface deforms and so does the gauge. The dimensions of the gauge change, and this deformation is what causes the change in electric resistance.

Strains generally range from 1 microstrain to 50,000 microstrains. To better understand this concept, let us use as an example a piece of aluminum

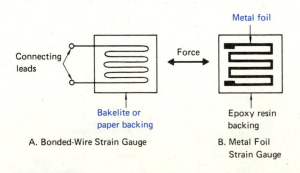

FIGURE 8.40 Pictorial Representations of Wire and Foil Strain Gauges

2 m long with 5000 microstrains applied. What is the change in length of the bar? The strain is given by

$$\text{strain} = \frac{\Delta l}{l_o} \tag{8.12}$$

where

Δl = change in length of bar
l_o = original length of bar

Solving for the change in length, we get

$$\Delta l = (\text{strain})\,(l_o) \tag{8.13}$$

Therefore, for our example, we have

$$\Delta l = (5000\,\mu\text{m/m})\,(2\,\text{m})$$

$$= 0.01\,\text{m}\quad\text{or}\quad 1\,\text{cm}$$

• EXAMPLE 8.16

Given a steel bar compressed with a deformation of 2500 microstrains, whose length was 0.5 m, find the length after the force is applied.

Solution

We find the change in length by using Equation 8.13:

$$\Delta l = (\text{strain})\,(l_o)$$

$$= (2500\,\mu\text{m/m})\,(0.5\,\text{m})$$

$$= 0.00125\,\text{m}\quad\text{or}\quad 1.25\,\text{mm}$$

Since the bar was compressed, the change in length must be subtracted from the original length:

$$l = l_o - \Delta l$$

$$= 0.5\,\text{m} - 0.00125\,\text{m}$$

$$= 0.49875\,\text{m}$$

As mentioned in the introduction to this section, the change in an object's length compared with its original length is called strain ($\Delta l/l_o$). Strain is often represented by ε (Greek letter epsilon). How much the resistance of a strain gauge changes depends on several things. Any resistance change is directly proportional to the original resistance, the elasticity of the material, and the strain, or deformation. These factors are related in the following equation:

$$\Delta R = R_o G \varepsilon \tag{8.14}$$

where

ΔR = change in resistance
R_o = original resistance
G = gauge factor
ε = strain

The gauge factor contains a constant called *Poisson's ratio*, which is an index of elasticity.

We can use Equation 8.14 to find the change in the electric resistance of an object being strained. For example, suppose a metal has a resistance of 120 Ω, a gauge factor of 3, and a strain of 6000 microstrains. According to Equation 8.14, the change in resistance is

$$\Delta R = R_o G \varepsilon$$

$$= (120\,\Omega)\,(3)\,(6000 \times 10^{-6}) = 2.16\,\Omega$$

• EXAMPLE 8.17

Given a gauge with a resistance of 100 Ω when unstrained, a gauge factor of 2.5, and a strain of 1200 microstrains, find the change in resistance.

Solution

The change in resistance is found from Equation 8.14:

$$\Delta R = R_o G \varepsilon$$

$$= (100\,\Omega)\,(2.5)\,(1200 \times 10^{-6}) = 0.3\,\Omega$$

Dividing both sides of Equation 8.14 by the original resistance and the strain, we see that the *gauge factor G* is a measure of the fractional change in resistance per unit of strain:

$$G = \frac{\Delta R / R_o}{\varepsilon} \qquad (8.15)$$

Or we can say that the gauge factor tells us the percent change in resistance of the gauge compared with its percentage change in length. Most metals have a gauge factor between 1.8 and 5.0.

Stress and strain are related through a law called *Hooke's law*. Hooke pointed out that for many materials, there is a constant ratio between stress and strain. This constant of proportionality between stress and strain is called *Young's modulus*. The relationship is expressed mathematically as

$$E = \frac{\sigma}{\varepsilon} \qquad (8.16)$$

where

E = Young's modulus (in force per unit area)
σ = stress (in force per unit area)
ε = strain

As an example, suppose we have a gauge with a modulus of elasticity (Young's modulus) of 1.03×10^8 kPa (kilopascals) undergoing a stress of 206 kPa. Solving Equation 8.16 for the strain ε, we get

$$\varepsilon = \frac{\sigma}{E} \qquad (8.17)$$

which, after substitution, yields

$$\varepsilon = \frac{206 \text{ kPa}}{1.03 \times 10^8 \text{ kPa}} = 2 \text{ μcm/cm}$$

As another example, suppose we have a gauge made of steel with a modulus of elasticity (Young's modulus) of 20.7×10^{10} N/m² undergoing a stress of 1×10^6 N. From Equation 8.17, we get

$$\varepsilon = \frac{\sigma}{E} = \frac{1 \times 10^6 \text{ N/m}^2}{20.7 \times 10^{10} \text{ N/m}^2}$$

$$= 4.8 \text{ microstrains}$$

Looking at this problem from another angle, we may want to know the stress on an object. According to the result of Example 8.17, the resistance of the gauge changed by 0.3 Ω under a strain of 1200 microstrains. What is the stress applied, in newtons per square meter and in pounds per square inch (psi) if Young's modulus is 20.7×10^{10} N/m²? First, we solve Equation 8.17 for stress, and then we substitute appropriate values:

$$\sigma = \varepsilon E \qquad (8.18)$$

so

$$\sigma = (12 \times 10^{-6})(20.7 \times 10^{10} \text{ N/m}^2)$$

$$= 2.48 \times 10^8 \text{ N/m}^2$$

Since 1 N/m² = 1.45×10^{-4} psi, we multiply our result by 1.45×10^{-4} to get the stress in pounds per square inch:

$$\sigma = (2.48 \times 10^8 \text{ N/m}^2)$$

$$\times [1.45 \times 10^{-4} \text{ psi } (\text{N/m}^2)]$$

$$= 36,000 \text{ psi}$$

• EXAMPLE 8.18

Given a gauge with a resistance of 100 Ω when unstrained, a modulus of elasticity of 20.7×10^{10} N/m², a gauge factor of 2.5, and a change in resistance of 0.5 Ω, find the stress on the gauge in newtons per square meter and in pounds per square inch.

Solution

We must solve Equation 8.14 for strain, to obtain

$$\varepsilon = \frac{\Delta R}{R_o G}$$

Substituting into this equation, we get

$$\varepsilon = \frac{0.5 \ \Omega}{(100 \ \Omega)(2.5)} = 2000 \text{ microstrains}$$

If the modulus of elasticity is 20.7×10^{10} N/m², the stress is, from Equation 8.18,

$$\sigma = \varepsilon E = (2000 \times 10^{-6})(20.7 \times 10^{10} \text{ N/m}^2)$$

$$= 414 \times 10^6 \text{ N/m}^2$$

Since 1 N/m² = 1.45×10^{-4} psi, we multiply this result by 1.45×10^{-4} to get the stress in pounds per square inch:

$$\sigma = (414 \times 10^6 \text{ N/m}^2)$$

$$\times [1.45 \times 10^{-4} \text{ psi (N/m}^2)]$$

$$= 60,030 \text{ psi}$$

Using the preceding equations, we can find how much stress a material is receiving. We will use these equations for the system shown in Figure 8.41.

Strain gauge A in Figure 8.41 is the sensor. As a force is applied down at point X, the metal bar bends. So does strain gauge A. In bending, the strain gauge wire increases its length and decreases its cross-sectional area. Thus, its resistance increases. (Strain gauge unstrained resistances range from 10 to 10,000 Ω.) This change in resistance is related to

strain and the stress (or force) that produced it. From the gauge factor equation (8.15), we can see that the important quantity is the change in the gauge resistance, not the actual resistance.

One problem with this type of strain gauge is its inherent lack of sensitivity. Let us see how great a resistance change we would expect with a gauge factor of 2, a strain of 1 microstrain, and a sensor with an unstrained resistance of 120 Ω. Solving Equation 8.14 for the change in resistance, we get

$$\Delta R = G \varepsilon R_o = (2)(1.0 \times 10^{-6})(120 \text{ }\Omega)$$

$$= 0.000240 \text{ }\Omega$$

Detecting this very small change in resistance means we must use instrumentation with microohm sensitivity. Measuring so small a resistance change usually requires a bridge, as shown in Figure 8.42. The unbalance in the bridge caused by the strain gauge resistance change is detected by a differential amplifier. Resistances R_A and R_B are designed to be greater than ten times the strain gauge resistance in order to make the circuit a constant-current source for the strain gauge. The design current is generally 1 mA in order to keep I^2R heat to a minimum and still have the strain gauge voltage high enough to be above the background noise level.

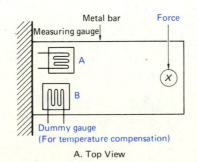

A. Top View

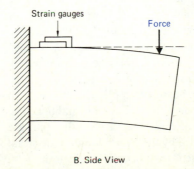

B. Side View

FIGURE 8.41   Strain Gauges Mounted on Bar and Subjected to Stress

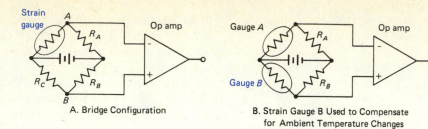

A. Bridge Configuration

B. Strain Gauge B Used to Compensate for Ambient Temperature Changes

FIGURE 8.42 Using Bridge to Measure Change in Strain Gauge Resistance

Now that we have a basic understanding of the bridge, let us turn our attention to the bridge's voltage behavior. Using the voltage divider equation, we can write an equation that tells us the output voltage of a bridge if the bridge supply voltage and resistances are known:

$$V_o = V_i \left(\frac{R_A}{R_A + R_g} - \frac{R_B}{R_C + R_B} \right)$$ (8.19)

where

R_g = resistance of strain gauge

The configuration in Figure 8.42A is called a *quarter bridge* since only one of the resistances is an active gauge. The other three are fixed resistors. When the ratio $R_B/R_C = R_A/R_g$, the voltage difference between the two legs of the bridge is zero. The output voltage is then 0 V, which can be seen by examining Equation 8.19. At this time, the bridge is said to be in balance.

Referring to Equation 8.15, we see that the quantity we want to measure is the gauge resistance change when a strain is applied. If we adjust resistor R_B with no strain applied to the gauge and then apply strain, the output voltage will be some value other than zero. The output voltage is produced by the unbalance caused by the change in the gauge resistance. With the strain applied, we again can adjust the bridge to a balanced or null condition. The amount of resistance change in R_B will equal the change in R_g caused by the strain. Some static strain indicators calibrate R_B and directly read out strain on a dial with a pointer fixed to the shaft of a variable resistor in the position of R_B.

Suppose we used a gauge with an unstrained resistance of 1000 Ω that changed to 1002 Ω under strain. If $R_A = R_B = 2000\ \Omega$, $R_C = 1000\ \Omega$, and $V_i = 10$ V, what is the output voltage? Using Equation 8.19, we can solve for the output voltage of the bridge:

$$V_o = V_i \left(\frac{R_A}{R_A + R_g} - \frac{R_B}{R_C + R_B} \right)$$

$$= 10\ \text{V} \left(\frac{2000}{2000 + 1002} - \frac{2000}{1000 + 2000} \right)$$

$$= 4.44\ \text{mV}$$

Balancing the bridge in a strain measurement is not always convenient because it would need to be done manually. Fortunately, we do not need to balance the bridge to get a strain measurement, as we did in the preceding example. By rearranging Equation 8.19, we get the following equation:

$$\frac{V_o}{V_i} = \frac{R_A}{R_A + R_g} - \frac{R_B}{R_C + R_B}$$

This equation is true for both strained and unstrained measurements; it also holds whether or not the bridge is balanced. At this point, it may be helpful to define a new term, V_r:

$$V_r = \left(\frac{V_o}{V_i} \right)_s - \left(\frac{V_o}{V_i} \right)_{us}$$ (8.20)

(strained)   (unstrained)

This term V_r is the difference between the ratios of the input and output voltages of the bridge under strained and unstrained conditions. By substitution, we can derive a formula relating strain, gauge factor, and this new variable V_r:

$$\varepsilon = \frac{-4V_r}{G\,(1 + 2V_r)} \qquad (8.21)$$

By measuring the output voltage V_o under strained and unstrained conditions, and having a knowledge of the gauge factor, we can calculate the strain an object undergoes.

An example at this point will help you understand this concept. Suppose we are working with an unbalanced bridge with an input voltage that remains constant at 1.980 V. In the unstrained condition, the output voltage is 0.000128 V. The unstrained output-to-input ratio then is

$$\left(\frac{V_o}{V_i}\right)_{us} = \frac{0.000128\text{ V}}{1.98\text{ V}} = 0.000065$$

In the strained condition, we measure an output voltage of 0.000418 V. The strained output-to-input ratio is

$$\left(\frac{V_o}{V_i}\right)_{s} = \frac{0.000418\text{ V}}{1.98\text{ V}} = 0.000211$$

We can now compute V_r, the difference between the two ratios:

$$V_r = \left(\frac{V_o}{V_i}\right)_{s} - \left(\frac{V_o}{V_i}\right)_{us}$$

$$= 0.000211 - 0.000065 = 0.000146$$

Using Equation 8.21, we can now calculate the strain, assuming a gauge with $G = 2$:

$$\varepsilon = \frac{-4V_r}{G\,(1 + 2V_r)} = \frac{-4\,(0.000146)}{2\,[1 + 2\,(0.000146)]}$$

$$= 291 \ \mu\text{in./in.}$$

The strain on this object, then, is 291 μin./in. or 291 μcm/cm.

Up to this point, we have not talked about the other strain gauge in Figure 8.41, strain gauge B. Another problem in strain gauge measurements results from resistance changes due to ambient temperature fluctuations. Strain gauge B, called the *dummy gauge*, is used for temperature compensation. Figure 8.42B shows both strain gauges in a balanced bridge. If temperature rises, both strain gauges change resistance by the same amount, creating no imbalance. Only the change in strain gauge A is proportional to strain.

Strain gauges are also found in load cells, which are commonly used to measure force. The *load cell* consists of one or more strain gauges mounted on some form of metal beam or bar, as shown in Figure 8.43. The gauges usually are wired in a bridge configuration. The imbalance is then taken from the load cell directly.

Semiconductor Strain Gauge. A more sensitive device is the *silicon semiconductor strain gauge*. Semiconductor strain gauges exhibit the piezoresistive effect. Materials that display this effect change their electric resistance with applied stress. All materials produce this effect, but in certain semiconductors, the effect is very large. The semiconductor crystals from which these sensors are made are grown with a controlled impurity content. Different

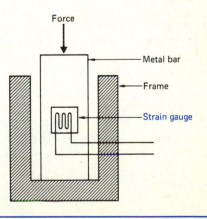

FIGURE 8.43 Load Cell

amounts of impurities produce different sensor characteristics. Advantages of the semiconductor strain gauge over the wire or foil device are as follows:

1. *Higher Sensitivity*: Gauge factors are 20 to 90 times higher than those for foil gauges.
2. *Smaller Sizes*: Sensors as small as 0.5 mm in length and 0.25 mm in width are possible with semiconductor strain gauges.
3. *Wider Resistance Ranges*: Resistances can range from 60 to 10,000 Ω.
4. *Higher Fatigue Life*: Semiconductor strain gauges can be stressed more than 10^7 times without damage.
5. *Negative Gauge Factors*: Both positive and negative gauge factors are available.

To show how the higher sensitivity of the semiconductor gauge can make a difference, let us compare a semiconductor strain gauge with a wire gauge. Recall the data from Example 8.17, which was based on a wire gauge. The gauge had a resistance of 100 Ω when unstrained, a gauge factor of 2.5, and a strain of 1200 microstrains. The change in resistance was 0.3 Ω. Let's assume that the wire gauge is replaced with a semiconductor gauge with a gauge factor of 250. What will the change in resistance be? From Equation 8.14, the change in resistance is

$$\Delta R = R_o G \varepsilon$$

$$= (100 \ \Omega)(250)\left(1200 \times 10^{-6}\right) = 30 \ \Omega$$

The semiconductor strain gauge typically is used in a bridge (with temperature compensation), as wire or foil strain gauges are. It is 30 to 50 times more sensitive than the comparable bonded-wire or foil strain gauge. Over wide ranges, linearity sometimes is a problem. So these devices commonly are used over narrow ranges. Where the application calls for wide ranges and high sensitivity, these devices are combined with linearity-compensating devices or circuits.

Semiconductor strain gauges have gauge factors as high as 200. Some manufacturers of semiconductor strain gauges state that the linearity of these gauges is as good as that of the best foil or wire gauges. When assessing linearity, we must consider the linearity of the circuit as well as that of the gauge itself. Semiconductor gauge linearities of $\pm 1\%$ are not unusual. Gauge resistance values range from under 100 to 10,000 Ω. Gauge resistance is determined by length, cross-sectional area, and resistivity of the crystal.

Another device used for strain or stress measurement is the *piezoelectric crystal*. Such crystals (commonly quartz, Rochelle salts, or tourmaline) exhibit a potential difference when compressed, as illustrated in Figure 8.44. The voltage that appears at the edges of the crystal can be high but is normally low in the amount of current it provides. Also, the charge leaks off very quickly under static conditions. Therefore, this device usually is used to measure forces or stresses that change rapidly, such as vibrations.

Many natural and artificial crystals exhibit a property called the *piezoelectric effect*. When force is applied to a piezoelectric crystal, the crystal lattice structure is strained or deformed. This deformation produces an electric charge on the edges of the crystal, as shown in Figure 8.44. When the force causing the deformation is taken away, the charge dissipates quickly. The transduction principle involved here is pressure to voltage. Generally, the more pressure applied to the crystal, the larger the voltage is. The piezoelectric effect is not exhibited in every crystal, nor is every crystal equally efficient

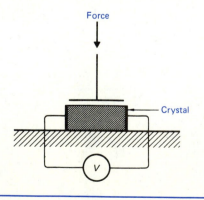

FIGURE 8.44 Piezoelectric Crystal Generating Voltage

at producing voltage. The amount of voltage produced depends on the crystal type, its thickness, and the amount of force exerted on it. Table 8.6 lists some crystals that exhibit the piezoelectric effect and the amount of voltage they produce.

We can use this parameter to calculate the output voltage of a crystal by multiplying the appropriate constant in Table 8.6 times the force (pressure) applied to the crystal times the thickness of the crystal. An example will help at this point. We would like to know how much voltage to expect from a 1 mm thick Rochelle salt crystal with a pressure of 1×10^6 N/m^2 applied to it. We will substitute values in the following formula:

$$V = KFd \qquad (8.22)$$

We obtain

$$V = (0.098 \text{ Vm/N}) \left(1 \times 10^6 \text{ N/m}^2\right) (0.001 \text{ m})$$
$$= 98 \text{ V}$$

• EXAMPLE 8.19

Given a lithium sulfate crystal 1.5 mm thick, find the voltage when the pressure is 10,000 psi.

Solution

First, we must convert the psi pressure into newtons per square meter. Since 1 psi = 6.895×10^3 N/m^2, we multiply 10,000 psi by 6.895×10^3:

$$(10,000 \text{ psi}) \left[6.895 \times 10^3 \text{ (N/m}^2)/\text{psi}\right]$$
$$= 68.95 \times 10^6 \text{ N/m}^2$$

TABLE 8.6 Crystals Exhibiting Piezoelectric Effect

Crystal	K (Vm/N) (Voltage per Unit Thickness per Unit Applied Pressure)
Quartz	0.055
Rochelle salt	0.098
Lithium sulfate	0.165
Tourmaline	0.0275

We now use Equation 8.22 to solve for the output voltage:

$$V = KFd$$
$$= (0.165 \text{ Vm/N}) \left(68.95 \times 10^6 \text{ N/m}^2\right) (0.001 \text{ m})$$
$$= 11,377 \text{ V}$$

Acceleration Transducers

We have seen how displacement transducers can be used to measure stress and strain. The displacement transducer can also be used to measure acceleration. Such a device is called an *accelerometer*.

Most accelerometers use the same principle, which is illustrated in Figure 8.45. As acceleration starts, the mass, which is free to move, goes backward. The mass moves backward because all mass has inertia, or resistance to motion. The higher the acceleration, the more backward movement that results. Any displacement, force, strain, or pressure transducers can be used to detect the motion or backward force.

Many accelerometers use a piezoelectric crystal to convert the force to electricity. Measuring the charge on the edges of a piezoelectric crystal gives the most accurate estimate of acceleration. Because measuring charge is both difficult and costly, however, voltage is measured instead. Acceleration in these cases is calculated by using the following equation:

$$a = \frac{V}{s}$$

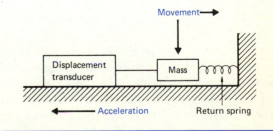

FIGURE 8.45 Basic Accelerometer

where

a = acceleration
V = peak voltage (in millivolts)
s = sensitivity of the sensor (in millivolts per *g*)

Recall that 1 *g* is the acceleration due to the earth's gravity, which is about 981 cm/s². Using this equation, we can calculate the acceleration of a system with a sensitivity of 20 mV/*g* and an output of 45 mV peak. We find the acceleration to be

$$a = \frac{V}{s} = \frac{45 \, mV}{20 \, mV/g} = 2.25 \, g$$

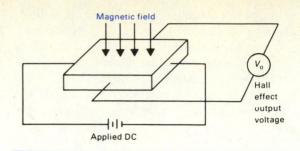

FIGURE 8.46 Hall Effect

MAGNETISM

Recall that magnetic fields are distributed spatially around an object as lines of force. Sensors that detect the strength of magnetic fields are used in measuring devices called *gauss meters*. Other common applications of magnetic field sensors are in ammeters. Two devices in particular are used in industry to detect the presence of magnetic fields: Hall effect devices and magnetoresistors. We will describe each type in turn.

Hall Effect Devices

The most popular device for detecting magnetic fields is the *Hall effect sensor*. The transduction principle used in the Hall effect device was discovered in 1879 by Edward H. Hall at Johns Hopkins University. He found that when a magnetic field was brought close to a gold strip in which a current was flowing, a voltage was produced. This effect is named the *Hall effect*.

Today, semiconductors are used in Hall devices instead of the gold with which Hall worked. Semiconductors produce the highest Hall voltages of any solid material.

The Hall effect, illustrated in Figure 8.46, is the generation of a voltage across opposite edges of an electric conductor carrying current and placed in a magnetic field. This effect is based on a force called the *Lorentz force*, which deflects charged carriers in a semiconductor material. The deflection of the charged carriers produces the difference in potential

shown in Figure 8.46. The Lorentz force is exerted in a direction perpendicular to the path of the particle's movement and the magnetic field direction.

The Hall effect is basically a majority carrier effect, depending on the bulk-material properties of a semiconductor material. Unlike transistors and diodes, the Hall effect device is free of surface effects, junction leakage currents, and junction threshold voltages. The most common material used in Hall effect devices is indium arsenide (InAs).

Hall effect devices are typically four-terminal devices. Two terminals are used for excitation and two for output voltage. Hall effect devices come in two basic functional classifications: linear and digital.

Linear Hall effect devices are used for gauss meters, with a sensitivity of 1.5 mV/*g*, and for DC current probes. Normally, to measure DC current, we must break the circuit and insert an ammeter. But breaking the circuit is not necessary with the Hall probe. The Hall effect DC current probe senses the magnetic field strength around a wire, which is proportional to current flow. Other applications include cover interlocks and ribbon speed monitoring in computer printers. Hall effect devices also find uses in brushless DC motors in computer disc and tape drives. When a PM rotor and a stator of coil windings are used, Hall sensors can replace brushes.

Two further applications of linear Hall effect devices are illustrated in Figures 8.47A and 8.47B. The diagram in Figure 8.47A shows a means of

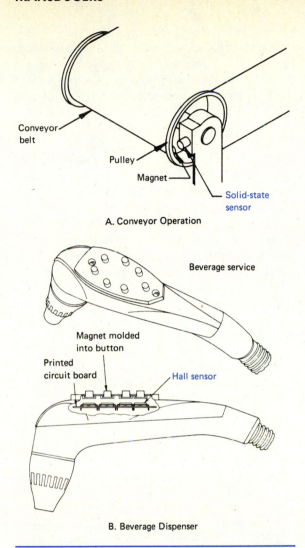

A. Conveyor Operation

Beverage service

B. Beverage Dispenser

FIGURE 8.47 Applications of Hall Effect Sensors

keeping tabs on a conveyor operation. The Hall effect sensor is mounted on the frame. A magnet is mounted on the drum. The Hall effect sensor actuates every time the magnet passes by. This system is used for speed control or for informing the operator that the conveyor is still in operation. Figure 8.47B shows a beverage dispenser. Magnets are attached to the buttons. When the buttons are pushed, the magnets actuate the sensors, dispensing the correct drink. The unit can be completely sealed for ease of cleaning.

Digital Hall effect devices are used in counting, switching, and proximity sensors. Proximity is sensed by placing a magnet on the object whose proximity is to be detected.

Hall effect devices typically produce about 10 mV potentials at the output terminals. Therefore, many manufacturers include an amplifier in the package with a Hall effect device.

Magnetoresistors

Magnetoresistors, although as popular as Hall effect devices, are also used to sense the presence of a magnetic field. Magnetoresistors look like photoresistors in structure. In these devices, indium antimonide is deposited in a serpentine pattern on a substrate, as shown in Figure 8.48. When a magnetic field impinges on the semiconductor, the carrier paths are distorted, as in the Hall effect device. However, in the magnetoresistor, carrier path distortion effectively narrows the cross-sectional area of the conductor, thereby increasing resistance.

Note that the magnetoresistor is a two-terminal device whose resistance varies with magnetic field proximity. As field strength increases, magnetoresistor resistance increases.

Magnetoresistors, being two-terminal devices, are used to replace resistors in low-voltage circuits, which Hall effect devices cannot be used for. Magnetoresistors also are used to sense mechanical position (proximity), current, and magnetic fields, as Hall effect devices are. An advantage of the magnetoresistor is its sensitivity. In a bridge circuit, magnetoresistors produce up to 1 V output, whereas the Hall effect devices typically produce less than 10 mV.

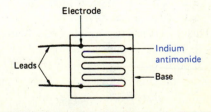

FIGURE 8.48 Magnetoresistor Construction

PRESSURE

Pneumatic and hydraulic systems abound in industry, even in the electronics age. There are many advantages to such systems, advantages that are discussed in Chapter 10. Suffice it to say here that these systems are still used; and one must often measure the pressures exerted at different points in these systems.

Pressure can be defined as the action of one force acting against an opposing force. Pressure normally is measured as a force per unit area, such as pounds per square inch (commonly abbreviated as psi, rather than lb/in.^{2}, in industry). In the English system, if the pressure is referenced to ambient pressure, the measurement is called *gauge pressure* (symbolized by psig). If the pressure measurement is referenced to a vacuum, the measurement is called *absolute pressure* (symbolized by psia). The metric system unit for absolute pressure is the pascal (Pa) and is equal to 1 N/m^2. Also, 1 Pa = 1.45×10^{-4} psi. Because the pascal is so small, the kilopascal (kPa) is often used.

Pressure transducers are classified into two basic types: gauge and absolute pressure transducers. The *gauge transducer* measures pressure with respect to the local atmospheric pressure, which, you will recall from physics, is 14.7 psi or 75 cm of mercury. This transducer is vented to the outside atmosphere. When the pressure-measuring port is exposed to the atmospheric pressure, the gauge should read 0 psig. The *absolute pressure transducer* measures pressure with reference to an internal chamber that is evacuated of air. When the pressure-measuring port is exposed to atmospheric pressure, the transducer indicates atmospheric pressure of 14.7 psia since the reference chamber exerts no pressure, being evacuated of all gas. Another type of pressure transducer you may see is called the *differential pressure transducer*. This device has two input ports and measures the difference (psid) between the pressures applied to the two ports.

Pressure-measuring devices will be broken down into two basic classifications: pressure-to-position transducers, such as the manometer, Bourdon tube, bellows, and diaphragm; and pressure-to-electrical transducers, such as the piezoresistive transducer.

Although the transducers discussed in this section primarily are pressure-to-position transducers, they can be converted easily to produce an electric output. As we proceed within this section, we will point out methods to achieve this conversion.

Manometers

The U tube manometer, shown in Figure 8.49, is representative of the liquid-column pressure transducers. The *U tube manometer* is a differential pressure-measuring device constructed from a glass or plastic tube bent in a U shape. The tube contains an amount of liquid, whose makeup depends on the application. Note that pressure p_1 is greater than pressure p_2 in the figure. The difference in pressure causes the liquid to move into the right leg until all pressures are in equilibrium. The difference in the height (Δh) of the liquid is proportional to the difference in pressures.

When no pressure is applied, $\Delta h = 0$. The difference in height is equal to the difference in pressure ($p_1 - p_2$) divided by the weight density (w_m) of the material in the manometer:

$$\Delta h = \frac{p_1 - p_2}{w_m} \qquad (8.23)$$

where

Δh = difference in height between two levels
p_1 = pressure in left limb of manometer
p_2 = pressure in right limb
w_m = weight density of liquid in manometer

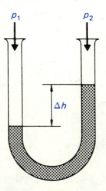

FIGURE 8.49   U Tube Manometer

An example may help at this point. Suppose p_1 is 8 psi, p_2 is 6 psi, and $w_m = 0.488$ lb/in.^{3}. The value of Δh can be found by substituting the appropriate values into Equation 8.23:

$$\Delta h = \frac{p_1 - p_2}{w_m}$$

$$= \frac{8 \text{ psi} - 6 \text{ psi}}{0.488 \text{ lb/in.}^3} = 4.1 \text{ in.}$$

• **EXAMPLE 8.20**

Given a manometer with $p_1 = 10$ psi, $p_2 = 5$ psi, and $w_m = 0.036$ lb/in.^{3}, find the difference in height in the two legs of the manometer.

Solution

The difference in height Δh can be found by substituting the appropriate values into Equation 8.23:

$$\Delta h = \frac{p_1 - p_2}{w_m}$$

$$= \frac{10 \text{ psi} - 5 \text{ psi}}{0.036 \text{ lb/in.}^3} = 138 \text{ in.}$$

The major disadvantage of manometers is that they are strictly mechanical. They can, however, be converted to electrical systems with a float device attached to a displacement transducer, such as the LVDT.

Elastic Deformation Transducers

All *elastic deformation pressure transducers* use the same transduction principle. Pressure causes a bending, or deformation, of the transducer material, usually a metal. This deformation results in a deflection or displacement. We might say, then, that these transducers are pressure-to-position converters. As we will see, the indicators can be purely mechanical, or they can use electric transducers to convert the displacement to an electrical parameter. We will discuss three types of elastic deformation transducers, all based on the same principle but each different in physical shape and size. The transducers we will consider are the Bourdon tube, the bellows, and the diaphragm.

Bourdon Tube. The *Bourdon tube* is one of the oldest and most popular (even today) pressure transducers. It was invented in 1851 by Eugene Bourdon and has been used in industry ever since. Bourdon tubes come in three basic shapes: C tube, spiral tube, and helical tube, as illustrated in Figures 8.50A, 8.50B, and 8.50C.

All three devices work in essentially the same fashion. They are all composed of a flattened metal tube, usually made of brass, phosphor bronze, or steel; see the cross section in Figure 8.50D. The pressure at the input is communicated to the inside of the device. Since there is more area on the outside of the tube than on the inside, the tube unwinds, much like a paper-coil party toy. The unwinding of the Bourdon tube is not linear. Therefore, mechanical pointers fastened to the moving tips have displacement applied through a gearing system. This gearing system compensates for the nonlinearity of movement. Nonlinearity may also be compensated for by using a nonlinear pointer scale.

The Bourdon tube is simple to manufacture, inexpensive, and accurate; and it measures pressures up to 100,000 psi (700,000 kPa). The C-shaped Bourdon tube is the least sensitive to pressure changes, and the helical tube is highest in sensitivity. These devices do not work well with pressures under 50 psi (750 kPa). Bourdon tubes are converted very simply to electric transducers by using electric displacement transducers such as potentiometers and LVDTs.

Bellows. The *pressure bellows* is a cylindrical-shaped device with corrugations along the edges, as shown in Figure 8.51A. As pressure increases, the bellows expands, moving the shaft upwards. The shaft may be attached to a pointer or an electric displacement transducer like an LVDT.

In Figure 8.51B, the bellows is connected to indicate differential pressure. If p_1 is greater than p_2, the pointer moves up. The amount of displacement is proportional to the difference in pressures.

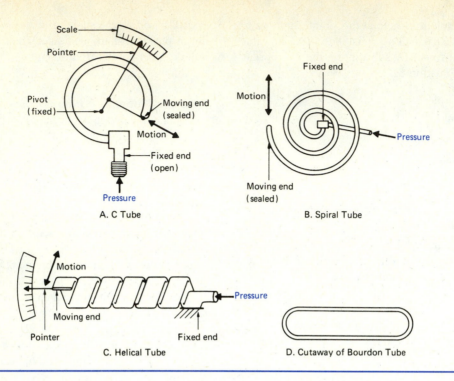

FIGURE 8.50 Bourdon Tubes

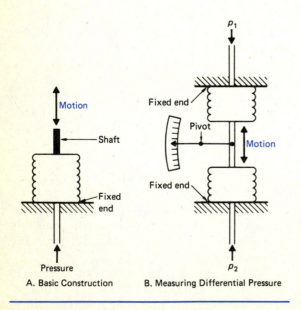

FIGURE 8.51 Pressure Bellows

Bellows are more sensitive to pressures in the 0–30 psi (0–210 kPa) range than are Bourdon tubes. Thus, they are more useful at lower pressures.

Diaphragm. The *diaphragm* is nothing more than a flexible plate, as illustrated in Figure 8.52. The plates are usually made of metal or rubber. The flat (or curved) metal plates generally are less flexible (and less sensitive) than the corrugated type. Increases in pressure will cause the shafts to move up. Diaphragms are used in the 0–15 psi (0–105 kPa) pressure range.

Piezoresistive Transducers

Both National Semiconductor and the Micro Switch Division of Honeywell manufacture very sensitive and linear piezoresistive pressure transducers. These sensors use the piezoresistive effect and have their

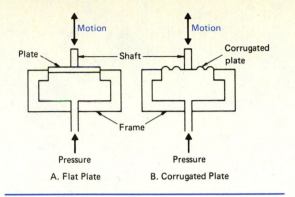

FIGURE 8.52 Pressure Diaphragms

own built-in amplifiers and temperature compensation networks. National's line of products can sense pressures ranging from 10–5000 psi (70–35,000 kPa)

with linearity less than 1%. Figure 8.53 shows one of National's transducers.

FLUID FLOW

Along with temperature, fluid flow is a very important measure. Many industrial processes, especially biomedical applications, depend on accurate assessment of fluid flow. Fluid flow measurements are an integral part of chemical, steam, gas, and water treatment plants. The usual approach to measuring the rate of a flowing fluid is to convert the kinetic energy that the fluid has to some other measurable form. This process is carried out by the fluid flow transducer.

There are many different transducers available to measure fluid flow. As with the other transducers we have discussed, we limit our treatment to only the more common fluid flow measuring devices: differential pressure flowmeters, variable-area flowmeters, positive-displacement flowmeters, velocity flowmeters, and thermal heat mass flowmeters.

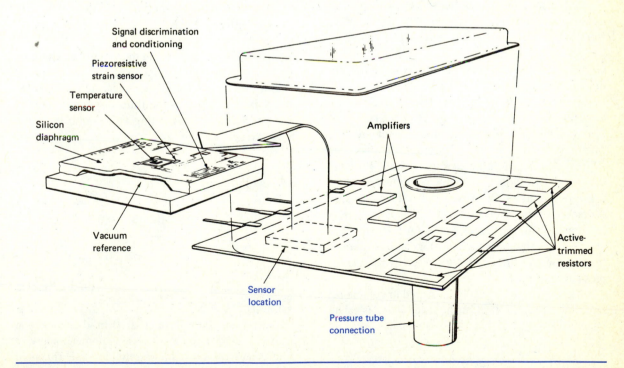

FIGURE 8.53 Piezoresistive Pressure Transducer

Differential Pressure Flowmeters

The most common industrial flowmeters come from this group. Most *differential pressure flowmeters* use a pressure difference caused by a restriction as a transduction principle. Each flowmeter has a different method of producing this pressure differential. The principle involved here is called the Bernoulli effect. The *Bernoulli effect* states that as the velocity of a fluid (liquid or gas) increases, the pressure decreases, and vice versa. We say that an inverse relationship exists between the pressure and velocity of a fluid.

If the fluid is incompressible, the same volume of fluid must pass through a given area in the same time interval. According to the laws of conservation of mass and energy, the mass entering any portion of the pipe must equal the mass leaving that portion of the pipe during some time interval. The same is true for the kinetic energy of the fluid. The fluid flow velocity is, therefore, increased in the smaller section of the pipe, causing an increase in kinetic energy per unit of mass.

In this section, we consider five of the commonly used differential pressure flowmeters: orifice plate, Venturi tube, flow nozzle, Dall tube, and elbow.

Orifice Plate. We will use the most common of the differential pressure flowmeters, the *orifice plate flowmeter*, as an example. It is shown in Figure 8.54. This flowmeter consists of a plate with a hole in it. If we place liquid columns all along the tube to indicate pressure, we see a *pressure gradient*, or pressure distribution, as pictured in Figure 8.54. Pressure increases slightly just before the restriction and falls off quickly after the restriction. The pressure decreases sharply because of the increase in velocity at and just after the restriction. A manometer or other differential pressure indicator can be calibrated to read the flow rate. As fluid flow increases, the pressure difference increases also.

The flow rate, normally measured in volume per unit time, is symbolized by the letter Q and is calculated as

$$Q = \left(\frac{\pi}{4}\right)\left(\frac{d_b}{d_a}\right)^2 (K)(2gd)^{1/2} \qquad \text{(8.24)}$$

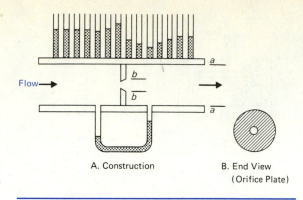

A. Construction

B. End View (Orifice Plate)

FIGURE 8.54 Static Pressure Gradient in Orifice Plate

where

d_b = diameter at b–b in orifice plate
d_a = diameter at a–a
K = constant, called *flow coefficient*
g = acceleration of gravity (32 ft/s^2)
d = difference between heights in manometer

An example may help at this point. Let's assume we have water flowing through a 12 in. diameter pipe with a 6 in. orifice. A manometer shows $d = 0.88$, and K is 0.98. What is the fluid flow rate, in cubic feet per second? Substituting values into Equation 8.24, we get

$$Q = \left(\frac{\pi}{4}\right)\left(\frac{d_b}{d_a}\right)^2 (K)(2gd)^{1/2}$$

$$= \left(\frac{\pi}{4}\right)\left(\frac{6}{12}\right)^2 (0.98)[(2)(32.2)(0.88)]^{1/2}$$

$$= 1.45 \text{ ft}^3/\text{s}$$

• EXAMPLE 8.21

Given a water flow through a 10 in. diameter pipe with an 8 in. orifice for which a manometer shows $d = 1.2$, and with $K = 0.98$, find the flow rate (in cubic feet per second).

Solution

Substituting values into Equation 8.24, we obtain

$$Q = \left(\frac{\pi}{4}\right)\left(\frac{d_b}{d_a}\right)^2 (K)(2gd)^{1/2}$$

$$= \left(\frac{\pi}{4}\right)\left(\frac{8}{10}\right)^2 (0.98)\,[(2)(32.2)(1.2)]^{1/2}$$

$$= 4.33 \text{ ft}^3/\text{s}$$

The orifice plate flowmeter is the most widely used of all the flowmeters. It is simple in structure, low in cost, and easy to install. It suffers from the disadvantages of poor accuracy, limited range, and inability to be used with slurries (liquids with suspended particles of solid matter).

Venturi Tube. Another popular flowmeter is the *Venturi tube*, shown in Figure 8.55. The Venturi tube employs the same principle as the orifice: differential pressure. The pressure difference is caused by the restriction in the tube.

The Venturi has two major advantages over the orifice. First, it creates less turbulence than the orifice because of the gently sloping sides. Thus, the final pressure loss is less than that in the orifice. This loss is called *insertion loss*. Up to 90% of the initial pressure can be recovered with the Venturi. Second, the Venturi can be used for slurries because there are no sharp edges or projections for solids to catch on. Its major disadvantage is cost. It is expensive to buy and install.

Flow Nozzle. The *flow nozzle*, shown in Figure 8.56, falls somewhere between the Venturi and the orifice plate in cost and pressure loss. In the flow nozzle, the edges of the restriction are somewhat rounded. Thus, the flow nozzle permits up to 50% more flow rate than the orifice. An additional advantage is its ability to be welded into place, which cannot be done with other differential pressure flowmeters.

Dall Tube. Of all the differential pressure flowmeters, the *Dall tube* has the least insertion loss. The Dall tube, shown in Figure 8.57, consists of two cones, the shorter one upstream of the restriction, the larger cone downstream. It is generally cheaper to purchase and install than the Venturi tube. However, it cannot be used with solids and slurries, as the Venturi can.

Elbow. The *elbow tube flowmeter* is a differential pressure flowmeter that does not use the Bernoulli effect. It is shown in Figure 8.58. The differential pressure is developed by the centrifugal force exerted

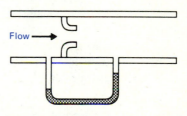

FIGURE 8.56 Flow Nozzle Flowmeter

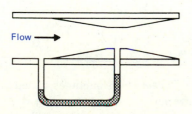

FIGURE 8.55 Venturi Tube Flowmeter

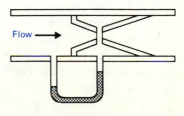

FIGURE 8.57 Dall Tube Flowmeter

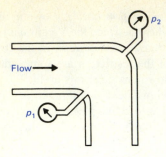

FIGURE 8.58   Elbow Tube Flowmeter

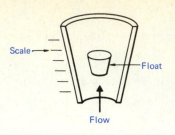

FIGURE 8.59   Rotameter

by the fluid flowing around the corner. In this case, p_2 is greater than p_1. Since elbows (corners) usually already exist in pipes, installation of elbows poses no problems. There is no additional pressure loss because there is no restriction. However, the elbow flowmeter has only ±5% accuracy at best.

The devices we have discussed thus far are called *primary elements*. They are so named because, in a primary sense, they convert fluid flow to pressure differential. *Secondary elements* are needed to convert this pressure difference into a usable parameter. This conversion can be accomplished with any of the pressure transducers discussed previously. In addition, National Semiconductor manufactures a differential pressure transducer that is used as a secondary element to measure flow.

Variable-Area Flowmeters

The *variable-area flowmeter* is a flowmeter that controls the effective flow area and, therefore, the flow rate within the meter. An example of the variable-area flowmeter is the rotameter, also called the float or rising ball. Shown in Figure 8.59, the *rotameter* consists of a float that is free to move up and down in a tapered tube. The tube is wider at the top and narrows toward the bottom. Fluid flows from the bottom to the top, carrying the float with it. The float rises until the area between it and the tube wall is just large enough to pass the amount of fluid flowing. The higher the float rises, the higher is the fluid flow rate.

Normally, the rotameter is used for visual flow measurement. It cannot be used for automatic process control.

Positive-Displacement Flowmeters

Positive-displacement flowmeters divide the flowing stream into individually known volumes. Each volume is then counted. Adding each volume increment gives an accurate measurement of how much fluid volume is passing through the meter. Positive-displacement meters are accurate to less than 1%, simple, inexpensive, and reliable. Furthermore, pressure losses generally are lower compared with those of other flowmeters.

The *nutating-disc flowmeter*, shown in Figure 8.60, is a common example of a positive-displacement flowmeter. The nutating-disc meter is used in residential water metering as well as in industrial metering. As the liquid enters the left chamber, it fills quickly. Because the disc is off center, it has unequal pressures on it, causing it to wobble, or nutate. This action empties the volume of liquid from the left chamber into the right. The process repeats itself, driven by the force of the flow stream. Each nutation of the disc is counted by a mechanical or an electromagnetic counting mechanism.

Because of the mechanical structure of the nutating-disc flowmeter, it is not suitable for measuring the flow of slurries.

Velocity Flowmeters

Velocity flowmeters (sometimes called *volumetric rate flowmeters*) give an output that is proportional to the velocity of fluid flow. We will consider three of the most commonly used velocity flowmeters: turbine, electromagnetic, and target.

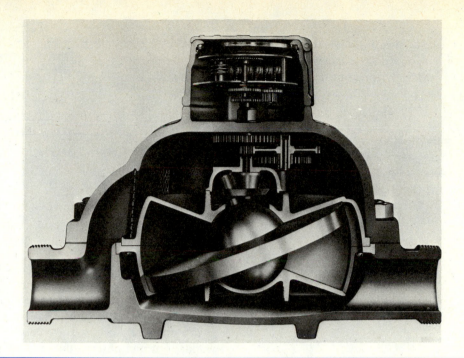

FIGURE 8.60 Nutating-Disc Meter

Turbine Flowmeter. The *turbine flowmeter* is shown in Figure 8.61. The fluid flowing past the turbine exerts a force on the blades. The blades then turn at a rate directly proportional to the fluid velocity. The turbine blades can be magnetized to produce pulse as voltage is induced in the coil. Turbine meters are useful because of their good linearity, range, and accuracy. They are among the most accurate flowmeters available.

Electromagnetic Flowmeter. The *electromagnetic flowmeter* operates on the principle of electromagnetic induction. The *principle of electromagnetic induction*, known as Faraday's law of electromagnetic induction, states that a conductor passing through a magnetic field (at right angles) produces a voltage. The voltage produced is proportional to the relative velocity of fluid flow.

The electromagnetic flowmeter, shown in Figure 8.62, generates a magnetic field through an electromagnet. The conductor is the flowing fluid. As we

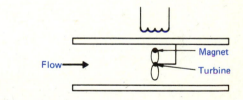

FIGURE 8.61 Turbine Flowmeter

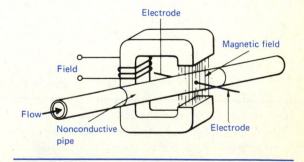

FIGURE 8.62 Electromagnetic Flowmeter

might expect, the fluid must be conductive. A potential is then produced between the electrodes, which are imbedded in a piece of nonconductive pipe. Since no obstructions exist, the pressure loss is small. Additionally, the lack of obstruction makes the measurement of slurries easy to do.

Target Flowmeter. The *target flowmeter*, shown in Figure 8.63, presents an obstruction to fluid flow. The fluid exerts a pressure on the target that is directly proportional to the fluid velocity. The force on the target is sensed by a strain gauge.

Thermal Heat Mass Flowmeters

The *thermal heat mass flowmeter* is an accurate device for measuring fluid flow. In this device, heat is applied to the flowing fluid. A temperature sensor is then used to measure the rate at which the fluid conducts heat away. This rate is directly proportional to the flow rate. One of the foremost advantages of the thermal heat mass system is its lack of obstructions to fluid flow.

Table 8.7 summarizes the fluid flow measuring devices discussed in this section.

LIQUID LEVEL

The methods of measuring liquid level are as numerous and varied as those for fluid flow. Also, there are as many reasons for wanting to know the level of a liquid (or solid) as there are reasons for wanting to know the rate of fluid flow. Liquid level is an important part of process control.

Liquid-level measurement is divided into two basic types: point and continuous. The *point level* is just what its name implies—measurement of a level at a point. *Continuous-level measurement* is analog in nature. The output of a continuous-level sensing instrument is directly proportional to the level of the liquid. Some sensors are more suited to one or the other of these types of measurement. In our description of the sensors, we point out which application is more appropriate. For purposes of discussion, we classify liquid-level sensors into sight, force, pressure, electric, and radiation sensors.

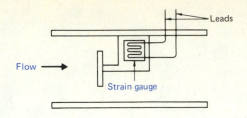

FIGURE 8.63 Target Flowmeter

Sight Sensors

Sight-level sensors are among the oldest and simplest. The *dipstick* is possibly the oldest, and it is still used in some parts of industry. To make a measurement, you insert a stick (sometimes calibrated with lines, as in a car oil dipstick) into the container. You then withdraw it and observe the level of fluid on the dipstick.

Related to the dipstick method are the sight glass and the gauge glass. The *sight glass* is similar to coffee urns found in cafeterias. The sight glass allows you to see the liquid level in the tube, as shown in Figure 8.64A. A *gauge glass* is similar (Figure 8.64B). It has windows in its sides to allow you to see the liquid level.

The *displacer* works on a different principle entirely. It uses Archimedes' principle. Archimedes found that a body immersed in water loses weight equal to the amount of water displaced. In Figure 8.64C, the weight of the displacer object is proportional to the liquid level. As the liquid rises, the displacer object has less weight. The weight measurement is made mechanically and displayed on a pointer, or electric force sensor. A strain gauge works nicely here.

The *tape float* (Figure 8.64D) works in a similar fashion. The float rides on top of the liquid. It is attached to a tape, at the end of which is an indicator. As the level rises, the tape moves down the scale.

All the sight methods have the same disadvantage: They require human operators to see and record the levels. However, these methods are very inexpensive, and all are suited to continuous measurement.

TABLE 8.7   Summary of Flowmeters

Type	Name	Range	Operating Principle	Accuracy	Advantages	Disadvantages
Differential pressure	Orifice plate	3:1	Variable head	1%–2%	Lowest in cost Easy installation	Highest permanent-head loss Cannot be used with slurries
	Flow nozzle	3:1	Variable head	1%–2%	Can be used at high velocity Can be used with slurries	Expensive Difficult to remove Difficult to remove and install
	Venturi tube	3:1	Variable head	1%–2%	Most accurate of these meter types Can be used with slurries Low permanent-head loss	Difficult to remove and install Costly
	Dall tube	3:1	Variable head	1%–2%	Lowest permanent-head loss	Costly
	Elbow tube	3:1	Centrifugal force	5%–10%	Saves space No head loss No obstructions	Low accuracy Small differential pressure created
Positive displacement	Nutating disc	5:1	Measured volume	0.1%	High accuracy Simple to install	Costly to install and maintain Cannot be used with slurries
Velocity	Turbine	10:1	Spinning turbine or propeller	0.5%	High accuracy Low head loss Wide temperature range	Cannot be used with slurries Hard to install
	Electro-magnetic	30:1	Electromagnetic voltage generation	1%	Linear No obstructions No head loss	Must have conductive fluid High cost
	Target	3:1	Fluid pressure on target	0.5%–3%	Compact Easy to install Low cost	Subject to temperature variations
Variable area	Rotameter	10:1	Balance between gravity and fluid pressure	1%–2%	Little head loss Linear response	Expensive Fragile
Mass flow	Thermal heat mass flowmeter	100:1	Temperature rise to mass flow rate	2%	Measures low flow rates Low head loss Fast response	Useful only for gases Costly and complex

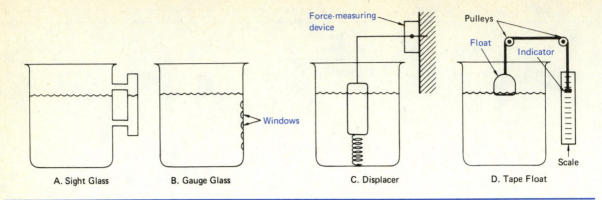

FIGURE 8.64 Liquid-Level Measurement by Sight

Force Sensors

The weight of a container (force) is often used to indicate liquid levels. When more liquid (or solid) enters the container, it weighs more. The weight is detected by *force* or *strain sensors*, as in the load cell of Figure 8.65A. Alternatively, the buoyant force of a float may move a rigid rod or flexible tape, cable, or chain, as illustrated in Figures 8.65B and 8.65C.

All these methods are suited to continuous measurement of liquid level. The float methods shown may be converted to an electrical output by placing a potentiometer at the pivot point. Changes in the linear displacement of the float are then changed to angular displacement by the pulley or pivot. The potentiometer changes angular displacement into an electric resistance change.

Pressure Sensors

Pressure is another variable that is affected by liquid level. You will recall from physics that pressure in a liquid increases with the depth of the liquid. For instance, if more water is added to a tank, there will be more water above the point at which the pressure is measured. The increased weight of the water will, therefore, create a greater pressure.

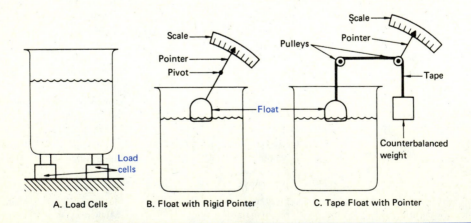

FIGURE 8.65 Liquid-Level Measurement by Force

The pressure at the bottom of the tank is related to the height of the liquid and the weight density of the liquid by the following equation:

$$p_b = hw \qquad (8.25)$$

where

p_b = pressure at bottom of tank
h = height of liquid
w = weight density of liquid

Let's consider an example. Suppose we have a tank full of oil with 4 psi pressure at the bottom of the tank. We wish to find the height of the oil. The specific gravity of the oil is 0.9. Recall that the specific gravity of a liquid is the ratio of the mass or weight density of the liquid to the mass or weight density of a selected reference material. The reference liquid is usually water at specified temperature and volume. The weight density of water is 62.4 lb/ft^3. The oil in our example has a weight density 0.9 times that of water, according to the equation

$$w_{oil} = (spg)\left(w_{H_2O}\right) \qquad (8.26)$$

Thus, we have

$$w_{oil} = (0.9)\left(62.4 \ lb/ft^3\right)$$

$$= 56.16 \ lb/ft^3$$

Since the weight density is in pounds per cubic foot, we should change the 4 psi measurement to pounds per square foot:

$$p_b = \left(4 \ lb/in.^2\right)\left(144 \ in.^2/ft^2\right) = 576 \ lb/ft^2$$

Now solve Equation 8.25 for h:

$$h = \frac{p_b}{w} \qquad (8.27)$$

So we obtain

$$h = \frac{576 \ lb/ft^2}{56.16 \ lb/ft^3} = 10.26 \ ft$$

The height of the oil is 10 ft.

• **EXAMPLE 8.22**

Given a bellows at the bottom of a tank of gasoline with a specific gravity of 0.7 and a pressure of 9 psi, find the height of the gasoline (in feet).

Solution

First, we find the weight density of the gasoline, using Equation 8.26:

$$w_{gas} = (spg)\left(w_{H_2O}\right)$$

$$= (0.7)\left(62.4 \ lb/ft^3\right)$$

$$= 43.68 \ lb/ft^3$$

Since the weight density is in pounds per cubic foot, we should change the 9 psi measurement to pounds per square foot:

$$p_b = \left(9 \ lb/in.^2\right)\left(144 \ in.^2/ft^2\right) = 1296 \ lb/ft^2$$

We may now use Equation 8.27:

$$h = \frac{p_b}{w} = \frac{1296 \ lb/ft^2}{43.68 \ lb/ft^3}$$

$$= 29.67 \ ft$$

The pressure exerted on the bottom of the tank is called the *hydrostatic head*, which is defined as the pressure a column of liquid exerts on the bottom of a container. As the liquid level increases, more pressure is exerted at the bottom of the tank. Devices that use this method of sensing level are called *hydrostatic-head devices*. In the *diaphragm type*, shown in Figure 8.66A, more pressure is exerted on the diaphragm as the level increases. The air pressure inside the tube then increases and registers a pressure on the meter. Figure 8.66B shows a similar device mounted on the bottom of the tank.

Other hydrostatic-head devices use differential pressure, as shown in Figure 8.66C. The pressure difference between the bottom and top of the tank

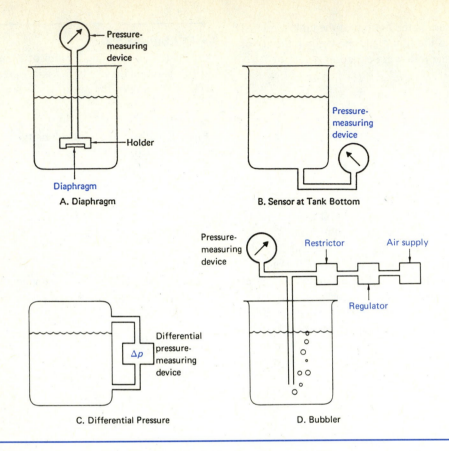

FIGURE 8.66 Liquid-Level Measurement by Pressure

depends on the liquid level. The difference is sensed by a differential pressure–measuring device, such as bellows or National Semiconductor's IC differential pressure transducers.

The *bubbler system*, shown in Figure 8.66D, is one of the oldest and simplest devices for measuring pressure. Air flows in through the tube placed in the tank. If enough pressure is exerted in the tube, bubbles are forced out of the tube. The pressure required to force the bubbles out varies proportionally with the liquid level. This pressure is indicated by the pressure-measuring device.

Electric Sensors

Beyond the mechanical-to-electrical conversions mentioned previously, certain sensors give direct electrical outputs with changing liquid levels. One interesting device, manufactured by Metritape, changes the resistance of a wire helix as pressure is applied. This device, illustrated in Figure 8.67, is suited to continuous-level measurement.

Electrodes are also used to sense electrical changes. In Figures 8.68A and 8.68B, the capacitance changes as the liquid level fluctuates. The change in capacitance is caused by the changing dielectric constant between the liquid and the air. The changes in capacitance usually are sensed with an oscillator or an AC bridge.

Note the resistive electrode sensor in Figure 8.68C. With conductive liquids, the electrodes in Figure 8.68C provide a path for current flow through the series circuit. This type of system is suited to a point-sensing application.

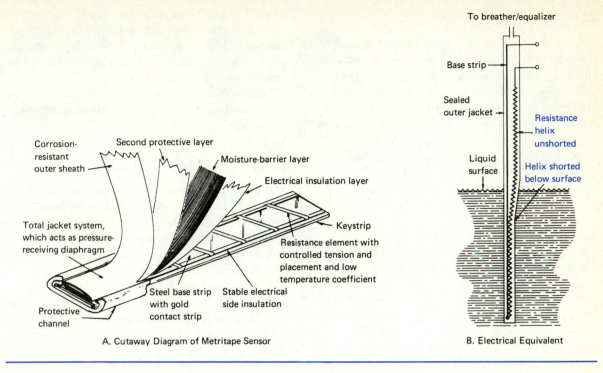

Corrosion-
resistant
outer sheath

Second protective layer

Moisture-barrier layer

Electrical insulation layer

Total jacket system,
which acts as pressure-
receiving diaphragm

Keystrip

Resistance element with
controlled tension and
placement and low
temperature coefficient

Protective
channel

Steel base strip
with gold
contact strip

Stable electrical
side insulation

A. Cutaway Diagram of Metritape Sensor

To breather/equalizer

Base strip

Sealed
outer jacket

Resistance
helix
unshorted

Liquid
surface

Helix shorted
below surface

B. Electrical Equivalent

FIGURE 8.67 Resistance Changes with Changes in Liquid Level

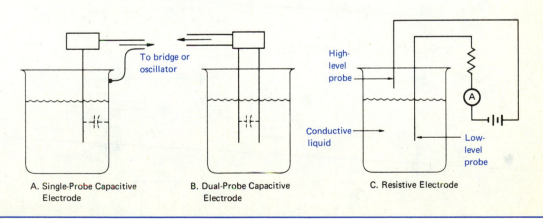

To bridge or
oscillator

A. Single-Probe Capacitive
Electrode

B. Dual-Probe Capacitive
Electrode

High-
level
probe

Conductive
liquid

Low-
level
probe

C. Resistive Electrode

FIGURE 8.68 Liquid-Level Measurement by Electrical Means

Radiation Sensors

Most of the level measurement devices discussed previously have elements that are in contact with the liquid. In some industrial measurements, however, the level-sensing device must be used with corrosive liquids or liquids under high pressures. Hence, contact with the liquid might destroy the device. In this section, we will discuss one class of sensors that do not physically contact the liquid at all: radiation sensors. We divide these sensors into two classifications: sonic (including ultrasonic) and nuclear radiation sensing devices.

Sonic. Sonic and ultrasonic beams (7.5–600 kHz) use the *echo principle* to measure liquid level. In Figure 8.69A, the sonic beam is transmitted from the transducer to the liquid surface. It is then reflected back to the transducer from the surface. The time it takes for the beam to travel from the transducer to the surface and back to the transducer depends on the liquid level. The lower the level, the more time it takes for the beam to travel the distance. The time interval, then, is inversely proportional to the liquid level.

Nuclear. Nuclear radiation is sometimes used to make liquid-level measurement for point-sensing applications. In Figure 8.69B, radiation is emitted from the transmitter in the form of gamma rays. The receiver is usually a Geiger-Müller tube. The radiation is absorbed by the liquid or solid to be measured. The output of the receiver, therefore, is dependent on

whether or not the liquid is between the sensor and the transmitter. Nuclear sensors are expensive compared with other electrical methods of sensing level. They also require extensive in-plant safety precautions.

Table 8.8 summarizes the different types of liquid-level sensors and some pertinent information about them.

MEASUREMENT WITH BRIDGES

Many of the transducers discussed previously use bridges to transform their resistance or impedance change to a voltage change. The *bridge* can transform a small change in resistance or impedance to a large change in voltage. The bridge circuit is capable of detecting changes on the order of 0.1%.

The basic four-arm DC bridge is shown in Figure 8.70. No current flows through the meter when the following resistance ratio is met:

$$\frac{R_1}{R_3} = \frac{R_2}{R_4} \tag{8.28}$$

When no current flows through the meter, the bridge is said to be in a *null state*.

In transducer applications, one of the resistors of Figure 8.70 is replaced by the transducer. Since the three remaining resistors are constant, any change in the bridge null state is caused by the transducer. Furthermore, the resistance of the transducer can be calculated at null by solving Equation 8.28 for the unknown resistance.

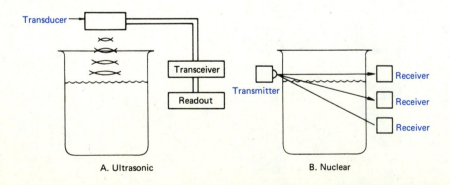

A. Ultrasonic B. Nuclear

FIGURE 8.69 Liquid-Level Measurement by Radiation

TABLE 8.8 Summary of Liquid-Level Sensors

Type	Name	Use*	Temperature Range	Pressure (lb/in.²)	Accuracy	Range
Sight	Tape float	C	To 149°C	300	±⅛ in.	To 60 ft
	Displacer	P, C	To 260°C	300	±¼ in.	10 ft
	Sight glass	P, C	To 260°C	10,000	±½%–1%	6-8 ft
Pressure	Diaphragm	P, C	To 459°C	Atmospheric	±1 in.	Unlimited
	Differential pressure	P, C	To 649°C	To 6000	±½%–1%	—
	Bubbler	C	Dew point	Atmospheric	±1%–2%	Unlimited
Electrical	Resistive electrode	P	–26°–82°C	To 5000	±⅛ in.	To 66 ft
	Capacitance	P, C	–26°–982°C	To 5000	±1%	To 20 ft
Radiation	Sonic	P, C	–40°–149°C	To 150	±1%–2%	To 125 ft
	Ultrasonic	P, C	–26°–60°C	To 320	±1%–2%	To 12 ft
	Nuclear	P, C	To 1648°C	Unlimited	±1%–2%	Unlimited

*P = point; C = continuous.

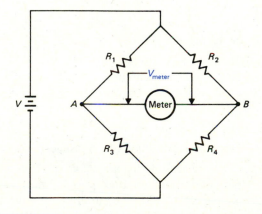

FIGURE 8.70 Four-Arm DC (Wheatstone) Bridge

If R_4 is the unknown resistance, it is related to the other resistances in the bridge by the following equation:

$$R_4 = \frac{R_2 R_3}{R_1}$$

For example, if $R_1 = 75 \ \Omega$, $R_2 = 100 \ \Omega$, and $R_3 = 150 \ \Omega$, then R_4 is 200 Ω.

A resistance change causes a bridge to become unbalanced. The imbalance is in the form of a volt-age between points A and B. This difference in potential causes current to flow through the meter. The potential difference (V_{meter}) between A and B can be approximated by

$$V_{meter} = V \left[\frac{R_2 \Delta R_4}{(R_2 + R_4)^2} \right]$$

where
$\quad V$ = supply voltage
$\quad \Delta R_4$ = change in resistance of R_4
$\quad R_4$ = original resistance of R_4

Let us suppose that the bridge is balanced with $R_1 = 100 \ \Omega$, $R_2 = 200 \ \Omega$, $R_3 = 400 \ \Omega$, and $R_4 = 800 \ \Omega$. The supply voltage is 100 V. If R_4 changes by 10 Ω, then V_{meter}, previously at 0 V, will be at about 0.2 V, as shown in the following calculation:

$$V_{meter} = V \left[\frac{R_2 \Delta R_4}{(R_2 + R_4)^2} \right]$$

$$= (100 \ \text{V}) \left[\frac{(200 \ \Omega)(10 \ \Omega)}{(200 \ \Omega + 800 \ \Omega)^2} \right]$$

$$= 0.2 \ \text{V}$$

Another helpful relationship to know is the bridge's sensitivity. The *sensitivity* of a bridge is the change in output voltage compared with the transducer's change in resistance. That is, the sensitivity (*s*) of a bridge (in volts per ohm) is equal to the change in voltage between points *A* and *B* (V_{meter}) divided by the change in the transducer's resistance (ΔR_4):

$$s = \frac{\Delta V_{meter}}{\Delta R_4} = V\left[\frac{R_2}{(R_2 + R_4)^2}\right]$$

The sensitivity of the bridge in our example is 0.2 V/10 Ω, or 20 mV/Ω. In other words, for every ohm that R_4 changes, the voltage V_{meter} changes about 20 mV.

CONCLUSION

We have covered a great many sensors in this chapter. All sensors convert the variable to be measured to an analog of that variable. As we have seen, the most common conversion is to an electrical quantity because electric control systems predominate in industry.

You should have a firm grasp of the transduction principles behind each sensor. Although you probably will not be asked to choose a sensor for a system, you will have to work with existing sensors in control systems. Thus, you will need a knowledge of the operating principles of the sensor.

In Chapter 10, the sensor is assumed to be an integral part of a process control system.

■ QUESTIONS •

1. A transducer is a device that converts a _____ into an output that facilitates measurement. The mercury thermometer converts temperature to _____.

2. The mechanical temperature transducers operate on the principle that gases and liquids change _____ when heated. The potential produced by the junction of two dissimilar methods is used in the _____ temperature sensor.

3. The psychrometer assesses relative humidity by comparing the _____ of a wet and dry bulb. The hair hygrometer element changes _____ with varying relative humidity.

4. The displacement sensor with a primary wound between two secondary windings is called a(n) _____. The bonded-wire strain gauge changes its _____ with strain.

5. Two devices used to measure magnetic field strength are the Hall device and the _____.

6. The _____ pressure transducer is more suited to measuring high pressures. A bellows converts pressure to _____.

7. The most commonly used differential pressure flowmeter is the _____ _____. Differential

pressure flowmeters use the _____ effect to generate a pressure difference.

8. The two types of liquid-level measurement are called continuous and _____. The electronic level probe is more suited to _____ measurement.

9. Define temperature.

10. Describe the function of a transducer.

11. Compare sensors that are active (give an output voltage with input energy) and those that are passive (give no output voltage). Give an example of each of the two types of sensors from among the temperature transducers.

12. Name the parts of a filled-system, temperature-measuring device.

13. Describe the behavior of a thermistor when it is heated. What type of temperature coefficient does it have? Is its response linear?

14. Describe how a thermistor can be used in (a) a fluid flow system, (b) a liquid-level system, and (c) time delay. *Hint*: Does the thermistor change its resistance immediately when heated?

15. How can a thermistor be made more linear? Name two methods.

16. Of the temperature sensors discussed in this chapter, which is the most sensitive?

17. Describe how a thermocouple is used to measure temperature.

18. Describe the operational principle behind the bimetallic strip.

19. Describe how the following humidity sensors work: (a) hair hygrometer, (b) resistance hygrometer, and (c) microwave transceiver.

20. Define the gauge factor for a strain gauge.

21. Describe two types of differential pressure transducers.

22. Describe measurement of liquid level and fluid flow using differential pressure transducers.

23. Compare the operation of the Bourdon tube with that of the bellows.

24. List the maximum ranges of the bellows, diaphragm, and Bourdon tube.

■ PROBLEMS ●

1. A type J (iron-constantan) thermocouple measures 6.907 mV across the device. Find the temperature of the test junction if the reference junction is at (a) 0°C and (b) 25°C.

2. The temperature of a type J thermocouple is 400°C. Find the voltage across the thermocouple if the reference junction is at 0°C.

3. A straight bimetal strip is 20 cm long, 0.1 cm thick, and clamped at one end. How much would it move if it was made of brass-invar and heated to 15°C?

4. A spiral bimetal strip has a radius of 10 cm and a thickness of 0.2 cm, and it is made of metals with coefficients of 17 millionths/°C and 5 millionths/°C. The spiral is 50 cm long, and it deflects by 1 cm. What is the heat change that produced this deflection?

5. A straight brass-kovar bimetal strip is 20 cm long and 0.2 cm thick. If the strip is straight at 25°C and the temperature increases to 100°C, find the deflection of the strip.

6. A spiral brass-kovar bimetal strip is 55 cm long and 0.2 cm thick. If the spiral has a radius of 10 cm at 25°C and if the temperature increases to 80°C, find the deflection of the strip.

7. Given a type J thermocouple, with a reference junction at 0°C and an output voltage of 5.485 mV, find the temperature of the reference junction.

8. Given a type J thermocouple with a reference junction at 0°C, find the voltage output you would expect from the thermocouple at 250°C.

9. Given a type J thermocouple with a reference junction at 85°C and a measuring junction producing 14.502 mV, find the temperature of the measuring junction.

10. An iron-constantan thermocouple produces an output voltage of 10.055 mV with the reference junction at 20°C. Find (a) the output voltage you would expect if the thermocouple reference junction was at 0°C and (b) the temperature of the measuring junction.

11. Given a type J thermocouple with a reference junction at 0°C and a measuring junction at 450°C, find the estimated output voltage from the thermocouple.

12. A thermocouple has a mass of 0.008 g, a specific heat of 0.08 cal/g°C, a heat transfer coefficient of 0.05 cal/cm-s°C, and an area of 0.05 cm². What is its thermal time constant?

13. A strain gauge has an initial resistance of 200 Ω. Under a strain of 10^{-4}, resistance increases to 202 W. What is the gauge factor?

14. A spiral bimetal strip has a radius of 6 cm, a length of 40 cm, and a diameter of 0.075 cm. If the metals are aluminum and invar, what will the deflection be for a 25°C change?

15. A thermistor at room temperature (25°C) is placed in an oven with a temperature of 100°C. With a time constant of 0.5 s, what will the thermistor's temperature be after 1.5 s?

16. A thermistor has a mass of 0.015 g, a specific heat of 0.05 cal/g°C, a heat transfer coefficient

of 0.028 cal/cm-s°C, and an area of 0.2 cm^2. Find the thermal time constant.   ,133 sec

17. A thermistor with a time constant of 2.5 s at a temperature of 20°C is immersed in boiling water ($T = 100$°C). Find the temperature of the sensor (a) after 1 s and (b) after 2 s.

18. A thermistor has a time constant of 1.0 s. If the thermistor starts at 30°C and is immersed in boiling water ($T = 100$°C), how long will it take the thermistor to heat to 75°C?

19. Given the op amp thermistor thermometer in Figure 8.17, find the output voltage with the thermistor at 75°C.

20. An RTD has a resistance of 150 Ω at 0°C. If the RTD has a temperature coefficient of resistance of 0.00412, what is the resistance of the RTD at (a) 25°C and (b) 50°C?

21. If a transistor has a V_{BE} of 0.600 V at 25°C, what is the base-emitter voltage at 100°C?

22. Given a platinum RTD with a temperature coefficient of 0.00392 and an original resistance of 125 Ω at 0°C, find the resistance at 85°C.

23. For the RTD thermometer circuit in Figure 8.21, with the sensor at a temperature of 45°C and with $R_0 = 100$ Ω, find the circuit output voltage.

24. For the RTD thermometer circuit in Figure 8.21, with the sensor at a temperature of 55°C and with $R_0 = 100$ Ω, find the circuit output voltage.

25. Given an LM335 at a temperature of 85°C, find the output voltage you would expect at that temperature.

26. Given an LM335 with an output voltage of 2.28 V, find the temperature of the sensor (in degrees Celsius).

27. The LM335 has a voltage out that rises 10 mV for every degree Celsius of temperature change. If the device is calibrated so that it has an output voltage of 2.98 V at 25°C, what will the output of the device be at (a) 55°C and (b) 75°C?

28. A strain gauge has an original, unstrained resistance of 150 Ω and a gauge factor of 1.8. If the gauge undergoes a strain of 2 μcm/cm, and assuming an increase in resistance occurs, what is the new resistance?

29. In Problem 28, if Young's modulus is 1.0×10^8 kPa, what is the stress (in kilopascals)?

30. A steel bar was compressed with a deformation of 5000 microstrains. If the original length of the bar was 0.75 m, what is the length after the force is applied?

31. A gauge has a resistance of 150 Ω when unstrained, a gauge factor of 2.0, and a strain of 2000 microstrains. Find the change in resistance.

32. A gauge has a resistance of 150 Ω when unstrained, a modulus of elasticity of 20.7×10^{10} N/m^2, a gauge factor of 3.0, and a change in resistance of 0.25 Ω. Find the stress on the gauge in newtons per square meter and in pounds per square inch.

33. What is the output voltage of the bridge in Figure 8.42A if the gauge is the one referred to in Problems 28 and 29 and is strained with a 2 μcm/cm strain? Assume $R_A = R_B = 100$ Ω, $R_C = 150$ Ω, and the bridge starts out balanced. ($V_i = 10$ V)

34. A crystal made of lithium sulfate has a pressure of 5.0×10^5 N over a surface area of 0.005 m^2, and the crystal thickness is 0.001 m. What is the output voltage of the crystal?

35. A pressure sensor at the bottom of a tank filled with liquid is used to measure the depth of the fluid. If the tank contains water, what would the pressure be if the tank was filled to a level of 10 m? How deep would the tank be if the sensor measured a pressure of 1.0×10^5 N/m^2?

36. A tourmaline crystal is 1.0 mm thick. Find the voltage when the pressure is 15,000 psi.

37. Given a bellows at the bottom of a tank of gasoline, with a specific gravity of 0.7 and a pressure of 15 psi, find the height of the gasoline (in feet).

Optoelectronics

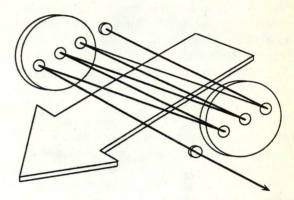

OBJECTIVES

On completion of this chapter, you should be able to:

- Describe the nature of light.
- Define radiometry and photometry.
- Contrast spontaneous emission and stimulated emission.
- Calculate the wavelength of a photon.
- Calculate the series resistance of an LED.
- Calculate the following laser parameters: frequency difference between longitudinal modes, coherence length and number of coherent wavelengths, divergence, focus beam spot radius, and peak and average power in a laser pulse train.
- List the four different types of lasers.
- Classify at least one laser from each type as either CW or pulsed and give the wavelength of that laser.
- Identify the schematic symbols for all optoelectronic receivers.

INTRODUCTION

Light, like sound, is a wave that can transport energy without transporting mass. Unlike sound, however, light does not have a basis in mechanics. Sound is the vibration of the molecules of a material medium produced by mechanical forces, so its properties can be derived from the Newtonian laws of mechanics. A light wave, on the other hand, is not the vibration of a material substance. Light is fundamentally different from sound, and its properties cannot be derived from the laws of mechanics. Nevertheless, a wave is a wave and light wave behavior applies as well to light as to sound.

In this chapter, we will discuss some important properties of light that can be understood without a detailed knowledge of optics, photometry, and stress analysis. You already have a good knowledge of electricity and its measurement. To understand optoelectronic devices, you must also have a working knowledge of light, its characteristics, and how it is measured.

THE NATURE OF LIGHT

Aristotle (384–322 B.C.) knew that sound is caused by vibrations in the air. He probably based this knowledge on the observation that music is produced by vibrating strings. In fact, the study of the relation of musical tones to the length of the vibrating string was well developed in the ancient world. Of course, an adequate explanation of sound waves was not possible until the time of Newton. Though nothing was known about the fundamental nature of light in Newton's time, it was natural to speculate that light was a wave, similar to sound. Christian Huygens (1629–1695), who lived at the same time as Newton, developed a wave theory of light.

Newton himself favored a theory according to which light is composed of massless particles (corpuscles). Newton supposed that these corpuscles travel through space at constant speed and that there is a different type of corpuscle for each color. His main objection to the wave theory was that light, unlike sound, does not appear to bend around corners. Newton's great reputation and the absence

of any hard evidence one way or another led to the general acceptance of the corpuscular theory during the eighteenth century.

The wave nature of light was finally established by a series of experiments that demonstrated that light obeys the superposition principle. These experiments were first performed by Thomas Young (1773–1829), the great Egyptologist, and later and more clearly by Augustin Fresnel (1788–1827). (You may wish to refer to a text on physics for a summary of the experiments these two men performed.)

At the time of these advances in the study of light, important discoveries were being made in the fields of electricity and magnetism. The basic laws of electricity and magnetism had been discovered in the first half of the nineteenth century. These laws were made into a comprehensive mathematical theory by James Clerk Maxwell (1831–1879). From this theory, Maxwell deduced that there should exist electromagnetic waves consisting essentially of oscillating electric and magnetic fields that propagate through space with a definite speed. According to Maxwell's theory, this speed is given in terms of certain well-known electric constants that enter into the theory. When Maxwell calculated the speed of electromagnetic waves from these constants, he found it equal to the speed of light. This great success of Maxwell's theory established that light is a form of electromagnetic radiation.

Today, physicists are familiar with electromagnetic waves with wavelengths ranging from less than 10^{-17} m to more than 10^4 m. Only waves with wavelengths in the narrow range between 4×10^{-7} and 7×10^{-7} m are detected by the human eye and thus constitute visible light. Waves with longer and shorter wavelengths have special names, such as radio waves, microwaves, infrared waves, ultraviolet waves, and X-rays.

Figure 9.1 shows the range of electromagnetic radiation that has been "light," together with the names given to different regions. Except for the visible region (400–700 nm), the boundaries between the regions are not sharply defined. Note that in the visible light spectrum, the wavelengths are measured in nanometers. One nanometer is 1×10^{-9} m and is equal to 100 μm. In older texts, the nanometer is called the *millimicron*. This term comes from the

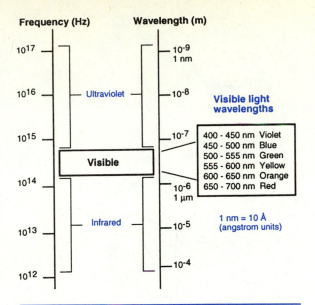

FIGURE 9.1 Electromagnetic Spectrum— Visible and Invisible Light

combination of the prefix milli- (10^{-3}) with the micron (10^{-6}). Another older unit, the angstrom unit (Å), is sometimes used and represents 1×10^{-10} m. For example, a 690 nm red wavelength would equal 6900 Å. You should be able to convert between the different units used for the measurement of the wavelengths of light. A summary is shown in Table 9.1.

TABLE 9.1 Units of Measurement for Light Wavelengths

Unit of Measurement	Abbreviation	Quantity
Angstrom unit	Å	1×10^{-10} m
Nanometer	nm	1×10^{-9} m
Millimicron	—	1×10^{-9} m
Micron	μ	1×10^{-6} m

• EXAMPLE 9.1

Given a light with a wavelength of 4000 Å, find the wavelength in nanometers.

Solution

Since one angstrom unit is equal to 10^{-10} m and one nanometer is equal to 10^{-9} m, the angstrom units must be divided by a factor of 10:

$$\text{wavelength (in nanometers)} = \frac{\text{angstrom units}}{10}$$

$$= \frac{4000 \text{ Å}}{10}$$

$$= 400 \text{ nm}$$

• EXAMPLE 9.2

Given a light with a wavelength of 750 nm, find the wavelength in microns.

Solution

Since one nanometer is equal to 10^{-9} m and one micron is equal to 10^{-6} m, the wavelength in nanometers must be divided by a factor of 1000:

$$\text{wavelength (in microns)} = \frac{\text{nanometers}}{1000}$$

$$= \frac{750 \text{ nm}}{1000}$$

$$= 0.75 \text{ μ}$$

RADIOMETRY AND PHOTOMETRY: MEASURING LIGHT

There are two basic systems in common use for describing and making measurements in optical systems: photometric and radiometric. The radiometric system was used for infrared applications by engineers more familiar with metric (or MKS) terminology. The basic units of the radiometric system are the watt and the centimeter; all other units within the system are derived from these units.

Radiometry

Radiometry is the measurement of the electromagnetic energy (radiant energy) emitted by a source or the energy falling on a detector. The radiation can be in the infrared and ultraviolet regions of the electromagnetic spectrum (Figure 9.1) as well as in the visible region. Figure 9.2 shows a typical radiometric system, which includes an emitter, a medium of transmission, and a detector.

The total output power of the emitter is called *radiant flux*, Φ_e, and is expressed in watts (W). Remember that one watt equals one joule per second. The radiant power, P_e, is the rate at which electromagnetic energy is emitted by a source. The radiant flux parameter tells us nothing about the distribution of the power emitted. We need to find out how much power is in each unit of solid angle. *Radiant intensity*, I_e, is the amount of power given off by an emitter per unit of solid angle and is expressed in watts per steradian. A *steradian* (ψ) is a dimensionless unit much like a radian. A graphic representation of a steradian is shown in Figure 9.3. For a solid angle with the apex at the center of a sphere, the size of the solid angle, in steradians, is the ratio of the area cut on the surface of the sphere divided by the square of the radius of that sphere, see Figure 9.3A. A sphere thus contains 4π steradians, analogous to a circle having 2π radians.

We can visualize the concept of the steradian by taking the radius of the sphere, say 10 cm, and squaring that value (100 cm²). To visualize the concept of a steradian, refer to Figure 9.3B. Imagine that we cut a cloth in the shape of a circle with an area of 100 cm². Let's assume that the sphere is made out of foam rubber. If we laid that cloth on the surface of the sphere and cut down to the center of the sphere, what shape object would we have? We would have cut an object in the shape of a cone. That cone contains one steradian (one unit) of solid angle.

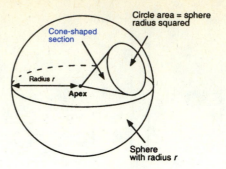

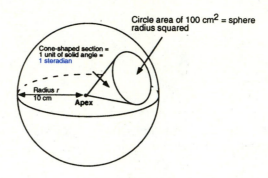

A. General Example

B. Specific Example

FIGURE 9.3 Pictoral Representation of a Steradian

With the concept of the steradian, we can now express the amount of power directed toward any area in space. When we know the amount of power a source emits in one steradian, we call this concept *intensity*. For example, if we measure 10 W of power coming out of the cone in Figure 9.3B, we say that the radiant intensity of the source is 10 W/sr (10 watts per steradian).

Radiant intensity is a light measurement that refers to what is called a *point source*. A point source is an emitter that is assumed to have no surface area. With a point source, we consider that the light comes from a single infinitely small point. A point source radiates equally in all directions. Many sources, like LEDs (covered later in this chapter), are considered an area source. An *area source* is a source that emits light from a clearly defined area. If the source of light is a flat surface (without any lenses), this type of area source is also called a *Lambertian* source. It

FIGURE 9.2 Radiometric System

is a property of a flat Lambertian source that the maximum radiant intensity occurs along the normal to the surface. (Recall that the normal is an imaginary line perpendicular to the surface.) If a Lambertian surface is viewed at angles other than head-on, the radiant intensity decreases as the cosine of the angle from the normal. The intensity of the source when viewed from path 1 (the normal) is greater than that of path 2, as shown in Figure 9.4.

It should be obvious at this point that we cannot use the radiant intensity parameter of a point source to describe the intensity of an area source. The intensity measurement for an area source is called the *radiant sterance* or *radiance*, L_e. Radiant sterance is defined as the radiant intensity of a source per unit of surface area of that source. It is measured in watts per steradian–square meter. It is interesting to note that for a Lambertian emitter, the radiance is constant at all viewing angles. Although the radiant intensity decreases as the cosine of the viewing angle, the apparent area of the source also decreases in the same way; thus, the ratio is constant. You can observe how the apparent area decreases by viewing a piece of paper, as shown in Figure 9.5. As the paper is tilted, the apparent or projected area decreases. If we continue to tilt the paper until we view the edge, the apparent area is zero. The final emitter parameter in common use is the term *radiant exitance*, M_e, formerly called *radiant emittance*. Radiant exitance is

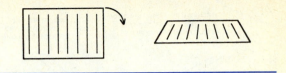

FIGURE 9.5 Apparent or Projected Area of an Area Source

defined as the density of the radiant power emitted from a surface. Its unit of measurement is in watts per square meter (W/m^2).

All of the parameters we have discussed to this point have related to emitters. When working with detectors, we must be able to measure the amount of power falling on a surface. The *irradiance* or *radiant incidence*, E_e, is defined as the amount of power arriving at the detector per unit of the detector's surface area. Radiant incidence is generally measured in either watts per square meter (W/m^2) or watts per square centimeter (W/cm^2). Irradiance is the radiometric term for what we call intensity. Radiant power and irradiance are related by the inverse square law. As the separation between the source and detector increase, the incidence decreases as the inverse of the square of the separation distance. Care must be taken here, as the radiant intensity of the source does not change as the detector moves further away.

Irradiance is best measured by a thermal detector, such as a *bolometer*, which absorbs all the radiation incident on a black nonreflecting element and converts it into internal thermal energy, thus increasing the temperature of the element. In a bolometer, the element is a blackened strip of platinum, the electrical resistance of which varies with temperature. The change in temperature of the strip is measured by measuring its resistance, and this, together with knowledge of the strip's heat capacity, determines the total energy absorbed. Bolometers and other thermal detectors are accurate devices for measuring irradiance because their response is independent of the wavelength of the incident radiation. Other electrical detectors, such as photoconductors, are more convenient to use, but their responses are highly sensitive to wavelength.

All the terms just discussed are related pictorially in Figure 9.6.

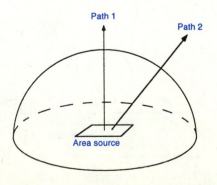

FIGURE 9.4 An Area Source (Lambertian)

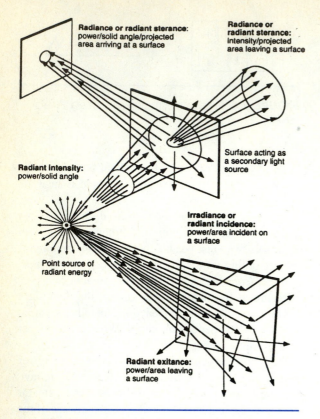

Radiance or radiant sterance: power/solid angle/projected area arriving at a surface

Radiance or radiant sterance: intensity/projected area leaving a surface

Surface acting as a secondary light source

Radiant intensity: power/solid angle

Irradiance or radiant incidence: power/area incident on a surface

Point source of radiant energy

Radiant exitance: power/area leaving a surface

FIGURE 9.6 Radiometric System with Terminology

Photometry

Photometry is the measurement of visible light as it appears to a human being with normal vision. The same physical concepts used in the radiometric system are used in the photometric system. Concepts such as intensity, optical power, and sterance all have photometric counterparts. The photometric system was originally designed for use with optical systems involving visible radiation and the human eye.

The response of the human eye to the brightness of an illuminated surface depends on the wavelength of the light coming from the surface. That is, for the same amount of illumination, different spectral colors appear to have different degrees of brightness. In 1924, after psychophysical measurements on many

subjects in different countries, the CIE (Commission Internationale d'Eclairage) adopted standard values for the relative efficiency of the human eye for light of different wavelengths. These values, plotted in Figure 9.7, officially define the *standard* observer. The efficiency curve of the standard observer is used to convert radiometric measurements into photometric measurements, just as the chromaticity diagram is used to convert a measurement of the spectral composition of a light into a measurement of its hue and saturation.

The radiometric power measurement is the watt. In the photometric system, the output power of the emitter is called *luminous flux*, Φ_v. (Note that the subscript v indicates the measurement is photometric.) The unit of measurement of luminous flux is the *lumen* (lm). Spectral luminous power, P_v, is the luminous power per wavelength at a particular wavelength. It is measured in lumens per nanometer (lm/nm).

The luminous power of a source depends on its radiant power—that is, on the amount of electromagnetic radiation emitted—and on the efficiency of this radiation to produce the sensation of brightness in the human eye. Figure 9.7 shows that light at 555 nm is most efficient. There is no practical light source that emits all its radiant power as visible light. Certain light sources produce more visible light than others as a proportion of the total radiant power emitted. For

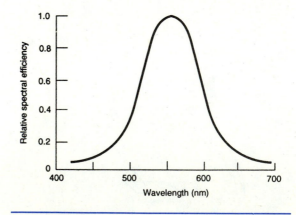

FIGURE 9.7 Relative Spectral Efficiency of the Standard Observer—CIE Curve or "Standard Eyeball"

example, only 7.5% of the radiant power of a 60 W incandescent light bulb is in the visible region of the spectrum. A 40 W fluorescent lamp, on the other hand, emits 20% of its radiant power as visible light.

Again, we need to know something about an emitter's optical power distribution into space. The unit of measurement describing this quantity in the radiometric system is the watt per steradian and is a measure of radiant intensity. The photometric term is *luminous intensity*, I_v, and is expressed in lumens per steradian, or candelas. One candela is equal to one lumen per steradian. You can visualize this concept by referring back to Figure 9.3. The intensity of the source is one candela when a total optical power equal to one lumen emerges from the cone (one steradian). The lumen is the SI unit that replaces the older unit called the *candela*.

As in the radiometric system, we need a way to express an area source emitter's intensity. The photometric unit expressing this concept is called *luminous sterance* or *luminance*, L_v. Luminance is defined as the luminous intensity per unit area leaving an emitter in a given direction. The surface area is the projected area, as viewed from the specified direction. Luminous sterance is measured in lumens per steradian–square meter, or candelas per square meter. Sometimes called *photometric brightness*, luminous sterance has two common units: the *nit*, equaling one candela per square meter, and the *stilb*, equaling one candela per square centimeter. The other important area source parameter is the source's *luminous exitance*, M_v. Luminous exitance is the luminous power leaving a unit area of surface. It is measured in lumens per square meter and was formerly called *luminous emittance*.

Turning from the source to the detector, we have the measurement called *illuminance*, E_v. Illuminance, more commonly called *luminous incidence*, is the luminous power per unit area incident on a surface (detector). Its SI unit is the lumen per square meter (lm/m^2), which is called a *lux* (lx). When one lumen falls on a surface with an area of one square meter, we say that the incidence is one lux. Other units of illuminance in common use are the lumen per square foot (lm/ft^2), called a *foot-candle* (fc), and the lumen per square centimeter (lm/cm^2), called a *phot* (ph).

Accurate electrical measurements of illuminance

are difficult because no electric device has the same response to radiation as the human eye. Some photoconductors, however, have response characteristics similar enough to that of the eye to make them valuable for monitoring illumination. A photoresistor (discussed later in this chapter) is made of a material whose electric resistance increases when exposed to radiation of certain wavelengths. Cadmium sulfide (CdS) is a widely used photoresistor, often used in photographic light meters. Its response characteristics are similar to those of the standard observer (Figure 9.7), except for tails in the ultraviolet and infrared regions. With the use of appropriate filters, these regions can be largely eliminated, and the response of the photoconductor can approximate that of the standard observer better.

All the terms just discussed are related pictorially in Figure 9.8.

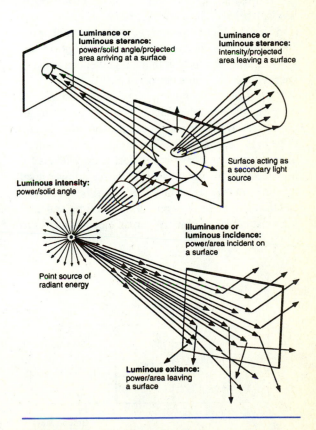

FIGURE 9.8 Photometric System with Terminology

EMITTERS

This section will introduce you to two popular optoelectronic emitters: the light-emitting diode (LED) and the laser. Before discussing these devices, we must have a clear understanding of how light is created at the atomic level.

Light Physics

Radiation and matter interact in several different ways. Recall that the atom consists of a small, dense nucleus and one or more electrons in motion around the nucleus. Electrons can only exist in certain predetermined orbits around the nucleus. Each of these individual orbits can be thought of as an energy level. Thus, an electron in a certain orbit has a certain discrete amount of energy.

In Figure 9.9, for example, note that four energy levels, labeled E_1 through E_4, exist in this atom. Note also that the higher the energy level, the higher the amount of energy an electron in this orbital possesses. For example, an electron in the E_2 level has a higher energy level than an electron in E_1.

Electrons normally occupy the lowest energy levels available. When electrons in an atom fill all the lowest available energy levels, the atom is said to be in its "ground state." Some electrons, however, may be forced to leave the lowest energy levels for higher levels. When an electron leaves a lower level for a higher one, a vacant place is left behind. Each electron can be in only one orbit at a time and cannot exist between orbits. When an electron moves to a level in an atom higher than the ground state, the atom is said to be "excited."

Absorption. Electrons may travel from one orbital to another. If an electron moves from a lower energy level to a higher energy level, energy must be added to cause that movement. When we add energy to an atom, causing an electron transition, we say that the atom has "absorbed" the energy. The electron's increased energy causes it to jump to a higher energy level. When an electron in an atom moves from the ground state to a higher level, we then say that the atom is in an "excited state." It is important to realize that the electron will only accept the precise amount of energy needed to move it from one energy level to a higher energy level. In Figure 9.9, for example, an electron in the ground state (E_1) must have a total of 12.1 eV added to make a jump to the E_3 orbital. Energies greater or less than 12.1 eV will leave the electron stationary.

Spontaneous Emission. Electrons in the atomic structure try to place themselves in the lowest possible energy state or level. An electron in an excited state will try to "de-excite" itself by one of several means. Just as the atom absorbs energy when an electron moves to a higher level, an atom with an electron in a higher level gives off energy moving from a higher state to a lower one. An electron may spontaneously move from a higher level to a lower one. The energy is emitted in the form of a photon of light. The photon released by the atom as it is de-excited will have a total energy exactly equal to the difference between the excited and lower energy levels.

For example, let us suppose that an electron moved from level E_3 to level E_2, as shown in Figure 9.10. The difference between energy levels is equal to Planck's constant times the speed of light, divided by the wavelength:

$$E_3 - E_2 = \frac{hc}{\lambda} \qquad \text{(9.1)}$$

where

  h = Planck's constant (6.63×10^{-34} J-s)
  c = speed of light (3×10^8 m/s)
  λ = wavelength of electromagnetic energy

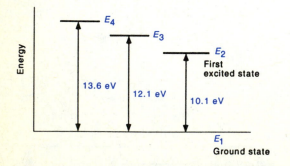

FIGURE 9.9 Hydrogen Atom Energy Levels

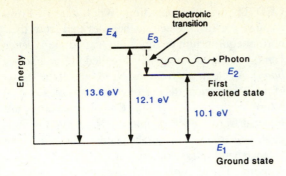

FIGURE 9.10 Electron Moving from Higher Energy Level to Lower Level

The wavelength of the photon energy can be found by rearranging Equation 9.1 to solve for wavelength:

$$\lambda = \frac{hc}{E_3 - E_2} \qquad (9.2)$$

$$= \frac{(6.63 \times 10^{-34} \text{ J-s})(3 \times 10^8 \text{ m/s})}{(12.1 \text{ eV} - 10.1 \text{ eV})(1.6 \times 10^{-19} \text{ J/eV})}$$

$$= 622 \text{ nm}$$

Note that since Planck's constant is given in joule-seconds and the energy difference is given in electron volts, the energy difference must be converted into joules. One electron volt is equal to 1.6×10^{-19} joules. The wavelength for the energy in this example is given as 622 nm, which falls within the orange spectrum. The release of a photon with this energy is called *spontaneous emission* or, sometimes, *fluorescence*. A familiar example of absorption and spontaneous emission can be seen in a phosphorescent material. Atoms of this type are excited by photons of just the right energy from a lamp or from the sun. Later, in the dark, the electrons in the excited atom spontaneously de-excite themselves, emitting photons of light. In this example, the exciting force is not of a unique energy, so electrons may be excited to many different levels. The photons released in de-excitation will then have many different discrete wavelengths. If enough discrete wavelengths are present, the light given off may appear to the eye as white.

EXAMPLE 9.3

Given that an electron falls from E_4 (13.6 eV) to E_3 (12.1 eV), find the wavelength of the photon given off from this transition.

Solution

Using Equation 9.2, we get

$$\lambda = \frac{hc}{E_4 - E_3}$$

$$= \frac{(6.63 \times 10^{-34} \text{ J-s})(3 \times 10^8 \text{ m/s})}{(13.6 \text{ eV} - 12.1 \text{ eV})(1.6 \times 10^{-19} \text{ J/eV})}$$

$$= 828.75 \text{ nm}$$

Light-Emitting Diodes (LEDs)

The *light-emitting diode* (LED) is one of the most common light sources used in industry today. Basically, LEDs are semiconductor diodes that when forward-biased, radiate optical energy. The energy may be visible, as in the V-LED (visible light-emitting diode), or invisible, as in the IR-LED (infrared light-emitting diode). The V-LED is useful as an indicator or warning light because it can be seen with the eye. The IR-LED produces radiation that is not visible to the eye but that can be detected by an infrared detector. Note the schematic diagram for the LED shown in Figure 9.11.

LEDs differ from other diodes because they emit light when current flows in the forward-bias direction. Conventional diodes do not emit light. In addition, the forward voltage for GaAsP LEDs is approximately 1.6 V while for conventional silicon diodes it is approximately 0.7 V.

FIGURE 9.11 LED Schematic Symbol

LED Structure. An LED is simply a PN junction in gallium arsenide (GaAs), gallium phosphide (GaP), or a combination of the two in a solid solution known as gallium arsenide phosphide (GaAsP). GaAsP is the most widely used today. In addition, zinc selenide (ZnSe) and silicon carbide (SiC) have been used recently for special colors.

The cross-sectional view of a typical GaAsP LED is shown in Figure 9.12. An epitaxial layer of GaAsP is grown on a wafer of monocrystalline GaAs. During the first 125 µm (1 mil) of growth, the concentration of GaP is increased from zero until the proper concentration of GaAs and GaP is achieved. This gradual increase is required to preserve the monocrystalline structure of the GaAs substrate. The final alloy contains approximately 60% GaAs and 40% GaP ($x = 0.4$ in the formula $GaAs_{(1-x)}P_x$). A more detailed description of the manufacturing processes involved can be found in most texts on semiconductors.

During the growth period, a material such as tellurium is added to provide N-type characteristics. The surface is then coated with Nitrox®, which is a layer of silicon dioxide over a layer of silicon nitride. Windows of the desired shape, either round, square, or rectangular, are next etched through this Nitrox®, using standard semiconductor photolithography techniques. Zinc is diffused into the N-type material, forming a thin layer of P-type material. Finally, an aluminum contact, with a "finger" pattern, is deposited on the P-type material. This pattern helps distribute the current evenly through the PN junction. The bottom of the chip is then coated with gold and germanium. These two substances become the electrical contacts for the LED. At this point, the structure is complete and ready for testing.

LED Theory of Operation. When current flows through the LED, electrons are elevated from their normal equilibrium energy state (called the *valence band*) to a higher energy level (called the *conduction band*). As electrons from the N side reach the P side of the PN junction, they recombine with holes as they return to the valence band. Conversely, holes are injected from the P side into the N side and recombine. This transition from the conduction band to the valence band releases energy in the form of heat and photons. These energy changes occur in discrete amounts called *quanta*.

A quantum change from the conduction to the valence band is called a *bandgap transition* and is accompanied by the release of a photon. A quantum change from the conduction to the valence band can also occur via intermediate levels. This situation, where the electron does not go directly from one band to another, is called a *nonradiative recombination*.

The energy of a photon released by a bandgap transition is equal to the bandgap energy in electron volts (eV). The wavelength of the photon determines the color of the emitted light. The wavelength of a photon released by a bandgap transition is given by the equation

$$\lambda = \frac{1240}{\varepsilon \, (eV)} \qquad (9.3)$$

where

λ = photon wavelength (in nanometers)
ε = photon energy (in electron volts)

Pure GaAs has a bandgap energy that produces photons in the infrared region at a wavelength of approximately 900 nm. Pure GaP produces photons in the visible green region around 550 nm. The alloy of 60% GaAs with 40% GaP used by some LED manufacturers produces photons in the visible red region at 655 nm. Table 9.2 shows the different types of LED materials and the wavelengths of the photons that they produce.

High-efficiency red LEDs can be made by using GaP material that is doped with oxygen. For the same amount of power in, oxygen-doped red GaP will produce more light than red GaAsP. GaP is transparent to its own light, unlike GaAsP or GaAs. This means that photons have greater probability of

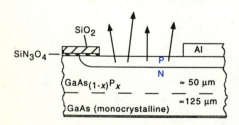

FIGURE 9.12   Cross Section of an LED

TABLE 9.2 LED Semiconductor Characteristics

Material	Dopant	Peak Emission Wavelength (nm)	Color
GaAs	Zn	900	IR
GaAs	Si	900–1020	IR
GaP	N	570	Green
GaP	N, N	590	Yellow
GaP	Zn, O	700	Red
GaAsP	—	650	Red
GaAsP	N	632	Orange
GaAsP	N	589	Yellow
SiC	—	490	Blue
ZnSe	—	490	Blue

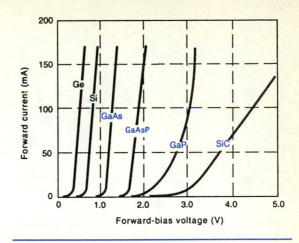

FIGURE 9.13 Forward Voltage-Current Curves for the LED and Other Diodes

escaping to the outside world in a GaP LED, thereby increasing the total light output. The light that normally would be lost in GaAsP due to absorption can now escape via internal reflections. This is why GaP LEDs are more efficient.

Keep one important thing in mind when high-efficiency GaP is used. GaP will "saturate"—that is, reach the point of constant light output for increased current input—sooner than GaAsP.

LED Voltage-Current Curves. The LED voltage-current curves are considerably different in comparison to those of other diodes. The forward-current curves are shown in Figure 9.13. Note the higher forward voltages for the gallium LEDs. The forward voltages for silicon and germanium are 0.7 and 0.3, respectively—much lower than those for the LEDs. The reverse-voltage behavior (not shown in the diagram) is also different for LEDs. Most Si diodes have peak inverse-voltage ratings greater than 100 V. Most LEDs have reverse-voltage maximums of between 3 and 5 V. You must take care to make sure that the LED is not damaged by applying a large reverse voltage.

At 20 mA average forward current, the brightness of GaAsP LEDs is usually adequate for visual indicators. The actual visibility of the lighted chip on a dark background is excellent under most viewing conditions. Remember, however, that LEDs are not designed for large-area illumination.

The colors presently available for LEDs range from invisible infrared to visible green. For GaAsP LEDs to be viewed by human observers, the color red has proven to be the most effective, even though the colors orange, yellow, and green can also be obtained. Red is used because it provides the best coupling between the relative efficiency of the human eye and the relative efficiency of gallium arsenide phosphide. If the human eye is to be the "detector" for an LED, we must consider the eye sensitivity as one of the factors to obtain maximum coupling efficiency. The diagram in Figure 9.14 shows simultaneous plots of the eye response and the efficiency of GaAsP. From this figure, we can see that optimum coupling with human vision occurs around 655 nm at a 40% concentration of GaP, which is why most LED manufacturers use this wavelength for red GaAsP LEDs. To get other colors, we must increase the concentration of GaP in GaAsP, even though the efficiency will be lower.

LED Circuits. Like other semiconductor diodes, the LED requires a series-limiting resistor, as shown in Figure 9.15. If we assume that $+V = 5$ V and that a

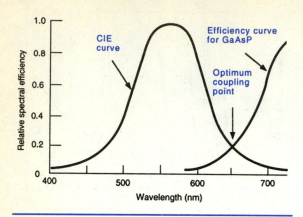

FIGURE 9.14   Relative Eye Efficiency and GaAsP versus Wavelength

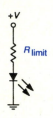

FIGURE 9.15   LED Current Limiting

GaAs LED has a 1.2 V drop across it when a current of 50 mA is flowing through it, the resistor R_{limit} will equal

$$R_{\text{limit}} = \frac{+V - V_F}{I_F} \qquad (9.4)$$

$$= \frac{5\text{ V} - 1.2\text{ V}}{50\text{ mA}} = 76\ \Omega$$

where
    $+V$ = supply voltage
    V_F = LED forward voltage
    I_F = LED forward current

• EXAMPLE 9.4

Given an LED with a forward voltage (V_F) of 1.3 V at 20 mA and a supply of +15 V, find the size of the series-limiting resistor.

Solution

Substituting into Equation 9.4, we obtain

$$R_{\text{limit}} = \frac{15\text{ V} - 1.3\text{ V}}{20\text{ mA}} = 685\ \Omega$$

Be careful when +V becomes small. When +V approaches V_F, small changes in the LED forward voltage can cause I_F to change drastically. A large increase in forward current could damage the diode. Damaging reverse voltage can be prevented by placing a silicon diode across the LED, as shown in Figure 9.16. When the supply goes negative, the silicon diode is forward-biased, clamping the voltage across the LED to 0.6 V.

LED Packages. Light-emitting diodes come in several popular packages. When LEDs were first brought to market, the package designs were borrowed from the incandescent lamp. The two most common sizes for LEDs are the T-1 and T-1¾ sizes. In this standard, called the *T-X designation*, the X stands for the base diameter in eighths of an inch. The T-1¾ size package is shown in Figure 9.17. Note

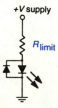

FIGURE 9.16   LED Reverse-Voltage Protection

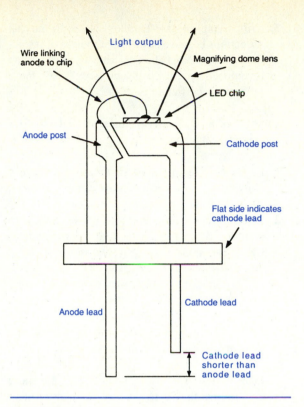

FIGURE 9.17 Plastic-Dome LED Package Construction (T-1¾)

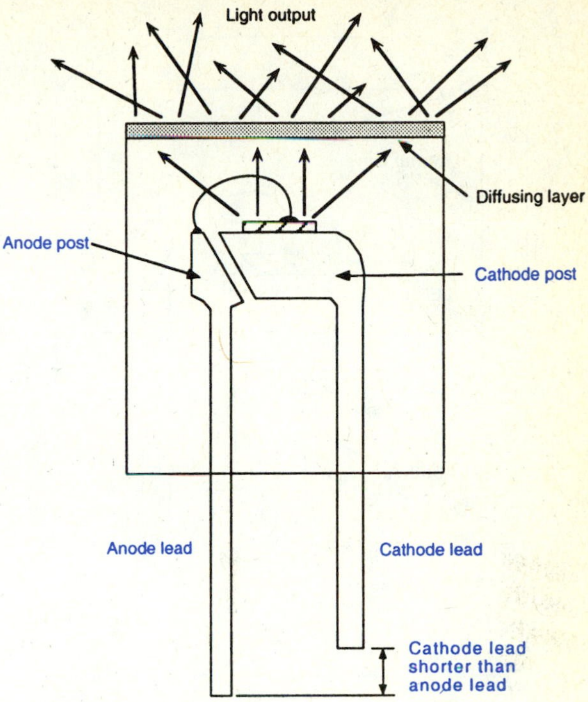

FIGURE 9.18 Rectangular LED Package Construction

the convention used to identify the cathode lead, which is shorter than the anode. Alternately, the cathode lead is the lead closest to the flat side of the dome. The T-1 package is similar to the T-1¾ size, only smaller.

Another popular LED package, as shown in Figure 9.18, is the rectangular plastic LED. It is designed for use in those applications where a cylindrically shaped LED is not effective. The rectangularly shaped LED is often used to illuminate a letter from behind, a process called *backlighting*. Designers also use this LED in bar graphs and in 7-segment displays. Note the similarity in construction to the standard LED. The rectangular-shaped

plastic case acts as a light pipe with a diffusing layer at the top. The light travels from the chip to the diffusing layer, where it exits the package.

Most LEDs use plastic packages. The plastic package, however, does not stand up well to adverse environments. For example, temperature fluctuations can cause small cracks in the plastic, which, in turn, can allow moisture into the LED. Extreme temperature changes can cause stresses on the wire that attaches the anode lead to the chip, causing separation. The hermetically sealed LED (Figure 9.19) can stand up to more severe temperature changes and will not allow moisture into the LED. Many military and industrial applications demand such a package.

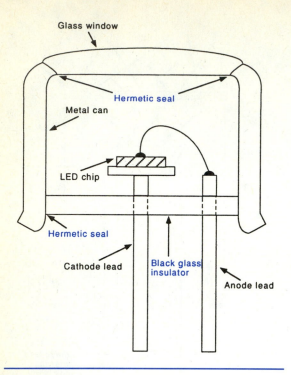

FIGURE 9.19 Hermetically Sealed LED Package

5. Critical angle loss—This loss is due to internal reflections because of different indices of refraction of GaAsP and air. The critical angle loss can be reduced by using a plastic lens. Plastic has a refraction coefficient of 1.55.

Advantages and Disadvantages of LEDs. LEDs offer significant advantages:

1. LEDs can be mounted in locations where a conventional socket/indicator cannot be used, such as inside a molded block of plastic.
2. LEDs dissipate less heat, thus allowing them to be mounted near heat-sensitive components.
3. LEDs can stand more physical abuse because they are solid-state devices.
4. LEDs require relatively low voltage and current from the circuit.
5. LEDs have a longer life expectancy.
6. LEDs do not radiate RF noise, so they can be used near RF-sensitive devices.

One limitation is the fact that LEDs are more sensitive to heat than conventional filament or neon indicators. With an increase in heat, the light output will go down slightly due to the LEDs temperature coefficient of light output ($-1\%/°C$). When the LED is cooled, the light output will return to normal.

LED Efficiency. Several factors determine LED efficiency:

1. Quantum efficiency, or percentage of recombinations that produce photons—This percentage of recombinations may range from 10% to 50% for GaAs and from 1% to 10% for GaAsP.
2. I^2R loss, or contact resistance and bulk resistance loss—This loss is significant only at high current levels.
3. Absorption of photons within the chip—The P layer (shown in Figure 9.12) is very thin to decrease the absorption of the semiconductor material.
4. Fresnel, or optical mismatch loss—This loss is due to internal reflections. In this case, part of the light generated within the chip is not transmitted. This loss may range from 15% to 30%.

Laser Physics

Since its discovery in 1960, the laser has seen increasing use in scientific, educational, and industrial applications. Seldom has a discovery in the area of applied physics had so great an impact on so many areas of life. Since the laser's discovery, thousands of laboratories have discovered thousands of applications for this versatile device. Its versatility has led to applications in the fields of medicine, biology, chemistry, and industrial material processing. In many cases, the applications discovered have been difficult or impossible to achieve with other technologies.

This section will discuss the basic physics behind laser operation. Later, the four basic types of lasers in common use today will be discussed.

Stimulated Emission. In 1917, Albert Einstein proved in theory a concept of great importance to the theory of the laser. Scientists at that time knew that an atom in an excited state would eventually decay to a state of lower energy. During the decay, a photon of energy would be released. As we have seen, this natural decay is called *spontaneous emission*. Einstein showed that emission could be caused by a photon striking an excited atom. The photon released by the excited atom is identical in frequency, energy, direction, and phase to the triggering photon. This type of emission, called *stimulated emission*, causes two photons to leave the atom—the original photon and the stimulated one. Where there was one photon, now there are two. Stimulated emission is illustrated in Figure 9.20.

In Figure 9.20A, we see a photon striking an excited atom. In Figure 9.20B, we see the result of the interaction between the incoming photon and the excited atom. Two photons leave where only one

entered. These two photons leave and trigger nearby atoms into stimulated emission. It is from this process that we get the acronym *laser*: light amplification by stimulated emission of radiation. Since one photon enters and two photons leave, stimulated emission is really light amplification, as suggested by the laser name.

Stimulated emission will only occur if certain conditions are met. First, the incoming photon must have the same energy as the photon it is stimulating. Second, the electron must stay in its excited, high-energy state long enough for a photon to come along and stimulate it to decay down to a lower level. In other words, large numbers of atoms must not spontaneously emit photons. Large numbers of atoms must remain excited long enough to be stimulated to emit photons.

A summary of the three different processes involving the interaction of photons with atoms is shown in Figure 9.21.

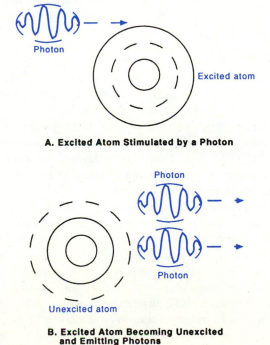

FIGURE 9.20 Stimulated Emission

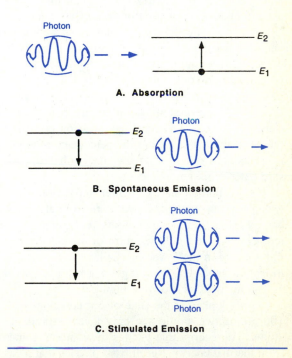

FIGURE 9.21 Interaction of Photons with Atoms

In 1958, A. L. Shawlow and C. H. Townes published a paper proposing that light could be produced by stimulated emission the way that microwave radiation had been produced a few years earlier. In 1960, T. H. Maiman built the first optical laser. It was a very simple device—a ruby rod, silvered on both ends, surrounded by a spiral-wound flash lamp similar to those found in modern photographic flashes. Soon after Maiman produced the ruby laser, other materials that could act as lasing media were discovered. Not long after the invention of the ruby laser, the helium-neon (HeNe) laser was developed as well as the carbon dioxide (CO_2) laser.

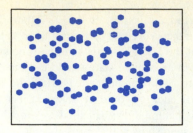

FIGURE 9.22 One Hundred Atoms in a Container

Requirements for a Laser

All lasers must have three basic elements: a lasing medium, an energy source, and an optical resonator. Each one of these fundamental components will be discussed in detail.

Lasing Media. Only certain materials serve well as lasing media. Only materials that can undergo a situation called *population inversion* can qualify as lasing media. After Einstein predicted that stimulated emission was possible, little interest was aroused in the scientific community. It was thought that stimulated emission could not be controlled because the process occurs in less than one-millionth of a second. If only two of several million atoms are in an excited state, the chances of stimulated emission occurring are very small. The greater the percentage of atoms in an excited state, the greater the probability for stimulated emission to occur.

In the normal state of matter, most atoms in a group are not excited. If we were to analyze the energy levels of a group of atoms kept at a constant temperature, we would see most of the electrons in those atoms in the lower energy levels. To see this more clearly, let us image a container with 100 atoms in it, as shown in Figure 9.22.

If we looked at the state of our container with 100 atoms, we would see only a few stimulated atoms. The energy that excites these atoms comes from thermal energy. Recall from physics that a gas in a container is constantly in motion. The amount of motion depends on the temperature; the higher the temperature, the more motion in the atoms. As the atoms energetically bounce off each other and the walls, certain numbers of atoms will get enough energy from the collisions to excite electrons to higher orbitals. The number of excited atoms in a group of atoms depends on the temperature. The number of atoms in an excited state is predicted by Boltzmann's law.

We can see clearly how many atoms would be in an excited state by looking at the chart in Figure 9.23A. The chart shows us that the number of atoms with electrons in high energy levels is less than the number in low energy levels. Atoms with electrons in lower energy levels are more numerous. This kind of distribution is called an *equilibrium distribution*. It is the normal distribution of atoms with the temperature constant. If we raise the temperature, the shape of the curve would be similar, as we can see in Figure 9.23B. Note that the highest numbers of atoms are still in the ground state. The numbers of atoms in higher states decrease as we go to higher energy levels. To make a laser, we must have a population of atoms different from the ones we have just described. Stimulated emission will not produce significant amplification of light unless a condition called *population inversion* occurs. We must have an inverted population. In the normal equilibrium population, we have more atoms in an unexcited state than in an excited state. Most electrons reside in the ground state or lowest energy levels, and the higher energy levels are depopulated. We can change this distribution by adding energy, but we must add energy in very precise, discrete amounts.

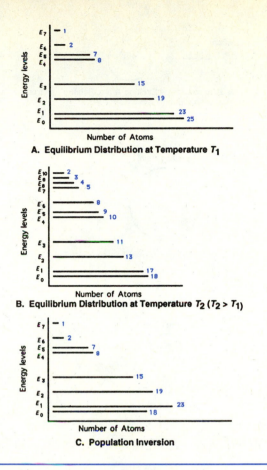

FIGURE 9.23 Energy Distribution in a Population of 100 Atoms

tion causes electrons in an atom to jump to a higher energy level. The excited atom is now ready to give off a photon by stimulated emission.

One of the most common optical pumps is the xenon flashtube, similar to the kind used in photographic flashes and strobe lights. Xenon gas is placed in quartz tubes at low pressures. A high voltage is placed across the tube, causing current to flow through the gas. The current produces a high-temperature plasma, which emits a great number of photons with many different wavelengths. Large numbers of photons with different wavelengths and energies ensure that some photons of the correct energies will be present. Designers of lasers make sure that the gas used in the flash produces enough photons that have the right energy and wavelength to produce population inversion in the lasing medium.

Flash lamps are fashioned into one of three types—linear, helical, and U-shaped—as shown in Figure 9.24. All lamps have two electrodes, one at either end. The high-voltage potential is applied across these two electrodes. The current produced by the potential flows from one electrode through the gas to the other electrode. Although xenon gas is most commonly used in laser flashtubes, sometimes krypton and helium gases are used instead. Krypton and helium gases provide a brighter flash.

How is population inversion achieved? We must add a lot of energy to the lasing medium. This energy, furthermore, must be in discrete amounts. Energy in discrete amounts ensures that electrons in the ground state are raised to the proper levels within the atom.

Energy Sources. The addition of energy to a laser that causes population inversion is called *pumping.* Laser designers use three common pumping methods. A brief discussion of each method follows.

Optical Pumping: In this method, a light source illuminates the lasing medium, usually a gas or a solid. The electrons from the lasing medium absorb incident photons of the correct energy. This absorp-

A. Straight Flashtube

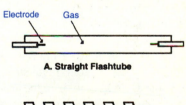

B. Helical Flashtube

C. U-shaped Flashtube

FIGURE 9.24 Flashtube Construction

Electron Collision Pumping: Modern lasers use two types of collision pumping—electron and atomic pumping. Consider the container of atoms shown in Figure 9.22. If we focus an electron gun at the container, we can strike the atoms in it with electrons. If the electrons have enough energy and strike atoms in the ground state, the atoms will be excited. Electrons within each excited atom will have absorbed energy from the moving electrons. The moving electrons transfer their energy to the atoms by colliding with them. The kinetic energy (energy of motion) of the electrons is converted to the excitational energy of the atoms in the container. If enough electrons are fired at the container, a population inversion will occur.

Electron collision pumping is most often used in gas lasers. Some experimental lasers use free electrons. A free-electron laser uses electrons that are not bound to an atom or a molecule. More often, electrons for electron collision pumping come from an electrical discharge. Since an electrical discharge can be sustained continuously, the output of the laser using this type of pumping can be a continuous beam.

Atomic Collision Pumping: This type of collision pumping uses excited atoms or molecules. Instead of electrons colliding with atoms, two atoms collide. The sequence of events is shown in Figure 9.25. First, an electron, usually from an electrical discharge, collides with an atom, exciting the atom (Figure 9.25A). Next, the excited atom crashes into the atom from the lasing medium, transferring energy from the exciting atom to the lasing atom (Figure 9.25B). After the collision, the exciting atom is in the ground state, and the lasing atom is in the excited state (Figure 9.25C). If enough excited atoms are created, a population inversion occurs.

Optical Cavities. The last requirement for a laser is an optical or resonant cavity. After we pump the lasing medium and produce a population inversion, lasing will begin. To get light amplification, however, we must have a resonant cavity. We can see what happens in a nonresonant cavity in Figure 9.26. In Figure 9.26A, we see a collection of atoms in the ground state. The black dots are excited atoms, and the white circles are unexcited atoms. We can see that most of the atoms in Figure 9.26A are in the ground

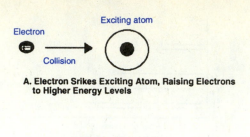

A. Electron Srikes Exciting Atom, Raising Electrons to Higher Energy Levels

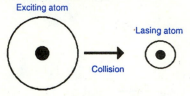

B. Exciting Atom Strikes Lasing Atom, Transferring Energy

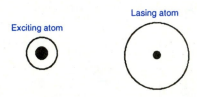

C. End Product is an Excited Lasing Atom

FIGURE 9.25 Atomic Collision Pumping

state. The diagram in Figure 9.26B shows the lasing medium being pumped. The pumping of the lasing medium raises electrons in atoms to higher energy levels. Figure 9.26C shows electrons in excited atoms dropping down to lower energy levels and emitting photons. The photons traveling in the horizontal direction strike other atoms in excited states and so emit photons. This process increases until the photons reach the far right wall of the container. At this point, if the container is glass, all the photons pass out of the container. No further amplification of photons occurs. Any photons not traveling in a horizontal direction will also pass out of the container. Amplification of the photons can be increased by adding two parallel mirrors, as shown in Figure 9.26D.

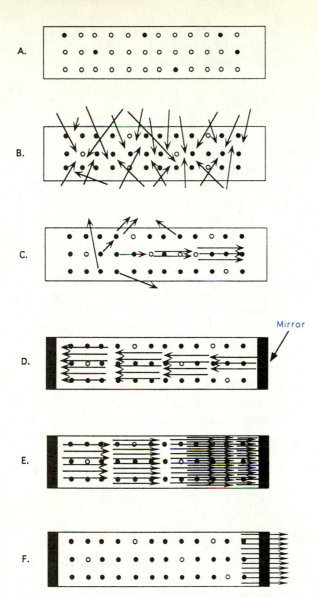

FIGURE 9.26 Laser Stimulated Emission

The amplification process of Figure 9.26C will continue as the photons strike the right mirror and are reflected back. The reflected photons cause more stimulated emission, and the number of photons continue to grow (Figure 9.26E). The photons will bounce back and forth between the mirrors and will

continue to increase in number. If the mirrors at either end of the cavity were 100% reflective, no light would escape the laser. The laser would, therefore, have no output. In a practical laser, one of the mirrors is made 100% reflective. The other mirror, called the *output mirror*, is made only partially reflective. The amount of reflection depends on the type of laser. In general, the output mirror is made reflective enough to sustain stimulated emission, while transmitting the rest of the photons to the outside world. If the pumping is continuous, the laser will reach a state of equilibrium. The number of photons produced by the atoms raised to the excited state and the number of photons emitted and lost will come into a balance. The photons emitted through the output mirror emerge as a powerful beam of parallel rays of light (Figure 9.26F). The other photons remaining inside the resonant cavity continue to generate more photons. The emission will continue as long as a population inversion is maintained among the lasing-medium atoms or molecules.

Laser Beam Characteristics

As we have suggested, the light leaving the laser is different from other sources of light, such as the LED and the fluorescent light bulb. Laser light is different from other light sources in four ways. Laser light is monochromatic, coherent, highly directional, and very intense.

Monochromaticity. Laser light is very close to being monochromatic. The term *monochromatic* means one color. All other conventional light sources produce more than one wavelength of light. For example, a halogen lamp produces a series of spectral lines. If we closely examined the spectral lines, we would find that they are not sharp. The lines are spread over a band of wavelengths. The energy-level characteristics of the hydrogen atom produce this spread. The narrow spread of wavelengths in the laser beam results from the special way in which the light is generated.

A white light source, say from a xenon flashtube, produces light over the whole visible spectrum, from 390 to 770 nm. The bandwidth of this spectrum is about 380 nm. A red LED has a broad spectrum output, perhaps about 50 nm wide. The bandwidth of a laser is very low because the light created by the laser is almost monochromatic and because the numbers of photons of different wavelengths are few. Most lasers generate only a few different frequencies. The number of frequencies generated depends on the characteristics of the lasing medium and the geometry of the resonant cavity. These different frequencies of oscillation are called *longitudinal modes*. The diagram in Figure 9.27 shows two waves of different frequencies oscillating in a laser tube. We have shown only two modes oscillating in this resonant cavity. In some lasers, the number may be high as 5000. Some special lasers may be tuned so that only one mode oscillates.

Most lasers have between 6 and 5000 longitudinal modes oscillating. But since the difference between the mode frequencies is small, the bandwidth of the laser output is small. The actual difference in frequency is defined by the equation

$$\Delta f = \frac{c}{2d} \qquad (9.5)$$

where

Δf = difference in frequencies between adjacent modes (in hertz)
c = speed of light (in meters per second)
d = distance between mirrors at either end of optical cavity (in meters)

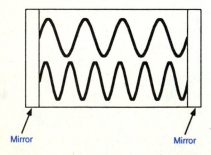

FIGURE 9.27 Two Longitudinal Modes Oscillating in a Laser

As an example, let us say that we have a laser cavity 10 cm long. Using Equation 9.5, the difference between frequencies is then

$$\Delta f = \frac{c}{2d} = \frac{3 \times 10^8 \text{ m/s}}{2\,(10 \text{ cm})} = 1.5 \text{ GHz}$$

At first glance, we might think that this is a large difference in frequency. It is, but not compared to the frequency of light. Suppose that we are working with a helium-neon laser, which produces photons with a wavelength of 632.8 nm. The frequency of the light at this wavelength is

$$f = \frac{c}{\lambda} = \frac{3 \times 10^8 \text{ m/s}}{632.8 \text{ nm}} = 474{,}000 \text{ GHz} \qquad (9.6)$$

A difference in frequencies of 1.5 GHz seems large. Compared to the frequencies of light lasers used, however, it is only a small fraction—in this case, about three-thousandths of a percent.

• EXAMPLE 9.5
Given a laser with a length of 1 m, find the frequency difference between longitudinal modes.

Solution
Using Equation 9.5, we get

$$\Delta f = \frac{c}{2d} = \frac{3 \times 10^8 \text{ m/s}}{2\,(1 \text{ m})} = 150 \text{ MHz}$$

Coherence. Laser light is said to be coherent. The term *coherent* describes a particular relationship between two waveforms. Two waves with exactly the same frequency, phase amplitude, and direction are said to be coherent. Such waves are "in step" with each other, as shown in Figure 9.28.

Each of these waves has certain characteristics, such as amplitude and wavelength. The amplitude of each of the sinusoidal waves varies with time, constantly changing its rate of change sinusoidally. The maximum amplitudes and wavelengths remain the same as the wave moves outward. If we were to

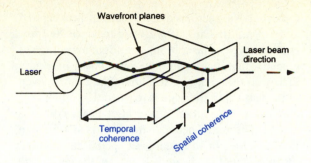

FIGURE 9.28   Coherence in a Laser

in phase both along the direction of propagation (temporal) and along any direction perpendicular to the direction of propagation (lateral).

How long will a wave stay coherent? We can make an approximation of the distance over which a coherent wave will stay coherent. The length or distance over which a wave will remain coherent is equal to the velocity of the wave divided by the bandwidth:

$$L_c = \frac{c}{BW} \qquad (9.7)$$

where

    c = velocity of light (in meters per second)
    BW = bandwidth (in MHz)
    L_c = coherence length (in meters)

Let's take a helium-neon (HeNe) laser as an example. Most HeNe lasers have a bandwidth of about 1500 MHz. If the speed of light is about 3×10^8 m/s, the coherence length is, by Equation 9.7,

$$L_c = \frac{c}{BW} = \frac{3 \times 10^8 \text{ m/s}}{1500 \text{ MHz}} = 0.2 \text{ m}$$

To determine the number of wavelengths over which the wave train remains coherent, we can divide the coherence length by the wavelength n, so

$$n = \frac{c}{(BW)\,\lambda} \qquad (9.8)$$

where

    n = number of coherent wavelengths
    c = speed of light (in meters per second)
    λ = wavelength of coherent radiation (in nanometers)

freeze the waveforms in time, each of the two coherent waveforms would be at the same point in their sinusoidal variation. For example, if we stopped one waveform in time and it was exactly at its peak, the same would be true of the other waveform. This particular type of coherence is called *temporal coherence*. Another type of coherence, called *lateral coherence*, can be seen in Figure 9.29. Imagine that we are looking at the front of the laser, seeing the sine waves propagating toward us. We would see two vertical lines, perfectly perpendicular. If the waves are laterally coherent, they will not turn away from the perpendicular. At any point, the distance between the two lateral points on the waves will be the same. If we combine these two types of coherence, lateral and temporal, we get the concept of *spatial coherence*. Two waves are spatially coherent when they are

Going back to the example of the HeNe laser, which produces waves with a wavelength of 632.8 nm, the number of coherent wavelengths is, by Equation 9.8,

$$n = \frac{c}{(BW)\,\lambda} = \frac{3 \times 10^8 \text{ m/s}}{(1500 \text{ MHz})\,(632.8 \text{ nm})}$$

$$= 316,056 \text{ wavelengths}$$

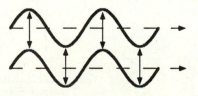

FIGURE 9.29   Lateral Coherence

• EXAMPLE 9.6

Given an argon laser with a wavelength of 488.0 nm and a bandwidth of 1000 MHz, find the coherence length and the number of coherent wavelengths.

Solution

By Equation 9.7,

$$L_c = \frac{c}{BW} = \frac{3 \times 10^8 \text{ m/s}}{1000 \text{ MHz}} = 0.3 \text{ m}$$

By Equation 9.8,

$$n = \frac{c}{(BW)\,\lambda} = \frac{3 \times 10^8 \text{ m/s}}{(1000 \text{ MHz})\,(488.0 \text{ nm})}$$

$$= 614{,}754 \text{ wavelengths}$$

Divergence. Light coming out of a laser tends to stay together in a beam. It does not spread out, or *diverge*, as much as do other sources of light. Laser light does not spread out as much as other sources because only beams that are parallel to each other leave the laser. We need to bear in mind that the beams of light produced by the laser are not perfectly parallel. There is always some bending of the light, a phenomenon known as *diffraction*. All sources of light will, therefore, have some divergence, or beam spread.

The beam spreading of the laser is usually measured in milliradians, or 1×10^{-3} radians. Since there are 2π radians in a circle, one milliradian equals about three minutes of arc, or 0.05°. A

typical HeNe laser has a divergence of between 0.5 and 1.5 milliradians. If we were to view the most common type of laser beam, we might expect to see a spatial distribution like the one shown in Figure 9.30A. In this energy distribution, the beam is even in intensity until the edge is reached. Unfortunately, the power distribution of the laser beam is not so simple. The actual beam distribution is as shown in Figure 9.30B. In an actual laser spatial distribution, the intensity of the beam is highest at the center and decreases in intensity away from the center. The actual intensity is described by the equation

$$I = I_o e^{-2x/w} \qquad (9.9)$$

where
   I = intensity of beam at some distance from center
   I_o = intensity at center
   x = distance from center
   w = beam radius or waist

The *beam waist* is the measurement across the beam at its narrowest point, as shown in Figure 9.31.

For illustration, let's take an example of a beam with a waist of 1 cm and an intensity of 10 W/sr. What will the intensity be 0.5 cm out from the center? Using Equation 9.9, we obtain

$$I = I_o e^{-2x/w}$$

$$= (10 \text{ W/sr})e^{-1}$$

$$= 3.68 \text{ W/sr}$$

The intensity at a distance of 0.5 cm would be about 0.37 of what it was at the center.

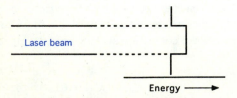

A. Apparent Energy Distribution in Laser Beam

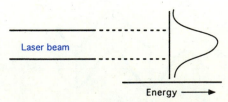

B. Actual Energy Distribution in Laser Beam

FIGURE 9.30   Energy Distribution in a Laser Beam

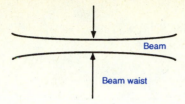

FIGURE 9.31   Measurement of Beam Waist

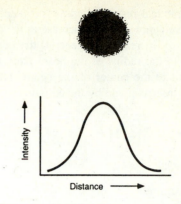

FIGURE 9.32   Gaussian Energy Distribution of a Laser Beam

• EXAMPLE 9.7

Given a laser with a beam waist of 2.5 cm and an intensity of 100 W/sr, find the intensity 1 cm out from the center of the laser.

Solution

By Equation 9.9, we get

$$I = I_0 e^{-2x/w}$$

$$= (100 \text{ W/sr})e^{-0.8}$$

$$= 44.9 \text{ W/sr}$$

A plot of the intensity with respect to distance is shown in Figure 9.32. The shape of this curve is sometimes called *Gaussian*, or bell-shaped. Our eyes see the edge of the beam at the point where the intensity drops to $1/e^2$ or about 14%. At this point, we think we see the edge of the beam. The beam divergence angle of a Gaussian laser beam is given by the equation

$$q = \frac{4\lambda}{\pi w_0} \qquad (9.10)$$

where

    q = angle of divergence (in radians)
    λ = wavelength of laser
    w_0 = beam waist

An example at this point may clarify this equation. An HeNe laser has a wavelength of 632.8 nm and a beam waist of 1 mm. What is the divergence of the beam in radians? Using Equation 9.10, we find

$$q = \frac{4\lambda}{\pi w_0}$$

$$= \frac{4 \ (632.8 \text{ nm})}{\pi \ (1 \text{ mm})}$$

$$= 0.0008 \text{ rad} \quad \text{or} \quad 0.8 \text{ millirad}$$

We can see that the laser beam diverges at an angle determined by several factors.

• EXAMPLE 9.8

Given an argon laser with a beam waist of 2.5 mm and a wavelength of 514.5 nm, find the divergence of the beam in radians.

Solution

By Equation 9.10, we have

$$q = \frac{4\lambda}{\pi w_0}$$

$$= \frac{4 \ (514.5 \text{ nm})}{\pi \ (2.5 \text{ mm})}$$

$$= 0.00026 \text{ rad} \quad \text{or} \quad 0.262 \text{ millirad}$$

One practical result of laser divergence is that the laser beam spot size will increase as the distance from the laser increases. In general, the beam spot size refers to the radius of the beam from the $1/e^2$ power point to the center of the beam. The beam radius of a laser is given by the equation

$$w = w_o \sqrt{1 + \left(\frac{x\lambda}{\pi w_o^2}\right)^2} \qquad (9.11)$$

where

w = beam radius at distance x
λ = wavelength
w_o = beam waist at its smallest point

An example at this point will be useful. Suppose that we are designing a reflector for our laser. We need to know the size of the beam to be reflected. If the beam waist is 1 mm, the distance is 1 km, and the laser beam is an HeNe laser with a wavelength of 632.8 nm, what will the size of the reflector need to be to reflect most of the energy? Using Equation 9.11, we find

$$w = w_o \sqrt{1 + \left(\frac{x\lambda}{\pi w_o^2}\right)^2}$$

$$= (1 \times 10^{-3} \text{ m})$$
$$\sqrt{1 + \left[\frac{(1 \times 10^3 \text{ m})(632.8 \times 10^{-9} \text{ m})}{\pi(1 \times 10^{-3} \text{ m})^2}\right]^2}$$

$$= 0.2 \text{ m}$$

Our reflector should have a diameter of about 0.4 m at a distance of 1 km from the laser.

• EXAMPLE 9.9
Given a laser with a beam waist of 2 mm, a distance of 10 km, and a wavelength of 514.5 nm, find the size of the reflector needed to reflect most of the energy.

Solution
By Equation 9.11, we find

$$w = w_o \sqrt{1 + \left(\frac{x\lambda}{\pi w_o^2}\right)^2}$$

$$= w_o \sqrt{1 + \left[\frac{(10 \times 10^3 \text{ m})(514.5 \times 10^{-9} \text{ m})}{\pi(2 \times 10^{-3} \text{ m})^2}\right]^2}$$

$$= 0.82 \text{ m}$$

A laser beam may be focused by a lens. A lens can either expand the beam spot size or contract it, depending on the type of lens used. The diverging concave lens expands the beam spot size, while the converging convex lens allows the beam size to decrease. The focus spot size for a Gaussian laser beam focused by a lens is given by the equation

$$w = \frac{2\lambda}{\pi w_o} \qquad (9.12)$$

where

w = beam spot radius when in focus
λ = wavelength of laser
w_o = beam waist diameter before it strikes lens

As an example, take an HeNe laser with a beam waist size of 1 cm. What will the beam spot radius be? Using Equation 9.12, we find

$$w = \frac{2\lambda}{\pi w_o}$$

$$= \frac{2(632.8 \times 10^{-9} \text{ m})}{\pi(1 \times 10^{-2} \text{ m})}$$

$$= 40.3 \text{ μm}$$

The focused spot size is approximately 40.3 μm.

• EXAMPLE 9.10

Given a laser with a beam waist of 2 mm and a wavelength of 514.5 nm, find the in-focus beam spot radius.

Solution

By Equation 9.12, we get

$$w = \frac{2\lambda}{\pi w_0}$$

$$= \frac{2(514.5 \times 10^{-9} \text{ m})}{\pi(2 \times 10^{-3} \text{ m})}$$

$$= 163.8 \text{ } \mu\text{m}$$

Intensity. The laser beam produces very intense light. Although the total amount of energy produced by a laser is small, the stimulated emission process concentrates the light produced. Each atom giving up a photon synchronizes with photons from other atoms. Coherence makes the amplitude of the beam very high. In a normal tungsten-filament light source, individual tungsten atoms produce photons randomly. Electric heating excites atoms of tungsten. Each tungsten atom decays spontaneously to lower energy levels, each emitting a photon. An atom emitting a photon must then be re-excited. The re-excitation process may take some time, or it may occur quickly. The tungsten-filament source produces an irregular stream of photons, all randomly emitted. Contrast that with the laser, which emits photons by stimulated emission.

Another reason the laser beam is so intense relates to its optical cavity. Only those photons that are emitted in the horizontal direction make more than one pass down the optical cavity.

As we have seen earlier in this chapter, the intensity of a point source is the power emitted per unit of solid angle. For an area source, the intensity of an area source, called *sterance*, is the power emitted per unit area per unit of solid angle. Since the laser concentrates all its energy into a very narrow beam, the laser source appears very bright. For

purposes of comparison, the sun has a sterance of approximately 130 W/cm^2-sr. HeNe lasers produce a sterance of about 10^6 W/cm^2-sr. Bear in mind that some pulsed lasers can have a sterance of greater than 10^{18} W/cm^2-sr.

In calculating these intensity figures, we need to keep in mind that energy is a measure of the capacity of a device to do work. We measure energy in joules. Power is the rate at which work is being done and is commonly measured in watts. One watt equals one joule per second, and one joule equals one watt-second. A laser, therefore, that emits 10 J in 1 s can be classified as a 10 W laser. If those same 10 J are emitted as a single pulse with a duration of a hundredth of a second, the laser would then be a 1000 W laser. Sometimes, the output of pulsed lasers are indicated in terms of the number of joules per second.

Up to this point, we have been assuming that the power from a laser beam distributes itself in a Gaussian, or bell-shaped, curve. As shown in Figure 9.32, the power of the Gaussian beam is highest at the center of the beam and decreases as we go away from the center. The actual beam shape depends on the wavelength of the photons; the curvature, spacing, and alignment of mirrors; and the diameter of the laser tube. The electromagnetic fields that the laser emits can have many different cross-sectional shapes. The different cross-sectional shapes a beam may have are called *transverse electromagnetic modes*, abbreviated TEM. The most common TEM is the Gaussian shape, TEM$_{00}$. The TEM$_{00}$ shape is shown in Figure 9.32. The subscripts in the TEM nomenclature relate to the number of nulls (low-power areas) in the beam's spatial distribution. In this nomenclature, the first subscript refers to a null in the horizontal direction; the second subscript refers to a null in the vertical direction. Refer to Figure 9.33 to see this concept applied.

Note that the beam labeled TEM$_{01}$ (Figure 9.33A) has a null in the horizontal direction but not in the vertical. The beam labeled TEM$_{10}$ (Figure 9.33B) shows a null in the vertical but not in the horizontal direction. The TEM$_{11}$ mode of operation (Figure 9.33C) has nulls in both

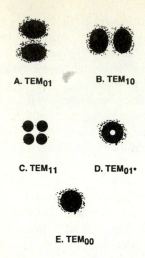

A. TEM$_{01}$

B. TEM$_{10}$

C. TEM$_{11}$

D. TEM$_{01*}$

E. TEM$_{00}$

FIGURE 9.33 Laser Transverse Excitational Modes

horizontal and vertical directions. Another mode seen sometimes in laser outputs is the TEM$_{01*}$ mode (pronounced T-E-M zero one star). Note that it has a donut-shaped pattern (Figure 9.33D). Most of the energy in the TEM$_{01*}$ mode is concentrated at the edge of the beam.

Laser Energy Delivery

Lasers deliver their energy in one of two basic methods: continuous-wave and pulsed. In *continuous-wave* (CW) output, the laser delivers its energy in a continuous stream of photons. The beam's power density is constant with time. Some lasers, like the HeNe laser, normally operate in the CW mode. In the *pulsed* mode of operation, the laser emits energy in a single burst. The laser usually repeats the burst at a certain rate, called the *pulse repetition rate* (PRR). Some lasers, like the ruby laser, normally operate in the pulsed mode.

As we have seen earlier in this chapter, laser output can be described in terms of power or in terms of energy. Energy is the ability to do work, while power is the rate of expending energy. Energy is measured in joules. The watt measures energy expended, with one watt equal to one joule per second. If a 10 W CW laser is turned on, it will

expend energy at a rate of 10 J/s. If left on for one minute, the energy expended will equal 60 s × 10 J/s, or 600 J. In the 10 W CW laser, the power measurement of 10 W refers to an average power measurement. When measuring pulsed lasers, we can measure power in two ways: We can measure *average* power, or we can measure *peak* power.

Look at the diagram in Figure 9.34. The peak power is measured over time interval t, while the average power is measured over the entire waveform period T. The average power is then

$$P_{avg} = \frac{E}{T} \qquad (9.13)$$

where

P_{avg} = average power
E = energy in the pulse
T = waveform period

The peak period is described by the equation

$$P_{pk} = \frac{E}{t} \qquad (9.14)$$

where

P_{pk} = peak power
E = energy in the pulse
t = time of pulse duration or pulse width

To illustrate the use of these equations, let's focus on an example of a pulsed laser. The laser we are working with has a pulse repetition rate (PRR)

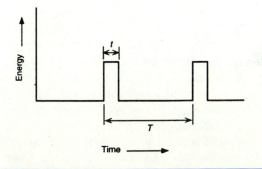

FIGURE 9.34 Pulsed Laser Output with Respect to Time

of 10 Hz and produces a 250 mJ pulse in 10 ns. The peak power in the pulse will be

$$P_{pk} = \frac{E}{t}$$

$$= \frac{250 \times 10^{-3} \text{ J}}{10 \times 10^{-9} \text{ s}}$$

$$= 25 \times 10^6 \text{ J/s} \quad \text{or} \quad 25 \text{ MW}$$

The average power over the entire period is

$$P_{avg} = \frac{E}{T}$$

$$= \frac{250 \times 10^{-3} \text{ J}}{0.1 \text{ s}}$$

$$= 2.5 \text{ W}$$

• EXAMPLE 9.11

Given a pulsed laser with a PRR = 1 kHz, producing a 100 mJ pulse in 1 μs, find the peak power of the pulse and the average power of the pulse train.

Solution

The peak power in the pulse will be, using Equation 9.14,

$$P_{pk} = \frac{E}{t}$$

$$= \frac{100 \times 10^{-3} \text{ J}}{1 \times 10^{-6} \text{ s}}$$

$$= 100 \times 10^3 \text{ J/s} \quad \text{or} \quad 100 \text{ kW}$$

The average power over the entire period is, using Equation 9.13,

$$P_{avg} = \frac{E}{T}$$

$$= \frac{100 \times 10^3 \text{ J}}{0.001 \text{ s}}$$

$$= 100 \text{ W}$$

Another way to calculate the average power is to multiply the peak power by the duty cycle:

$$P_{avg} = P_{pk} \times D \tag{9.15}$$

where
P_{avg} = average power
P_{pk} = peak power
D = duty cycle

The duty cycle (D) is the ratio of the time of the peak power to the entire period:

$$D = \frac{t}{T} \tag{9.16}$$

where
D = duty cycle
t = pulse width
T = waveform period

In our example, t is 10 ns and T is 0.1 s. The duty cycle is

$$D = \frac{t}{T} = \frac{10 \times 10^{-9} \text{ s}}{0.1 \text{ s}} = 0.1 \times 10^{-6}$$

Average power may then be calculated by using Equation 9.15:

$$P_{avg} = P_{pk} \times D$$

$$= (25 \times 10^6 \text{ W}) \times (0.1 \times 10^{-6})$$

$$= 2.5 \text{ W}$$

We can see from these examples that laser power can be increased by operating a laser in a pulsed mode rather than in a CW mode.

Types of Lasers

Earlier in this section we discussed basic laser principles. Now it is time to consider those lasers you will be likely to see in industry. We will discuss the output characteristics and how each laser works.

Four different types of lasers are in common use today. The four types are the solid-state, gas, liquid, and semiconductor lasers. We will consider representative lasers from each of these classifications.

Solid-State Lasers. Solid-state lasers use a rod made out of a solid material, usually a crystal or glass. The rod is cylindrically shaped with its ends polished. Normally, the term *solid-state* refers to semiconductor material. With lasers, however, it takes on a different meaning. In lasers, solid-state means that a lasing material has been added to a solid host. The glass or crystal material is the host material. The actual lasing material is a small amount of impurity added to the glass or crystal. The ions added to this host come from a transition metal such as chromium, nickel, or cobalt or from rare-earth materials such as neodymium or erbium.

T. H. Maiman, working at Hughes Aircraft, constructed and operated the first laser, a *ruby laser*, in 1960. The ruby, commonly known as a gem, is made up of aluminum oxide (Al_2O_3). Pure aluminum oxide is as clear as glass. To get a reddish color, the aluminum oxide is doped with *chromium oxide*. Ruby lasers contain chromium dioxide in a concentration equal to 0.05% by weight, which gives the crystal a pink color. Aluminum oxide doped with 0.05% of chromum contains about 1.6×10^{19} chromium atoms per cubic centimeter. The familiar ruby red color comes from a higher concentration of chromium, about 0.5% by weight.

Ruby crystal rods are seldom made in lengths greater than 6 in. This limitation in length is due to the difficulty in growing long crystals of good optical quality. Rods have a diameter of up to 1 in. Although rods can be made with a greater diameter, pumping rods with a large diameter is difficult. Rods with a diameter of greater than 1 in. tend to be pumped around the outside of the rod and not at the center. The ends of the ruby rods are polished, and mirrors are deposited on the ends of the rods. One of the mirrors will be close to 100% reflective. This means that nearly 100% of the light striking that mirror will be reflected back. The other mirror, the output mirror, has a reflectivity from 65% to 80%.

The chromium atom in the aluminum oxide crystal lattice structure is responsible for lasing. The chromium atoms absorb energy from the flashtube in a band centered around 545.1 nm. Electrons are raised from ground level E_1 to excited level E_3, as shown in Figure 9.35. From level E_3, electrons drop quickly to level E_2 in a radiationless phonon transition.

In a phonon transition, energy is given up in the form of heat. Recall from physics that heat is simply vibrations between atoms and molecules. Phonon transitions produce no photons. The electrons reside now in the E_2 level (metastable state). The electrons in E_2 stay there for a considerable amount of time, about 3 ms. A time of 3 ms may not seem to be a long time, but to an electron it is. A xenon flash lamp placed beside or around the ruby causes a population inversion, in which more electrons are in E_2 than at ground level.

The excited atoms in the E_2 level begin to de-excite spontaneously. During spontaneous de-excitation, electrons drop from E_2 to E_1 (ground level). Stimulated emission will take place since a population inversion is in effect. In any lasing medium, stimulated emission will occur in all directions. No particular direction of propagation is favored. As stated earlier, the lasing medium is placed within an optical cavity. The optical cavity controls the emission direction and increases the amount of energy in the pulse.

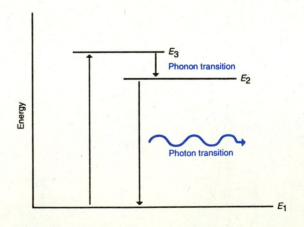

FIGURE 9.35 Atomic Energy Levels

Photons not emitted along the axis of the optical cavity will pass out of the cavity and be lost. If photons are aligned along the horizontal axis of the rod, they will eventually strike a mirror. Photons striking a mirror will be reflected back, passing again through the lasing medium. More passes through the medium trigger more excited electrons to undergo spontaneous emission. The pulse grows in size with more reflections. On each encounter with the less reflective mirror, a number of photons emerge from the laser. The ruby laser then emits high-intensity, coherent light through the output mirror. The wavelength of the emitted coherent light is 694.3 nm; this light is within the red spectrum.

The energy-level structure of the ruby laser is different from most other material used for lasers. It has a three-level structure, as shown in Figure 9.36. Electrons normally reside in the pump band, level E_3; the metastable state, level E_2; or the ground state, E_1. All electrons destined for the pump band in level E_3 come from the ground state. Almost all other lasers have a four-level system. The four-level systems take some electrons from a level higher than the ground state. Four-level systems take less energy to pump, therefore, than three-level systems. It is interesting to note, in light of this fact, that the first material to lase was the difficult-to-pump ruby.

Pulses from the ruby laser can go as high as 150 J but normally are in the tens of joules. The pulse from the ruby laser lasts only a few milliseconds since the pumping is not continuous. Continuous pumping is not possible in the ruby laser because the rod tends to overheat and crack. The xenon flashtube must provide several joules of energy per cubic centimeter to cause a population inversion in ruby crystals. The flash must be so intense that the flashtube heats the rod. Since ruby has poor thermal conductivity, the ruby cannot be cooled easily. Pulses from the ruby can be shortened to several nanoseconds by Q switching. Q-switching ruby lasers can produce pulses with power in the tens of megawatts. Ruby lasers may also be mode-locked, producing peak powers in the low-gigawatt range with pulse durations of 10 ps (picoseconds).

A schematic of a ruby laser system is shown in Figure 9.37. A charge is stored in a capacitor, which discharges quickly through the flash lamp when triggered. A bright wide-spectrum pulse of light from the flash lamp illuminates the nearby ruby rod, causing a population inversion. After triggering, the capacitor is recharged through the power supply. Pumping cannot be done at a rate of greater than 1–2 pulses per second. During the early 1960s, ruby lasers were very popular. After that time, several other lasers were invented that replaced the ruby laser in many applications. The popularity of the ruby laser, therefore, declined somewhat in the intervening years.

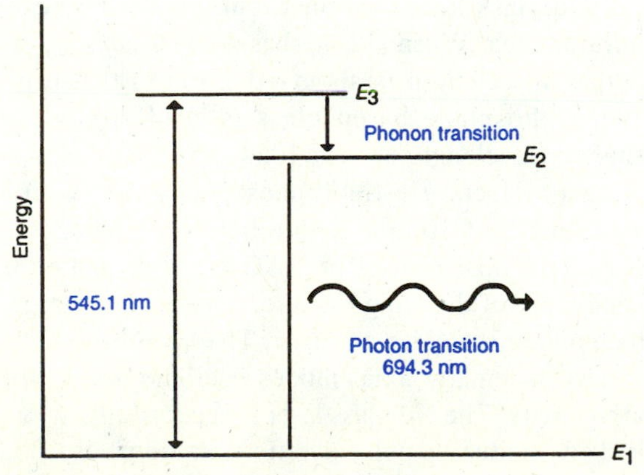

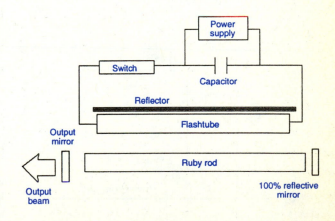

FIGURE 9.36 Ruby Laser Chromium Atom Energy Levels

FIGURE 9.37 Ruby Laser Schematic Diagram

The *YAG laser* is similar to the ruby laser in many ways. Like the ruby, it is pumped by a flashtube and uses a crystalline host. The host in the YAG laser, however, is yttrium (pronounced IT-tree-um) aluminum garnet ($Y_3A_{15}O_{12}$), not aluminum oxide. YAG, a much better host than aluminum oxide, is very hard and can be polished easily. Its most important advantage over aluminum oxide is its better thermal conductivity. The high thermal conductivity makes it possible to operate the YAG laser in the CW mode since the heat can be more easily removed. Like the ruby laser, the size of the rod is limited to short lengths because of difficulties in fabrication. The YAG rod takes much longer to grow than the ruby crystal and is also much more expensive. It is difficult to make YAG rods more than 6 in. long.

Like the ruby laser, the YAG rod is doped with an impurity that serves as the lasing material. The impurity in the YAG host is called *neodymium* (pronounced nee-oh-DIM-ee-um), symbolized by the abbreviation Nd. Neodymium, a rare-earth material, is added to the YAG crystal in a concentration of about 3% by weight. Its energy-level structure is shown in Figure 9.38. Note the four-level structure. In this system, electrons are pumped from level E_2

to E_4. In the E_4 state, electrons make a phonon transition to E_3. Stimulated emission takes place from the E_3 level. When electrons fall from the E_3 level to the E_2 level, a photon of light is produced, with a wavelength of 1.06 μm. This wavelength is in the invisible near-infrared area.

Nd:YAG lasers can be operated in the CW mode with power levels up to several hundred watts. The Nd:YAG laser can also be operated in the pulsed mode with PRRs up to several kilohertz. *Q* switching yields pulse energies from 100 to 150 mJ with pulse widths from 10 to 20 ns. Mode locking can produce pulse widths down to 30 ps.

In a *neodymium-glass laser*, glass serves as a host to the active neodymium lasant. Glass can be made in many different sizes and shapes, all with good optical quality at reasonable prices. Nd:glass lasers may have rods with a diameter as small as several micrometers to rods with lengths greater than 2 m. Most Nd:glass rods have diameters of about 1 cm. Since glass can be fabricated in large sizes, the Nd:glass laser is chosen where a burst of large power is needed. An Nd:glass laser can deliver an energy level greater than 5000 J. It has one major disadvantage—poor thermal conductivity. Glass has a thermal conductivity ten times lower than YAG. This makes CW operation of Nd:glass difficult since it is hard to remove the heat. Nd:glass lasers are, therefore, used in a pulse mode. They usually produce pulses of several joules at PRRs of up to 100 pulses per minute.

Nd:glass lasers produce outputs in the near-infrared area. When silicate-based glass is used, the output wavelength is about 1062 nm (1.062 μm). When phosphate-based glass is used, a slightly shorter wavelength around 1054 nm results. Newer Nd:glass lasers use fluorophosphate glass, with a wavelength of 1060 nm, as a host material. Using glass as a host instead of YAG tends to make the bandwidth of the Nd:glass laser several times larger than that of the Nd:YAG laser. The large bandwidth means that many axial modes oscillate within the laser cavity. The Nd:glass laser is successfully modelocked, producing pulse durations in the picosecond range. Dyes specially developed by Kodak have been used to mode-lock these glass lasers. When *Q*-switched, the Nd:glass laser produces output pulse energies comparable to the ruby laser.

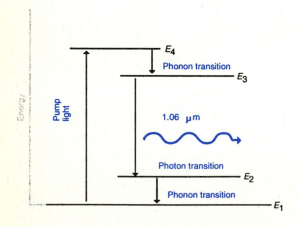

FIGURE 9.38 Four-Level Structure of Neodymium

Several materials have been developed in recent years to challenge the popularity of the three solid-state lasers just discussed. The crystal called *alexandrite* seems to lack many of the disadvantages of ruby, Nd:glass, and Nd:YAG. Alexandrite, also called *chrysoberyl*, has a $BeAl_2O_4$ chemical structure. It is a high-melting-point, hard crystal with excellent thermal properties (about twice as good as those of YAG). Good thermal properties make the alexandrite laser well suited to high average power applications. Since the host is doped with chromium, this laser can be operated as a three-level system. This laser can also be operated in a four-level state, in which it may be tuned by adjusting the height of the ground level. Alexandrite lasers can be operated CW, *Q*-switched, or mode-locked.

Table 9.3 shows a comparison of the important features of the four solid-state lasers discussed. In Table 9.3, the pumping threshold refers to the amount of energy needed from the flash lamp to produce a population inversion. Note the large amount of energy needed to pump the three-level ruby laser. The efficiency parameter refers to the output power from the laser compared to the input power from the lamp.

New developments in lasers occur frequently as corporations invest heavily in laser research and development. Recently, Swartz Electro-Optics developed a pulsed Er:YAG laser. This new laser, operating in the mid-IR at a wavelength of 2.94 μm, uses the rare-earth element *erbium*, abbreviated Er. The pulse energy of this laser is about 0.2 J with a pulse width of 200 μs and a PRR of 2 pulses per second. This unit can be *Q*-switched or mode-locked.

Gas Lasers. Gas lasers use a gas or a mixture of gases within a glass tube as a lasing medium. Gas lasers can be classified into several groups, depending on the type of gas used. Types of gases used include neutral atoms such as helium-neon, ionized atoms such as argon and krypton, and molecular gases such as CO_2 and water vapor. Since most gas lasers are similar in construction, a representative of each type will be discussed.

Neutral-Atom Gas Lasers: The *helium-neon laser*, a representative gas laser of the neutral-atom type, is the laser most commonly used in industry and education. Invented in 1961 by Ali Javan, it proved to be the forerunner of the entire family of gas lasers.

The lasing medium in the helium-neon laser is a mixture of gases made up of about 90% helium (He) and 10% neon (Ne) under a pressure of 1 torr (about 1 mm of mercury). The neon is the lasing material. We can get a picture of the lasing process by looking at Figure 9.39, a schematic of an HeNe system. Electrons from a cathode are accelerated down the tube. Since the He atoms have a large cross-sectional area compared to the Ne atoms, the electrons strike the He atoms. The electrons excite the He atoms, which collide with the Ne atoms. The He atoms give up their energy when they strike and transfer their energy to the Ne atoms. The final stage of the excitation process leaves an unexcited He atom and an excited Ne atom. If enough Ne atoms are excited, a population inversion will occur with stimulated emission following.

TABLE 9.3 Comparison of Four Solid-State Lasers

	Laser			
Feature	Alexandrite	Ruby	YAG	Nd:Glass (Phosphate)
Level type	4	3	4	4
Wavelength (nm)	700–815	694	1064	1056
Pumping threshold (J)	10	200	4	4
Efficiency (%)	2.5	0.5	2.5	3.5
Q-switched pulse duration	30–250	20	10–20	10–20
Thermal conductivity (W/cm/K)	0.23	0.35	0.13	0.0074

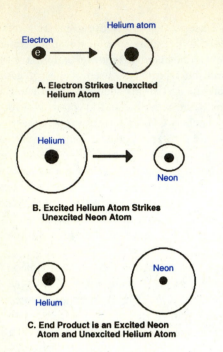

A. Electron Strikes Unexcited Helium Atom

B. Excited Helium Atom Strikes Unexcited Neon Atom

C. End Product is an Excited Neon Atom and Unexcited Helium Atom

FIGURE 9.39 HeNe Collision Pumping

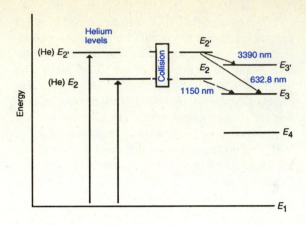

FIGURE 9.40 HeNe Laser Energy Levels

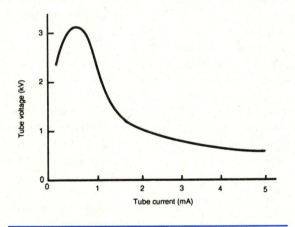

FIGURE 9.41 Current versus Voltage in a Standard HeNe Laser Tube

We can see why He was chosen to transfer energy to the Ne atom by examining the HeNe energy-level diagram in Figure 9.40. The two gases have electron excitation levels that are almost identical. When the He atom collides with an Ne atom, the He atom transfers the correct amount of energy to the Ne atom. The Ne spontaneous emission process produces photons with three wavelengths—632.8 nm, 3390 nm, and 1150 nm. Although HeNe lasers could emit any of these three wavelengths, most HeNe lasers are tuned to emit 632.8 nm, the only visible output wavelength.

The voltage-current characteristic curve of an HeNe gas laser, shown in Figure 9.41, gives important information about this laser. Note that as we increase voltage from zero, very little current flows. Significant current flows only when we reach a voltage called the *breakdown voltage*, usually around 3–4 kV for most small HeNe lasers. After breakdown voltage is exceeded, the current starts to increase and the voltage across the tube decreases. Exceeding the breakdown voltage on the tube produces what is known as a *plasma*. Creating the plasma state increases the current in the tube. Most HeNe lasers operate with a voltage drop of 1000–1500 V dropped across the tube. As in the fluorescent tube, a ballast resistor is needed to drop the remainder of the voltage that the power supply produces. Note that the power supply must produce a voltage at least as great as the breakdown voltage of the tube.

Output powers of HeNe lasers range from less than a milliwatt to 50 mW.

Ionized-Atom Gas Lasers: The *argon laser* is a gas laser. Unlike the HeNe laser, which is classified as a neutral-atom gas laser, the argon laser is an ionized-atom gas laser. A laser is considered an ionic laser when stimulated emission occurs in the levels of an ion instead of a complete neutral atom. The design and operating conditions of ionic and neutral gas lasers are different.

In CW ionic lasers, the main difference is in the excitation process and power supply requirements. The ionic gas laser uses a high-current, low-voltage power supply. This is exactly the opposite of the requirement of the neutral gas laser. The voltage drop across the electrodes in the ionic laser usually ranges from 200 to 300 V, with current ranging from 5 to 50 A. Note the large power supply in the argon ionic gas laser shown in Figure 9.42. The large power supply adds cost to this type of laser. Pressure in the tube is about 0.5 torr with a bore diameter between 1 and 3 mm.

The schematic diagram of an argon ionic gas laser in Figure 9.42 shows one of the consequences of the high-current, low-voltage mode of operation. Such tubes give off great amounts of heat and must be cooled. The tube in Figure 9.42 shows that water is used as a coolant. Since the bore diameter is small and the current is high, most ionic tubes have a current density of several hundred amperes per square centimeter. These high current densities tend to erode most materials used for bores in other lasers. Ionic lasers have bores built out of material that can withstand the high current densities. Current ionic lasers use graphite or beryllium oxide. Both of these materials have good thermal conductivities. High thermal conductivity makes it easier to remove the heat generated in the bore.

Most ionic gas lasers have power outputs between 1 and 20 W. The energy-level diagram in Figure 9.43 shows the transitions in an argon gas laser. Argon gas produces light of many different frequencies. When all these frequencies are produced by the argon laser, the laser is said to be operating in the *multiline mode*. Although the argon laser could be operated in the multiline mode, usually the laser is tuned to a particular wavelength. Tuning can be accomplished by using a prism, as shown in Figure 9.44. The prism is mounted so that it can be rotated. The different wavelengths take different paths out of the prism due to dispersion. In this case, the prism is adjusted so that the 488.0 nm beam strikes the mirror. All other beams will miss the 100% reflective mirror. Only the 488.0 nm beam sees the optical cavity. All other wavelengths will not be amplified. The output mirror emits 488.0 nm only. Most argon lasers use the 514.5 nm line and the 488 nm line. These two lines have about two-thirds of the total output of the laser. The other one-third of the power is split between the other wavelengths.

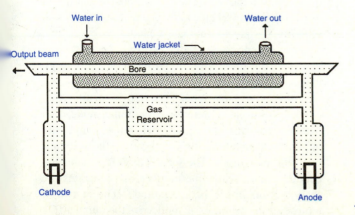

FIGURE 9.42 Argon Gas Laser Tube

FIGURE 9.43 Argon Gas Laser Energy Levels

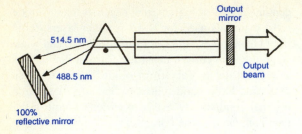

FIGURE 9.44 Tuning the Argon Gas Laser

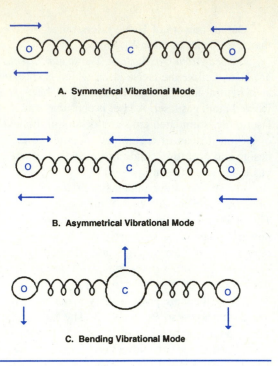

A. **Symmetrical Vibrational Mode**

B. **Asymmetrical Vibrational Mode**

C. **Bending Vibrational Mode**

FIGURE 9.46 Carbon Dioxide Molecule Vibrational Modes

Other ionic lasers use different gases. One of the most popular is the *krypton laser*. The krypton laser uses the noble gas krypton as a lasing medium. Like argon, it, too, has a large number of lines. The highest power is in the 647.1 nm line, but the laser can be tuned over several other lines. The *xenon laser*, often operated in the pulsed mode, is tunable over the 539.5–995 nm range.

Molecular Gas Lasers: The *carbon dioxide* (CO_2) *laser* is a good example of a molecular gas laser. The CO_2 laser uses molecular vibrational states of CO_2 to produce lasing action. All the lasers discussed up to this point have used the transitions of electrons to produce stimulated emission. If we were to look at a CO_2 molecule, we would see a structure like that shown in Figure 9.45. The carbon atom is in the middle, with the oxygen molecules on either end. The molecular bonds are shown as "springs."

The CO_2 molecule vibrates in three different ways, as shown in Figure 9.46. The first vibrational mode is the *symmetrical mode* (Figure 9.46A). Note that the carbon atom is stationary, while the oxygen atoms move in and out. In the *asymmetrical mode* (Figure 9.46B), one oxygen atom moves out, while the carbon atom and the other oxygen atom move toward each other. The final mode is the *bending*

mode (Figure 9.46C). Note that the molecule vibrates by bending—hence, its name. The molecule may vibrate in any of the three modes in any combination. Each vibrational mode has a certain resonant frequency. Like electronic energy levels, energy may be added or removed from a vibrational state only in discrete packets. When energy is transferred from one vibrational state to another, a photon is given off. Refer to the energy-level diagram in Figure 9.47 and note that the three vibrational modes have energy levels associated with them. Levels are abbreviated by a three-digit sequence. The first digit represents a level within the symmetric vibrational mode. The number indicates the actual level. For example, the number 100 indicates the first level of the symmetric vibrational mode. The second digit represents the bending mode. Thus, the number 020 indicates the second level in the bending mode. The last digit represents the asymmetric mode. So, the number 001 tells us that the first level of the asymmetric vibrational mode is in view. Two transitions are

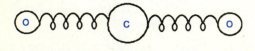

FIGURE 9.45 Carbon Dioxide Molecule

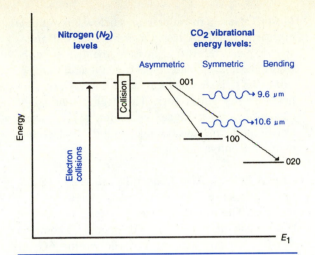

FIGURE 9.47 Carbon Dioxide Molecule Energy Levels

important in the CO_2 laser. First, the transition from levels 001 to 100 produces a strong line with a wavelength of 10.6 μm. Another strong line occurs from a transition from levels 001 to 020. The output wavelength from this transition centers around 9.6 μm. Most CO_2 lasers use the 10.6 μm output wavelength.

It was soon found by those working with CO_2 lasers that CO_2 gas by itself did not lase effectively. CO_2 lasers work best with nitrogen gas (N_2) and helium gas (He) mixed with the CO_2. The N_2 gas molecule, since it is diatomic, has only one vibrational state.

As you can see from Figure 9.47, the energy level of N_2 corresponds almost exactly with the 001 asymmetric vibrational level. The N_2 gas serves the same purpose as the He gas in the HeNe laser. The N_2 gas is excited by electron collision to a vibrational level. The energy stored in the vibrating N_2 atoms is transferred to the CO_2 atom by collision. The addition of He further increases the CO_2 laser output and also helps to keep the gases cool.

Another factor discovered by early CO_2 laser experimenters concerned the breakup of the gas. In a sealed tube, the CO_2 gas decomposes, making the laser lose power. As a result, fresh CO_2 gas is normally added to the tube while the old gas is

discharged out of the tube. In small lasers, the used gas is discharged out into the atmosphere. In larger lasers, the gas is recirculated and replenished.

The first type of CO_2 laser developed is shown in Figure 9.48. The gas flows down the length of the tube. The excitation is applied at either end of the tube. Because the gas flows down the length or axis of the tube, this type of CO_2 laser is called an *axial-flow laser*. The flow rates of early lasers were slow, around 50 ft³/hr. Slow axial-flow CO_2 lasers give 50–70 W/m of length, with total average power outputs to 1000 W. Power higher than 500–1000 W makes the tube lengths excessively long. One early laser of this type produced about 8000 W of power, but the tube was 750 feet long. Besides the length, the other factor that limits the power output of the slow axial-flow laser is heat buildup.

As we increase the gas flow rate, output power increases. The main difference between the slow axial-flow CO_2 laser and the fast axial-flow CO_2 laser is the gas flow rate. Gas flows up to 1000 ft³/hr, producing outputs up to 5000 W. Fast axial-flow CO_2 lasers produce about 600 W/m of tube length. Since the gas flows at a high rate, the gas is collected, cooled through a heat exchanger, and recycled.

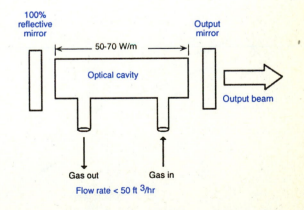

FIGURE 9.48 Slow Axial-Flow CO_2 Laser Construction

A type of CO_2 laser called a *transverse-flow laser* allows greater gas flow rates to be achieved by flowing the gas transversely and by folding the beam with mirrors, as shown in Figure 9.49. Folding allows maximum usage of the excited gas, any molecule of which only stays in the lasing region for about 2 ms. Large mirrors are used in this laser to reflect the beam through the lasing region several times. After reflecting a number of times, the photons escape through the output coupler. Using this folded construction, the gas-transport CO_2 laser can generate a large amount of power in a small space. Chemical catalysts recombine the gas molecules after excitation and are recirculated. The gas-transport laser can give CW power outputs up to 20 kW.

All of the CO_2 lasers discussed may be operated in pulsed mode. To get pulsed-mode operation, the excitation must be applied in short pulses. Pulsed operation is preferred in several industrial processes. A special type of CO_2 laser gives very high pulsed power operation. This laser is called the *transverse-excited atmospheric* (TEA) *laser*. Electrical excitation of the gas in a CO_2 laser can be done more efficiently by applying the voltage across the tube. Transverse excitation reduces the voltage needed to excite the N_2 gas. Some axially excited systems use voltage up to a million volts. High field strengths can be maintained at low voltages, from 20 to 50 kV.

Most CO_2 lasers run at pressures below atmospheric pressure. The TEA laser, however, runs at atmospheric pressure. At this pressure, the density of the CO_2 molecules is much higher than that of conventional CO_2 lasers, which have a CO_2 pressure between 0.5 and 1 torr. The higher density of CO_2 molecules produces very high peak powers, usually in the megawatt range, with pulses less than a microsecond in duration.

Liquid Lasers. Also frequently called *dye lasers*, liquid lasers use a complex organic dye as a lasing medium. The organic dye is dissolved in a liquid host, usually methyl alcohol. For example, a common dye used in the dye laser is rhodamine 6G, or xanthene dye. Rhodamine 6G has a $C_{26}H_{27}N_2O_3Cl$ chemical composition and has a molecular weight of 450. It dissolves in water and methyl alcohol. It is used in the garment industry to color silk pink.

The most important characteristic of this laser is its ability to be tuned. Because of its large molecular structure, it has many electron excitation levels. Within each excitation level are many vibrational levels. All these levels tend to overlap and therefore blend together. A wavelength-selecting instrument placed in the optical cavity allows tuning to a particular wavelength out of the many possible wavelengths. By changing the amount and concentration of the dyes, the entire visible spectrum may be covered. Rhodamine 6G, for example, may be tuned over a bandwidth of about 200 nm in the red spectral area.

Semiconductor (Injection) Lasers. Unlike the solid-state lasers, which are made of a crystal or glass, the semiconductor lasers are made from semiconductor materials. The diagram in Figure 9.50A shows a normal semiconductor energy-level system. As in most solid materials, electron energy levels are grouped in bands. Each band represents a large number of closely spaced levels. The lower level, the valence band, is separated from the conduction band by a space, the forbidden band. The valence band is the highest filled band. The conduction band is the lowest empty band. No electron can exist in the forbidden band. If the semiconductor material were cooled to absolute zero, all electrons would be

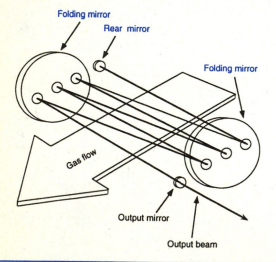

FIGURE 9.49 Transverse-Flow CO_2 Laser with Folded Beam

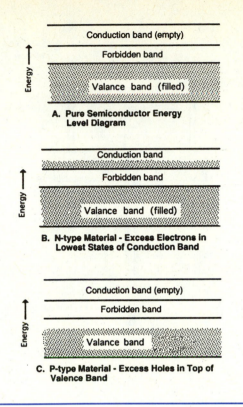

A. Pure Semiconductor Energy Level Diagram

B. N-type Material - Excess Electrons in Lowest States of Conduction Band

C. P-type Material - Excess Holes in Top of Valence Band

FIGURE 9.50 Semiconductor Energy Levels

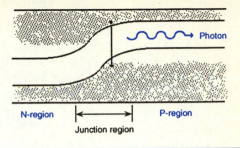

FIGURE 9.51 Semiconductor PN Junction Energy Levels

in the valence band. As the temperature rises, some electrons will gain enough energy to leave the valence band and populate the conduction band. This is the common excitation process we have described in other lasers. Every electron leaving the valence band for the conduction band leaves behind a "hole" in that band. In a pure semiconductor material, the numbers of electrons and holes are exactly equal.

Adding a small amount of impurities, called *dopants*, produces a semiconductor material that has more electrons in the conduction band than there are holes in the valence band. This type of material, called *N-type material*, is shown in Figure 9.50B. Addition of a different type of impurity produces a semiconductor material with exactly the opposite characteristic. Called *P-type material*, this semiconductor has an excess of holes in the valence band and a deficiency of holes in the conduction band, as

shown in Figure 9.50C. When we place P-type and N-type materials together, we have a semiconductor diode. The diagram in Figure 9.51 shows the energy-band diagram of the junction area.

Operation: Application of a forward-bias potential to the diode drives electrons and holes to the junction. The electrons and holes recombine at the junction area. This process is called *carrier injection*. When recombination occurs, electrons fall from the conduction band to the holes to the valence band. The diode gives off a photon of radiation for every transition. The process where radiation is given off by the diode is called *recombination radiation*. The principle of recombination radiation is used in the light-emitting diode, discussed earlier in this chapter.

To make a semiconductor laser, we need all the characteristics we have come to associate with laser operation. First, we need an optical cavity. In the semiconductor laser diode, shown in Figure 9.52, both ends of the semiconductor crystal are cut. The ends are cut so that they are perpendicular to the plane of the junction and parallel to each other. The crystal is either cleaved along one of the crystal faces or cleaved and polished to flatness. The ends of the crystal serve as mirrors in other lasers and reflect most of the radiation due to the change in refractive indices. Photons will, therefore, be reinforced between the flat ends of the crystal resonant cavity.

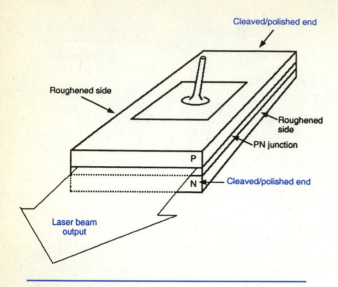

Cleaved/polished end

Roughened side

Roughened side

PN junction

P

N

Cleaved/polished end

Laser beam output

FIGURE 9.52 Semiconductor PN Junction Injection Laser

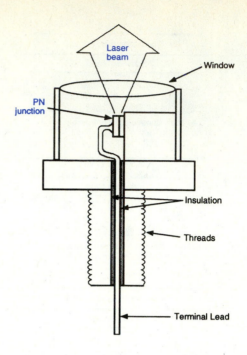

Laser beam

Window

PN junction

Insulation

Threads

Terminal Lead

FIGURE 9.53 Injection Laser Construction

Since the semiconductor material serves as the lasing medium, we need a population inversion to produce lasing action. A population inversion exists when we place a forward-biased potential across the junction and inject enough carriers into the junction. The population inversion must be great enough so that the optical gain exceeds the optical losses in the cavity. The laser crystal is then put into a package like the one shown in Figure 9.53. The beam leaves the edge of the laser diode and goes through the window at the top of the package.

Types of Junctions: The earliest semiconductor lasers were called *homojunction* (HJ) lasers. As the name implies, the laser had only one junction made of gallium arsenide (GaAs). The actual light-emitting area in the semiconductor laser diode was only a few microns thick, and the entire crystal was no more than 1 mm long. This early laser had some serious problems. First, the output tended to decrease rapidly with time because of the high power density at the junction, which was likely to be greater than 20,000 A/cm^2. Gallium arsenide has an index of refraction of about 3.6, so the reflectivity of the

crystal ends is about 36%. Because the power density for lasing was so high, the simple GaAs injection laser had a high lasing threshold. The *lasing threshold* is defined as that anode current necessary to cause lasing. The simple homojunction GaAs laser needed from 2 to 30 A to lase, depending on the diode. Since the junction dropped about 2 V at this current, the power was high. With so small a device, it was necessary to operate it in the pulsed mode at a very low duty cycle to keep the laser from burning up.

The efficiency of the homojunction was improved by adding a P+-region on top of the normal P-region, as shown in Figure 9.54A. This figure shows the structure of the improved GaAs homojunction with the variation in refractive index. The heavily doped P+-region had a slightly lower refractive index than the normally doped P-region. Light is confined in the normally doped P-region by a light piping effect. The wave will tend to

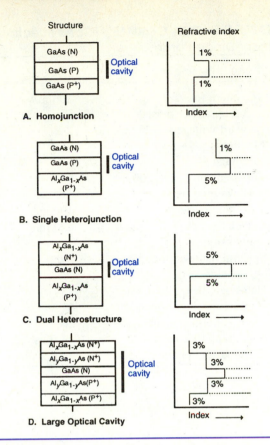

FIGURE 9.54 PN Junction Injection Laser Construction

TABLE 9.4 RCA SG2007 GaAs Injection Laser Diode Parameters

Parameter	Value
Peak forward current	40 A
Duty factor	0.1%
Pulse duration	200 ns
Peak forward voltage at 40 A forward current	8 V
Minimum total radiant flux at 40 A forward current	10 W
Threshold current	11 A
Radiant efficiency	3.5%
Wavelength of peak radiant intensity	904 nm
Spectral bandwidth at 50% intensity	3.5 nm
Half-angle beam spread	
Plane parallel to junction	7.5°
Plane normal to junction	9°
Rise time	<1.0 ns
Emitting-region dimensions	228.5×2 μm

travel more slowly in the region having a higher refractive index. Because the differences in the index of refraction are not great, the confinement of the beam is poor.

Table 9.4 summarizes some information on an RCA SG2007 GaAs injection laser. Note the high lasing threshold. Driving circuitry must produce a current of 11 A before the diode will lase. Note also the peak forward current and voltage, 40 A and 8 V. The peak power dissipation will be

$$P_{pk} = I_{pk}V_{pk} \qquad (9.17)$$

$$= (40 \text{ A}) (8 \text{ V}) = 320 \text{ W}$$

What will the average power be? Recall that the average power is equal to the peak power multiplied by the duty cycle D (Equation 9.15). With a duty cycle of 0.1%,

$$P_{avg} = P_{pk} \times D$$

$$= 320 \text{ W} \times 0.001$$

$$= 0.320 \text{ W} \quad \text{or} \quad 320 \text{ mW}$$

We can easily see that this laser may only be operated in a pulsed mode. Another interesting feature of this device is the small emitting area. The small emitting area is well suited to fiber optics, an application that we will discuss in detail in later chapters. In the fiber optics application, light must be launched down a glass fiber, in some cases smaller than a human hair. The semiconductor injection laser is well suited to this task due to its small emitting area. The small emitting area also produces a wide beam divergence angle. The divergence is larger in this laser than in any other laser. To cover long distances, the injection laser must be collimated by a lens.

Poor confinement in the homojunction laser meant high losses and a high lasing threshold. Researchers developed the *single heterojunction* (SH) structure to combat this lack of confinement. Instead of the P+-layer being heavily doped GaAs, the heterojunction used a different material entirely for the P+-region. In the SH injection laser, the P+-region was made of aluminum gallium arsenide (AlGaAs), as shown in Figure 9.54B. Note the decrease in the refractive index on the P+ side of the junction. The greater difference in the refractive index made the light pipe more effective. The lasing threshold was reduced to about 0.1 of the homojunction. Because the N side of the junction did not have a large change in refractive index, the losses were heavier on this side. This lead to the invention of the *dual heterostructure* (DH), shown in Figure 9.54C. Note that the AlGaAs is now on both sides, causing a corresponding decrease in refractive index on the N side of the junction. The lasing threshold decreases again by a factor of 10 over the single heterojunction. The lasing threshold for this device is usually about 1 A but can be much lower in modern devices, such as the C86046E injection laser manufactured by RCA. The parameters for this laser are listed in Table 9.5.

Because of the addition of the second heterojunction, the device gained a problem not seen in the HJ, SH, or earlier DH lasers. Because the DH structure confines the flux to the junction area so well, the flux density becomes very high. The high-density flux tends to damage the active area. The *large optical cavity* (LOC) structure, also sometimes called the *separate confinement heterostructure* (SCH) laser, shown in Figure 9.54D, overcomes this difficulty. Note that the LOC structure adds an $Al_yGa_{1-y}As$ layer on either side of the $Al_xGa_{1-x}As$. The difference between these two layers is in the proportion of Al and Ga, as indicated by the x and y. The y concentration is typically much lower than the x. Changes in the proportion of Al and Ga change the refractive index. The diagram in Figure 9.55 shows the change

TABLE 9.5 RCA C86046E AlGaAs Injection Laser Diode Parameters

Parameter	Value
Maximum forward current, continuous or pulsed	200 mA
Forward voltage at 125 mA	2 V
Maximum peak reverse voltage	2 V
Total radiant flux at 200 mA forward current	10 mW
Threshold current	100 mA
Wavelength of peak radiant intensity	820 nm
Spectral bandwidth at 50% intensity	2 nm
Half-angle beam spread	
Plane parallel to junction	5°
Plane normal to junction	20°
Rise time (10% to 90%)	<1.0 ns
Emitting-region dimensions	13×2 μm

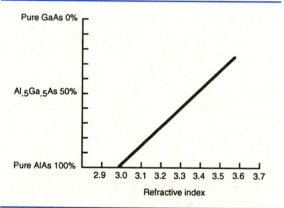

FIGURE 9.55 Change in Refractive Index versus Proportion of Al to Ga in AlGaAs

in refractive index versus the proportion of Al to Ga in the AlGaAs compound. Note that pure GaAs has a refractive index of about 3.58, while AlAs has a refractive index of about 2.98. The changes in the material composition make the optical cavity much wider than the area to which the carriers are confined (Figure 9.54D). Because of the wider optical cavity, the diffraction of the beam is lowered. Beam divergence in this LOC laser is about 2°.

The DH injection laser has been further refined into the *stripe-geometry* DH injection laser, shown in Figure 9.56A. The normal DH injection laser confines the carriers and light well vertically but not as well from side to side or laterally. The stripe geometry helps confine the beam laterally. Note the presence of a gap in the SiO_2 insulating layer running down the length of the laser. The insulator confines the electrical excitation to a narrow stripe running down the length of the laser. The carriers are then injected only over the narrow width of the stripe, usually about 10–20 μm thick. The carriers do not tend to spread out over the entire junction region. This improves the spatial distribution pattern of the beam. Instead of being somewhat flat, the beam from a stripe-geometry laser tends to be rounded. The

diagram in Figure 9.56B shows the radiation pattern of the stripe-geometry DH injection laser.

Another improvement on the DH design is shown in Figure 9.57. This design is called the *buried heterostructure* (BH) injection laser. This laser has a heterostructure buried within the N-type AlGaAs. The narrow width (as low as 2–3 μm) in the BH laser reduces carrier spreading. The threshold current is reduced to 5–50 mA in this type of laser with peak power from 1 to 5 mW.

One of the most advanced injection laser structures is called the *distributed feedback* (DFB) laser. In this laser, one of the sides of the active layer is corrugated. This corrugation produces a feedback mechanism for the laser oscillations. This structure differs from the others in that the wavelength of the output is determined by the spacing of the corrugation and not the bandgaps of the semiconductor material. Although the output frequency is determined in a different way, the DFB laser is made of AlGaAs and GaAs and has a DH structure. The corrugated structure is known as a *phase grating* and acts as a diffraction grating.

DFB lasers have the following advantages. They produce a single wavelength, high radiance, and low noise. They also have a small active area and are easy to use. At present, however, the lasers are extremely expensive primarily because they are so difficult to make. Most analysts believe that the price will come

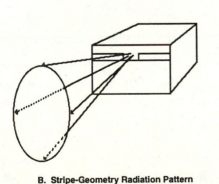

A. Stripe-Geometry Structure

Metal contact
SiO_2 Insulating layer
$Al_xGa_{1-x}As$ (P)
p-GaAs
GaAs (P)
$Al_xGa_{1-x}As$ (N)
GaAs (N) (substrate)
Metal contact

B. Stripe-Geometry Radiation Pattern

FIGURE 9.56 Stripe-Geometry DH Injection Laser

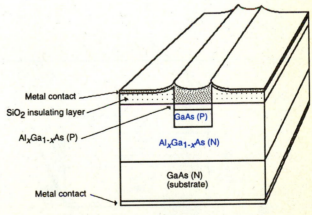

Metal contact
SiO_2 insulating layer
$Al_xGa_{1-x}As$ (P)
GaAs (P)
$Al_xGa_{1-x}As$ (N)
GaAs (N) (substrate)
Metal contact

FIGURE 9.57 Buried Heterostructure Injection Laser

down as more demand increases the production of these devices. The wavelengths of the GaAs lasers we have discussed depends on their chemical composition, except for the DFB laser.

In general, lasers using GaAs or the ternary compound AlGaAs can radiate wavelengths between 800 and 1000 nm. Above 1000 nm, we must turn to different materials. Particularly in fiber optics applications, it is necessary to have emitters that produce wavelengths between 1000 and 1600 nm. The most popular semiconductors for this range of wavelengths are the indium gallium arsenide phosphide/indium phosphide (InGaAsP/InP) compounds. Note the stripe geometry of the laser pictured in Figure 9.58. A 0.2 μm thick N-type InGaAsP active layer is grown on top of the N-type InP layer. Data for the RCA C86045E injection laser, which uses this structure, is shown in Table 9.6. As in other semiconductor lasers, the threshold current and the output power both depend on the case temperature of the laser. A diagram of the temperature dependency of the injection laser is shown in Figure 9.59. Note the almost linear change in output for a given change in drive current. Note also how temperature changes this curve. Increases in temperature increase the threshold current of the laser and decrease the output power. To keep the temperature stable, electronic circuits keep the temperature from changing under load conditions.

TABLE 9.6 RCA C86045E InGaAsP Injection Laser Diode Parameters

Parameter	Value
Peak forward current	2.5 A
Duty factor	1%
Pulse duration	1 μs
Peak forward voltage at 2.5 A forward current	2.5 V
Minimum total radiant flux at 2.5 A forward current	500 mW
Threshold current	0.7 A
Wavelength of peak radiant intensity	1300 nm
Spectral bandwidth at 50% intensity	7.0 nm
Half-angle beam spread	
Plane parallel to junction	12°
Plane normal to junction	24°
Rise time	<1.0 ns
Emitting-region dimensions	2 × 37 μm

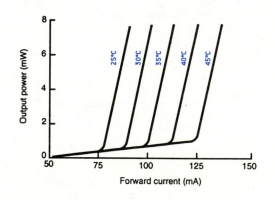

FIGURE 9.59 Effects of Temperature on Threshold Current in an Injection Laser

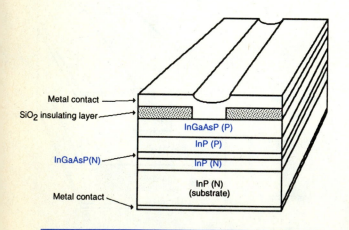

FIGURE 9.58 Long-Wavelength DH Injection Laser

InGaAsP and other indium-based compounds can take the wavelengths of injection lasers up to about 1300 μm. Longer-wavelength lasers require different lasing mediums. The lead-salt injection lasers allow access to wavelengths between 2.7 and 30 μm. At wavelengths below 4 μm, lead cadmium selenide (PbCdSe) diodes are used. Diodes of lead sulfide selenide serve at intermediate wavelengths between 4 and 8 μm. PbSnSe and PbSnTe provide output between 6 and 30 μm. A typical lead-salt laser system includes a crystal mount and refrigerator with a temperature stabilizer and power supply. The temperature of the mount is kept very closely controlled to within 10^{-4} degrees Celsius. These lead-salt devices may be tuned by adjusting the temperature.

Injection lasers are normally used as single devices. Several manufacturers have produced high-power arrays of injection lasers called *phased arrays*. Current arrays have achieved power outputs of over 200 mW and at room temperature with efficiencies of greater than 25%. One of the most common uses of this type of laser array is to pump solid-state Nd:YAG lasers. Why pump a YAG laser rather than use the output of the laser diode itself? First, the laser diode may not emit light at the wavelength necessary for a particular application. Second, the laser diode does not have great frequency stability. Third, the spatial properties of the beam are inferior to the TEM_{00} beam output from a YAG laser. Fourth, the laser diode has a limited power output. Last, high power outputs can be achieved by Q switching YAG lasers. This is possible in solid-state lasers because energy can be stored in the upper-lasing level to be released by Q switching. Q switching is a method used to decrease the pulse width and increase the pulse power in lasers. Q switching is not possible with the injection laser diode.

The injection laser diode has several advantages over conventional flash lamps. First, since the flash lamp's power density is high, water cooling is usually required. Also, since the YAG laser rod absorbs much of the lamp's energy, the laser rod may need to be cooled. Second, flash lamps have a lifetime of only several hundred hours. Third, flash lamp–pumped YAG lasers typically have efficiencies of about 0.5% with a TEM_{00} output. Nd:YAG lasers sometimes use tungsten-halogen lamps as pumps. The main advantage of the tungsten lamp is that the lamp presents a simple resistive load to the power supply. This simplifies the power supply design and reduces cost. Table 9.7 shows a comparison of the three types of pump sources for YAG lasers.

TABLE 9.7 Comparison of Nd:YAG Laser Pump Sources

Feature	Pump Source		
	Inert Gas Arc Lamp	Tungsten Lamp	Injection Laser
Input power to pump source	2 kW	500 W	1 W
Useful pump power	100 W	5 W	0.2 W
Laser power (TEM_{00})	8 W	0.2 W	0.06 W
Electrical efficiency	0.4%	0.04%	5000 h

The most common laser diodes used to pump YAG lasers are made of GaAs. They typically emit wavelengths in the 800 nm region. These GaAs lasers are placed in arrays 10 deep, producing about 200 mW of power from the array. Although the lifetime of the laser diode array is longer, the main advantage of the diode array over conventional sources is in the beam. Unlike conventional sources, the laser diode array output beam is collimated and focused. Most of the power from the array goes into the rod, thus improving efficiency. Another advantage of the diode array occurs in tuning. Unlike conventional sources, the laser diode can be temperature-tuned to put out a wavelength exactly in the center of the YAG laser's pump band. Most of the energy from the array goes into raising electrons to higher levels in the neodymium. Little of the energy goes into heating up the YAG crystal. This makes cooling of the diode-pumped YAG laser unnecessary and simplifies the design of the laser.

Another interesting application for the diode-pumped YAG laser involves frequency doubling. In this application, the YAG laser produces 532 nm, instead of the normal 1064 nm. The frequency doubling is done by passing the beam through a potassium titanyl phosphate (KTP) crystal.

Other Lasers. Some lasers in experimental stages and those just being introduced to industry defy the classifications we have just used. These lasers, however, need some discussion since the experimental lasers of today usually become the workhorses of tomorrow.

Chemical Lasers: All of the lasers discussed up to this point have some sort of chemical input. Although electricity may not cause the population inversion directly, the electrical input has an indirect effect. In the chemical laser, a chemical reaction liberates the energy needed to cause a population inversion. Let's take an example. A series of chemical reactions can take place where a vibrationally excited molecule of hydrogen fluoride (HF*) is created. (The asterisk symbolizes an atom or a molecule in an excited state.) The chemical reaction automatically produces a population inversion when HF* is produced. The HF* molecule decays back to unexcited HF by giving up a photon. Stimulated emission may take place if enough HF* is created by the chemical reaction and an optical cavity exists. In theory, all you need to do is get the reaction going. From then on, all you need to do is supply chemicals to keep it going. In fact, most chemical lasers need a source of electrical power to sustain their operation. Chemical lasers are powerful but not very common. Most research in chemical lasers is occurring in military applications. HF chemical lasers produce wavelengths between 2.6 and 3.6 μm.

Excimer Lasers: The excimer laser has achieved much more popularity than the chemical laser. Classified by some as a molecular laser, the excimer laser combines inert gases into molecules as the lasing medium. We may wonder about how an inert gas may form into a molecule. If we learned anything from chemistry, we learned that inert gases are exactly that—inert. The word *inert* as applied to chemical reactions means that these gases do not react with any other substance. This meaning of inert is only partially true. Inert gases will react with other substances to form molecules. The molecules, however, are not stable.

The excimer laser, then, makes an unstable molecule containing an inert gas as one of the ingredients. An *excimer* is a molecule that is bound together when the molecule is in an excited state but that is not bound when in the ground state. An example at this point may help. One of the more popular excimer lasers uses krypton fluoride (KF) as a lasing medium. The excimer laser starts out with a mixture of krypton and fluorine gases. A high-power electron beam irradiates these two gases. A complex process results in the creation of many KF* molecules. If an optical cavity exists, the KF* molecule quickly decays into K, F, and a photon, and stimulated emission will take place. Excimer lasers emit wavelengths in the ultraviolet region, where biological tissues, plastics, and semiconductors absorb very strongly. Excimers supply average powers in the range of 50–150 W with pulse durations between 10 and 30 ns. PRFs (pulse repetition frequency) range from 1 to 500 Hz. The output wavelength depends on the gases used. Table 9.8 shows excimer molecule compounds and the wavelengths produced.

RECEIVERS

In the previous section, we discussed two emitters of light—the LED and the laser. In this section, we will discuss receivers of light. In some cases, these receivers act as sensors or transducers. (See Chapter 8 to review the concept of the sensor.) Light receivers may also be used as a part of a communications system, such as fiber optics. In either case, the basic device is the same.

Photoreceivers may be classified into three basic groups: bulk photoconductors (such as the photoresistor), the PN junction group (which includes the photodiode), and phototubes (such as the photomultiplier). We will discuss the major devices in each classification.

TABLE 9.8 Excimer Laser Wavelengths

Compound	Wavelength (nm)
ArF*	193
KrF*	248
XeCl*	308
XeF*	351

Bulk Photoconductors

Photoresistors. Several years before Hertz discovered the photoelectric effect, Willoughby Smith noticed that the resistance of a block of selenium decreased when exposed to light. This same effect is used in the *photoresistor*. The photoresistor, shown in Figure 9.60, is made of either cadmium sulfide (CdS) or cadmium selenide (CdSe). These substances are vapor-deposited on a glass or ceramic substrate and then hermetically sealed in plastic or glass. When light strikes the photoresistive material, it liberates electrons. These electrons are then available as current flow, and the resistance decreases. More light falling on the photoresistor causes resistance to decrease further.

The spectral response (how the device reacts to different light wavelengths) depends on the type of material used. Notice in Figure 9.61 that the spectral response of the CdS photoresistor closely matches the response of the human eye. The CdSe spectral response is shifted toward the infrared.

The specific material used determines the ratio of dark-to-light resistance, which can range from 100:1 to 10,000:1. The photoresistor, then, is very sensitive to changes in light. It is generally easy to use and inexpensive. Disadvantages include narrow spectral response, poor temperature stability, and light history effect. The light history (hysteresis)

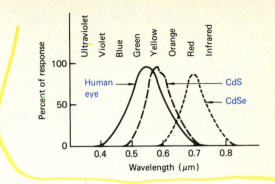

FIGURE 9.61 Spectral Response of CdS and CdSe Photoresistors Compared with Response of Human Eye

effect is particularly annoying since it causes the resistance to depend on past light intensity. In addition, photoresistors are slow to react to light intensity changes. The CdS takes about 100 ms to respond, while the faster CdSe takes about 10 ms.

Because the photoresistor is so sensitive to light changes, it does not normally need amplification. A simple potentiometric circuit, like the ones shown in Figure 9.62, is sufficient. Note the different schematic symbols in common use. The Greek letter λ (lambda), as indicated in Figure 9.62, is often used to symbolize light.

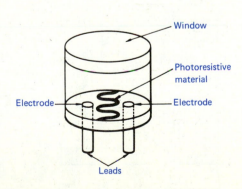

FIGURE 9.60 Cutaway View of Photoresistor

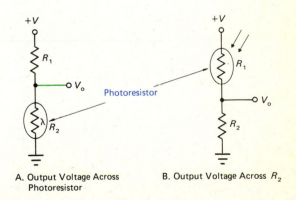

FIGURE 9.62 Photoresistors Used in Voltage Divider Circuits

An example may help at this point. Suppose we have a CL704L photoresistor in series with a 10 kΩ resistor (Figure 9.62B). If the cell is illuminated by 0.2 footcandle (fc), what is the output voltage with respect to ground? We see from the cell resistance curves in Figure 9.63 that the resistance is approximately 3 kΩ. The output voltage may be found from the voltage divider equation:

$$V_o = \left(\frac{R_2}{R_1 + R_2}\right)(+V) \tag{9.18}$$

$$= \left(\frac{10 \text{ k}\Omega}{10 \text{ k}\Omega + 3 \text{ k}\Omega}\right)(10 \text{ V})$$

$$= 7.7 \text{ V}$$

• **EXAMPLE 9.12**

Given a CL707HL with an illumination of 2 fc, find the output for the circuit of Figure 9.62B.

Solution

From Figure 9.63, we see that the resistance of the photoresistor is 10 kΩ. Again, using Equation 9.17, we get

$$V_o = \left(\frac{R_2}{R_1 + R_2}\right)(+V)$$

$$= \left(\frac{10 \text{ k}\Omega}{10 \text{ k}\Omega + 10 \text{ k}\Omega}\right)(10 \text{ V})$$

$$= 5.0 \text{ V}$$

PN Junction Photoconductors

Photodiodes. Light levels are also detected by PN junctions, like those found in photodiodes and phototransistors. *Photodiodes* are simply PN diodes with their junctions exposed to light.

Early experimenters with PN junctions discovered that minority carrier current flow was influenced

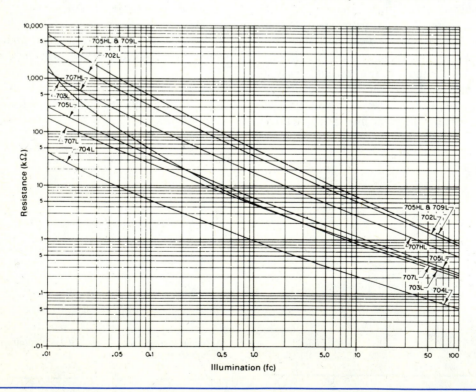

FIGURE 9.63 Cell Resistance Curves

by temperature and light levels. Increasing light falling on the junction creates more electron-hole pairs, which increases minority carrier current flow. Since we are dealing solely with minority carrier current, the photodiode is operated with reverse bias.

Note the spectral response curve for the photodiode shown in Figure 9.64. This curve is much wider than that of the photoresistor, covering the visible spectrum as well as the infrared.

Photodiode construction is illustrated in Figure 9.65. The simple PN device (Figure 9.65A) is nothing more than a PN junction whose structure is optimized for light reception. The PIN device (Figure 9.65B) adds a layer of undoped semiconductor material. (I stands for "intrinsic".) The layer of undoped material increases the size of the depletion region. Widening the depletion region effectively decreases junction capacitance. This decrease makes the PIN diode respond much faster to light-level changes than the PN photodiode does. PIN photodiode dark currents are much smaller, also.

Photodiodes produce relatively small currents even under fully illuminated conditions. Some manufacturers add an amplifier in the same package to increase sensitivity. Packages for photodiodes without amplifiers are shown in Figure 9.66, along with common schematic symbols for photodiodes.

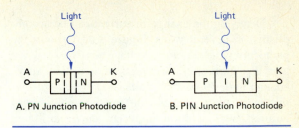

FIGURE 9.65 Two Types of Photodiodes

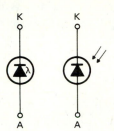

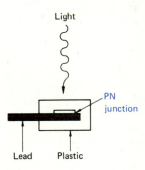

A. Schematic Symbols

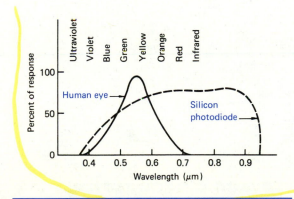

FIGURE 9.64 Spectral Response of Silicon Photodiode Compared with Response of Human Eye

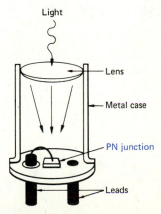

B. Packaged Photodiodes

FIGURE 9.66 Photodiodes

Detecting the change in current can be accomplished in several ways. Figure 9.67A shows the photocurrent developing a voltage across an external load R_L. Normally, this method is not sensitive enough to develop small voltages. For better development of small voltages, a differential amplifier may be used, or the Norton amplifier, shown in Figure 9.67B, may be used. In the Norton amplifier, as the light level increases, diode D_1 increases its conduction. The increase in current is mirrored in the current flowing through the feedback resistor R_F. Hence, more voltage is produced at the output.

The best mode of operation of the photodiode is shown in Figure 9.67C. In this op amp mode of operation, a constant voltage is applied across the photodiode, and the current changes with variations in the amount of light that strikes the detector. The amount of change in current is based on the sensitivity of the device. A data sheet for an SFH205 is shown in Figure 9.68. Note that it has a sensitivity 0.57 A/W when illuminated with a source that has a wavelength of 950 nm. (Wavelengths longer than 770 nm are considered to be in the infrared spectrum.)

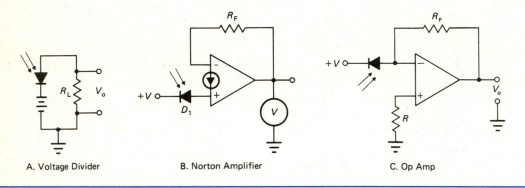

A. Voltage Divider   B. Norton Amplifier   C. Op Amp

FIGURE 9.67   Applications of Photodiodes

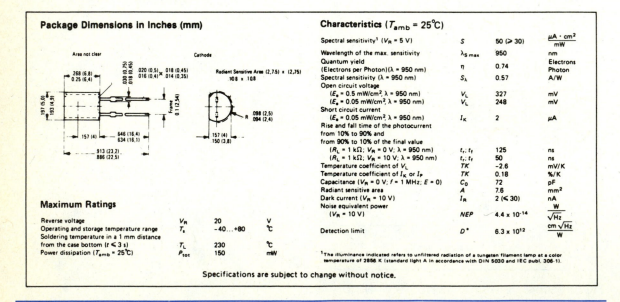

FIGURE 9.68   PIN Photodiode Data Sheet for SFH205

The output current of the SFH205 photodiode can be calculated as follows: Let us assume that the photodetector is illuminated by 0.5 mW/cm² of light radiation from a 950 nm source. From the data sheet, we find that the detector has an area of 7.6 mm². To find the total amount of power striking the radiant-sensitive area, we multiply the illumination (E), in units of power per square unit of area, by the area (A) of the detector. To get the photocurrent (I_{photo}), we multiply the total amount of power striking the detector by the sensitivity (S) of the detector. This relationship is summarized in Equation 9.19:

$$I_{photo} = ESA \qquad\qquad (9.19)$$

where

E = amount of optical power falling on detector per unit detector area (sometimes called the incidence or illumination)
S = sensitivity of detector (in amperes per watt)
A = radiant-sensitive area of detector

Note that A and S are available from the data sheet. From Equation 9.19, the photocurrent is then

$$I_{photo} = ESA$$
$$= (0.5 \text{ mW/cm}^2)(0.57 \text{ A/W})(7.6 \text{ mm}^2)$$
$$= 21.6 \text{ } \mu A$$

If we placed the SFH205 in the circuit in Figure 9.67C with a feedback resistance of 50 kΩ, the output voltage would be

$$V_o = (I_{photo})(R_F)$$
$$= (21.6 \text{ } \mu A)(50,000 \text{ } \Omega)$$
$$= 1.083 \text{ V}$$

• **EXAMPLE 9.13**
Given an SFH205 diode with an incidence of 0.25 mW/cm², find the output of the circuit in Figure 9.67B with a 50,000 Ω feedback resistance.

Solution
The photocurrent is found by substituting the appropriate values into Equation 9.19:

$$I_{photo} = ESA$$
$$= (0.25 \text{ mW/cm}^2)(0.57 \text{ A/W})(7.6 \text{ mm}^2)$$
$$= 10.8 \text{ } \mu A$$

If we place the SFH205 in the CDA circuit of Figure 9.67B with a feedback resistance of 50 kΩ, the output voltage is

$$V_o = (I_{photo})(R_F)$$
$$= (10.8 \text{ } \mu A)(50,000 \text{ } \Omega)$$
$$= 0.54 \text{ V}$$

Photodiodes, with their extremely fast response, are ideal for laser detection. They also find applications in ultrahigh-speed demodulation, switching, and decoding.

Phototransistors. As we might expect, if diodes can be made photosensitive, transistors can, too. The *phototransistor* is electrically similar to a small-signal silicon transistor. Structurally, the only difference is that in the phototransistor, the collector-base junction is larger and is exposed to light.

A diagram of the structure of the phototransistor is shown in Figure 9.69A. The base material is thin so that impinging light can strike the collector-base (CB) junction. When light strikes the collector-base junction, electron-hole pairs are created, as in the diode. In this way, base current is created and amplified by the current gain of the transistor. The CB junction then acts like a current source, as indicated in Figure 9.70. Although the photodiode needs amplification, the phototransistor does not: The current generated in the phototransistor is automatically amplified in the collector.

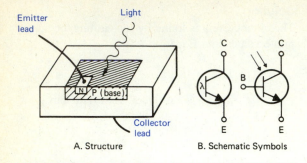

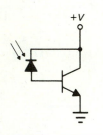

FIGURE 9.69   Phototransistor

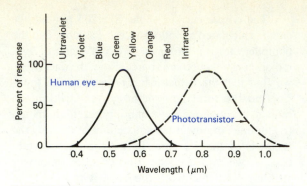

FIGURE 9.71   Spectral Response of Phototransistor Compared with Response of Human Eye

FIGURE 9.70   Equivalent Circuit for Phototransistor

The phototransistor is much more sensitive than the photodiode. However, it has higher junction capacitances, which give it poor frequency response compared with the photodiode.

Phototransistors can be either two-lead or three-lead devices (see Figure 9.69B). In the two-lead package, only the emitter and collector are connected; the base is not electrically available for biasing. The only drive for the two-lead device is the light falling on the CB junction. Two-lead packages remain the most common form of phototransistor. The three-lead form allows electrical connection to the base for purposes of biasing and decreasing device sensitivity.

Spectral response characteristics of the phototransistor are shown in Figure 9.71. Notice that the response characteristics of the phototransistor extend well into the infrared. For this reason, tungsten lamps, which emit radiation in this area, are often used to illuminate phototransistors.

Because its response is slower than that of the photodiode and because it is nonlinear, the phototransistor most often is used for computer punch and paper tape readers. When the phototransistor is employed in a digital application, linearity considerations are not important.

The phototransistor produces a photocurrent in much the same way as does the photodiode. Recall that the transistor has a reverse-biased, collector-base junction diode. The photocurrent is produced by the light radiation falling on the junction. The photocurrent is then multiplied by the current gain of the transistor. The phototransistor is thus more sensitive than the photodiode. The frequency response of the photodiode is better than that of the phototransistor, however.

For purposes of comparison, Figure 9.72A shows a data sheet for the CLT2165 phototransistor. Note that the current flow is 1.2 mA (approximately). The SFH205 photodiode has a current flow of 216 µA at the same incident radiation of 0.5 mW/cm². Using this information, we can calculate the output voltage produced by the circuit shown in Figure 9.72B. A 1.2 mA current flow through a 1.0 kΩ resistor will drop 1.2 V. With a +10 V supply, the output voltage will be 8.8 V.

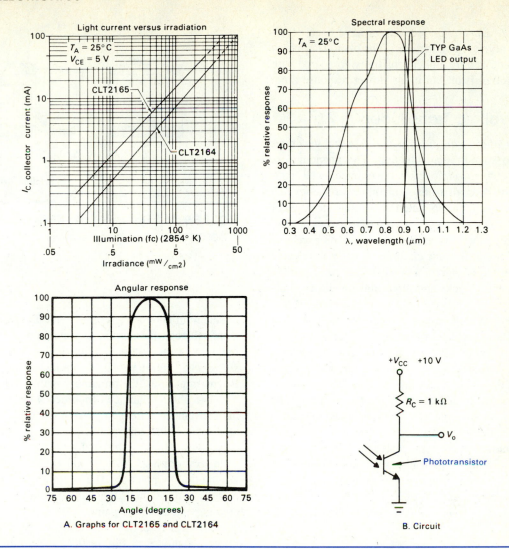

FIGURE 9.72   Phototransistor Data Sheet and Circuit for CLT2165/CLT2164

• EXAMPLE 9.14

Given a CLT2164 phototransistor in the circuit shown in Figure 9.72B, find the output voltage if the incident radiation is 0.5 mW/cm².

Solution

The chart in Figure 9.72A shows a collector current of 0.5 mA with incident radiation of 0.5 mW/cm². A current of 0.5 mA through a 1.0 kΩ resistor will drop 0.5 V. The output voltage will be +10 V −0.5 V, or 9.5 V.

Photo Darlington transistors have been developed by several manufacturers. The photo Darlington is characterized by higher sensitivity compared with the phototransistor, but it has slower response time.

Photovoltaic Transducers

A *photovoltaic transducer* is a device that generates a voltage when irradiated by light. This phenomenon, called the *photovoltaic effect*, is exhibited by all semiconductors. Recall from semiconductor theory that the area around the PN junction becomes depleted of carriers. The P-type region receives an excess of electrons, while the N-type region receives an excess of holes. An equilibrium condition finally is reached where no more migration of carriers occurs. The PN junction then blocks any more movement of majority carriers.

The photovoltaic cell is so constructed that when light falls on it, electron-hole pairs are formed, as shown in Figure 9.73. These pairs become minority carriers. Unlike majority carriers, minority carriers can and do move across the junction. Thus, a negative potential is developed at the terminal attached to the N-type material. If a load is attached to the two terminals, current flows; the current is proportional to the amount of carriers created by light.

Two basic types of photovoltaic devices are in general use today: silicon cells and selenium cells. Pictorial representations of both are shown in Figure 9.74. In the *selenium cell* (Figure 9.74A), a thin layer

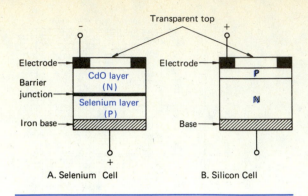

FIGURE 9.74 Photovoltaic Cells (Photocells)

of selenium is deposited on a metal base. Cadmium is then deposited on top of the selenium. A PN junction forms between the cadmium and selenium. In the *silicon cell* (Figure 9.74B), a thin layer is doped with P-type impurities, the remainder with N-type.

The photovoltaic cell (photocell) has low internal resistance in order to prevent the voltage generated from being lost within the device under load conditions. This low internal resistance is achieved by heavy doping. A depletion region then forms around the PN junction. The operation of both types of cells is the same: Light falls on the PN junction, creating electron-hole pairs.

Spectral responses of these two types of cells differ, as shown in Figure 9.75. Under high light intensities, silicon cells produce as much as 0.5 V between their terminals and supply as much as 100 mA of current. Selenium cells, although possessing better spectral qualities, are much less efficient as power converters. For this reason, silicon cells are more useful as solar cells. The purpose of a solar cell is to convert radiant energy from the sun into electric energy. Solar cells are often found in series, parallel, or series-parallel combinations to increase current and voltage output.

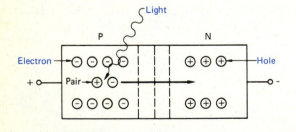

FIGURE 9.73 Construction of Photovoltaic Cell

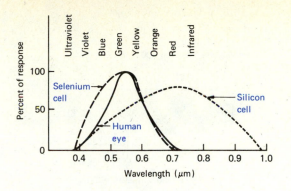

FIGURE 9.75 Spectral Response of Selenium and Silicon Photovoltaic Cells Compared with Response of Human Eye

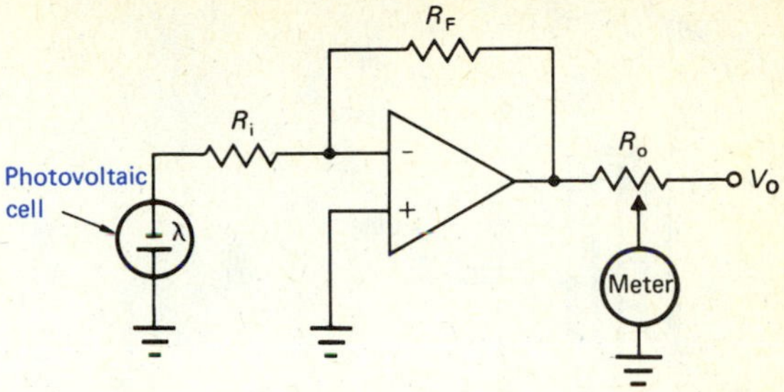

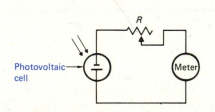

FIGURE 9.77 Op Amp Combined with Photovoltaic Cell

Selenium cells are often used in light meters, an application illustrated in Figure 9.76. Such circuits are very simple, requiring only a sensitive meter movement, a variable resistor, and the photovoltaic cell.

A more sensitive meter may be obtained by using an op amp, as shown in Figure 9.77. The op amp amplifies the small voltage change from the photovoltaic cell. Note the different schematic symbols used for photovoltaic cells in Figures 9.76 and 9.77. Both symbols are used interchangeably.

Photoemissive Transducers

Photoemissive transducers emit electrons when struck by light. Most of the transducers using this principle are vacuum tubes. Although most vacuum tube devices have been replaced in industry, two are still being used as photosensors: the phototube and the photomultiplier.

Phototubes. The *phototube*, shown in Figure 9.78A, has a peculiar construction. The rod-shaped device is the anode. The cathode, shaped as an open cylinder, is coated with a *photoemissive material*. When light strikes this material, it emits electrons. Increasing light levels increase the anode current.

Although the phototube has been replaced by the photodiode in many applications, it is still used. Its spectral response is very linear, and it has excellent frequency response characteristics. Because current flows are only a few microamperes, the phototube usually needs amplification.

Photomultipliers. The *photomultiplier* operates on a principle similar to that of the phototube. It has a photosensitive cathode that emits electrons when light

FIGURE 9.76 Light Meter Using Photovoltaic Cell

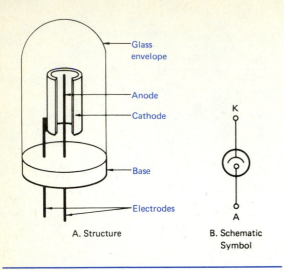

A. Structure

B. Schematic Symbol

K

A

FIGURE 9.78 Phototube

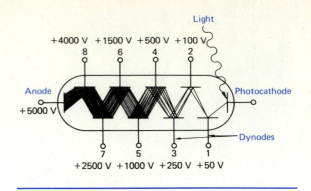

FIGURE 9.79 Structure and Operation of Photomultiplier Tube

strikes it. In addition to one anode, the photomultiplier has between 10 and 15 electrodes, which are called *dynodes*. The structure of the photomultiplier is shown in Figure 9.79.

Starting with the dynode closest to the cathode, each dynode has an increasingly high positive potential applied to it. When a photon strikes the cathode, electrons are emitted. The emitted electrons are attracted to dynode 1, striking it. Secondary emission occurs from dynode 1, and more electrons are emitted. This cascading process occurs until a great many electrons are emitted from dynode 8. Photomultipliers have gain factors up to 10^9, which makes them one of the most sensitive photodetectors available.

Photosensors have gained a wide following in industry in recent years. Their popularity is due in large measure to their versatility. Figure 9.80 illustrates the photosensor's wide area of application.

Web breaks—that is, the breaks that sometimes occur in a roll of paper or cloth—can be detected with the arrangement shown in Figure 9.80C. Any break will let the light from a source pass through to a detector, thus setting off an alarm or turning off the drive motor. In Figure 9.80E, boxes on a conveyor may be merged without colliding. The box on the right will be stopped until the box on the left passes by. In Figure 9.80F, information from registration marks (reflected back to a sensor) can initiate related operations, such as printing, cutting, or folding. Gluing can be accomplished, as shown in Figure 9.80G, by using the photodetector to detect the presence of a product. Obviously, the manufacturer would not want glue applied when there is no product present to receive it. Guillotine blades can be controlled by photosensors in cutting operations, as shown in Figure 9.80H. A common and important industrial application of photosensors is in detecting fill levels, as shown in Figure 9.80L. The fill detectors (above the box) are turned on by the box detectors (on opposite sides of the box). Without this circuit, the fill inspection pair might mistake the space between the boxes as an improper fill.

A summary of optoelectronic sensors is given in Table 9.9. A summary of their schematic symbols is given in Figure 9.81.

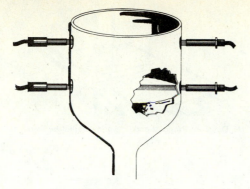

A. Keeping Hopper Fill Level between High and Low Limits

B. Counting Products

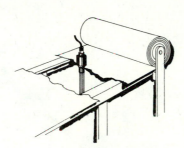

C. Detecting Web Break

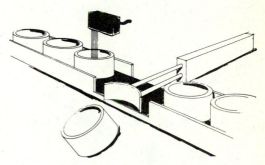

D. Checking Dark Caps for White Liners

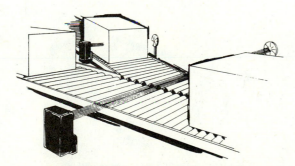

E. Preventing Collisions Where Two Conveyors Merge

F. Detecting Registration Marks

FIGURE 9.80 Applications of Optoelectronic Sensors

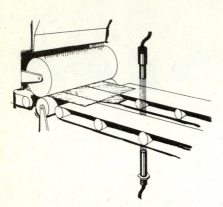

G. Gluing, Buffing, or Flattening

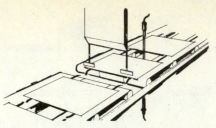

H. Detecting Products and Operating Guillotine Blade

I. Detecting Thread Breaks

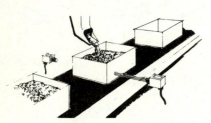

J. Slowing Conveyor and Filling Carton

K. Controlling Size of Paper or Fabric Roll

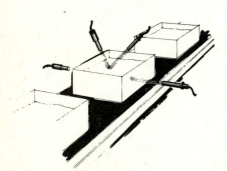

L. Checking Fill Level

M. Turning Glue Nozzle On and Off

FIGURE 9.80 (Continued)

TABLE 9.9 Summary of Optoelectronic Sensors

Type	Name	Spectral Response (µm)	Response Time	Advantages	Disadvantages
Photoconductive	Photorcsistor			Small	Slow
	CdS	0.5–0.7	100 ms	High sensitivity	Hysteresis
	CdSe	0.6–0.9	10 ms	Low cost	Temperature
				Visual range	range limited
	Photodiode	0.4–0.9	1 ns	Very fast	Low-level output
				Good linearity	(current)
				Low noise	
				Wide spectral range	
	Phototransistor	0.25–1.1	1 µs	High current gain	Low frequency
				Can drive TTL	response (500
				Small size	kHz)
					Nonlinear
Photovoltaic	Photovoltaic (solar) cell	0.35–0.75	20 µs	Linear	Slow
				Self-powered	Low-level output
				Visual range	(voltage)
Photoemissive	Phototube	0.15–1.2	1 µs	Good stability	Fragile
				Fast response	
				High impedance	
	Photomultiplier	0.2–1.0	1–10 ns	Fast response	Fragile
				High sensitivity	Need high-voltage
				Good linearity	power supply

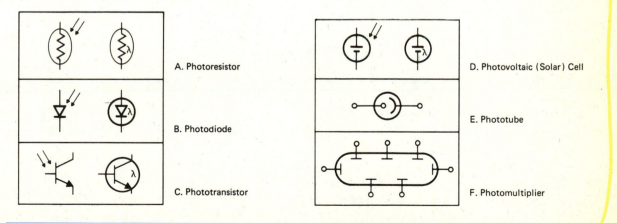

A. Photoresistor

B. Photodiode

C. Phototransistor

D. Photovoltaic (Solar) Cell

E. Phototube

F. Photomultiplier

FIGURE 9.81 Schematic Symbols for Optoelectronic Sensors

CONCLUSION

We have seen in this chapter that optoelectronic devices are widely used in industry. In the case of the laser, the applications of these devices are so varied that the inventors could not have imagined the uses to which these devices are put. You should now have a firm grasp of the basic principles of optoelectronic receivers and transmitters. Although you will probably not have to specify these devices, a good understanding of how they work will help you in troubleshooting equipment malfunctions.

■ QUESTIONS ●

1. A wavelength of 7000 Å is equal to _____ nanometers.

2. Differentiate between wavelength measurements of nanometers, millimicrons, microns, and angstrom units.

3. Define the following terms: (a) radiometry and (b) photometry.

4. Define the following terms and list their abbreviations and units of measurement: (a) radiant flux, (b) radiant intensity, (c) steradian, (d) radiant sterance, (e) radiant incidence, and (f) radiant exitance.

5. Define the following terms and list their abbreviations and units of measurement: (a) luminous flux, (b) luminous intensity, (c) luminous sterance, (d) luminous incidence, and (e) luminous exitance.

6. Describe and contrast the following concepts: absorption, spontaneous emission, and stimulated emission.

7. Which color LED is best for viewing by the human eye?

8. In the LED package, the longer lead is usually the _____ lead.

9. List three factors that determine LED efficiency.

10. List three advantages of LEDs and one disadvantage when compared to filament light sources.

11. State the requirements for a laser.

12. Describe population inversion in a laser.

13. List three characteristics of a laser.

14. List the wavelengths of the following lasers: (a) HeNe, (b) ruby, (c) Nd:YAG, (d) Nd:glass, (e) CO_2, and (f) argon.

15. Injection lasers are made out of _____ material.

■ PROBLEMS ●

1. Given a light with a wavelength of 5650 Å, find the wavelength in nanometers.

2. Given a light with a wavelength of 800 nm, find the wavelength in microns.

3. Given that an electron falls from E_3 (5.75 eV) to E_1 (4.75 eV), find the wavelength of the photon given off from this transition.

4. Given an LED with a forward voltage (V_F) of 1.25 V at 25 mA and a supply of +12 V, find the size of the series-limiting resistor.

5. Given a laser with a length of 0.75 m, find the frequency difference between longitudinal modes.

6. Given a krypton laser with a wavelength of 752.5 nm and a bandwidth of 1200 MHz, find the coherence length and the number of coherent wavelengths.

7. Given a laser with a beam waist of 1.75 cm and an intensity of 120 W/sr, find the intensity 10 cm out from the center of the laser.

8. Given an argon laser with a beam waist of 2.5 mm and a wavelength of 514.5 nm, find the divergence of the beam in radians.

9. Given a laser with a beam waist of 1.75 mm, a distance of 7 km, and a wavelength of 441.6 nm, find the size of the reflector needed to reflect most of the energy.

10. Given a laser with a beam waist of 2.5 mm and a wavelength of 752.5 nm, find the in-focus beam spot radius.

11. Given a pulsed laser with a PRR = 2 kHz, producing a 150 mJ pulse in 2.4 μs, find the peak power of the pulse and the average power of the pulse train.

12. Given a CL707HL with an illumination of 1.5 fc, find the output voltage for the circuit in Figure 9.62B.

13. Given an SFH205 diode with an incidence of 0.25 mW/cm^2, find the output of the circuit in Figure 9.65B with a 25,000 Ω feedback resistance.

14. A CLT2164 phototransistor is used in the circuit shown in Figure 9.72B. If the incidence is 0.25 mW/cm^2, find the output voltage.

Industrial Process Control

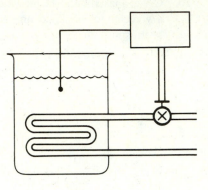

OBJECTIVES

On completion of this chapter, you should be able to:

- Define process control.
- List the elements of a process control system.
- Define process load, process lag, and stability, and describe how they relate to a process control system.
- List and give examples of two types of process control systems.
- Distinguish between feedback and feedforward control.
- Describe the following controller modes: on-off, proportional, integral, and derivative.
- Contrast electric controllers and pneumatic controllers in terms of the advantages and disadvantages of each.
- Distinguish between servos and other types of automatic controllers.
- Describe the operation of a synchro.

INTRODUCTION

A *process control system* can be defined as the functions and operations necessary to change a material either physically or chemically. Process control normally refers to the manufacturing or processing of products in industry.

Every process has one or more controlled, or dynamic, variables. The *controlled* (or *dynamic*) *variable* is a variable we wish to keep constant. Processes also have one or more manipulated variables, or control agents. A *manipulated variable* is a variable that we change to regulate the process. Specifically, the manipulated variable enables us to keep the controlled variable constant. Examples of controlled and manipulated variables are pressure, temperature, fluid flow, and liquid level. Finally, each process control system has one or more disturbances; *disturbances* tend to change the controlled variable. The *function* of the process control system is to regulate the value of the controlled variable when the disturbance changes it.

An example may help to identify the components of a process control system. Figure 10.1 illustrates a system in which milk is heated to pasteurizing temperature. The temperature of the milk is the controlled variable. Steam flows through the pipes, transferring its heat to the milk. The flow of steam controls the temperature of the milk. Therefore, the steam flow is the manipulated variable. The ambient temperature surrounding the tank can be considered a disturbance. For instance, if the ambient temperature decreases, the temperature of the milk will eventually decrease.

In an automatic-control situation, the temperature of the milk would be monitored. If it decreased beyond allowable limits, the controller would make adjustments to bring the temperature under control. The *controller* is that part of a process control system that decides how much adjustment the system needs and implements the results of that system. For example, the controller might open the steam control valve, causing the milk to be heated and returning it to the required temperature. This valve is the final control element in a process control system.

In this chapter, we will present a basic outline of process control systems. Since this subject is rather complex, we will present only a general overview of these systems. We will first discuss the characteristics that are found in most process control systems. Next, we will consider general types of process control and the ways in which controllers operate. Finally, we will give a short summary of those final control elements we have not discussed in previous chapters. Recall that we have already covered motors in Chapters 3, 4, and 5.

Well-designed process control systems can save an enormous amount of time and money, reduce error (which improves the quality of a product) and can provide greater operator safety. Because of these advantages, sophisticated process control systems are frequently found in the manufacturing industry. The tasks of preventive and corrective maintenance vary directly with the complexity of the system. Therefore, technicians should have a firm grasp of the concepts behind process control systems.

CHARACTERISTICS OF PROCESS CONTROL SYSTEMS

All process control systems have three characteristics in common. First, each automatic process control system makes a *measurement* of the controlled variable. This measurement usually is made by a sensor or transducer. As we saw in Chapter 8, transducers change the controlled variable into another form, usually electrical. In our milk example, a thermistor

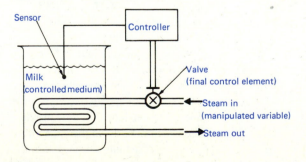

FIGURE 10.1 Process Control System for Pasteurizing Milk

could have been used to change the temperature of the milk into a corresponding electric resistance. As the temperature of the milk changed, so would the sensor's resistance. Since the sensor's output must be evaluated by the controller, the output must be in a suitable form. For example, if the controller is a computer, the sensor's output must be digital. Thus, the sensor's output may need to be converted, or conditioned, to an appropriate form for evaluation.

Second, as mentioned, the controller needs to *evaluate* the information from the sensor. It does so by comparing the sensor measurement with a reference called the *set point*. If the sensor measurement differs from the set point, an *error condition* occurs. The controller then decides whether or not this amount of difference is acceptable. If it is not acceptable, the controller initiates action to reduce the error.

Finally, each process control system must have a final control element. A *final control element* makes those adjustments that are necessary to bring the controlled variable back to the set point value.

PROCESS CHARACTERISTICS

Every process exhibits three basic characteristics that are important in the understanding of process control systems. These characteristics are process load, process lag, and stability.

Process Load

Process load can be defined as the total amount of control agent needed to keep the process in a balanced condition. In the milk pasteurization example, we need a certain amount of steam (the control agent) to keep the milk at the correct temperature. Suppose the ambient temperature around the tank decreases. Then we need a greater rate of steam flow to keep the milk at a constant temperature. Thus, a decrease in the ambient temperature constitutes a change in the process load.

The process load is directly related to the setting of the final control element. Any process load change will cause a change in the state of the final control element. The final control element's adjustment is what keeps the process balanced.

Process Lag

Process lag is the time it takes the controlled variable to reach a new value after a process load change. This time lag is a function of the process and not the control system. The control system has time lags of its own. Process lags are caused by three properties of the process: capacitance, resistance, and transportation time.

Capacitance can be defined as the ability of a system to store a quantity of energy or material per unit quantity of a reference. For example, a large volume of water has the ability to store heat energy. As a result, we could say that the water has a large thermal capacitance.

A large capacitance in a process means that it takes more time for process load changes to occur. From one viewpoint, large capacitance is a desirable characteristic because keeping the controlled variable constant is then easier. Small disturbances do not have very much effect on the process load. On the other hand, large capacitance also means that it is more difficult to change the controlled variable back to the desired point once it has been changed.

Resistance in a process can be defined as opposition to flow. In the pasteurization example, we would notice thermal resistance in the walls of the steam pipes. In other words, the material of the pipes slows down the transfer of heat to the liquid. Sometimes, gases and liquids surround the pipes in layers or films, increasing the thermal resistance compared with pipes without the layers or films. These layers have a slowing effect on the transfer of heat energy and thus have resistance. Large resistances will, therefore, increase the process lag by opposing the change in the controlled variable.

Often, the capacitance and resistance of a system are combined into a factor called *RC delay*, or the *RC time constant*. Recall that, in an electric system, the time it takes to charge a capacitor to 63% of its final voltage is one time constant. The rate at which the capacitor charges will depend on the capacitance of the capacitor and the resistance through which it must charge. The product of the resistance and the capacitance (*RC*) gives one time constant (in seconds).

Process control systems have the same RC lag. In the pasteurization example, the product of the thermal capacitance (in joules per degree Celsius) and the thermal resistance (in degrees Celsius per joule per second) will give one time constant (in seconds). In practice, one system may have many RC lags caused by many individual resistances and capacitances.

The third component of process lag, *transportation time*, or *dead time*, can be defined as the time it takes for a change to move from one place to another in a process. Or we can define transportation time as the time between the application of the disturbance and the changing of the process load. Dead time is most easily illustrated in applications where fluids or solids are moving, as in a flowing fluid or a conveyor belt application. In a fluid flow application, for example, let us say that a temperature change is made some distance upstream from the sensor. It will take some time for the temperature change to be transported downstream to the sensor. In the conveyor illustration, the same principle applies. Any change in the condition of the product (weight, for example) will not be detected until that change is communicated to the sensor. This time delay is the dead time, or transportation time.

Note that any controller action is delayed by the amount of time delay present. To find the time delay, we use the equation $d = rt$, which can be rewritten as follows:

$$t = \frac{d}{r} \qquad (10.1)$$

where

t = transportation (delay) time (in minutes)
d = distance (in centimeters)
r = flow rate (in centimeters per minute)

In the fluid flow example, suppose that fluid is flowing at a rate of 100 cm/min, and say the temperature sensor is located 50 cm from the temperature change. Then transportation (delay) time is

$$t = \frac{d}{r} = \frac{50 \text{ cm}}{100 \text{ cm/min}} = 0.5 \text{ min} \quad \text{or} \quad 30 \text{ s}$$

In other words, it will take 30 s for the temperature change to reach the sensor. Thus, there will be a delay of 30 s *before* the controller can even start to react.

• EXAMPLE 10.1

Given a hopper that discharges a solid on a 5 ft long conveyor belt at a rate of 10 lb/min, with the rate at which solids fall off the end of the belt also 10 lb/min and the conveyor belt moving at a rate of 1 ft/min, find the amount of time it will take for the rate at which solids fall from the belt to increase if the rate at which the solids are discharged to the belt is increased to 20 lb/min.

Solution

The change in the flow rate will not be seen at the end of the belt until the material has been transported the entire length of the conveyor. The same change in the flow rate of the solid moving off the end of the conveyor will be seen after a delay equal to the transportation time. The transportation time (t) can be found by using Equation 10.1:

$$t = \frac{d}{r} = \frac{5 \text{ ft}}{1 \text{ ft/min}} = 5 \text{ min}$$

Lags occur in other areas also, such as the sensor, the controller, and the final control element. As we saw in Chapter 8, a temperature sensor takes some amount of time to respond to a temperature change. The sensor must be heated to the same temperature as the dynamic variable. In the case of the RTD, which may be enclosed in a protective sheath, some delay will occur while the temperature change is transmitted through the sheath to the sensor. You will also recall that the sensor itself has a thermal mass, which will slow down the sensor temperature change. This delay is called the *transducer lag*, or *measurement delay*.

Sometimes, the controller takes a significant amount of time to evaluate the input data and make a decision. Since most controllers are chosen for their speed, this factor is not usually a problem unless the process conditions change quickly. Sometimes, a *controller lag* is introduced into the system by the process engineer to meet certain system requirements.

Many final control elements have significant lags. Large motors, for example, may take a minute or so to come up to full speed after a load change.

Electrically operated valves are notoriously slow in opening and closing. Some valves may take 2 min to fully open from a closed position.

Process engineers are especially interested in keeping dead time to a minimum. Long dead times make accurate process control difficult.

Stability

An important consideration when examining control systems using feedback is the stability factor. We say that a process control system is *stable* if it can return the controlled variable to a steady-state value. Typically, an unstable system will cause the controlled variable to oscillate above and below the desired value.

If the controlled variable oscillates above and below the desired value, three things will happen. First, the strength of the oscillations will increase in amplitude as time increases. This result occurs when the feedback is in phase and the loop gain is greater than 1. (Recall the Barkhausen criteria for oscillators discussed in Chapter 1.) Second, if the feedback is in phase with the oscillations and the loop gain is 1, the oscillations will have a constant amplitude. Third, if the loop gain is less than 1 and the feedback is out of phase with the oscillations, then the oscillations will gradually die out. The amount of time that it takes for the oscillations to die out, or damp, is called the *settling time*. A good process control system will reduce settling time to a minimum.

Important information can be gained from analysis of the system response to a load change. If too much control action is present, corresponding to a loop gain larger than 1, the system responds with an unstable and possibly dangerous response (Figure 10.2A). If the amount of control action or feedback is decreased to a loop gain of 1, we will see the system responding as in Figure 10.2B, with a stable, constant-amplitude response. As the amount of control is decreased to a loop gain of less than 1, we see a type of response that is called *underdamped*.

This system response is so named because it does not have enough damping or resistance to keep the system from oscillating, as shown in Figure 10.2C. Damping in a control system is a force that opposes the change in the manipulated variable. A bit less control (less positive feedback) will give a waveform known as a *critically damped* response (Figure 10.2D). Even less control action and more damping produce the waveform known as *overdamped* (Figure 10.2E).

Which system response is best? We can rule out the first response, the one with the increasing amplitude, as generally undesirable. The oscillating, constant-amplitude waveform, typical of one particular type of controller, may not be particularly desirable but may be acceptable if the system does not require close control of the dynamic variable. The choice usually is between the critically damped, underdamped, or overdamped responses. The overdamped response usually is chosen if close control is necessary and no oscillations are tolerable. The underdamped response is chosen where fast dynamic variable response is required. In general, however, the critically damped response is most desired, offering a compromise between speed of response and oscillations.

TYPES OF PROCESS CONTROL

Generally speaking, process control can be classified into two types: open-loop and closed-loop control. Closed-loop control can be further divided into feedback and feedforward control.

Open-Loop Control

Open-loop control involves a prediction of how much action is necessary to accomplish a process. That is, in an open-loop system, no check is made during the process to see whether corrective action is necessary to accomplish the end result.

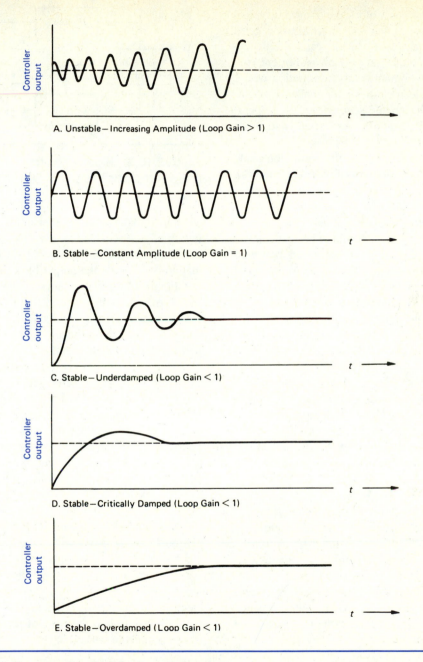

A. Unstable—Increasing Amplitude (Loop Gain > 1)

B. Stable—Constant Amplitude (Loop Gain = 1)

C. Stable—Underdamped (Loop Gain < 1)

D. Stable—Critically Damped (Loop Gain < 1)

E. Stable—Overdamped (Loop Gain < 1)

FIGURE 10.2   Process Control System Responses

Figure 10.3 shows a block diagram of a typical open-loop system. A washing machine is a good example of such a system. In this case, the person operating the machine takes a measurement (the clothes are dirty), compares this information with a reference level (clean clothes), and makes a prediction (a cycle is chosen and soap added). The operator then starts the machine and attends to other business.

The operator assumes that the prediction made will accomplish the desired objective (clean clothes). If the prediction is absolutely correct in all aspects (amount of soap, amount of hot and cold water, and so forth), the clothes will be absolutely clean. Therefore, open-loop control is capable of perfect control. If the prediction was wrong for any reason, however, the objective will not be reached. Open-loop control does not always provide perfect control. Since the washing machine does not make any measurements or comparisons, any mistakes in the prediction will produce an undesirable outcome.

Closed-Loop Control

In a *closed-loop system*, a measurement is made of the variable to be controlled. This measurement is then compared with a reference point, or set point. If a difference or error exists between the actual and desired levels, an automatic controller will take the necessary corrective action.

Feedback Control. A block diagram of a typical closed-loop control system is shown in Figure 10.4. A comparison of Figures 10.3 and 10.4 shows the difference between open-loop and closed-loop systems. In the open-loop system, no actual measurement is made on the process. In contrast, in closed-loop control, the variable to be controlled is measured and compared with a reference (the set point), and corrective action is taken.

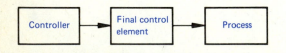

FIGURE 10.3 Block Diagram of Open-Loop Control System

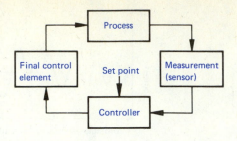

FIGURE 10.4 Block Diagram of Closed-Loop Feedback Control System

The milk pasteurization example is a good illustration of a closed-loop system. This system is often called a *feedback control system* since sensor information is fed from the output back to the input.

Figure 10.4 shows a block diagram for a closed-loop feedback control system. The measurement of the dynamic variable is taken by the sensor. The controller compares the information from the set point and the measurement. It then decides whether or not to take corrective action on the basis of its evaluation. The controller can be any circuit or system capable of evaluation and decision making, from a simple op amp comparator to a complex digital computer. In the block diagram, the controller also performs any signal conditioning needed to make the sensor measurement compatible with the input to the evaluation circuit. The controller, if it decides action is to be taken, adjusts the final control element to bring the dynamic variable back under control.

We can relate this block diagram to the pasteurization example discussed at the beginning of the chapter. The measurement is taken by a temperature sensor. The controller, on the basis of this measurement, decides whether to adjust the final control element (a valve) or to leave it unchanged. Recall that the valve controlled the flow of steam in the pipes (the manipulated variable).

Feedforward Control. Closed-loop feedback control has disadvantages in two areas of process control systems. Feedback control is not satisfactory when there are large disturbances in the process load and when there are large process lags. These disadvantages can be dealt with by using a feedforward (or predictive) type of control. *Feedforward control* is defined as a closed-loop system that feeds a correction signal forward to the controller based on a measurement of the disturbance. A simple feedforward control system is shown in Figure 10.5.

There is one basic difference between feedforward control and open-loop control: Open-loop control makes an assumption about the variables in the system. In contrast, feedforward control makes a measurement of a disturbance. From this measurement, the controller estimates what action needs to be taken to keep the process from changing. In other words, the controller decides what change to make in the manipulated variable so that the change, when combined with the disturbance, will produce no change in the controlled variable. The controller anticipates the effect the disturbance will have on the process.

Feedforward control, like open-loop control, relies on a prediction. It differs from open-loop control in that it does not rely on a fixed program. Feedforward control is a type of closed-loop control.

Note the difference between feedback control and feedforward control. In feedback control, a measurement of the dynamic variable is taken. In feedforward control, a measurement of the disturbance is taken.

Like feedback control, feedforward control is not without its problems. The prediction made in feedforward control assumes that all significant disturbances are known. It also assumes that there are sensors present to measure these disturbances and that we know exactly how the disturbances will affect the process. However, the engineering and technical know-how needed to accomplish this type of control is immense. Thus, feedforward control is used only where the process load and the disturbances to it are well understood and can be accurately predicted.

BASIC CONTROL MODES

As we discussed previously, the controller performs evaluation and decision-making operations in a control system. It takes the difference or error signal from the comparator and determines the amount of change the manipulated variable needs to bring the controlled variable back to normal. The controller may contain the *error detector* (or comparator), a signal-conditioning device, a recording device, and a telemetering device. The entire system with all these possible combinations is shown in Figure 10.6.

Controllers may be classified in several different ways. For instance, they may be classified according to the type of power they use. Two common types in this category are electric and pneumatic controllers. *Pneumatic controllers* are decision-making devices that operate on air pressure.

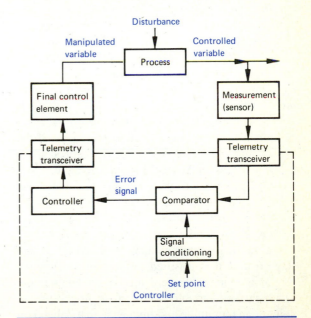

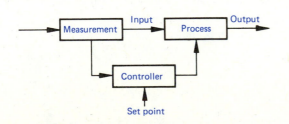

FIGURE 10.5 Block Diagram of Closed-Loop Feedforward Control System

FIGURE 10.6 Expanded Block Diagram of Closed-Loop Feedback Control System

Electric (or *electronic*) *controllers* operate on electric signals. You will see both types used widely in industry. Pneumatic controllers are used in the chemical and petrochemical industries for reasons of safety. They are less expensive and simpler than comparable electric controllers. However, pneumatic controllers are difficult to interface with digital computers, a great disadvantage in light of the increasing popularity of computers in process control. Electric controllers suffer no such disadvantage.

Controllers are also classified according to the type of control they provide. In this section, we will discuss five types of control in this category: on-off, proportional, proportional plus integral, proportional plus derivative, and proportional plus integral plus derivative.

On-Off Control

In *on-off control*, the final control element is either on or off. This controller is also called "bang-bang" from the speed of the response of the on-off state. In the pasteurizing example, if the controller were an on-off controller, the valve would be either open or closed. This type of control is sometimes also called *two-position control* since the final control element is either in the open or the closed position. That is, the controller will never keep the final control element in an intermediate position.

On-off control is undoubtedly the most popular mode in industry today. On-off control also has domestic applications. For example, most home heating systems use the on-off control mode. If the room temperature goes below a predetermined point, the heater turns on. When the room temperature goes above that point, the heater shuts off. Thus, the control is on-off.

On-off control is illustrated in Figure 10.7. As soon as the measured variable goes above the set point, the final control element is turned off. It will stay off until the measured variable goes below the set point. Then the final control element will turn on. The measured variable will oscillate around the set point at an amplitude and frequency that depend on the process's capacity and time response. This oscillatory response is typical of the on-off controller. For processes that require that the dynamic variable be

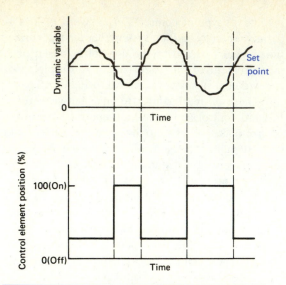

FIGURE 10.7 Plot of Measured Variable and Final Control Element Position versus Time in On-Off Controller

held relatively constant, this type of controller is not a wise choice. The oscillations may be reduced in amplitude by increasing the sensitivity of the controller, but this would cause the controller to turn on and off more frequently, a possibly undesirable consequence.

An example of an on-off controller is shown in Figure 10.8. In this simple circuit, the sensor is an LM335, a temperature-to-voltage converter. (A data sheet for this device is included at the back of this book.) The sensor is calibrated to give 2.98 V at 25°C and gives an output of 10 mV/K. Thus, if the temperature is 30°C, the output of the LM335 is

$$V_o = (\text{temperature, in K}) (10 \, \text{mV/K}) \qquad (10.2)$$

or

$$V_o = (303 \, \text{K}) (10 \, \text{mV/K}) = 3.03 \, \text{V}$$

The output of the LM335 is applied to the comparator, which in this case is the controller. Note that the comparator has a reference voltage of 3.23 V. In process control terminology, this voltage

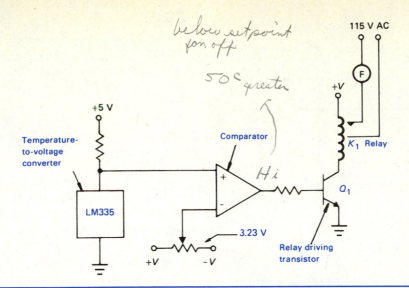

FIGURE 10.8   Op Amp On-Off Controller

represents the set point. It is called the set point because it sets the point at which the controller switches from on to off or off to on. With a temperature of 30°C and an output of 3.03 V from the LM335, the comparator output will be low. Recall that, in a noninverting comparator, the output will go high only when the input voltage goes above the reference voltage. In this case, the comparator output will go high only when the input goes above the reference of 3.23 V. This voltage will occur when the LM335 reaches a temperature of 323 K, or 50°C. When the output of the comparator goes high, it turns on the transistor Q_1, and relay K_1 energizes. The relay in this case controls a fan, which brings the ambient temperature down. The fan will stay on until the ambient temperature goes below 50°C. When the temperature goes below this set point, the comparator output will go low, turning off the transistor and the relay. The relay contacts then open, removing power from the fan.

This simple on-off controller has one main disadvantage. If the fan cools the sensor too quickly, the comparator and the relay will turn on and off quickly, possibly damaging the relay. One way to

stop this action is to add a dead zone or a differential gap, to be discussed later.

Despite its disadvantages, on-off control is useful. The on-off control mode is chosen by the process engineer under the following conditions:

1. Precise control must not be needed.

2. The process must have sufficient capacity to allow the final control element to keep up with the measurement cycle.

3. The energy coming in must be small compared with the energy already existing in the process.

On-off control is most often found in air-conditioning and refrigeration systems. It is also widely used in safety shutdown systems to protect equipment and people. In this application, on-off control is called a *shutdown* or *cutback alarm.*

Differential-Gap Control. *Differential-gap control,* shown in Figure 10.9, is similar to on-off control. Notice that a band, or gap, exists around the control point. When the measured variable goes above the

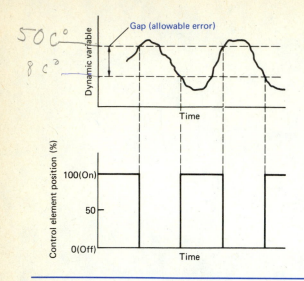

50 C°

8 C°

FIGURE 10.9 Plot of Measured Variable and Final Control Element Position versus Time in Differential-Gap Controller

upper boundary of the gap, the final control element closes. It will stay closed until the measured variable drops below the lower boundary. Some home heating systems use this type of control mode rather than on-off.

Differential-gap control does not work as well as on-off, but it does save wear and tear on the final control element. In industry, differential-gap control is often found in noncritical-level control applications, such as keeping a tank from running dry or from flooding.

The gap in differential-gap control is often called a *dead zone*. Engineers normally choose a dead zone of about 0.5% to 2.0% of the range of the final control element.

Let us consider an example of a differential-gap controller, shown in Figure 10.10. This circuit uses the LM335 as a temperature sensor, and control circuitry is provided by the versatile LM3900 CDA. The first CDA (1IC) is used as a buffer amp, isolating the sensor from the controller. The second CDA (2IC) is the controller, a circuit you may recall from Chapter 1 of this text. The 1IC is connected as a noninverting comparator, with a reference voltage of +5 V connected to the inverting input. The output of the comparator drives a transistor, which, in turn, drives a relay that controls 115 V AC power to a fan. When the comparator output goes high, the transistor is forward-biased, turning on relay K_1. The NO contacts of the relay then close, applying power to the fan.

As you know from studying this circuit in Chapter 1, the comparator output is either 0 V or +5 V,

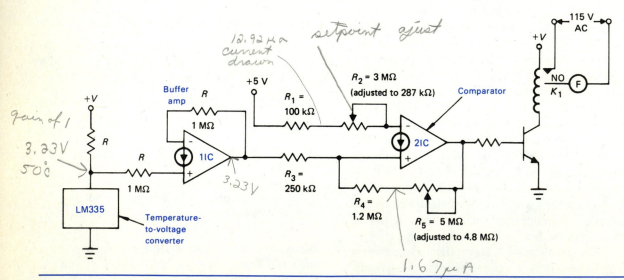

gain of 1

3.23 V

50°C

12.92 μa current drawn

setpoint adjust

3.23 V

1.67 μA

FIGURE 10.10 Op Amp Differential-Gap On-Off Controller = *simple, Less expensive*

setpoint = center of gap

depending on the state of the input. The output will go high when the input current from the buffer amp exceeds the current drawn in the reference circuit, which consists of the +5 V potential, R_1, and R_2. With the output at 0 V, the only current drawn from the noninverting terminal results from the input voltage. The current drawn from the reference circuit is

$$I_{ref} = \frac{V_{ref}}{R_1 + R_2} \qquad (10.3)$$

which, for our example, is

$$I_{ref} = \frac{+5 \text{ V}}{100 \text{ k}\Omega + 287 \text{ k}\Omega} = 12.92 \text{ μA}$$

For the output voltage to go high, then, the input must draw an amount of current greater than 12.92 μA. The input will draw this current at the following input voltage:

$$V_i = (I_{ref})(R_3) \qquad (10.4)$$

or

$$V_i = (12.92 \text{ μA})(250 \text{ k}\Omega) = 3.23 \text{ V}$$

Combining these two equations, we can find the input voltage that will make the output go high:

$$V_i = \left(\frac{V_{ref}}{R_1 + R_2}\right)(R_3) \qquad (10.5)$$

If the LM335 has been properly calibrated, this voltage should be attained when the temperature of the sensor is 50°C. This set point can be changed by varying the size of R_2. Increasing R_2 will decrease the reference current. The output of the comparator goes low when a lower voltage is applied to 1IC by the LM335.

When the output goes high, current is drawn from the noninverting terminal to the output. This current drain reduces the amount of current needed to trip the comparator to make it go low. Specifically, the input voltage needed to make the comparator go low is

$$V_i = \left(\frac{V_{ref}}{R_1 + R_2} - \frac{V_{sat}}{R_4 + R_5}\right)(R_3) \qquad (10.6)$$

Using a value of 287 kΩ for R_2 and 4.8 MΩ for R_5 and solving Equation 10.6, we have

$$\begin{aligned}
V_i &= \left(\frac{V_{ref}}{R_1 + R_2} - \frac{V_{sat}}{R_4 + R_5}\right)(R_3) \\
&= \left(\frac{5 \text{ V}}{387 \text{ k}\Omega} - \frac{5 \text{ V}}{6 \text{ M}\Omega}\right)(250 \text{ k}\Omega) \\
&= 3.02 \text{ V}
\end{aligned}$$

When the input voltage goes below 3.02 V, the output will go low again. This voltage (3.02 V) corresponds to a temperature of 29°C at the LM335. In this application, the fan switches on when the temperature reaches about 50°C. This value, the set point, can be varied by changing R_2. The fan will not go off until the temperature is brought down to about 29°C. The difference between these two temperatures is the dead or neutral zone, or differential gap. In this circuit, the differential gap is changed by adjusting resistor R_5.

• **EXAMPLE 10.2**
Given the CDA differential-gap controller in Figure 10.10, with R_2 set to 250 kΩ and R_5 set to 4 MΩ, find (a) the upper and lower trip points and (b) the temperatures that trip the comparator and the differential gap in temperature and voltage.

Solution
(a) The upper trip point is

$$UTP = \left(\frac{V_{ref}}{R_1 + R_2}\right)(R_3) \qquad (10.7)$$

which gives us

$$\begin{aligned}
UTP &= \left(\frac{+5 \text{ V}}{100 \text{ k}\Omega + 250 \text{ k}\Omega}\right)(250 \text{ k}\Omega) \\
&= +3.57 \text{ V}
\end{aligned}$$

The lower trip point is

$$LTP = \left(\frac{V_{ref}}{R_1 + R_2} - \frac{V_{sat}}{R_4 + R_5}\right)(R_3) \qquad (10.8)$$

which gives us

$$LTP = \left(\frac{+5\ V}{100\ k\Omega + 250\ k\Omega} - \frac{+5\ V}{1.2\ M\Omega + 4\ M\Omega}\right)$$
$$\times (250\ k\Omega)$$
$$= +3.33\ V$$

(b) A voltage of 3.57 V from an LM335 corresponds
to a temperature of 357 K, or 84°C. The voltage
equal to 3.33 V corresponds to a temperature of
333 K, or 60°C. The differential gap in voltage
is 3.57 V – 3.33 V = 0.24 V. In temperature, the
neutral zone is 84°C – 60°C, or 24°C.

An op amp differential-gap controller is shown
in Figure 10.11. Let us assume that we are using ideal
op amps with a grounded negative supply and a
+15 V potential for the positive supply. The output
from op amp A_3 will be high (+15 V) when the input
to the inverting terminal goes below the set point. The
set point voltage is applied to the noninverting
terminal of op amp A_3 through resistor R_5 and op amp

A_2. The output of A_3 will go low when the input
voltage (TP$_2$) goes above the set point input.

When the output of op amp A_3 goes high (to
+15 V), a +15 V potential is fed to the input of the
noninverting input of A_3. If the wiper of R_3 is
adjusted for +9 V, a +9 V potential will appear at the
output of A_2 at TP$_3$. The wiper of R_6 is set at 90 kΩ.
The voltage at TP$_4$, called the *upper trip point*, can
be found by using the following equation:

$$UTP = \frac{(V_{TP5} - V_{TP3})(R_5)}{R_5 + R_6} + V_{TP3} \qquad (10.9)$$

For the circuit of Figure 10.11, we obtain

$$UTP = \frac{[(+15\ V) - (+9\ V)](4700\ \Omega)}{4700\ \Omega + 90,000\ \Omega} + (+9\ V)$$
$$= 9.3\ V$$

When the output voltage of A_3 is low, the voltage at
TP$_4$, called the *lower trip point*, can be found from
the following equation:

$$LTP = \frac{(V_{TP3})(R_6)}{R_5 + R_6} \qquad (10.10)$$

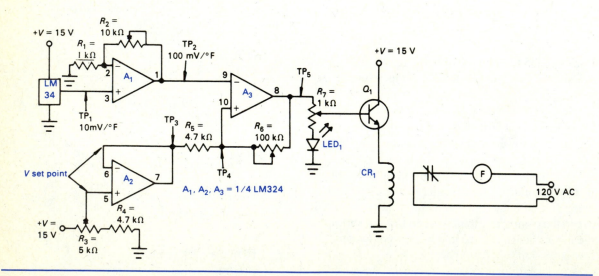

FIGURE 10.11 Op Amp Differential-Gap Controller

For our example, we get

$$LTP = \frac{(+9 \text{ V}) (90{,}000 \text{ }\Omega)}{4700 \text{ }\Omega + 90{,}000 \text{ }\Omega}$$

$$= 8.55 \text{ V}$$

Next, we need to find the temperatures at which the controller will react. The LM34 produces an output equal to 10 mV/°F. If we adjust R_2 to 9 kΩ, the output of A_1 will be equal to 100 mV/°F. With an 8.55 V potential at TP_4, the output of A_3 will go high when the sensor goes below 85.5°F. A high from A_3 will turn on Q_1 and turn on the relay coil, turning off the fan. As the temperature increases, the potential at TP_2 will rise at a rate of 100 mV/°F. When the output of A_1 goes above 93°F, the voltage at TP_2 will rise above the voltage at TP_4, driving the A_3 output low. Note that the neutral zone, or gap, voltage is 9.3 V − 8.55 V = 0.75 V. In temperature, the neutral zone is 93°F − 85.5°F, or 7.5°F.

• EXAMPLE 10.3

Given the circuit shown in Figure 10.11, with R_6 set to 80 kΩ and the wiper of R_3 set for 8 V, find (a) the upper and lower trip points in voltage and (b) the temperatures that trip the device and the differential gap in temperature and voltage.

Solution

(a) The upper trip point can be found by using Equation 10.9:

$$UTP = \frac{(V_{o3} - V_{TP3}) (R_5)}{R_5 + R_6} + V_{TP3}$$

$$= \frac{[(+15 \text{ V}) - (+8 \text{ V})] (4700 \text{ }\Omega)}{4700 \text{ }\Omega + 80{,}000 \text{ }\Omega} + (+8 \text{ V})$$

$$= +8.39 \text{ V}$$

The voltage at TP_4, the lower trip point, can be found from Equation 10.10:

$$LTP = \frac{(V_{TP3}) (R_6)}{R_5 + R_6}$$

$$= \frac{[(+8 \text{ V})] (80{,}000 \text{ }\Omega)}{4700 \text{ }\Omega + 80{,}000 \text{ }\Omega}$$

$$= 7.56 \text{ V}$$

(b) With a 7.56 V potential at TP_4, the output of A_3 will go high when the sensor goes below 75.6°F. When the output of A_1 goes above 83.9°F, the voltage at TP_2 will rise above the voltage at TP_4, driving the A_3 output low. The voltage neutral zone, or gap, is 8.39 V − 7.56 V = 0.83 V. In temperature, the neutral zone is 83.9°F − 75.6°F, or 8.3°F.

Another example of a differential-gap on-off controller, this time in a level control application, is shown in Figure 10.12. This application uses conductive probes as a transducer. The application requires the use of a liquid that is conductive but not flammable. When the circuit is energized, current flows through the primary and the NC relay contacts, applying power to the pump. Note that, at this time, no current is flowing in the secondary circuit. The tank starts to fill as the pump transfers fluid into it. When the fluid level reaches the high-level probe, current flows from the secondary into the fluid via the ground probe, through the high-level probe, R_4, R_1, and the diode 1CR. A negative potential is developed on the Q_2 gate with respect to its anode. Since this circuit is a PUT (discussed in a previous chapter), the gate-to-anode voltage turns it on. The PUT conducts through resistor R_3, placing a positive potential on the gate of Q_1, turning it on. Relay K_1 turns on when Q_1 conducts, changing the state of the relay contacts. The NC contact opens, removing power from the pump. Note that, during the process, when the liquid falls below the high-level probe, current can still flow in the control circuit through the low-level probe and the closed relay contacts. When the level falls below the low-level probe, however, the control circuit current flow is interrupted since it can no longer flow through the fluid. When current can no longer flow through R_1, Q_2 will not turn on, keeping Q_1 and relay K_1 off. Power is then applied to the pump, filling the tank.

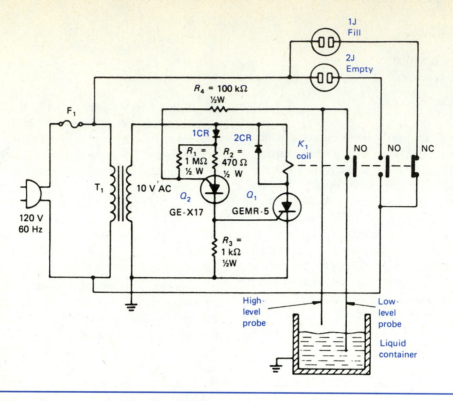

FIGURE 10.12 Liquid-Level On-Off Controller

We can clearly see that this device, too, is an on-off controller. Note that there is a neutral zone established between the two liquid-level probes. In the previous example, the differential gap was temperature; in this case, it is liquid level.

Let's assume that the tank has a capacity of 200 gallons (gal). We will take out the liquid at a rate of 40 gal/min and replace it while the pump is on at a rate of 120 gal/min. The tank will fill at a rate of (120 − 40) gal/min = 80 gal/min. We will place the low-level probe at the 25% high level. The 25% high level corresponds to 0.25 times 200 gal, or 50 gal. The pump will go on when the liquid level goes below the 50 gal mark. If we place the high-level probe at the 75% full point, the pump will go off when the level goes above 150 gal. How long will it take for the tank to fill when the level goes below the 25% limit? We assume that the pump starts when the liquid is at the 50 gal level, and the tank fills at a rate of 80 gal/min. To find the fill time, we divide

the distance the liquid level must reach (100 gal, from 50 to 150) by the fill rate of 80 gal/min, according to Equation 10.1:

$$t = \frac{d}{r} = \frac{\text{distance}}{\text{fill rate}} = \frac{100 \text{ gal}}{80 \text{ gal/min}} = 1.25 \text{ min}$$

The fill time is 1.25 min.

When the fill pump goes off, the liquid will be taken out at a rate equal to 40 gal/min. How long will it take to empty the tank? We can use the same relationship (Equation 10.1) we used to find the fill time in order to find the emptying time:

$$t = \frac{d}{r} = \frac{\text{distance}}{\text{empty rate}} = \frac{100 \text{ gal}}{40 \text{ gal/min}} = 2.5 \text{ min}$$

The time to empty the tank to the 50 gal level is about 2.5 min.

Together, the fill time and the emptying time make up what is called the *process reaction time*. The process reaction time (also called the *cycling period*) is the fill time plus the emptying time, 1.25 min + 2.5 min, or 3.75 min. As the capacitance of the tank increases, the cycling time will increase also.

• EXAMPLE 10.4

Given a 100 gal tank that begins filling when the level goes below 20 gal and stops when the level goes above 80 gal, find the fill and emptying times and the cycling time if the tank fills at 50 gal/min and empties at 15 gal/min.

Solution

To find the fill time, we divide the distance the liquid level must reach (60 gal, from 20 to 80) by the fill rate of 50 gal/min, according to Equation 10.1:

$$t = \frac{d}{r} = \frac{\text{distance}}{\text{fill rate}} = \frac{60 \text{ gal}}{50 \text{ gal/min}} = 1.2 \text{ min}$$

The fill time is 1.2 min.

When the fill pump goes off, the liquid will be taken out at a rate equal to 15 gal/min. The emptying time is then

$$t = \frac{d}{r} = \frac{\text{distance}}{\text{empty rate}} = \frac{60 \text{ gal}}{15 \text{ gal/min}} = 4.0 \text{ min}$$

The time to empty the tank is about 4.0 min.

The cycling period is the fill time plus the emptying time, or 1.2 min + 4.0 min = 5.2 min.

The diagram in Figure 10.13 shows the time relationship between filling and emptying for the tank we described in Figure 10.12. The pump turns on at time t_1, when we empty the tank and the level drops below 50 gal. Note the time delay between the time when the level goes below the low-level probe and the time when the liquid actually starts to flow into the tank. This delay is the dead time. In this example, dead time may be due to such factors as the time it

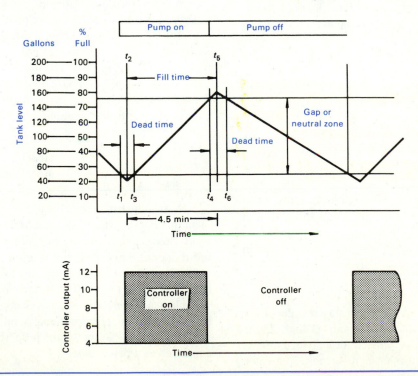

FIGURE 10.13 Controller Action in Tank-Filling Operation

takes for the pump to turn on and the length of the pipe between the pump and the tank. The delay in this example is about 15 s. Note that the delay causes an undershoot between times t_1 and t_3. When the pump drives the liquid level to the high-level probe, the pump turns off. As with most pumps, the device continues to rotate as it slows down. The liquid in the line will continue to fill the tank until the pump actually stops rotating. Note that this delay causes an overshoot between times t_4 and t_6. The liquid level varies between the 20% and 80% levels, with a period of about 3.75 min. In the liquid-level example, the position of the high- and low-level probes determines the gap or neutral zone.

On-off control works best when the dynamic variable changes slowly with respect to time. The rate at which the dynamic variable changes is called the *process reaction rate*.

On-off control is almost always the simplest and least expensive controller mode. This type of control works equally well with electric systems and with pneumatic systems. However, many industrial processes need better, more sophisticated control. Proportional control was developed to meet this need.

Proportional Control

In *proportional control*, the final control element is purposely kept in some intermediate position between on and off. Proportional control is a term usually applied to any type of control system where the position of the final control element is determined by the relationship between the measured variable and the set point.

An example of a proportional-control mode using a filled-system temperature sensor is shown in Figure 10.14. The bulb of the filled system senses the temperature of the fluid in the line. Any change in the temperature causes the valve position to change.

The *gain* of the controller is determined by the relationship of change in temperature to change in valve position. If the valve moves a great deal for a given temperature change, the gain is high. If the valve does not move very much for a given temperature change, the controller gain is small. The valve normally is set so that when the temperature is in the center of its range, the valve is one-half open. In this

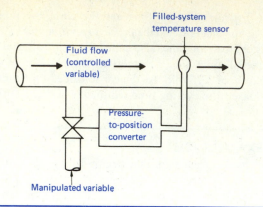

FIGURE 10.14 Proportional-Control System Using Filled-System Temperature Sensor

system, the set point is determined by the initial setting of the valve (50%). Once this system is installed, the controller gain is also fixed. More sophisticated proportional-control systems have variable gain or set points.

The final control element setting when the measured variable equals the set point is called the *bias*. In our example, the bias is 50%. As the measurement deviates from the set point, the final control element setting will change from 50%. The amount of change in output (final control element) for a given change in input is a function of the gain of the controller's amplifier, as shown in the following equation:

$$\text{controller gain} = \frac{\Delta \text{output}}{\text{set point} - \text{measurement}} \quad \textbf{(10.11)}$$

Most industrial controllers have a gain adjustment that is expressed in percent of the proportional band. *Proportional band* is the amount of change in the dynamic variable that causes a full range of controller output. Or we can say that the proportional band is equal to the range of values of the dynamic variable that corresponds to a full or complete change in controller output. Proportional band is also called *throttling range*. Normally, the proportional band is expressed as a percentage:

$$\% \text{ proportional band} = \frac{1}{\text{gain}} \times 100 \quad \textbf{(10.12)}$$

Then we have

$$\text{gain} = \frac{100}{\% \text{ proportional band}} \qquad (10.13)$$

Note the relationship between gain (or sensitivity) and proportional band. Systems with large proportional bands are less sensitive (have less gain) than narrow-band systems.

The equation for determining output for the proportional controller in Figure 10.14 is

$$\text{output} = \frac{100}{\% \text{ proportional band}}$$

$$\times (\text{set point} - \text{measurement})$$

$$+ \text{bias} \qquad (10.14)$$

Thus, for a fixed gain (proportional-band setting) and a fixed bias, we can calculate the output. To do so, we must know the measurement and the set point. For example, let us say the proportional band is 100% (gain = 1) and the bias is 50%. When the measurement equals the set point, the output (final control element setting) is 50%. When the measurement exceeds the set point by 10%, the output is 40%. When the measurement is below the set point by 10%, the output is 60%.

The higher the gain, the more the output will change for a given change in either measurement or set point. Figure 10.15A shows the effect on the input-output relationship for a gain change.

Another way of showing the effect of changing the proportional band is illustrated in Figure 10.15B. Each position in the proportional band produces a controller output. The wider the band, the more the input signal (set point – measurement) must change to cause the output to swing from 0% to 100%. Changing the bias shifts the proportional band so that a given input signal will cause a different output level.

Let's assume that we have a temperature controller with a set point of 150°F. The proportional band is 20°F, and the extremes of temperature during operation range from 130° to 170°F. So the input range is 40°F. The proportional band is, therefore, 50% of the input range. If we assume that the controller will produce 100% of its output between 130°

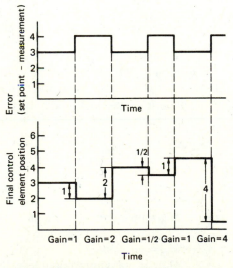

A. Effect of Gain Change on Input-Output Relationships

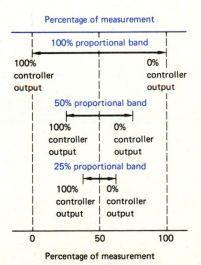

B. Effect of Proportional Band Change on Controller Output and Measurement

FIGURE 10.15 Plots of Controller Gain and Proportional-Band Changes

and 140°F, it will produce a zero output between 160° and 170°F. The output between 140° and 160°F will be proportional to the difference between the set point and the measurement. We can see this result in the diagram in Figure 10.16. In this case, the controller gain (symbolized by the letter K) is 2.

A change in the gain will change the proportional band. We can also see the effect of changing the gain in Figure 10.15. Note that the proportional band is larger for smaller K values and smaller for larger K values. When the gain (K) is less than 1, the controller range is reduced. For example, with a gain of 0.5, the controller output is limited to an output swing of 25% to 75%. This curve is based on a bias of 50%.

An example may help at this point. Let us go back to the example of the pasteurization process presented in the introduction to this chapter. Assume that the set point is 65°C. Since the pasteurization process requires that the milk stay between 60° and 70°C for a half hour, we will assume that the extremes of our temperature range will be limited to these two points. If the input range is 10°C, the proportional band is 50% of the input range, or 5°C (±2.5°C). The controller will be at 100% when the

sensor is measuring 62.5°C. If the sensor moves down to 60°, no further control action is possible. On the other hand, the controller will be operating at 0% when the temperature reaches 67.5°C. If the temperature goes above that value, say to 70°C, no further control action will occur. When the input is between 62.5° and 67.5°C, the controller output will be proportional to the difference between the set point and the measurement. The gain of this system is, according to Equation 10.13,

$$\text{gain} = \frac{100}{\% \text{ proportional band}} = \frac{100}{50} = 2$$

Since a 50% change in the measurement produces a 100% change in the controller output, the gain is 2.

Rewriting Equation 10.14, we can calculate the controller output with the following formula:

$$C_o = Ke + C_b \qquad (10.15)$$

where

C_o = controller output (in percent)
K = gain of controller (inverse of proportional band)
e = error term (percent drift away from set point)
C_b = controller output bias (in percent when measurement equals set point or error equals 0)

The error term in this equation is, as suggested in Equation 10.4, a function of the difference between set point and measurement but referenced to the full input error range—in this case, 10°C. The error term may be calculated by the following formula:

$$e = \frac{M_{dv} - M_{sp}}{M_{max} - M_{min}} \times 100 \qquad (10.16)$$

where

M_{dv} = measurement of dynamic (controlled) variable
M_{sp} = set point
M_{max} = maximum possible value of dynamic variable
M_{min} = minimum possible value of dynamic variable

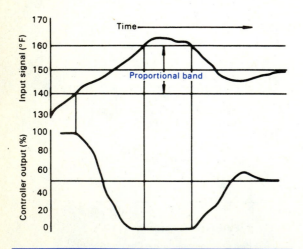

FIGURE 10.16 Input-Output Relationships for Proportional Controller

We can use Equations 10.15 and 10.16 to help us solve for the controller output. For example, if the measurement is 64°C, solving Equation 10.16 for e, we get

$$e = \frac{M_{dv} - M_{sp}}{M_{max} - M_{min}} \times 100$$

$$= \frac{64 - 65}{70 - 60} \times 100 = -10.0\%$$

Substituting into the controller output equation (10.15) and assuming a controller bias setting of 50%, we get

$$C_o = Ke + C_b = [(-2)(-10\%)] + 50\%$$

$$= 70\%$$

From this example, we can see that the controller output would be 70% with a dynamic-variable measurement of 64°C.

• EXAMPLE 10.5

Given a process control system in which the gain is 1.6, the controller output range from 0% to 100% is 4 to 20 mA, the final control element is set at 50% when the controller produces a 12 mA output, a sudden change of load causes an error signal change of 0.5 V, and the error range is 10 V (±5 V), find the controller output (in milliamperes).

Solution
Using Equation 10.16, we obtain

$$e = \frac{M_{dv} - M_{sp}}{M_{max} - M_{min}} \times 100 = \frac{\text{error}}{\text{error range}} \times 100$$

$$= \frac{0.5}{10 \text{ V}} \times 100 = 5\%$$

With a controller bias setting of 50%, we can use Equation 10.15 to find the controller output:

$$C_o = Ke + C_b = [(1.6)(5\%)] + 50\% = 58\%$$

To find the actual controller output in milliamperes, we take 58% of 16 mA (the total controller range) and add that value to 4 mA (the lowest value of the range). Thus, the controller will provide 13.28 mA to the final control element.

In Example 10.5, the condition that caused the error may have been a temporary condition, say the addition of a new batch of cold liquid. If the system returns to normal, the controller will settle back down to 12 mA. However, if the condition is permanent and not temporary, a constant error value of 0.5 V may be needed to keep the material at a constant temperature. The difference between the 12 mA original controller setting and the new 13.28 mA controller setting is called the *offset*. It is the difference between the measurement and the set point after a new controller level has been reached. In the case of Example 10.5, the offset is 1.28 mA, as indicated in Figure 10.17.

Proportional control tries to return a measurement to the set point after a disturbance has occurred. However, it is impossible for a proportional controller to return the measurement so that it equals the set point. By definition, the output must equal the bias setting (normally 50%) when the measurement equals

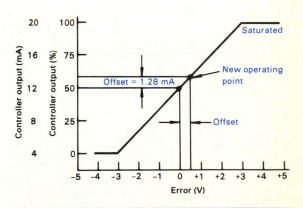

FIGURE 10.17 Input-Output for Proportional Controller

the set point. If the loading conditions require a different output, a difference between measurement and set point must exist for this output level. Proportional control may reduce the effect of a load change, but it can never eliminate it.

The resulting difference between measurement and set point, after a new balance level has been reached, is called *offset*. The amount of offset may be calculated from the following equation:

$$\Delta\text{offset} = \frac{\%\ \text{proportional band}}{100} \times \Delta\text{measurement} \qquad (10.17)$$

The change in measurement is the change required by load upset.

From this equation, we see that as the proportional band goes toward zero (gain approaches infinity), offset will approach zero. This result seems logical because a controller with infinite gain is by definition an on-off controller. We know that we cannot have an offset in an on-off controller. On the other hand, as proportional band increases (gain decreases), proportionately more offset will exist.

Many proportional-control circuits use the op amp, like the one illustrated in Figure 10.18. This particular amplifier has a standard industrial control signal in the form of a current that varies between 4 and 20 mA. Resistor R_4 converts the 4–20 mA signal

to a voltage between 1 and 5 V. If the resistor R_1 is properly adjusted, the output from 1IC and 2IC will be zero. Since 1IC is just a summing amp, an equal positive and negative voltage at the two inputs will cancel each other. For example, if the system is to give a zero output voltage with a 4 mA input, the zero-adjust must be set to +1 V. This voltage will cancel the –1 V potential developed across R_1. The span-adjust sets the gain of the system. Let us further specify in our system that the output should go from 0 to 10 when the input goes from 4 to 20 mA. The zero-adjust would be adjusted to produce +1 V to counteract the –1 V developed across the resistor R_4 by the 4 mA signal. A 20 mA current signal would develop a –5 V potential across R_4. The output of 1IC would be +4 V with a 20 mA signal. Adjusting R_8 to a resistance of 5 kΩ will give the 2IC a gain of 2.5 and an output voltage of 10 V with an input of 20 mA. We can see, therefore, that our requirements have been met.

Using this system, let us see how it responds to a change in the process. The full measurement range is 20 mA – 4 mA = 16 mA. Suppose that our set point is 12 mA, which corresponds to a 50% controller output (+5 V). What will the controller output be with an increase of 1 mA in the measurement? A 13 mA signal will develop a –3.25 V potential across R_4, giving an output of 5.625 V at the output of 2IC.

Note in this example that a full, 100% output swing of the controller was caused by the full, 100%,

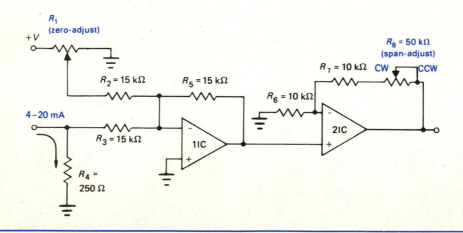

FIGURE 10.18   Op Amp Proportional-Control System

4–20 mA control signal. The proportional band for this system is, therefore, 100%. If we limit the input to 8–16 mA to produce 100% swing in output voltage (0–10 V), we have a 50% proportional band. If the output is set to +5 V when the error is zero with a set point of 12 mA, an increase of 25% (4 mA) will cause the output to go to 100% (+10 V). Conversely, a decrease of the input by 25% (to 8 mA) will cause the output to go to 0% or 0 V. Note also that the inverse of the proportional band does not equal the gain of the op amp circuit. The relationship between gain and proportional band holds only when both measurement and output are in the same units.

• EXAMPLE 10.6

Given the 4–20 mA converter in Figure 10.18, find the setting of resistors R_1 and R_8 that will give an output voltage of 0–7.5 V with a 4–20 mA input signal.

Solution

The setting of R_1 is unchanged since we want 0 V out when the 4 mA signal is applied.

We also want an output voltage of +7.5 V when the 20 mA signal is applied. The 20 mA signal will drop a –5 V potential across R_4. This voltage will be added to the +1 V at the wiper of R_1, giving an output of +4 V from 1IC. (Remember that 1IC is an inverting summer.) What 2IC gain will take a +4 V input and give an output of +7.5 V? The gain A_V is equal to the output voltage divided by the input voltage, or

$$A_V = \frac{V_{out}}{V_{in}}$$

In this case, the gain is

$$A_V = \frac{+7.5 \text{ V}}{+4 \text{ V}} = 1.875$$

Since this amplifier is a noninverting amplifier, the gain is

$$A_V = \frac{R_f}{R_i} + 1 \qquad \qquad \textbf{(10.18)}$$

For this example, we can modify Equation 10.18 to suit our needs by substituting circuit values into it, which yields

$$A_V = \frac{R_7 + R_8}{R_6} + 1$$

To solve our problem, we must solve for the value of R_8, our unknown:

$$R_8 = [(A_V - 1)(R_6)] - R_7 \qquad \qquad \textbf{(10.19)}$$

Substituting appropriate values gives

$$R_8 = [(1.875 - 1)(10 \text{ k}\Omega)] - 5 \text{ k}\Omega$$

$$= 3750 \, \Omega$$

Another example of proportional control is shown in Figure 10.19. This circuit is a precision proportional temperature controller that uses a thermistor as a temperature sensor. The load is a heating element. You will recognize this circuit as the ramp-and-pedestal circuit presented in Chapter 7. Suppose that the temperature detected by the thermistor in the process decreases. A decreased temperature will increase the resistance of the thermistor R_4, increasing the voltage at the top of R_4. This action raises the pedestal. Since the capacitor charge now takes less time to reach the firing voltage of the UJT, the SCR will fire earlier in the cycle, thus applying more average power to the load. Finally, more power to the heating element will increase the amount of heat it gives off.

The gain of this circuit can be adjusted over a wide range by changing the size of the charging resistor, R_6. When a ramp amplitude of 1 V is selected with a 20 V zener, a 22% change in the thermistor resistance will give a linear, full-range change in the controller output. The controller in this case is the heater.

Recall the example presented at the beginning of this chapter concerning the pasteurization process. That example also uses proportional control. A simplified electric circuit for such a control is presented in Figure 10.20. The sensor in this case is a thermistor, which, as you will recall

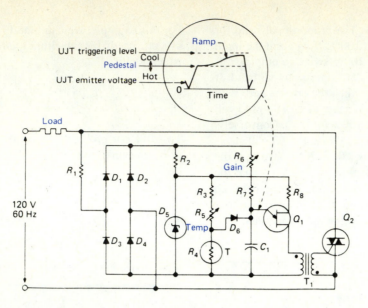

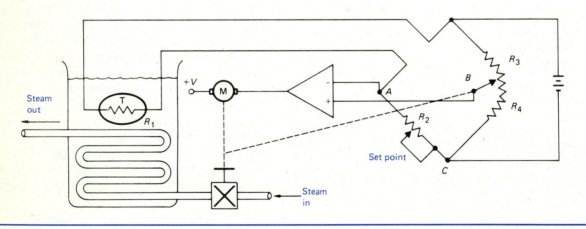

FIGURE 10.19 Proportional Controller for Precision Temperature Control

FIGURE 10.20 Proportional Control Using Liquid Temperature

from Chapter 8, has a negative temperature coefficient. The final control element is a motor-driven valve that controls the rate of steam flow through the pipes in the liquid. The motor in the valve is driven by an amplifier that is connected to a bridge. Note the dotted line in the illustration. This line is connected to the valve, indicating that the motor drives

the valve. But note also that the motor drives a potentiometer in the bridge. In the schematic diagram, the upper half of the potentiometer is labeled R_3, and the lower half is labeled R_4. When the bridge is in balance, no difference in potential exists between points A and B on the bridge. The output of the amplifier is 0 V and the motor is at rest.

Let us suppose that at a temperature of 65°C each resistor, including the thermistor, is at a resistance of 2000 Ω. The voltage at each point A and B then is +5 V. Let us further say that the temperature of the liquid drifts downward. This change will be detected by the thermistor and reflected in an increase in resistance, say, to 3000 Ω. If R_2 stayed at 2000 Ω, the potential at point A with respect to C would decrease to +4.44 V. The potential at B would be +5 V. So a difference in potential of about 0.56 V appears between A and B. This difference in potential is amplified by the amplifier and used to drive the motor on the valve. In this case, the valve will open, letting in more steam, thus adding more heat to the system. The motor, which drives the valve, also drives the bridge. As the motor is driving the valve open, the motor is readjusting the bridge to a balanced condition again. When the null condition is reached on the bridge, the motor will stop because the voltage to the amplifier is zero again. How much the valve opens before the motor stops turning depends on the gain of the system.

An increase in the temperature of the liquid will have the opposite effect. The thermistor resistance will decrease, unbalancing the bridge in the opposite direction. The voltage at point A will be greater than that at point B. This voltage, amplified by the amplifier circuitry, will move the motor in the opposite direction, closing the valve slightly. Closing the valve will decrease the flow of steam to the system, reducing the amount of heat that the system receives.

The bridge may be unbalanced manually by an operator, who can adjust resistor R_2. Manipulating resistor R_2 adjusts the set point. The dial on the potentiometer R_2 in this case is calibrated in degrees. The operator can adjust the temperature of the liquid by varying this potentiometer. The system will treat this adjusted value as the new set point and try to maintain that temperature as disturbances tend to change it.

Another proportional-control circuit is shown in Figure 10.21. This circuit also controls temperature. The CA3059 is a zero-voltage switch (discussed in

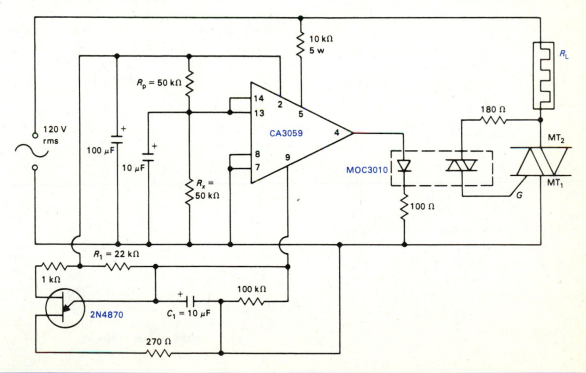

FIGURE 10.21 Proportional Temperature Control Using ZVS

Chapter 7). The UJT relaxation oscillator provides a sawtooth waveform at the input to pin 9. If the supply voltage at pin 2 is 6.5 V and the intrinsic standoff ratio is 0.65, the period of the sawtooth is approximately

$$T = (R_1 C_1)\left(\ln \frac{1}{1-\eta}\right) \tag{10.20}$$

or

$$T = (22\ \text{k}\Omega)\,(10\ \mu\text{F})\left(\ln \frac{1}{1-0.65}\right)$$

$$= 0.23\ \text{s} \quad \text{or} \quad 230\ \text{ms}$$

The frequency of the oscillator can be calculated by taking the reciprocal of the period:

$$f = \frac{1}{T} \tag{10.21}$$

or

$$f = \frac{1}{230\ \text{ms}} = 4.3\ \text{Hz}$$

If we adjust R_p and R_x to 10 kΩ, the voltage at pin 13 will be half the supply (6.5 V), or 3.25 V. The sawtooth output from the capacitor is applied to pin 9 of the CA3059 ZVS. During the time that the voltage on pin 9 is less than the voltage on pin 13, the output of the CA3059 at pin 4 will be high, turning on the power triac. The MOC3010 is an optoisolator, which isolates the low-power ZVS circuit from the high-power triac circuit. When the sawtooth at pin 9 ramps above the pin 13 potential—in this case, 3.25 V—the CA3059 output is inhibited and the triac is off. The voltage at pin 13 determines the on-off ratio (or duty cycle) at the output. In this case, the oscillator period is 230 ms.

When does the voltage at pin 9 reach the 3.25 V potential at pin 13? We can find this period by modifying the Equation 10.20, which is

$$T = (R_1 C_1)\left(\ln \frac{1}{1-\eta}\right)$$

We modify this equation as follows:

$$T = (R_1 C_1)\left[\ln \frac{1}{1-(V_{13}/V_2)}\right] \tag{10.22}$$

For our example, we find

$$T = (22\ \text{k}\Omega)\,(10\ \mu\text{F})\left[\ln \frac{1}{1-(3.25\ \text{V}/6.5\ \text{V})}\right]$$

$$= 0.152\ \text{s} \quad \text{or} \quad 152\ \text{ms}$$

The duty cycle can be calculated by dividing the time the device is on by the sawtooth period:

$$\text{duty cycle} = \frac{t_{\text{on}}}{T_{\text{saw}}} \tag{10.23}$$

or

$$\text{duty cycle} = \frac{152\ \text{ms}}{230\ \text{ms}} = 0.66 \quad \text{or} \quad 66\%$$

Suppose the load is resistive and 100 W of power is consumed when the power is applied continuously. How much average power will be dissipated at a duty cycle of 66%? To answer this question, we multiply the decimal value of the duty cycle by the 100% power value—in this case 100 W:

$$P_{\text{load}} = \text{duty cycle} \times \text{full power} \tag{10.24}$$

or

$$P_{\text{load}} = 0.66 \times 100\ \text{W}$$

$$= 66\ \text{W}$$

• EXAMPLE 10.7

Given the ZVS UJT power control circuit in Figure 10.21, with R_p = 10 kΩ and R_x = 4 kΩ, find (a) the duty cycle of the triac and (b) the power delivered to the load if 100% power is 100 W.

Solution

(a) Since the oscillator values have not been changed, the period of the oscillator is still 230 ms. We can find the time it is on by using Equation 10.22:

$$T = (R_1 C_1) \left[\ln \frac{1}{1 - (V_{13}/V_2)} \right]$$

From the voltage divider equation, we see that the voltage across the resistor R_x is 0.286 times 6.5 V, or 1.86 V. We can now substitute that value into Equation 10.22 to get

$$T = (22 \text{ k}\Omega)(10 \text{ μF}) \left[\ln \frac{1}{1 - (1.86 \text{ V}/6.5 \text{ V})} \right]$$

$$= 0.074 \text{ s} \quad \text{or} \quad 74 \text{ ms}$$

The duty cycle can be calculated from Equation 10.23:

$$\text{duty cycle} = \frac{t_{\text{on}}}{T_{\text{saw}}} = \frac{74 \text{ ms}}{230 \text{ ms}}$$

$$= 0.32 \quad \text{or} \quad 32\%$$

(b) To find the power applied to the load, we use Equation 10.24 and multiply the decimal value of the duty cycle by the 100% power value—in this case, 100 W:

$$P_{\text{load}} = \text{duty cycle} \times \text{full power}$$

$$= 0.32 \times 100 \text{ W}$$

$$= 32 \text{ W}$$

A proportional temperature controller may be constructed by replacing R_x with a thermistor and replacing the load with a heating element. When the temperature is low, say around room temperature, and power is applied to the ZVS circuit, the voltage at pin 13 will be high. A high voltage will be present because the thermistor has a high resistance at room temperature. As the temperature starts to rise, the potential at pin 13 drops, decreasing the number of cycles of AC applied to the heater element. The voltage will continue to decrease until an equilibrium condition exists. The system will be in equilibrium when the power to the heater equals the losses of heat in the system, as shown in the timing diagram in Figure 10.22B. Note that about 50% of the available

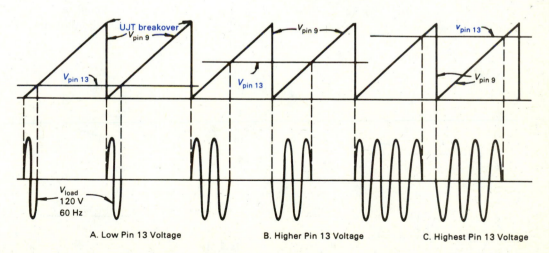

A. Low Pin 13 Voltage   B. Higher Pin 13 Voltage   C. Highest Pin 13 Voltage

FIGURE 10.22 ZVS Proportional-Control Timing Diagram

power is applied to the heater at this time since the duty cycle is about 50%.

If the heat losses increase, the temperature will decrease, raising the resistance of the thermistor. An increase in the resistance of the thermistor will increase the voltage at pin 13, applying more power to the heating element to make up for the increased losses. The result of increased heat losses is shown in Figure 10.22C. On the other hand, if the heat losses decrease, the temperature will increase, lowering the resistance of the thermistor. A decrease in the resistance of the thermistor will decrease the voltage at pin 13, applying less power to the heating element to compensate for the decreased losses. The result of decreased heat losses is shown in Figure 10.22A.

Proportional Plus Integral Control

Often, in industrial applications, the offset caused by proportional control cannot be tolerated. Process engineers solve this problem by adding another control mode. Recall from Chapter 1 that the op amp integrator integrates any voltage present at the input. *Integral control* (sometimes called *reset action*) in a process control system will integrate any difference between the measurement and the set point. The controller's output will change until the difference between the measurement and the set point is zero.

The response of a pure integral controller is shown in Figure 10.23A. Note that the controller output changes until it reaches 0% or 100% of scale (or until the measurement is returned to the set point). This figure also assumes that an open-loop condition exists where the controller's output is not connected to the process.

Figure 10.23B shows an open-loop, proportional plus integral controller's response to a step change. *Proportional plus integral* (PI) *control* combines the characteristics of both types of control. Reset time is the amount of time required to treat the amount of change caused by proportional action. In Figure 10.23B, reset time t is the amount of time required to repeat the amount of output change y.

Another way to visualize PI action is shown in Figure 10.24. In this example, a 50% proportional band is centered around the set point. If a disturbance comes into the system, the measurement will deviate from the set point. Proportional response will be seen immediately in the output, followed by integral action.

We may think of integral action as forcing the proportional band to shift. This shift causes a new controller output for a given difference between measurement and set point. Integral control will continue to shift the proportional band as long as there is a difference between the measurement and the set point.

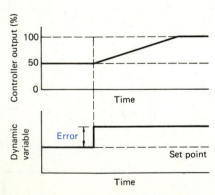

A. Pure Integral Controller (Open Loop)

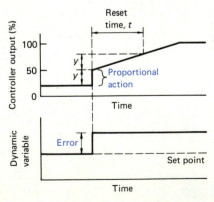

B. PI Controller (Open Loop)

FIGURE 10.23 Plots of Controller Output and Dynamic-Variable Measurement versus Time

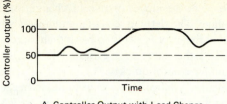

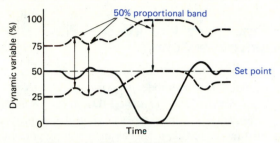

A. Controller Output with Load Change

50% proportional band

Set point

Time

B. Dynamic Variable with Shift of Proportional Band

FIGURE 10.24 Shifting of Proportional Band in PI Controller When Error Changes

A circuit for an electronic PI controller is shown in Figure 10.25. When the error signal increases (V_i in Figure 10.25), the increased rate of change of the integral output will be added to the proportional output. The input to the summing amplifier is the sum of the input voltage, the set point voltage, and the integral response. The integral control gradually reduces the error to zero, reducing the offset. The gain of the integral side of the controller (K_I) is $1/RC$, and the gain of the proportional section (K_P) is R/R_1.

Integral-control action does exactly what an operator would do by manually adjusting the bias in the proportional controller. The width of the proportional band stays constant. It is shifted in a direction opposite to that of the measurement change. Thus, an increasing measurement signal results in a decreasing output, and vice versa. Because integral control acts only on long-term or steady-state errors, it is seldom used alone.

The PI control mode is used in situations where changes in the process load do not happen very often; but when they do happen, changes are small. These small changes may occur for a long time before they finally go beyond the allowable error limits. In many industrial processes, even small amounts of error that persist for long periods of time are undesirable. Consequently, PI control is a very popular mode of control in industry today.

Proportional Plus Derivative Control

Some process control systems have errors that change very rapidly. This situation is especially true in processes that have a small capacitance. Neither proportional control nor PI control responds well to errors that change rapidly. Recall from Chapter 1 that the differentiator circuit responded to the input rate of change, or slope. The same concept applies to the type of controller mode called *derivative*, or *rate*, *control*. By adding derivative control to proportional control, we get a controller output that responds to the measurement's rate of change as well as to its size.

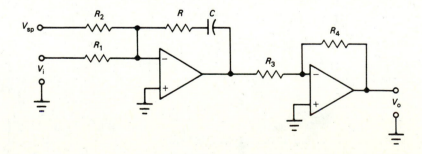

FIGURE 10.25 PI Controller

Proportional plus derivative (PD) *control* is illustrated in Figure 10.26. When a measurement changes, derivative action differentiates the change and maintains a level as long as the measurement continues to change at a given rate. Note that the step decrease in the controller output graph is proportional to the slope of the dynamic-variable change.

An example of a circuit for a PD controller is shown in Figure 10.27. The first amplifier takes the derivative of the input, which gives an output

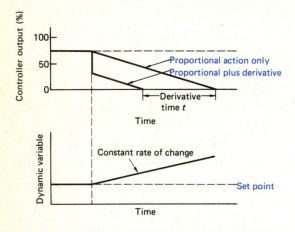

FIGURE 10.26 Plot of Measurement and Controller Output versus Time in PD Controller

proportional to the slope, or rate of change, of the input voltage. The derivative signal anticipates what the value of the error is likely to be. It then increases or decreases the value of the output to bring the process under control quickly.

Derivative control is never used alone because it can react to measurements only when they are changing. It cannot react to steady-state errors.

Proportional Plus Integral Plus Derivative Control

All three controller modes we have discussed can be combined into one mode, *proportional plus integral plus derivative* (PID) *control*. PID control has all the advantages of the three types of control. The proportional-mode portion produces an output proportional to the difference between the measurement and the set point. The integral-control action produces an output proportional to the amount and the length of time the error is present. The derivative section produces an output proportional to the error rate of change. The PID mode of control applies to systems that have transient as well as steady-state errors.

A circuit for an electronic PID controller is presented in Figure 10.28. Each of the three types of control is present—integral, derivative, and proportional.

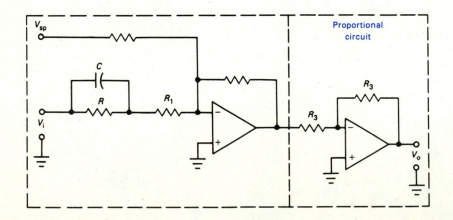

FIGURE 10.27 PD Controller

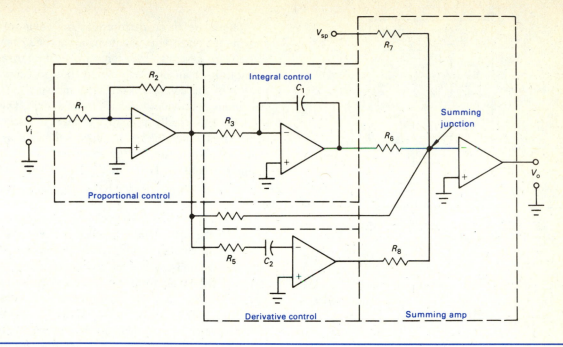

FIGURE 10.28 PID Electronic Controller

CONTROLLERS

As we have seen, the controller is the part of a control system that compares the measurement of the controlled variable with the set point. The controller also directs the action of the final control element, which corrects or limits the deviation.

Controllers in industry may be classified in several ways. For example, they may be classified as either electric (electronic) or pneumatic. Also, electric and pneumatic controllers can both be broken down into two popular types: analog and digital. In this section, we will discuss both analog and digital types of electric and pneumatic controllers, including synchros and servos, which are considered analog electric controllers.

Electric Controllers

Many electric (or electronic) analog controllers use the op amp as a basic building block. Recall that the op amp can perform the processes of comparison, summation, integration, and differentiation. These are precisely the building blocks we need to put together the types of controller modes we have just discussed. Process engineers choose the types of control to suit the particular process application. All the controller modes we have discussed can be accomplished with the op amp.

Although each manufacturer of electronic analog process controllers produces a different system, many of these systems have common factors. For example, many electronic controllers have similar front panels, with metering systems similar to the one shown in Figure 10.29. This meter displays two types of information: the set point and the measured variable values. The set point normally is adjustable from the front panel by means of a potentiometer. Thus, to change the set point for different applications is simple. The scale usually is calibrated to read from 0% to 100% of signal variation. In addition, many controllers give operators the ability to switch back and forth from manual to automatic mode. The manual mode often is used when the process is started up and when it is shut down.

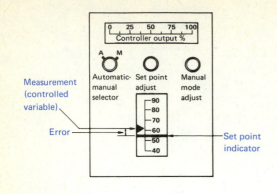

FIGURE 10.29 Front Panel of Typical Electronic Controller

Digital electronic control is also found in industry. The computer, of course, can be classified as a digital controller. Because of the ability of the computer to make decisions and initiate action, it is ideally suited to process control applications. The most popular digital controller today is the programmable controller, which is a microprocessor-based system capable of a wide range of tasks. The programmable controller will be discussed in Chapter 14.

Synchros and Servos

Automatic control systems can be as simple as an on-off heat control for an oven or as complex as the guidance system for a missile or a space vehicle. As the field of automatic control has become wider and more complex, the controlling equipment has likewise become more complex and sophisticated. During the course of this change, the field of servomechanisms has emerged to become a separate field of study and application.

Although the general theory of operation for automatic control systems and servomechanisms is the same, there are some differences between the two. A servomechanism is a special type of automatic control system. A *servomechanism*, sometimes called a *servo*, is an automatic control system that controls the physical motion or position of an object. A servo system has four important features: (1) an input command, (2) feedback, (3) power amplification, and (4) an output that turns or positions an object.

There are three differences between servo systems and other types of automatic control systems. First, servo systems exert control over the position or motion of an object. Second, the position or motion of the controlled object is measured. This measurement provides feedback to the system. Third, servo systems usually respond to changes in fractions of a second, rather than in seconds or minutes, as in other forms of control. You may recognize these requirements as the simple feedback control system discussed earlier in this chapter.

The basic block diagram of a servo system is shown in Figure 10.30. The command, or reference input, sets the position the controlled device is to seek and hold. In process control terminology, the reference corresponds to the set point. The controlled device may be a platform, a missile, a ship's rudder, or some kind of tool. The controlled device usually receives power from a power amplifier, either electric, hydraulic, or pneumatic. This power amplifier is part of the control system, which amplifies the low-power error signal and produces the power needed to drive the controlled device. The error detector takes the position or velocity information from the feedback device and produces an error signal proportional to the difference between the command input and the feedback. The error detector may be a transistor, a bridge circuit, a differential gear, an op amp, a synchro (discussed below), or a thermostat.

Notice that the system includes both a feedback device and an error detector. Without these components, the system would not be a closed-loop system. Recall that a system without feedback is an open-loop system. An open-loop system is not considered a true

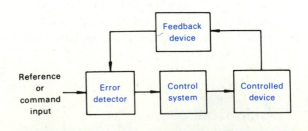

FIGURE 10.30 Block Diagram of Servo System

servo system. There are many examples of an open-loop control system in everyday life. One of the most common is the automobile accelerator. We control the speed of the car by foot pressure on the accelerator. By varying the foot pressure, we reach our desired speed. When we reach the desired speed, we may keep the same pressure on the accelerator in order to maintain that speed. But in most instances, the speed will not remain constant. Many outside factors will intervene to change the speed, such as the level of the terrain, the road surface condition, and air pressure on the car. As drivers, we note the effect of the changes and regulate the pressure on the accelerator pedal accordingly.

Contrast this action with the automatic-cruise-control feature of some automobiles. With cruise control, we set the desired speed. The control continually measures the actual speed by a tachometer attached to the automobile. Changes in speed, whatever the cause, are detected automatically by the system. The system responds to the change by automatically regulating the pressure on the accelerator to maintain a constant speed. So cruise control is a true servo system.

Velocity Servos. As its name implies, the *velocity servo* controls the speed of a rotating device. An example of a velocity servo is shown in Figure 10.31. The DC motor in this servo system has an AC tachometer output. Motors of this type can be found on many 5.25 in. computer floppy disc drives. Combining these motors with an LM2907 F/V converter (discussed in Chapter 2), we can construct an efficient, low-cost speed control. The AC tachometer output is connected to the noninverting input of the LM2907 comparator. The charge pump converts the AC tachometer frequency to a voltage, which appears at pins 3 and 4. The voltage at pins 3 and 4 is compared with the preset voltage that appears at the wiper of the speed control potentiometer. The setting of the speed control potentiometer can be considered as the command input—or in process control terminology, the set point.

When the voltage from the charge pump is lower than the voltage at the speed control potentiometer, the output of the comparator goes low, applying power to the motor. That is, the speed of the motor is less than the preset value, requiring that we apply power to the motor to increase its speed. The LED

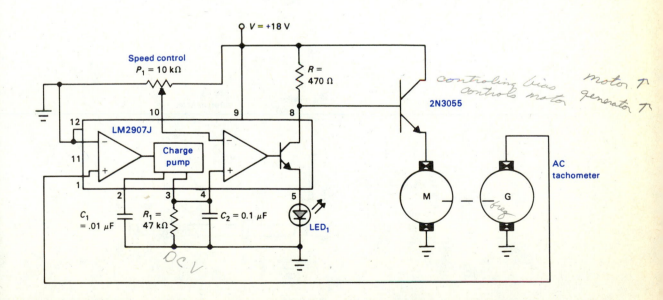

FIGURE 10.31   Velocity Servo Control System

will be off at this time since no emitter current flows. When the motor speed increases, the AC tachometer frequency increases, increasing the voltage at pins 3 and 4. When this voltage goes above the set point, the comparator output goes high, turning on the LM2907 output transistor, which turns off the power transistor. The power is then removed from the motor. The proper speed is maintained by operating the transistor in a switching mode. The average power applied to the motor will increase as the duty cycle increases.

An advantage of the velocity servo loop is its improved response to changes in the command signal input. The system can regulate speed as long as a signal proportional to the motor speed is available. In our example, this signal is provided by a tachometer. The output of the tachometer is called the *velocity feedback signal*. In this tachometer, the output frequency is directly proportional to rotational speed. Many types of feedback devices are used in industry. When velocity feedback is required, a tachometer is generally used.

Positioning Servos. The positioning servo finds many more applications than the velocity servo system. The *positioning servo system* controls the physical position of an object. The control of a ship's

rudder is a common application of a positioning servo; this application is illustrated in Figure 10.32. The objective of this system is to move the rudder a specified distance. Note the difference between the goal of the positioning servo and the goal of the velocity servo. The command potentiometer is connected across the positive and negative supplies. When we place the command potentiometer in the exact center of its span, the wiper arm of the potentiometer will be at 0 V and the motor will not turn. If we turn the shaft of the potentiometer in either direction, a voltage will appear at the wiper with respect to ground. This voltage is the command signal. We are telling the system that we want to turn the rudder. The command signal is applied to the amplifier input. The motor will rotate in the corresponding direction.

As an example, suppose we wish to turn the motor rudder clockwise (CW). So we turn the shaft of the potentiometer clockwise, placing a positive potential at the wiper. This potential will be amplified by the amplifier and will cause the motor attached to the rudder to turn. The motor will turn the rudder in a clockwise direction. The size of the voltage that appears at the amplifier input will determine how fast the rudder will move to its new position.

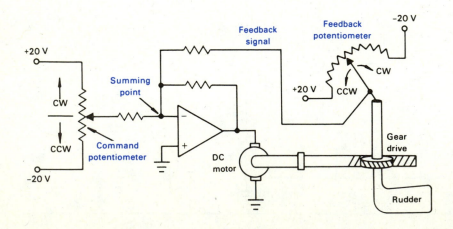

FIGURE 10.32 Positioning Servo Control System

If you recall the original definition of a servo system, you will see that we need a device that will tell us about the position of the rudder. If we do not have this feedback information, the system will not know when (or if) the rudder reaches its new position. The feedback device we will use in this system is a potentiometer. The potentiometer's output—in this case, electric resistance—is a function of the angular position of its shaft. Note that we connect the wiper arm to the rudder shaft and the arms of the potentiometer to a positive and negative supply. The voltage at the wiper of the potentiometer will then be proportional to the shaft position. This potentiometer is the feedback potentiometer. The voltages from the feedback potentiometer and the command input potentiometer will be fed through resistors to the input of a summing amplifier.

Let's start by assuming that the wipers of both potentiometers are at center, or 0 V, positions. Let's assume further that the rudder is straight out at its 0° mechanical position. We will call this position the *zero-error condition*. We turn our command input potentiometer clockwise so that we get a value of +2 V at the wiper. The motor will turn clockwise, thus turning the rudder. As the motor turns clockwise, it turns the feedback potentiometer attached to the rudder. Note that when the feedback potentiometer turns clockwise, it turns the wiper in the negative voltage direction. As the motion of the rudder continues, the feedback potentiometer will produce a negative voltage that will add to the +2 V from the command potentiometer. The amplifier output will then be driven toward zero as the rudder moves in response to the command input. Motion will continue until the feedback potentiometer reaches a point where the wiper is at a –2 V potential. The output of the amplifier will be zero, and the motor will stop turning.

The summing of the command potentiometer and feedback potentiometer voltages causes a net input of 0 V, a zero-error condition. But note the difference in the rudder position. In our starting position, the zero-error condition was when the rudder was straight out at 0 mechanical degrees. After the command input was processed, the system returned to a zero-error condition, but the rudder was turned to a new, nonzero mechanical position. If we turn the command potentiometer back to 0 V, the rudder will return to its straight position at 0 mechanical degrees.

We see from this discussion that the output is a position rather than a velocity. The voltage at the input of the amplifier is the difference between the command input and the feedback. The servo loop always responds to this input by reducing the error to 0 V. This voltage is called the *servo error signal*. If the servo error voltage is zero, the motor has turned the rudder to the position indicated by the command input.

Both velocity and positioning servo systems are similar in that they both are feedback systems. Both systems make constant, continuous corrections to the object being controlled.

Servo System Gain. In servo systems, as in many automatic control systems, the input and output measurement units are always different. Let's use our rudder-positioning system as an example, and let's assume that the gain of our amplifier is 1. Let's further assume that the rudder may span an arc of 180°. A rudder movement of 180° requires a voltage change of 40 V (–20 V to +20 V). Each volt of change causes a 4.5° shift in the rudder position.

We can see that the higher the gain of this system, the faster the rudder will move to its new position. The accepted terms for measuring *servo gain* are output velocity and position error. These terms are related in the following equation:

$$\text{positioning servo gain} = \frac{\text{output velocity}}{\text{position error}} \quad \textbf{(10.25)}$$

In our rudder example, we can express the velocity of the rudder in degrees per second. So for a rudder velocity of 9°/s, the servo gain is

$$\text{positioning servo gain} = \frac{\text{output velocity}}{\text{position error}}$$

$$= \frac{9}{4.5} = 2$$

Thus, for each degree of error signal within the servo loop, the servo will produce a correction velocity of 2°/s.

Synchros. The term *synchro* is a generic term for a family of instruments that have been in the past and are now used as position transducers. Synchros have been available for about forty years in various forms as part of electromechanical servo and shaft angle positioning systems. It was not until recently, though, that the value of the synchro was fully realized, even in the midst of an electronics industry that has been literally overwhelmed by the microprocessor. With appropriate interfacing, the synchro can form the heart of a digital shaft angle measurement system. In terms of reliability and cost, the synchro-digital system cannot be surpassed.

Synchro Operation. All synchro devices work on essentially the same principle, that of a transformer with a rotating winding. In appearance, synchros are

cylindrical and look like small AC motors. They vary in diameter from 0.5 to about 4 in. Internally, almost all synchros are similar in construction. Each has a rotor, with either one or three windings (depending on the type of synchro). The synchro rotor is free to rotate inside the fixed stator. The stator has three windings, which are connected in a star or wye arrangement. Each of the three windings are wound so that they are located 120° apart on the stator. The diagram in Figure 10.33A illustrates a basic synchro; the schematic diagram is shown in Figure 10.33B. Note that the three stator windings are each brought directly to the terminals marked S_1, S_2, and S_3. For a rotor with a single winding, the windings from the rotor are normally connected by slip rings and brushes to terminals marked R_1 and R_2. The letters CX in Figure 10.33B refer to a special type of synchro called a *control transmitter*. The letters RX

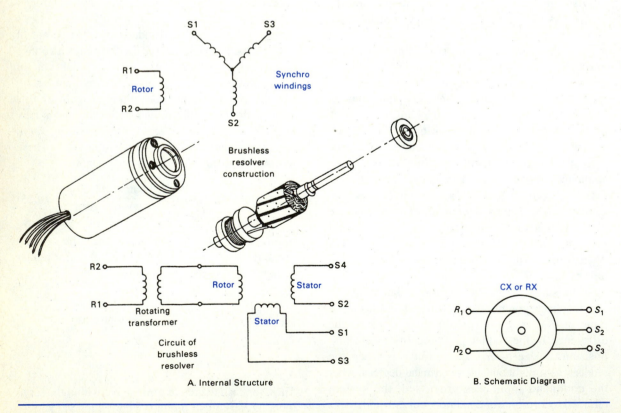

A. Internal Structure   B. Schematic Diagram

FIGURE 10.33 Synchro Construction

refer to a device called a *resolver transmitter*. A resolver is a type of synchro in which the windings are displaced 90° in relation to each other instead of 120°, as in the synchro.

In general, if the rotor winding of a synchro is connected to an AC voltage (called the *reference voltage*), the voltage induced in any stator winding will be proportional to the cosine of the angle theta (θ) between the rotor coil axis and the stator coil axis. The voltage induced across any pair of stator terminals will be the sum or difference, depending on the phase, of the voltage across the two coils concerned.

As an example, consider the two connected synchros in Figure 10.34. The synchro on the left is the transmitter, and the device on the right is the receiver. In this illustration, the magnetic polarities apply only to a half cycle of the line voltage. On the opposite half cycle, all polarities are reversed, and the field set up by the rotor coil changes polarity at the same time as the stator field changes its polarity.

In Figure 10.34, the transmitter rotor is set to the vertical position shown. The receiver rotor is shown held momentarily in 30° counterclockwise position. The induced voltages have the values and polarities

shown. In this unbalanced position, instantaneous current flows in the direction indicated by the arrows on the diagram. This current sets up a stator field in a direction that will exert a clockwise torque on the receiver rotor. The current continues until the receiver rotor is in alignment with the stator field. When alignment occurs, the voltages induced in the stator windings of the receiver are equal and opposite to the voltages induced in the transmitter. In the balanced condition, no current flows in the stator windings. The only current that flows in the system is the excitation current drawn by the two rotors. We see that the transmitter supplies current to establish a field in the receiver only when the receiver rotor is out of alignment. As a result, the system draws little power from the line.

Synchro Applications. Beyond the obvious applications of remote positioning, the synchro has uses in the digital and microprocessor fields. The diagrams in Figure 10.35 show synchro-to-digital and digital-to-synchro converters. In the *synchro-to-digital converter* (SDC) of Figure 10.35A, the synchro transmitter is connected to an SDC IC instead of a synchro

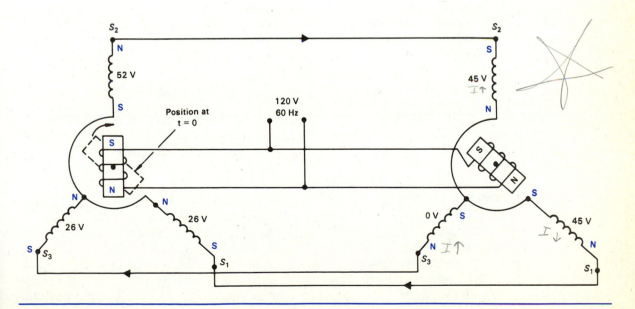

FIGURE 10.34 Synchro Transmitter-Receiver

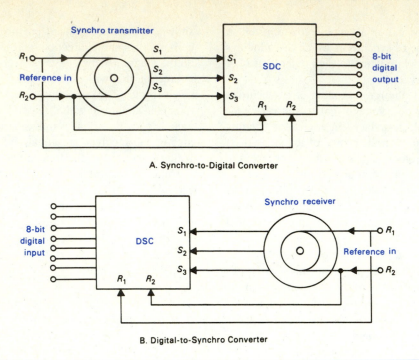

A. Synchro-to-Digital Converter

B. Digital-to-Synchro Converter

FIGURE 10.35 SDC and DSC Conversion Systems

receiver. The synchro signal is converted to a digital signal, which has a binary value proportional to the synchro shaft's position. This binary number may be evaluated by a microprocessor for processing. The other general application is illustrated in Figure 10.35B, a *digital-to-synchro converter* (DSC). The digital input to the DSC IC produces a synchro signal that positions the shaft of the synchro receiver.

Both of these devices are used in the application shown in Figure 10.36. To assist in a ship's navigation, the chart of the area through which the ship is sailing is placed on a flat, plotting table. A spot of light is generated on the table to show the ship's position on the chart. Many plotting tables take inputs in synchro form from the ship's navigation equipment to drive the equipment that positions the light spot. Because the latest plotting tables have microprocessor-controlled x and y drives for the light spot, SDCs are necessary to convert the synchro

information into digital form. The mechanisms that drive the spot are sometimes DC servomotors, which need DSC so that the microprocessor can control them.

Pneumatic Controllers

Because of their low cost, low maintenance, and safety advantages, pneumatic controllers are still popular in industry. Pneumatic controllers, like electric controllers, come in both analog and digital versions.

The analog *pneumatic controller* uses the pressure bellows and the flapper-nozzle combination to effect its control. A simple *flapper-nozzle amplifier* is shown in Figure 10.37A. Constant air pressure is supplied to the nozzle through a restrictor whose diameter is about 0.010 in. (0.254 mm). The nozzle itself has a diameter of about 0.020 in. (0.5 mm). The

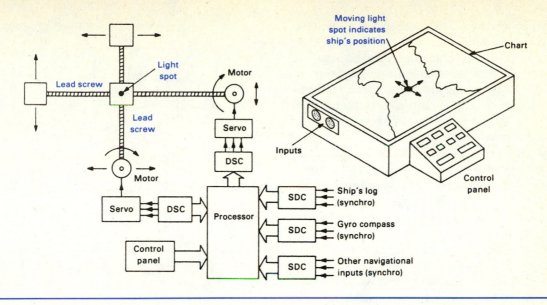

FIGURE 10.36   Microprocessor-Controlled Plotting Table with Synchro Inputs

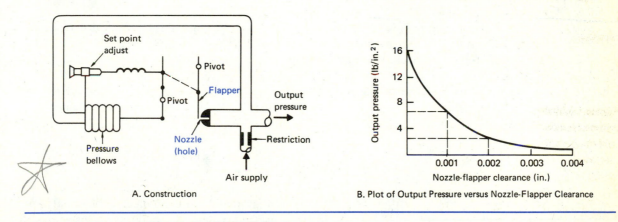

A. Construction

B. Plot of Output Pressure versus Nozzle-Flapper Clearance

FIGURE 10.37   Pneumatic Flapper-Nozzle Amplifier

flapper is positioned against the nozzle opening. The position of the flapper is determined by the transducer output and the set point. The nozzle's back pressure is inversely proportional to the distance between the nozzle opening and the flapper. As shown in Figure 10.37B, a flapper motion of about 0.002 in. (0.05 mm) is sufficient to provide a full range of output. The sensitivity of this system can be adjusted by moving the flapper pivot point.

The flapper-nozzle amplifier shown in Figure 10.37A can be used as a proportional controller. When the controlled variable (output pressure) increases, the bellows expand, causing the flapper to move away from the nozzle. Air then escapes through the nozzle into the atmosphere. This action reduces the output pressure to the set point value. Conversely, a decrease in output pressure causes the output pressure to increase to the set point value.

A proportional plus derivative pneumatic controller is shown in Figure 10.38. The addition of a variable restrictor results in a delayed negative feedback. If the controlled variable increases or decreases suddenly, the transducer causes the flapper to open or close. This change in flapper position causes the output pressure to change suddenly, but the pressure in the bellows can only change slowly. The slower change in the bellows is due to the variable restriction and the capacity of the bellows. This slower change delays and reduces the amount of negative feedback. Since the feedback is negative, the output pressure is higher and leads the transducer signal. Therefore, the delayed negative feedback produces a derivative response.

These examples show that control can be accomplished by analog pneumatic means. However, digital pneumatic control also is available through fluidic devices. A *fluidic device* is a digital logic element that operates by airflow and pressure.

One of the most popular fluidic devices uses the *Coanda*, or *wall attachment*, *effect*. This effect is illustrated in Figure 10.39. A fluid flowing past a wall has a tendency to attach itself to that wall. When air pressure flows from the supply, the stream of air will attach to one of the two walls. The stream will stay attached until a jet of air from the control port detaches it from the wall.

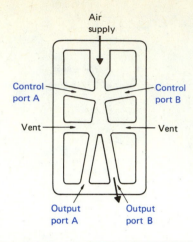

FIGURE 10.39 Wall Attachment Effect in Fluidic Device

You may recognize this fluidic device, functionally, as a bistable flip-flop. By a rearrangement of the input ports and the addition of a restriction to one of the output legs, both AND/NAND and OR/NOR gates may be formed. The logic symbols are shown in Figure 10.40. With the three logic elements, a simple digital pneumatic computer may be made. It can be used for simple controlling and decision-making functions in the same way an electronic digital computer is used.

Fluidic systems have several advantages over electric systems. Since no electric potentials are present, there is less likelihood of explosive gases being ignited. This advantage is especially important in the chemical and petrochemical industries. Also, the fluidic system can be completely flushed out with cleaning solvents at regular intervals. Such cleaning is difficult to do in electric equipment. This advantage is particularly important in the food-processing industry, where bacterial contamination must be kept low.

There are some disadvantages to fluidics, however. Generally, fluidic systems are much slower than comparable electric systems. Also, fluidic systems take up much more space since large-scale integration is not available with pneumatic systems.

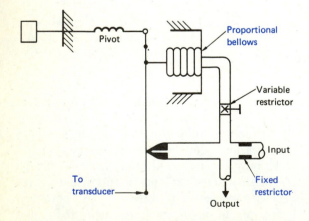

FIGURE 10.38 Proportional Plus Derivative Pneumatic Controller

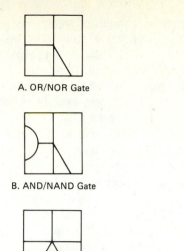

A. OR/NOR Gate

B. AND/NAND Gate

C. Bistable Flip-Flop

FIGURE 10.40 Schematic Diagrams of Fluidic Logic

FINAL CONTROL ELEMENTS

Previously, we defined the *final control element* as a device that corrects the value of the controlled variable. For instance, in an assembly line, the controlled variable might be conveyor belt speed. The conveyor belt is driven by a motor, which is the final control element. In a proportional-control system, an increase in the load on the belt would tend to slow the belt down. This speed change would be sensed and compared with a set point. The controller would increase the belt speed to normal by varying the speed of the motor, the final control element. In the pasteurization example, recall that the steam heated the milk to the proper temperature. Therefore, the valve controlling the flow of steam is the final control element. If the liquid were heated by a resistance element, that element would be the final control element.

In many process control applications, the final control element is a control valve. The *control valve* is a mechanism that regulates the flow rate of a fluid, which can be a gas, a liquid, or a vapor. The control valve can be classified on the basis of its flow characteristics or on the basis of its body style.

In electrical terms, a valve can be considered analogous to a resistance. Flow through a valve is then proportional to two things: the area of the valve opening and pressure drop across the valve. The following formula expresses this relationship:

$$Q = KA \sqrt{\Delta p} \qquad \textbf{(10.26)}$$

where

Q = quantity of fluid flow
K = constant of proportionality for conditions of flow
A = area of valve opening
Δp = pressure drop across valve

Figure 10.41 is a diagram of the percent of valve travel, or position, plotted against the percent of flow. Note that the *quick-opening valve* shows a large increase in flow rate with a small change in valve opening. This type of valve is used in on-off control. The *linear valve* has a linear response curve. That is, the valve position is linearly related to the flow rate.

The most commonly used valve type is the *equal-percentage valve*. Note that a change in the valve position produces an equal-percentage change in flow. If you were to plot this relationship on a logarithmic scale, the relationship would be a linear one. The equal-percentage valve is designed to be used between a certain minimum and maximum flow rate. The maximum flow rate (Q_{max}) divided by the

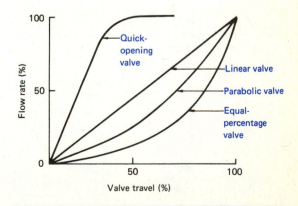

FIGURE 10.41 Plot of Valve Travel versus Flow Rate for Four Control Valve Designs

minimum flow rate (Q_{min}) is called the *rangibility* (R) of the valve. The following formula expresses the rangibility of the valve:

$$R = \frac{Q_{max}}{Q_{min}} \qquad \textbf{(10.27)}$$

Most commercial valves have a rangibility between 30 and 50.

Generally speaking, valves are controlled in industry in two ways: electrically and pneumatically. Electrical actuation of valves may be done by a solenoid, such as the one illustrated in Figure 10.42A. When current flows through the coil, a magnetic field is generated, which moves the plunger. Such a device is used in conjunction with on-off controllers.

Another method of controlling valve actuation electrically is by motors or servomechanisms, as illustrated in Figure 10.42B. The DC signal from the controller is amplified and operates a geared motor, driving the valve stem. The valve position is converted to an electric signal that feeds back to the amplifier circuit. Here it is compared with the DC signal. Any difference is amplified to drive the valve to a position exactly proportional to the original DC signal.

Valves may also be actuated pneumatically. Pneumatic actuation is shown in Figure 10.43. In this case, the diaphragm moves proportionally to the amount of air pressure at the inlet. Air pressure may be used to either open or close the valve.

Although control valves are the most common final control elements used in industry, there are others in use. In Chapters 3, 4, and 5, we discussed DC and AC motors. These devices are used in industrial velocity and position control systems, such as conveyor belts and mixing operations. The speed of DC motors is easily controlled and is capable of reversible operation. Speed control is possible in AC motors as well, although not as conveniently.

CONCLUSION

At this point, we have discussed most of the components of a process control system. These parts should be starting to fit together into a coherent whole for you. All the previous chapters have been building up to this point—a clear understanding of process control systems and their components.

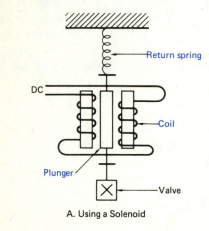

A. Using a Solenoid

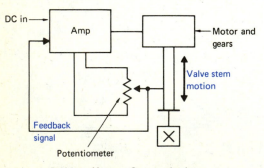

B. Using a Motor or Servomechanism

FIGURE 10.42   Electric Valve Control

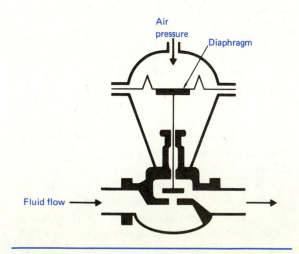

FIGURE 10.43   Pneumatic Valve Control

■ QUESTIONS ●

1. The function of a process control is to keep the controlled variable constant when it tends to be changed by a _____.

2. The process of comparing the information from a sensor to a reference is called _____. The difference between the sensor and the set point reference is called the _____.

3. _____ control has no feedback from output to input. Feedback control is sometimes unsuited to control applications since large _____ or _____ cannot be dealt with effectively by feedback control.

4. _____ control is sometimes called anticipatory control because it anticipates the effect of a disturbance and tries to correct it before it happens. _____ control is suited to systems that have large capacitances.

5. In proportional control, the final control element action is directly related to the difference between the measurement and the _____. The amount of final control element change for a given input change is called _____.

6. Integral control provides controller action that changes as a function of _____ _____ the error is present. Integral control is also called _____ control.

7. Derivative control provides controller action that varies as a function of the _____ at which the error changes. Derivative control is also called _____ control.

8. Define the following terms: (a) dynamic variable and (b) manipulated variable.

9. Draw a block diagram of a closed-loop control system with blocks for measurement, evaluation, and control.

10. Define process load.

11. Define process lag, and list the three properties of a process control system that cause it.

12. Give an example of an open-loop control system.

13. Draw a block diagram of a feedforward control system, and describe the system's operation.

14. Is it necessary to have a sensor in an open-loop control system? Why or why not?

15. Is a sensor necessary in a closed-loop control system? Why or why not?

16. What is an error signal? How is it generated?

17. Describe and draw response curves for the following control modes (open loop): (a) on-off, (b) proportional, (c) integral, and (d) derivative.

18. As the controller gain increases, what happens to the controller output for a given change in the measurement?

19. Define offset. How may offset be calculated?

20. Explain the advantages and disadvantages of (a) proportional control, (b) integral control, and (c) derivative control.

21. Can integral and derivative control be used alone? Why or why not?

22. Explain how control can be achieved by pneumatic means. Describe digital and analog pneumatic controllers and how they work.

23. Define final control element.

24. What is the relationship between controller gain and proportional band? Is it ever possible to have a 0% proportional band? Why?

25. Define the following terms: (a) velocity servo, (b) positioning servo, (c) servo error signal, and (d) synchro.

26. List three differences between a servo system and other types of automatic control.

27. Draw and label the block diagram of a servo system.

■ PROBLEMS ●

1. A process control system with a DC motor as a final control element has a proportional band of 100%. What is the gain of the controller? If the proportional band were 50%, what would the gain be?

2. In Problem 1, suppose a system had a gain of 3, a set point of 70°C, and a measurement of 50°C. How much would the controller output change over this difference?

3. Calculate the proportional band of a pneumatic controller that has a total range of 0–200 psi with its set point at 50 psi. A deviation of ±10 psi will cause the output of the controller to vary from 0 to 100 psi.

4. A pneumatic-to-electric transducer provides the following information:

Pounds per Square Inch	V_o
0.5	1.00
7.5	1.50
10.0	1.75

What is the gain of the transducer, using the values in the first and second rows?

5. In the on-off controller in Figure 10.8, we would like the fan to turn on when the temperature goes to 60°C.

   a. Calculate the output of the LM335 at 60°C.

   b. Determine the reference voltage needed at the noninverting terminal.

6. The circuit in Figure 10.10 has the following resistor value changes: $R_2 = 250$ kΩ and $R_5 = 4.5$ MΩ. Calculate the temperatures at which the circuit will trip.

7. For the controller in Figure 10.10, calculate the resistance changes in R_2 and R_5 to cause switching at temperatures of 55° and 35°C.

8. The entire measurement range of a proportional-control system is 100°C. The controller is a fan motor with an adjustable speed. Its bias point is 50%, which is equal to a speed of 1000 r/min. When temperature increases, the fan speed increases. The proportional band is 25%.

   a. Calculate the gain or sensitivity of the system.

   b. Calculate the fan speed if the measurement is 110°C.

9. In the circuit in Figure 10.18, what would the output voltage of the circuit be if the input were 18 mA?

10. In the proportional-control system shown in Figure 10.18, what values of resistance would be needed to change this controller's proportional band to 25% with a bias point of 3.5 V at the output with a 13 mA signal in?

11. A hopper discharges a solid on a moving conveyor belt at a rate of 20 lb/min. The rate at which solids fall from the end of the belt is also 20 lb/min. The rate at which the solids are discharged to the belt is increased to 35 lb/min. The conveyor is 10 ft long and is moving at a rate of 1.5 ft/min. Find the amount of time it will take for the rate at which solids fall off the belt to increase.

12. A 200 gal tank is filled when the level goes below 10 gal and stops when the level goes above 190 gal. If the tank fills at 70 gal/min and empties at 10 gal/min, find the fill and emptying rates and the cycling time.

13. For the CDA differential-gap controller in Figure 10.10, with R_2 set to 235 kΩ and R_5 set to 3.5 MΩ, find (a) the upper and lower trip points and (b) the temperatures that trip the comparator and the differential gap in temperature and voltage.

14. The circuit shown in Figure 10.11 has R_6 set to 70 kΩ and the wiper of R_3 set for 7.5 V. Find (a) the upper and lower trip points in voltage and (b) the temperatures that trip the device and the differential gap in temperature and voltage.

15. Given a process control system in which the gain is 1.2, the controller output range from 0% to 100% is 4 to 20 mA, the final control element is set at 50% when the controller produces a 12 mA output, a sudden change of load causes an error signal change of 0.25 V, and the error range is 10 V (±5 V), find the controller output (in milliamperes).

16. For the 4–20 mA converter in Figure 10.18, find the setting of resistors R_1 and R_8 that will give an output voltage of 0–15 V with a 4–20 mA input signal.

17. For the ZVS UJT power control circuit in Figure 10.21, with $R_p = 15$ kΩ and $R_x = 8$ kΩ, find the duty cycle of the triac and the power delivered to the load if 100% power is 150 W.

CHAPTER 11

Pulse Modulation

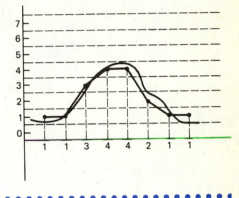

OBJECTIVES

On completion of this chapter, you should be able to:

- Define pulse modulation.
- Determine the minimum sample rate for pulse modulation.
- Modulate and demodulate the four types of analog pulse modulation.
- List the advantages and disadvantages of the pulse modulation types.
- Generate and demodulate pulse-code modulation.
- Determine quantizing noise values for pulse-code modulation.
- Use the 555 timer in circuit applications.

INTRODUCTION

Pulse modulation has been known to humanity for a long time. Modulated drum beats and smoke signals are two examples of this process. However, in this chapter, we will be concerned with a more recent development: human-generated electric pulse modulation.

Pulse modulation is a process in which an analog signal is sampled at regular periods of time. Information contained in the analog signal is transmitted only at the sampling time and may also have synchronizing and calibrating pulses. At the receiver, the original waveform may be reconstructed from the information contained in the samples. If the original samples are taken frequently enough, the analog signal can be reproduced with minimal error or distortion.

This chapter will deal with the various types of pulse modulation and demodulation and some examples of circuits using these pulse modulation methods. Pulse modulation is fundamental to the understanding of telemetry, the subject of Chapter 12.

ELECTRIC PULSE COMMUNICATION

Electric pulse communication had its origin in 1837 when Samuel Morse invented the telegraph. *Telegraphy* is the process of sending written messages from one point to another in the form of code. A few years later, in 1845, a Russian general, K. I. Konstantinov, and Dr. Poulié constructed a telemetry system. A *telemetry system* performs measurements on distant objects. The original system recorded and analyzed the flight of a cannonball. This system automatically reported the course of the cannonball as it passed through screens, and it recorded the electric impulses and a timing impulse on a manually turned recording drum loaded with graph paper. This system was one of the first written records concerning telemetry.

More recently, with the development of television and radar, an additional type of pulse communication—data communication—has come into widespread use. *Data communication* is the process of transmitting pulses, which are the output of some data source, from one point to another. Examples of data communication are computer-to-computer transmission, data collection, or telemetry and alarm systems. Other commercial uses are financial/credit information, travel and accommodation booking services, and inventory control for stores.

Facsimile is also considered to be a form of data communication. *Facsimile* is the process whereby fixed graphic material, such as pictures, drawings, or written material, is scanned and converted into electric pulses, transmitted, and, after reception, used to produce a likeness (facsimile) of the original. This process can be considered similar to the transmission of a single television picture, but in facsimile, the picture is recorded on paper. Facsimile has been used by news services to transmit newspaper photos and by ships for reception of up-to-date weather charts and maps.

In recent years, the volume of pulse communication has increased greatly. There are at least three factors contributing to this increase in pulse communication: (1) Much of the information to be transmitted is in pulse form to begin with, such as computer data or, to some extent, picture elements; (2) a more error-free transmission process is possible as, for example, in distant space probe picture transmission; and (3) the advent of large-scale integration has greatly simplified the necessary electronic circuitry. In addition, large-scale integration has permitted the use of complex coding systems that take the best advantage of channel capacities.

Telegraphy and telemetry can properly be considered subsets of data communication, but historically, they have been kept separate, as indicated in Figure 11.1. Telegraphy has been considered a separate branch because it is the transmission of messages, which are not judged to be data. Telemetry has been considered a separate branch because its first major use employed pulse modulation and radiotelemetry and did not use public transmission networks, such as the telephone line, for its transmission. Since the inception of data communication, however, these distinctions have become less clear.

Technicians should understand the relationships among telegraphy, telemetry, and data communication. This background will be beneficial when you are studying Chapter 12, which is concerned with telemetry.

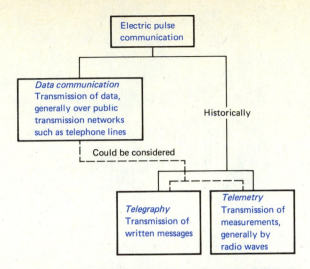

FIGURE 11.1 Subsets of Electric Pulse Communication

PULSE MODULATION TYPES

Pulse modulation can be divided into two major categories: analog and digital. In *analog pulse modulation*, a characteristic of the pulse, such as height or width, may be infinitely variable and varies proportionally to the amplitude of the original waveform. In *digital pulse modulation*, a code is transmitted; the *code* indicates the sample amplitude to the nearest discrete level.

All modulation systems sample the information waveform to be telemetered, but they all have different ways of indicating the sampled amplitude. Each modulation system also has its advantages and disadvantages, as will be discussed in the following sections.

ANALOG PULSE MODULATION

There are four major types of analog pulse modulation: pulse-amplitude modulation, pulse-width (or duration) modulation, pulse-position modulation, and pulse-frequency modulation. We will discuss each type in detail in this section.

Pulse-Amplitude Modulation (PAM)

Pulse-amplitude modulation (PAM) is illustrated in Figure 11.2. PAM is a process in which the signal is sampled at regular intervals, and each sample is made proportional to the amplitude of the signal at the instant of sampling (Figure 11.2A). As shown in Figure 11.2, there are two types of PAM. *Double-polarity* PAM can have pulse excursions both above and below the reference level (Figure 11.2B). *Single-polarity* PAM has a fixed DC level added to the pulses so that the pulse excursions are always positive (Figure 11.2C).

Noise. Since the amplitude of the pulse contains the information, it is very important that the amplitude be a true representation of the original level. However, in the transmission of the signal, noise can be added to the pulse, causing reconstruction errors. *Reconstruction errors* are errors that cause the reconstructed waveform to differ from the original waveform. These errors may be caused by noise, or they may result from other phenomena.

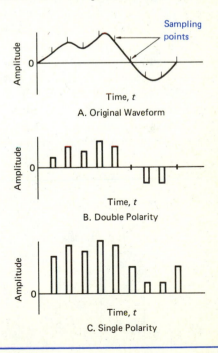

FIGURE 11.2 Pulse-Amplitude Modulation (PAM)

Figure 11.3 illustrates reconstruction errors due to noise. Noise and other reasons for error are discussed in more detail in Chapter 12. The added noise depends on many factors, such as the strength of the signal, the distance the signal must travel, and the environment the signal must pass through. Because PAM is greatly affected by noise, it is not as often as other forms of pulse modulation.

Aliasing. Another problem that can exist with any form of sampled telemetering is aliasing. *Aliasing* is the reconstruction of an entirely different signal from the original signal. The different signal is a result of the sampling rate being low in comparison with the rate of change of the signal sampled. If, for example, the signal to be sampled varies at a rate of 10 Hz, the signal may be sampled ten times during its period (sampling rate of 100 Hz). This sampling rate will not result in severe aliasing. However, if the sampling rate were reduced to 9 Hz, or less than once each signal period, a certain amount of aliasing would occur. Figures 11.4 and 11.5 illustrate these ideas.

In Figure 11.4, the sampling rate is ten times the frequency of the waveform to be sampled. As shown, the resulting, or recovered, waveform is approximately equal to the original signal. In Figure 11.5, the original waveform frequency remains the same as in Figure 11.4, but the sampling rate is reduced by more than ten times the sample rate of Figure 11.4. The resulting, or recovered, waveform in Figure 11.5 no longer resembles the original waveform. Therefore, considerable aliasing is produced.

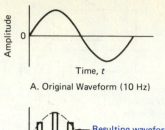

A. Original Waveform (10 Hz)

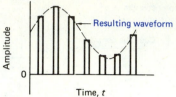

B. Pulses Used to Reconstruct Original Waveform

FIGURE 11.4 Sample Rate Resulting in Minimum Aliasing

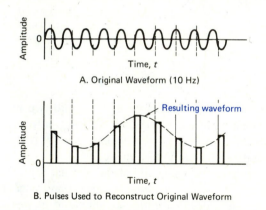

FIGURE 11.5 Sample Rate Resulting in Considerable Aliasing

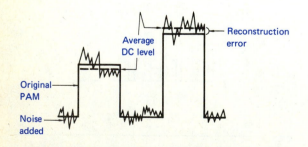

FIGURE 11.3 Effect of Noise on PAM

Sampling Theorem. The aliasing problem associated with sample rate has been investigated by many researchers and has resulted in a theorem known as the *sampling theorem*:

If the sampling rate in any pulse modulation system exceeds twice the maximum signal frequency (or Nyquist rate), the original signal can be reconstructed in the receiver with vanishingly small distortion.

As a result of this theorem, most pulse modulation sampling rates for speech over standard telephone channels, whose bandwidth is 300–3400 Hz, are standardized at 8000 samples per second. The 8000 samples per second is slightly more than twice 3400 Hz, and therefore the sampling theorem is satisfied.

To illustrate the theorem, we will assume Figure 11.6A represents a small portion of a speech transmission. The portion of the waveform from point A to point B represents the steepest slope, or greatest rate of change, of the wave. Since it is telephonic speech, this rate of change corresponds to the frequency of 3400 Hz. A 3400 Hz sine wave (the dotted line) whose peaks are at points A and B is shown for comparison. The period of the 3400 Hz sine wave is also shown. According to the sampling theorem, the samples are taken 8000 times per second and therefore have a pulse period of 0.125 ms.

The waveform in Figure 11.6B is the same as the waveform in Figure 11.6A, but it shows how far apart the samples are taken to indicate 8000 pulses per second. This rate is the slowest pulse rate required to reproduce the original signal with "vanishingly small distortion."

Generation and Demodulation of PAM. In most electronic systems, there are many different circuits that can be used to produce the same outcome. So it is here. The circuits in this chapter represent only a few of the circuits that can be used to produce pulse modulation. PAM can be produced by first generating a pulse, then varying the gain of the amplifier to which the pulse is going. A block diagram for producing PAM is shown in Figure 11.7. A National Semiconductor LM555 timer (discussed later in this chapter) connected in the astable configuration or CD4047 low-power, monostable/astable multivibrator in the astable mode can be used as the pulse generator. The CD4047 is CMOS (complementary metal-oxide semiconductor).

The variable-gain amplifier employed here is often used to generate analog amplitude modulation (AM). The basic circuitry is that of a differential amplifier, shown in Figure 11.8. The pulses are input into one of the differential transistors, while the modulating signal is applied to the normally constant current transistor. The modulating signal thus causes the amplifier gain to vary. The output is now pulse-amplitude-modulated. This result is essentially what the transconductance amplifier, discussed in detail in Chapter 1, produces.

Other classes of circuits that can be used for the variable-gain amplifier are Motorola Semiconductor's MC1595L linear four-quadrant multiplier, National Semiconductor's LM1596 balanced modulator-demodulator, and RCA's CA3080 operational transconductance amplifier (OTA).

The complete circuit for PAM is shown in Figure 11.9.

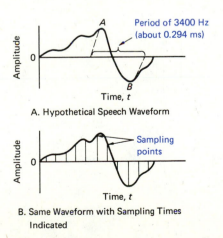

A. Hypothetical Speech Waveform

B. Same Waveform with Sampling Times Indicated

FIGURE 11.6 Waveform Illustrating Sampling Theorem

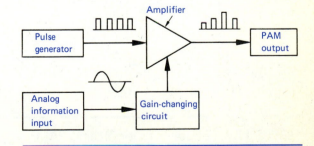

FIGURE 11.7 Block Diagram for Producing PAM

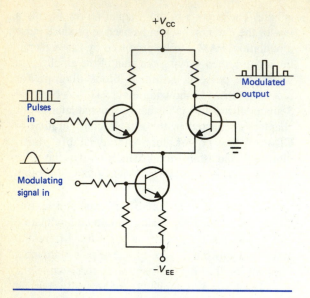

FIGURE 11.8 Variable-Gain Differential Amp

Demodulation, or signal recovery, of PAM can be accomplished with a low-pass filter, as shown in Figure 11.10. Additional filtering or signal smoothing may be required.

One important question that sometimes needs to be addressed is, "How much bandwidth is needed to transmit the PAM signal?" The required bandwidth has been found experimentally as

$$BW = \frac{K}{t} \tag{11.1}$$

where

BW = bandwidth (in hertz)

K = constant that depends on spacing between adjacent pulses and other factors

t = pulse width (in seconds)

In theoretical systems, K is equal to 0.5; in practical systems, a constant equal to 1 is more often used.

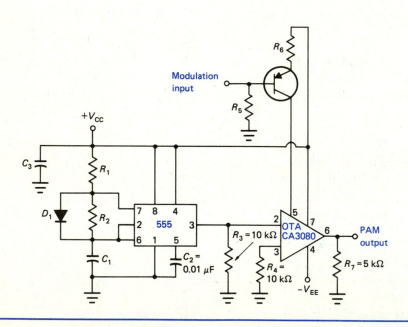

FIGURE 11.9 Circuit for Generating PAM

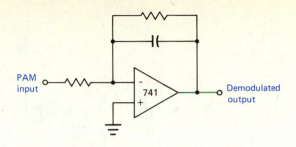

FIGURE 11.10 Low-Pass Filter for Demodulating PAM

Notice that when $K = 1$ in Equation 11.1, the standard equation $f = 1/t$ results, where f is the frequency to be determined and t is the period of the sine wave. Therefore, when $K = 1$ in Equation 11.1, we are approximating the PAM signal by substituting a sine wave of the same period.

An example may help here. Suppose we are sampling an analog signal at a rate of 5000 Hz and our pulses have a width of 40 μs. What bandwidth would we need to transmit the signal? Using Equation 11.1, we get

$$BW = \frac{K}{t} = \frac{0.5}{40\ \mu s} = 12.5\ \text{kHz}$$

for $K = 0.5$. With a constant K equal to 1, BW is then equal to the reciprocal of the pulse width of the pulse, or

$$BW = \frac{K}{t} = \frac{1}{t}\frac{1}{40\ \mu s} = 25\ \text{kHz}$$

• EXAMPLE 11.1

Given an analog signal that is sampled at a rate of 7500 Hz and that pulses with a width of 50 μs, find the bandwidth needed to transmit the signal at $K = 0.5$ and $K = 1$.

Solution

Using Equation 11.1 and substituting, we get

$$BW = \frac{K}{t} = \frac{0.5}{50\ \mu s} = 10,000\ \text{Hz}$$

for $K = 0.5$. With a constant K equal to 1, BW is equal to the reciprocal of the pulse width of the pulse, or

$$BW = \frac{K}{t} = \frac{1}{t} = \frac{1}{50\ \mu s} = 20\ \text{kHz}$$

Pulse-Width Modulation (PWM)

Pulse-width modulation (PWM), sometimes referred to as *pulse-duration modulation* (PDM), is shown in Figure 11.11. In PWM, the signal is sampled at regular intervals, but the pulse width is made proportional to the amplitude of the signal at the instant of sampling (Figure 11.11A). As shown in Figure 11.11B, the pulses are of equal amplitude, and the leading edges of the pulses are the same time apart. However, the trailing edges of the pulses are not.

One concern in PWM is that the pulse-width variations do have practical limits. Obviously, the pulse width cannot exceed the spacing between pulses. If it did, there would be overlapping, and the proportional relationship between pulse width and signal amplitude would no longer be true. And since

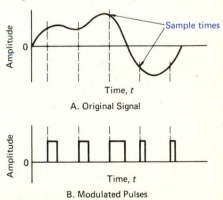

FIGURE 11.11 Pulse-Width Modulation (PWM)

a negative pulse width makes no sense, the most negative swing of the signal to be sampled can, at most, produce zero pulse width. In practice, these limits are more restricted. As a rule of thumb, the widest pulse width should not exceed 80% of maximum; the narrowest pulse width should not be less than 20% of maximum. Figure 11.12 illustrates these limitations.

PWM is less affected by noise than PAM. However, it is not completely immune. Figure 11.13 shows the possible error due to noise for PWM.

Generation of PWM can be produced by again employing the LM555. National Semiconductor's LM555 specification sheet contains a circuit similar to the one shown in Figure 11.14. In this circuit, two LM555s are used. However, the circuit can be made more compact by using an LM556, which is a dual LM555.

Demodulation of PWM can be accomplished in the same manner as for PAM (Figure 11.10). A low-pass filter will demodulate pulses that vary in either amplitude or width.

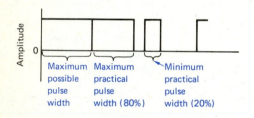

FIGURE 11.12 Practical Pulse-Width Limitations

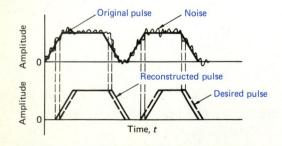

FIGURE 11.13 Exaggerated Effect of Noise on PWM

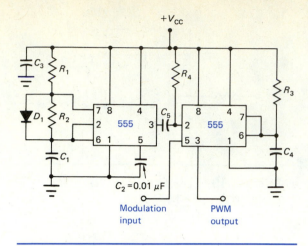

FIGURE 11.14 Circuit for Generating PWM

Another PWM circuit, this one using the LM566, is shown in Figure 11.15A. The modulation process occurs when the comparator compares the triangle wave (f_c) from the 566 and the sine wave input (which represents the input signal intelligence). When the sine wave input goes below the triangle wave input, the output of the comparator goes low, as shown in Figure 11.15B. Note that the comparator output is high when the sine wave input goes above the triangle wave input. Note also that the output waveform of the comparator is pulse-width–modulated.

Precautions to consider for this circuit are:

1. The triangle wave frequency should not go lower than twice the intelligence signal frequency (sine wave generator frequency in Figure 11.15A) in order to satisfy the sampling theorem requirements.

2. The intelligence signal amplitude should not exceed 80% of the triangle wave amplitude in order to satisfy practical pulse-width requirements.

Also note that the pulse-width–modulated waveform of Figure 11.15 has a variable pulse period as well as pulse width. The output of the circuit of Figure 11.14 varies only in pulse width, not in pulse period.

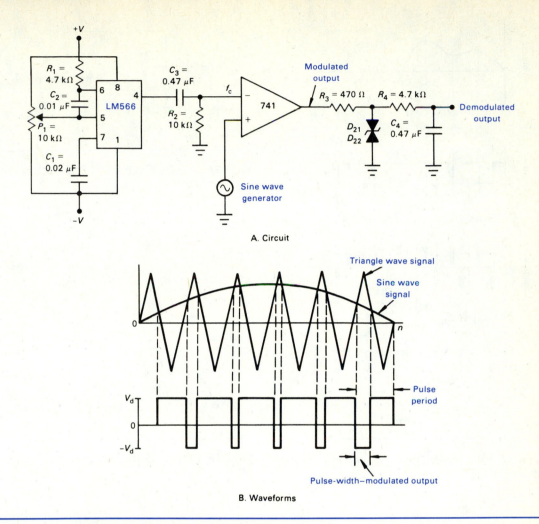

A. Circuit

B. Waveforms

FIGURE 11.15 Pulse-Width Modulation with LM566

Pulse-Position Modulation (PPM)

Pulse-position modulation (PPM) and a method of generating it are shown in Figure 11.16. The original signal is first pulse-width–modulated, then differentiated, and clipped. The resulting pulses vary their position about a reference line, which is established by the zero-signal pulse position. As shown in Figure 11.16D, the PPM pulses are on the reference line, or

lag the reference, or lead it. The amount of lead or lag from the reference line is proportional to the amplitude of the original signal above or below the zero line. Note that the leftmost pulse of Figure 11.16D is on the reference line; the second, third, and fourth pulses from the left lag the reference lines; the fifth and sixth pulses lead the reference lines; and the last pulse is on the reference line.

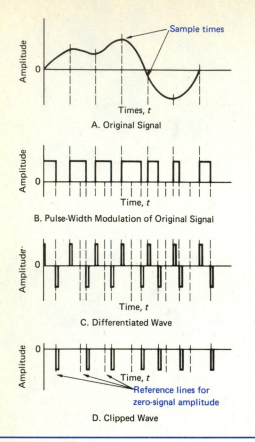

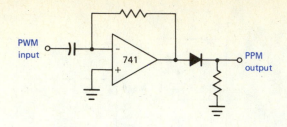

FIGURE 11.17 Circuit for Producing PPM from PWM

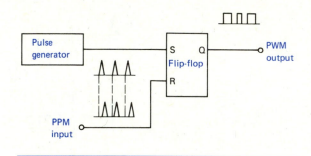

FIGURE 11.16 Generation of Pulse-Position Modulation (PPM)

FIGURE 11.18 Converting PPM to PWM

PPM is as susceptible to noise on the pulse as PWM is, but it is not as bad as PAM. The same cause of error shown in Figure 11.13 can affect PPM. But, of course, for PPM the position of the pulse is affected; its width is not important.

As shown in Figure 11.16, in the generation of PPM, the signal is first converted into PWM. The circuit of Figure 11.14 shows generation of PWM. The output from the circuit of Figure 11.14 is then input to the circuit of Figure 11.17. The PWM is thus differentiated, and the negative pulses are clipped. Because the differentiator is an inverter, the PPM output is now a positive-going pulse.

PPM can be demodulated in at least two different ways. In the first method, PPM is converted back to PWM, and then the output is filtered. Figure 11.18

illustrates the conversion to PWM. Here, a flip-flop is used for the conversion to PWM, which then can be filtered to the original signal. This circuit has a drawback, however: The pulse generator input must somehow be synchronized with the PPM input. This synchronization calls for more circuitry and is more difficult than necessary. Therefore, it will not be discussed further.

The second method for demodulating PPM does not require synchronization and is easier to do than the first. This method requires the use of a PLL circuit, which can be used to demodulate any frequency-modulated signal. The circuit is shown in Figure 11.19. The two diodes, D_1 and D_2, are used to limit the input signal to a safe range in case it is large. The input of the LM565 responds to zero or

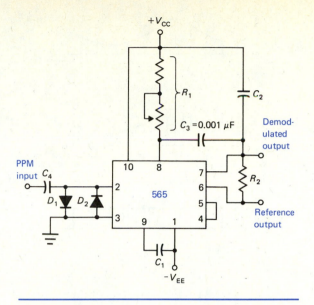

FIGURE 11.19 Demodulation of PPM Using PLL

reference crossings only, and therefore the input signal does not have to be a sine wave but can be any waveform.

The VCO free-running frequency f_o of the 565 is given by the following equation:

$$f_o = \frac{1.2}{4R_1C_1}$$

This free-running frequency should be adjusted to be near the center frequency of the input PPM frequency range. In the equation, R_1 should be between 2 and 20 kΩ, with 4 kΩ being the optimum value. The value of C_1 (other than conforming to the equation above) is not critical. Resistor R_2 in Figure 11.19 is not necessary but can be used to decrease the lock range of the 565. The output of the 565 (pin 7) will need filtering and amplification. (The PLL and its use are discussed in detail in Chapter 2.)

Pulse-Frequency Modulation (PFM)

Pulse-frequency modulation (PFM) may look very much like PPM, but it is generated in a different way. PPM has only one pulse in each sample space, but

PFM can have more than one pulse in each sample space. (See Figure 11.22E.)

Of the four pulse modulation systems discussed up to this point, PFM is the least affected by noise. PFM depends not on pulse width or pulse location but, rather, on the number of pulses per second. Thus, noise spikes would have to be very large in order to affect frequency.

One method for generating PFM uses a single LM555 connected as shown in Figure 11.20. This circuit is shown in National Semiconductor's specifications for the LM555 as PPM. However, according to the definitions in this chapter, Figure 11.20 is the circuit for PFM, not PPM.

Another circuit for generating PFM very simply is that of Figure 11.21. Here, the LM741 op amp is used as a relaxation oscillator. The R_1C_1 combination determines the rate of change of voltage at the inverting input to the op amp. The R_2R_3 combination provides positive feedback to the op amp and is the comparator-with-hysteresis portion of the circuit.

Some means of varying the value of R_1 would cause the output square wave to change its frequency. Therefore, if the modulation signal is applied such that R_1 varies its resistance at the modulation rate, then the output square wave will vary its frequency at the modulation rate. Figure 11.21 shows an LED (D_1) and a photoresistor (R_1) enclosed in a lightproof

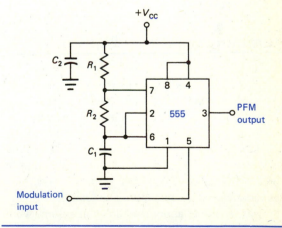

FIGURE 11.20 Generation of PFM Using LM555

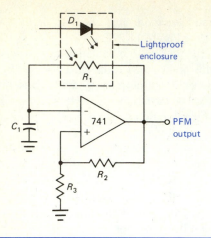

FIGURE 11.21 Generation of PFM Using LM741

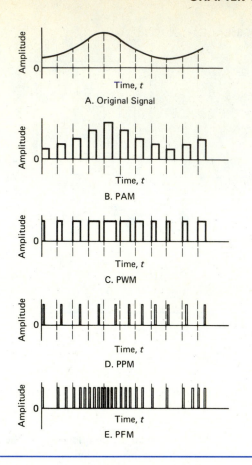

FIGURE 11.22 Analog Pulse Modulation Types

enclosure. The circuitry associated with the LED causes it to vary its intensity, which, in turn, varies the resistance of R_1. The output of the 741 is now PFM.

Demodulation of PFM can be done in the same way that demodulation of PPM is done—that is, with the PLL of Figure 11.19. The PLL *chip* (the semiconductor material that makes up the integrated circuit) is a very handy circuit. You should get to know it well.

Summary of Analog Pulse Modulation

Figure 11.22 gives a visual summary of the analog pulse modulation types. As we have seen, these pulse modulation types can be implemented easily by using the simple circuits illustrated in this chapter. These circuits have been chosen to facilitate understanding of the modulation types and also to illustrate the problems associated with each modulation type, such as modulation, demodulation, and noise. An industrial-grade modulation/demodulation system would necessarily be more complex, but the principles involved would be the same.

THE LM555 TIMER

As you have seen in the implementation of the analog pulse modulation circuits, the LM555 timer was used extensively. Since its introduction, the 555 timer has had a significant impact and enjoyed great popularity in the electronics industry. Its success can be attributed to at least four characteristics: versatility, simplicity of use, stability, and low cost. These attributes have made the 555 timer almost as popular as the IC operational amplifier. In the following presentation, we will discuss the 555's basic specifications, chip operation, and component value calculations.

Basic Specifications

We cannot emphasize too strongly how important device specifications are. As a user of electronic devices, you should know that company data books and application notes have a significantly higher level of accuracy of information than *any* textbook has. The accuracy of data books impacts directly in a company's survival; therefore, a company will do everything in its power to keep its data books accurate. Whenever you see circuit schematics that you wish to construct that are given in magazine or journal articles or in books, be sure to check them against the manufacturers' data books in order to save yourself troubleshooting time.

Most manufacturers use standardized device specifications, as can be seen by comparing various data books for the same device. The specification information is generally divided into a number of helpful categories: description, important features and general applications, pin configurations, absolute maximum ratings, block diagrams, equivalent schematics, electrical characteristics, typical performance curves, and some typical applications to answer many questions the user may have. Examples of specifications are given in the data sheets presented at the end of this text. The order of information may vary among manufacturers, but all the categories just listed are almost always given.

The description of the device is often the first item of information. The description is a short and precise summary of the device and some important considerations in its use. This information will help the user quickly locate the right device for the intended use.

To illustrate, the data sheets for the LM555 timer at the back of the book list these description features: (1) It is a stable device; (2) it can generate time delays or oscillations; (3) it can be externally triggered or reset; (4) one external resistor and one external capacitor are used for time delays; (5) two external resistors and one capacitor are used to generate oscillations; (6) it may be triggered or reset on falling waveforms; and (7) it can source or sink up to 200 mA or drive TTL circuits. (Sourcing and sinking are discussed in Chapter 2.)

The next section of the specifications, the features and applications section, gives single-phrase descriptions of the important features of the device and general application categories. The features section is quite detailed and may be an expansion of the description paragraph. The "selling points" of the device would be located there.

For example, the data sheets at the end of the book list the following major features of the LM555: The timer can be set from microseconds through hours (and even days with two 555s); (2) it can operate in the astable (no stable state) or monostable (one stable state) modes; (3) the pulse duty cycle can be adjusted (this feature will be discussed later in this chapter); and (4) the timer's output voltage changes no more than 0.005% per degree Celsius.

Notice that the applications section lists pulse-width and pulse-position modulation as well as the expected timer applications.

The information necessary for applying the device in a circuit is given next. The pin configuration or connection diagrams show the "pinout," or IC package pin numbers, in numerical order and the pin functions. Also shown are the different types of packages available. Data books list, in a separate section, all the information about chip packaging that the company provides. This section lists package outlines, dimensions, lead spacing, tolerances, and sometimes thermal resistance values for the different packages. The package construction material, such as plastic or ceramic, may also be given. This packaging section is most helpful in determining circuit board layout and orientation.

Absolute maximum ratings are the values of supply voltage, power dissipation, and temperature that must not be exceeded; if they are, the manufacturer cannot guarantee safe operation. These values may be exceeded accidentally, as when the device is used with an inductive load or in an enclosed automobile in the summer. Therefore, environmental conditions must always be considered and taken into account.

Inside the connection diagram is the block diagram representation of the schematic. The 555 block diagram is easier to see in Figure 11.23 than

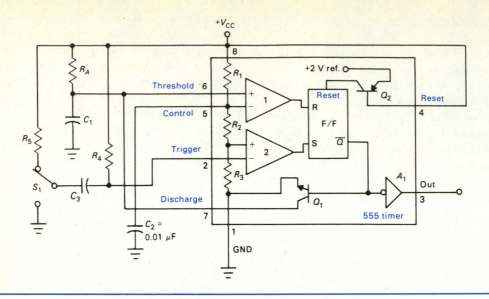

FIGURE 11.23 555 Timer Connected as Monostable Multivibrator

in the data sheets, but they are the same. The block diagram shows the major operation blocks and how they are connected. This diagram is very helpful in visualizing the device's principles of operation. Knowing a device's internal operation helps you to apply the device in new situations. Many students neglect this diagram and are thus mystified by the device's operation in a circuit. We will deal more with the block diagram later.

The equivalent circuit, or schematic diagram, is often an approximation of the actual circuit, but it can be very useful. This diagram shows the circuit concepts and how the block diagram is actually implemented. It can also be used to determine approximate input impedance, output impedance, frequency response, and other circuit characteristics through circuit analysis.

The major features shown on the schematic diagram in the data sheets for the 555 are the input and output circuits. Differential amplifiers make up the input circuit, and a "push-pull" or "totem pole" amplifier constitutes the output. Each type of

amplifier has its own characteristics and advantages. The differential amplifiers provide common-mode rejection, among other things, as discussed in Chapter 1; the totem pole output amplifier provides greater power gain.

The electrical characteristics and typical performance curves complement each other. The electrical characteristics list important parameters of the device and give the manufacturer's guaranteed minimums and maximums as well as typical values. One or more of these values may be missing either because no limit is possible or because values are so inconsistent that a limit is not practical. Typical values are given more often than maximums and minimums are. Typical values are *averages* for a large number of devices, so one particular device measured may not give the exact values listed in the tables. A slight deviation does not mean that the device is bad, however. If the measurement is "close," it is probably acceptable. Pay particular attention to the units of the parameter, which can help you understand the parameter more fully. For example, ppm (parts per million) is a short way of representing a very small number, and it indicates a fraction with one million in the denominator. That is, 50 ppm is 50 divided by 1,000,000, or 0.00005.

Typical performance curves amplify and expand upon the electrical characteristics table. The curves show more precisely how some parameters vary with others over a large range; the values in the characteristics table may be single points on the curves, specified by the test conditions.

The typical performance curves may be used for design as well as for troubleshooting. For example, the minimum pulse width required for triggering, as shown on the triggering curve for the 555, can be used to set a minimum limit on the design. If the measured value goes below the minimum, you can be almost certain that there is one reason the timer is not working correctly. The supply current versus supply voltage curve can be used to determine the maximum power supply requirements for the system in which the timer is located. And the source and sink current curves can be used to ensure that maximum loads for the timer are not exceeded. These illustrations for the 555 help explain why manufacturers include typical performance curves in the specifications.

Lastly, the typical applications show a few examples that have been worked out by the manufacturer. Generally, you will understand many other applications once you understand the examples provided by the manufacturer. Most data books have additional application notes at the end of the book, or an entire book may be devoted to nothing but applications. These applications are very useful and are recommended reading.

Using the 555

The 555 timer is shown in Figure 11.23 within the rectangle. This block diagram is arranged somewhat differently than the diagram given in the data sheets at the end of this text, but close inspection will show that the features are the same. The timer consists of two voltage comparators, a bistable flip-flop, a discharge transistor, a reset transistor, and a resistor divider network. The resistor divider network is used to set the comparator levels. Since all three resistors

(R_1, R_2, and R_3) are of equal value, the threshold comparator is referenced internally at $\frac{2}{3}$ of the supply voltage level and the trigger comparator is referenced at $\frac{1}{3}$ of the supply voltage. The outputs of the comparators are connected to the bistable flip-flop.

The output of the flip-flop is initially set high (see Figure 11.24E). The flip-flop output goes to two places: the power amplifier, which inverts the signal and drives the output load (see Figure 11.24F), and the discharge transistor, which is held on by the flip-flop and causes capacitor C_1 to remain discharged.

The timing diagram in Figure 11.24A shows that the voltage at pin 2 of the 555 is held at V_{CC} because of the pull-up resistor R_4. (This feature is not shown in the data sheets but is the only way pin 2 would stay high.) As the switch S_1 goes from V_{CC} to the ground position, the voltage on pin 2 is driven in the negative direction. When the trigger voltage is driven below $\frac{1}{3}$ of the supply, the comparator changes state and sets the flip-flop, driving the output of the flip-flop to a low state, as shown in Figure 11.24C. With the flip-flop output low, the output power amplifier drives the output high, and the discharge transistor is turned off, allowing capacitor C_1 to start charging; see Figures 11.24B, 11.24E, and 11.24F. Capacitor C_1's charging now controls the circuit. When the voltage across C_1—and, therefore, the positive input to the comparator—reaches $\frac{2}{3}$ of the supply, the threshold comparator resets the flip-flop (11.24D), which, in turn, drives the power amplifier output low and also turns on transistor Q_1, discharging capacitor C_1. Once the capacitor is discharged, the timer will await another trigger pulse, and the timing cycle is completed.

An important point can be made here. In the discussion concerning the pull-up resistor R_4, we noted that the data sheets did not show this feature. But by observing the block diagram and understanding the timer's operation, we can see that the trigger must be held high for the circuit to operate as specified. The point is that the block diagram is an essential piece of information for understanding and using the chip.

In the operation of the circuit in Figure 11.23, note that once the device has triggered and the bistable flip-flop is set, continued triggering will not interfere with the timing cycle. However, there may

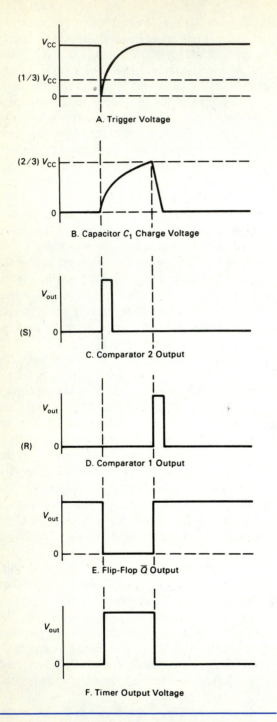

FIGURE 11.24 Timing Diagram for Monostable Multivibrator

come a time when it is necessary to interrupt or halt a timing cycle. This operation is the function of the reset.

In the normal operating mode, the reset transistor Q_2 is off, with its base held high. When the base of Q_2 is grounded, it resets the flip-flop such that its output is driven high, turning on Q_1 and discharging the timing capacitor C_1. The reset overrides all other functions within the timer.

Due to the nature of the trigger circuitry, the timer will trigger on the negative-going edge of the input pulse. For the device to time-out properly, the trigger voltage level must be returned to some voltage greater than $\frac{1}{8}$ of supply voltage before the time-out period. If the trigger is not AC-coupled, as in Figure 11.23, the output of the timer will remain high as long as the trigger voltage remains below $\frac{1}{8}$ of supply and will override any timing done by the timing capacitor C_1. This override is a common problem encountered by anyone who does not understand the timer's operating principles.

For applications where the control voltage function (pin 5) is not used, as in Figure 11.23, it is strongly recommended that a bypass capacitor (0.01 μF) be placed across the control voltage pin and ground. This capacitor will increase the noise immunity of the timer, preventing noise voltages from altering the timing capacitor's discharge point and thereby causing timing error.

The supply voltage may be provided by any number of sources; however, several precautions should be taken. The most important—and the one that causes the most headaches if not practiced—is good power supply filtering and adequate bypassing. Ripple on the supply line can cause loss of timing accuracy. A capacitor across V_{CC} and the ground—ideally, directly across the device—is necessary. The size of the capacitor will depend on the specific application. Values of capacitance from 0.01 to 10 μF are most common. Note that the bypass capacitor should be as close to the device as physically possible.

In selection of the timing resistor and capacitor, several considerations should be taken into account. Stable external components are necessary for the RC network if good timing accuracy is to be maintained.

The timing resistor(s) should be of the metal film variety if timing accuracy and repeatability are important. A good-quality trim potentiometer, placed in series with the timing resistor, will allow for best adjustability and performance. The timing capacitor should be a high-quality, stable component with very low leakage characteristics. Do not use ceramic disc capacitors since their capacitance values are not sufficiently stable. Several acceptable capacitor types are silver mica, mylar, polycarbonate, polystyrene, or tantulum. If timer accuracy over a temperature range is a consideration, timing components with a small positive temperature coefficient should be chosen. This combination will tend to cancel timing drift due to temperature.

Component Value Calculations

In the selection of the values for the timing resistor(s) and capacitor, several points should be considered. A minimum value of about 0.25 μA for the threshold current is necessary to trip the threshold comparator. In calculating the maximum value of resistance, keep in mind that at the time the threshold current is required, the voltage potential on the threshold pin is ⅔ of the supply. Therefore, maximum resistance is defined as

$$R_{max} = \frac{V_{CC} - V_C}{I_{thresh}} \qquad (11.2)$$

where V_C is control voltage (generally ⅔ V_{CC}).

• EXAMPLE 11.2
Given the 555 timer with supply voltages of 15 and 5 V, find R_{max}.

Solution
Using Equation 11.2 with $V_{CC} = 15$ V, $V_C = 10$ V, and $I_{thresh} = 0.25$ μA, we get

$$R_{max} = \frac{V_{CC} - V_C}{I_{thresh}} = \frac{15\ V - 10\ V}{0.25 \times 10^{-6}\ A}$$

$$= 20\ M\Omega$$

For $V_{CC} = 5$ V, we get

$$R_{max} = \frac{V_{CC} - V_C}{I_{thresh}} = \frac{5\ V - 3.33\ V}{0.25 \times 10^{-6}\ A}$$

$$= 6.6\ M\Omega$$

Note: If you are using a large value for the timing resistor, be certain that the capacitor leakage is significantly lower than the charging current available in order to minimize timing error. (If the timing capacitor has very much leakage, the 555 may never produce an output.)

Certain minimum values of resistance should also be observed. Remember that the discharge transistor Q_1, when turned on, will be carrying two current loads. The first is the constant current through timing resistor R_A. The second is the varying discharge current from the timing capacitor. For best operation, the current contributed by the R_A path should be minimized so that the majority of discharge current can be used to reset the capacitor voltage. Hence a 5 kΩ value is recommended as the minimum feasible value for R_A. Lower values can be used successfully in certain cases but should be avoided if at all possible.

To determine the external component values for the circuit of Figure 11.23, let us look at the timing circuit, R_A and C_1. When the discharge transistor Q_1 is turned off and capacitor C_1 starts to charge, it does so through resistor R_A. The voltage on the capacitor increases exponentially with a time constant $T = R_A C_1$. Ignoring capacitor leakage, the capacitor will reach the ⅔ of supply voltage in 1.1 time constants, or

$$T_P = 1.1(R_A C_1) \qquad (11.3)$$

where

    T_P = pulse period of timer output pulse

• EXAMPLE 11.3
For the circuit of Figure 11.23, with a 0.1 μF silver mica capacitor, find the external circuit values necessary to generate a 50 ms pulse.

Solution

Solving Equation 11.3 for R_A gives

$$R_A = \frac{T_P}{1.1 C_1} = \frac{50 \times 10^{-3}\text{ s}}{(1.1)(0.1 \times 10^{-6}\text{ F})}$$

$$= 455\text{ k}\Omega$$

For generation of pulse-width modulation, the timer is connected in the monostable mode, as shown on the right in Figure 11.14. The monostable circuit is triggered with a continuous pulse train, as shown in Figure 11.25A. The threshold voltage is modulated by the signal applied to the control voltage terminal, pin 5, causing it to vary about the ⅔ supply bias set by the internal voltage divider string. See Figure 11.25C. As the control voltage varies, the charge time of the timing capacitor C_4 (Figure 11.14) varies proportionally, as shown in Figure 11.25D. This variation also causes the output pulse width to vary in accordance with the modulation signal, as indicated in Figure 11.25H. Pulse-width modulation results.

For the 555 to be used in an astable multi-vibrator, the circuit of Figure 11.26 is used. In the astable (free-running) mode, only one additional component, R_B in Figure 11.26, is necessary. The trigger is now connected to the threshold pin. At power-up, the timing capacitor C_1 is discharged, holding the trigger low. This low triggers the timer, which establishes the capacitor charge path through R_A and R_B. When the capacitor reaches the threshold level of ⅔ of the supply, the output of the timer drops low and the discharge transistor turns on. The timing capacitor now discharges through R_B alone. When the capacitor voltage drops to ⅓ of the supply, the trigger comparator trips, automatically retriggering the timer, creating an oscillator whose frequency is given by

$$f = \frac{1.49}{(R_A + 2R_B)(C_1)} \tag{11.4}$$

The duty cycle is

$$\text{duty cycle} = \frac{R_A + R_B}{R_A + 2R_B} \tag{11.5}$$

Note that this equation differs from the data sheet equation. Equation 11.5 defines the duty cycle as the time the device is high divided by the total period of the waveform. Also, in different data books, some constants in the equations differ.

The timing diagram for the circuit of Figure 11.26 is shown in Figure 11.27. Different ratios of R_A and R_B will vary the duty cycle according to Equation 11.5. If a duty cycle less than 50% is required, this circuit needs some adjustment because the charge path is $R_A + R_B$ while the discharge path is R_B alone. In this case, you must insert a diode in parallel with R_B, as shown on the left in Figure 11.14. Now the charge path becomes R_A and the parallel diode into C_1. Discharge is through R_B and the discharge transistor Q_1. This arrangement will allow a duty cycle range from less than 5% to greater than 95%. Note that for reliable operation, a minimum value of 3 kΩ for R_B is recommended to ensure that oscillation begins.

To determine the pulse period for the circuit on the left in Figure 11.14, use the following equations:

$$t_1\text{ (output high)} = 0.67 R_A C_1 \tag{11.6}$$

$$t_2\text{ (output low)} = 0.67 R_B C_1 \tag{11.7}$$

$$T_P\text{ (total period)} = t_1 + t_2$$
$$= 0.67(R_A + R_B)(C_1) \tag{11.8}$$

• EXAMPLE 11.4

Given an astable multivibrator, with $R_A = 10$ kΩ, $R_B = 10$ kΩ, and $C_1 = 0.1$ μF, find the total pulse period and duty cycle for (a) the circuit in Figure 11.26 and (b) the circuit in Figure 11.14.

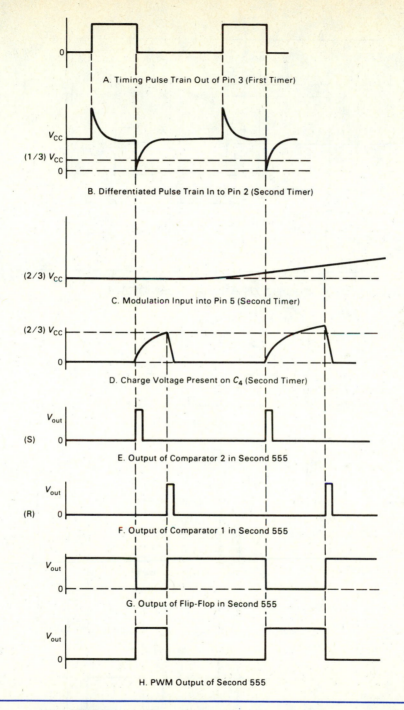

A. Timing Pulse Train Out of Pin 3 (First Timer)

B. Differentiated Pulse Train In to Pin 2 (Second Timer)

C. Modulation Input into Pin 5 (Second Timer)

D. Charge Voltage Present on C_4 (Second Timer)

E. Output of Comparator 2 in Second 555

F. Output of Comparator 1 in Second 555

G. Output of Flip-Flop in Second 555

H. PWM Output of Second 555

FIGURE 11.25 Timing Diagram for PWM of Figure 11.14

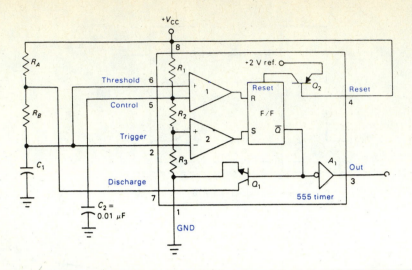

FIGURE 11.26   555 Timer Connected as Astable Multivibrator

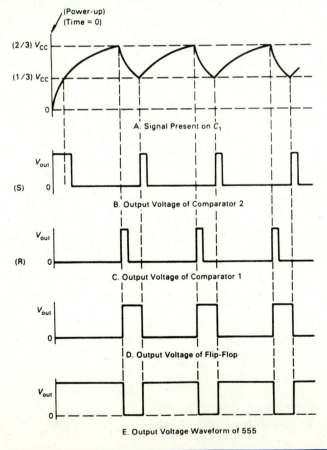

FIGURE 11.27   Timing Diagram for Astable Multivibrator of Figure 11.26

Solution

(a) For the circuit of Figure 11.26, we use Equations 11.4 and 11.5:

$$f = \frac{1.49}{(R_A + 2R_B)(C_1)}$$

$$= \frac{1.49}{\left[(10 \times 10^3) + 2(10 \times 10^3)\right](0.1 \times 10^{-6}\,\text{F})}$$

$$= 497\,\text{Hz}$$

Since $T_P = 1/f$, then $T_P = 1/497\,\text{Hz} = 2.01\,\text{ms}$. The duty cycle is

$$\text{duty cycle} = \frac{R_A + R_B}{R_A + 2R_B}$$

$$= \frac{10 \times 10^3\,\Omega + 10 \times 10^3\,\Omega}{10 \times 10^3\,\Omega + 2(10 \times 10^3\,\Omega)}$$

$$= 66.7\%$$

The pulse is high for ⅔ of the period and low during the remaining ⅓ of the period.

(b) For the circuit of Figure 11.14, we use Equations 11.6, 11.7, and 11.8. From Equation 11.8, we get

$$T_P = 0.67(R_A + R_B)(C_1)$$

$$= (0.67)\left(10 \times 10^3\,\Omega + 10 \times 10^3\,\Omega\right)$$

$$\times \left(0.1 \times 10^{-6}\,\text{F}\right)$$

$$= 1.34\,\text{ms}$$

The duty cycle is found by dividing the time for a high output by the total pulse period. Since

$$t_1\,(\text{output high}) = 0.67 R_A C_1$$

$$t_2\,(\text{output low}) = 0.67 R_B C_1$$

then

$$\text{duty cycle} = \frac{0.67 R_A C_1}{0.67(R_A + R_B)(C_1)}$$

$$= \frac{R_A}{R_A + R_B}$$

$$= \frac{10 \times 10^3\,\Omega}{10 \times 10^3 + 10 \times 10^3\,\Omega} = 50\%$$

The pulse is high for half of the period and low for the other half.

DIGITAL PULSE MODULATION

Pulse-code modulation is the major type of digital pulse modulation in use today. In this section, we will describe pulse-code modulation in detail and briefly discuss some other digital systems found in industry.

Pulse-Code Modulation (PCM)

Pulse-code modulation (PCM), a form of digital pulse modulation, is very different from analog pulse modulation. Analog pulse modulation requires that some characteristic of the pulse, such as amplitude, width, position, or frequency, vary in accordance with the amplitude of the original signal. Pulse-code modulation, in contrast, transforms the amplitude of the original signal into its binary equivalent. This binary equivalent represents the approximate amplitude of the signal sampled at that instant. The approximation can be made very close, but it still is an approximation.

To illustrate PCM, let us suppose a signal could vary between 0 and 7 V. This voltage range is divided into a number of equally spaced levels, or *quanta*, which can be a representation of the integer values of voltages, as shown in Figure 11.28A.

When the signal is sampled (the vertical lines in Figure 11.28A), the digit produced depends on the level, or quantum, the signal is in at that instant. For instance, the leftmost sample time in Figure 11.28A

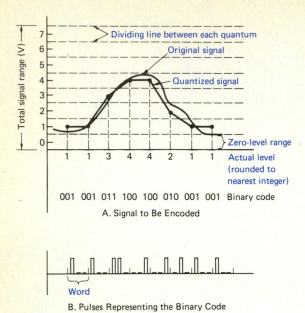

A. Signal to Be Encoded

B. Pulses Representing the Binary Code

FIGURE 11.28 Encoding a Signal for Pulse-Code Modulation (PCM)

finds the signal in the range between ½ and 1½ V. This range corresponds to the decimal number 1 and the binary number 001. The process of determining which level the signal is in is called *quantization*. This quantized digit is represented in binary code by ones and zeros (Figure 11.28A) and sent as pulses (Figure 11.28B), generally in reverse order to make decoding easier.

As shown in Figure 11.28B, it may be difficult to determine when one group of pulses, called a *word*, ends and the next group or word starts. For this reason, a *supervisory*, or *synchronizing*, *bit* is generally added to each coded word to separate them. This bit is distinguishable from the other pulses in some way, such as having a longer pulse duration or a greater pulse amplitude. This bit allows easier decoding.

You probably immediately notice in Figure 11.28 that the quantized signal does not look like the original signal. This difference is called *quantizing noise*. It is called noise because the errors are randomly produced. The largest quantizing error that

can occur is equal to one-half the amplitude of the levels into which the signal is divided. In Figure 11.28, the largest quantizing error is ½ V.

For reduction of this quantizing noise, the obvious solution is to increase the number of quantizing levels used to encode the original signal. For example, ½ V intervals instead of 1 V intervals could have been used. In the case of Figure 11.28, then, four bits (16 voltage levels) instead of three bits (8 voltage levels) would have to be used to represent each level. The increased number of levels results in more pulses per second at the output of the PCM encoder and therefore an increase in the frequency of the signal to be transmitted. The advantage of increasing quantizing levels is that quantizing noise is reduced; the disadvantage is that a greater bandwidth is required to transmit the signal. In practical systems, 128 levels for speech are considered adequate.

The actual number of levels is equal to 2^n, where n is the number of bits in the system. A six-bit system would have 2^6 levels, or 64 levels. To find the voltage each bit is worth, we divided 2^n into the full-voltage scale. For example, a 7 V full-scale range in a four-bit system will have one bit worth

$$\frac{7 \text{ V}}{2^4} = 0.4375 \text{ V}$$

Each bit will be worth 0.4375 V, a value close to ½ V. The quantizing error value will be (½)(0.4375) V, or 0.21875 V, since this is how much an analog signal can be off and still be represented by the same bit code. See Figure 11.28. The quantizing error in Figure 11.28 is ½ V.

• EXAMPLE 11.5
Given a six-bit PCM system with a 10 V full-scale voltage, find the voltage each bit is worth and the quantizing error.

Solution
A 10 V full-scale range in a six-bit system will have one bit worth

$$\frac{10 \text{ V}}{2^6} = 0.15625 \text{ V}$$

Each bit will be worth 0.15625 V. The quantizing error value will be (½)(0.15625 V), or 0.078125 V, or 78.125 mV.

PCM Bandwidth Requirements. Using the Nyquist rate, we should be able (in theory) to transmit a pulse train with a frequency f over a channel with a bandwidth of $(\frac{1}{2})f$. For example, a pulse train may have a frequency of 8000 Hz. We should be able to transmit this frequency over a channel with a bandwidth of 4000 Hz. If we sample the audio signal twice during its period and convert the audio signal to an eight-bit digital output, the bit rate will be

$$\text{bit rate} = (2)\,(4000\ \text{Hz})\,(8\ \text{bits})$$

$$= 64,000\ \text{bits/s (bps)}$$

The minimum bandwidth will be half that value, or 32,000 Hz. This will be covered in more detail in Chapter 12.

• EXAMPLE 11.6

Given an audio frequency spectrum with a highest frequency of 5000 Hz, find the minimum bandwidth required for a PCM signal with six-bit accuracy.

Solution

If we sample the audio signal twice during its period and convert the audio signal to a six-bit digital output, the bit rate will be

$$\text{bit rate} = (2)\,(5000\ \text{Hz})\,(6\ \text{bits})$$

$$= 60,000\ \text{bps}$$

The minimum bandwidth will be half that value, or 30,000 Hz.

Generation and Demodulation of PCM. Generation and demodulation of PCM are much more complex than they are for analog pulse modulation. The block diagrams of Figures 11.29A and 11.29C give some idea of the complexity involved. Figure 11.29B shows the original input analog signal at the top, the analog signal transformed into PCM in the middle, and the demodulated PCM on the bottom line. The middle diagram of Figure 11.29B also shows when the sample-and-hold (S/H) circuit is sampling and holding (to the end of the S/H block). The dotted lines indicate the precise time the input signal is held to be converted to digital code, which follows the S/H block.

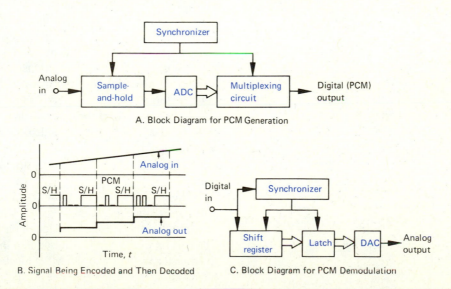

A. Block Diagram for PCM Generation

B. Signal Being Encoded and Then Decoded

C. Block Diagram for PCM Demodulation

FIGURE 11.29 PCM Generation and Demodulation

Figure 11.30 shows the encoding process for generating PCM. For encoding of the incoming analog signal into a serial pulse train, the signal is first sampled at regular intervals, as shown in Figure 11.30A. This sampled voltage is held at a constant level for a short period of time by an S/H circuit (Figure 11.29A). An example of an S/H circuit is shown in the bottom left corner of Figure 11.31. The components involved are the CD4016 bidirectional switch (the *B* part), the 1 μF capacitor, and the 741 voltage follower.

The next step in the process is to convert the sampled signal into a digital representation. This conversion is done by the ADC (or A/D) shown in Figure 11.29A. The ADC in Figure 11.31 is the ADC0804. The specifications and description of this device are available from the manufacturer. Because the ADC does not operate instantaneously, the input signal must be sampled and held at a steady-state voltage long enough for the digital conversion to occur. In addition, the ADC converts the single input into a number of parallel outputs, each output representing a bit location. The thick arrow out of the ADC in Figure 11.29A represents this parallel output. Figure 11.30B illustrates the sampled intervals, and Figure 11.30C shows the output of the ADC.

Lastly, the outputs of the ADC are multiplexed onto the output line (Figure 11.29A), one at a time and each in turn. Multiplexing starts with the LSB (least significant bit) and ends with the MSB (most significant bit); see Figures 11.30C and 11.30D. At the same time, a synchronizing pulse (S) is added in order to separate each word and allow for easier demodulation. In Figure 11.31, the multiplexing circuit is on the right side and is composed of the CD4017 and two CD4016s. There may or may not be a space between bits, depending on the system used. The resulting signal is shown in Figure 11.30D.

All these operations must occur repeatedly and within a given time frame. This job belongs to the synchronizer shown in Figure 11.29A. In Figure 11.31, the synchronizer is the pair of 555s at the top.

The encoded signal is now ready to be sent to the transmitter. The transmitter will convert the pulses into AM, FM, single sideband (SSB), or whatever type of modulation is desired. The signal does not necessarily have to be sent by radio, however. Light, sound, or telephone line transmission may be employed, depending on transmission distance and other factors.

At the receiver, the transmitted signals are converted back into pulses, then demodulated into the original analog signal (or a close approximation). Figures 11.29 and 11.32 show the PCM demodulation process.

The PCM signal is first input into a *shift register* (Figures 11.29C and 11.32B), which converts the serial pulses of each word into a parallel output. The first bit of the word is shifted into the portion of the register we will call the MSB location. As each bit, in turn, is shifted into the MSB location, the preceding bit in that location is shifted into the next bit location. Finally, when the last bit of the word is

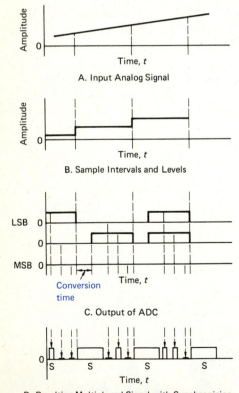

A. Input Analog Signal

B. Sample Intervals and Levels

C. Output of ADC

D. Resulting Multiplexed Signal with Synchronizing Pulse Added

FIGURE 11.30 Signal Process for Generating PCM

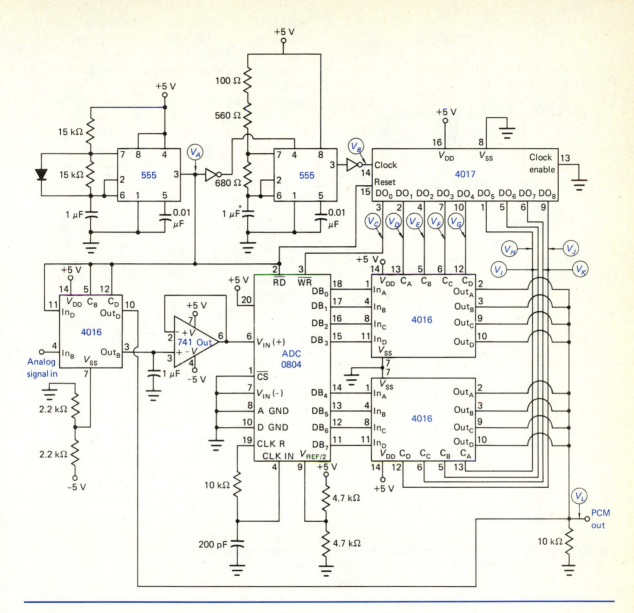

FIGURE 11.31 Schematic of PCM Modulator

shifted in, the register contains the serial pulse code displayed in parallel fashion. This process is illustrated in Figures 11.32A and 11.32B. The shift register is shown in Figure 11.33; it is the 74164.

Also, at this time, the output of the shift register is latched into the latch (Figures 11.29C and 11.32C) to preserve the pulse code and present it to the DAC

(or D/A) shown in Figure 11.29C. *Latching* is the process of locking the input signal onto the output and holding it there regardless of what the input signal does following the latching process. The shift register is cleared during each synchronizing pulse time. The pulse is shown in Figure 11.33; it is the 74LS377.

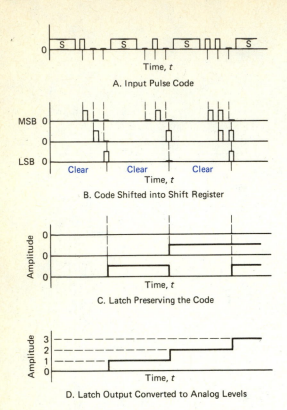

0   S   S   S   S

Time, *t*

A. Input Pulse Code

MSB 0

0

LSB 0

Clear   Clear   Clear

Time, *t*

B. Code Shifted into Shift Register

0

Amplitude   0

0

Time, *t*

C. Latch Preserving the Code

3
2
Amplitude 1
0

Time, *t*

D. Latch Output Converted to Analog Levels

FIGURE 11.32   Signal Process for Demodulating PCM

The DAC (Figure 11.29C) converts the code into the appropriate analog level, and the process is complete. Figure 11.32C shows the latched digital code, and Figure 11.32D shows the resulting analog output. The DAC in Figure 11.33 is the DAC0808. If this signal is now applied to a low-pass filter, the high-frequency components of the steps between voltage levels can be reduced, and the waveform will look more like the original analog signal.

If we look again at the block diagram in Figure 11.29C, we see that the PCM signal was input into the synchronizer as well as the shift register. The synchronizer detects the synchronizing pulse and synchronizes the shifting and latching process just described. In Figure 11.33, the entire top half of the circuit is the synchronizing pulse detector and synchronizer. This part of the circuit is a very critical part. If the synchronizer is off slightly, the latching process can occur too early or too late to give correct information. Most of the problems associated with this circuit will involve synchronizer timing.

We have gone through the process of generation and demodulation of PCM in detail. Certainly, there are many other ways to do the same thing. However, the ideas presented here are valid. Detailed schematics of a working PCM modulator and demodulator were presented in Figures 11.31 and 11.33. These circuits are designed not for efficiency but as teaching aids to illustrate PCM modulation and demodulation. With the ideas presented here, you can design your own system or improve on the circuits given.

Example of PCM. Figure 11.34 shows several *synchrograms*, which are diagrams showing waveforms synchronized in time. (Another name for synchrogram is *timing diagram*.) The diagrams in Figure 11.34 are synchrograms of the voltage waveforms for the voltages specified by the letters in the PCM modulator schematic of Figure 11.31. The data word represented in Figure 11.34 is 00000010, which corresponds to an analog input of 0.05 V.

The photograph in Figure 11.35A presents the same information as the synchrograms in Figure 11.34. The voltage waveforms for V_A through V_L (the letters in the schematic in Figure 11.31) are consecutive in each photograph in Figure 11.35 from top to bottom.

Figure 11.35B shows one data word of 01010101, or an analog voltage input of 1.57 V. Notice that the bottom line in each photograph in Figure 11.35 shows the data word as it comes from the PCM modulator. Figures 11.35C and 11.35D have an expanded time base in order to show more than one data word at a time; these photographs also show that the data words are varying with time.

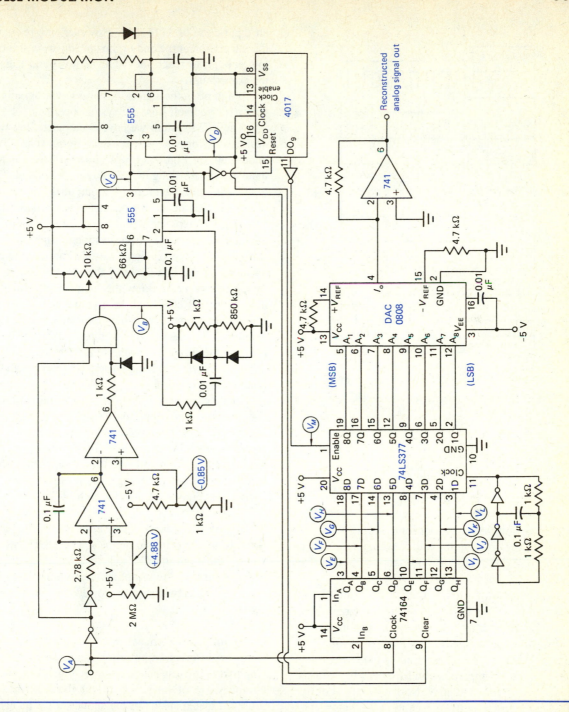

FIGURE 11.33 Schematic of PCM Demodulator

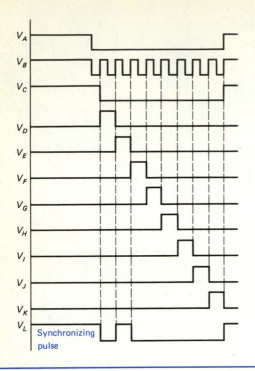

V_A
V_B
V_C
V_D
V_E
V_F
V_G
V_H
V_I
V_J
V_K
V_L

Synchronizing pulse

FIGURE 11.34 Synchrograms for PCM Modulation

Figure 11.36 shows synchrograms for the PCM demodulator whose schematic was given in Figure 11.33. The voltage waveforms for V_A through V_M (the letters in the schematic) are consecutive in each photograph of Figure 11.36, from top to bottom. The last line of the photographs is unused.

Each photograph in Figure 11.37 shows the PCM signal (V_A at the top of each photograph) and the output of the integrator (bottom). (The integrator is also called the *synchronizing pulse detector*.) The bottom signal in each photograph is the input to the next 741 (the comparator) and triggers it when its input signal goes to –0.85 V. Notice the effect of the PCM code on the bottom trace in each photograph. The worst-case condition shown in Figure 11.37C has

an input code of 11111111. However, the result of the integration at the end of the coded part is still not low enough to reach the –0.85 V level and trigger the comparator.

Go through the schematics and synchrograms to verify for yourself how the system operates.

The PCM system we have been discussing was designed to operate at very low frequencies. The photograph in Figure 11.38A shows a 1 Hz analog input and the resulting PCM demodulated output. The input is the top waveform; the output is the bottom waveform. As the input frequency increases (Figure 11.38B), the steps in the output become much more visible, and the output becomes increasingly phase-shifted. The output is being sampled approximately 44 times during one cycle of the input signal. This sampling rate is 22 times the Nyquist rate for this input signal. (Recall that the Nyquist sampling rate is twice the highest frequency expected. See Figure 11.38D.) The phase shift of Figure 11.38B is a direct result of the *frame time*—the time from synchronizing pulse to synchronizing pulse—compared with the input signal frequency.

In the circuit for Figure 11.38, the frame time is about 23 ms. The input analog signal is sampled at the beginning of the synchronizing pulse, but it is not converted back into an analog signal until the end of the frame in the demodulator. Therefore, a phase shift results as the input analog period approaches the frame time. Figures 11.38A through 11.38C illustrate this progression.

Figure 11.38D shows the highest input frequency possible, resulting in two samples per period (Nyquist rate) of the input in order to be within the specifications set forth by the sampling theorem. In this case, filtering would be necessary to recover the original signal.

Companding. Often, we must reduce quantizing error for different-amplitude signals. To do so, we employ the process called *companding*, which is the compression of large signals at the transmitter, followed by expansion of the signals at the receiver. This process is called *companding* because it

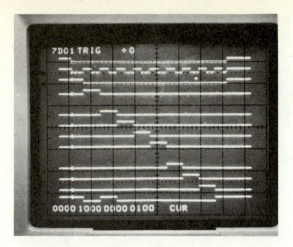

A. Data Word 00000010, Representing Analog Input of 0.05 V

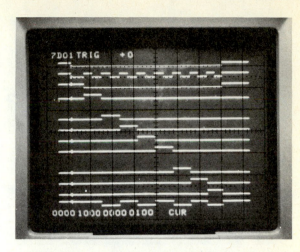

B. Data Word 01010101, Representing Analog Input of 1.57 V

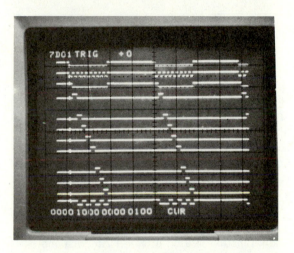

C. Two Data Words Representing Same Information as in Part B

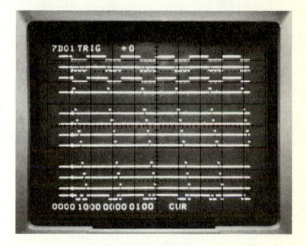

D. Five Data Words Representing a Time-Varying Signal

FIGURE 11.35 Synchrograms for PCM Modulator

*com*presses the large signals (at the transmitter) and then ex*pand*s those that were compressed (at the receiver). The following discussion demonstrates why this process is helpful.

As was previously shown, the maximum quantizing error for a PCM system is one-half the amplitude of a quantizing level. The PCM system of Figure 11.31 has a signal amplitude range of 5 V and uses eight bits for each word. Thus, the maximum quantizing error in this system is as follows:

$$\frac{1}{2^8} \times 5 \text{ V} = \frac{1}{256} \times 5 \text{ V} = 19.5 \text{ mV}$$

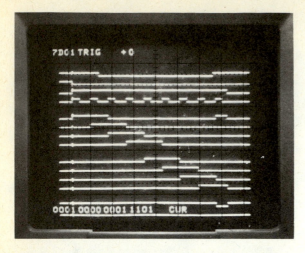

A. Data Word 00000001, Representing Analog Input of 0.03 V

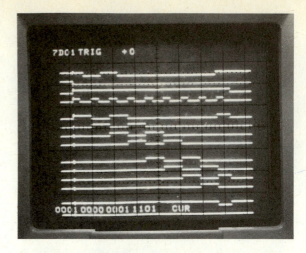

B. Data Word 00000010, Representing Analog Input of 0.05 V

FIGURE 11.36   Synchrograms for PCM Demodulator

For an input analog signal of 5 $V_{(p-p)}$, 19.5 mV represents about a 0.4% error. However, for an input signal of 0.1 $V_{(p-p)}$, 19.5 mV is almost a 20% error. Thus, small-signal inputs can produce large errors at the output because of quantizing.

One way to reduce this problem is to have more quantizing levels for small signals, or, in other words, to *taper* the quantizing levels. Figure 11.39 illustrates a nonlinear quantizing-level arrangement. At the receiver, the levels would be corrected by a corresponding amount so that the amplitude of the original signal is returned.

The circuitry necessary to produce nonlinear quantizing levels is much more complex than that required for linear levels. However, a suitable alternative can be employed. The large signals can be attenuated proportionally more than small signals before they are encoded. Then at the receiver, the large signals can be amplified more than the small signals. The results here would be the same as those for the nonlinear circuitry, without greatly complicating the circuitry.

Integrated Circuit CODEC. At this point, we can appreciate the complexity of designing a PCM modulator and demodulator system. National Semiconductor now has two ICs that perform these functions. They are called *pulse-code modulation coders/decoders* (PCM CODECs). They are the TP3001 μ-law CODEC and the TP3002 A-law CODEC. The *μ-law* and *A-law* designations refer to the two different transfer functions used in companding. The μ-law function is used in North America and parts of the Far East. The A-law function is used in most of the remaining countries. For more applications and information, see National Semiconductor's application note 215. A reprint of the application note is given in National Semiconductor's *Special Functions Databook*.

Advantages of PCM. Unless the noise signal completely obliterates the PCM pulse or pulses, PCM is not affected by amplitude noise, as PAM is. Nor is PCM affected by rising or falling edge noise, as PWM, PPM, and PFM are. Thus, PCM is less

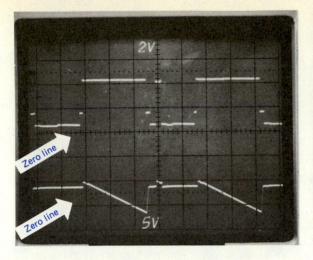

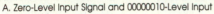

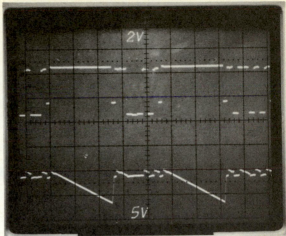

A. Zero-Level Input Signal and 00000010-Level Input

B. Time-Varying Input

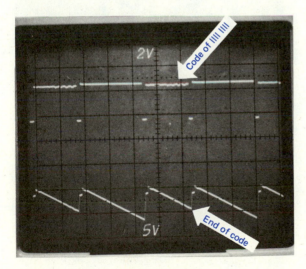

C. Worst-Case Condition

FIGURE 11.37 Results of Processing Input Signal in Order to Synchronize Demodulator (Figure 11.33)

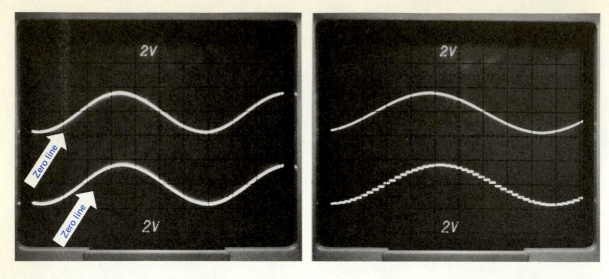

A. Well-Reconstructed Waveform

B. Approximately 22 Times the Nyquist Sampling Rate

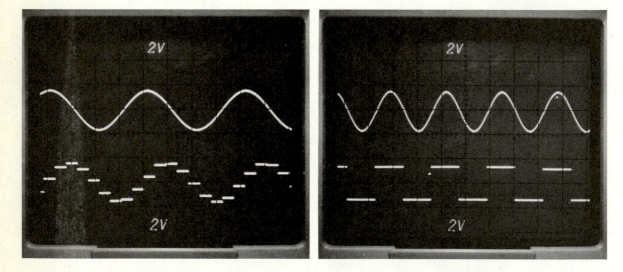

C. Approximately 5 Times the Nyquist Sampling Rate

D. Highest Input Frequency (Nyquist Sampling Rate)

FIGURE 11.38 Waveforms of Input and Output Signals of Circuits in Figures 11.31 and 11.33

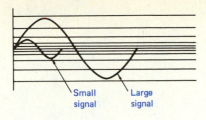

Small   Large
signal   signal

FIGURE 11.39 Tapered Quantizing Level

affected by noise than any of the other pulse modulation methods so far presented. Insensitivity to noise is a great advantage, especially in systems where the signal must be relayed from point to point before it reaches its destination. Other signals are degraded slightly after each relay process; PCM is not.

If noise rejection is such an advantage, why is PCM not the only system in use today? Other systems are used because they were developed first, because PCM circuitry is much more complex, and because PCM requires a much larger bandwidth than the other systems.

PCM was patented in 1937 by Alex H. Reeves of Great Britain. Even though PCM was well documented, it did not come into its own until the early 1960s, when ICs could be used to implement the circuitry. As discussed, ICs make implementation of complex PCM much easier.

The bandwidth requirement for PCM will also become less of a problem as technology advances. With the steady increase in the use of lasers and fiber optics, noise-free transmission of wide-bandwidth signals can be realized.

PCM is becoming more popular as circuit complexity is solved. In 1965, the *Mariner IV* space probe transmitted pictures by using PCM. The system required 30 min to transmit each picture, but the transmitter was over 200,000,000 km away, and the transmitter power was only 10 W! Today, PCM is being used to produce noise-free recordings on phonograph records (digital recordings) and on video recorder systems (laser discs). PCM will become more widespread as technology continues to develop.

Other Digital Systems

Other digital pulse modulation systems have been proposed, but none is as popular as PCM. We will discuss briefly two of the other digital modulation systems.

One digital system in use is called *differential PCM*. Differential PCM is very similar to regular PCM. The difference is that regular PCM quantizes the absolute amplitude of the signal, whereas differential PCM quantizes the difference in amplitude, positive or negative, between one sample and the previous sample. The idea here is that most signals do not change much from their previous level. Thus, it would take fewer pulses to represent a change in amplitude than it would take to represent an absolute amplitude. If this technique could be implemented, then a bandwidth reduction could be realized. Again, complex encoding and decoding circuits have kept differential PCM from becoming widely accepted.

The second and more popular digital system is called *delta modulation*. Delta modulation in its simplest form is similar to differential PCM. However, delta modulation changes only one bit, positive or negative, per sample. Figure 11.40 illustrates delta modulation.

One of the problems associated with delta modulation is *slope clipping*. Here, the input signal is changing so rapidly that the delta modulator cannot keep up with it. Slope clipping is illustrated in Figure 11.41.

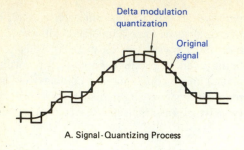

A. Signal-Quantizing Process

B. Pulse Code to Be Transmitted

FIGURE 11.40 Delta Modulation

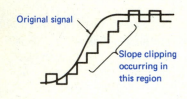

FIGURE 11.41 Slope Clipping in Delta Modulation

CONCLUSION

This chapter has covered pulse modulation, the basis for telemetry. Of all the pulse modulation types, PCM is becoming the most widely used. PCM's advantage of excellent noise rejection is beginning to overshadow its disadvantage of complex circuitry, especially now that single ICs are acting as encoders and decoders. Already, the industrial, consumer, and government markets have made extensive use of the process.

■ QUESTIONS •

1. _____ is the process for performing measurements on distant objects.

2. News services transmit photographs electronically by a process known as _____.

3. The _____ measurement technique is easier to use over short distances. The _____ technique is more noise-free over long distances.

4. The four types of analog pulse modulation are _____, _____, _____, and _____.

5. The subclassification of PAM in which pulse excursions occur both above and below the reference level is called _____.

6. In PWM, the widest pulse width should not exceed _____% of maximum, and the narrowest pulse width should not be less than _____%.

7. Pulse-_____ modulation is differentiated and clipped PWM.

8. Pulse-_____ modulation is a form of digital pulse modulation.

9. In PCM, the analog signal voltage is divided into levels called _____.

10. Name one of the five modulation types that is most affected by noise.

11. Identify the reconstruction phenomenon that may occur when the sampling rate of the analog signal is too slow.

12. Name the type of modulation produced by an amplifier whose gain can be controlled by an analog signal.

13. Name the type of filter with which both PAM and PWM can be demodulated.

14. The PLL can be used to demodulate what type of modulation?

15. Does the input to the PLL have to be a sine wave for the PLL to work properly?

16. Name the term used for a group of pulses in PCM representing an analog voltage level.

17. Discuss why quantizing noise is random noise.

18. Discuss why the largest quantizing error is one-half the amplitude of the levels into which the analog signal is divided.

19. Discuss the use of a synchronizing pulse in PCM.

20. Define a synchrogram.

21. Discuss the need for companding.

22. Explain what CODEC means.

23. Discuss two advantages that are keeping PCM from becoming more common.

24. Define slope clipping.

■ PROBLEMS •

1. Observing the rules of the sampling theorem, determine the minimum sampling rate for high-fidelity voice (20 kHz).

2. Draw to scale the PPM waveform for a straight line that has a slope of 0.2 and a length of 20 cm and goes through zero at the 10 cm length point. The pulse width (of PWM) at the lower end of the straight line is 20% of its maximum, and the pulse width (of PWM) at the upper end of the straight line is 80% of its maximum pulse width. Use 11 equally spaced sample points from one end of the 20 cm point to the other.

3. Determine the VCO free-running frequency of the 565 given in Figure 11.19. The value of R_1 is 4 kΩ, and the value of C_1 is 0.1 μF.

4. In a PCM system with six bits in each word, the maximum input signal swing is 10 V. For quanta that are equally spaced, determine the maximum quantizing error.

5. In Figure 11.21, suppose $C_1 = 0.01$ μF, $R_1 = 1$ kΩ, $R_2 = 4.6$ kΩ, and $R_3 = 5.4$ kΩ. What is the approximate output frequency of the circuit if the power connections on the 741 are ±10 V?

6. If R_1 in Problem 5 changes to 1.5 kΩ, what is the new output frequency?

7. What are the digital bit values of the two data words shown in Figure 11.35C, given in the order of transmission (that is, given as in Figure 11.35A, but two data words instead of one)?

8. An analog signal is sampled in a PAM system at a rate of 10,000 Hz, and the pulses have a width of 30 μs. Find the bandwidth needed to transmit the signal at $K = 1$ and $K = 0.5$.

9. With a supply voltage of +12 V, find the maximum value for the timing resistor for a 555 timer.

10. Using the 555 timer circuit in Figure 11.23, find the timing resistor value needed to produce a 20 ms pulse with a 0.2 μF capacitor.

11. A 555 timer has a +6 V supply. Find the timing capacitor needed to give a 25 ms pulse with the maximum-value timing resistor.

12. Find the total pulse period and the duty cycle for the circuit in Figure 11.26 and on the left in Figure 11.14 with the following values: $R_A = 12$ kΩ, $R_B = 15$ kΩ, and $C_1 = 0.2$ μF.

13. A 12-bit PCM system has a 5 V full-scale voltage. Find the voltage each bit is worth and the quantizing error.

14. Given an audio frequency spectrum with a highest frequency of 10,000 Hz, find the minimum bandwidth required for a PCM signal with 12-bit accuracy.

Industrial Telemetry and Data Communication

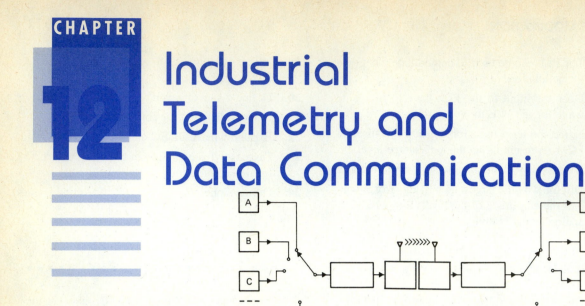

OBJECTIVES

On completion of this chapter, you should be able to:

- Describe frequency-division and time-division multiplexing.
- Determine the relationship among data bandwidth, frequency deviation, and modulation index.
- Relate pulse rise time to data bandwidth.
- List the sources of error in multiplexed transmission.
- Determine the classification of multiplexing systems.
- List methods used for data communication and recording on magnetic media.
- Identify the most important circuits in the RS-232C serial interface standard.
- Calculate frame time, frame rate, sample time, and commutation rate for pulse modulation systems.

INTRODUCTION

Telemetry was discussed briefly at the beginning of Chapter 11 because it relies heavily on pulse modulation. Now that pulse modulation has been presented, telemetry can be discussed in more detail.

Telemetry can be defined as the science involved in performing a measurement at a remote location and transmitting the data to a central location for processing or storage. The telemetry system can be divided functionally into three parts: transducers, transmission system, and display and interpretation system. See Figure 12.1.

The transducers are located at the remote station and transform the physical quantities to be monitored into electric signals. The transmission system consists of a device for transforming the electric signals from the transducers into signals suitable for transmission to the receiving station; the transmission and receiving devices; and a device for transforming the output of the receiver to forms suitable for display and interpretation. The display and interpretation system consists of the devices that calculate the desired parameters and display them for final interpretation by automatic or human means.

Transducers were discussed in detail in Chapter 8 and that discussion need not be repeated here. Our attention in this chapter will focus on the transmission system portion of the telemetry system and, in particular, on the remote and receiving processor portions shown in Figure 12.1. These devices transform the signals for transmission and later transform them back to signals suitable for display.

APPLICATIONS OF TELEMETRY

Telemetry is considered a necessary topic in a text on industrial electronics because of the need in industry to transmit measurements over a distance. Typical applications, to name only a few, have been the monitoring and automatic adjustments of electric power transmission grids or networks and other utility systems; the monitoring of meteorological (weather) conditions in atmospheric, subsurface, or severe-climate locations; the monitoring of seismic (earthquake) conditions; the monitoring or control of satellite, space probe, missile, and ordnance systems; easier monitoring of vital signs of animals or humans who are ambulatory (able to move about); and the monitoring and control of rotating machines, such as the blades of a turbine.

Of the telemetry applications listed, most of the measurements to be telemetered are analog (or continuously variable) and may be transmitted by analog methods. A simple example of an analog method is a voltmeter (or ammeter) with long leads. Many factors must be considered when telemetering analog measurements, such as ground loops, line loss, and noise. Nevertheless, most of these applications are relatively easy to realize with telemetry when compared with pulse modulation, especially over short distances.

If analog transmission is easy and pulse modulation is more complicated, why is pulse modulation used so much? Some of the advantages have already been listed in Chapter 11. To summarize: (1) Some of the information is in pulse form to start with;

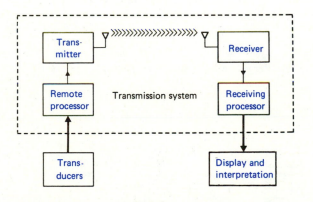

FIGURE 12.1 Functional Diagram of Telemetry System

(2) pulse transmission is a more error-free transmission process; and (3) large-scale integration has reduced circuit complexity. In addition to these advantages are two more: Transmitters can operate on very low duty cycles (percent of time that the transmitter is on), and the time intervals between pulses can be filled with samples of other data. This last advantage illustrates how a number of different sets of data can be *multiplexed*—that is, transmitted over the same channel. We will discuss multiplexing in detail in this chapter.

MULTIPLEXING

In most telemetry applications, a number of different measurements must be performed. Although a separate transmission line or link could be used for each measurement, the problems of efficiency in terms of power, bandwidth, weight, size, and cost normally prevent us from doing so. Therefore, we commonly send many measurements over a single transmission channel. This process, as mentioned, is called *multiplexing*. Multiplexing can be used in systems for purposes other than telemetering, but it is so useful in telemetering that it will be the subject of much of this chapter.

There are two types of multiplexing in common use today: frequency-division multiplexing (FDM) and time-division multiplexing (TDM). These two types of multiplexing are illustrated in Figure 12.2.

Figure 12.2A is a representation of the space available for communication. This communication

space can be divided into different frequency channels at the same time for *frequency-division multiplexing* (Figure 12.2B) or into different time slots for the same frequency for *time-division multiplexing* (Figure 12.2C). Of course, combinations of these two methods could also be used. Notice that each type has a *guard space* to separate channels. Generally, the guard space is used to reduce cross talk (interference between channels) and therefore reduce circuit cost and complexity.

FREQUENCY-DIVISION MULTIPLEXING (FDM)

A familiar example of FDM is radio broadcasting. The signals received by a radio antenna contain many different programs traveling together but occupying different frequencies on the radio spectrum. The tuning circuits in the radio receiver allow the selection of one station from all the others.

The use of the radio frequency spectrum for telemetry and other applications is regulated by various government agencies. The Inter-Range Instrumentation Group (IRIG) publishes a set of telemetry standards that are revised periodically and that specify all pertinent details of telemetry applications.* Table 12.1 is an excerpt from these telemetry standards.

*A copy of the standards may be requested from Secretariat, Inter-Range Instrumentation Group, White Sands Missile Range, New Mexico 88002.

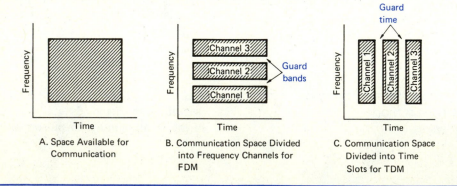

FIGURE 12.2 Frequency-Division and Time-Division Multiplexing

TABLE 12.1   Proportional Bandwidth FM Subcarrier Channels

			±7.5% Channels				
Channel	Center Frequencies (Hz)	Lower Deviation Limit* (Hz)	Upper Deviation Limit* (Hz)	Nominal Frequency Response (Hz)	Nominal Rise Time (ms)	Maximum Frequency Response† (Hz)*	Minimum Rise Time† (ms)
1	400	370	430	6	58.00	30	11.700
2	560	518	602	8	42.00	42	8.330
3	730	675	785	11	32.00	55	6.400
4	960	886	1,032	14	24.00	72	4.860
5	1,300	1,202	1,398	20	18.00	98	3.600
6	1,700	1,572	1,828	25	14.00	128	2.740
7	2,300	2,127	2,473	35	10.00	173	2.030
8	3,000	2,775	3,225	45	7.80	225	1.560
9	3,900	3,607	4,193	59	6.00	293	1.200
10	5,400	4,995	5,805	81	4.30	405	0.864
11	7,350	6,799	7,901	110	3.20	551	0.635
12	10,500	9,712	11,288	160	2.20	788	0.444
13	14,500	13,412	15,588	220	1.60	1,088	0.322
14	22,000	20,350	23,650	330	1.10	1,650	0.212
15	30,000	27,750	32,250	450	0.78	2,250	0.156
16	40,000	37,000	43,000	600	0.58	3,000	0.117
17	52,500	48,562	56,438	790	0.44	3,938	0.089
18	70,000	64,750	75,250	1050	0.33	5,250	0.067
19	93,000	86,025	99,975	1395	0.25	6,975	0.050
20	124,000	114,700	133,300	1860	0.19	9,300	0.038
21	165,000	152,624	177,375	2475	0.14	12,375	0.029
			±15% Channels‡				
A	22,000	18,700	25,300	660	0.53	3,330	0.106
B	30,000	25,500	34,500	900	0.39	4,500	0.078
C	40,000	34,000	46,000	1200	0.29	6,000	0.058
D	52,500	44,625	60,375	1575	0.22	7,875	0.044
E	70,000	59,500	80,500	2100	0.17	10,500	0.033
F	93,000	79,050	106,950	2790	0.13	13,950	0.025
G	124,000	105,400	142,600	3720	0.09	18,600	0.018
H	165,000	140,250	189,750	4950	0.07	24,750	0.014

*Rounded off to nearest hertz.

†The indicated data for maximum frequency response and minimum rise time are based on the maximum theoretical response that can be obtained in a bandwidth between the upper and lower frequency limits specified for the channels.

‡Channels A through H may be used by omitting adjacent lettered and numbered channels. Channels 13 and A may be used together with some increase in adjacent channel interference.

Twenty-one data channels with subcarrier center frequencies ranging from 400 Hz to 165 kHz (column 2 of Table 12.1) have been designated for telemetry use. *Subcarriers* are carrier frequencies in telemetry systems that are modulated by the measurement or intelligence signal. These center frequencies are designed with a ±7.5% frequency deviation. (For channel 1, a 7.5% frequency deviation is 7.5% of 400 Hz, or ±30 Hz.) Thus, Table 12.1 is used for FM purposes.

A chart of the frequencies of channels 1–4 is shown in Figure 12.3; it is based on the information in Table 12.1. Note the presence of the guard bands between the channels. The guard band between channels 1 and 2 extends from 430 to 518 Hz. The guard bands help keep adjacent channels from interfering with each other.

The *nominal frequency response*, or data bandwidth (column 5 of Table 12.1), designates the highest frequencies at which the data can vary for each channel. For example, channel 1's data cannot vary any faster than 6 Hz in order to stay in channel 1, assuming a modulation index of 5. The *modulation index* for FM is defined as the maximum frequency deviation of the signal divided by the modulating

frequency. (For channel 1, the modulation index, M_f, is 30/6, or 5.) While modulation indices of 5 are recommended, indices as low as 1 or less may be used. For these low indices, low signal-to-noise ratios, increased harmonic distortion, and cross talk must be expected.

The relationship among the data bandwidth (data BW), the frequency deviation of the center frequency (400 ±30 for channel 1), and the modulation index is expressed by the following equation:

$$\text{data } BW = \frac{\text{frequency deviation}}{\text{modulation index}}$$
$$= \frac{(\Delta f)}{(M_f)} \quad \textbf{(12.1)}$$

For channel 1, we get

$$\text{data } BW = \frac{30}{5} = 6 \text{ Hz}$$

The bandwidth of the subcarrier band is actually limitless (infinite), but for practical purposes, it is considered finite.

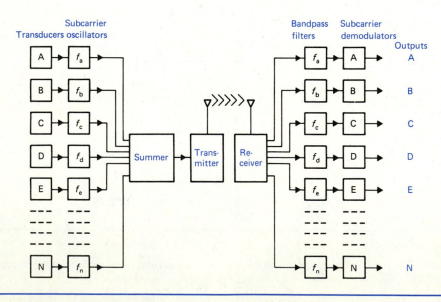

FIGURE 12.3 FDM System Using Channels 1, 2, 3, and 4 of Table 12.1

The effective bandwidth of channel 1, when modulated with a 6 Hz data signal, is

$$BW_{chx} = 2(\Delta f + \text{data } BW) \qquad (12.2)$$

or

$$BW_{ch1} = 2(30 \text{ Hz} + 6 \text{ Hz}) = 72 \text{ Hz}$$

We will assume that the frequency spectrum around the center frequency of 400 Hz is symmetrical. The lower limit of the channel 1 spectrum is 400 Hz − [(72 Hz)/2], or 364 Hz. The upper limit of the frequency spectrum is 400 Hz + [(72 Hz)/2], or 436 Hz. Referring to Figure 12.3 again, we see that the upper limit of channel 1 of 430 Hz and the lower limit of channel 2 of 518 Hz still leaves a significant guard band.

The rise time (t_r) for channel 1 (see column 6 of Table 12.1) is

$$t_r = \frac{0.35}{\text{data } BW} \qquad (12.3)$$

or

$$t_r = \frac{0.35}{6 \text{ Hz}} = 0.05833 \text{ s} \quad \text{or} \quad 58.33 \text{ ms}$$

• EXAMPLE 12.1

Given an FM frequency-division multiplexing system that uses channel 2, find (a) the data bandwidth (the highest modulating frequency), (b) the bandwidth of the frequency spectrum when channel 2 is modulated by the highest modulating frequency, (c) the upper and lower limits of the carrier frequency, and (d) the nominal rise time (t_r).

Solution

(a) The data bandwidth can be found by using Equation 12.1:

$$\text{data } BW = \frac{\text{frequency deviation}}{\text{modulation index}}$$

$$= \frac{42 \text{ Hz}}{5} = 8 \text{ Hz}$$

(rounded to the lowest whole number).

(b) The effective bandwidth of channel 2, when modulated with an 8 Hz data signal, is given by Equation 12.2:

$$BW_{ch2} = 2(\Delta f + \text{data } BW)$$

$$= 2(42 \text{ Hz} + 8 \text{ Hz}) = 100 \text{ Hz}$$

(c) We will assume that the frequency spectrum around the center frequency of 560 Hz is symmetrical. The lower limit of the channel 2 spectrum is then 560 Hz − [(100 Hz)/2], or 510 Hz. The upper limit of the frequency spectrum is 560 Hz + [(100 Hz)/2], or 610 Hz.

(d) The rise time (t_r) for the highest-frequency signal for channel 2 is, from Equation 12.3,

$$t_r = \frac{0.35}{\text{data } BW} = \frac{0.35}{8 \text{ Hz}}$$

$$= 0.04375 \text{ s} \quad \text{or} \quad 43.75 \text{ ms}$$

Note that in part (a), we rounded the data BW from 8.4 to 8 Hz. If 8.4 Hz were used to calculate t_r, a time of 41.67 ms, or 42 ms, would be obtained. This answer corresponds to the value stated in Table 12.1.

In Table 12.1, the center frequency deviation and the maximum frequency response are identical since the maximum frequency response column is based on a modulation index of 1. The *rise times* t_r (nominal and minimum) given in Table 12.1 are related to the data BW by Equation 12.3 given earlier:

$$t_r = \frac{0.35}{\text{data } BW}$$

Again, nominal rise time is based on a modulation index of 5, and minimum rise time is for an index of 1. The rise time equation can now relate pulse rise time to data bandwidth and therefore expands the use of Table 12.1 beyond FM applications to pulse applications.

The 21 bands were chosen to make the best use of present equipment and the frequency spectrum. There is a ratio of approximately 1.3:1 between center frequencies of adjacent bands, except between 14.5 and 22 kHz, where a larger gap was left to provide for a compensation tone for magnetic tape recording. The deviation has been kept at ±7.5% for all bands, with the option of ±15% deviation on the five higher bands to provide for transmission of higher-frequency data. When this option is used on any of these five bands, certain adjacent bands cannot be used, as noted in a footnote to Table 12.1.

The basic operation of an FDM system is illustrated in Figure 12.4. The measurement signals from the transducers are used to modulate subcarrier oscillators tuned to different frequencies. The outputs of the subcarrier oscillators are then linearly summed (linearly mixed), and the resulting composite signal is used to modulate the main transmitter. At the receiving site, the composite signal is obtained from the receiver demodulator and input to a number of bandpass filters, which are tuned to the center frequencies of the subcarrier oscillators.

All types of modulation can be used for both the subcarrier and the main carrier. The types of modulation are FM (frequency modulation), PM (phase modulation), AM (amplitude modulation), SC (suppressed-carrier amplitude modulation), and SS or SSB (single-sideband amplitude modulation). FDM systems normally are designated by listing the type of modulation used by the subcarriers, followed by the type of modulation used by the main carrier. Thus, FM/AM indicates an FDM system in which the subcarriers are frequency-modulated by the measurement, and the main carrier is amplitude-modulated by the composite subcarrier signals. Almost all combinations of subcarrier and main-carrier modulation techniques have been used in the past. However, FM/FM is by far the most common technique in use today.

Let us assume that we are going to use an FM/FM system employing the first four channels of the IRIG chart, and let's calculate the transmission bandwidth needed to transmit channels 1–4. The highest modulating frequency for channel 4 can be found by using Equation 12.1:

$$\text{data } BW = \frac{\text{frequency deviation}}{\text{modulation index}}$$

$$= \frac{72 \text{ Hz}}{5} = 14 \text{ Hz}$$

The effective bandwidth of channel 4, when modulated with a 14 Hz data signal, is, according to Equation 12.2,

$$BW_{ch4} = 2(\Delta f + \text{data } BW)$$

$$= 2(72 \text{ Hz} + 14 \text{ Hz}) = 172 \text{ Hz}$$

The highest frequency when the four channels are put together is 960 Hz + [(172 Hz)/2], or 1046 Hz. Assuming a modulation index of 5 for the carrier frequency, the frequency deviation will be five times the highest modulating frequency:

$$\Delta f = (\text{modulation index}) (\text{total data } BW) \quad \textbf{(12.4)}$$

or

$$\Delta f = (5) (1046 \text{ Hz}) = 5230 \text{ Hz}$$

The transmission bandwidth of the composite signal is then 5.23 kHz.

• EXAMPLE 12.2

Given an FM/FM (subcarrier/carrier) system, using the first three channels of the IRIG chart, find the transmission bandwidth needed to transmit channels 1–3.

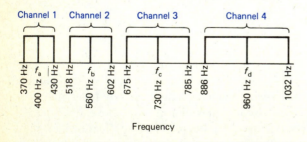

FIGURE 12.4 FDM System

Solution

The highest modulating frequency for channel 3 can be found by using Equation 12.1:

$$\text{data } BW = \frac{\text{frequency deviation}}{\text{modulation index}}$$

$$= \frac{55 \text{ Hz}}{5} = 11 \text{ Hz}$$

From Equation 12.2, the effective bandwidth of channel 3, when modulated with an 11 Hz data signal, is

$$BW_{\text{ch1–3}} = 2(\Delta f + \text{data } BW)$$

$$= 2(55 \text{ Hz} + 11 \text{ Hz}) = 132 \text{ Hz}$$

The highest frequency when the three channels are put together is 730 Hz + [(132 Hz)/2], or 796 Hz. Assuming a modulation index of 5 for the carrier, the frequency deviation will be five times the highest modulating frequency, according to Equation 12.4:

$$\Delta f = (\text{modulation index}) (\text{data } BW)$$

$$= (5) (796 \text{ Hz}) = 3980 \text{ Hz}$$

The transmission bandwidth of the composite signal is then 3.98 kHz.

The principal sources of errors in an FDM transmission system are drift, bandlimiting, cross talk, distortion, and radio frequency (RF) link noise.

The errors due to drifts (slowly changing circuit parameters caused by heat, age, and so on) are those associated with modulation and demodulation of the subcarriers. If the drifts are slow, their effect can be greatly diminished by use of calibration signals. Their reduction depends on the circuit design.

Errors due to bandlimiting (limited frequency spectrum) occur throughout the system whenever the data are dynamic (changing). This error source cannot be eliminated regardless of the circuit design.

Cross talk errors are of two kinds. The first is caused by nonlinearities in the summing amplifier before the main transmitter and in the modulation or demodulation process of the main carrier. This type of cross talk error can be eliminated. The second type of cross talk error is due to overlap of frequencies from the bandpass filters of adjacent channels. Since no filter is perfect, some overlap will always be present in other channels. This error cannot be eliminated.

Distortion error is due to nonlinearities in the subcarrier modulator and demodulator (not to be confused with cross talk errors due to nonlinearities in the main-carrier modulator or demodulator). Calibration signals can help reduce or eliminate this error.

RF link noise (the RF link is the transmitter, receiver, and medium in between) causes errors that are generally random and come from three sources: receiving station front-end noise, noise from space, and technological interference. These types of noise will always be present but can be reduced by proper circuit design.

TIME-DIVISION MULTIPLEXING (TDM)

The major alternative to FDM is time-division multiplexing (TDM). Here, the time available is divided into small slots, and each of these slots is occupied by a piece of one of the signals to be sent. The multiplexing apparatus scans the input signals sequentially. Only one signal occupies the channel at one instant. TDM is thus quite different from FDM, in which all the signals are sent at the same time but each occupies a different frequency band.

The operation of a TDM system from a functional standpoint is illustrated in Figure 12.5. The signals from the transducers are input to a *commutator* (a switching device), which samples the channels sequentially. Thus, the output of the commutator (also referred to as the *multiplexer*) is a series of pulses, the amplitudes of which correspond to the sampled values of the input channels from the transducers. This train of pulses is then passed through a device that converts it to a form suitable for modulating at the transmitter.

At the receiving station, the process is reversed. The demodulated output from the receiver is passed

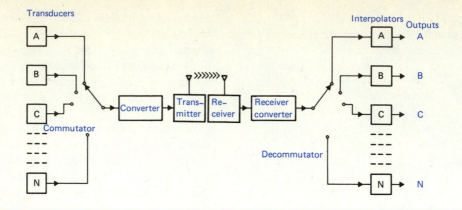

FIGURE 12.5 TDM System

through a converter, which reproduces the pulse train that existed at the commutator at the transmission site. This pulse train can then be decommutated to produce pulses with values corresponding to samples of the original measurement signals.

Let us look at the commutator section a little more closely. Figure 12.6A is a diagram of a mechanical commutator connected to three transducers. As the commutator arm or rotor rotates, it makes contact with conductors A, B, C, and sync.

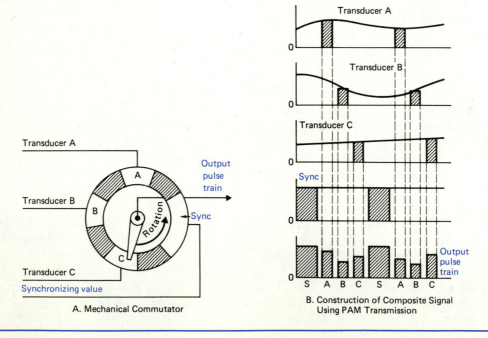

A. Mechanical Commutator

B. Construction of Composite Signal Using PAM Transmission

FIGURE 12.6 TDM Commutation

Each time a conductor is contacted, a voltage is output. Between each conductor is an insulator. When the commutator rotor is on an insulator, there is no output. The insulator contact corresponds to the space between signals in the output pulse train diagram of Figure 12.6B.

Figure 12.6B shows a synchrogram of the operation occurring in Figure 12.6A. If the values of the voltages from the instruments are not varying too rapidly compared with the rotation time of the rotor, the individual inputs can be reconstructed from the composite signal.

For separation of the signals when they are received, a commutator similar to that illustrated in Figure 12.6A might be used, but with the input and output reversed. The receiving commutator must be exactly synchronized with the transmitting commutator. That is exactly what the sync pulse is for—to synchronize the receiver commutator with the transmitter commutator. Modern commutators operate at very high speeds and are electronic instead of mechanical.

As in FDM, TDM modulation of the transmitter may take any form—that is, AM, FM, PM, and so on. However, the principal distinction among TDM systems lies in the form of the processor used (see Figure 12.1). For the processor, any of the pulse modulation methods discussed in Chapter 11, such as PAM, PWM, PPM, PCM, and delta modulation, will work. The system designation normally lists the type of processor (or converter) first, followed by the type of main-carrier modulation. For example, a PWM/PM system has a pulse-width modulation processor, the output of which is used to phase-modulate the main carrier.

Submultiplexing

FDM transmission systems commonly submultiplex some of the wider-band subcarrier channels. *Submultiplexing* usually involves the use of a TDM waveform as the modulation of one of the subcarrier channels, as shown in Figure 12.7. The use of submultiplexed channels of this nature allows the

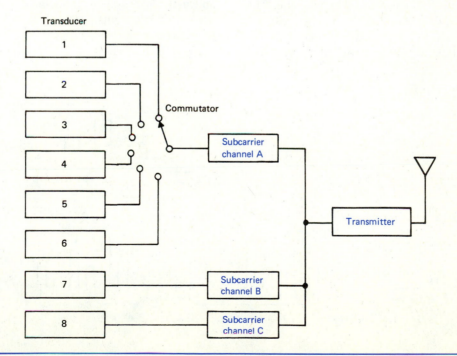

FIGURE 12.7 Combination of TDM and FDM in One System

total number of measurements handled by the transmission system to be increased considerably. The frequency content of the submultiplexed measurements must be low relative to the frequency content that normally could be transmitted over the subcarrier channel.

The most common types of submultiplexing are PAM/FM/FM and PWM/FM/FM, although almost all varieties have been used for some applications. Combined telemetry systems have not been widely used to date. However, there is some indication that they may receive increased attention in the future.

Pulse-Amplitude TDM

Let us return our attention now to pure TDM. The first modulation type we will consider is PAM. A PAM system is one in which the output of the commutator is used to modulate directly the main transmitter. The processor is simply a pair of wires, so PAM is the simplest of the TDM systems.

The PAM wave trains may take several forms, two of which are illustrated in Figure 12.8. The principal difference here lies in the duty cycle system (50% and 100%). The length of time necessary to sample all channels (including frame sync and

calibration pulses) is normally referred to as *frame time*. For identification of the sample at the receiving station, frame synchronization must be inserted. Several different methods can be used to designate a frame. The one illustrated in Figure 12.8 consists of forcing several consecutive channels to a level below the minimum allowable data value. Since drifts and nonlinearities in the system cause error directly, another common technique is to transmit calibration pulses (zero, half, and full scale) as shown. The data are offset in such a fashion that channels with signal outputs that are capable of both positive and negative polarities are centered about the half-scale value. The frame sync pulses and calibration pulses are sometimes referred to as *housekeeping pulses* since they are not part of the data but are used to synchronize and calibrate the receiver.

In general, the 100% duty cycle system (Figure 12.8A) requires a smaller frequency spectrum than the 50% duty cycle system (Figure 12.8B). However, the 50% system was used to a greater extent in the past because of the relative ease of synchronizing at the receiving station. In addition, the dead time available in the 50% system allows the use of circuitry to get rid of transients (undesirable short-duration signals) and therefore reduce cross talk between successive channels.

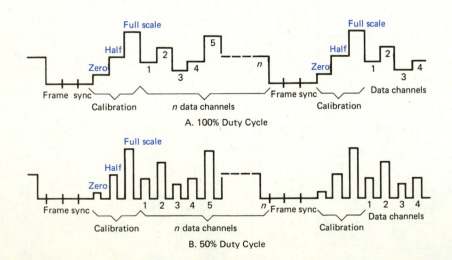

FIGURE 12.8   PAM Waveforms in a TDM System

Pulse-Width TDM

In PWM, the width of pulses is varied in proportion to the modulation signal. The PWM pulse train, then, consists of a string of pulses with different widths, as illustrated in Figure 12.9.

Guard time is allowed at both the beginning and the end of each pulse to reduce the difficulties associated with interchannel cross talk. Since the information is carried in terms of pulse width, drifts and nonlinearities in the system do not have as great an effect on data accuracy as they do in an equivalent PAM system.

Pulse-Position TDM

PPM is similar to PWM. Here, though, only the trailing edge of the pulses, rather than the entire duration of the pulse, is transmitted to identify the pulse width. See Figure 12.10. PPM is used principally in connection with an amplitude-modulated main carrier since its greatest advantage is in the small percentage of time that pulses are present. Thus, the transmitter operates at relatively high peak powers and low average powers. The wider system bandwidths required (as compared with PWM and PAM) makes its use in connection with a frequency-modulated main carrier undesirable because the reduction in average power for the FM system could not be realized.

A typical PPM waveform is shown in Figure 12.10 along with the equivalent PWM waveform. The synchronization pulses shown may vary from system to system. Pulses corresponding to the leading edge may be transmitted in order to reduce synchronization problems at the receiving station, although Figure 12.10 does not illustrate this possibility.

PFM systems will not be discussed here since they are so similar to PPM systems.

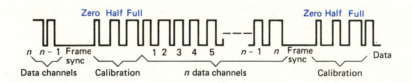

FIGURE 12.9 PWM Waveform in a TDM System

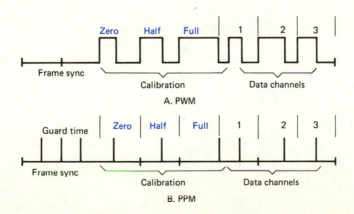

FIGURE 12.10 PPM Waveform Compared with PWM Waveform in TDM System

Analog TDM Errors

The principal sources of error in analog TDM systems are closely related to those in FDM systems. These error sources are drift and nonlinearity, bandlimiting, cross talk, interpolation errors, and RF link noise.

With the exception of PAM, only drifts and nonlinearities associated with equipment prior to and including the remote converters (processors), and after and including the receiver converter, are of primary concern to analog TDM systems. As in FDM systems, the drift and nonlinearity are entirely dependent on circuit design. Calibration signals can be of considerable help here.

Bandlimiting errors in TDM systems are produced only in the equipment preceding the multiplexer. Errors due to bandwidth restrictions in other parts of the system usually are classified under different names. Bandlimiting errors depend on the phase and the amplitude response of the filters involved, as well as on the character of the data.

Cross talk in a TDM system normally occurs from bandlimiting of the pulse signal, which causes transients from channel pulses to affect the pulses of the channels following. The relationship between bandlimiting and cross talk varies from system to system.

Interpolation error comes about when an attempt is made to reconstruct a continuous waveform from the sampled values of the original waveform available to the receiving station. The amount of interpolation error depends on the characteristic of the data that are sampled, the method of interpolation, and the sampling rate.

The RF link errors cannot be eliminated. They depend on the received signal power at the receiving station and many other system parameters.

Pulse-Code TDM

PCM comes in many varieties. The most common is binary modulation, the type shown in Figures 11.32 through 11.38 in Chapter 11. In *binary modulation*, the transmission of a 1 or 0 value during a bit interval is done by the presence or absence of a pulse. In actual practice, this waveform can have many different forms. Figure 12.11 shows a number of the different waveforms that can be used.

The nonreturn-to-zero (NRZ) waveform is probably the most commonly used of those shown. Notice there are two types of NRZ signals. The NRZ (change) transitions only when the next bit of data changes value. NRZ (mark) transitions with each 1 bit and does not change when a 0 bit is present. However, all waveforms shown in the figure either have been used or are to be used in modern systems. When the NRZ or return-to-zero (RZ) waveforms are used, it is common practice to employ a premodulation filter to round off the corners of the modulating waveform to limit the bandwidth requirement.

Even though the NRZ waveform is probably the most commonly used binary modulation, it is seriously lacking in at least one important area—clocking. As shown in Figure 12.11, a long string of ones or zeros would result in a lengthy data level without transitions. Now a very important point is to know *when* to examine the NRZ signal to determine the logic level of the signal in order to reconstruct accurately the transmitted data. For this task, a separate clock signal is transmitted along with the NRZ data.

This additional clock signal has at least two drawbacks: Two channels instead of one are needed to transmit the data and reconstruct it, and the phasing of the clock signal with the data can get out of synchronization. This phase shift is most likely to occur when the data are transmitted over long distances and the data and clock signals are on different channels.

So that a clock signal need not be transmitted with the data, a self-clocking format may be used. A self-clocking code, such as the RZ type in Figure 12.11, is one in which one type of data bit is a frequency multiple of the other. A clock signal as well as data can be recovered from this type of modulation by using a PLL.

The arrangement of the code sequence in a binary transmission system is called the *format*. In general, the format is defined by the arrangement of *bits* (binary digits) in each code group representing a sample value (sometimes called a *word*) and the arrangement of these code groups with respect to one another.

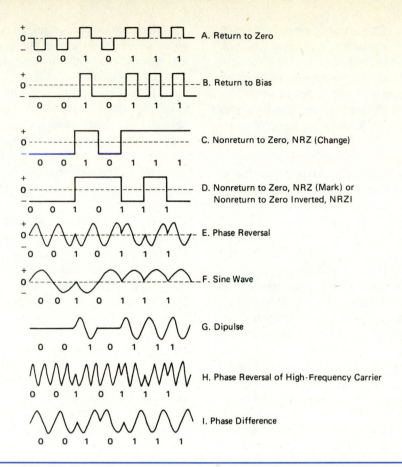

FIGURE 12.11   Different Ways to Represent 1 and 0 in Binary Modulation

The bits within a particular coded word may consist of information bits, parity bits, or synchronization (sync) bits. *Information bits* appear in a binary code that represents the sample value. A *parity bit* is either present or not present in order to make the total number of 1 bits per word either even or odd (depending on whether even or odd parity is to be used). Parity bits allow the detection of words with a single bit in error. Parity bits generally are not used in radio transmission systems since it is much more likely that two or more bits will be missing. However, in ground data-handling systems, parity bits are nearly always used since single-bit errors are most likely. The *sync bits* are timing bits. They may appear after every word or after a certain small sequence of words, and they usually are called *word synchronization bits*. Either the parity bits, the sync bits, or both may be missing, depending on the requirements of the system. Today, virtually all operational binary PCM systems include some form of word synchronization, although some systems presently under development do not.

In the simplest cases of TDM or PCM, the information channels are sampled sequentially (as in Figure 12.5) with fixed word lengths. In this case, each word represents the code for a separate information channel, appearing sequentially in time. The sequence is repeated each time the commutator completes a cycle. This entire cycle is called a *frame*.

In most recent PCM systems, the programming of channels within a frame has been quite complex, with some channels being sampled more often than others. This system can be best visualized by considering a commutator with a very large number of terminals and with some of the terminals tied together so that some channels are sampled more often than others.

In most PCM systems, one must also supply timing information to designate the start of the frame. Timing information usually is supplied by transmitting a unique or identifiable code pattern in one or more word positions.

Digital TDM Errors

Errors in a digital system can be categorized as analog errors (drifts, bandlimiting, nonlinearity, and so on, in analog portions of the system), digital dropouts, quantization errors, interpolation errors, and RF link errors.

Analog errors due to drifts, bandlimiting, and so on, are the same as those types discussed previously. Since most digital systems contain analog circuitry, they also have analog errors.

The errors associated with data after it has been digitized are called *dropouts*. In a binary-coded system, the principal sources of dropouts are in the tape and disc recording processes and are due to tape or disc imperfections.

Recall from Chapter 11 that the largest quantization error occurring is equal to one-half the amplitude of the levels into which the signal is divided. Interpolation and RF link errors are the same as those previously discussed.

Cassette Recording Methods

Cassette magnetic tape recording is an inexpensive method for storing telemetry data and computer programs and is used extensively in industry. The cassette tape and cartridge are relatively rugged, have a large storage space, and are very economical. A disadvantage is that the data or programs are, of necessity, stored serially. Thus, accessing a particular piece of data may take seconds, which may or may not be a problem, depending upon the application.

Many cassette recording techniques place data bytes sequentially along the length of the tape, leaving no blanks. Other methods leave blank spaces between blocks of data to facilitate searches, starts, and stops. Figure 12.12 illustrates ten popular recording methods. Note that these methods show basically two different waveshapes: sine and square waves. The square waves represent complete magnetic saturation, first in one direction and then in the other. The sine waves do not represent saturation but, rather, the normal recording of a frequency.

DFR, double-frequency recording (Figure 12.12A), is used for recording data on fixed-head discs at high transfer rates. A 1 bit is represented by a reversal of magnetization and a 0 bit by the absence of a reversal. An additional reversal is inserted between each bit to provide a timing signal. This encoding requires a maximum of two reversals per bit. Unfortunately, trying to record on cassette tape with this method poses a major problem: The cassette tape bandwidth is too narrow. DFR requires a bandwidth obtainable on cassette tape only at low speeds. Thus, data rates greater than 800 bits per second (bps) produce unreliable recordings. Tape speed changes do not affect this method because of its self-clocking characteristic. In DFR, however, noise and signal strength variations can cause loss of information.

FSK, frequency shift keying (Figure 12.12B), is the digitizing technique used in most modems used for telecommunications. Initially, FSK would appear to be an ideal method for recording data onto cassettes; however, there are a few drawbacks. Even though FSK is resistant to noise and amplitude variation—qualities important to data reliability—it is more costly to implement than DFR, and it cannot tolerate more than a ±6% tape speed variation. Since many cassette systems cannot maintain speed accuracies within this ±6% limit, tape recording using FSK generally has been limited to high-quality, expensive systems.

NRZ and NRZI (Figures 12.12C and 12.12D) find greatest use in seven-track tape drives that meet IBM compatibility standards. With NRZ and NRZI methods, current through the write head of the tape recorder does not return to zero after a write, so each state of the magnetic recording corresponds to a

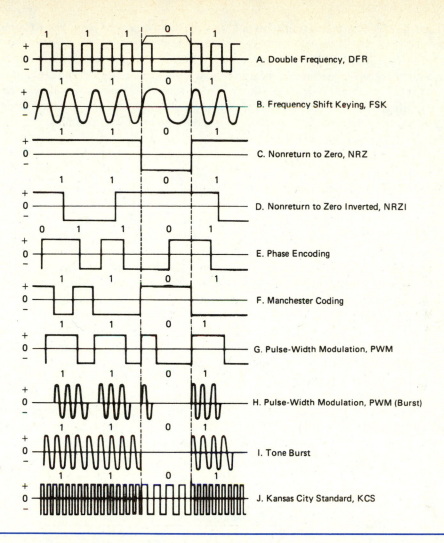

FIGURE 12.12   Cassette Tape Recording Methods

particular binary bit, either 1 or 0. NRZ and NRZI recording methods are resistant to noise and amplitude variation because of magnetic saturation. Data rates of 1500 bps and greater can produce accurate recordings. NRZ and NRZI techniques are not self-clocking since the transitions from one flux direction to the other depend upon the data and therefore may not occur at regular intervals necessary for generating the self-clocking signals. Thus, most NRZ/NRZI

recorders require a separate clock track to maintain a sync signal and consequently need a separate read/write head, which usually adds expense and mechanical complexity.

Phase encoding, also known as biphase-L (Figure 12.12E), is one of two phase-recording techniques that remain popular. This method is resistant to noise, amplitude variation, and frequency change, and it is also self-clocking. If one clock cycle is defined as a bit cell, a 1 or 0 is determined by both a transition in the middle of the cell (like NRZI) *and* by the polarity of that transition (not like NRZI). Equipment necessary to generate this recording technique is easy to implement and use.

Manchester coding, also called biphase-M (Figure 12.12F), is similar to biphase-L. The difference is in the location of the transition in the bit cell, which is at the beginning of each cell. Manchester coding also is easy to generate and decode.

PWM (Figure 12.12G) for magnetic tape recording is a patented process. It is self-clocking and resistant to frequency, noise, and amplitude variation. Its major disadvantage is the strict bandwidth requirement, making it only slightly less desirable than the phase-encoding methods.

PWM burst (Figure 12.12H) is similar to standard PWM with one exception. A tone burst replaces the magnetic saturation technique required to generate a 1 or 0 pulse. Tone burst (Figure 12.12I), also called continuous-wave recording, is one of the simplest techniques to generate and decode but probably the worst method for recording data on magnetic tape. It is nonclocking and is not resistant to noise or speed variations. Reliable recording data rates over 110 bps are almost impossible.

Kansas City Standard, KCS (Figure 12.12J), is a variation on the Manchester code and is quite popular among hobbyists. The KCS method uses an 8-cycle burst of a 2400 Hz saturation frequency to represent a 1 and a 4-cycle burst of a 1200 Hz saturation frequency to represent a 0.

Table 12.2 gives a summary of cassette recording techniques and a very general evaluation of the more important specifications.

IC TDM

Most of the TDM systems we have studied to this point have been assumed to be made up of discrete devices. As in many other areas of electronics, a manufacturer—in this case, National Semiconductor—has produced an integrated circuit that has all the required multiplexing circuitry in one chip. The LM604, shown in Figure 12.13, is a four-channel TDM amplifier. It contains four op amp input stages and one output stage. The input stage is selected and connected to the output stage by the digital control circuitry. This IC operates from DC power ranging from 4 to 32 V.

The LM604 is programmed by five different logic control pins. Two pins, *A* and *B*, select the desired input, and the enable (EN) controls the output stage. Inputs to *A* and *B* and enable are accepted

TABLE 12.2   A Comparison of Cassette Tape Recording Methods

Recording Method	Performance under Speed Variations	Self-Clocking Code (Yes or No)	Performance under Noise Variations	Ease of Implementation	Maximum Transfer (bps)
DFR	Good	Yes	Fair	Good	800
FSK	Poor	No	Excellent	Good	450
NRZ	Excellent	No	Excellent	Excellent	1500+
NRZI	Excellent	No	Excellent	Excellent	1500+
Phase encoding	Excellent	Yes	Excellent	Good	1500
Manchester coding	Excellent	Yes	Excellent	Good	1500
PWM	Excellent	Yes	Good	Excellent	1500
PWM (burst)	Excellent	Yes	Good	Excellent	1500
Tone burst	Poor	No	Poor	Excellent	450
Kansas City Standard (KCS)	Excellent	Yes	Excellent	Excellent	1200

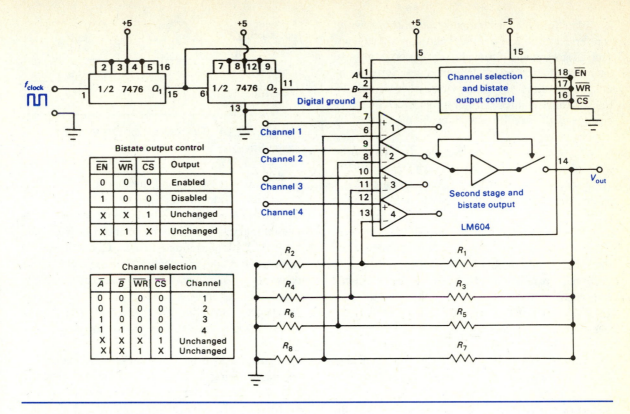

FIGURE 12.13 IC TDM

only if both the write (WR) and chip-select (CS) lines are held low. A high to either the write line or the chip-select line will latch the data at the other inputs and stop any other changes. The logic thresholds are referenced to the digital ground pin 4. This pin may be tied to any potential from $-V$ to $(+V - 4\text{ V})$, regardless of whether the IC is operating on single or split supplies. Resistors R_1 through R_8 determine the gain of the individual op amps. The 7476 is a flip-flop that selects each of the four inputs in turn.

DATA COMMUNICATION

As discussed early in Chapter 11, *data communication* is the process of transmitting pulses, which are the output of some data source, from one point to another. Telemetry deals mainly with the transmission of measurements. With the rapid incursion of the microcomputer (µC) into the industrial workplace, however, data communications other than telemetry

are becoming commonplace. Communication between computers in the same plant, communication between computers many miles apart, and storage of computer programs are just a few of the applications that will require technical expertise in data communication systems in order to keep them operational. Because of the pervasiveness of data communication in the industrial workplace, some of the more common and important concepts will be presented here.

Transmission Classifications

There are basically two ways in which data can be moved from one place to another: serial and parallel transmission. *Parallel transmission* is used for movement of data within the microcomputer and for short distances outside the computer. (Microcomputers generally deal with eight bits of data at a time.) This method is efficient and quick—important considerations for the computer. *Serial transmission* is reserved for longer distances and slower data rates, such as storage and retrieval of data or programs and data transmission over telephone lines. Serial transmission requires only two wires, whereas parallel transmission requires eight or more, which increases costs and the likelihood of problems caused by broken wires.

If serial transmission is used, the bits can be transmitted synchronously or asynchronously. In *synchronous transmission*, all characters are sent at a constant rate, and timing is synchronized between the transmitter and receiver so that the receiver can keep in step with the transmitter. (This concept was discussed in the pulse modulation chapter.) The synchronization can be done by sending a clock pulse train on a separate line or by recovering a clock signal from the data stream itself. One of the principal characteristics of synchronous data transfer is that data streams must be continuous, or one unit of data must immediately follow another, until the entire communication has been completed. Blank data generally terminate transmission. In *asynchronous transmission*, characters can be sent at any time, and the receiver is provided with a means for recognizing the start and end of each character. As a result, each character is sent as a unit unto itself, and there can be any amount of time between characters. Both synchronous and asynchronous methods will be discussed in the following sections.

Synchronous Serial Transmission. As stated earlier, the most important characteristic of synchronous data transmission is that data must be transmitted continuously. Both transmitter and receiver must agree upon a bits-per-second transfer rate. Thereafter, the data stream will be interpreted as one bit per recovered clock signal transition, as shown in Figure 12.14. Note that this data stream uses the NRZ transmission method.

Some problems now arise. How can the receiving device determine the beginning and end of each data unit, or character? And we know that the binary digits 1 and 0 by themselves cannot represent all the letter and number possibilities. How many bits are needed to represent these letters and numbers?

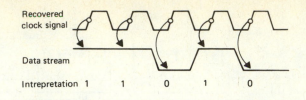

FIGURE 12.14 Recovery of Data from Data Stream

To answer the second question first, there are two popular code formats used for communications: the ASCII (American Standards Committee on Information Interchange) code (pronounced ASK-ee) and the EBCDIC (Extended Binary Coded Decimal Interchange Code) (pronounced EB-see-dick). These codes are given in Appendixes B and C, respectively. The EBCDIC code is used by IBM and systems attached to IBM equipment, and ASCII is used in most other applications. We will use the ASCII code in this text.

Now, back to the question posed earlier: How can the receiving device determine the beginning and end of each character? The answer is that synchronizing characters are first sent to alert the receiver that data are to follow. Thereafter, each specific number of bits in the string represents a character.

As you can see, many rules must be imposed upon the transmitter and receiver operation in order to ensure that information will be transferred correctly. The set of rules for each method of transmission is called its *communications protocol*.

The data unit in a synchronous serial data stream may contain 5, 6, 7, or 8 data bits that may be followed by a parity bit. Seven bits plus odd parity will be used here. Odd parity requires that the parity bit be a 1 or 0, whatever is required to make the data group have an odd number of ones. Of course, even parity will have an even number of ones, including the parity bit. An example of a transmission message is

Message␢to␢follow

where

␢ represents a blank, or space character

The middle part of the message would appear as

sync	char	char	sync	sync	char
-- 00101100	01000000	11101001	00101100	00101100	11011111 --
	space	t			o

Since the data characters may not be available when the transmitter needs to send them, sync characters are inserted to keep the transmission flowing. Eight bits per character make up part of the communications protocol for this particular synchronous serial transmission system.

Other types of synchronous serial data transmission protocol that are in present use are Bisync, Serial Data Link Control (SDLC), and High-Level Data Link Control (HDLC). These protocols are IBM standards that are fast becoming world standards for synchronous serial data.

Compared with asynchronous serial transmission, synchronous transmission has somewhat better characteristics, even though it is more complex to implement. Synchronous transmission has the advantages of a good speed-to-distance ratio (a minimum of 1000 ft/333,000 bps), reliability (error checking), expandability (able to communicate with over 200 stations), low cost (two conductor cables for connections), and efficiency (async uses up to 20% of each word transmission in start and stop bits).

Asynchronous Serial Transmission. In asynchronous transmission, characters can be sent at any time, with any amount of time between characters. The protocol provides a means for enabling the receiver to recognize the start and end of each character. A standard format such as that shown in Figure 12.15 generally is used. This collection of pulses, including the start and stop bits, is called a *frame*.

The frame begins with a start bit and ends with one or more stop bits. These bits enable the receiver to recognize the beginning and end of the character. The transition from a high in the idle state to a low start bit informs the receiver that the next eight bits are data. The stop bit (or bits) marks the end of the character and gives the receiver time to reset its hardware and prepare for the next character. There may be one, one and one-half, or two stop bits. (Fractional bits are permissible since their period is just a resting time for the receiver.) Hardware that is completely electronic needs only one stop bit, but the early electromechanical machines needed two stop bits before the next character. If noise pulses or other errors cause the receiver not to detect the stop pulse,

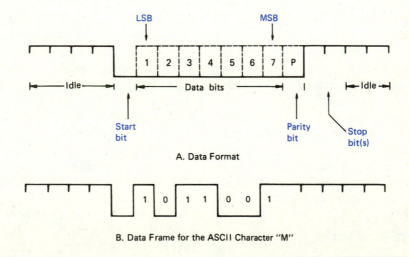

A. Data Format

B. Data Frame for the ASCII Character "M"

FIGURE 12.15 Data Frame Format

a *framing error* is indicated. The transmitter then may retransmit the frame that had the error.

The bit time determines the maximum rate at which bits can be transmitted and thus defines the *bit rate* at which transfer of data can occur. The bit rate is measured in bits per second (bps). The terms *baud* and *baud rate* are often used, somewhat inaccurately, in place of bps or data rate. Baud is a term used accurately in telegraphy and has been carried over into Teletype-like communications. The term *baud* comes from the early Teletype standard of 10 characters per second, and it involved 11 bits. The 11 bits were classified as 1 start bit, 2 stop bits and 8 data bits. Baud is not used accurately with the type of computer-based serial data communication for which it is generally applied. The correct term, bits per second, will be used in this text. Some common communication data rates are 110, 150, 300, 600, 1200, 2400, 4800, and 9600 bps.

Serial transmission also must consider the direction of signal travel on the data line. Figure 12.16 shows three techniques for dealing with data transmission direction. Simplex transmission (Figure 12.16A) is the simplest technique but is limited; it allows communication in only one direction. Duplex transmission allows bidirectional communication and comes in two varieties. Full duplex (Figure 12.16B) allows two-way communication and also allows both users to talk at the same time. The telephone is an example of such a process. The subtleties of human communication sometimes require the capabilities of full duplex transmission. Thus, the telephone must have that capability. Full duplex, however, is somewhat inefficient because it generally requires four wires in order to be implemented. Figure 12.16C shows the half duplex method, in which one user must listen while the other talks. Most radio voice communication uses this "press-to-talk" procedure.

Having discussed the rules, procedures, and protocol of data communication, we will examine the hardware implementation in the next section.

RS-232C Serial Interface

The RS-232C is an electrical interface standard for connecting system components such as modems, printers, terminals, and computers. This standard was

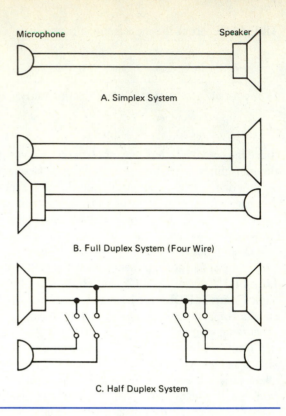

FIGURE 12.16 Simplex and Duplex Transmission

established by the Electronic Industries Association (EIA), an industry trade organization, to be an interface between some form of data terminal equipment (DTE) and some form of data communication equipment (DCE). The *RS* in RS-232C stands for *recommended standard*; the *C* indicates that this is the third change in the standard. The standard defines 20 distinct signals (Table 12.3) to be used to interface terminals and communication equipment.

What is a standard? A *standard* is a model or pattern laid down by some authority by which the quality or correctness of a thing may be determined. Authorities for interface standards may be some independent agency, such as the Institute of Electrical and Electronics Engineers (IEEE), or a manufacturer, such as IBM. Two things are important for the establishment of a standard: An authority is involved, and the standard is published and widely available.

TABLE 12.3   Similar Signals and Functions of the Two Serial Standards: RS-232C and RS-449

RS-232C			RS-449*		
Pin	Circuit	Description	Pin	Circuit	Description
7	AB	Signal ground (common return)	19	SG	Signal ground
			37	SC	Send common
			20	RC	Receive common
22	CE	Ring indicator	28	IS	Terminal in service
20	CD	Data terminal ready (DTR)	15	IC	Incoming call
			12	TR	Terminal ready
6	CC	Data set ready (DSR)	11	DM	Data mode
2	BA	Transmitted data (TxD)	4	SD	Send data
3	BB	Received data (RxD)	6	RD	Receive data
24	DA	Transmit signal element timing (DTE source)	7	TT	Terminal timing
			5	ST	Send timing
15	DB	Transmission signal element timing (DCE source)	8	RT	Receive timing
17	DD	Receiver signal element timing (DCE source)			
4	CA	Request to send (RTS)	7	RS	Request to send
5	CB	Clear to send (CTS)	9	CS	Clear to send
8	CF	Received line signal detector (DCD)	13	RR	Receiver ready
			33	SQ	Signal ready
21	CG	Signal quality detector	34	NS	New signal
23	CH	Data signal rate selector (DTE source)	16	SF	Select frequency
			16	SR	Signaling rate selector
23	CI	Data signal rate selector (DCE source)	2	SI	Signaling rate indicator
14	SBA	Secondary transmitted data	3†	SSD	Secondary send data
16	SBB	Secondary received data	4†	SRD	Secondary receive data
19	SCA	Secondary request to send	7†	SRS	Secondary request to send
13	SCB	Secondary clear to send	8†	SCS	Secondary clear to send
12	SCF	Secondary received line signal detector	2†	SRR	Secondary receiver ready
			10	LL	Local loopback
			14	RL	Remote loopback
			18	TM	Test mode
			32	SS	Select standby
			36	SB	Standby indicator

*37-pin connector except where noted.

†9-pin connector.

Some so-called standards are not standards at all because one or both of these elements are missing. These procedures are generally known as *de facto* standards because of their widespread use in industry.

Standards are an aid to the user, not the manufacturer. The larger the manufacturer, the less advantage is received from standardization. Many companies keep their standards and systems as private as possible. In addition, a standard does not have to be perfect, merely adequate. No matter what standard is produced, someone will disagree with the details; it is difficult to satisfy an entire industry.

The RS-232C standard also defines the voltages, the ranges for a logical 1 and a logical 0, used in all circuits. Figure 12.17A shows an RS-232C signal at the transmitter end of a cable just as it is being transmitted down the cable. A high, or 1, is represented by –12 V, and a low, or 0, by +12 V. Figure 12.17B shows the same signal after it has traveled the length of the cable and been received at the receiver end. Noise and transmission line parameters cause the signal to be degraded. However, the signal need only exceed –3 V at the receiver for a 1 and +3 V for a 0. Figure 12.17C shows the reconstructed TTL signal. TTL signals vary from 0 to 0.8 V for a 0 and from 2.4 to 5 V for a 1.

Some devices do not use all of the RS-232C standard signals, so not all devices will work together without modification. The most important pins in the RS-232C standard are transmit data (TxD) and receive data (RxD), pins 2 and 3, respectively. These pins are the two wires over which serial data are simultaneously sent and received. The remaining pins, except for ground (pin 7) and pins 14 and 16 of the secondary group, are control circuits. The original RS-232C interface was designed to connect some type of data terminal equipment, such as a time-sharing terminal, to some type of data communication equipment, such as a modem. Figure 12.18A shows the minimum configuration for connecting a DTE and a DCE. The DTE in this case is a microcomputer.

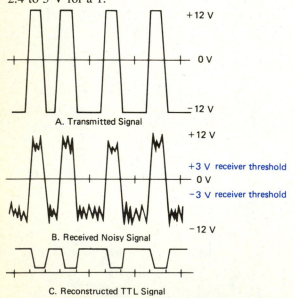

A. Transmitted Signal

B. Received Noisy Signal

C. Reconstructed TTL Signal

FIGURE 12.17 Reduced Effects of Noise and Transmission-Wire Length

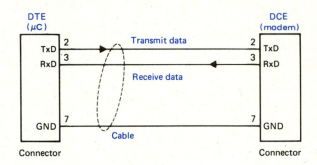

A. Minimum DTE-to-DCE Configuration

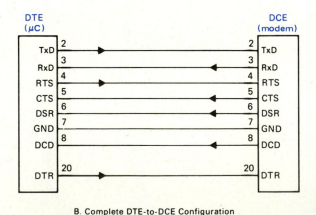

B. Complete DTE-to-DCE Configuration

FIGURE 12.18 DTE-to-DCE Connection Configuration

Notice in Figure 12.18A that the two transmit and two receive pins are connected. Which transmits and which receives? Just as directions in a computer system are from the computer's point of view, all the names of RS-232C signals are from the perspective of the DTE. Thus, the DTE transmits on the TxD line, and the DCE receives on it. Similarly, the DTE receives data on the RxD line, and the DCE transmits on it. Figure 12.18B shows the complete DTE-to-DCE connection.

The following list is a summary of the eight important pins on the RS-232C interface. Table 12.3 gives the pin description, and the following summary gives the general use of the listed pins.

- *Pin 2, Transmitted Data (TxD)*: This pin transmits data from the DTE to the DCE.

- *Pin 3, Received Data (RxD)*: This pin transmits data from the DCE to the DTE.

- *Pin 4, Request to Send (RTS)*: This pin must be at logic zero to enable the DCE to transmit data.

- *Pin 5, Clear to Send (CTS)*: A logic zero on this pin indicates that the DCE is ready to transmit.

- *Pin 6, Data Set Ready (DSR)*: A logic zero on this pin signals the DTE that the DCE is powered and ready to operate.

- *Pin 7, Signal Ground or Common Return*: This line *must* be connected.

- *Pin 8, Data Carrier Detect or Received Line Signal Detector (DCD)*: This input to the DTE may be used to disable data reception if, for example, the originating data channel is disconnected.

- *Pin 20, Data Terminal Ready (DTR)*: A logic low to the DCE indicates that the DTE is powered and ready to operate.

The designations of DTE and DCE are sometimes referred to as the "electronic sex" of the equipment because the RS-232C standard requires that the female connector be associated with the DCE. Connecting a DTE and DCE would be termed as connecting devices of complementary sex.

What happens when the RS-232C interface is used to connect two DTEs, a use for which it was not intended? It turns out that by switching certain wires, the connection and operation can be completed successfully. Figure 12.19A is the simple null modem arrangement that is expanded into the general-purpose or full null modem arrangement of Figure 12.19B.

Even though the RS-232C standard specifies pin numbers, it does not specify a connector. The commonly used connector for this standard is the DB25 connector. It is available in male and female "genders": the DB25P (male) and the DB25S (female). The connector's gender, however, has nothing to do with the device being a DTE or a DCE. Figure 12.20 shows these two connectors.

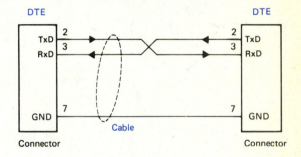

A. Simple Null Modem Connection

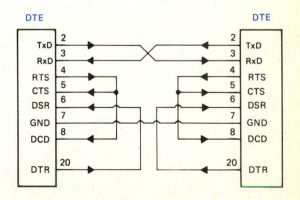

B. Full Null or General Purpose Modem Connection

FIGURE 12.19 DTE-to-DTE Connection

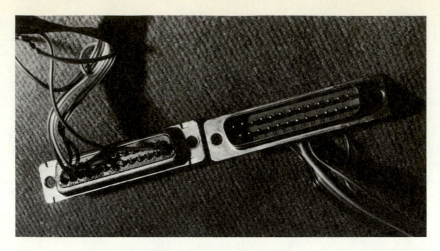

A. Male (Front and Back)

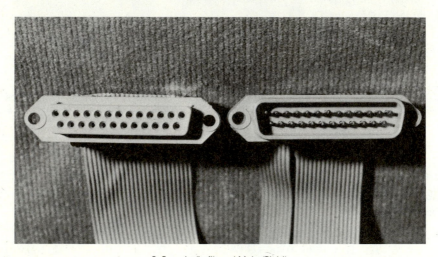

B. Female (Left) and Male (Right)

FIGURE 12.20   DB25 Connectors

One of the problem areas for the RS-232C interface standard is the electric signal level used. This standard was developed before the TTL levels (0–5 V) became popular. TTL can use a single 5 V DC supply for its operation, but the RS-232C standard requires two equal voltages of opposite polarity in the 5–25 V range. To resolve this problem, the EIA introduced two standards, the RS-422 and the RS-423, that specify TTL-compatible voltage levels. The new standards alter only the electrical characteristics of the interface signals; their functions remain the same.

The RS-232C standard has another problem. Its speed is limited to about 20,000 bps and about 50 ft (15 m) of length. For this reason, the EIA published another standard, designated RS-449, that allows data rates as high as 2 million bps. In addition, a few more pins and signals have been added. Refer again to

Table 12.3 for the pin designations. The RS-449 also uses two connectors, a 37-pin plug that carries the most frequently used signals and a 9-pin plug that contains the seldom-used secondary channel. The RS-449 standard is intended to gradually replace the RS-232C. In the meantime, the two standards are bound to cause confusion.

MULTIPLEXING SAMPLE PROBLEMS

We begin this section by presenting a few definitions that were given previously or were implied.

A *frame* is one set of data. The *frame time* is the time from the start of one sync pulse to the start of the next. Its unit of measure is the second. The *frame rate* is the reciprocal of frame time; its unit of measure is hertz (Hz). A *sample* is the individual data input into the frame. The *sample time* is the time allotted for each sample. *Commutation* is the process of changing from one sample to the next. The *commutation rate* is the rate of changing from sample to sample; it is also the reciprocal of sample time. Its unit of measure is hertz (Hz).

Note that the circuits in this section are only proposed circuits. Some have been constructed, and some have not.

Now, let us work some problems. A certain PAM TDM system has 14 data inputs, 3 calibration pulses, and a sync pulse that takes 3 sample times. The sample time divisions are all equal and are at a 100% duty cycle (see Figure 12.8). All data inputs are sampled only once per frame. The fastest data input changes no faster than 10 Hz, and it is sampled five times during its period. Determine the following: frame time, frame rate, sample time, and commutation rate.

The first thing to do is draw a diagram to clarify the problem. The diagram is shown in Figure 12.21.

If the fastest data input changes no faster than 10 Hz, then this sample will determine the frame time. The period of the fastest-changing data input is then 0.1 s. Five frames occur during this time, so frame time is 0.1 s divided by 5, or 0.02 s. Therefore, the frame rate is 50 Hz. Since there are 14 samples, 3 calibration pulses, and a sync pulse that takes 3 sample times, then the sum of these yields 20 sample times occurring within the frame. Sample

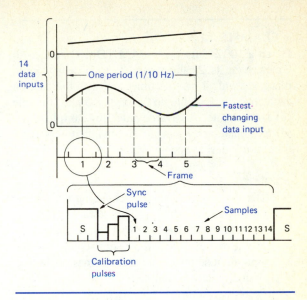

FIGURE 12.21   Diagram for First Problem (PAM TDM System)

time is the number of samples divided into the frame time, or 0.001 s. Commutation rate is then 1 kHz (the reciprocal of sample time).

The pulse width (t) with a 100% duty cycle is frame time divided by the number of sample times per frame, or

$$t = \frac{T_f}{20} \tag{12.5}$$

where
T_f = frame time

Thus, we obtain

$$T_f = \frac{0.02\ \text{s}}{20} = 1\ \text{ms}$$

Using Equation 11.1 from Chapter 11, we can calculate the composite bandwidth with $K = 0.5$:

$$BW = \frac{K}{t} = \frac{0.5}{1\ \text{ms}} = 500\ \text{Hz}$$

If the PAM signal modulates a carrier, the composite bandwidth would be at least twice the PAM bandwidth, or 1 kHz.

• EXAMPLE 12.3

Given a 100% duty cycle PAM TDM system, with 9 data channels, 2 calibration pulses, and a sync pulse that takes only 1 sample time, and with the data on each channel changing no faster than 5 kHz, find the (a) frame time, (b) frame rate, (c) sample time, (d) commutation rate, and (e) bandwidth if we sample three times during each waveform's period.

Solution

(a) The period of the fastest-changing data is the reciprocal of the highest data frequency, 1/5000 Hz = 200 µs. The frame time (T_f) is equal to the period (t_p) of the fastest-changing data input divided by the number of times the fastest-changing data is sampled:

$$T_f = \frac{t_p}{\text{number of samples}} \qquad (12.6)$$

or

$$T_f = \frac{200 \text{ µs}}{3} = 66.7 \text{ µs}$$

(b) The frame rate is the reciprocal of the frame time, so the frame rate is

$$\frac{1}{T_f} = \frac{1}{66.7 \text{ µs}} = 15 \text{ kHz}$$

(c) The sample time (T_s) is the frame time divided by the number of sample spaces:

$$T_s = \frac{T_f}{\text{number of sample spaces}} \qquad (12.7)$$

In this case, we have 9 data channels, 2 calibration pulses, and 1 sync pulse, for a total of 12 sample spaces. So we have

$$T_s = \frac{66.7 \text{ µs}}{12} = 5.56 \text{ µs}$$

Note that T_s, the sample time, is identical to t in Equation 11.1.

(d) The commutation rate (f_c) is equal to the reciprocal of the sample time:

$$f_c = \frac{1}{T_s} \qquad (12.8)$$

or

$$f_c = \frac{1}{5.56 \text{ µs}} = 180 \text{ kHz}$$

(e) The pulse width (t) with a 100% duty cycle in this case is equal to the sample time of 5.56 µs. Using Equation 11.1 from Chapter 11, we can calculate the composite bandwidth with $K = 0.5$:

$$BW = \frac{K}{t} = \frac{0.5}{5.56 \text{ µs}} = 90 \text{ kHz}$$

If the PAM signal modulates a carrier, the composite bandwidth would be twice the PAM bandwidth, or 180 kHz.

Now, let's see how we calculate the bandwidth requirements of the PAM TDM signal. As an example, suppose we need to sample seven signals, whose frequency will be no higher than 1000 Hz, and one sync pulse. Let us also assume that we want to sample at a rate 25% higher than the minimum Nyquist rate. The frame rate (f_f) is

$$f_f = 2f \qquad (12.9)$$

where
f = frequency of fastest-changing signal

For our example, we get

$$f_f = (2)(1000 \text{ Hz})(1.25) = 2.5 \text{ kHz}$$

The frame time is equal to the reciprocal of the frame rate:

$$T_f = \frac{1}{f_f} \qquad (12.10)$$

or

$$T_f = \frac{1}{2.5 \text{ kHz}} = 0.4 \text{ ms}$$

The pulse width (t) with a 50% duty cycle is the frame time divided by the number of samples per frame (8) divided by the duty cycle, or

$$t = \frac{T_f}{16} \qquad (12.11)$$

or

$$t = \frac{0.4 \text{ ms}}{16} = 25 \text{ μs}$$

Using Equation 11.1 from Chapter 11, we can calculate the composite bandwidth with $K = 0.5$:

$$BW = \frac{K}{t} = \frac{0.5}{25 \text{ μs}} = 20 \text{ kHz}$$

If the PAM signal modulates a carrier, the composite bandwidth would be twice the PAM bandwidth, or 40 kHz.

• EXAMPLE 12.4
Given a 50% duty cycle PAM TDM system that must sample 11 signals, whose frequency will be no higher than 500 Hz, with one sync pulse and a sampling rate 50% higher than the minimum Nyquist rate, find the (a) frame rate, (b) frame time, and (c) PAM bandwidth and composite BW if the PAM signal were to modulate a carrier.

Solution
(a) The frame rate (f_f) is, according to Equation 12.9,

$$f_f = (2)\,(500 \text{ Hz})\,(1.5) = 1.5 \text{ kHz}$$

(b) From Equation 12.10, the frame time is equal to the reciprocal of the frame rate:

$$T_t = \frac{1}{f_f} = \frac{1}{1.5 \text{ kHz}} = 0.00066 \text{ s} \quad \text{or} \quad 0.66 \text{ ms}$$

(c) The pulse width (t) with a 50% duty cycle is

$$t = \frac{T_f}{24} \qquad (12.12)$$

or

$$t = \frac{0.66 \text{ ms}}{24} = 0.0000277 \text{ s} \quad \text{or} \quad 27.7 \text{ μs}$$

Using Equation 11.1 from Chapter 11, we can calculate the composite bandwidth with $K = 0.5$:

$$BW = \frac{K}{t} = \frac{0.5}{27.7 \text{ μs}} = 18 \text{ kHz}$$

If the PAM signal modulates a carrier, the composite bandwidth would be twice the PAM bandwidth, or 36 kHz.

• EXAMPLE 12.5
Given a PWM system with five channels, each having a frequency that goes no higher than 1.5 kHz, a sampling rate twice the minimum Nyquist rate, and a rise time not exceeding 1% of the time allotted to each channel sample, find the minimum transmission bandwidth.

Solution
The minimum frame rate is equal to four times the 1.5 kHz frequency, or 6 kHz. The frame time is the reciprocal of the frame rate, or 1/6000 Hz = 166.7 μs. Each of the frames is divided into five slots for each of the five channels, so the sample time, according to Equation 12.7, is

$$T_s = \frac{T_f}{\text{number of sample spaces}}$$

$$= \frac{166.7 \text{ μs}}{5} = 33.3 \text{ μs}$$

Each of the data pulses will change its width depending on the size of the modulating signal. The pulse width cannot exceed 33 μs, however. The maximum rise time (t_r) is specified to be no greater than 1% of the time allotted to each channel—in this case, 33.3 μs. So the maximum rise time is 0.01 times 33.3 μs, or 0.333 μs. The approximate transmission *BW*, then, is approximately

$$BW = \frac{0.5}{t_r} = \frac{0.5}{0.333 \ \mu s} = 1.5 \ \text{MHz}$$

For the next problem, we want to design a single-channel, voice PCM system. Voice sample rate generally is set at 8000 samples per second. We will use eight bits for each word, a 50% duty cycle, and a sync pulse whose amplitude is twice that of a data bit but has the same sample time. Determine the following: pulse width, commutator frequency, frame time, and frame rate.

The diagram for this problem is shown in Figure 12.22. Since there are 8000 samples per second, frame time, which is the reciprocal of 8000 samples per second, is 125 μs, and frame rate is 8 kHz. A 50% duty cycle will require the equivalent of 18 sample spaces per frame because there is one sample space between bits and one-half sample space at each end of the frame. Thus, pulse width is frame time divided by 18, or 6.94 μs. Commutator frequency is 144.1 kHz.

If the circuit of Figure 11.31 (Chapter 11) were to generate the output desired in this problem, the first 555 would be set to 8 kHz and the second 555 to 144.1 kHz.

As another problem, let us design a physiological monitor of some type. It should be lightweight, portable, and inexpensive. As an example, we will make a heartbeat monitor. TDM is generally used for slowly changing data, so we will use PFM for this problem.

The PFM generator of Figure 11.21 (Chapter 11) would seem to be a prime candidate for this problem. And if we want an inexpensive system, we could use an FM radio as the receiver. The bandwidth for FM radio is from 88 to 108 MHz, but the frequency limit of the LM741 is 1 MHz, so this device won't do. However, other op amps may be used if they can operate at high frequencies.

An example of an op amp that might meet the frequency requirements is the LM733 differential video amp. It has a bandwidth of 120 MHz and selectable gains of 10, 100, and 400; and no frequency compensation is required. Figure 12.23 shows how the transmitter might be connected.

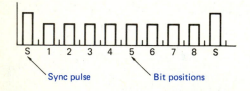

FIGURE 12.22 Diagram for Second Problem (Single-Channel, Voice PCM System)

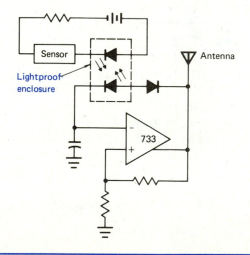

FIGURE 12.23 Proposed Transmitter for Heartbeat Monitor

An alternative might be to use a lower-frequency transmitter of the same design and a PLL IC as the receiver/demodulator. Since the IC is a low-power device, the transmission distances would be necessarily short.

For our next problem, we will make a single-channel oscilloscope into a multichannel oscilloscope. To do so, we will use TDM principles but no transmitter or receiver. (The oscilloscope will be the receiver.)

The idea in this problem is to divide the inputs (called *channels* here) into PAM and give each input channel a different DC offset. These PAM TDM signals will be input to the single-channel oscilloscope, which will display all the inputs on the screen, multiplexed in time. If the multiplexing is done fast enough, the display will appear to be that of a multichannel oscilloscope.

The schematic of Figure 12.24 can be used to produce the desired results. The input resistors, R_1

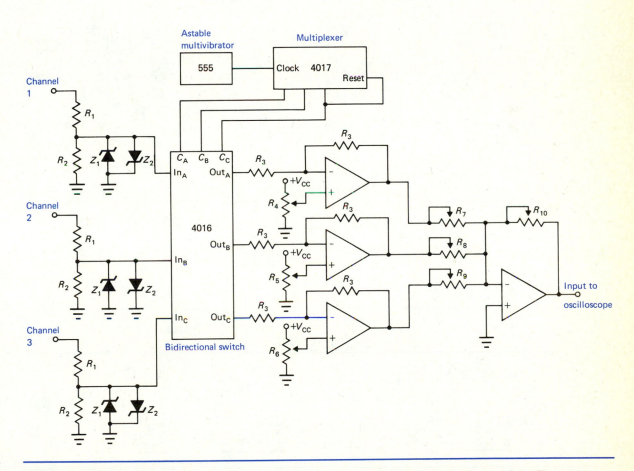

FIGURE 12.24 Basic Schematic for Changing Single-Channel Oscilloscope into Three-Channel Oscilloscope

and R_2, are used to attenuate the input signals to keep them within the voltage limits of the circuitry. Notice that the input resistors for the three channels have the same designation (R_1 and R_2). Thus, all R_1's are the same resistance, and all R_2's are the same resistance. The input zener diodes, Z_1 and Z_2, protect the ICs if the maximum voltages are still exceeded. The first

line of op amps gives each channel a different DC level offset so that each input channel will be displayed on a different level on the screen. Resistors R_7, R_8, and R_9 adjust the individual channel gains, and R_{10} adjusts the overall gain. The final op amp sums all the signals and prevents each channel from interfering with the others.

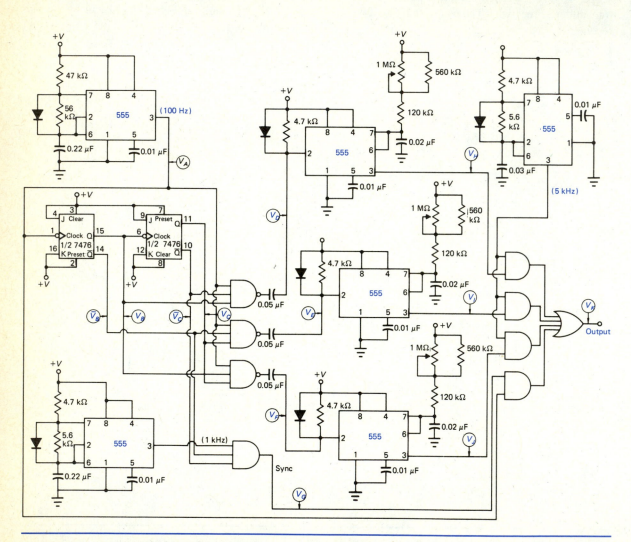

FIGURE 12.25 Three-Channel Multiplexer Using PWM

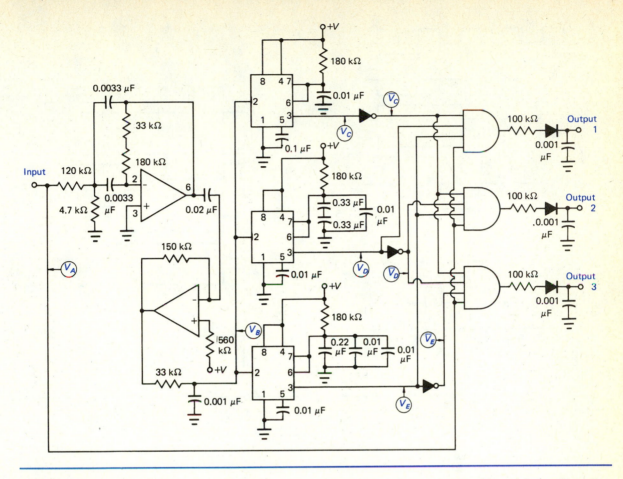

FIGURE 12.26 Three-Channel Demultiplexer Using PWM to Demultiplex Signal from Circuit of Figure 12.25

The final problem is to make a three-channel multiplexer suitable for a radio-controlled airplane. The solution is represented in Figures 12.25 through 12.28.

Figures 12.25 and 12.26 are the schematics for the modulator and demodulator. Figures 12.27 and 12.28 are the synchrograms for the schematics. The receiver in the schematic of Figure 12.27 is producing pulses whose pulse widths correspond to the pulse widths being transmitted. The output from each channel can be attached to appropriate mechanical actuators in the airplane.

CONCLUSION

As shown in this chapter, the principles of telemetry have many industrial applications. And as in the case of most modern electronics applications, only the competence and imagination of the investigator will limit the results.

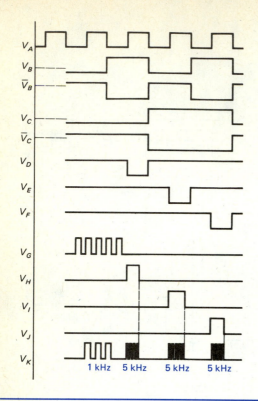

FIGURE 12.27 Waveforms for Schematic of Figure 12.25

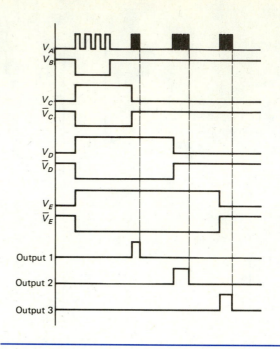

FIGURE 12.28 Waveforms for Schematic of Figure 12.26

■ QUESTIONS •

1. A telemetry system can be divided into three parts: The _____, which are at remote locations and transform the physical quantities to be monitored into electric signals; the _____ system, which transforms the transducer signals, transmits and receives signals, and transforms receiver signals for display; and the _____ and _____ system, which displays received signals for interpretation.

2. Communication using different frequency channels at the same time is called _____.

3. Communication using the same frequency channel but at different times is called _____.

4. According to Table 12.1, in order to keep a low signal-to-noise ratio while transmitting telemetry signals, we need a modulation index of _____.

5. An FM/AM multiplexed telemetry system designation indicates that the subcarriers are _____-modulated and the main carrier is _____-modulated.

6. _____ /_____ is by far the most common modulated telemetry technique in use today.

7. Communication errors caused by equipment aging can be reduced by using _____ signals.

8. A 100% duty cycle telemetry system uses a _____ frequency spectrum than the 50% duty cycle system uses.

9. In Question 8, the 50% duty cycle system _____ cross talk between successive channels.

10. Pulse-position TDM uses the _____ edge of the pulse to carry the information.

11. Pulse-code TDM can transmit 1 and 0 information only by the _____ or _____ of a pulse.

12. Define multiplexing.

13. Name two types of multiplexing in common use.

14. Define subcarrier.

15. Define frame time.

16. What are housekeeping pulses?

17. Define binary transmission system format.

18. Define a pulse-code TDM word.

19. Discuss when parity bits are and are not used.

20. If a data signal varies at a rate of 75 Hz, what would the lowest standard frequency band be that could be used to carry this information?

21. Band 14 would use a subcarrier of what frequency?

22. To demodulate an FM data signal on band 10, what discriminator center frequency would you use?

23. What is the lowest frequency the subcarrier can deviate to and still be in band 8?

24. Measurement of a signal indicates that its minimum rise time is 7 ms. What channels could be used to carry this information?

25. Sketch the two types of NRZ waveforms, and explain how they differ.

26. What are two disadvantages of non-self-clocking binary modulation methods compared with self-clocking methods?

27. List ten magnetic medium recording methods.

28. Which magnetic recording format has the best characteristics other than the transfer rate?

29. Which two magnetic recording formats have the best characteristics that include the fastest transfer rates?

30. The two major classifications of data transmission are _____ and _____.

31. The two major classifications of serial data transmission are _____ and _____. Explain the differences between these classifications.

32. List two character code formats in common use today.

33. Define simplex, half duplex, and full duplex transmission.

34. What are the requirements for a proposed model to become a true standard?

35. What are the data speed and cable length limitations for the RS-232C serial interface standard?

36. What is the data speed for the RS-449 serial interface standard?

■ PROBLEMS

1. Suppose you wish to transmit voice (20 kHz) on an FM system that has a modulation index of 5. What will be the maximum frequency deviation of this transmitter?

2. A telemetry system has seven different sensors to sample. Two additional sample spaces are set aside for the high-level and low-level references, and one sample is set aside for a sync pulse. In this telemetry system, the highest-frequency input variable has a maximum frequency of 10 Hz. You wish to sample this variable ten times in each waveform period. Each of the seven inputs is sampled only once during each frame. Determine (a) sample time, (b) frame time, (c) frame rate, and (d) commutation rate.

3. An FM frequency-division multiplexing system uses channel 3 on the IRIG chart. Find (a) the minimum rise time, (b) the bandwidth of the frequency spectrum when channel 3 is modulated by the highest modulating frequency, and (c) the upper and lower limits of the carrier frequency when modulated by the highest modulating frequency.

4. You would like to use the first five channels on the IRIG telemetry chart. Calculate the transmission bandwidth needed to transmit channels 1–5.

5. A 100% duty cycle PAM TDM system, with five data channels (maximum frequency of 2 kHz), three calibration channels, and one sync pulse (taking up only one sample space), must be sampled four times during each of the channel's periods. Find the (a) frame rate, (b) frame time, (c) commutation rate, (d) sample time, and (e) composite bandwidth.

6. A 50% duty cycle PAM TDM system, with ten data channels (maximum frequency of 1 kHz), three calibration channels, and one sync pulse (taking up only one sample space), must be sampled seven times during each of the channel's periods. Find the (a) frame rate, (b) frame time, (c) commutation rate, (d) sample time, and (e) composite bandwidth of the modulation and the carrier.

7. A PWM system has ten channels (including housekeeping pulses), and each channel's data rate is no higher than 1 kHz. If we sample at four times the minimum Nyquist rate and specify that the rise time cannot exceed 1% of the time allotted to each channel sample, find the minimum transmission bandwidth.

Sequential Process Control

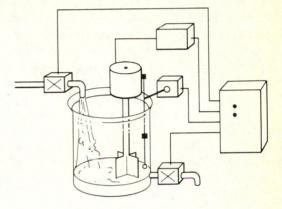

OBJECTIVES

On completion of this chapter, you should be able to:

- Define sequential control.
- Distinguish a batch process from a continuous process.
- Identify batch process applications.
- Recognize and explain the parts of a ladder logic diagram.
- Generate a ladder logic diagram from a pictorial diagram and sequential description of a sequential process.
- Determine that a ladder logic diagram is accurate by "hand running the logic."
- Draw relay logic symbols for the following digital functions: AND, OR, NAND, NOR.
- Convert relay logic ladder diagrams to equivalent semiconductor logic diagrams.
- Convert semiconductor logic circuit diagrams to relay logic ladder diagrams.

INTRODUCTION

Traditionally, industrial processes and control systems used relays, timers, and counters as well as sensors and actuators. The first three types of devices constitute a class of control devices that are used to control processes known as sequential control processes (as opposed to continuous control processes). A *sequential process* is a process in which one event follows another until a job is completed. Sequential process control can be subdivided into two major classes: control of event-based sequential processes and control of time-based sequential processes. Of course, many sequential processes may contain elements of both of these classes.

An *event-based* sequential process is one in which the occurrence of an event causes a certain action to take place. For example, a limit switch closes to reverse a motor moving a part because the part has reached the furthest limits of safe operation. Or the inputs to a logic device become true, such as the input connections to an AND gate going high, and cause the output of that device to activate some other device. In an event-based sequential process, the action of the controller is being governed by events external to the controller. These processes are of the type that can be controlled by a *programmable controller*.

A *time-based* sequential process is one in which the events occur in a specific sequence, not the nearly random closure of switches external to the controller. The action of this controller basically is determined by the program in the controller and not by external occurrences. The controlling system for this type of process is called a *programmable sequence controller* or *programmable process controller*. An extensive treatment of time-based control processes is beyond the scope of this book. This chapter and the next will deal mainly with event-based control.

Another name for a sequential process is a *batch process*. A batch process consists of a sequence of one or more *steps* or *phases* that must be performed in a defined order. The completion of this sequence of steps creates a finite quantity of finished product. If more of the product is to be created, the sequence must be repeated.

This chapter covers sequential control and how it can be implemented on automatic machines. The ladder diagram is the traditional method of choice for representing the logic necessary to help design and document these automatic processes. Sequential control and ladder diagrams form the base upon which the programming of the programmable controller, the subject of Chapter 14, is founded.

CHARACTERISTICS

The following list outlines some key concepts that distinguish a batch process from a continuous process.

1. *Discrete* loads of raw materials normally are introduced into the line for processing, rather than the continuous flow of materials found in a continuous process.

2. Each load of material being processed normally can be identified all the way through the processing because each is kept separate from all others being processed. In most continuous processes, raw materials cannot be uniquely tracked.

3. Each load of raw material can be processed differently at the various equipment areas in the line. In continuous processes, raw materials usually are processed in an identical fashion.

4. Movement of a load from one equipment area or step to the next can occur only when the step has been completed and the next equipment area has been vacated. In continuous processes, materials flow steadily from one area to the next.

5. Batch processes generally have *recipes* associated with each load of raw product to be processed. These recipes direct the processing of the raw product. *Note*: Even though the word

recipe is used here, it does not specifically refer to food preparation, although many foodstuffs are made by batch processing.

6. Batch processes usually require more *sequential logic* than continuous processes.

7. Batch processes often include processing steps that can fail and include other special processing steps to be taken in that event.

APPLICATIONS

Batch control systems generally are configured to control four different types of applications. They are used for the control of solids, which normally use input parameters from weighing equipment; for the control of liquids, with input parameters from flowmetering equipment; for the control of the environment, such as temperature, dew point, and other variables; and for the control of "universal" systems, which accept inputs from many different types of sensors.

When we weigh solids, two parameters of the system need to be considered. The first, called *dribble*, occurs when the loading chute discharge rate changes from high- to low-speed discharge so that the scale can have enough reaction time to measure the solid accurately and not overshoot the intended weight. The other parameter, sometimes called *preactuate*, considers the amount of solid still in the loading chute but not yet fallen to the scale after the discharge has been cut off. These two parameters will vary with each system but must be accounted for if accurate weighing is desired.

The accurate measurement of fluid will depend upon the type of flowmeter used and the density of the fluid. Because most fluids change density in a consistent manner with temperature, density can often be inferred from a temperature measurement.

Environmental control applications include controlling heat for sintering, carburizing, hardening, tempering, brazing, annealing, and gaseous diffusion processes such as that used in IC manufacturing.

Other applications are freeze-drying vegetables and baking out moisture from large rolls of paper to reduce their shipping weight.

Universal control system applications require special functions such as floating-point math, exponential functions, and various other functions. This class of processes covers such a wide range of applications that special sensors and devices may be required.

DEVICE SYMBOLS

Documentation of industrial processes is continuing to grow in importance particularly because systems are becoming more complex and interdependent. Logical and easy-to-read documentation is important not only for the design and installation of a system but also for continued system maintenance and troubleshooting. For these reasons, standardized symbols and formats need to be understood and used. The symbols that follow generally are accepted by industry personnel when documenting sequential control systems using relay logic. The standardized format for documenting sequential control systems is the ladder logic diagram, the topic of the next section.

Some mention should be made at this point about the general differences between electrical and electronic symbol designations. These two industries matured somewhat independently of each other, with different ruling bodies, and therefore some differences exist in their respective symbols for the same component. For example, the electronic symbol for a resistor is a zigzag line with an alphanumeric designation of R_1. The same symbol in the electrical or industrial world is a rectangle with lines coming from the ends and with an alphanumeric designation of 1R. These differences can sometimes be confusing. In this chapter, we will use industrial symbols because designing and documenting the sequential process is predominantly an industrial process for industrial machines.

Switch Symbols

One of the most common symbols used in sequential control documentation is for switches. Two switches are particularly common: push-button and limit switches (Figure 13.1).

The two categories for each switch shown in Figure 13.1 are normally open (NO), shown on the left, and normally closed (NC), shown on the right. There are variations to these symbols, such as those in Figure 13.2, but the symbols shown are the ones in general use. For more information concerning the push-button and limit switches, refer to Chapter 6 on control devices. Push-button switches generally are actuated by the operator, whereas limit switches are actuated by the machine or system being operated.

A newer type of switch that is becoming more popular is the proximity switch. A proximity switch senses the presence or absence of a target without physical contact. The six basic types are magnetic, capacitive, ultrasonic, inductive, air jet stream, and photoelectric. The symbols for proximity switches are the same as those for limit switches.

Temperature switches, shown in Figure 13.3A, open or close when a designated temperature is reached. You now know from the symbols which switches are NO and which are NC; therefore, they will not be labeled as such except for clarity. Level switches (Figure 13.3B) sense the height of a liquid and open or close their contacts when the predetermined height is reached. Pressure switches (Figure 13.3C) detect pressure and actuate (open or close) their contacts when the specified pressure is reached.

The most common type of flowmeter used in sequential control is one that produces pulses. These pulses can be counted by a counter, which, in turn, can cause an action when a specified count is reached. Counters will be discussed later, but the flowmeter symbol and its output are shown in Figure 13.4. Notice that the flowmeter output is a series of pulses with amplitudes of 24 V since this particular flowmeter operates on 24 V. Other voltages can be used, depending upon the requirements of the flowmeter.

The proximity, temperature, level, and pressure switches and the flowmeter are more likely to be called *sensors* and, as such, are discussed in Chapter 8 on transducers.

Relay Symbols

Another very commonly used device in sequential control is the electromagnetic relay. The symbol for the relay (Figure 13.5) is a circle and parallel lines. The circle represents the coil of the relay, and the parallel lines represent the relay contacts. A relay will have only one coil but may have any number of contacts, either NO, NC, or a mix of both. The relay

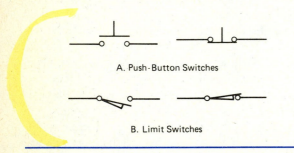

A. Push-Button Switches

B. Limit Switches

FIGURE 13.1 Common Switch Symbols

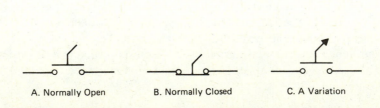

A. Normally Open   B. Normally Closed   C. A Variation

FIGURE 13.2 A Variation of Push-Button Switch Symbols

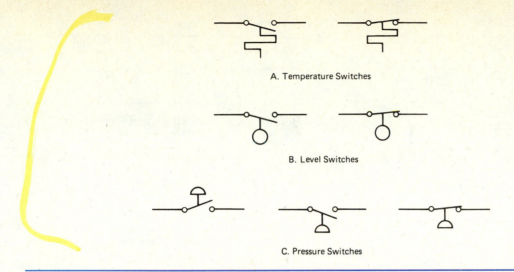

FIGURE 13.3   Temperature, Level, and Pressure Switch Symbols

A. Flowmeter   B. Flowmeter Output

FIGURE 13.4   Flowmeter Symbol and Output Pulses

A. Control Relay Coil   B. NO Contacts   C. NC Contacts

FIGURE 13.5   Control Relay and Contacts Symbols

coil generally has an alphanumeric designation, such as 1CR, inside the coil circle, and the same designation next to the contacts that are controlled by that coil. The letters CR represent *control relay*, and the number distinguishes that relay from all others in the schematic. Electromagnetic relays are discussed in Chapter 6 on control devices.

Other devices whose symbols are similar to the control relay are the timer relay and the counter. The timer relay symbol (Figure 13.6A) is also a circle, but there are two variations. The three-connection timer relay coil has the third, or bottom, connection permanently connected to the power in order to operate the timer motor. This connection may or may not be shown. Current flow through the other two connections initiates the timing sequence. The timer relay switch contacts will actuate upon completion of the timing period.

The contacts in Figure 13.6B are known as *on-delay* timer contacts. The energizing of coil 1TR causes the NO contacts to time close after a period of time and the NC contacts to time open. De-energizing timer coil 1TR instantaneously opens the NO contacts and closes the NC contacts.

The contacts in Figure 13.6C are known as *off-delay* timer contacts. Energizing the coil of the off-delay timer instantaneously closes the NO con-

A. Timer Relay Coils

B. On-Delay Timer Contacts

C. Off-Delay Timer Contacts

FIGURE 13.6   Timer Relay and Contacts Symbols

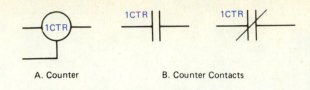

A. Counter   B. Counter Contacts

FIGURE 13.7   Counter and Contacts Symbols

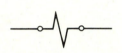

FIGURE 13.8   Solenoid Symbol

tacts and opens the NC contacts. De-energizing the off-delay timer coil causes the NO contacts to open and the NC contacts to close *after a time delay*.

There are various types of timers, but generally, the timer is reset and ready to start timing again whenever the current flow to the timer is interrupted.

The counter symbol (Figure 13.7A) generally has three connections. The third, or bottom, connection is used to enable (allow it to operate) and then to reset the counter when desired. The remaining two connections carry the signal or pulses to be counted. Notice that the counter contacts (Figure 13.7B) are no different from the control relay contacts, except for their alphanumeric designation. These contacts actuate when the predetermined, or preset, count is reached and reset when the counter is reset.

Solenoid Symbol

Another useful sequential control device is the solenoid (Figure 13.8). A solenoid provides electrical control over materials such as liquids and gases. An electric signal to the solenoid causes a valve to open and allow the liquid or gas to flow. The valve stays open until the electric signal ceases. A solenoid is somewhat different from a relay. A solenoid

generally has a coil of wire with a moving metal core in the coil, and the relay has a fixed metal core in the coil that, when activated, pulls an armature that makes the contact. Solenoids, as well as relays, can be used to switch heavy-current circuits.

Miscellaneous Symbols

Some other device symbols are shown in Figure 13.9. These symbols are self-explanatory.

Dotted lines connecting symbols indicate that those devices are physically connected and are actuated together. Figure 13.10 illustrates push buttons and limit switches that are activated together.

LADDER LOGIC DIAGRAM

The control logic for a sequential process traditionally has been depicted in the language of a ladder logic diagram. A *ladder logic diagram* is a diagram with a vertical line (the power line) on each side. All the components are placed between these two lines, called *rails* or *legs*, connecting the two power lines with what look like rungs of a ladder—thus the name, ladder logic diagram.

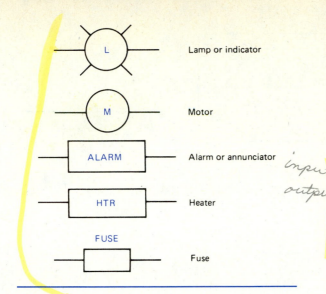

L — Lamp or indicator

M — Motor

ALARM — Alarm or annunciator

HTR — Heater

FUSE — Fuse

FIGURE 13.9 Miscellaneous Symbols

Ladder logic diagrams are universally used and understood in industry, whether in the process industry, in manufacturing, on the assembly line, or inside electrical appliances and products. The programmable controller (PC), which first appeared in 1969, increased its chances of success because it capitalized on well-established concepts and practices. Thus, the ladder logic diagram was the language of choice for programming the PC.

Power connections for ladder logic diagrams can be either AC or DC, as shown in Figure 13.11. The transformer (Figure 13.11A) and battery (Figure 13.11B) are the means by which power is introduced into their respective circuits. These power connections usually are not shown in ladder logic

diagrams but are understood to be there. The alphanumeric designations on the transformer symbol in Figure 13.11A are also located on the transformer wires or on the transformer case next to the wires on larger transformers. The H's designate the primary connections; the X's are the secondary connections. The standard features of a ladder logic diagram that are shown in the figure include the following:

inputs
outputs

- Left-hand rails (rails 1 and 3 in Figure 13.11)
- Right-hand rails (rails 2 and 4 in Figure 13.11)
- Ladder rungs that show the device symbol wiring
- Wired-in device symbols

Notice that if a circuit is completed on any rung, current will flow between the rails on that rung. If there is no resistance or impedance on that rung, the power supply will be shorted. For example, in Figure 13.11B, if the heating device were left out of rung 2, and the 1CR contact closed, the 24 V DC supply would be shorted. This condition should never occur.

There are a number of additional important points about ladder logic diagrams as illustrated in Figure 13.11.

1. By convention, all of the devices that represent a resistive or inductive load to the circuit are shown on the right, and the devices that make or break electric contact are shown on the left of the diagram.

2. The circuits connected in parallel, such as the 1CR contacts in parallel with the start push button 2PB, are sometimes referred to as *branches*. Some manufacturers, however, refer

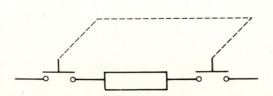

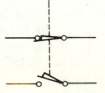

FIGURE 13.10 Symbols for Devices Acting Together

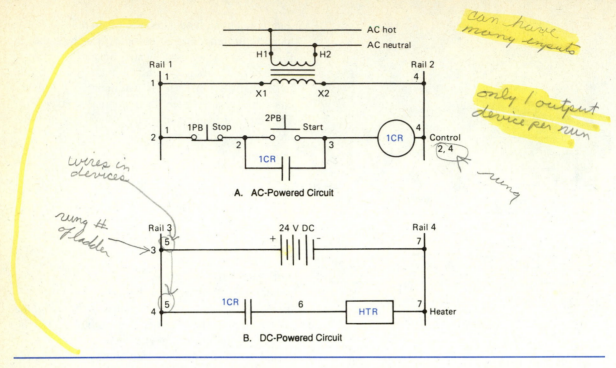

Handwritten annotations: *can have many inputs*; *only 1 output device per run*; *rung*; *wires in devices*; *rung # of ladder*

FIGURE 13.11 Basic Ladder Logic Diagrams

to each different horizontal line in the diagram as a rung. The rung concept will be used in this text because identifying and referring to each line in the diagram is then easier.

3. The numbers to the right of rails 1 and 3, to the left of rails 2 and 4, and at each point in the circuit where devices connect indicate *wiring labels*. These labels on the physical wires are numbered according to the diagram and are used by the installation and maintenance personnel to help them trace wires and prevent wiring mistakes. Figure 13.12 shows an example of wiring labels.

4. The words to the right of the diagrams in Figure 13.11 indicate the major function or component in that rung. For example, rung 2 has a control relay in it and thus the word *control* is to the right. The load in rung 4 is the *heater*.

5. Immediately under these words are another set of numbers that indicate the rung in which the contacts for that load device is located. Under the word *control* are the numbers "2, 4." These

numbers show that the contacts for the control relay 1CR are located in rungs labeled 2 and 4. If these numbers were underlined, they would refer to NC contacts. Since these contacts are NO, the numbers are not underlined. The word *heater* at the end of rung 4 does not have numbers under it because the heater does not control other contacts. The practice of indicating the rung in which the contacts are located is very helpful in large-system diagrams where the contacts may be spread throughout the diagram.

6. The 1CR contacts in parallel with the start switch in rung 2 illustrate a common industrial practice in ladder logic diagrams. The 1CR contacts are in the same rung as the control relay that activates these contacts. Thus, when the start button is pushed and the 1CR relay is energized (or pulled in or picked up), the 1CR contacts activate and keep the 1CR relay active even though the start button is released. This action *latches*, or *seals*, the relay on; therefore, the circuit is called a latch, or seal, circuit.

FIGURE 13.12 Example of Wiring Labels

The ladder logic diagrams of Figure 13.11 do nothing more than turn on a heater with no provision for controlling the heater automatically. This system is impractical, but it does illustrate the elements in ladder logic diagrams. The following sections will show examples that are more practical.

Writing a Ladder Logic Diagram

There are as many ways to start writing ladder logic diagrams as there are manufacturers who sell systems that use these diagrams. In general, the following suggestions apply to writing ladder logic diagrams.

1. Look at the system to be analyzed. Draw a picture of the system, making sure all of the components of the system are present in the drawing. Accuracy is important. If anything is left out here, it will most likely be forgotten later also.

2. Determine the sequence of operations to be performed. Write these operation sequences in

sentences or put them in a table, whatever is best suited to the person making the diagram.

3. Write the ladder logic diagram from this sequence of desired operations.

4. Check the operation of the diagram in the following manner: Go through the logic process as if you were the machine, generally starting with the pressing of the start button. As each load is activated, write its designation on a piece of paper. When that device is deactivated, put a line through its designation. When the sequence of operations is completed and ready to start over, all the load designations should have been written down and crossed off. This procedure is sometimes called *hand running the logic*. If any loads were not used or were used but not crossed off, recheck the diagram and system to see whether that operation was necessary or incomplete. If step 4 is successfully completed, an accurate ladder logic diagram should be generated.

Automatic-Mixer Example. These steps will be used to produce a ladder logic diagram for a simplified automatic-mixing process. Step 1, a picture or diagram of the system, is shown in Figure 13.13A. For step 2, the sequence of operations, the following short description is given: The tank in Figure 13.13A is filled with a fluid, agitated for a specified length of time, and then emptied.

Another way of examining the order of operation of the process is illustrated in Figure 13.13B and is called a *state description* of the process. A state description is similar to a flowchart in computer programming. It shows each action in a circle, and arrows indicate the order in which these actions occur.

Step 3, the ladder logic diagram, is shown in Figure 13.14. Now we will follow the series of events for the full control cycle of the automatic-mixer process and check the logic (step 4) at the same time. Mark down each device when it is activated and cross it off when it is deactivated. Refer to Figures 13.13C and 13.14. The process is initiated by pressing the start push button 2PB. Activating the start button energizes the control relay 1CR located in the start/

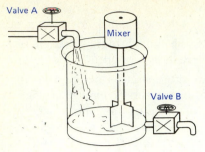

A. Simplified Mixer Process

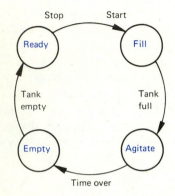

B. State Description of Process

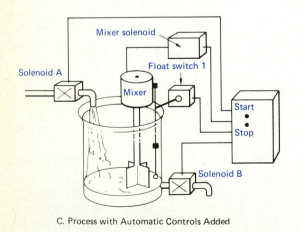

C. Process with Automatic Controls Added

FIGURE 13.13 Sequential Control Process

stop switch box of Figure 13.13C. (Write 1CR on a piece of paper or check sheet since it was energized.)

When the control relay 1CR is energized, the 1CR relay contacts change state; in this case, since they are both NO, they close. When the 1CR contacts in rung 1 close, current is allowed to continue through the coil of the 1CR control relay, even though the start push button 2PB has been released. This circuit holds the 1CR relay in as long as the power line power is applied, the stop button 1PB is not pushed, and the timing relay 1TR has not timed out.

This latch circuit has another useful purpose. If the main power to the system is interrupted for any reason, the process cannot start up by itself. This safety feature is important to an operator, unaware that the main power is off, investigating the cause of a stoppage when the main power resumes. Without this safety feature, the operator could be injured.

When the 1CR contacts in rung 2 close, current can flow through solenoid A. Solenoid A is an electromechanical device that is electrically activated to open a valve mechanically, which allows fluid to flow into the tank. Fluid flows because the float switch 1FS in rung 2 is closed. This situation is indicated in Figure 13.13C by the empty position for the float switch. (Write solenoid A on the check sheet.)

When the tank has filled, the float switch 1FS changes to the filled position. This tank-full condition de-energizes solenoid A by opening 1FS in rung 2, starts the timer relay 1TR in rung 3 by closing 1FS in rung 3, and energizes the mixer relay MR. (On the check sheet, put a line through solenoid A since it is de-energized, and add 1TR and MR to the sheet since they are energized.)

Notice here that the two limit switches do not operate in the same way for the same conditions. For example, 1FS in rung 2 opens when the tank is full but does not close until the tank is empty, not just below the full level as the tank is emptying. Limit switch 1FS in rung 3 closes when the tank is full but does not open until the tank is completely empty. In electronics, this type of operation would

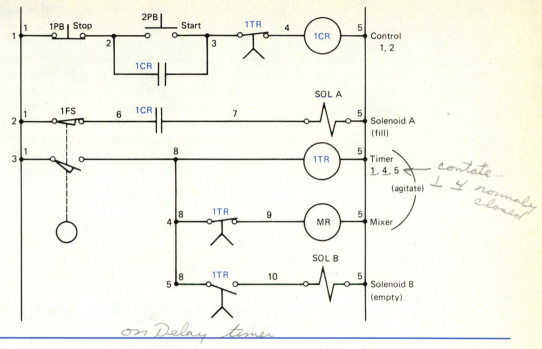

FIGURE 13.14 Ladder Logic Diagram for Control of Automatic Mixer of Figure 13.13C

be considered switching with hysteresis. If, in practice, this type of switch cannot be found, additional logic could be added to the ladder diagram to accomplish the desired result. However, a sump pump float switch will do the job.

After the timer has timed out, relay 1TR switches off the mixer in rung 4, energizes solenoid B in rung 5, which starts emptying the tank, and de-energizes the control relay 1CR in rung 1. (On the check sheet, put a line through MR and 1CR, and add solenoid B to the sheet.) Because the control relay 1CR was de-energized by the timer, the 1CR contacts in rungs 1 and 2 also opened at this time. This action does not change anything on the check sheet, but it does prevent solenoid A from turning on while the tank is being emptied.

When the tank is empty, float switch 1FS in rung 3 shuts off solenoid B and resets the timer, placing the system in the ready position for the next manual start. (Now cross off solenoid B and 1TR on the check sheet. All devices have now been turned on and then turned off; thus, step 4 has been successfully completed.)

The ladder logic diagram in Figure 13.14 is a basic configuration that will work properly even if the start button is pressed and held down. The 1TR contacts in rung 1 make sure that the control relay 1CR is not activated until the correct time in the cycle. Pressing the stop button will stop the fill process, which can be started again later, but the stop button will not stop the process once the timer and mixer are started. The only way to stop the timer and mixer in an emergency is to throw the main power. The operating personnel must be made aware of this situation, or the ladder logic diagram must be changed to accommodate this emergency condition.

Another helpful procedure that can be used to comprehend what is happening in the process is the timing diagram shown in Figure 13.15. This diagram shows, along the left side, all the ladder devices grouped by rung number; along the bottom, it shows time as the independent variable. The start of the process is indicated at the top left, and completion of the process is shown at the top right of the diagram. The circled numbers in each row of the diagram indicate the order in which events happen. For example, number 1 in the second row shows that

pressing the start button 2PB is the initiating event that causes the 1CR coil to energize. This coil, in turn, causes the two 1CR contacts to change state—in this case, close and conduct current. Closing the 1CR contacts in rung 2 causes solenoid A to energize, as shown in rows 6 and 8. The numbers in circles along with the arrows are called *cause-effect arrows*. In other words, the action produced where the circled number is located produces an action where the arrow points. It is not always easy to visualize these actions in the ladder diagram.

In general, there are three states in which the ladder logic devices can exist: contacts open or circuit not complete, contacts closed or circuit complete but no current flow, and current flow. These conditions

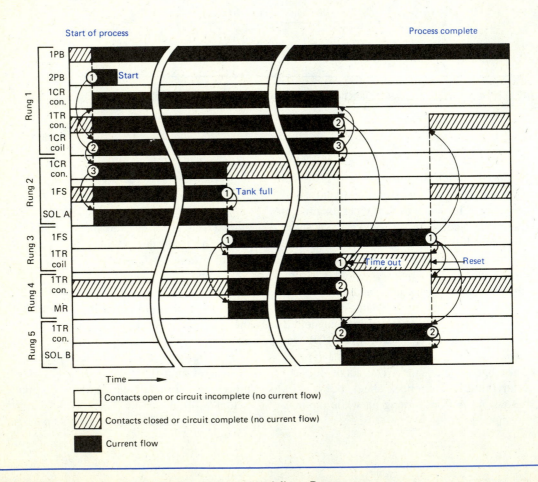

FIGURE 13.15 Timing Diagram for Automatic-Mixer Process

are shown in the key for the timing diagram in Figure 13.15. Follow through the ladder logic operation again, but this time follow the events as they occur in the timing diagram. From the timing diagram and the state description (Figure 13.13B), we see that there are only four places where device conditions may change state. These changes in state are not at all apparent in the ladder logic diagram.

Oscillatory-Motion Example. Becoming proficient in writing ladder logic diagrams takes practice, so here is another example. Figure 13.16 shows an example of a reciprocating-motion system. The workpiece starts on the left and moves to the right when the start button is pressed. When it reaches the rightmost limit, the drive motor reverses and brings the workpiece back to the leftmost position again, and the process repeats. This relatively uncomplicated process can be represented by the ladder logic diagram of Figure 13.17 and timing diagram of Figure 13.18.

The operation of the ladder logic is as follows (verification of the logic by means of the check sheet is left to the reader): While the process is stopped,

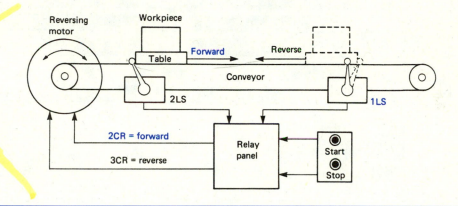

FIGURE 13.16 Reciprocating (Oscillatory) Motion Process

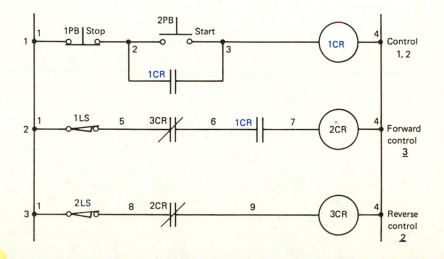

FIGURE 13.17 Ladder Logic Diagram for Oscillatory-Motion Process

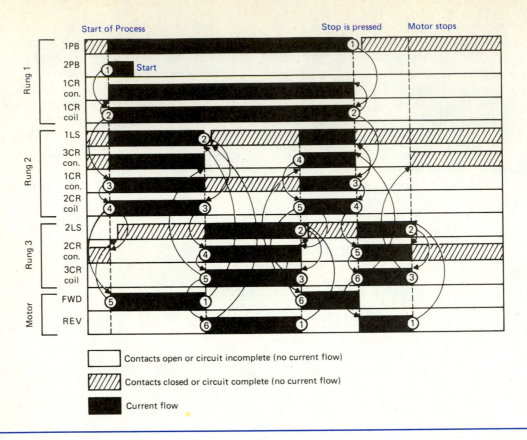

FIGURE 13.18 Timing Diagram for Oscillatory-Motion Process

limit switch 2LS is kept open by the table holding the workpiece. Limit switch 1LS is an NC switch and is in its normal position (closed). After the start button is pressed, the control relay 1CR is actuated, closing the two 1CR contacts in rungs 1 and 2. The 1CR contacts in rung 1 latch the 1CR relay on. In rung 2, the 1LS switch is closed since the workpiece is not all the way to the right, the 3CR contacts are closed since the 3CR control relay in rung 3 is not energized, and the 1CR contacts in rung 2 are closed. Thus, the 2CR control relay in rung 2 is energized, opening the 2CR contacts in rung 3. In addition, the 2CR control relay activates the contacts to the belt motor (not shown) that causes the motor to turn in the forward direction. When the table leaves the 2LS switch, the switch closes in rung 3. As the belt moves

the workpiece along, the rightmost position is reached, causing switch 1LS to open. When the 1LS switch opens, the power to the 2CR relay in rung 2 is broken, allowing the 2CR contacts in rung 3 to close. Closing of the 2CR contacts energizes the 3CR relay in rung 3, opens the 3CR contacts in rung 2, and activates the contacts to the belt motor (not shown) that causes the motor to turn in the reverse direction. The workpiece now travels to the left, eventually opening switch 2LS, and the process repeats. (It is left to the reader to determine that when the stop button is pressed, the workpiece will travel to the leftmost position and stop.)

In Figure 13.16, all of the switch connections have arrowheads that are pointing to the relay panel; the connections necessary to drive the motor are pointing out of the panel. The control relays are contained in the panel. Thus, in relay logic, the switches are considered inputs, and the motor control signals are considered outputs. Figure 13.19 shows the categories that are considered input and output devices. Information coming from the input devices goes into the control logic that, in turn, drives the output devices.

Conveyor Belt Example. One more example of writing a ladder logic diagram is given here. Try to write the ladder logic diagram from the information given; then check your answer. (One possible solution to the ladder logic diagram is given later in this chapter in Figure 13.26.)

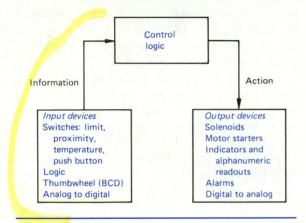

FIGURE 13.19 Input and Output Devices

This conveyor belt process example is illustrated in Figure 13.20. In this process, 2LS is set at 0.9 in. and 3LS is set at 1.1 in. to accommodate a part height between or equal to these values. If the part trips 2LS but not 3LS, the part is greater than or equal to 0.9 in. and less than or equal to 1.1 in. tall. This part is considered a "good" part. When this part trips 4LS, 5CR is actuated, dropping the part onto the good-part conveyor.

If the part trips both 2LS and 3LS, or neither, the part is too large or too small and is considered "bad." When either of these conditions occurs, the part travels down the conveyor, not falling onto the good-part conveyor when (or if) it trips 4LS but falling into the bad-part bin when it trips 5LS. Each time a part enters the bad-part bin, a counter is incremented. When the bin is full (count complete), solenoid 3 (SOL 3) is energized, opening the bottom of the bin and emptying it. The counter will then reset automatically. If a part comes on the conveyor, does not fall into either chute, and one minute elapses after 1LS is tripped, then an alarm sounds and the conveyor belt stops, indicating a malfunction.

Fail-Safe

The term *fail-safe* is used to denote the design of a system in which a failure (in a fail-safe system) would cause no or minimal damage and inconvenience. A perfect fail-safe device would never fail for any reason. Of course, a perfect device is not practical, but a low probability for failure is acceptable.

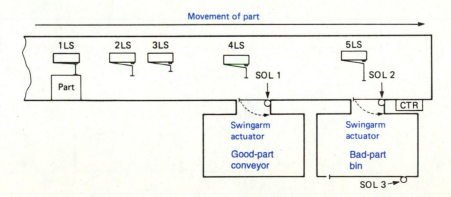

FIGURE 13.20 Conveyor Belt Process

In general, the best fail-safe circuit practice for input devices such as start and stop buttons, limit switches, float switches, and pressure switches is to adhere to the following procedure: *Start a sequence by closing NO circuit contacts, and stop a sequence by opening NC contacts.* The validity of this statement can be seen from examining the most common types of circuit failure:

1. Dirty or oxidized contacts close but fail to conduct.

2. A broken wire or loose connection opens the circuit.

3. A different power source used in the input device is interrupted, causing the input device to go dead.

If one or more of these common circuit failures occur, either the circuitry wired with the fail-safe procedure would fail to start or, if started, the system would stop. These two failure consequences usually are safe, but each case must be analyzed to see what the consequences are. Use the type of contacts on the input device that minimize danger.

A less likely but possible mode of failure could occur if the contacts were to weld shut or short in some other way. If NO contacts on the input device were selected to protect the system against dirty contacts or an open-circuit condition, then a failure could occur if the contacts welded closed. One way to overcome this difficulty would be to use two contacts in series in a redundant circuit, such that both contacts must work correctly to allow the sequence to operate. If the probability of one contact welding shut is 1 in 2 million, then the redundant circuit results in a probability of failure due to contacts welding of 1 in 4×10^{12}. The following examples will help clarify these concepts.

In a certain process, a chemical starts to fill a tank when contact 3CR closes, as shown in Figure 13.21. The tank fills until float switch 1FS closes, which energizes 5CR, opening NC contacts 5CR and closing the chemical valve. The system stays off until the reset button 2PB is pressed.

The problem with this circuit is that, although simple, it is not safe. If the contacts in the float switch failed or a wire broke, the system could not shut off and the chemical would overflow. For correction of the fault, the float switch could use NC contacts instead of NO contacts. An improved version of the circuit is shown in Figure 13.22.

An even more efficient circuit is illustrated in Figure 13.23. Here, the reset button is used to turn the control relay 5CR on, but 3CR is still used to turn on the solenoid. This circuit will protect against a loose or broken wire, bad contacts, and loss of power from the float switch; but it would not be safe if the float switch contacts were to become welded shut. In that case, a redundant circuit with two NC contacts on the float switch, as shown in Figure 13.24, would reduce the probability of failure of the system. *Note*: The line between 2PB and 1FS in Figures 13.23 and 13.24 connecting rungs 1 and 2 is called a *crossover*.

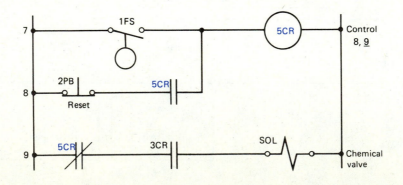

FIGURE 13.21 Partial Ladder Diagram for Chemical Tank-Filling System

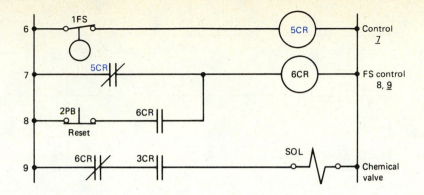

FIGURE 13.22 Increased Fail-Safe Capabilities for Chemical Tank-Filling System

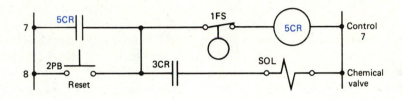

FIGURE 13.23 More Efficient Fail-Safe System

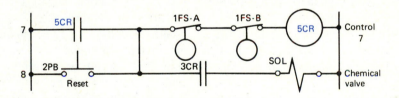

FIGURE 13.24 Redundant Fail-Safe System

The problem in Figure 13.21 could have been solved by using contacts of opposite type—NO and NC in a *parity* circuit, as shown in Figure 13.25. The redundancy solution in Figure 13.24 could have been defeated if both contacts had shorted simultaneously. The parity solution of Figure 13.25 could be defeated if the NC contacts were shorted and the NO contacts failed to close. The better solution depends upon which situation is less likely to occur.

Figure 13.26 is a possible solution to the conveyor belt problem of Figure 13.20.

RELAY LOGIC AND LOGIC GATES

At this point, you may have observed that relay logic and semiconductor digital logic are functionally equivalent. This observation is an important one since the relay must be understood as a logic element in a programmable controller. (Programmable controllers will be discussed in Chapter 14.) This point about relay logic is demonstrated in Figure 13.27. In Figure 13.27A, current will flow through the relay coil (C) if and only if switches A and B are closed.

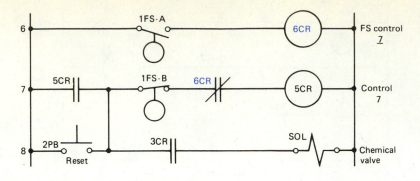

FIGURE 13.25   Fail-Safe System with Parity

The switches are electrically in series. Current can flow through the relay only if there is a complete path for current through the series circuit. This situation corresponds to the AND digital logic function. Both *A* and *B* are required for *C*. The corresponding semiconductor digital logic gate is shown in Figure 13.27B, and its truth table is given in Figure 13.27C. As indicated in the truth table, highs at both *A* and *B* will give a high out at *C*. Any other combination of inputs results in a low out. Both IEEE (Institute of Electrical and Electronics Engineers) and NEMA (National Electrical Manufacturers Association) have attempted to standardize logic symbols. The new IEEE symbol (IEEE standard 91-1984) for the AND function is shown in Figure 13.27D; the NEMA symbol is given in Figure 13.27E. Since some of these new logic function symbols now appear in industrial diagrams, you should be familiar with them.

The OR function, in relay logic form, is illustrated in Figure 13.28A. The relay coil (*C*) will energize if either *A* or *B* closes. Note that the switches are electrically in parallel. Closing one of the switches allows current to flow, completing the circuit. The equivalent semiconductor gate is shown in Figure 13.28B. As the truth table in Figure 13.28C shows, a high at either *A* or *B* will give a high output at *C*. The output is low only when both inputs are low. The new IEEE symbol for the OR function is shown in Figure 13.28D; the NEMA symbol is given in Figure 13.28E.

The final logic expression needed to complete the three basic logic classifications is the NOT circuit. The relay logic NOT circuit appears in Figure 13.29A. Recall that the NOT operation is a complement operation in Boolean algebra. That is, the output of this circuit will be the complement of the input. Note that the *A* coil contacts are normally closed. Thus, the *A* coil (not shown) is NOT energized when the *B* coil is energized. Conversely, when the *A* coil is energized, the *A* contacts will open and the current will not flow through *B*. The corresponding NOT semiconductor function is shown in Figure 13.29B, and the truth table is given in Figure 13.29C. The semiconductor NOT gate is also called an *inverter*. The new IEEE symbol for the NOT function is shown in Figure 13.29D; the NEMA symbol is shown in Figure 13.29E.

Most of the logic functions can be made with these three functions. For example, the NOT function can be combined with the AND function to give the NAND function. The relay logic equivalent of the NAND function is presented in Figure 13.30A. The top part of the relay circuit you will recognize as the AND circuit. The bottom part represents the NOT. The combination of these two relay circuits is the NAND. When both switches *A* and *B* are closed, current flows through *C*. Energizing coil *C* opens the *C* contact, removing current from *D*, which then de-energizes. The semiconductor version of the NAND circuit is shown in Figure 13.30B and the accompanying truth table is given in Figure 13.30C.

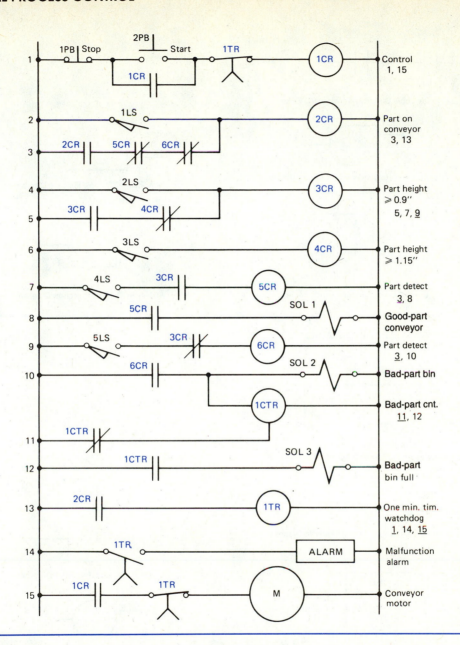

FIGURE 13.26 One Solution to Conveyor Belt Problem

Note the circle at the output of the NAND symbol. This circle at the output (or input) indicates a complementary function, or inversion. The new IEEE symbol for the NAND function is shown in Figure 13.30D; the NEMA symbol is presented in Figure 13.30E.

The NOR function shown in Figure 13.31A is the combination of the OR and NOT digital functions. As in the NAND function, the relay logic equivalent of the NOR function is the combination of the OR and NOT functions. The semiconductor logic function of this relay logic circuit is shown in

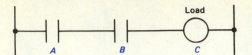

A. Relay Logic Circuit Diagram AND Function

B. Old IEEE AND Logic Gate Schematic Diagram

D. New IEEE AND Logic Function Diagram

A	B	C
1	1	1
1	0	0
0	1	0
0	0	0

C. AND Logic Function Truth Table

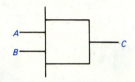

E. NEMA AND Logic Function Diagram

FIGURE 13.27   Various AND Logic Symbols

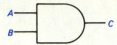

A. Relay Logic Circuit Diagram OR Function

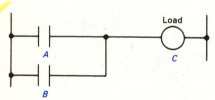

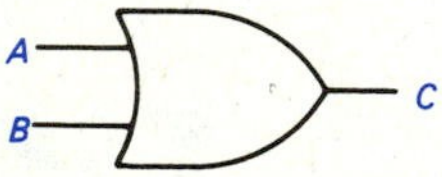

B. Old IEEE OR Logic Gate Schematic Diagram

A	B	C
1	1	1
1	0	1
0	1	1
0	0	0

C. OR Logic Function Truth Table

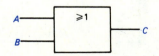

D. New IEEE OR Logic Function Diagram

E. NEMA OR Logic Function Diagram

FIGURE 13.28   Various OR Logic Symbols

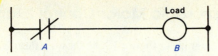

A. Relay Logic Circuit Diagram NOT Function

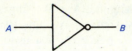

B. Old IEEE NOT Logic Function Schematic Diagram (Inverter)

C. NOT Logic Function Truth Table

D. New IEEE NOT Logic Function Diagram

E. NEMA NOT Logic Function Diagram

FIGURE 13.29 Various NOT Logic Symbols

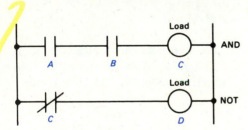

A. Relay Logic Circuit Diagram NAND Function

B. Old IEEE NAND Logic Gate Schematic Diagram

A	B	D
1	1	0
1	0	1
0	1	1
0	0	1

C. NAND Logic Function Truth Table

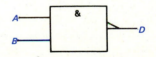

D. New IEEE NAND Logic Function Diagram

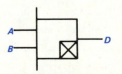

E. NEMA NAND Logic Function Diagram

FIGURE 13.30 Various NAND Logic Symbols

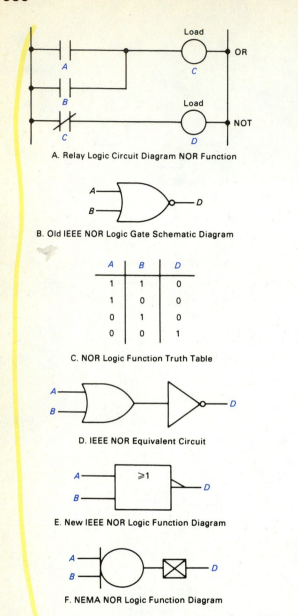

A. Relay Logic Circuit Diagram NOR Function

B. Old IEEE NOR Logic Gate Schematic Diagram

A	B	D
1	1	0
1	0	0
0	1	0
0	0	1

C. NOR Logic Function Truth Table

D. IEEE NOR Equivalent Circuit

E. New IEEE NOR Logic Function Diagram

F. NEMA NOR Logic Function Diagram

FIGURE 13.31 Various NOR Logic Symbols

Figure 13.31B, and the accompanying truth table is given in Figure 13.31C. Note that this circuit is equivalent to the diagram shown in Figure 13.31D. The new IEEE symbol for the NOR function is shown in Figure 13.31E; the NEMA symbol is given in Figure 13.31F.

Relay Logic Motor Control Example

The diagram in Figure 13.32 shows a familiar motor control circuit. Pushing the start momentary contact button allows current to flow through the 1M relay coil. Energizing the 1M coil changes the state of the three NO 1M contacts. AC power is then applied to the three-phase motor. Current also flows through the 1M1 contacts when the relay picks up. When the start button is released, current continues to flow through its own contacts and through the 1M coil. You will recognize this configuration as a latching relay. The relay coil will de-energize under three conditions:

1. If AC power is lost, all power will be removed from the circuit.
2. If the overload relay (OL) trips, the OL contact will open.
3. If the stop button is pushed, current will not flow in the 1M coil.

We can duplicate this logical condition with semiconductor digital logic circuits, as shown in Figure 13.33. Device A is an OR gate; devices B and

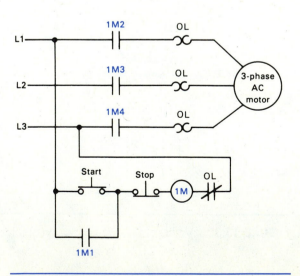

FIGURE 13.32 Three-Phase AC Motor Starting Circuit Using Relay Logic Symbols

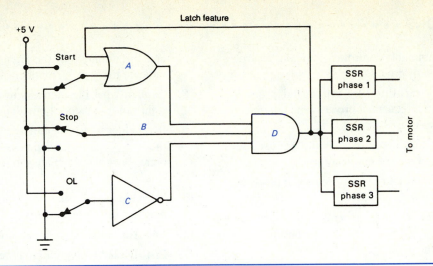

FIGURE 13.33 TTL Digital Logic Motor Starter

C are inverters. Device D is a three-input AND gate. If the start switch is closed, a high is applied to the lower OR gate input. Since both inverters have a low input, three highs are present at the input to the AND gate. Three highs at the input to the AND gate will cause the output of gate D to go high, turning on the three SSRs connected to the output. The three SSRs apply three-phase power to the motor as well as providing isolation between the power circuit and the control circuit. Changing the state of either the stop or OL switches will cause the output of the AND gate D to go low, removing power from the motor. Note also that a high is fed back from the output of the AND gate to the input of the OR gate. If the start switch changes state, one high will still be present at the input to the OR gate from the AND output. The output of the AND gate is then kept high by its own output.

This example shows how semiconductor digital logic can perform the same function as relay logic. Semiconductor digital logic has the advantages of being more reliable and cheaper to build and operate, and it takes up less space than equivalent relay logic circuitry. These advantages come from the basic structural differences between electromagnetic relays and semiconductor digital logic devices.

Both digital and relay logic systems have the same disadvantage: They are both difficult to change. When the task they must perform changes, the circuit structure must be changed. Modification means rewiring and troubleshooting the new circuit, which can be a time-consuming and frustrating task, especially with a complex circuit. The programmable controller, discussed in the next chapter, can do the same job as either the relay logic circuit or the semiconductor digital logic circuit. It has the added advantage of being able to easily change the tasks it performs in software.

CONCLUSION

The basis of the programmable controller's programming language has been, by popular demand, the ladder logic diagram with standard relay symbology. The reason for the ladder diagram's popularity is that plant personnel are very familiar with relay logic from their previous experience with sequential controls. Understanding how to generate a ladder logic diagram from a sequential process, the subject of this chapter, is the first step in being able to program most programmable controllers, the subject of the next chapter.

■ QUESTIONS •

1. A programmable controller generally controls what type of process?

2. Define the following: sequential process, event-based sequential process, and time-based sequential process.

3. Batch control systems generally are used for what four types of applications?

4. List seven differences between batch processes and continuous processes.

5. Push-button switches generally are actuated by the _____, whereas limit switches are actuated by the _____.

6. List at least six general types of proximity switches.

7. In a ladder logic diagram, the numbers to the left of the left rail give the _____ numbers, and the numbers to the right of the left rail indicate the _____ numbers.

8. Explain what the numbers to the right of the right rail in a ladder logic diagram indicate.

9. In a ladder logic diagram, loads are shown on the _____ side, and devices that make or break electric contact are shown on the _____ side.

10. What are the four steps in writing a ladder logic diagram for a process?

11. List five ladder logic input devices.

12. List five ladder logic output devices.

13. In the design of a fail-safe circuit, a sequence is started by activating _____ contacts and stopped by activating _____ contacts.

14. Define the term *fail-safe* in your own words.

■ PROBLEMS •

1. Draw a hierarchical flowchart using the following terms: (a) time-based sequential process, (b) sequential process, (c) event-based sequential process, (d) programmable sequence controller, (e) programmable controller, (f) batch process, and (g) programmable process controller. Start with sequential process at the top of the chart. Show how each process is broken down from that point and which controller is used with which process.

2. A life science laboratory technician wishes to automate the feeding of laboratory animals somewhat by automatically measuring the feed going onto a scale. After the first batch of feed is weighed automatically, the technician will remove the feed from the scale and distribute it while the next batch of feed is being automatically weighed. Write an accurate ladder logic diagram for the process described, taking into consideration the following constraints:

• The process must have a start and a stop button that can start or stop the process at any point and resume the process where it left off.

• When the process is started, food is augered onto the scale at regular speed. As the scale gets close to balance, the auger speed changes to half speed (or trickle). When the scale is almost balanced, the auger stops and turns on only for one second in every ten or so seconds until the scale is balanced. (This step is called *jogging* the motor.)

• If no jogging occurs for one minute, a light comes on to indicate the scale is in balance.

• The technician locks the scale, removes the feed, and gently releases the scale to start the next batch.

3. Draw the relay logic symbols for the following digital logic functions: (a) AND, (b) OR, (c) NAND, and (d) NOR.

4. A part is placed on a conveyor that is not moving. After the part is placed on the conveyor, the conveyor starts and the part moves. In the middle of the conveyor, the part moves through a 3 ft area where it is painted. The part does not stop during the time it is

being painted. When the part reaches the end of the conveyor, the conveyor stops and the part is removed. The process then starts all over again. Draw the relay logic necessary to accomplish this task. Use limit switches to detect the presence or absence of the part.

5. Use the same information as in Problem 4, but have the part stop and be stamped before it is painted.

6. Convert the semiconductor logic circuits in Figures 13.34A and 13.34B to the equivalent relay logic circuits.

7. Convert the relay logic circuits in Figures 13.35A and 13.35B to the equivalent semiconductor logic circuits.

8. A fan must be turned on and off from three different locations. Draw the relay logic circuit that will accomplish this task.

9. A milling machine (M) has a lubrication pump (P) that must be turned on before the milling machine can be turned on. If the pump stops for any reason, the milling machine must be shut off. Draw the relay logic circuit that will accomplish this task.

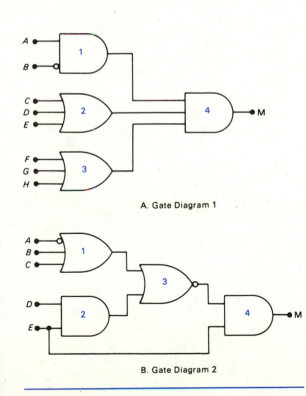

A. Gate Diagram 1

B. Gate Diagram 2

FIGURE 13.34 TTL Gate Diagrams for Problem 6

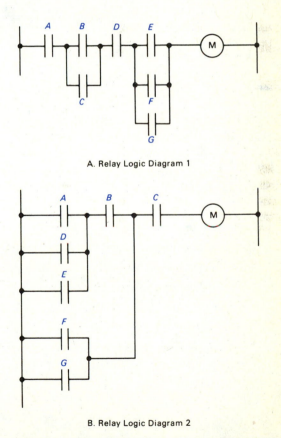

A. Relay Logic Diagram 1

B. Relay Logic Diagram 2

FIGURE 13.35 Relay Logic Diagrams for Problem 7

Programmable Controllers

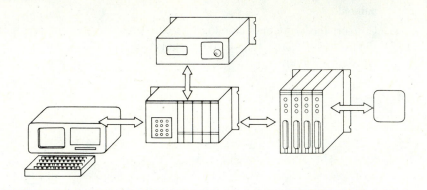

OBJECTIVES

On completion of this chapter, you should be able to:

- State the historical definition for the programmable controller (PC).
- State the parts and functional operation of a PC.
- Determine the important factors to consider when selecting a PC for an application.
- Identify the approximate size of a PC.
- Identify the symbols for a PC ladder diagram.
- Write a PC ladder diagram.

INTRODUCTION

In the 1960s, computers were considered by many in industry as the ultimate way to achieve increased efficiency, reliability, productivity, and automation of industrial processes. Computers possessed the ability to acquire and analyze data at extremely high speeds, make decisions, and then disseminate information back to the control process. However, there were disadvantages associated with computer control, such as high cost, complexity of programs, hesitancy on the part of industry personnel to rely on a machine, and lack of personnel trained in computer technology. Thus, computer applications throughout the 1960s were mostly in the area of data collection, on-line monitoring, and open-loop advisory control.

During the mid-1960s, however, a new concept in electronic controllers evolved: the *programmable controllers*, PCs. (Note that the abbreviation PC will be used in this text to denote the programmable controller and not the personal computer, which also is popularly represented by the abbreviation.) The concept of the PC developed from a mix of solid-state computer technology and traditional sequential controllers, such as the stepping drum (a mechanical, rotating switching device) and the solid-state programmer with plug-in modules. The PC first came about as a result of problems faced by the auto industry, which had to scrap costly assembly line controls each time a new model went into production. The first PCs were installed in 1969 as electronic replacements of electromechanical relay controls. The PC presented the best compromise of existing relay ladder logic techniques and expanding solid-state technology. It increased the efficiency of the auto industry's manufacturing system by eliminating the costly job of rewiring relay controls used in the assembly line process. The PC reduced the changeover downtime, increased flexibility, and considerably reduced the space requirements formerly used by the relay controls.

Since the introduction of the PC into the *manufacturing* industry in 1969, the PC has become widely used in the *process* industry as well. A process industry generally performs the functions and operations necessary to change a material either physically or chemically. In this chapter, we first will discuss the basic concepts of a PC.

DEFINING A PC

In 1978, the National Electrical Manufacturers Association (NEMA) released a standard for PCs. This standard, NEMA Standard ICS3–1978, was the result of four years of work by a committee made up of representatives from PC manufacturers. Part ICS3–304 of the standard defines a PC as "a digitally operating electronic apparatus which uses a programmable memory for the internal storage of instructions for implementing specific functions such as logic, sequencing, timing, counting, and arithmetic to control, through digital or analog input/output modules, various types of machines or processes. A digital computer which is used to perform the functions of a programmable controller is considered to be within this scope. Excluded are drum and similar mechanical type sequencing controllers."

The official NEMA definition allows almost any computer-based controller to be considered a PC, including single-board computers, numerical controllers, and programmable process controllers or sequential controllers. Although PCs can be used to perform the jobs of numerical and sequential controllers, numerical and sequential controllers typically cannot perform the event-based control functions discussed in the previous chapter. Because of this official definition confusion, we will use the historical, or *de facto*, definition for the PC that is developed in the following text.

Regardless of size, cost, or complexity, all PCs share the same basic parts and functional characteristics, as shown in Figure 14.1. A PC will always consist of input and output interfaces, memory, a processor, a programming language and device, a power supply, and housings.

Functionally, a PC examines the status of input interfaces and, in response, controls something through output interfaces. Several logic combinations or instructions usually are needed to carry out a control plan, or *program*, as it is commonly called. This control plan is stored in memory by using a programming device. All logic combinations stored in memory are periodically evaluated by the processor in a predetermined order. The period of time required to evaluate the status of input devices, output devices, and the control plan is called a *scan*. The input and output devices, such as switches, motors,

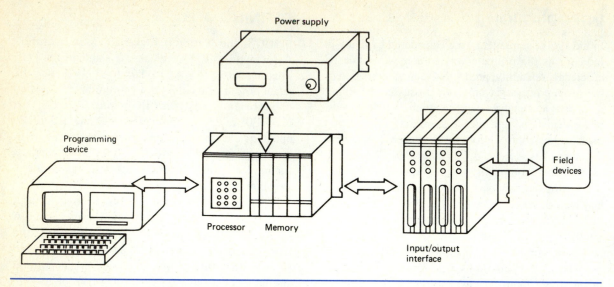

FIGURE 14.1 Basic Programmable Controller

lights, and so on, that attach to the input/output (I/O) interfaces are called *field devices*. During a scan, all the inputs are examined, the control plan is evaluated, and outputs are updated.

PCs employ combinational logic and frequently are programmed by using (but not restricted to) relay ladder logic. Conventional PC inputs and outputs are made at industrial line voltages—for example, 120 V AC, 240 V AC, 24 V DC, and 5 V DC. PCs are designed to operate in industrial environments that are dirty, are electrically noisy, have a wide fluctuation in temperatures (0°–60°C), and have relative humidities of from 0% to 95% (noncondensing). Air conditioning, which generally is required for computers, is not required for PCs. PCs can be maintained by the plant electrician or technician with minimal training. Most of the maintenance is done by replacing modules rather than components. Often, the PC has a diagnostic program that assists the technician in locating bad modules.

Mainframes and minicomputers, on the other hand, require highly trained electronics specialists to maintain and troubleshoot them. Note that computers are generally classified into four categories of increasing size and power: microcomputers, minicomputers, mainframes, and supercomputers.

However, the boundary between one type of computer and the next is no longer well defined. Generally, though, a microcomputer is a computer that is implemented by using a microprocessor chip (or chips) as the main control unit. A programmable controller is an example of a microcomputer. Supercomputers, because of their extreme complexity, high cost, and limited (for now) application, are not used in the manufacturing or process industries.

Microcomputers (personal computers) are increasingly being used in industry. General-purpose microcomputers are being designed to withstand the industrial environment of dirt, noise, temperature, and humidity in order to be used as stand-alone controllers or in conjunction with other devices such as programmable controllers. A listing of environmentally hardened (protected) personal computers is given in Appendix E. Microcomputers are relatively easy to troubleshoot and maintain, and they have added greatly to the flexibility and ease of use of the programmable controller. A subscription to a personal computer journal, such as *Personal Engineering & Instrumentation News*, Box 300, Brookline, MA 02146, may help you to stay current in the rapidly changing area of industrial personal computers.

In summary, the historical definition for the PC basically describes it as a special-purpose computer whose special features set it apart from other controllers. A PC consists of I/O interfaces, memory, a processor, a programming device, and a power supply. The PC inputs and outputs operate at one of the industrial line voltages of 240 V AC, 120 V AC, 24 V DC, or 5 V DC. The PC also is designed to operate in the harsh industrial workplace and be maintained by plant personnel. Most PCs have additional capabilities, but these are the qualities that set it apart from other controllers.

Appendix F provides a listing of most of the PCs that are available today, together with some of their more important specifications. Of these, manufacturers caution that scan rate data is, at most, an estimate of typical performance. For any given program, the scan rate will vary depending on the types of actions called for by the program.

We will discuss each basic part of a PC in the following sections.

INPUT/OUTPUT INTERFACES

Input interfaces, which are modular because they can be plugged into and unplugged from the system, accept signals from the machine or process devices (115 V AC) and convert them into signals (5 V DC) that can be used by the controller. *Output interfaces* or modules convert controller signals (5 V DC) into external signals (115 V AC) used to control the machine or process. Other voltage arrangements may be used, but the ones just listed are the most common. There are usually 1, 2, 4, 8, or 16 circuits on an I/O interface.

Typically, input and output interfaces are housed in a rigid card rack that can be mounted in an enclosure, either next to or thousands of feet away from the processor. In addition, there is some means of connecting field wiring on the I/O interface or I/O housing. Connecting the field wiring to the I/O housing allows easier disconnection and reconnection of the wiring in order to change modules. Lights also are added to interface modules to indicate the on or off status of each I/O circuit. Each circuit usually is isolated and fused. Some I/O interface modules have blown-fuse indicators. The following items are common I/O interface modules:

- AC voltage inputs and outputs
- DC voltage inputs and outputs
- Pulse inputs
- BCD (binary-coded decimal) inputs and outputs
- Low-level analog inputs (such as thermocouples)
- High-level analog inputs and outputs (in the 4–20 mA range)
- Special-purpose interfaces

The first four types of modules listed—AC, DC, pulse, and BCD—are often referred to as *discrete modules*. These modules come in many different varieties and combinations, but they are typically used to interface to on-off devices, such as limit switches, push buttons, motor starters, pilot lights, and annunciators having different voltage ranges.

The low-level input and high-level input/output modules perform signal conditioning, if necessary, and A/D and D/A conversions to directly interface analog signals to the programmable controller. A/D and D/A conversion was discussed in detail in Chapter 2.

Special-purpose interface modules cover a very broad range of applications, some of which are listed next.

1. The proportional/integral/derivative (PID) control module performs closed-loop PID control. It monitors the process variable, compares it with the desired set point, and calculates the required analog output based on its internal control algorithm. Generally, the PID module can operate independently. Or the module and processor can perform adaptive control, in which the processor continually adjusts the module's control algorithm on the basis of process changes monitored by the processor. PID control was discussed in detail in Chapter 10.

2. Motion control modules are used with different motion control devices such as servomotor positioning, stepper motor positioning, absolute-position encoding, and encoder counting. Stepper motors were discussed in Chapter 4.

These two types of modules, PID and motion control, are relatively common to most programmable controller manufacturers, as evidenced by the listing in Appendix F. The following special-purpose interface modules are not as common.

1. A fiber optics converter module converts electric signals to light signals and transmits these signals through a fiber optics cable. At the other end of the cable, a second fiber optics module converts light signals back to electric signals. Communication through fiber optics can be very useful in industry where electromagnetic or radio frequency noise could disrupt low-level electrical communication signals.

2. An ASCII I/O interface module provides an interface between a programmable controller and a peripheral device that generates and/or receives ASCII characters, such as a personal computer, a bar code reader, or an ASCII display terminal. Applications such as report writing, video display of machine function, menu selection, and soft-function key selection are not uncommon.

3. A new module in use now is the vision input module. The vision module provides precise visual data input for improved automatic control and verification. Applications include dimensional gauging of extruded metals, hole verification and inspection, limited machine/robot guidance, and machine tool and workpiece alignment and verification.

MEMORY

Memory is the location where the control plan or program is held or stored in the PC. The information stored in the memory relates to how the input and output data should be processed.

The complexity of the control plan determines the amount of memory required. Most PC memories are expandable in fixed increments. Memory elements store individual pieces of information called *bits* (for *bi*nary dig*its*). These memory modules are mounted on *printed-circuit* boards (another use for the abbreviation PC). Memory is specified in thousand or "K" increments, where 1K is 1024 bytes of memory storage (a *byte* is 8 bits). PC memory capacity may vary from less than 100 bits to over 256K bytes, depending on the manufacturer.

Memory is categorized as either *read-only memory* (ROM) or *random-access memory* (RAM). Both terms are somewhat misleading. ROM has to be written into at some time in order to enter the program, so it should probably be termed "read-mostly" memory. And both RAM and ROM can be accessed randomly, but RAM is the historical term that indicates memory that can be altered easily. Some ROM may be altered physically (by ultraviolet light) or electronically to change its contents. The most popular ROM used in PCs today is UV PROM (ultraviolet programmable read-only memory). The UV PROM ICs have a clear quartz window in the middle of the device for erasing the memory contents. They then may be reprogrammed.

EEPROM, or E^{2}PROM, is electrically erasable PROM. This type of memory uses an electric signal to erase the memory, and it can erase one memory location at a time. This memory has an advantage over UV PROM, which must erase all locations and then must be entirely reprogrammed. Some problems with EEPROM became evident recently, and it has not proven to be cost-effective. Intel Corporation has developed what it calls "flash" memory that can be electrically erased but, again, will erase the entire chip. The memory used in a PC is listed in the "Type of memory" column in Appendix F.

In the past, the most common RAM used in PCs was a magnetic core. The main advantages of core memory are that it can be easily altered, it is convenient for changing set points that must be changed often, and yet it is nonvolatile (does not lose its memory when power is lost). Today, most PCs use MOS IC memory that is more compact and less expensive. However, since this memory loses its stored data when power is removed, a backup battery supply is required to retain the memory contents.

PROCESSOR

The *processor*, sometimes called the *central processing unit* (CPU), is the heart of the PC and organizes all controller activity. The CPU causes the control plan logic stored in memory to be evaluated, along with the status of the inputs, and issues a specified command to the appropriate output.

A number or a combination of letters and numbers, such as 110/15, is used to code data locations referred to as *addresses* (locations in memory). The user selects the I/O address by assigning a specific input or output interface circuit to a specific field device. All programming of the field device is referenced to the number assigned.

In addition to doing straight logic processing, the processor may perform other functions such as timing, counting, latching, comparing, and retentive storage. It also can emulate stepping switches and shift registers. These additional processor functions may be either special hardware units that are part of the PC or software programs integrated into the memory.

Most PC manufacturers use a microprocessor as the CPU, thus decreasing the PC's size and increasing its decision-making capabilities. Some controllers have the ability to perform complicated math well beyond the basic four functions of add, subtract, multiply, and divide.

PROGRAMMING LANGUAGE AND PROGRAMMING DEVICE

The *programming language* allows the user to communicate with the PC via a programming device. PC manufacturers use slightly different programming languages, but all languages are designed to convey to the PC, by means of instructions, how to carry out the control plan.

Figure 14.2 illustrates some of the more common PC programming languages available. The PC ladder diagram (Figure 14.2B), based on the relay ladder diagram (Figure 14.2A), is by far the most common. Boolean statements (Figure 14.2C) relate

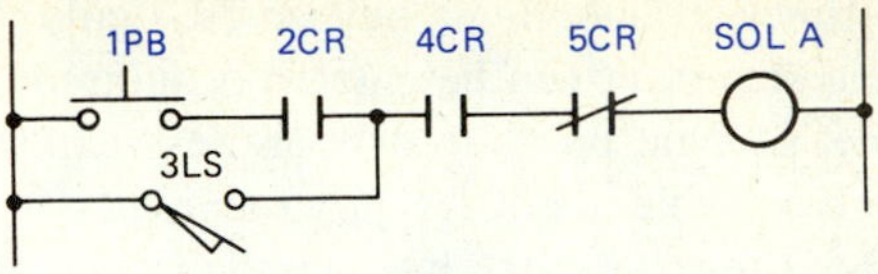

A. Relay Ladder Diagram

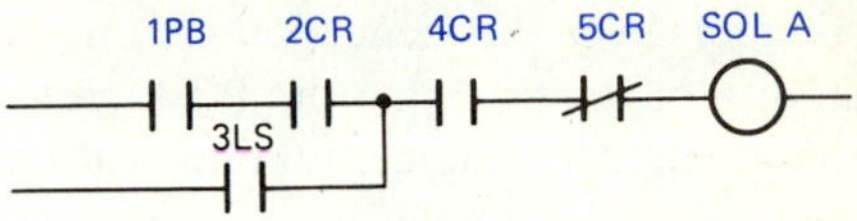

B. Free-Format-Equivalent PC Diagram

$$[(\,1PB \cdot 2CR) + 3LS] \cdot 4CR \cdot \overline{5CR} = SOL\ A$$

C. Boolean Statement

LOAD	1PB
AND	2CR
OR	3LS
AND	4CR
CAND	5CR
STORE	SOLA

D. Code or Mnemonic Language

FIGURE 14.2 Comparison of Programming Languages Used with Various PCs

logical inputs such as AND, OR, and INVERT to a single statement output—in this case, solenoid A (SOL A). Another type of PC language is the code or mnemonic (pronounced nee-MON-ic) language, shown in Figure 14.2D. This language is very similar to computer assembly language.

The *programming device* is used to load the program into the PC memory. The most common programming device is the CRT (cathode-ray tube) terminal, which provides a full alphanumeric and/or special-function keyboard. The CRT allows the user to see the ladder diagram or coded language as interpreted by the PC.

At the other end of the programming device spectrum is the small manual programmer. (All programming is manual, but this type of programming device has been referred to traditionally as a manual programmer.) In this case, data and instructions are entered through thumbwheels and special-function buttons or keypads. Usually, only one logic statement can be entered or monitored at one time. Manual programmers are less expensive and more portable than CRT programming devices.

After program entry has been completed, the programming device continues to be used as a diagnostic tool. Even the manual programmers have the capability of interrogating the PC to determine I/O, memory, and CPU status. With the CRT terminal, "living pictures" of the PC's operation make troubleshooting easier and faster.

POWER SUPPLY

The PC power supply may be integrated with the CPU, memory, and I/Os into a single housing, or it may be a separate unit connected to the main housing through a cable. As a system expands to include more I/O modules or special-function modules, most PCs require an auxiliary power supply to meet the power demand. Power supplies are also the first line of defense against electrical noise generated over the power lines.

HOUSINGS

One of the most popular PC features is its modularity. Modularity makes repairs easier and reduces downtime. Most major PC components are mounted on printed-circuit boards that can be inserted into a housing or card rack. One or more housings make up a PC system.

Housings may contain the CPU, memory, I/O modules, special interface modules, and a power supply—or in some cases, just I/O modules. Housings may be rack-mounted in a control console or mounted to a subpanel in an enclosure. Most housings are designed to protect the PC control circuits from dirt, dust, electrical noise, and vibration.

PC SIZE

There is much variation in size identification of PCs, but they can be roughly divided into three sizes: small, medium, and large. The small-size category covers units with up to 128 I/Os and memories up to 2K bytes. These PCs are capable of providing versatility and sophistication ranging from simple to advanced levels of machine control.

Medium-sized PCs generally have from 256 to 512 I/O modules and memories ranging from 4K to 7K bytes. Intelligent I/O cards make medium-sized PCs adaptable to temperature, pressure, flow, weight, position, or any type of analog function commonly encountered in process control applications.

Large PCs, of course, are the most sophisticated units of the PC family. In general, large units have 1024 to 4096 I/Os and memories from 8K to 192K bytes. The large PC has virtually unlimited applications. Large PCs can control individual production processes or entire plants. Modular design permits systems to be expanded to control thousands of analog and digital points (input or output connections).

SELECTING A PC

The key factor in selecting a PC is establishing exactly what the unit is supposed to do. Current designs cover a broad range of sizes and capabilities. At the small end, PCs are primarily used as relay replacements to provide standard relay logic, timing, counting, and shift register functions. At the large end, analog I/O capability has made it possible for the units to become an integral part of process control systems.

Probably the most important step in selecting the proper PC system is determining what the I/O requirements are, including types, location, and quantity. If the application involves the replacement of relays, the user can determine I/O needs quickly. Establishing analog I/O needs is much more complicated and may require expert assistance. Other requirements to be evaluated include memory type and capacity, programming procedures, and peripheral equipment needs. Normally, a 10%–20% spare or expansion capability should be allowed for each application.

Expense of the PC is another factor to consider. Determining the cost of a PC system is not easy, however. Many intangibles must be taken into consideration. System requirements dictate costs to a certain degree, but the true cost also depends on the value of increased production, improved quality, greater flexibility, and reduced downtime. Installation, operation, and maintenance costs are important economic factors to consider. Servicing the PC can be expensive. However, if in-house repair capability is available, servicing costs may be reduced.

In general, buying a PC system that is larger than current needs dictate is not advisable. All phases of the project must be considered, however, and future conditions must be accurately anticipated to ensure that the system is the proper size to fill the current, and possibly future, requirements of an application.

EXAMPLE OF A SMALL SYSTEM

Our main purpose in this chapter is to acquaint you with real programmable controllers and how they can be used. To this end, some examples have been chosen that are among the most popular systems in the industry. The Allen-Bradley SLC® 100 has been chosen as a representative small PC system that has most of the capabilities desired and is one of the easier ones to learn to use. Industrial-quality programmable controllers tend to be expensive, but the SLC® 100 is affordable and is an excellent choice

for educational purposes. Its industrial quality and ruggedness are appropriate for the classroom/laboratory environment.

The following information has been kindly provided by Allen-Bradley and its distributors. (Local distributors can be found by contacting the Allen-Bradley Company. The two we contacted were Kiefer Electric Supply Co. in Benton, Illinois, and French Gerleman Electric Co. in St. Louis, Missouri.)

The SLC® 100 is shown in Figure 14.3. This unit is basically self-contained and has all of the parts of a PC built into a small (95 mm × 124 mm × 250 mm) package. As we discussed earlier, a PC consists of I/O interfaces, memory, a processor, a programming device, and a power supply. We will now examine each of these features for the SLC® 100.

I/O Interfaces

As shown in Figure 14.3, the SLC® 100 PC has 10 input connections and 6 output connections. The wiring terminals have hinged covers that attach and prevent accidental contact: They also have write-on areas to identify external circuits. In addition, up to

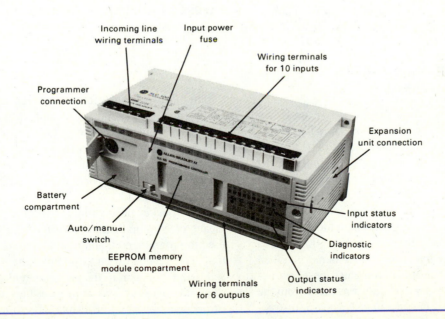

FIGURE 14.3 Features of the SLC® 100 PC

6 expansion units may be attached, which provide a total of 112 I/O (70 input and 42 output) connections. Input status indicators (red LEDs) and output status indicators on the front panel will light when an input or output circuit is energized. These indicators can be very helpful as troubleshooting aids.

Memory

The memory consists of 885 words of CMOS RAM. Battery backup is provided to retain the memory contents when processor power is removed. The lithium battery provides backup power for approximately 2–3 years. Programs can also be loaded from or stored in an EEPROM memory module.

Processor

The processor is the "brains" behind the device and integrates the I/O interface circuit memory that will process and manipulate programmed information. It causes output devices to be energized and de-energized in response to the on-off status of input devices. A CPU fault light on the front panel goes on when the processor has detected an error in either the CPU or memory. Operation of the PC is automatically stopped to prevent undesired results or damage to the field devices.

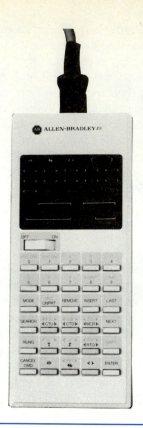

FIGURE 14.4 SLC® 100 Pocket Programmer

Programming Language and Programming Device

The programming format for this PC is the ladder diagram, which uses symbology similar to hard-wired relay ladder circuits. The relay ladder rung requires *electrical* continuity to energize the output, but the PC ladder rung requires *logical* continuity to energize the output. We will discuss this feature later in the chapter.

In the PC ladder rung, the individual symbols represent instructions; the numbers 001, 003, and so on, are the instruction addresses in memory. When you program the controller, you enter these instructions one by one into the processor memory from the pocket programmer (Figure 14.4) keyboard. Instructions are stored in the user program portion of the processor memory.

Some of the instructions entered are used to represent the external input and output devices connected to the processor unit. Other instructions are "internal"; that is, they are used to establish the exact conditions under which the processor will energize or de-energize output devices in response to the status of input devices.

The pocket programmer of Figure 14.4 is the means of communicating with the controller. It is used to program, edit, and monitor controller operation. The keyboard is used to enter the instructions and data that make up the program, and the display shows these instructions and data as they are entered. Error codes appear on the display if instructions have been entered incorrectly or if internal processor problems occur.

Power Supply

The power line connections to the PC are shown in Figure 14.3. A green LED shows when power is applied to the processor unit. A battery-low red LED lights when the battery voltage level falls below a threshold level.

Housing

The PC housing is industrial grade with protection against noise, dirt, and other environmental effects. Notice in Figure 14.3 that there are very few exposed parts.

Allen-Bradley also provides simplified training information for this PC. Publications 1745–800 and 1745–800A are the SLC® 100 *User's Manual* and *Self-Teaching Manual.*

LARGER PCs

Larger PCs are generally modular in design. The user must mix and match (with assistance from the PC manufacturer) the modules needed to control the process.

PC LADDER INSTRUCTIONS

The instructions for the SLC® 100 are representative of and similar to the PC logic used in other PCs. We will find that there are some possibly unexpected differences between PC logic and relay ladder logic. A small group of instructions and an explanation of what they mean will be given first. Then examples and tips will be provided to show you how to use those instructions.

Basic Instructions

The three basic instructions for the Allen-Bradley PC are shown in Figures 14.5A and 14.5B. Even though these instructions are symbols, they are also programming *instructions.* As in any computer language, instructions tell the computer what to do. As you type these symbols on the CRT or programming device, you are programming the PC.

Figure 14.5A shows the two instructions that are called *conditional input instructions*, examine-ON and examine-OFF. Each of these instructions refers to an area of the PC called the *data table.* Since the PC is just a specialized computer, you can think of the data table as a section of the PC's memory. Memory in a computer can be thought of as a place to store binary information. Each individual location is called a *status bit.* Each status bit has a location called an *address.* When an input device turns on or closes, as in the limit switch LS1 in Figure 14.6A, the status bit associated with that input location goes "on" or "high." The SLC® 100 has 10 inputs (001 through 010) and 6 outputs (011 through 016). Note that LS1 (refer to Figure 14.6B) is connected to PC input 001 and CR4 is associated with PC input 003. The output coil CR2 is connected to PC output 011. For the circuit in Figure 14.6A to function, both the limit switch and the CR4 contact must close. When they close, current flows through the CR2 coil, energizing that relay coil.

External Input Device	Input Terminal	Examine - ON Instruction	Examine - OFF Instruction
	OFF	FALSE Status bit OFF	TRUE Status bit OFF
	ON	TRUE Status bit ON	FALSE Status bit ON

A. Conditional Input Instructions

Output-Energize Instruction	Output Contacts	External Output Device
Status bit ON TRUE rung conditions	Closed	M Energized
Status bit OFF FALSE rung conditions	Open	M Not energized

B. Output Instructions

FIGURE 14.5 Basic Instructions

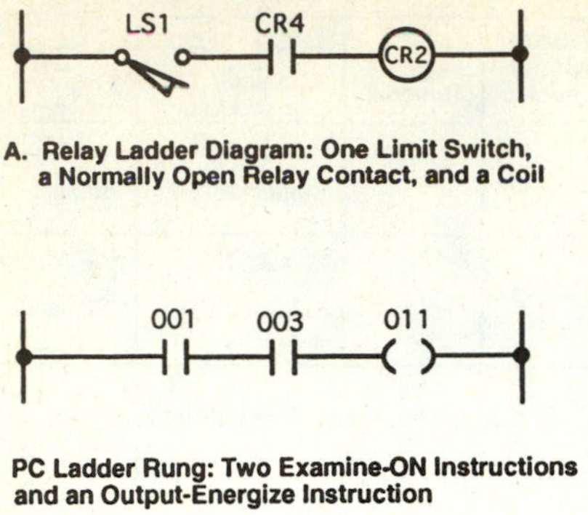

A. Relay Ladder Diagram: One Limit Switch, a Normally Open Relay Contact, and a Coil

B. PC Ladder Rung: Two Examine-ON Instructions and an Output-Energize Instruction

FIGURE 14.6 Relay Ladder Diagram with Corresponding PC Ladder Diagram

The PC does not work this way. The PC uses *logical continuity* rather than electrical continuity. The rung in Figure 14.6B must be "true" for the output 011 to be energized. For a rung to be true, all conditional input instructions must be true. In Figure 14.6B, for example, the examine-ON instructions at 001 and 003 must both be true to make the rung true. Looking at the chart in Figure 14.5A, we see that the examine-ON instruction is true when the switch attached to the input address turns on.

Let's trace the sequence of events when both LS1 and CR4 turn on. After they turn on, the PC detects this state and sets the status bits at 001 and 003 addresses on. The PC then evaluates each conditional input instruction. The examine-ON instruction asks the question, "Is the switch attached to input 001 ON?" If the answer is yes, the statement is true. If, however, the answer to that question is no, the statement is false. In this case, both input conditions are true, and the rung is therefore true. When the rung is true, the *output-energize instruction* tells the controller to "set the status bit of the output-energize instruction in this rung to ON." After this status bit is set on, the controller applies power to the device attached to output 011. When both switches close, the output attached to 011 goes on.

As you can see, the key to understanding this logic rests with a good understanding of Figure 14.5. To clarify the use of this chart, let's take an example.

• EXAMPLE 14.1

Given an SLC® 100 PC where a lamp connected to output 11 comes on when switch 7 is open and switch 5 is closed (Figure 14.7A), draw the PC ladder diagram.

Solution

To make the instruction true with switch 007 open, we must choose an examine-OFF instruction (Figure 14.5A). To make the instruction true with switch 005 closed, we must choose an examine-ON instruction (Figure 14.5A). The proper PC ladder diagram solution is shown in Figure 14.7B.

The circuit shown in Figure 14.8 is the familiar start-stop station common in industry. When the start button is pressed, current flows through the NC stop button and the coil 1CR. The coil 1CR energizes, picking up the NO 1CR contacts, which are now sealed or latched. Pressing the stop button interrupts current flow, unsealing or unlatching the circuit, de-energizing the 1CR coil. Note the PC solution to this circuit shown in Figure 14.8C. The stop switch 009 is represented by an examine-ON instruction, not an examine-OFF instruction. We must choose an

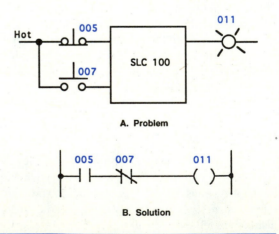

A. Problem

B. Solution

FIGURE 14.7 Example 14.1—Lamp Must Come on when Switch 5 is Closed and Switch 7 is Open

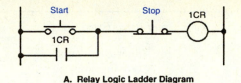

A. Relay Logic Ladder Diagram

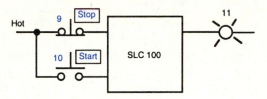

B. Wiring Diagram-PC

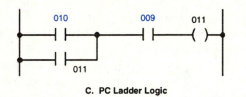

C. PC Ladder Logic

FIGURE 14.8 Start-Stop Station

instruction for the NC stop switch that will be true in its current state. In other words, the only action that will make the entire rung true is pressing the start switch. The instruction referring to the stop switch must be true already. The proper choice of instructions, according to Figure 14.5, is the examine-ON instruction.

An example may help to clarify this circuit. Suppose we want to use an NC start switch and an NO stop switch. Electrically, this would be impossible, but with the PC such a task is easy.

• EXAMPLE 14.2

Given an SLC® 100 PC where a lamp connected to output 11 comes on when an NC switch 9 is pressed and turns off when an NO switch 10 is pressed (Figure 14.9A), draw the PC ladder diagram.

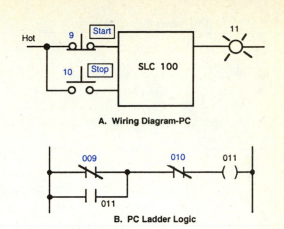

A. Wiring Diagram-PC

B. PC Ladder Logic

FIGURE 14.9 Example 14.2—Start-Stop Station

Solution

To make the instruction false with switch 009 closed and true when it is open, we must choose an examine-OFF instruction (Figure 14.5A). To make the instruction true with switch 010 open and false when it is closed, we must choose an examine-OFF instruction (Figure 14.5A). The proper PC ladder diagram solution is shown in Figure 14.9B.

TIMERS

After coils and contacts, the timer is the most commonly used device in a ladder logic system. Of the timing functions used, the most common is the on-delay timer. Recall from Chapter 6 that the on-delay timer initiates an action after a specified delay has occurred. The SLC® 100 has a retentive timer on-delay (RTO) and a retentive timer off-delay (RTF). The timer functions as an internal clock, counting in 0.1 s intervals. The number of 0.1 s intervals counted is called the *accumulated value*. Counting takes place under the conditions given in Table 14.1.

The time delay is set by programming a preset (PR) value. Both timers also have a reset accumulated (RAC) value. The RAC value is automatically set when you program the instruction to reset the

TABLE 14.1 Counting Conditions

RTO Timer Rung Conditions		RTF Timer Rung Conditions	
True	Timer is counting	False	Timer is counting
False	Counting stops	True	Counting stops
True	Counting resumes	False	Counting resumes
AC value represents the cumulative time during which the rung is true		AC value represents the cumulative time during which the rung is false	

counter. In most cases, the RAC value is left at 0000, but you can change it to any value up to 9999. If you program the RAC value, the time delay will equal the preset (PR) value minus the reset accumulated (RAC) value.

A simple on-delay timer is shown in Figure 14.10. When switch 001 closes, power is applied to the timer. At this time, the examine-ON instruction 901 is false. When the accumulated value reaches the PR value minus the RAC value (40 s), the examine-ON instruction at 901 will become true. The timer is reset by closing the normally open 002 contact. With RAC = 0000, the timer will be reset to 0 s.

COUNTERS

The PC counter is very similar in function to the PC timer. The counter is an event-driven function, while the timer is a time-driven function. The SLC® 100 has two counting functions, the *up-counter* and the *down-counter*. As its name suggests, the up-counter increments or adds to the count in a register when

the rung with the counter instruction goes from false to true. In other words, the counter counts false-to-true transitions. As in the timer, when the accumulator reaches the preset value, the status bit for the counter instruction goes on. Any examine-ON instruction with that address will then go true. Any examine-OFF instruction will go false.

An example may help at this point. Let's program an up-counter that counts every time an NO switch 010 is pressed and turns on output 011 after 20 counts. We will reset the counter with an NC switch 009 and reset the counter to 10. The program to accomplish this is found in Figure 14.11. Note that depressing switch 010 makes rung 1 go from false to true. Each time the rung goes true, one count is added to the accumulator. When the accumulator reaches a count of 20, the status bit at 901 goes on. The examine-ON instruction in rung 2 then becomes true, making the entire rung true. This action applies power to whatever output device is connected to output 011. The accumulator is reset to a count of 10 when switch 009 is pressed.

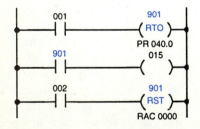

FIGURE 14.10 Simple On-Delay Timer Function

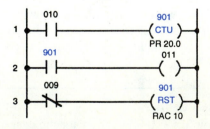

FIGURE 14.11 PC Ladder Diagram for Up-Counter

ALLEN-BRADLEY PLC-2® MEMORY STRUCTURE

As we have seen, each symbol in the PC diagram represents an instruction. In addition, each instruction has a reversed location in memory that contains the status bit of the instruction. The status bit is shown by a binary 1 or 0, the 1 representing a true condition and the 0 representing a false condition. When the input instructions on a rung are logically true (or the bits in memory for those input instructions are all ones because its input devices have electrical continuity), logic continuity is established. In turn, the output instruction is true, and the output device is turned on.

Each instruction status bit location in memory has a bit address represented by a five-digit octal number. The *octal number system* uses the digits 0 through 7 to represent all number combinations. (The decimal system uses the digits 0 through 9.) The numbers 0 through 7 are the same in octal as they are in decimal. However, the decimal number 8 is 10 in octal, the decimal 9 is octal 11, and so on. (A subscript of 8 beside an octal number distinguishes it from a decimal number.) In Figure 14.12, instruc-

tion status bit locations are shown in octal within each 16-bit *word* in memory. A word in this case is determined by the number of lines or connections on the data bus in the microcomputer system. This microcomputer has a 16-bit data bus. There are six different word sizes for PCs: 1, 4, 8, 12, 16, and 24. An 8- or 16-bit word size is the most common. A specific bit location can be identified by combining the three-digit word address and the two-digit bit number to form the five-digit bit address. Figure 14.12 shows the bit address of 01514_8 identified. Each input and output device is associated with a bit address that is displayed next to the device in the PC logic diagram, as shown in Figure 14.13.

ALLEN-BRADLEY PLC-2® MEMORY ORGANIZATION

The organization of memory, or a memory map, can now be discussed. A *memory map*, as illustrated in Figure 14.14, is a picture of the memory space that shows the memory size and for what, if anything, each part of the memory is reserved. Every PC has a memory map, but it may not be like the one illustrated.

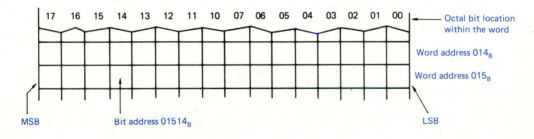

FIGURE 14.12 Memory Word and Bit Address Structure

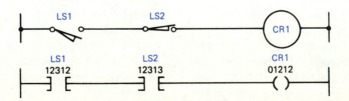

FIGURE 14.13 Bit Addresses on the PC Logic Diagram

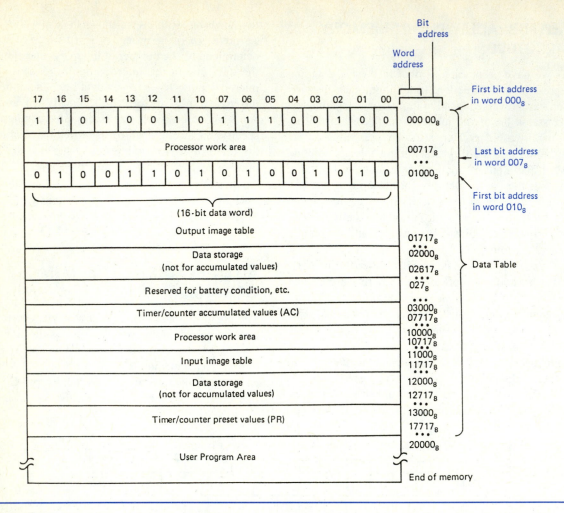

FIGURE 14.14 Memory Map of PC Memory

Figure 14.14 shows the memory space divided into two major areas: the data table and the user program area. The size of each area can be varied within limits to suit user needs, but the total cannot exceed the processor memory size. The data table stores the information needed in the execution of the user program, such as the status of input and output devices, timer/counter preset and accumulated values,

data storage, and so on. Any instruction in the user program can address any word or bit in the data table except in the processor work area. The user program is the logic that controls the machine operation. The logic consists of instructions that are programmed in the PC ladder logic format. Each instruction requires one word of memory.

HARDWARE-TO-MEMORY INTERFACE

The processor monitors input conditions and controls output devices according to a user-entered program. The interface between the hardware and the program occurs in the I/O image table (Figure 14.14). The purpose of the input image table is to duplicate the status of the input devices wired to input module terminals. If an input device is electrically closed, its corresponding input image table bit is a 1. If an input device is open, its corresponding input image table bit is a 0. The input image table bits are *monitored* in conjunction with the user program. The output image table contains the status of the output devices wired to the output module terminals. The output image table bits are *controlled* by the user program.

The instruction addresses in the I/O image table have a dual role: The five-digit bit address references both an I/O image table address and a hardware location, as shown in Figure 14.15. The most significant digit indicates an input or output device; the example in Figure 14.15B is for an output device. Referring to Figure 14.14, we can see why it refers

to an output. The output image table starts at address 01000_8 and ends at 01717_8. The most significant digit in these address locations is a 0. The most significant digit in the input image table memory area is a 1. Each five-digit bit address in the I/O image table directly relates to an I/O module terminal, as shown in Figure 14.16.

Figure 14.17 illustrates the hardware-to-memory interface. When an input device connected to terminal 11212_8 is closed, the input module circuitry senses a voltage. The true condition is entered into the input image table bit 11212_8. During the program execution or scan, the processor examines bit 11212_8 for a true condition. If the bit is true (in this case, it is), the examine-ON instruction is logically true. The rung is true because bit 11212_8 is true. The processor then sets output image table bit 01306_8 to true. The processor turns on terminal 01306_8 during the next I/O scan, and the output device wired to this terminal becomes energized. This process is repeated as long as the processor is in the run mode. If the input device were to open, a 0 would be placed in the input image table, causing the output image table to go to 0 and, in turn, causing the output device to turn off.

CONCLUSION

About 12 years ago, the U.S. automotive industry needed a control system that was easily programmed, highly reliable, small, able to communicate with a computer, and inexpensive. The programmable controller grew out of these requirements. The programmable controller has proved to be effective and has replaced the electromagnetic relay in most applications. Through their early years, programmable controllers gained a good reputation and were soon accepted by industries other than the automotive industry. Today, the programmable controller is more flexible and more reliable than earlier generations of controllers. Its uses have been more far-reaching than anyone ever imagined. It has become a major tool for solving control problems in most industrial plants, and its popularity is likely to increase in the future.

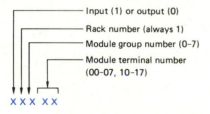

- Input (1) or output (0)
- Rack number (always 1)
- Module group number (0–7)
- Module terminal number (00–07, 10–17)

X X X X X

A. Generic Structure

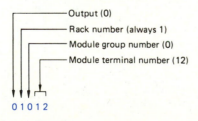

- Output (0)
- Rack number (always 1)
- Module group number (0)
- Module terminal number (12)

0 1 0 1 2

B. Specific Hardware Location

FIGURE 14.15 Hardware Address Structure

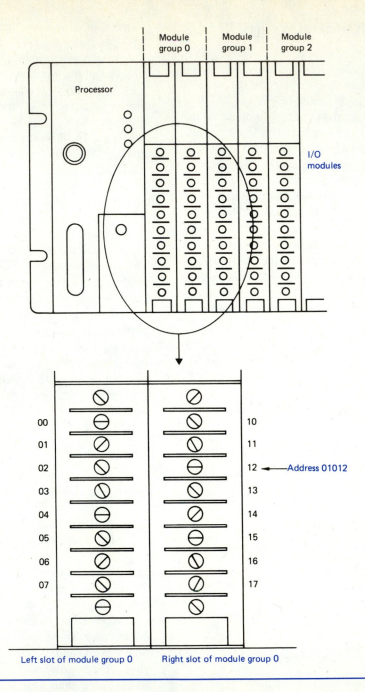

FIGURE 14.16 Address Bit Relationship to Hardware Location

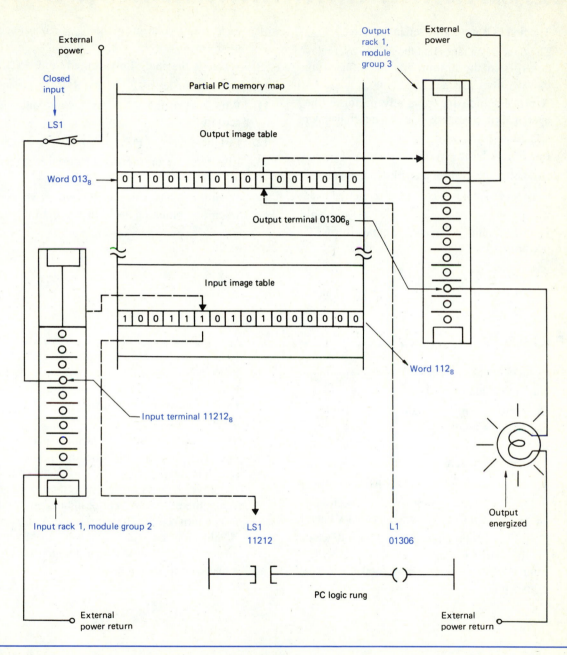

FIGURE 14.17 Hardware-to-Memory Interface

■ QUESTIONS •

1. The first PCs were installed in the year _____.

2. The three advantages that the PC has over relay controls in the automobile manufacturing industry are _____, _____, and _____.

3. A process industry generally performs the operations necessary to change a material either _____ or _____.

4. List the six basic physical parts of a PC.

5. How is the PC control plan placed in memory?

6. Input and output devices are also referred to as _____ devices.

7. Are scan rates an accurate measure of PC performance?

8. Field wiring connects to the _____ housing or modules.

9. Where is the control plan stored in the PC?

10. What do the letters RAM and ROM represent?

11. What part of the PC organizes all controller activity?

12. List three types of PC programming languages.

13. List two major factors to consider when selecting a PC.

14. What three questions should you ask yourself when programming each rung in the PC?

■ PROBLEMS •

1. Draw the PC logic diagram for the relay logic diagram of Figure 13.14.

2. Draw the PC logic diagram for the relay logic diagram of Figure 13.26.

3. Draw the PC logic diagram for the relay logic diagram of Problem 2 at the end of Chapter 13.

Write a PLC program for the following situations.

4. Relay A coil goes on immediately after pushing an NO switch. Coil B is energized 11 s later.

5. Coils C and D both turn on immediately when the limit switch is turned on. When the switch is turned off, D remains on for another 5 s; then it goes off.

6. Both E and F coils go on when a switch is turned on. Coil E turns off after 5 s, and F continues to run until the switch is de-energized.

7. Three mixers (A, B, C) are located on a process line. The process starts when the master start switch is pressed. When the master switch goes on, mixer A turns on for 30 s, then goes off. While mixer A is on, a pilot light is on. Mixer B then turns on for 20 s after mixer A turns off. A pilot light goes on when B is on. Mixer C goes on after B goes off and remains on until a master stop switch is pressed.

Introduction to Robotics

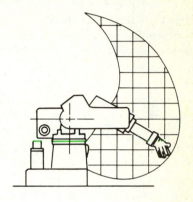

OBJECTIVES

On completion of this chapter, you should be able to:

- Define an industrial robot.
- List four requirements for an industrial robot.
- List and explain the function of the three basic parts of a robot.
- Explain the concept of degrees of freedom.
- Contrast the four different robot work envelopes.
- Differentiate between a servo and a non-servo robot.

INTRODUCTION

If you were to ask ten people to define a robot, you would probably get ten different answers. Each answer would correspond to that person's mental image of a robot. Most descriptions would be something like this: "A robot is a manlike mechanical device capable of performing human tasks." This type of definition might be sufficient for a robot in a science fiction movie, but it does not fit an industrial robot.

Karel Capek, a Czech novelist and dramatist, coined the term *robot* in his 1923 play entitled *R.U.R.* This play showed class struggle in a society with automated workers. Robot is the Czech word for "worker." The word was picked up by science fiction authors in the 1930s and 1940s. Isaac Asimov, an American writer, first used the term *robotics*. These authors inspired scientists and engineers such as Joseph Engelberger to develop industrial robots.

In the early 1950s, Robert C. Goertz developed simple manipulators to handle radioactive materials. George Devol, who worked with Engelberger, holds a 1961 patent on the first industrial robot. The first computer-controlled robot was developed by H. A. Ernst at MIT in 1961.

There is no universally accepted definition of a robot. Typical definitions include ideas such as mobility, programmability, and the use of sensory feedback in determining robot behavior. Perhaps the most comprehensive definition of *industrial robot* is the one established by the Robotic Institute of America (RIA) in 1979. It reads: "An industrial robot is a re-programmable multifunction manipulator designed to move materials, parts, tools, or specialized devices through variable programmed motions for the performance of a variety of tasks." The key word in this definition is "re-programmable." This word implies that a computer is associated with the robot as its "brain" and also that a new task for the robot can be made in software, much as we reprogram a programmable controller. The robot needs only small hardware changes for a new job.

Robot programmability provides major advantages over hard automation. If there are to be many models or options on a product, programmability allows the variations to be handled easily. If product models change frequently, as in the automotive industry, it is generally far less costly to reprogram a robot than to rework hard automation. A robot workstation may be programmed to perform several tasks in succession rather than just a single step on a line. This makes it easy to accommodate fluctuations in product volume by adding or removing workstations. Also, robots may be quickly reprogrammed to do different tasks. It is often possible to amortize or spread the cost of the robot over several products.

An earlier Japanese definition of an industrial robot is more general than the RIA definition. It states that an industrial robot is an all-purpose machine, equipped with a memory and appropriate mechanism to automatically perform motions in replacing human labor. This definition implies that robots are useful in many ways and can perform applications that are poorly suited to human abilities. In the past, for example, industry used robots to perform operations in hostile environments, such as picking up parts and placing them inside a radioactive chamber. Today, some of these applications involve work in unusual environments, like clean rooms, furnaces, and space. Robots also operate in undesirable environments and in applications that are dull and monotonous. These include manipulation of small and large objects like electronic parts and turbine blades.

Future use for industrial robots includes countless applications involving improved quality, increased productivity, and lower costs. Introducing robots into the workplace will change our economic and social structure. If this technology is properly applied, however, the transition to more robotic workers should be gradual, with time for human adjustment. Proponents of robotics point with pride to the benefits of mechanization within the U.S. agricultural industry. They point out that the efforts of just 3% of our population can now produce enough food to feed us well. Agricultural mechanization also allows us to share surplus food with others in our world family.

The largest single force driving robotics technology is the need to increase productivity and reduce costs in manufacturing. The circumstances under which robots can help achieve these goals depend on a number of factors. The nature of a manufacturing task may or may not lend itself to the current capabilities of robots. If the product volume is small, manual labor may be more cost-effective. If the volume is very large and uses similar repetitive tasks, greater throughput can typically be obtained by developing hard automation for the job. Robots are strongest in the middle ground.

The science of robotics is now in the same position computers were 30 years ago. No one is really expert in building or using them. You have probably seen pictures of the first computer. It filled a huge room and broke down every 7.5 minutes. Experience by the builders and users over a period of time has made computers practical. Robotics will, no doubt, follow the same pattern.

An interesting sidelight to robotic automation is the countless variety of parts that can be handled efficiently with robotics and automated devices. The automobile is a good example. We often think of an automobile as a large four-wheeled assembly weighing in excess of a ton. We do not realize that most of these parts start out as bits and pieces weighing less than a few pounds. When properly applied, robots can hasten the process from "inventory" to "salable units" by working long hours for low pay. Some sources report that robots work 24 hours a day for a few dollars per hour.

Japan has led the world in the use of robots in manufacturing. The two sectors making heaviest use of robots today are the automotive and electronics industries. It is not surprising, therefore, that Japan has had its greatest success in these two areas. Growth of industrial robotics in the United States has been steady in recent years. It has not been as rapid as popular magazine articles of the early 1980s led the public to expect, however. If U.S. industries are going to thrive in world markets, greater dependence on automated manufacturing is essential.

ROBOT REQUIREMENTS

A number of stringent requirements are imposed upon robots in order for them to be competitive in the world of manufacturing.

1. Reliability and durability are very important. An industrial robot must work every day, often all day, to pay for itself.

2. Robots are not usually as fast as hard automation or human workers doing the same job. Currently, the speed of robots is constrained by computational as well as mechanical factors.

3. Accuracy of robots is important for applications such as precise electronic testing and for assembly tasks.

4. The ability to conform to manufacturing requirements is important in assembly and machining applications. Precisely machined parts are usually expensive.

5. It is highly desirable that robots be able to use new sensors. If new sensors cannot be used, a very expensive machine may become obsolete.

6. Ease of programming is important. When robots are easy to program, robotic applications can be developed quickly. This saves time and money.

7. Versatility is needed to avoid the cost of special-purpose fixtures required for new robot workstations.

8. Cleanliness is of increasing importance in many electronics applications.

ROBOT STRUCTURE

Industrial robots typically consist of three parts: a *manipulator* (mechanical arm), a *controller*, and a *power supply* (Figure 15.1). The term *controller* is used in several different ways. In this context, a controller is the computer system used to control the robot and is sometimes called a *robot workstation controller*. The manipulator serves the same function as the torso, arm, wrist, and hand in the human body. The power supply provides the energy in the appropriate form for the manipulator to use in doing the work.

Robot Arms

Robot arms come in a variety of different types, one of which is shown in Figure 15.2. The hand of the robot is called its *end effector*. The types of joints (whether they revolve or slide), their arrangement, and the geometry of the links that connect them make up the kinematic structure of the robot. You can see from Figure 15.2 that the nomenclature of a robot is very similar to that of the human body.

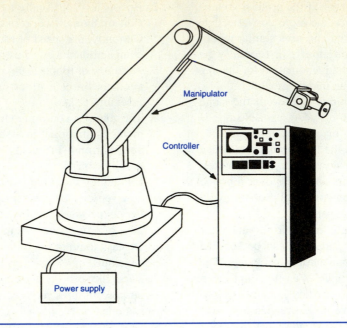

FIGURE 15.1 Parts of a Robot System

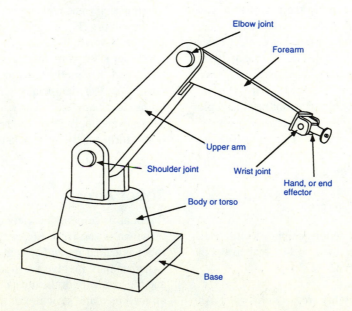

FIGURE 15.2 Parts of a Manipulator

Movement of manipulator arms is often described as having *degrees of freedom*. You may think of degrees of freedom as axes of movement. The first degree of freedom of arm movement is the movement from side to side; we call this the *x axis* in a Cartesian coordinate system. The second degree of freedom of arm movement is the movement of going up and down; we call this the *z axis* in a Cartesian coordinate system. The third degree of freedom of arm movement is the movement of reaching in and out; we call this the *y axis* in a Cartesian coordinate system. All three together are called the three *major axes* of manipulator movement, as shown in Figure 15.3.

Not all types of robots have all three major axes of movement. A robot, for example, that moves up and down (*z* axis) and in and out (*y* axis) is a two-axis robot. A two-axis robot has two degrees of freedom.

Three *minor axes* of manipulator movement are found in the three possibilities of wrist movement (Figure 15.4). Note, in Figure 15.2, that the hand is attached to the end of the forearm at the wrist joint. The diagram in Figure 15.4 shows the three possible

movements of the hand attached to the wrist: roll, pitch, and yaw. The *roll* motion (fourth degree of freedom) is performed by twisting the wrist, as in turning a key in a lock. The *pitch* movement (fifth degree of freedom) allows the hand to bend up and down. The *yaw* motion (sixth degree of freedom) is from side to side.

When major and minor axes of movement are combined, they give the robot six possible axes of movement, or six degrees of freedom. Again, not all robots are equipped with this many axes, but many are. For example, a robot may have three major and two minor axes of movement. Such a robot is said to have five degrees of freedom.

Robot Controllers

The controller may be programmed to operate the robot in a number of ways, thus distinguishing it from hard automation. *Hard automation* is controller machinery that must be reset for each job. Hard automation machinery has, for example, various cams and sequencing valves that must be readjusted for each new job.

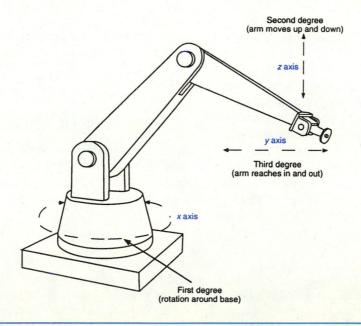

FIGURE 15.3 Three Major Axes (Degrees of Freedom) of Manipulator Movement

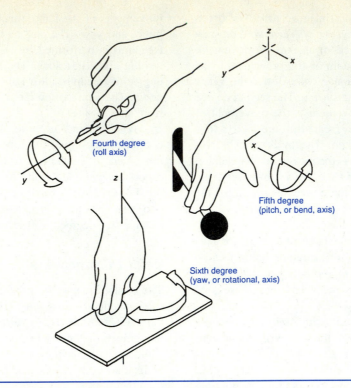

FIGURE 15.4   Three Minor Axes (Degrees of Freedom) of Manipulator Movement

The controller also monitors auxiliary sensors that detect the presence, distance, velocity, shape, weight, or other properties of objects. Robots may be equipped with vision systems, depending on the application for which they are used. Most often, industrial robots are stationary, and work is transported to them by conveyors or robot carts, which are often called *autonomous guided vehicles* (AGVs).

Autonomous guided vehicles are becoming popular in industry for materials transport. Most frequently, these vehicles use a sensor to follow a wire in the factory floor. Some systems employ an arm mounted on an AGV.

Robot Power Supplies

Some important criteria for the evaluation of manipulator power supplies are their dynamic range, the precision with which they may be controlled, the force or torque that they can generate, and their size, mass, and cleanliness. The most common types of power supplies for robots are electric, hydraulic, and pneumatic.

ROBOT CLASSIFICATIONS

Robots are classified by work envelope, by power supply, by power supply control, and by motion control. We will briefly consider each of these methods of robot classification.

Work Envelope

The *work envelope* of the robot is a description, usually graphic, of what area the robot can reach. All points within the reach of the robot are then said to be in the robot's work envelope. An accurate knowledge of where the robot can reach is essential in choosing the right robot and in applying it.

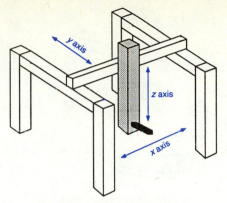

A. Rectangular Coordinate Robot

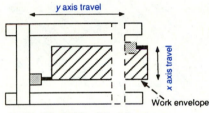

B. Top (Plan) View of Work Envelope

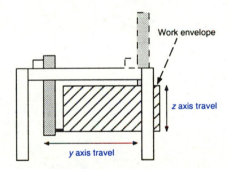

C. Side (Elevation) View of Work Envelope

FIGURE 15.5 Rectangular Coordinate Robot Work Envelope

Rectangular Coordinate Robot. Note that the *rectangular coordinate robot*, shown in Figure 15.5A, is a three-axis robot. It allows movement in each of the three major axes. From the top or plan view (Figure 15.5B), we see that the work envelope is a rectangle. This rectangle forms when the robot moves through the limits of its x and y axis movements.

From the side or elevation view (Figure 15.5C), we see that the work envelope here is also rectangularly shaped. This rectangle forms when the robot moves through the limits of its x and z axis movements. When we combine both plan view and elevation view, the composite work envelope is box shaped.

Cylindrical Coordinate Robot. A *cylindrical coordinate robot* is shown in Figure 15.6A. When comparing the elevation view of the cylindrical coordinate robot (Figure 15.6C) with that of the rectangular coordinate robot, we see very little difference in their work envelopes. Both work envelopes are rectangularly shaped because both robots have arms that can rise and fall in the vertical, or z, axis and also travel in and out in the y axis direction. Any differences that might appear in the work envelope would be due to one robot's having a greater reach (y axis) or height (z axis). It is when we look at the plan view that we see the difference in these two robots. The plan view of the cylindrical coordinate robot (Figure 15.6B) shows that it can pivot on its base, unlike the rectangular coordinate robot. When the base pivots, it forms a circular-shaped work envelope. By combining the plan and elevation views, we can see where the cylindrical coordinate robot gets its name. The entire work envelope is shaped in the form of a cylinder.

Spherical Coordinate Robot. A *spherical coordinate robot* is shown in Figure 15.7A. When we compare the plan view of the spherical coordinate robot (Figure 15.7B) with that of the cylindrical coordinate robot, the robots appear to have identical work envelopes. The cylindrical coordinate robot and spherical coordinate robot both can pivot on their bases, which accounts for the similarity in their plan views. When we view the spherical coordinate robot from the side (Figure 15.7C), we see the difference between the two robots. The spherical coordinate robot does not move straight up and down in the z axis plane. Instead, the spherical coordinate robot pivots up and down, rotating in the vertical axis. If we combine the plan and elevation views of this robot, we can visualize part of a sphere or ball. The spherical coordinate robot is also called the *polar coordinate robot*.

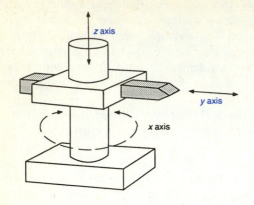

A. Cylindrical Coordinate Robot

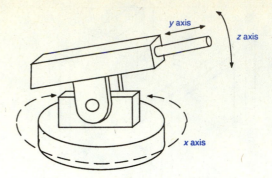

A. Spherical Coordinate Robot

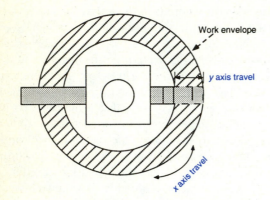

B. Top (Plan) View of Work Envelope

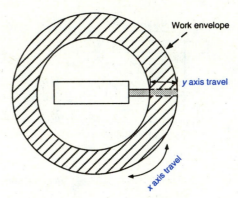

B. Top (Plan) View of Work Envelope

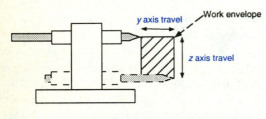

C. Side (Elevation) View of Work Envelope

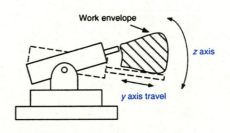

C. Side (Elevation) View of Work Envelope

FIGURE 15.6 Cylindrical Coordinate Robot Work Envelope

FIGURE 15.7 Spherical Coordinate Robot Work Envelope

Jointed-Arm Coordinate Robot. The *jointed-arm coordinate robot*, shown in Figure 15.8A, gets its name from its similarity in construction to the human arm. The human arm has a shoulder, an elbow, and a wrist joint. The manipulator arm shown in Figure 15.2 is also a jointed-arm coordinate robot.

This type of robot is sometimes called the *anthropomorphic robot*. The word "anthropomorphic" comes from two Greek terms, the first meaning "man" and the second meaning "shape." In this chapter, we will use the word "jointed-arm" in referring to this type of robot.

Since this robot can pivot on its base (sometimes called the *waist*), the work envelope plan view (Figure 15.8B) is similar to that of the spherical and cylindrical robots. The work envelope elevation view (Figure 15.8A) shows a significant difference, however. The jointed-arm robot's elevation view shows a complex teardrop-shaped work envelope. Because of its six degrees of freedom, the jointed-arm robot is the most versatile of the robots we have discussed. It can reach down and pick up a workpiece on the floor. It can place that workpiece anywhere in its work envelope by reaching up and over any obstacle. This combination of complex movements is not possible with any of the other coordinate robots.

The major disadvantage of the jointed-arm robot is its cost. Because of its six axes of movement, the manipulator arm is expensive. Also, the controller must have more complex hardware and software, which drives the cost of this robot even higher. The jointed-arm robot is used only in those situations in which other robots are ineffective.

Power Supply

Robots are also classified by the form of power that drives the manipulator—*pneumatic*, *hydraulic*, and *electric*. Pneumatic supplies provide a great deal of torque for the size of the manipulator. However, they are hard to control precisely because of the compressibility of air. They are often used in simple robots that merely move back and forth between hard stops, but they are not widely used in more programmable robots.

Most advanced robots are either hydraulically or electrically driven. A number of factors tend to favor hydraulic manipulators for robots that must carry heavy payloads. One reason is that the torque-to-mass ratio is currently better for hydraulic systems.

Environmental requirements are important. Hydraulic systems tend to drip hydraulic fluid or produce particles. This makes them unsuitable for the clean environments that are often needed for electronics manufacturing. On the other hand, some electric motors may be explosion hazards in volatile applications such as paint spraying.

Electric motors have become increasingly popular for powering small-to-medium-sized robots. The design of new types of motors has become an important topic in robotics research. Conventional motors spin fast and generate low torque, thus requiring a transmission for speed reduction. This may be done with gears or with a widely used device called a *harmonic drive* that provides speed reductions of the order of a hundred to one in a very small package. However, a transmission introduces a number of factors, including static friction, binding, wear, backlash, and cogging, that make it difficult to apply.

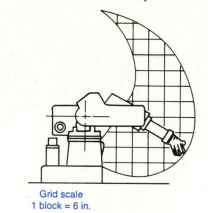

Grid scale
1 block = 6 in.

A. Teardrop Side View

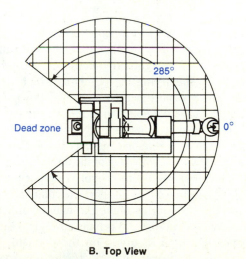

285°

Dead zone   0°

B. Top View

FIGURE 15.8 Jointed-Arm Coordinate Robot Work Envelope

A recent trend has been in the direction of direct-drive arms in which no speed reduction is necessary. At IBM and elsewhere, new types of motors have been invented that produce high torque at low speed. In conjunction with suitable mechanical robot designs, these motors show promise for reducing some of the unpredictable aspects of manipulator behavior that have made robot control something less than a science.

Power Supply Control

The manipulator arm must have a supply of power to use in doing work. As we have seen, the three types of power supplies are electric, hydraulic, and pneumatic. Robot systems can be classified by the way in which this power is controlled. These two types of robots are *servo controlled* and *non-servo controlled*.

The non-servo robots are the simplest. They are inexpensive, are easy to understand and easy to set up, and have good precision with high reliability. The non-servo robot controller sends two and only two signals to the power supply: ON and OFF. Operation is simple: When you want the arm to move, apply power; when you want it to stop, remove power. This type of robot usually uses mechanical stops to position the arm at the end of its travel. Since the robot literally bangs into these stops, the robot is sometimes called a *bang-bang robot*.

Non-servo robots are a logical choice for point-to-point transfer in manufacturing processes and assembly operations. Non-servo robots usually operate along the axes of a Cartesian coordinate system with one or more rotations, such as wrist motion. Motion is controlled through the use of a limited number of stops, and the actual path between points may be difficult to define. The control system can be as simple as a cam or drum timer that sequences valves or switches.

Servo-controlled robots have a wider range of capabilities. They can be programmed to stop at any point in their work envelope. They can perform multiple point-to-point transfers and move along a controlled path. Most servo-controlled robots use a jointed-arm mechanism and can be programmed to avoid an obstruction. Programming can be rather sophisticated, and the price tag for a complete servo-controlled robot system is relatively high.

One of the major things that differentiates a servo robot from a non-servo robot is the presence of feedback in the servo robot system. This is the same principle discussed in Chapter 10 on process control. The feedback in a robot system is usually position feedback. The feedback allows the controller to know where the robot arm is in space.

Motion Control

Robots are classified by the types of movements the manipulator makes: pick and place, point to point, controlled path, and continuous path.

Pick and Place. A *pick-and-place robot* is a good example of a non-servo robot. The name describes what it does. The pick-and-place robot picks up a part in one location and places it in another location. Of the four types of robots, the pick-and-place robot is the simplest and least expensive. The pick-and-place robot is also easier and less costly to maintain. It is used in many industrial applications, such as loading and unloading a conveyor. In this type of application, a more complicated robot is not necessary and would certainly be more costly.

Point to Point. The *point-to-point robot* is the most common of the types of robots. This type of robot is a servo-controlled robot. Note that the hand path of this robot is not straight. Programming of this device is done by specifying the points to which it will travel. This is done either by "teaching" it by moving its arm to the actual points in space or by describing the points in the computer program. Each point in space will have a reference number in the Cartesian coordinate system, with each of the three axes having one number. Usually, the robot is taught using a *teaching pendant*, which is a portable component that allows the operator to move the manipulator to a particular point in space. When the arm is in position, the point is then recorded on the pendant. The operator then moves the arm to the next point, recording it. The machine does not remember the path the operator used to get the arm to the points. It only remembers the points themselves. When the program is run, the robot will move its arm to each of the programmed positions. The path the manipulator hand takes is not predictable during programming, however. The arm usually travels in an arc, as shown in Figure 15.9A.

Controlled Path. The path the point-to-point robot takes to get from one point to another is difficult to predict. The path is usually an arc. When the manipulator hand must travel in a straight line between points, a *controlled-path robot* is used. The endpoints are entered into the robot program software, and the robot then moves in a straight line between the points, as shown in Figure 15.9B. This type of robot would be used, for example, in a drilling operation. The drill bit must move through the workpiece in a straight line. The controlled-path robot is also used in welding and assembly operations.

Continuous Path. As seen in Figure 15.9C, the *continuous-path robot* can be programmed to take any path between two points. Programming the continuous-path robot is different from that of the point-to-point and controlled-path robots. The point-to-point and controlled-path robots usually use a teach pendant to record the points to which the arm must move. The continuous-path robot is programmed by grabbing the robot and moving it through the path we want it to take. The controller not only records the path in memory, but also remembers the timing of the movement. This type of robot is used in spray painting. If you have ever tried to spray paint, you know how important the correct path and the correct timing are to a high-quality product.

ROBOT PROGRAMMING

The program in a robotic system tells the robot how to perform its task. In this section, we examine some of the tools and techniques that have been used and proposed for robot programming.

Guiding

Guiding is the process of moving a robot through a sequence of motions to "show it" what it must do. One guidance method is to physically drag around

the hand of the robot while it records joint positions at frequent intervals along the trajectory. The robot then plays back the motion just as it was recorded.

Guiding may also be applied using a teaching pendant, which has keys that are used to command the robot. Several modes of operation are often

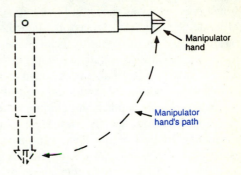

A. Point-to-Point Robot Hand Path

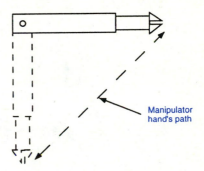

B. Controlled-Path Robot Hand Path

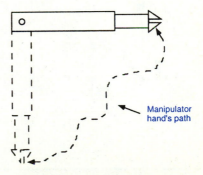

C. Continuous-Path Robot Hand Path

FIGURE 15.9 Classification of Robot by Motion

available on the teach pendant. In joint mode, a pair of buttons is used to move each joint back and forth. In addition to joint mode, one or more Cartesian modes may be provided. In Cartesian modes, buttons are associated with Cartesian axes in some three-dimensional coordinate system.

Two examples of coordinate systems are *world mode* and *hand mode*, as shown in Figure 15.10. In world mode, the coordinate system is aligned with the base or frame of the robot. In hand mode, the coordinate system is always aligned with the gripper. In the figure, the x, y, and z axes of the hand coordinate system are labeled x', y', and z', respectively.

For point-to-point applications, guiding systems usually allow the user to specify a few key positions to be recorded. The system fills in the points between adjacent pairs of these positions. It provides a path that does not play back every fumble and overshoot of the teacher.

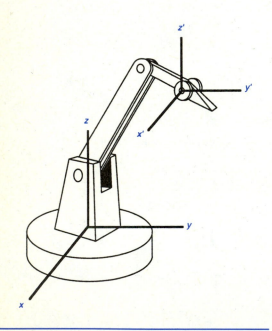

FIGURE 15.10 World Mode and Hand Mode

Programming Languages

It is no surprise that the well-established technology of programming languages should be adopted for programming robots. Many robot-programming languages have been developed. The first, MHI, was a limited language designed by Ernst at MIT in 1961. Two languages developed at Stanford University, WAVE and AL, were particularly influential in the field. Conventional computer programming languages like C, LISP, Pascal, and BASIC have been extended with subroutine libraries for robot control. Robot languages in use in industry today include AML (IBM), HELP (GE), KAREL (GMF), LM (Scemi, Inc.), MCL (McDonnell Douglas), RAIL (Automatix), and VAL-11 (Unimation, Adept).

Robot Programs

Keep in mind that robot programs are computer programs. Many of the advances in computer programming in the last couple of decades have applied to robots. Robot programs deal with a richer variety of I/O devices than conventional computer programs do.

Robot programs must command robots to move. The way in which we specify motion is important. Also, the programs use information from sensors. One way of using sensory information is to monitor sensors until a given condition occurs and then perform a specified action in response. Another use of sensory information is to use feedback from sensors to change the robot's behavior continuously.

No robot is an island. Industrial robots function in factory environments. These factories contain other robots, automation equipment, and computers. Effective communication among these elements is critical to computer-integrated manufacturing (CIM). Robot programs must also deal with passing data back and forth to the different places in the factory where they are needed.

Robot-programming languages must contain specifications of motion for the robot. Some of the commands used to specify motion are discussed here. The joint-level move is the most fundamental motion command. In the simplest case, the command might be written as follows:

move (joint, goal)

The joint is either a sliding or a revolving joint that is to be brought to some linear or angular position. Even in this simplest case, much is left unsaid. For a revolving joint without joint limits, for example, the path has not been fully specified since the goal could be reached in either a clockwise or a counterclockwise direction. Also, how fast should the motion be? How accurately must the goal be achieved? Is a little overshoot allowable? There are always trade-offs between these considerations. To keep things simple, robot programs have a number of options that allow these factors to be controlled. These options are specified as defaults. With large numbers of optional parameters, keyword parameters are an advantage. For a simple arm movement, goals are given in absolute terms. It is worth noting that it is often more convenient to specify all of these in terms relative to the current position.

Joint moves may involve more than one joint. In this case, the command might be written as follows:

move ($[j_1, j_2, j_3]$ $[goal_1, goal_2, goal_3]$)

where j_1, j_2, j_3 specifies the three joints and $goal_1$, $goal_2$, $goal_3$ indicates the three goals for each joint. In this example, we have used AML notation to specify one list of three joints and another of three goals. It is convenient to think of such a goal as a point in joint space. When this program is run, each of the three joints would move to each specified goal.

In most cases, it is better to specify Cartesian coordinates rather than joint motions. As in the case of guiding, Cartesian motion specifications are made in a Cartesian coordinate system. In a robot with six degrees of freedom, the position and orientation of an object in space can be described with six numbers. Each one of the numbers refers to a particular axis.

Regardless of the type of program used, each program has a similar function. The program controls the robot in each of its tasks.

END EFFECTORS

End effectors, also called *work-holding devices*, pick up, grasp, or handle objects for transfer or during processing. Robot hands or end effectors are designed to do a specific task such as painting, or they are considered general-purpose, such as grippers are. General-purpose end effectors are usually needed for material-handling tasks when it is not known where the robot will grip a product or which of several products will be moved.

Selecting an end effector early in a program involving robotics is usually not difficult since characteristics of the object's characteristics are well known. End effectors come in all types and sizes. Many must be custom made to match special handling requirements. Some typical examples of end effectors are shown in Figure 15.11.

Grippers are perhaps the most universal of all end effectors. In many cases, reprogramming involves only a change in finger style. Note the different types of grippers shown in Figure 15.12. In this diagram, the workpieces have been shaded. Most robot manufacturers have standard grippers in several sizes, in two- and three-finger configurations for external and internal gripping, as well as soft, blank fingers that can be easily modified for special shapes.

The smallest gripper is about twice the dimensions of a thimble, and the largest compares to a human hand. Two-finger models simulate the motions of the thumb and index finger for reaching into channels, grasping parts with closely spaced components, or picking and placing any small object. Three-finger models duplicate the motions of the thumb, index finger, and a third finger for grasping bodies of revolution and objects of spherical or cylindrical shape. Fingers are interchangeable in most sizes and provide for rapid changeover when robots are reprogrammed for another task.

A pilot diameter and bolt circle are provided on the rear face for mounting. Simple adapters will mount the grippers to almost any linear- or rotary-motion device to transport objects within the operating range of finger travel. Transport can be as simple as mounting a gripper to the rod of an air

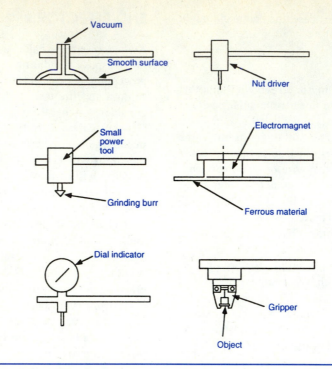

FIGURE 15.11   Various Types of End Effectors

cylinder for straight-line motion or mounting to a more complex device such as to the "business end" of a six-component manipulator to reach any point in a Cartesian coordinate system.

APPLICATIONS

In this section, we will consider several representative applications of robots.

Spot welding involves applying a welding tool to some object, such as a car body, at specified locations. This requires the robot to move its hand (end effector) to a sequence of positions with sufficient accuracy to perform the task properly. The robot should move at high speed to reduce cycle time, while avoiding collisions and excessive wear or damage.

Pick and place is the name commonly given to the operation of picking up a part and placing it appropriately for subsequent operations. Pick-and-place operations have some requirements in addition to those for spot welding. The part must not be

dropped. It must be held securely enough to prevent it from slipping in the gripper but gently enough to avoid damage. In addition, the robot must avoid disturbing the part during approach and departure.

Spot-welding and pick-and-place operations are characterized by their point-to-point nature. In such applications, what happens at the beginning and the end of the motion is the most important part. There is some latitude in choosing the motions between these two points.

Spray painting requires covering a surface with an even coat of paint. This is typically done by specifying the trajectory along which the arm will move. The trajectory should include both position and orientation of the nozzle as a function of time.

Seam welding requires that a welding torch continuously follow a seam on a surface. Unlike spray painting, seam welding typically requires real-time correction of the path to accommodate small deviations of the actual seam from the expected path. Spray painting and seam welding are both continuous-path applications. In such applications,

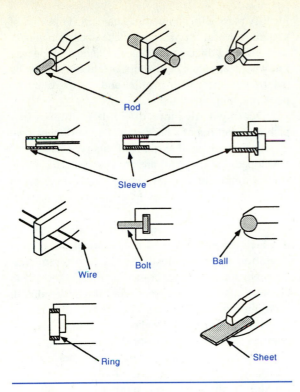

Rod

Sleeve

Wire Bolt Ball

Ring Sheet

FIGURE 15.12 Various Kinds of Robot Grippers

as inserting an electronic component into a printed-circuit board. The robot is programmed to perform a single operation as a single step in an assembly line. Each robot is fed parts of a single type from a part feeder, which presents them in the correct orientation. In this approach, the robot is used in the same way as hard automation is traditionally used.

An alternative method is to feed all parts directly into a robot workstation in which the entire assembly is to be completed. Part feeders and magazines may be arranged about the workstation, as may a variety of tools and fixtures required for the assembly. Another option is that the workstation is presented with a "kit" of preoriented parts containing all components required for the assembly. To have individual robot workstations do independent assembly of complete products is extremely advantageous for flexible production capacity.

Machining of mechanical parts is a growing application of robotics technology. Operations like grinding, deburring, and sanding parts require the ability to follow surfaces and to maintain the forces required to perform the specified operation.

An unusual but fascinating application is sheepshearing. Recently, several researchers constructed a robot system that would perform sheepshearing on live sheep. The robot was sufficiently adaptable to cut the wool without harming the sheep. Sheep appreciate compliance, which is defined as the quality of bending to stresses within the elastic limit.

SENSING

A variety of sensors and sensing techniques are used in robotics. In this short section, we will mention a few of the more important ones.

The most widely used sensors in robotics are *binary sensors*. The breaking of a beam of light or the depressing of a switch is used to detect the presence of parts in almost every application. Chapters 8 and 9 have an extensive treatment of these types of sensors. Many industrial applications require the monitoring of hundreds of such sensors in a robot workstation.

Strain gauges (Chapter 8) are commonly used to measure force. Absolute accuracy, dynamic range,

the robot's position and orientation as a function of time are important throughout the motion.

Electronic testing by robots is being used extensively in industry. One application is that of testing the electrical continuity between pins, which involves point-to-point operations. Another application is the detection of flaws in printed circuits by probing along metal traces on circuit boards.

Weighing and measuring are now often performed using automated coordinate measuring machines. These devices are very slow and accurate robots. They are used to weigh and measure dimensions of mechanical parts, usually by a sequence of point-to-point motions.

Assembly is an application of increasing importance. Robotic assembly may be done in different ways. One typical method is to equip a simple robot with a special end effector for a particular task, such

linearity, and hysteresis are some parameters by which the usefulness of these devices may be judged. *Force sensors* in the gripper may be used to sense collisions, the presence of an object, tightness of grasp, and the weight of objects.

Direct measurement of the relationship between the manipulator end effector and the workpiece is called *endpoint sensing*. Wrist-mounted force sensors, sensors on the end effector, and structured light projected from the end effector are all endpoint-sensing techniques. It is highly desirable to use endpoint sensing because it provides a direct measurement of the error to be corrected. However, as was noted earlier, effective control of endpoint sensing is complicated by the flexibility of robot structures.

A variety of types of *proximity sensors* are in use in robotics. Critical parameters are (1) the range of distances over which the sensor is useful, (2) accuracy, (3) linearity, and (4) sensitivity to environmental conditions. Ultrasonic sensors have been widely used in mobile robots. These sensors have, in the past, been somewhat limited by their inability to sense objects at close range accurately. However, a technique used by one manufacturer allows an ultrasonic sensor located in the base of a gripper to accurately sense objects as close as one inch. Ultrasonic sensors tend to be somewhat sensitive to temperature variations, atmospheric disturbances (humidity, turbulence), and extraneous reflections.

In the field of robotics, *computer vision* is often used for the identification and location of parts. The problem of binary two-dimensional vision in environments with controlled lighting is sufficiently well understood to be widely used in industry. A number of companies sell products that may be used to identify parts with an overhead camera. This kind of system is useful for pick-and-place operations under the following three conditions: (1) The part types are known in advance, (2) their orientation may be determined from their profile, and (3) there are no parts touching one another.

Dealing with more difficult problems like picking a part out of a bin requires three-dimensional vision. Experimental systems of this type have been developed, but they still are too slow and unreliable for commercial use. However, robot vision is promising and continues to be an active area of research in computer vision.

Another application of vision is visual servoing, in which the image is used to determine and correct deviation from the desired path. Using special-purpose hardware to compute image moments 60 times a second, one researcher has succeeded in servoing a robot to catch a Ping-Pong ball.

The strongest industrial economic incentives for computer vision comes from the area of inspection and measurement rather than from robot control. Two-dimensional vision is used widely in industry for the inspection of mechanical and electronic parts. Inspection systems may also use robots for positioning of parts.

CONCLUSION

Robots have been popular in industry for increasing the speed of manufacture and the quality of manufactured goods. Robots are flexible and fit into many factory automation systems. We will see more of these devices as the United States tries to meet the demands for high-quality and low-cost goods in the world marketplace,

■ QUESTIONS ....

1. Define an industrial robot.

2. List five requirements for a robot to be successful in the manufacturing industry.

3. The industrial robot consists of _____ parts.

4. The movement of a robot's manipulator arm is called _____ of freedom.

5. Differentiate between a robot and hard automation.

6. List three of the four types of robot classifications.

7. The spherical coordinate robot is also called the _____ _____ robot.

8. The robot that has a shoulder, an elbow, and a wrist joint is called a _____ coordinate robot.

9. The most common types of robot power supplies are electric, hydraulic, and _____.

10. Contrast the pick-and-place robot with the point-to-point robot.

11. State the definition of guiding, as it applies to robot programming.

12. The part of a robot used to grasp parts is called the _____ _____.

13. Describe three industrial applications where a robot might be used to increase industrial manufacturing productivity.

Data Sheets

- 741 Operational Amplifier
- 3900 Current-Differencing Amplifier
- SCR
- Triac
- 335 Temperature Sensor
- 555 Timer
- 565 Phase-Locked Loop
- ADC0801 Analog-to-Digital Converter
- DAC0808 Digital-to-Analog Converter

Operational Amplifiers/Buffers

LM741/LM741A/LM741C/LM741E Operational Amplifier

General Description

The LM741 series are general purpose operational amplifiers which feature improved performance over industry standards like the LM709. They are direct, plug-in replacements for the 709C, LM201, MC1439 and 748 in most applications.

The amplifiers offer many features which make their application nearly foolproof: overload pro-

tection on the input and output, no latch-up when the common mode range is exceeded, as well as freedom from oscillations.

The LM741C/LM741E are identical to the LM741/LM741A except that the LM741C/LM741E have their performance guaranteed over a 0°C to +70°C temperature range, instead of −55°C to +125°C.

Schematic and Connection Diagrams (Top Views)

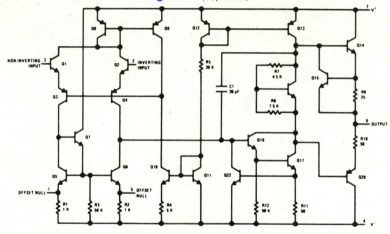

Metal Can Package

Order Number LM741H, LM741AH,
LM741CH or LM741EH
See NS Package H08C

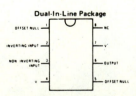

Dual-In-Line Package

Order Number LM741CN or LM741EN
See NS Package N08B
Order Number LM741CJ
See NS Package J08A

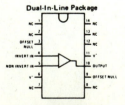

Dual-In-Line Package

Order Number LM741CN-14
See NS Package N14A
Order Number LM741J-14, LM741AJ-14
or LM741CJ-14
See NS Package J14A

Absolute Maximum Ratings

	LM741A	LM741E	LM741	LM741C
Supply Voltage	±22V	±22V	±22V	±18V
Power Dissipation (Note 1)	500 mW	500 mW	500 mW	500 mW
Differential Input Voltage	±30V	±30V	±30V	±30V
Input Voltage (Note 2)	±15V	±15V	±15V	±15V
Output Short Circuit Duration	Indefinite	Indefinite	Indefinite	Indefinite
Operating Temperature Range	−55°C to +125°C	0°C to +70°C	−55°C to +125°C	0°C to +70°C
Storage Temperature Range	−65°C to +150°C	−65°C to +150°C	−65°C to +150°C	−65°C to +150°C
Lead Temperature (Soldering, 10 seconds)	300°C	300°C	300°C	300°C

Electrical Characteristics (Note 3)

PARAMETER	CONDITIONS	LM741A/LM741E MIN	TYP	MAX	LM741 MIN	TYP	MAX	LM741C MIN	TYP	MAX	UNITS
Input Offset Voltage	$T_A = 25°C$										
	$R_S \leq 10\ k\Omega$					1.0	5.0		2.0	6.0	mV
	$R_S \leq 50\Omega$		0.8	3.0							mV
	$T_{AMIN} \leq T_A \leq T_{AMAX}$										
	$R_S \leq 50\Omega$			4.0							mV
	$R_S \leq 10\ k\Omega$						6.0			7.5	mV
Average Input Offset Voltage Drift			15								$\mu V/°C$
Input Offset Voltage Adjustment Range	$T_A = 25°C$, $V_S = \pm20V$	±10				±15			±15		mV
Input Offset Current	$T_A = 25°C$		3.0	30		20	200		20	200	nA
	$T_{AMIN} \leq T_A \leq T_{AMAX}$			70		85	500			300	nA
Average Input Offset Current Drift				0.5							$nA/°C$
Input Bias Current	$T_A = 25°C$		30	80		80	500		80	500	nA
	$T_{AMIN} \leq T_A \leq T_{AMAX}$			0.210			1.5			0.8	μA
Input Resistance	$T_A = 25°C$, $V_S = \pm20V$	1.0	6.0		0.3	2.0		0.3	2.0		$M\Omega$
	$T_{AMIN} \leq T_A \leq T_{AMAX}$, $V_S = \pm20V$	0.5									$M\Omega$
Input Voltage Range	$T_A = 25°C$							±12	±13		V
	$T_{AMIN} \leq T_A \leq T_{AMAX}$				±12	±13					V
Large Signal Voltage Gain	$T_A = 25°C$, $R_L > 2\ k\Omega$										
	$V_S = \pm20V$, $V_O = \pm15V$	50									V/mV
	$V_S = \pm15V$, $V_O = \pm10V$				50	200		20	200		V/mV
	$T_{AMIN} \leq T_A \leq T_{AMAX}$, $R_L > 2\ k\Omega$,										
	$V_S = \pm20V$, $V_O = \pm15V$	32									V/mV
	$V_S = \pm15V$, $V_O = \pm10V$				25			15			V/mV
	$V_S = \pm5V$, $V_O = \pm2V$	10									V/mV
Output Voltage Swing	$V_S = \pm20V$										
	$R_L > 10\ k\Omega$	±16									V
	$R_L > 2\ k\Omega$	±15									V
	$V_S = \pm15V$										
	$R_L \geq 10\ k\Omega$				±12	±14		±12	±14		V
	$R_L \geq 2\ k\Omega$				±10	±13		±10	±13		V
Output Short Circuit Current	$T_A = 25°C$	10	25	35		25			25		mA
	$T_{AMIN} \leq T_A \leq T_{AMAX}$	10		40							mA
Common Mode Rejection Ratio	$T_{AMIN} \leq T_A \leq T_{AMAX}$										
	$R_S \leq 10\ k\Omega$, $V_{CM} = \pm12V$				70	90		70	90		dB
	$R_S \leq 50\ k\Omega$, $V_{CM} = \pm12V$	80	95								dB

Electrical Characteristics (Continued)

PARAMETER	CONDITIONS	LM741A/LM741E			LM741			LM741C			UNITS
		MIN	TYP	MAX	MIN	TYP	MAX	MIN	TYP	MAX	
Supply Voltage Rejection Ratio	$T_{AMIN} \leq T_A \leq T_{AMAX}$, $V_S = \pm 20V$ to $V_S = \pm 5V$										
	$R_S \leq 50\Omega$	86	96								dB
	$R_S \leq 10\ k\Omega$				77	96		77	96		dB
Transient Response	$T_A = 25°C$, Unity Gain										
Rise Time			0.25	0.8		0.3			0.3		µs
Overshoot			6.0	20		5			5		%
Bandwidth (Note 4)	$T_A = 25°C$	0.437	1.5								MHz
Slew Rate	$T_A = 25°C$, Unity Gain	0.3	0.7			0.5			0.5		V/µs
Supply Current	$T_A = 25°C$					1.7	2.8		1.7	2.8	mA
Power Consumption	$T_A = 25°C$										
	$V_S = \pm 20V$		80	150							mW
	$V_S = \pm 15V$					50	85		50	85	mW
LM741A	$V_S = \pm 20V$										
	$T_A = T_{AMIN}$			165							mW
	$T_A = T_{AMAX}$			135							mW
LM741E	$V_S = \pm 20V$			150							mW
	$T_A = T_{AMIN}$			150							mW
	$T_A = T_{AMAX}$			150							mW
LM741	$V_S = \pm 15V$										
	$T_A = T_{AMIN}$					60	100				mW
	$T_A = T_{AMAX}$					45	75				mW

Note 1: The maximum junction temperature of the LM741/LM741A is 150°C, while that of the LM741C/LM741E is 100°C. For operation at elevated temperatures, devices in the TO-5 package must be derated based on a thermal resistance of 150°C/W junction to ambient, or 45°C/W junction to case. The thermal resistance of the dual-in-line package is 100°C/W junction to ambient.

Note 2: For supply voltages less than ±15V, the absolute maximum input voltage is equal to the supply voltage.

Note 3: Unless otherwise specified, these specifications apply for $V_S = \pm 15V$, $-55°C \leq T_A \leq +125°C$ (LM741/LM741A). For the LM741C/LM741E, these specifications are limited to $0°C \leq T_A \leq +70°C$.

Note 4: Calculated value from: BW (MHz) = 0.35/Rise Time(µs).

Operational Amplifiers/Buffers

LM2900/LM3900, LM3301, LM3401 Quad Amplifiers

General Description

The LM2900 series consists of four independent, dual input, internally compensated amplifiers which were designed specifically to operate off of a single power supply voltage and to provide a large output voltage swing. These amplifiers make use of a current mirror to achieve the non-inverting input function. Application areas include: ac amplifiers, RC active filters, low frequency triangle, squarewave and pulse waveform generation circuits, tachometers and low speed, high voltage digital logic gates.

Features

- Wide single supply voltage   4 V_{DC} to 36 V_{DC}
  range or dual supplies ± 2 V_{DC} to ± 18 V_{DC}
- Supply current drain independent of supply voltage
- Low input biasing current   30 nA
- High open-loop gain   70 dB
- Wide bandwidth   2.5 MHz (Unity Gain)
- Large output voltage swing   (V^+ -1) Vp-p
- Internally frequency compensated for unity gain
- Output short-circuit protection

Schematic and Connection Diagrams

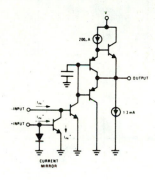

Order Number LM2900J
See NS Package J14A
Order Number LM2900N,
LM3900N, LM3301N
or LM3401N
See NS Package N14A

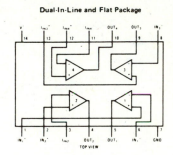

Dual-In-Line and Flat Package

TOP VIEW

Typical Applications   (V^+ = 15 V_{DC})

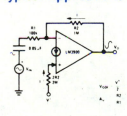

Inverting Amplifier

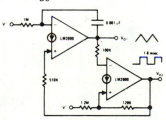

Triangle/Square Generator

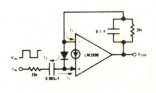

Frequency-Doubling Tachometer

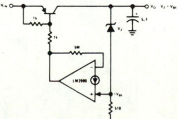

Low V_{IN} – V_{OUT} Voltage Regulator

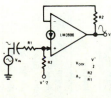

Non-Inverting Amplifier

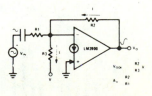

Negative Supply Biasing

Absolute Maximum Ratings

	LM2900/LM3900	LM3301	LM3401
Supply Voltage	32 VDC / ±16 VDC	28 VDC / ±14 VDC	18 VDC / ±9 VDC
Power Dissipation (TA = 25°C) (Note 1)			
Cavity DIP	900 mW		
Flat Pack	800 mW		
Molded DIP	570 mW	570 mW	570 mW
Input Currents, I_{IN}^+ or I_{IN}^-	20 mADC	20 mADC	20 mADC
Output Short-Circuit Duration – One Amplifier TA = 25°C (See Application Hints)	Continuous	Continuous	Continuous
Operating Temperature Range			
LM2900	-40°C to +85°C		
LM3900	0°C to +70°C	-40°C to +85°C	0°C to +75°C
Storage Temperature Range	-65°C to +150°C	-65°C to +150°C	-65°C to +150°C
Lead Temperature (Soldering, 10 seconds)	300°C	300°C	300°C

Electrical Characteristics (Note 6)

PARAMETER	CONDITIONS	LM2900			LM3900			LM3301			LM3401			UNITS
		MIN	TYP	MAX	MIN	TYP	MAX	MIN	TYP	MAX	MIN	TYP	MAX	
Open Loop														
Voltage Gain	TA = 25°C, f = 100 Hz										800			V/mV
Voltage Gain	TA = 25°C, Inverting Input	1.2	2.8		1.2	2.8		1.2	2.8		1.2	2.8		V/mV
Input Resistance			1			1			1		0.1	1		MΩ
Output Resistance			8			8			8			8		kΩ
Unity Gain Bandwidth	TA = 25°C, Inverting Input		2.5			2.5			2.5			2.5		MHz
Input Bias Current	TA = 25°C, Inverting Input		30	200		30	200		30	300		30	300	nA
	Inverting Input												500	nA
Slew Rate	TA = 25°C, Positive Output Swing		0.5			0.5			0.5			0.5		V/μs
	TA = 25°C, Negative Output Swing		20			20			20			20		V/μs
Supply Current	TA = 25°C, RL = ∞ On All Amplifiers		6.2	10		6.2	10		6.2	10		6.2	10	mADC
Output Voltage Swing	TA = 25°C, RL = 2k, VCC = 15.0 VDC													
VOUT High	$I_{IN}^- = 0$, $I_{IN}^+ = 0$	13.5			13.5			13.5			13.5			VDC
VOUT Low	$I_{IN}^- = 10\mu A$, $I_{IN}^+ = 0$		0.09	0.2		0.09	0.2		0.09	0.2		0.09	0.2	VDC
VOUT High	$I_{IN}^- = 0$, $I_{IN}^+ = 0$ $R_L = \infty$, VCC = Absolute Maximum Ratings		29.5			29.5			25.5			15.5		VDC
Output Current Capability	TA = 25°C													
Source		6	18		6	18		5	18		5	10		mADC
Sink	(Note 2)	0.5	1.3		0.5	1.3		0.5	1.3		0.5	1.3		mADC
ISINK	VOL = 1V, IIN = 5μA		5			5			5			5		mADC

Electrical Characteristics (Continued) (Note 6)

PARAMETER	CONDITIONS	LM2900			LM3900			LM3301			LM3401			UNITS
		MIN	TYP	MAX	MIN	TYP	MAX	MIN	TYP	MAX	MIN	TYP	MAX	
Power Supply Rejection	$T_A = 25°C$, f = 100 Hz		70			70			70			70		dB
Mirror Gain	@ 20μA (Note 3)	0.90	1.0	1.1	0.90	1.0	1.1	0.90	1	1.10	0.90	1	1.10	μA/μA
	@ 200μA (Note 3)	0.90	1.0	1.1	0.90	1.0	1.1	0.90	1	1.10	0.90	1	1.10	μA/μA
ΔMirror Gain	@ 20μA To 200μA (Note 3)	2		5	2		5	2		5	2		5	%
Mirror Current	(Note 4)		10	500		10	500		10	500		10	500	μADC
Negative Input Current	$T_A = 25°C$ (Note 5)		1.0			1.0			1.0			1.0		mADC
Input Bias Current	Inverting Input		300			300								nA

Note 1: For operating at high temperatures, the device must be derated based on a 125°C maximum junction temperature and a thermal resistance of 175°C/W which applies for the device soldered in a printed circuit board, operating in a still air ambient.

Note 2: The output current sink capability can be increased for large signal conditions by overdriving the inverting input. This is shown in the section on Typical Characteristics.

Note 3: This spec indicates the current gain of the current mirror which is used as the non-inverting input.

Note 4: Input V_{BE} match between the non-inverting and the inverting inputs occurs for a mirror current (non-inverting input current) of approximately 10μA. This is therefore a typical design center for many of the application circuits.

Note 5: Clamp transistors are included on the IC to prevent the input voltages from swinging below ground more than approximately −0.3 V_{DS}. The negative input currents which may result from large signal overdrive with capacitance input coupling need to be externally limited to values of approximately 1 mA. Negative input currents in excess of 4 mA will cause the output voltage to drop to a low voltage. This maximum current applies to any one of the input terminals. If more than one of the input terminals are simultaneously driven negative smaller maximum currents are allowed. Common-mode current biasing can be used to prevent negative input voltages; see for example, the "Differentiator Circuit" in the applications section.

Note 6: These specs apply for −55°C ≤ T_A ≤ +125°C, unless otherwise stated.

Typical Performance Characteristics

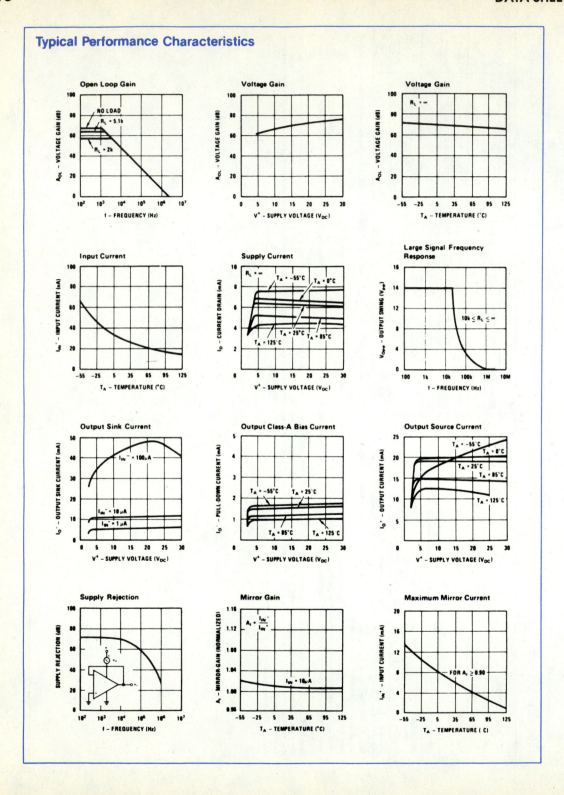

Application Hints

When driving either input from a low-impedance source, a limiting resistor should be placed in series with the input lead to limit the peak input current. Currents as large as 20 mA will not damage the device, but the current mirror on the non-inverting input will saturate and cause a loss of mirror gain at mA current levels—especially at high operating temperatures.

Precautions should be taken to insure that the power supply for the integrated circuit never becomes reversed in polarity or that the unit is not inadvertently installed backwards in a test socket as an unlimited current surge through the resulting forward diode within the IC could cause fuzing of the internal conductors and result in a destroyed unit.

Output short circuits either to ground or to the positive power supply should be of short time duration. Units can be destroyed, not as a result of the short circuit current causing metal fuzing, but rather due to the large increase in IC chip dissipation which will cause eventual failure due to excessive junction temperatures. For example, when operating from a well-regulated +5 V_{DC} power supply at T_A = 25°C with a 100 kΩ shunt-feedback resistor (from the output to the inverting input) a short directly to the power supply will not cause catastrophic failure but the current magnitude will be approximately 50 mA and the junction temperature will be above T_J max. Larger feedback resistors will reduce the current, 11 MΩ provides approximately 30 mA, an open circuit provides 1.3 mA, and a direct connection from the output to the non-inverting input will result in catastrophic failure when the output is shorted to V^+ as this then places the base-emitter junction of the input transistor directly across the power supply. Short-circuits to ground will have magnitudes of approximately 30 mA and will not cause catastrophic failure at T_A = 25°C.

Unintentional signal coupling from the output to the non-inverting input can cause oscillations. This is likely only in breadboard hook-ups with long component leads and can be prevented by a more careful lead dress or by locating the non-inverting input biasing resistor close to the IC. A quick check of this condition is to bypass the non-inverting input to ground with a capacitor. High impedance biasing resistors used in the non-inverting input circuit make this input lead highly susceptible to unintentional ac signal pickup.

Operation of this amplifier can be best understood by noticing that input currents are differenced at the inverting-input terminal and this difference current then flows through the external feedback resistor to produce the output voltage. Common-mode current biasing is generally useful to allow operating with signal levels near ground or even negative as this maintains the inputs biased at $+V_{BE}$. Internal clamp transistors (see note 5) catch negative input voltages at approximately -0.3 V_{DC} but the magnitude of current flow has to be limited by the external input network. For operation at high temperature, this limit should be approximately 100μA.

This new "Norton" current-differencing amplifier can be used in most of the applications of a standard IC op amp. Performance as a dc amplifier using only a single supply is not as precise as a standard IC op amp operating with split supplies but is adequate in many less critical applications. New functions are made possible with this amplifier which are useful in single power supply systems. For example, biasing can be designed separately from the ac gain as was shown in the "inverting amplifier," the "difference integrator" allows controlling the charging and the discharging of the integrating capacitor both with positive voltages, and the "frequency doubling tachometer" provides a simple circuit which reduces the ripple voltage on a tachometer output dc voltage.

Typical Applications (Continued)

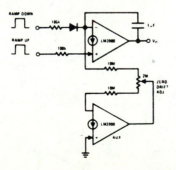

Low-Drift Ramp and Hold Circuit

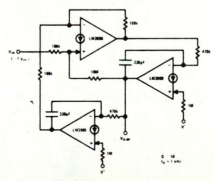

Bi-Quad Active Filter
(2nd Degree State-Variable Network)

Typical Applications (Continued)

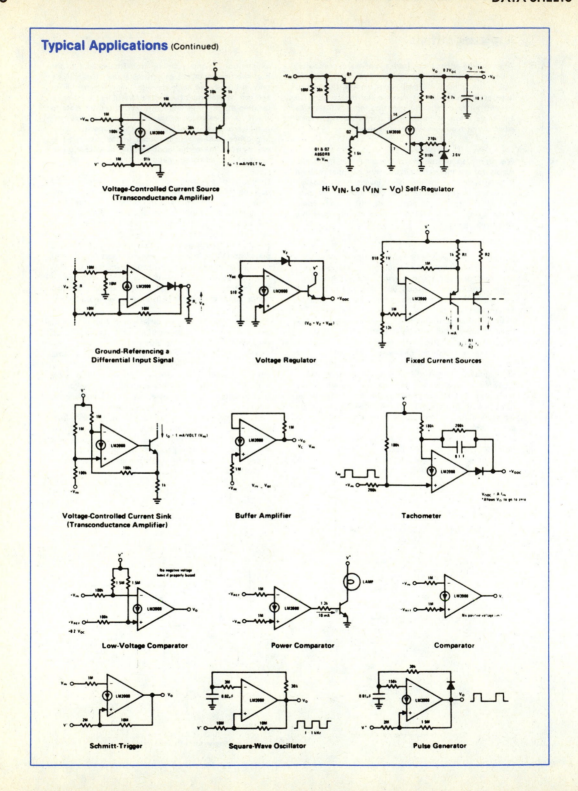

Voltage-Controlled Current Source
(Transconductance Amplifier)

Hi V$_{IN}$, Lo (V$_{IN}$ − V$_O$) Self-Regulator

Ground-Referencing a
Differential Input Signal

Voltage Regulator

Fixed Current Sources

Voltage-Controlled Current Sink
(Transconductance Amplifier)

Buffer Amplifier

Tachometer

Low-Voltage Comparator

Power Comparator

Comparator

Schmitt-Trigger

Square-Wave Oscillator

Pulse Generator

Typical Applications (Continued)

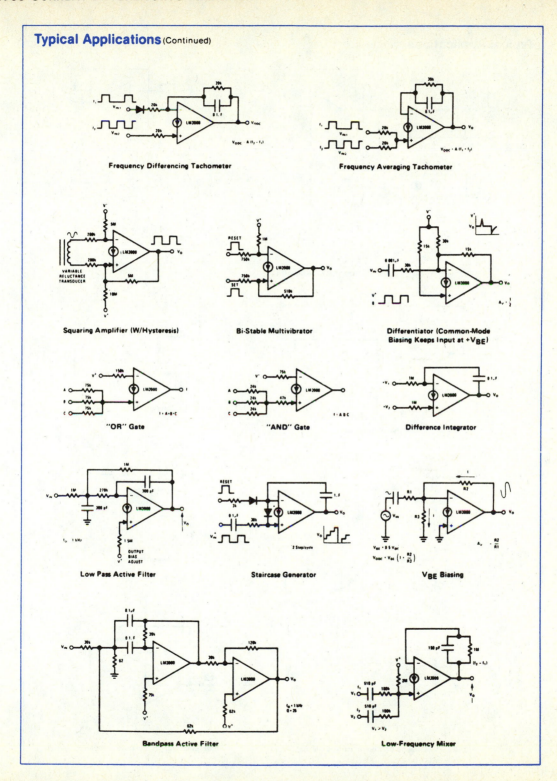

Frequency Differencing Tachometer

Frequency Averaging Tachometer

Squaring Amplifier (W/Hysteresis)

Bi-Stable Multivibrator

Differentiator (Common-Mode Biasing Keeps Input at $+V_{BE}$)

"OR" Gate

"AND" Gate

Difference Integrator

Low Pass Active Filter

Staircase Generator

V_{BE} Biasing

Bandpass Active Filter

Low-Frequency Mixer

Typical Applications (Continued)

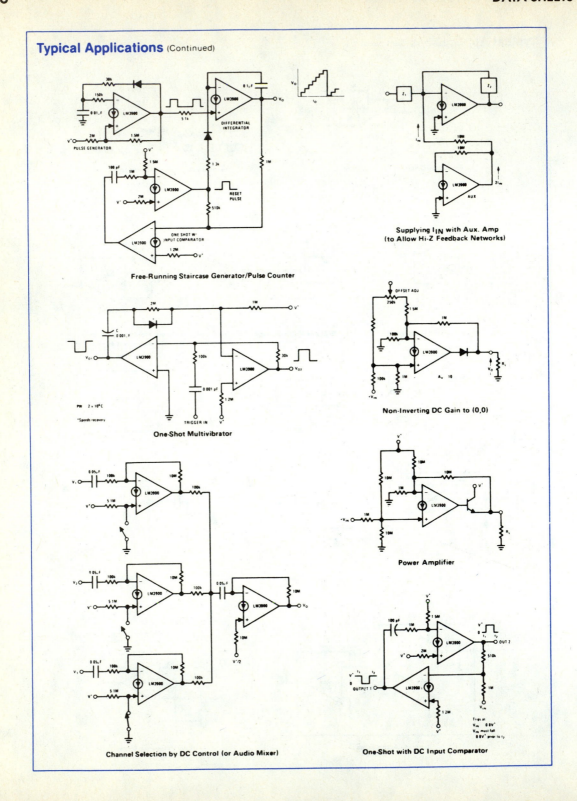

Free-Running Staircase Generator/Pulse Counter

Supplying I_{IN} with Aux. Amp
(to Allow Hi-Z Feedback Networks)

One-Shot Multivibrator

Non-Inverting DC Gain to (0,0)

Power Amplifier

Channel Selection by DC Control (or Audio Mixer)

One-Shot with DC Input Comparator

Typical Applications (Continued)

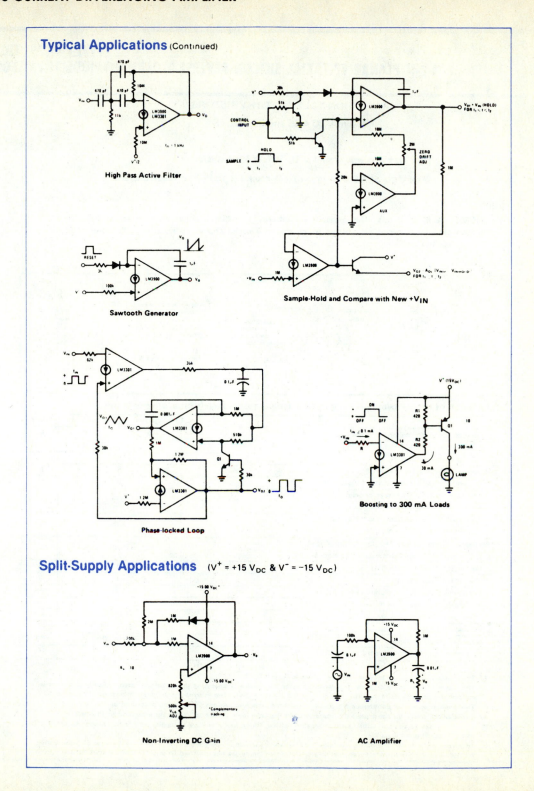

High Pass Active Filter

Sawtooth Generator

Sample-Hold and Compare with New +V$_{IN}$

Phase-locked Loop

Boosting to 300 mA Loads

Split-Supply Applications (V$^+$ = +15 V$_{DC}$ & V$^-$ = −15 V$_{DC}$)

Non-Inverting DC Gain

AC Amplifier

TYPES TIC35, TIC36
P-N-P-N PLANAR EPITAXIAL SILICON REVERSE-BLOCKING TRIODE THYRISTORS

RADIATION-TOLERANT THYRISTORS
400 mA DC ● 15 and 30 VOLTS

- Max I_{GT} of 5 mA after 1×10^{14} Fast Neutrons/cm^2
- Max V_{TM} of 1.6 V at I_{TM} of 1 A after 1×10^{14} Fast Neutrons/cm^2

description

The TIC35, TIC36 thyristors offer a significant advance in radiation-tolerant-device technology. Unique construction techniques produce thyristors which maintain useful characteristics after fast-neutron radiation fluences through 10^{15} n/cm^2.

mechanical data

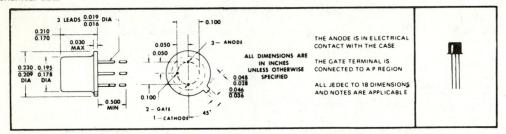

absolute maximum ratings over operating free-air temperature range (unless otherwise noted)

	TIC35	TIC36	UNIT
Continuous Off-State Voltage, V_D (See Note 1)	15	30	V
Repetitive Peak Off-State Voltage, V_{DRM} (See Note 1)	15	30	V
Continuous Reverse Voltage, V_R	5		V
Repetitive Peak Reverse Voltage, V_{RRM}	5		V
Nonrepetitive Peak Reverse Voltage, V_{RSM} (See Note 2)	5		V
Continuous On-State Current at (or below) 55°C Case Temperature (See Note 3)	400		mA
Continuous On-State Current at (or below) 25°C Free-Air Temperature (See Note 4)	225		mA
Average On-State Current (180° Conduction Angle) at (or below) 55°C Case Temperature (See Note 5)	320		mA
Surge On-State Current (See Note 6)	3		A
Peak Negative Gate Voltage	−4		V
Peak Positive Gate Current (Pulse Width ≤ 300 µs)	250		mA
Peak Gate Power Dissipation (Pulse Width ≤ 300 µs)	500		mW
Average Gate Power Dissipation	10		mW
Operating Free-Air or Case Temperature Range	−55 to 125		°C
Storage Temperature Range	−65 to 200		°C
Lead Temperature 1/16 Inch from Case for 10 Seconds	260		°C

NOTES: 1. These values apply when the gate-cathode resistance $R_{GK} = 1 \text{ k}\Omega$.
2. This value applies for a 5-ms rectangular pulse when the device is operating at (or below) rated values of peak reverse voltage and on-state current. Surge may be repeated after the device has returned to original thermal equilibrium.
3. These values apply for continuous d-c operation with resistive load. Above 55°C derate according to Figure 2.
4. These values apply for continuous d-c operation with resistive load. Above 25°C derate according to Figure 3.
5. This value may be applied continuously under single-phase, 60-Hz, half-sine-wave operation with resistive load. Above 55°C derate according to Figure 2.
6. This value applies for one 60-Hz half sine wave when the device is operating at (or below) rated values of peak reverse voltage and on-state current. Surge may be repeated after the device has returned to original thermal equilibrium.

TYPES TIC35, TIC36
P-N-P-N PLANAR EPITAXIAL SILICON REVERSE-BLOCKING TRIODE THYRISTORS

electrical characteristics at 25°C free-air temperature (unless otherwise noted)

	PARAMETER	TEST CONDITIONS		MIN	TYP	MAX	UNIT
I_D	Static Off-State Current	V_D = Rated V_D,	R_{GK} = 1 kΩ, T_A = 125°C			20	μA
I_R	Static Reverse Current	V_R = 5 V,	R_{GK} = 1 kΩ, T_A = 125°C			100	μA
I_{GR}	Gate Reverse Current	V_{KG} = 4 V,	I_A = 0			5	μA
I_{GT}	Gate Trigger Current	V_{AA} = 6 V, R_L = 100 Ω, V_{GG} = 6 V, $R_{G(source)}$ ⩾ 10 kΩ, $t_{p(g)}$ ⩾ 20 μs, T_A = −55°C				100	μA
		V_{AA} = 6 V, R_L = 100 Ω, V_{GG} = 6 V, $R_{G(source)}$ ⩾ 10 kΩ, $t_{p(g)}$ ⩾ 20 μs				20	
V_{GT}	Gate Trigger Voltage	V_{AA} = 6 V, R_L = 100 Ω, R_{GK} = 1 kΩ, $t_{p(g)}$ ⩾ 20 μs, T_A = −55°C				0.9	V
		V_{AA} = Rated V_D, R_L = 100 Ω, R_{GK} = 1 kΩ, $t_{p(g)}$ ⩾ 20 μs, T_A = 125°C		0.2			
		V_{AA} = 6 V, R_L = 100 Ω, R_{GK} = 1 kΩ, $t_{p(g)}$ ⩾ 20 μs				0.75	
I_H	Holding Current	V_{AA} = 6 V, R_{GK} = 1 kΩ, Initiating I_T = 10 mA, T_A = −55°C				4	mA
		V_{AA} = 6 V, R_{GK} = 1 kΩ, Initiating I_T = 10 mA				2	
V_{TM}	Peak On-State Voltage	I_{TM} = 1 A,	See Note 7			1.6	V
dv/dt	Critical Rate of Rise of Off-State Voltage	V_D = Rated V_D,	R_{GK} = 1 kΩ		12		V/μs

post-irradiation electrical characteristics at 25°C free-air temperature

	PARAMETER	TEST CONDITIONS	RADIATION FLUENCE†	MIN	TYP	MAX	UNIT
I_{GT}	Gate Trigger Current	V_{AA} = 6 V, R_L = 100 Ω	1 × 10^{14} n/cm^2			5	mA
V_{TM}	Peak On-State Voltage	I_{TM} = 1 A, See Note 7	1 × 10^{14} n/cm^2			1.6	V

† Radiation is fast neutrons (n) at E ⩾ 10 keV (reactor spectrum).

thermal characteristics

	PARAMETER	MIN	TYP	MAX	UNIT
θ_{J-C}	Junction-to-Case Thermal Resistance			124	°C/W
θ_{J-A}	Junction-to-Free-Air Thermal Resistance			345	

NOTE: 7. These parameters must be measured using pulse techniques. t_w = 300 μs, duty cycle ⩽ 2%. Voltage-sensing contacts, separate from the current-carrying contacts, are used.

TYPES TIC226B, TIC226D
SILICON BIDIRECTIONAL TRIODE THYRISTORS

8 A RMS • 200 V and 400 V
TRIACS
for
HIGH-TEMPERATURE, HIGH-CURRENT, and HIGH-VOLTAGE APPLICATIONS
• Typ dv/dt of 500 V/μs at 25°C

description

These devices are bidirectional triode thyristors (triacs) which may be triggered from the off-state to the on-state by either polarity of gate signal with Main Terminal 2 at either polarity.

mechanical data

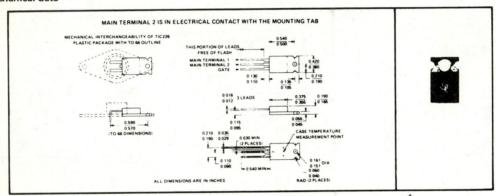

absolute maximum ratings over operating case temperature range (unless otherwise noted)[†]

			UNIT
Repetitive Peak Off-State Voltage, V_{DRM} (See Note 1)	TIC226B	200	V
	TIC226D	400	
Full-Cycle RMS On-State Current at (or below) 85°C Case Temperature, $I_{T(RMS)}$ (See Note 2)		8	A
Peak On-State Surge Current, Full-Sine-Wave, I_{TSM} (See Note 3)		70	A
Peak On-State Surge Current, Half-Sine-Wave, I_{TSM} (See Note 4)		80	A
Peak Gate Current, I_{GM}		1	A
Peak Gate Power Dissipation, P_{GM}, at (or below) 85°C Case Temperature (Pulse Width ≤ 200 μs)		2.2	W
Average Gate Power Dissipation, $P_{G(av)}$, at (or below) 85°C Case Temperature (See Note 5)		0.9	W
Operating Case Temperature Range		−40 to 110	°C
Storage Temperature Range		−40 to 125	°C
Lead Temperature 1/16 Inch from Case for 10 Seconds		230	°C

NOTES: 1. These values apply bidirectionally for any value of resistance between the gate and Main Terminal 1.
    2. This value applies for 50-Hz to 60-Hz full-sine-wave operation with resistive load. Above 85°C derate according to Figure 2.
    3. This value applies for one 60-Hz full sine wave when the device is operating at (or below) the rated value of on-state current. Surge may be repeated after the device has returned to original thermal equilibrium. During the surge, gate control may be lost.
    4. This value applies for one 60-Hz half sine wave when the device is operating at (or below) the rated value of on-state current. Surge may be repeated after the device has returned to original thermal equilibrium. During the surge, gate control may be lost.
    5. This value applies for a maximum averaging time of 16.6 ms.

[†]All voltage values are with respect to Main Terminal 1.

TYPES TIC226B, TIC226D
SILICON BIDIRECTIONAL TRIODE THYRISTORS

electrical characteristics at 25°C case temperature (unless otherwise noted)[†]

	PARAMETER	TEST CONDITIONS			MIN	TYP	MAX	UNIT
I_{DRM}	Repetitive Peak Off-State Current	V_{DRM} = Rated V_{DRM},	$I_G = 0$	$T_C = 110°C$			±2	mA
I_{GTM}	Peak Gate Trigger Current	V_{supply} = +12 V[†],	$R_L = 10\ \Omega$,	$t_{p(g)} \geqslant 20\ \mu s$		15	50	mA
		V_{supply} = +12 V[†],	$R_L = 10\ \Omega$,	$t_{p(g)} \geqslant 20\ \mu s$		−25	−50	
		V_{supply} = −12 V[†],	$R_L = 10\ \Omega$,	$t_{p(g)} \geqslant 20\ \mu s$		−30	−50	
		V_{supply} = −12 V[†],	$R_L = 10\ \Omega$,	$t_{p(g)} \geqslant 20\ \mu s$	75			
V_{GTM}	Peak Gate Trigger Voltage	V_{supply} = +12 V[†],	$R_L = 10\ \Omega$,	$t_{p(g)} \geqslant 20\ \mu s$		0.9	2.5	V
		V_{supply} = +12 V[†],	$R_L = 10\ \Omega$,	$t_{p(g)} \geqslant 20\ \mu s$		−1.2	−2.5	
		V_{supply} = −12 V[†],	$R_L = 10\ \Omega$,	$t_{p(g)} \geqslant 20\ \mu s$		−1.2	−2.5	
		V_{supply} = −12 V[†],	$R_L = 10\ \Omega$,	$t_{p(g)} \geqslant 20\ \mu s$	1.2			
V_{TM}	Peak On-State Voltage	$I_{TM} = \pm12$ A,	$I_G = 100$ mA, See Note 6				±2.1	V
I_H	Holding Current	V_{supply} = +12 V[†],	$I_G = 0$,	Initiating $I_{TM} = 500$ mA		20	60	mA
		V_{supply} = −12 V[†],	$I_G = 0$,	Initiating $I_{TM} = -500$ mA		−30	−60	
I_L	Latching Current	V_{supply} = +12 V[†],	See Note 7			30	70	mA
		V_{supply} = −12 V[†],	See Note 7			−40	−70	
dv/dt	Critical Rate of Rise of Off-State Voltage	V_{DRM} = Rated V_{DRM},	$I_G = 0$,	$T_C = 110°C$		500		V/μs
dv/dt	Critical Rate of Rise of Commutation Voltage	V_{DRM} = Rated V_{DRM},	$I_{TRM} = \pm12$ A, $T_C = 85°C$, See Figure 3		5			V/μs

[†]All voltage values are with respect to Main Terminal 1.

NOTES: 6. This parameter must be measured using pulse techniques. $t_w \leqslant 1$ ms, duty cycle $\leqslant 2\%$. Voltage-sensing contacts, separate from the current-carrying contacts, are located within 0.125 inch from the device body.

7. The triacs are triggered by a 15-V (open-circuit amplitude) pulse supplied by a generator with the following characteristics: $R_G = 100\ \Omega$, $t_w = 20\ \mu s$, $t_r \leqslant 15$ ns, $t_f \leqslant 15$ ns, $f = 1$ kHz.

thermal characteristics

	PARAMETER	MAX	UNIT
$R_{\theta JC}$	Junction-to-Case Thermal Resistance	1.8	°C/W
$R_{\theta JA}$	Junction-to-Free-Air Thermal Resistance	62.5	

 **National Semiconductor**

LM135/LM235/LM335, LM135A/LM235A/LM335A Precision Temperature Sensors

General Description

The LM135 series are precision, easily-calibrated, integrated circuit temperature sensors. Operating as a 2-terminal zener, the LM135 has a breakdown voltage directly proportional to absolute temperature at +10 mV/°K. With less than 1Ω dynamic impedance the device operates over a current range of 400 μA to 5 mA with virtually no change in performance. When calibrated at 25°C the LM135 has typically less than 1°C error over a 100°C temperature range. Unlike other sensors the LM135 has a linear output.

Applications for the LM135 include almost any type of temperature sensing over a −55°C to +150°C temperature range. The low impedance and linear output make interfacing to readout or control circuitry especially easy.

The LM135 operates over a −55°C to +150°C temperature range while the LM235 operates over a −40°C to +125°C temperature range. The LM335 operates from −40°C to +100°C. The LM135/LM235/LM335 are available packaged in hermetic TO-46 transistor packages while the LM335 is also available in plastic TO-92 packages.

Features

- Directly calibrated in °Kelvin
- 1°C initial accuracy available
- Operates from 400 μA to 5 mA
- Less than 1Ω dynamic impedance
- Easily calibrated
- Wide operating temperature range
- 200°C overrange
- Low cost

Schematic Diagram

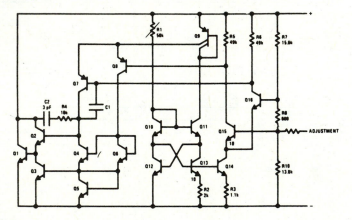

Typical Applications

Basic Temperature Sensor

Calibrated Sensor

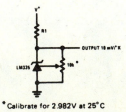

*Calibrate for 2.982V at 25°C

Wide Operating Supply

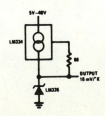

Absolute Maximum Ratings

Reverse Current	15 mA
Forward Current	10 mA
Storage Temperature	
TO-46 Package	$-60°C$ to $+180°C$
TO-92 Package	$-60°C$ to $+150°C$

Specified Operating Temperature Range

	Continuous	Intermittent (Note 2)
LM135, LM135A	$-55°C$ to $+150°C$	$150°C$ to $200°C$
LM235, LM235A	$-40°C$ to $+125°C$	$125°C$ to $150°C$
LM335, LM335A	$-40°C$ to $+100°C$	$100°C$ to $125°C$
Lead Temperature (Soldering, 10 seconds)		$300°C$

Temperature Accuracy LM135/LM235, LM135A/LM235A (Note 1)

PARAMETER	CONDITIONS	LM135A/LM235A			LM135/LM235			UNITS
		MIN	TYP	MAX	MIN	TYP	MAX	
Operating Output Voltage	$T_C = 25°C$, $I_R = 1$ mA	2.97	2.98	2.99	2.95	2.98	3.01	V
Uncalibrated Temperature Error	$T_C = 25°C$, $I_R = 1$ mA		0.5	1		1	3	°C
Uncalibrated Temperature Error	$T_{MIN} < T_C < T_{MAX}$, $I_R = 1$ mA		1.3	2.7		2	5	°C
Temperature Error with 25°C Calibration	$T_{MIN} < T_C < T_{MAX}$, $I_R = 1$ mA		0.3	1		0.5	1.5	°C
Calibrated Error at Extended Temperatures	$T_C = T_{MAX}$ (Intermittent)		2			2		°C
Non-Linearity	$I_R = 1$ mA		0.3	0.5		0.3	1	°C

Temperature Accuracy LM335, LM335A (Note 1)

PARAMETER	CONDITIONS	LM335A			LM335			UNITS
		MIN	TYP	MAX	MIN	TYP	MAX	
Operating Output Voltage	$T_C = 25°C$, $I_R = 1$ mA	2.95	2.98	3.01	2.92	2.98	3.04	V
Uncalibrated Temperature Error	$T_C = 25°C$, $I_R = 1$ mA		1	3		2	6	°C
Uncalibrated Temperature Error	$T_{MIN} < T_C < T_{MAX}$, $I_R = 1$ mA		2	5		4	9	°C
Temperature Error with 25°C Calibration	$T_{MIN} < T_C < T_{MAX}$, $I_R = 1$ mA		0.5	1		1	2	°C
Calibrated Error at Extended Temperatures	$T_C = T_{MAX}$ (Intermittent)		2			2		°C
Non-Linearity	$I_R = 1$ mA		0.3	1.5		0.3	1.5	°C

Electrical Characteristics (Note 1)

PARAMETER	CONDITIONS	LM135/LM235 LM135A/LM235A			LM335 LM335A			UNITS
		MIN	TYP	MAX	MIN	TYP	MAX	
Operating Output Voltage Change with Current	$400 \mu A < I_R < 5$ mA At Constant Temperature		2.5	10		3	14	mV
Dynamic Impedance	$I_R = 1$ mA		0.5			0.6		Ω
Output Voltage Temperature Drift			+10			+10		mV/°C
Time Constant	Still Air		80			80		sec
	100 ft/Min Air		10			10		sec
	Stirred Oil		1			1		sec
Time Stability	$T_C - 125°C$		0.2			0.2		°C/khr

Note 1: Accuracy measurements are made in a well-stirred oil bath. For other conditions, self heating must be considered.

Note 2: Continuous operation at these temperatures for 10,000 hours for H package and 5,000 hours for Z package may decrease life expectancy of the device.

Typical Performance Characteristics

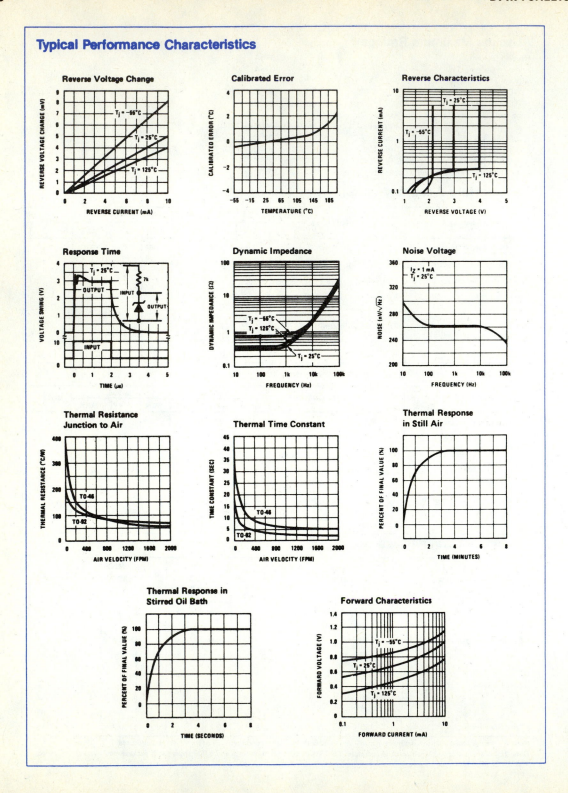

 National Semiconductor

Industrial Blocks

LM555/LM555C Timer

General Description

The LM555 is a highly stable device for generating accurate time delays or oscillation. Additional terminals are provided for triggering or resetting if desired. In the time delay mode of operation, the time is precisely controlled by one external resistor and capacitor. For astable operation as an oscillator, the free running frequency and duty cycle are accurately controlled with two external resistors and one capacitor. The circuit may be triggered and reset on falling waveforms, and the output circuit can source or sink up to 200 mA or drive TTL circuits.

Features

- Direct replacement for SE555/NE555
- Timing from microseconds through hours
- Operates in both astable and monostable modes

- Adjustable duty cycle
- Output can source or sink 200 mA
- Output and supply TTL compatible
- Temperature stability better than 0.005% per °C
- Normally on and normally off output

Applications

- Precision timing
- Pulse generation
- Sequential timing
- Time delay generation
- Pulse width modulation
- Pulse position modulation
- Linear ramp generator

Schematic Diagram

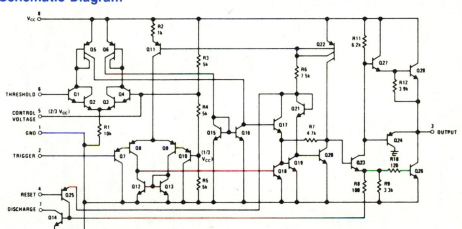

Connection Diagrams

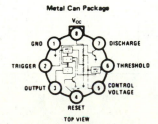

Metal Can Package

TOP VIEW

Order Number LM555H, LM555CH
See NS Package H08C

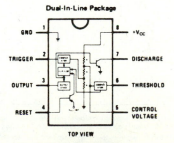

Dual-In-Line Package

TOP VIEW

Order Number LM555CN
See NS Package N08B
Order Number LM555J or LM555CJ
See NS Package J08A

Absolute Maximum Ratings

Supply Voltage	+18V
Power Dissipation (Note 1)	600 mW
Operating Temperature Ranges	
LM555C	0°C to +70°C
LM555	−55°C to +125°C
Storage Temperature Range	−65°C to +150°C
Lead Temperature (Soldering, 10 seconds)	300°C

Electrical Characteristics (T_A = 25°C, V_{CC} = +5V to +15V, unless otherwise specified)

PARAMETER	CONDITIONS	LIMITS						UNITS
		LM555			LM555C			
		MIN	TYP	MAX	MIN	TYP	MAX	
Supply Voltage		4.5		18	4.5		16	V
Supply Current	V_{CC} = 5V, R_L = ∞		3	5		3	6	mA
	V_{CC} = 15V, R_L = ∞		10	12		10	15	mA
	(Low State) (Note 2)							
Timing Error, Monostable								
Initial Accuracy			0.5			1		%
Drift with Temperature	R_A, R_B = 1k to 100 k,		30			50		ppm/°C
	C = 0.1μF, (Note 3)							
Accuracy over Temperature			1.5			1.5		%
Drift with Supply			0.05			0.1		%/V
Timing Error, Astable								
Initial Accuracy			1.5			2.25		%
Drift with Temperature			90			150		ppm/°C
Accuracy over Temperature			2.5			3.0		%
Drift with Supply			0.15			0.30		%/V
Threshold Voltage			-0.667			0.667		x V_{CC}
Trigger Voltage	V_{CC} = 15V	4.8	5	5.2		5		V
	V_{CC} = 5V	1.45	1.67	1.9		1.67		V
Trigger Current			0.01	0.5		0.5	0.9	μA
Reset Voltage		0.4	0.5	1	0.4	0.5	1	V
Reset Current			0.1	0.4		0.1	0.4	mA
Threshold Current	(Note 4)		0.1	0.25		0.1	0.25	μA
Control Voltage Level	V_{CC} = 15V	9.6	10	10.4	9	10	11	V
	V_{CC} = 5V	2.9	3.33	3.8	2.6	3.33	4	V
Pin 7 Leakage Output High			1	100		1	100	nA
Pin 7 Sat (Note 5)								
Output Low	V_{CC} = 15V, I_7 = 15 mA		150			180		mV
Output Low	V_{CC} = 4.5V, I_7 = 4.5 mA		70	100		80	200	mV
Output Voltage Drop (Low)	V_{CC} = 15V							
I_{SINK} = 10 mA			0.1	0.15		0.1	0.25	V
I_{SINK} = 50 mA			0.4	0.5		0.4	0.75	V
I_{SINK} = 100 mA			2	2.2		2	2.5	V
I_{SINK} = 200 mA			2.5			2.5		V
V_{CC} = 5V								
I_{SINK} = 8 mA			0.1	0.25				V
I_{SINK} = 5 mA						0.25	0.35	V
Output Voltage Drop (High)	I_{SOURCE} = 200 mA, V_{CC} = 15V		12.5			12.5		V
	I_{SOURCE} = 100 mA, V_{CC} = 15V	13	13.3		12.75	13.3		V
	V_{CC} = 5V	3	3.3		2.75	3.3		V
Rise Time of Output			100			100		ns
Fall Time of Output			100			100		ns

Note 1: For operating at elevated temperatures the device must be derated based on a +150°C maximum junction temperature and a thermal resistance of +45°C/W junction to case for TO-5 and +150°C/W junction to ambient for both packages.

Note 2: Supply current when output high typically 1 mA less at V_{CC} = 5V.

Note 3: Tested at V_{CC} = 5V and V_{CC} = 15V.

Note 4: This will determine the maximum value of R_A + R_B for 15V operation. The maximum total (R_A + R_B) is 20 MΩ.

Note 5: No protection against excessive pin 7 current is necessary providing the package dissipation rating will not be exceeded.

Typical Performance Characteristics

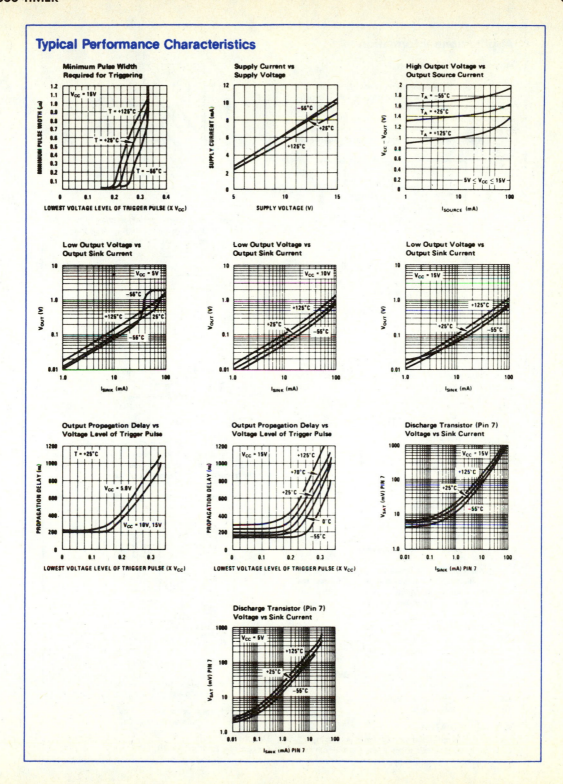

Applications Information

MONOSTABLE OPERATION

In this mode of operation, the timer functions as a one-shot (*Figure 1*). The external capacitor is initially held discharged by a transistor inside the timer. Upon application of a negative trigger pulse of less than 1/3 V_{CC} to pin 2, the flip-flop is set which both releases the short circuit across the capacitor and drives the output high.

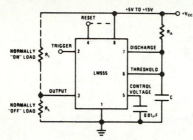

FIGURE 1. Monostable

The voltage across the capacitor then increases exponentially for a period of t = 1.1 R_AC, at the end of which time the voltage equals 2/3 V_{CC}. The comparator then resets the flip-flop which in turn discharges the capacitor and drives the output to its low state. *Figure 2* shows the waveforms generated in this mode of operation. Since the charge and the threshold level of the comparator are both directly proportional to supply voltage, the timing internal is independent of supply.

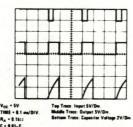

V_{CC} = 5V
TIME = 0.1 ms/DIV.
R_A = 9.1kΩ
C = 0.01μF

Top Trace: Input 5V/Div.
Middle Trace: Output 5V/Div.
Bottom Trace: Capacitor Voltage 2V/Div.

FIGURE 2. Monostable Waveforms

During the timing cycle when the output is high, the further application of a trigger pulse will not effect the circuit. However the circuit can be reset during this time by the application of a negative pulse to the reset terminal (pin 4). The output will then remain in the low state until a trigger pulse is again applied.

When the reset function is not in use, it is recommended that it be connected to V_{CC} to avoid any possibility of false triggering.

Figure 3 is a nomograph for easy determination of R, C values for various time delays.

NOTE: In monostable operation, the trigger should be driven high before the end of timing cycle.

ASTABLE OPERATION

If the circuit is connected as shown in *Figure 4* (pins 2 and 6 connected) it will trigger itself and free run as a

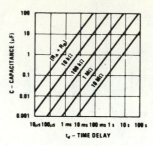

FIGURE 3. Time Delay

multivibrator. The external capacitor charges through R_A + R_B and discharges through R_B. Thus the duty cycle may be precisely set by the ratio of these two resistors.

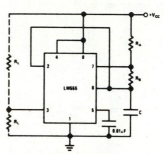

FIGURE 4. Astable

In this mode of operation, the capacitor charges and discharges between 1/3 V_{CC} and 2/3 V_{CC}. As in the triggered mode, the charge and discharge times, and therefore the frequency are independent of the supply voltage.

Figure 5 shows the waveforms generated in this mode of operation.

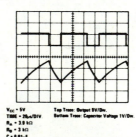

V_{CC} = 5V
TIME = 20μs/DIV.
R_A = 3.9 kΩ
R_B = 3 kΩ
C = 0.01μF

Top Trace: Output 5V/Div.
Bottom Trace: Capacitor Voltage 1V/Div.

FIGURE 5. Astable Waveforms

The charge time (output high) is given by:
$$t_1 = 0.693 (R_A + R_B) C$$

And the discharge time (output low) by:
$$t_2 = 0.693 (R_B) C$$

Thus the total period is:
$$T = t_1 + t_2 = 0.693 (R_A + 2R_B) C$$

Applications Information (Continued)

The frequency of oscillation is:

$$f = \frac{1}{T} = \frac{1.44}{(R_A + 2R_B)C}$$

Figure 6 may be used for quick determination of these RC values.

The duty cycle is:

$$D = \frac{R_B}{R_A + 2R_B}$$

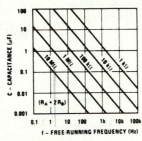

FIGURE 6. Free Running Frequency

FREQUENCY DIVIDER

The monostable circuit of *Figure 1* can be used as a frequency divider by adjusting the length of the timing cycle. *Figure 7* shows the waveforms generated in a divide by three circuit.

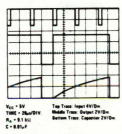

$V_{CC} = 5V$
TIME = 20μs/DIV.
$R_A = 0.1$ kΩ
$C = 0.01$μF

Top Trace: Input 4V/Div.
Middle Trace: Output 2V/Div.
Bottom Trace: Capacitor 2V/Div.

FIGURE 7. Frequency Divider

PULSE WIDTH MODULATOR

When the timer is connected in the monostable mode and triggered with a continuous pulse train, the output pulse width can be modulated by a signal applied to pin 5. *Figure 8* shows the circuit, and in *Figure 9* are some waveform examples.

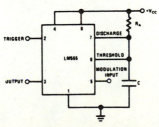

FIGURE 8. Pulse Width Modulator

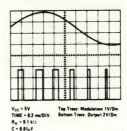

$V_{CC} = 5V$
TIME = 0.2 ms/DIV.
$R_A = 0.1$ kΩ
$C = 0.01$μF

Top Trace: Modulation 1V/Div.
Bottom Trace: Output 2V/Div.

FIGURE 9. Pulse Width Modulator

PULSE POSITION MODULATOR

This application uses the timer connected for astable operation, as in *Figure 10*, with a modulating signal again applied to the control voltage terminal. The pulse position varies with the modulating signal, since the threshold voltage and hence the time delay is varied. *Figure 11* shows the waveforms generated for a triangle wave modulation signal.

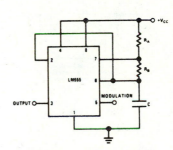

FIGURE 10. Pulse Position Modulator

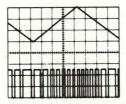

$V_{CC} = 5V$
TIME = 0.1 ms/DIV.
$R_A = 3.0$ kΩ
$R_B = 3$ kΩ
$C = 0.01$μF

Top Trace: Modulation Input 1V/Div.
Bottom Trace: Output 2V/Div.

FIGURE 11. Pulse Position Modulator

LINEAR RAMP

When the pullup resistor, R_A, in the monostable circuit is replaced by a constant current source, a linear ramp is

Applications Information (Continued)

generated. *Figure 12* shows a circuit configuration that will perform this function.

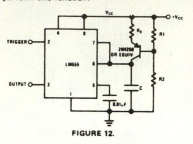

FIGURE 12.

Figure 13 shows waveforms generated by the linear ramp.

The time interval is given by:

$$T = \frac{2/3\ V_{CC}\ R_E\ (R_1 + R_2)\ C}{R_1\ V_{CC} - V_{BE}\ (R_1 + R_2)}$$

$$V_{BE} \simeq 0.6V$$

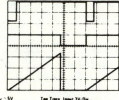

$V_{CC} = 5V$ Top Trace: Input 2V/Div.
TIME = 20µs/DIV.   Middle Trace: Output 5V/Div.
R_1 47 kΩ   Bottom Trace: Capacitor Voltage 1V/Div.
R_2 100 kΩ
R_E = 2.7 kΩ
C = 0.01µF

FIGURE 13. Linear Ramp

50% DUTY CYCLE OSCILLATOR

For a 50% duty cycle, the resistors R_A and R_B may be connected as in *Figure 14*. The time period for the out-put high is the same as previous, $t_1 = 0.693\ R_A\ C$. For the output low it is $t_2 =$

$$[(R_A\ R_B)/(R_A + R_B)]\ CLn\left[\frac{R_B - 2R_A}{2R_B - R_A}\right]$$

Thus the frequency of oscillation is $f = \dfrac{1}{t_1 + t_2}$

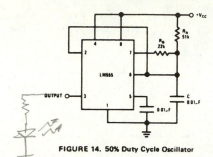

FIGURE 14. 50% Duty Cycle Oscillator

Note that this circuit will not oscillate if R_B is greater than 1/2 R_A because the junction of R_A and R_B cannot bring pin 2 down to 1/3 V_{CC} and trigger the lower comparator.

ADDITIONAL INFORMATION

Adequate power supply bypassing is necessary to protect associated circuitry. Minimum recommended is 0.1µF in parallel with 1µF electrolytic.

Lower comparator storage time can be as long as 10µs when pin 2 is driven fully to ground for triggering. This limits the monostable pulse width to 10µs minimum.

Delay time reset to output is 0.47µs typical. Minimum reset pulse width must be 0.3µs, typical.

Pin 7 current switches within 30 ns of the output (pin 3) voltage.

Industrial Blocks

LM565/LM565C Phase Locked Loop

General Description

The LM565 and LM565C are general purpose phase locked loops containing a stable, highly linear voltage controlled oscillator for low distortion FM demodulation, and a double balanced phase detector with good carrier suppression. The VCO frequency is set with an external resistor and capacitor, and a tuning range of 10:1 can be obtained with the same capacitor. The characteristics of the closed loop system—bandwidth, response speed, capture and pull in range—may be adjusted over a wide range with an external resistor and capacitor. The loop may be broken between the VCO and the phase detector for insertion of a digital frequency divider to obtain frequency multiplication.

The LM565H is specified for operation over the –55°C to +125°C military temperature range. The LM565CH and LM565CN are specified for operation over the 0°C to +70°C temperature range.

Features

- 200 ppm/°C frequency stability of the VCO

- Power supply range of ±5 to ±12 volts with 100 ppm/% typical
- 0.2% linearity of demodulated output
- Linear triangle wave with in phase zero crossings available
- TTL and DTL compatible phase detector input and square wave output
- Adjustable hold in range from ±1% to > ±60%.

Applications

- Data and tape synchronization
- Modems
- FSK demodulation
- FM demodulation
- Frequency synthesizer
- Tone decoding
- Frequency multiplication and division
- SCA demodulators
- Telemetry receivers
- Signal regeneration
- Coherent demodulators.

Schematic and Connection Diagrams

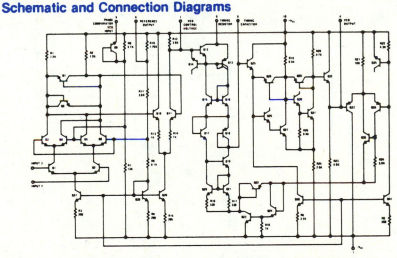

Metal Can Package

Order Number LM565H or LM565CH
See NS Package H10C

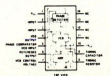

Dual-In-Line Package

Order Number LM565CN
See NS Package N14A

Absolute Maximum Ratings

Supply Voltage	±12V
Power Dissipation (Note 1)	300 mW
Differential Input Voltage	±1V
Operating Temperature Range LM565H	−55°C to +125°C
LM565CH, LM565CN	0°C to 70°C
Storage Temperature Range	−65°C to +150°C
Lead Temperature (Soldering, 10 sec)	300°C

Electrical Characteristics (AC Test Circuit, T_A = 25°C, V_C = ±6V)

PARAMETER	CONDITIONS	LM565			LM565C			UNITS
		MIN	TYP	MAX	MIN	TYP	MAX	
Power Supply Current			8.0	12.5		8.0	12.5	mA
Input Impedance (Pins 2, 3)	$-4V < V_2, V_3 < 0V$	7	10			5		kΩ
VCO Maximum Operating Frequency	C_o = 2.7 pF	300	500		250	500		kHz
Operating Frequency Temperature Coefficient			−100	300		−200	500	ppm/°C
Frequency Drift with Supply Voltage			0.01	0.1		0.05	0.2	%/V
Triangle Wave Output Voltage		2	2.4	3	2	2.4	3	V_{p-p}
Triangle Wave Output Linearity			0.2	0.75		0.5	1	%
Square Wave Output Level		4.7	5.4		4.7	5.4		V_{p-p}
Output Impedance (Pin 4)			5			5		kΩ
Square Wave Duty Cycle		45	50	55	40	50	60	%
Square Wave Rise Time			20	100		20		ns
Square Wave Fall Time			50	200		50		ns
Output Current Sink (Pin 4)		0.6	1		0.6	1		mA
VCO Sensitivity	f_o = 10 kHz	6400	6600	6800	6000	6600	7200	Hz/V
Demodulated Output Voltage (Pin 7)	±10% Frequency Deviation	250	300	350	200	300	400	mV_{pp}
Total Harmonic Distortion	±10% Frequency Deviation		0.2	0.75		0.2	1.5	%
Output Impedance (Pin 7)			3.5			3.5		kΩ
DC Level (Pin 7)		4.25	4.5	4.75	4.0	4.5	5.0	V
Output Offset Voltage $|V_7 - V_6|$			30	100		50	200	mV
Temperature Drift of $|V_7 - V_6|$			500			500		µV/°C
AM Rejection		30	40			40		dB
Phase Detector Sensitivity K_D		0.6	.68	0.9	0.55	.68	0.95	V/radian

Note 1: The maximum junction temperature of the LM565 is 150°C, while that of the LM565C and LM565CN is 100°C. For operation at elevated temperatures, devices in the TO-5 package must be derated based on a thermal resistance of 150°C/W junction to ambient or 45°C/W junction to case. Thermal resistance of the dual-in-line package is 100°C/W.

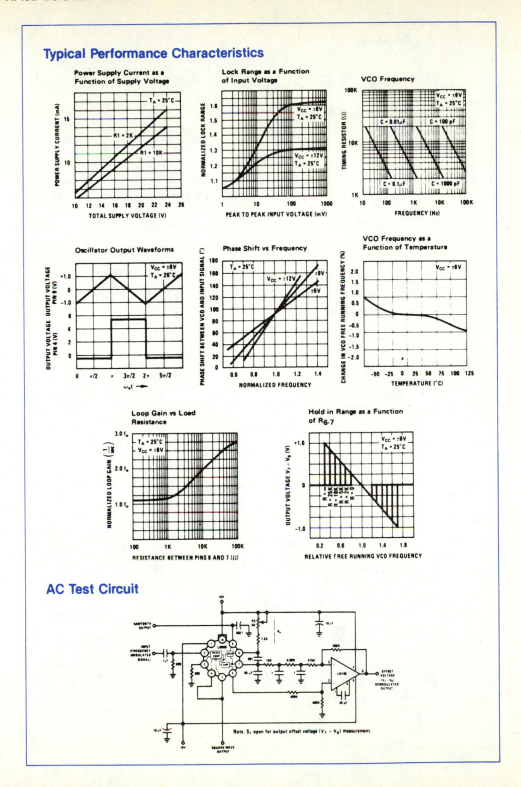

Typical Performance Characteristics

AC Test Circuit

Typical Applications

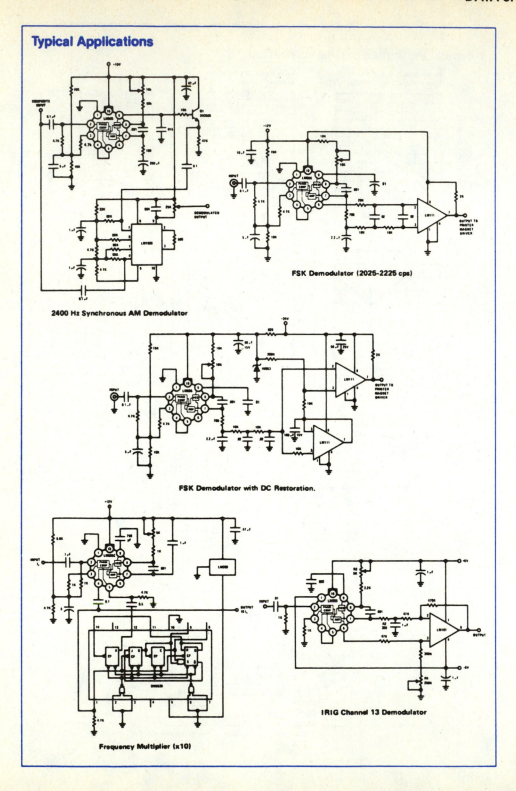

2400 Hz Synchronous AM Demodulator

FSK Demodulator (2025-2225 cps)

FSK Demodulator with DC Restoration.

Frequency Multiplier (x10)

IRIG Channel 13 Demodulator

Applications Information

In designing with phase locked loops such as the LM565, the important parameters of interest are:

FREE RUNNING FREQUENCY

$$f_o \cong \frac{1}{3.7\,R_oC_o}$$

LOOP GAIN: relates the amount of phase change between the input signal and the VCO signal for a shift in input signal frequency (assuming the loop remains in lock). In servo theory, this is called the "velocity error coefficient".

$$L \quad \rho \text{ gain} = K_o K_D \left(\frac{1}{\text{sec}}\right)$$

$$K_o = \text{oscillator sensitivity} \left(\frac{\text{radians/sec}}{\text{volt}}\right)$$

$$K_D = \text{phase detector sensitivity} \left(\frac{\text{volts}}{\text{radian}}\right)$$

The loop gain of the LM565 is dependent on supply voltage, and may be found from:

$$K_o K_D = \frac{33.6\,f_o}{V_c}$$

f_o = VCO frequency in Hz

V_c = total supply voltage to circuit.

Loop gain may be reduced by connecting a resistor between pins 6 and 7; this reduces the load impedance on the output amplifier and hence the loop gain.

HOLD IN RANGE: the range of frequencies that the loop will remain in lock after initially being locked.

$$f_H = \pm \frac{8\,f_o}{V_c}$$

f_o = free running frequency of VCO

V_c = total supply voltage to the circuit.

THE LOOP FILTER

In almost all applications, it will be desirable to filter the signal at the output of the phase detector (pin 7) this filter may take one of two forms:

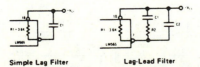

Simple Lag Filter   **Lag-Lead Filter**

A simple lag filter may be used for wide closed loop bandwidth applications such as modulation following where the frequency deviation of the carrier is fairly high (greater than 10%), or where wideband modulating signals must be followed.

The natural bandwidth of the closed loop response may be found from:

$$f_n = \frac{1}{2\pi} \sqrt{\frac{K_o K_D}{R_1 C_1}}$$

Associated with this is a damping factor:

$$\delta = \frac{1}{2} \sqrt{\frac{1}{R_1 C_1 K_o K_D}}$$

For narrow band applications where a narrow noise bandwidth is desired, such as applications involving tracking a slowly varying carrier, a lead lag filter should be used. In general, if $1/R_1 C_1 < K_o K_d$, the damping factor for the loop becomes quite small resulting in large overshoot and possible instability in the transient response of the loop. In this case, the natural frequency of the loop may be found from

$$f_n = \frac{1}{2\pi} \sqrt{\frac{K_o K_D}{\tau_1 + \tau_2}}$$

$$\tau_1 + \tau_2 = (R_1 + R_2)\,C_1$$

R_2 is selected to produce a desired damping factor δ, usually between 0.5 and 1.0. The damping factor is found from the approximation:

$$\delta \simeq \pi\,\tau_2 f_n$$

These two equations are plotted for convenience.

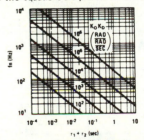

Filter Time Constant vs Natural Frequency

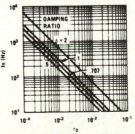

Damping Time Constant vs Natural Frequency

Capacitor C_2 should be much smaller than C_1 since its function is to provide filtering of carrier. In general $C_2 \leq 0.1\,C_1$.

A to D, D to A

ADC0801, ADC0802, ADC0803, ADC0804, ADC0805 8-Bit μP Compatible A/D Converters

General Description

The ADC0801, ADC0802, ADC0803, ADC0804 and ADC0805 are CMOS 8-bit successive approximation A/D converters which use a differential potentiometric ladder—similar to the 256R products. These converters are designed to allow operation with the NSC800 and INS8080A derivative control bus, and TRI-STATE® output latches directly drive the data bus. These A/Ds appear like memory locations or I/O ports to the microprocessor and no interfacing logic is needed.

A new differential analog voltage input allows increasing the common-mode rejection and offsetting the analog zero input voltage value. In addition, the voltage reference input can be adjusted to allow encoding any smaller analog voltage span to the full 8 bits of resolution.

- Differential analog voltage inputs
- Logic inputs and outputs meet both MOS and T^2L voltage level specifications
- Works with 2.5V (LM336) voltage reference
- On-chip clock generator
- 0V to 5V analog input voltage range with single 5V supply
- No zero adjust required
- 0.3" standard width 20-pin DIP package
- Operates ratiometrically or with 5 V_{DC}, 2.5 V_{DC}, or analog span adjusted voltage reference

Key Specifications

- Resolution   8 bits
- Total error   ±1/4 LSB, ±1/2 LSB and ±1 LSB
- Conversion time   100 μs

Features

- Compatible with 8080 μP derivatives—no interfacing logic needed – access time – 135 ns
- Easy interface to all microprocessors, or operates "stand alone"

Typical Applications

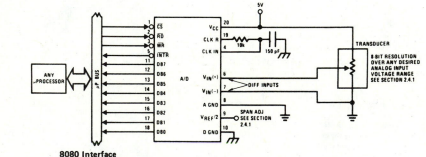

8080 Interface

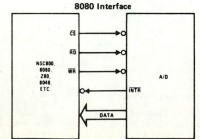

ERROR SPECIFICATION (INCLUDES FULL SCALE, ZERO ERROR, AND NON-LINEARITY)			
PART NUMBER	FULL-SCALE ADJUSTED	$V_{REF}/2$ = 2.500 V_{DC} (NO ADJUSTMENTS)	$V_{REF}/2$ = NO CONNECTION (NO ADJUSTMENTS)
ADC0801	±1/4 LSB		
ADC0802		±1/2 LSB	
ADC0803	±1/2 LSB		
ADC0804		±1 LSB	
ADC0805			±1 LSB

TRI-STATE® is a registered trademark of National Semiconductor Corp.

Absolute Maximum Ratings (Notes 1 and 2)

Supply Voltage (V_{CC}) (Note 3)	6.5V
Voltage	
Logic Control Inputs	−0.3V to +18V
At Other Input and Outputs	−0.3V to (V_{CC} + 0.3V)
Storage Temperature Range	−65°C to +150°C
Package Dissipation at T_A = 25°C	875 mW
Lead Temperature (Soldering, 10 seconds)	300°C

Operating Ratings (Notes 1 and 2)

Temperature Range	$T_{MIN} \leq T_A \leq T_{MAX}$
ADC0801/02LD	−55°C ≤ T_A ≤ +125°C
ADC0801/02/03/04LCD	−40°C ≤ T_A ≤ +85°C
ADC0801/02/03/05LCN	−40°C ≤ T_A ≤ +85°C
ADC0804LCN	0°C ≤ T_A ≤ +70°C
Range of V_{CC}	4.5 V_{DC} to 6.3 V_{DC}

Electrical Characteristics

The following specifications apply for V_{CC} = 5 V_{DC}, $T_{MIN} \leq T_A \leq T_{MAX}$ and f_{CLK} = 640 kHz unless otherwise specified.

PARAMETER	CONDITIONS	MIN	TYP	MAX	UNITS
ADC0801:					
Total Adjusted Error	With Full-Scale Adj.			±1/4	LSB
(Note 8)	(See Section 2.5.2)				
ADC0802:					
Total Unadjusted Error	V_{REF}/2 = 2.500 V_{DC}			±1/2	LSB
(Note 8)					
ADC0803:					
Total Adjusted Error	With Full-Scale Adj.			±1/2	LSB
(Note 8)	(See Section 2.5.2)				
ADC0804:					
Total Unadjusted Error	V_{REF}/2 = 2.500 V_{DC}			±1	LSB
(Note 8)					
ADC0805:					
Total Unadjusted Error	V_{REF}/2 − No Connection			±1	LSB
(Note 8)					
V_{REF}/2 Input Resistance (Pin 9)	ADC0801/02/03/05	2.5	8.0		kΩ
	ADC0804 (Note 9)	1.0	1.3		kΩ
Analog Input Voltage Range	(Note 4) V (+) or V (−)	Gnd−0.05		V_{CC}+0.05	V_{DC}
DC Common-Mode Error	Over Analog Input Voltage Range		±1/16	±1/8	LSB
Power Supply Sensitivity	V_{CC} = 5 V_{DC} ±10% Over Allowed V_{IN}(+) and V_{IN}(−) Voltage Range (Note 4)		±1/16	±1/8	LSB

AC Electrical Characteristics

The following specifications apply for V_{CC} = 5 V_{DC} and T_A = 25°C unless otherwise specified.

	PARAMETER	CONDITIONS	MIN	TYP	MAX	UNITS
T_c	Conversion Time	f_{CLK} = 640 kHz (Note 6)	103		114	µs
T_c	Conversion Time	(Note 5, 6)	66		73	1/f_{CLK}
f_{CLK}	Clock Frequency	V_{CC} = 5V, (Note 5)	100	640	1460	kHz
	Clock Duty Cycle	(Note 5)	40		60	%
CR	Conversion Rate In Free-Running Mode	$\overline{INTR}$ tied to $\overline{WR}$ with $\overline{CS}$ = 0 V_{DC}, f_{CLK} = 640 kHz			8770	conv/s
$t_W(\overline{WR})L$	Width of $\overline{WR}$ Input (Start Pulse Width)	$\overline{CS}$ = 0 V_{DC} (Note 7)	100			ns
t_{ACC}	Access Time (Delay from Falling Edge of $\overline{RD}$ to Output Data Valid)	C_L = 100 pF		135	200	ns
t_{1H}, t_{0H}	TRI-STATE Control (Delay from Rising Edge of $\overline{RD}$ to Hi-Z State)	C_L = 10 pF, R_L = 10k (See TRI-STATE Test Circuits)		125	200	ns
t_{WI}, t_{RI}	Delay from Falling Edge of $\overline{WR}$ or $\overline{RD}$ to Reset of $\overline{INTR}$			300	450	ns
C_{IN}	Input Capacitance of Logic Control Inputs			5	7.5	pF
C_{OUT}	TRI-STATE Output Capacitance (Data Buffers)			5	7.5	pF

Electrical Characteristics

The following specifications apply for V_{CC} = 5 V_{DC} and $T_{MIN} \leq T_A \leq T_{MAX}$, unless otherwise specified.

PARAMETER		CONDITIONS	MIN	TYP	MAX	UNITS
CONTROL INPUTS [Note: CLK IN (Pin 4) is the input of a Schmitt trigger circuit and is therefore specified separately]						
V_{IN} (1)	Logical "1" Input Voltage (Except Pin 4 CLK IN)	V_{CC} = 5.25 V_{DC}	2.0		15	V_{DC}
V_{IN} (0)	Logical "0" Input Voltage (Except Pin 4 CLK IN)	V_{CC} = 4.75 V_{DC}			0.8	V_{DC}
I_{IN} (1)	Logical "1" Input Current (All Inputs)	V_{IN} = 5 V_{DC}		0.005	1	μA_{DC}
I_{IN} (0)	Logical "0" Input Current (All Inputs)	V_{IN} = 0 V_{DC}	−1	−0.005		μA_{DC}
CLOCK IN AND CLOCK R						
V_T+	CLK IN (Pin 4) Positive Going Threshold Voltage		2.7	3.1	3.5	V_{DC}
V_T-	CLK IN (Pin 4) Negative Going Threshold Voltage		1.5	1.8	2.1	V_{DC}
V_H	CLK IN (Pin 4) Hysteresis $(V_T+) - (V_T-)$		0.6	1.3	2.0	V_{DC}
V_{OUT} (0)	Logical "0" CLK R Output Voltage	I_O = 360 μA V_{CC} = 4.75 V_{DC}			0.4	V_{DC}
V_{OUT} (1)	Logical "1" CLK R Output Voltage	I_O = −360 μA V_{CC} = 4.75 V_{DC}	2.4			V_{DC}
DATA OUTPUTS AND INTR						
V_{OUT}(0)	Logical "0" Output Voltage Data Outputs	I_{OUT} = 1.6 mA, V_{CC} = 4.75 V_{DC}			0.4	V_{DC}
	INTR Output	I_{OUT} = 1.0 mA, V_{CC} = 4.75 V_{DC}			0.4	V_{DC}
V_{OUT} (1)	Logical "1" Output Voltage	I_O = −360 μA, V_{CC} = 4.75 V_{DC}	2.4			V_{DC}
V_{OUT} (1)	Logical "1" Output Voltage	I_O = −10 μA, V_{CC} = 4.75 V_{DC}	4.5			V_{DC}
I_{OUT}	TRI-STATE Disabled Output Leakage (All Data Buffers)	V_{OUT} = 0 V_{DC}	−3			μA_{DC}
		V_{OUT} = 5 V_{DC}			3	μA_{DC}
I_{SOURCE}		V_{OUT} Short to Gnd, T_A = 25°C	4.5	6		mA_{DC}
I_{SINK}		V_{OUT} Short to V_{CC}, T_A = 25°C	9.0	16		mA_{DC}
POWER SUPPLY						
I_{CC}	Supply Current (Includes Ladder Current)	f_{CLK} = 640 kHz, $V_{REF}/2$ = NC, T_A = 25°C and $\overline{CS}$ = "1"				
		ADC0801/02/03/05		1.1	1.8	mA
		ADC0804 (Note 9)		1.9	2.5	mA

Note 1: Absolute maximum ratings are those values beyond which the life of the device may be impaired.

Note 2: All voltages are measured with respect to Gnd, unless otherwise specified. The separate A Gnd point should always be wired to the D Gnd.

Note 3: A zener diode exists, internally, from V_{CC} to Gnd and has a typical breakdown voltage of 7 V_{DC}.

Note 4: For $V_{IN}(-) \geq V_{IN}(+)$ the digital output code will be 0000 0000. Two on-chip diodes are tied to each analog input (see block diagram) which will forward conduct for analog input voltages one diode drop below ground or one diode drop greater than the V_{CC} supply. Be careful, during testing at low V_{CC} levels (4.5V), as high level analog inputs (5V) can cause this input diode to conduct—especially at elevated temperatures, and cause errors for analog inputs near full-scale. The spec allows 50 mV forward bias of either diode. This means that as long as the analog V_{IN} does not exceed the supply voltage by more than 50 mV, the output code will be correct. To achieve an absolute 0 V_{DC} to 5 V_{DC} input voltage range will therefore require a minimum supply voltage of 4.950 V_{DC} over temperature variations, initial tolerance and loading.

Note 5: Accuracy is guaranteed at f_{CLK} = 640 kHz. At higher clock frequencies accuracy can degrade. For lower clock frequencies, the duty cycle limits can be extended so long as the minimum clock high time interval or minimum clock low time interval is no less than 275 ns.

Note 6: With an asynchronous start pulse, up to 8 clock periods may be required before the internal clock phases are proper to start the conversion process. The start request is internally latched, see *Figure 2* and section 2.0.

Note 7: The $\overline{CS}$ input is assumed to bracket the $\overline{WR}$ strobe input and therefore timing is dependent on the $\overline{WR}$ pulse width. An arbitrarily wide pulse width will hold the converter in a reset mode and the start of conversion is initiated by the low to high transition of the $\overline{WR}$ pulse (see timing diagrams).

Note 8: None of these A/Ds requires a zero adjust (see section 2.5.1). To obtain zero code at other analog input voltages see section 2.5 and *Figure 5*.

Note 9: For ADC0804LCD typical value of $V_{REF}/2$ input resistance is 8 kΩ and of I_{CC} is 1.1 mA.

1.0 UNDERSTANDING A/D ERROR SPECS

A perfect A/D transfer characteristic (staircase waveform) is shown in *Figure 1a*. The horizontal scale is analog input voltage and the particular points labeled are in steps of 1 LSB (19.53 mV with 2.5V tied to the $V_{REF}/2$ pin). The digital output codes which correspond to these inputs are shown as D−1, D, and D+1. For the perfect A/D, not only will center-value (A−1, A, A+1, . . .) analog inputs produce the correct output digital codes, but also each riser (the transitions between adjacent output codes) will be located ±1/2 LSB away from each center-value. As shown, the risers are ideal and have no width. Correct digital output codes will be provided for a range of analog input voltages which extend ±1/2 LSB from the ideal center-values. Each tread (the range of analog input voltage which provides the same digital output code) is therefore 1 LSB wide.

Figure 1b shows a worst case error plot for the ADC0801. All center-valued inputs are guaranteed to produce the correct output codes and the adjacent risers are guaranteed to be no closer to the center-value points than

±1/4 LSB. In other words, if we apply an analog input equal to the center-value ±1/4 LSB, *we guarantee* that the A/D will produce the correct digital code. The maximum range of the position of the code transition is indicated by the horizontal arrow and it is guaranteed to be no more than 1/2 LSB.

The error curve of *Figure 1c* shows a worst case error plot for the ADC0802. Here we guarantee that if we apply an analog input equal to the LSB analog voltage center-value the A/D will produce the correct digital code.

Next to each transfer function is shown the corresponding error plot. Many people may be more familiar with error plots than transfer functions. The analog input voltage to the A/D is provided by either a linear ramp or by the discrete output steps of a high resolution DAC. Notice that the error is continuously displayed and includes the quantization uncertainty of the A/D. For example the error at point 1 of *Figure 1a* is +1/2 LSB because the digital code appeared 1/2 LSB in advance of the center-value of the tread. The error plots always have a constant negative slope and the abrupt upside steps are always 1 LSB in magnitude.

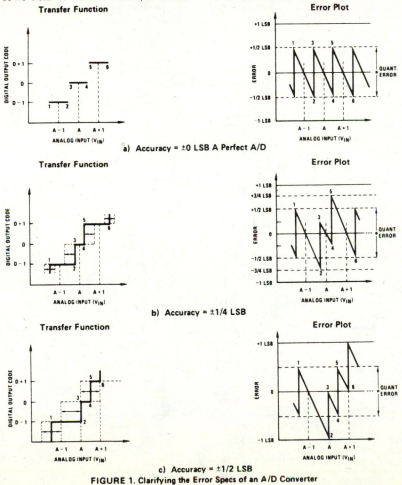

a) Accuracy = ±0 LSB A Perfect A/D

b) Accuracy = ±1/4 LSB

c) Accuracy = ±1/2 LSB

FIGURE 1. Clarifying the Error Specs of an A/D Converter

2.0 FUNCTIONAL DESCRIPTION

The ADC0801 series contains a circuit equivalent of the 256R network. Analog switches are sequenced by successive approximation logic to match the analog difference input voltage $[V_{IN}(+) - V_{IN}(-)]$ to a corresponding tap on the R network. The most significant bit is tested first and after 8 comparisons (64 clock cycles) a digital 8-bit binary code (1111 1111 = full-scale) is transferred to an output latch and then an interrupt is asserted ($\overline{INTR}$ makes a high-to-low transition). A conversion in process can be interrupted by issuing a second start command. The device may be operated in the free-running mode by connecting $\overline{INTR}$ to the $\overline{WR}$ input with $\overline{CS} = 0$. To insure start-up under all possible conditions, an external $\overline{WR}$ pulse is required during the first power-up cycle.

On the high-to-low transition of the $\overline{WR}$ input the internal SAR latches and the shift register stages are reset. As long as the $\overline{CS}$ input and $\overline{WR}$ input remain low, the A/D will remain in a reset state. *Conversion will start from 1 to 8 clock periods after at least one of these inputs makes a low-to-high transition.*

A functional diagram of the A/D converter is shown in *Figure 2*. All of the package pinouts are shown and the major logic control paths are drawn in heavier weight lines.

The converter is started by having $\overline{CS}$ and $\overline{WR}$ simultaneously low. This sets the start flip-flop (F/F) and the resulting "1" level resets the 8-bit shift register, resets the Interrupt (INTR) F/F and inputs a "1" to the D flop, F/F1, which is at the input end of the 8-bit shift register. Internal clock signals then transfer this "1" to the Q output of F/F1. The AND gate, G1, combines this "1" output with a clock signal to provide a reset signal to the start F/F. If the set signal is no longer present (either $\overline{WR}$ or $\overline{CS}$ is a "1") the start F/F is reset and the 8-bit shift register then can have the "1" clocked in, which starts the conversion process. If the set signal were to still be present, this reset pulse would have no effect (both outputs of the start F/F would momentarily be at a "1" level) and the 8-bit shift register would continue to be held in the reset mode. This logic therefore allows for wide $\overline{CS}$ and $\overline{WR}$ signals and the converter will start after at least one of these signals returns high and the internal clocks again provide a reset signal for the start F/F.

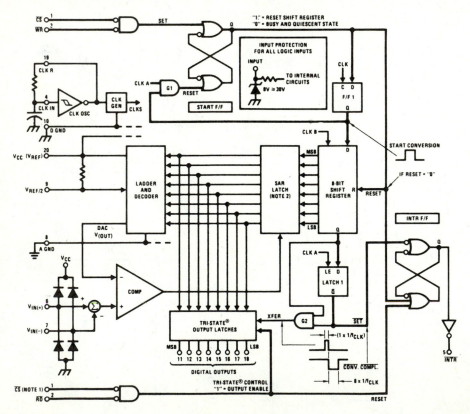

Note 1: $\overline{CS}$ shown twice for clarity.

Note 2: SAR = Successive Approximation Register.

FIGURE 2. Block Diagram

After the "1" is clocked through the 8-bit shift register (which completes the SAR search) it appears as the input to the D-type latch, LATCH 1. As soon as this "1" is output from the shift register, the AND gate, G2, causes the new digital word to transfer to the TRI-STATE output latches. When LATCH 1 is subsequently enabled, the Q output makes a high-to-low transition which causes the INTR F/F to set. An inverting buffer then supplies the $\overline{\text{INTR}}$ output signal.

Note that this $\overline{\text{SET}}$ control of the INTR F/F remains low for 8 of the external clock periods (as the internal clocks run at 1/8 of the frequency of the external clock). If the data output is continuously enabled ($\overline{\text{CS}}$ and $\overline{\text{RD}}$ both held low), the $\overline{\text{INTR}}$ output will still signal the end of conversion (by a high-to-low transition), because the $\overline{\text{SET}}$ input can control the Q output of the INTR F/F even though the RESET input is constantly at a "1" level in this operating mode. This $\overline{\text{INTR}}$ output will therefore stay low for the duration of the $\overline{\text{SET}}$ signal, which is 8 periods of the external clock frequency (assuming the A/D is not started during this interval).

When operating in the free-running or continuous conversion mode ($\overline{\text{INTR}}$ pin tied to $\overline{\text{WR}}$ and $\overline{\text{CS}}$ wired low—see also section 2.8), the START F/F is SET by the high-to-low transition of the $\overline{\text{INTR}}$ signal. This resets the SHIFT REGISTER which causes the input to the D-type latch, LATCH 1, to go low. As the latch enable input is still present, the $\overline{\text{Q}}$ output will go high, which then allows the INTR F/F to be RESET. This reduces the width of the resulting $\overline{\text{INTR}}$ output pulse to only a few propagation delays (approximately 300 ns).

When data is to be read, the combination of both $\overline{\text{CS}}$ and $\overline{\text{RD}}$ being low will cause the INTR F/F to be reset and the TRI-STATE output latches will be enabled to provide the 8-bit digital outputs.

2.1 Digital Control Inputs

The digital control inputs ($\overline{\text{CS}}$, $\overline{\text{RD}}$, and $\overline{\text{WR}}$) meet standard T^2L logic voltage levels. These signals have been renamed when compared to the standard A/D Start and Output Enable labels. In addition, these inputs are active low to allow an easy interface to microprocessor control busses. For non-microprocessor based applications, the $\overline{\text{CS}}$ input (pin 1) can be grounded and the standard A/D Start function is obtained by an active low pulse applied at the $\overline{\text{WR}}$ input (pin 3) and the Output Enable function is caused by an active low pulse at the $\overline{\text{RD}}$ input (pin 2).

2.2 Analog Differential Voltage Inputs and Common-Mode Rejection

This A/D has additional applications flexibility due to the analog differential voltage input. The $V_{IN}(-)$ input (pin 7) can be used to automatically subtract a fixed voltage value from the input reading (tare correction). This is also useful in 4 mA−20 mA current loop conversion. In addition, common-mode noise can be reduced by use of the differential input.

The time interval between sampling $V_{IN}(+)$ and $V_{IN}(-)$ is 4-1/2 clock periods. The maximum error voltage due

to this slight time difference between the input voltage samples is given by:

$$\Delta V_e(\text{MAX}) = (V_P)\,(2\pi f_{cm})\left(\frac{4.5}{f_{CLK}}\right)$$

where:

ΔV_e is the error voltage due to sampling delay

V_P is the peak value of the common-mode voltage

f_{cm} is the common-mode frequency

As an example, to keep this error to 1/4 LSB (~5 mV) when operating with a 60 Hz common-mode frequency, f_{cm}, and using a 640 kHz A/D clock, f_{CLK}, would allow a peak value of the common-mode voltage, V_P, which is given by:

$$V_P = \frac{[\Delta V_e(\text{MAX})\,(f_{CLK})]}{(2\pi f_{cm})\,(4.5)}$$

or

$$V_P = \frac{(5 \times 10^{-3})\,(640 \times 10^3)}{(6.28)\,(60)\,(4.5)}$$

which gives

$$V_P \cong 1.9\text{V}.$$

The allowed range of analog input voltages usually places more severe restrictions on input common-mode noise levels.

An analog input voltage with a reduced span and a relatively large zero offset can be easily handled by making use of the differential input (see section 2.4 Reference Voltage).

2.3 Analog Inputs

2.3.1 Input Current

Normal Mode

Due to the internal switching action, displacement currents will flow at the analog inputs. This is due to on-chip stray capacitance to ground as shown in *Figure 3*.

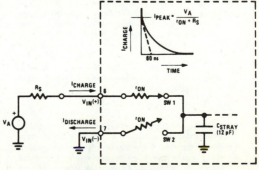

r_{ON} of SW 1 and SW 2 $\cong$ 5 kΩ

$\tau = r_{ON}\,C_{STRAY} \cong$ 5 kΩ x 12 pF $\cong$ 60 ns

FIGURE 3. Analog Input Impedance

The voltage on this capacitance is switched and will result in currents entering the $V_{IN}(+)$ input pin and leaving the $V_{IN}(-)$ input which will depend on the analog differential input voltage levels. These current transients occur at the leading edge of the internal clocks. They rapidly decay and *do not cause errors* as the on-chip comparator is strobed at the end of the clock period.

Fault Mode

If the voltage source which is applied to the $V_{IN}(+)$ pin exceeds the allowed operating range of V_{CC} + 50 mV, large input currents can flow through a parasitic diode to the V_{CC} pin. If these currents could exceed the 1 mA max allowed spec, an external diode (1N914) should be added to bypass this current to the V_{CC} pin (with the current bypassed with this diode, the voltage at the $V_{IN}(+)$ pin can exceed the V_{CC} voltage by the forward voltage of this diode).

2.3.2 Input Bypass Capacitors

Bypass capacitors at the inputs will average these charges and cause a DC current to flow through the output resistances of the analog signal sources. This charge pumping action is worse for continuous conversions with the $V_{IN}(+)$ input voltage at full-scale. For continuous conversions with a 640 kHz clock frequency with the $V_{IN}(+)$ input at 5V, this DC current is at a maximum of approximately 5 μA. Therefore, *bypass capacitors should not be used at the analog inputs or the $V_{REF}/2$ pin* for high resistance sources (> 1 kΩ). If input bypass capacitors are necessary for noise filtering and high source resistance is desirable to minimize capacitor size, the detrimental effects of the voltage drop across this input resistance, which is due to the average value of the input current, can be eliminated with a full-scale adjustment while the given source resistor and input bypass capacitor are both in place. This is possible because the average value of the input current is a precise linear function of the differential input voltage.

2.3.3 Input Source Resistance

Large values of source resistance where an input bypass capacitor is not used, *will not cause errors* as the input currents settle out prior to the comparison time. If a low pass filter is required in the system, use a low valued series resistor ($\leq$ 1 kΩ) for a passive RC section or add an op amp RC active low pass filter. For low source resistance applications, ($\leq$ 1 kΩ), a 0.1 μF bypass capacitor at the inputs will prevent pickup due to series lead inductance of a long wire. A 100Ω series resistor can be used to isolate this capacitor—both the R and C are placed outside the feedback loop—from the output of an op amp, if used.

2.3.4 Noise

The leads to the analog inputs (pins 6 and 7) should be kept as short as possible to minimize input noise coupling. Both noise and undesired digital clock coupling to these inputs can cause system errors. The source resistance for these inputs should, in general, be kept below 5 kΩ. Larger values of source resistance can cause undesired system noise pickup. Input bypass capacitors, placed from the analog inputs to ground, will eliminate

system noise pickup but can create analog scale errors as these capacitors will average the transient input switching currents of the A/D (see section 2.3.1). This scale error depends on both a large source resistance and the use of an input bypass capacitor. This error can be eliminated by doing a full-scale adjustment of the A/D (adjust $V_{REF}/2$ for a proper full-scale reading—see section 2.5.2 on Full-Scale Adjustment) with the source resistance and input bypass capacitor in place.

2.4 Reference Voltage

2.4.1 Span Adjust

For maximum applications flexibility, these A/Ds have been designed to accommodate a 5 V_{DC}, 2.5 V_{DC} or an adjusted voltage reference. This has been achieved in the design of the IC as shown in *Figure 4*.

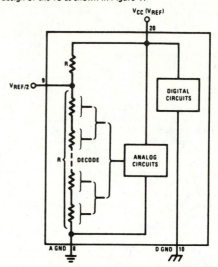

FIGURE 4. The $V_{REFERENCE}$ Design on the IC

Notice that the reference voltage for the IC is either 1/2 of the voltage which is applied to the V_{CC} supply pin, or is equal to the voltage which is externally forced at the $V_{REF}/2$ pin. This allows for a ratiometric voltage reference using the V_{CC} supply, a 5 V_{DC} reference voltage can be used for the V_{CC} supply or a voltage less than 2.5 V_{DC} can be applied to the $V_{REF}/2$ input for increased application flexibility. The internal gain to the $V_{REF}/2$ input is 2 making the full-scale differential input voltage twice the voltage at pin 9.

An example of the use of an adjusted reference voltage is to accommodate a reduced span—or dynamic voltage range of the analog input voltage. If the analog input voltage were to range from 0.5 V_{DC} to 3.5 V_{DC}, instead of 0V to 5 V_{DC}, the span would be 3V as shown in *Figure 5*. With 0.5 V_{DC} applied to the $V_{IN}(-)$ pin to absorb the offset, the reference voltage can be made equal to 1/2 of the 3V span or 1.5 V_{DC}. The A/D now will encode the $V_{IN}(+)$ signal from 0.5V to 3.5V with the 0.5V input corresponding to zero and the 3.5 V_{DC} input corresponding to full-scale. The full 8 bits of resolution are therefore applied over this reduced analog input voltage range.

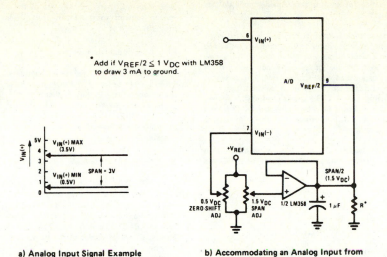

a) Analog Input Signal Example

b) Accommodating an Analog Input from
0.5V (Digital Out = 00$_{HEX}$) to 3.5V
(Digital Out = FF$_{HEX}$)

FIGURE 5. Adapting the A/D Analog Input Voltages to Match an Arbitrary Input Signal Range

2.4.2 Reference Accuracy Requirements

The converter can be operated in a ratiometric mode or an absolute mode. In ratiometric converter applications, the magnitude of the reference voltage is a factor in both the output of the source transducer and the output of the A/D converter and therefore cancels out in the final digital output code. The ADC0805 is specified particularly for use in ratiometric applications with no adjustments required. In absolute conversion applications, both the initial value and the temperature stability of the reference voltage are important accuracy factors in the operation of the A/D converter. For $V_{REF}/2$ voltages of 2.5 V_{DC} nominal value, initial errors of ±10 mV_{DC} will cause conversion errors of ±1 LSB due to the gain of 2 of the $V_{REF}/2$ input. In reduced span applications, the initial value and the stability of the $V_{REF}/2$ input voltage become even more important. For example, if the span is reduced to 2.5V, the analog input LSB voltage value is correspondingly reduced from 20 mV (5V span) to 10 mV and 1 LSB at the $V_{REF}/2$ input becomes 5 mV. As can be seen, this reduces the allowed initial tolerance of the reference voltage and requires correspondingly less absolute change with temperature variations. Note that spans smaller than 2.5V place even tighter requirements on the initial accuracy and stability of the reference source.

In general, the magnitude of the reference voltage will require an initial adjustment. Errors due to an improper value of reference voltage appear as full-scale errors in the A/D transfer function. IC voltage regulators may be used for references if the ambient temperature changes are not excessive. The LM336B 2.5V IC reference diode

(from National Semiconductor) is available which has a temperature stability of 1.8 mV typ (6 mV max) over 0°C $\leq$ T$_A$ $\leq$ +70°C. Other temperature range parts are also available.

2.5 Errors and Reference Voltage Adjustments

2.5.1 Zero Error

The zero of the A/D does not require adjustment. If the minimum analog input voltage value, $V_{IN(MIN)}$, is not ground, a zero offset can be done. The converter can be made to output 0000 0000 digital code for this minimum input voltage by biasing the A/D V_{IN} (−) input at this $V_{IN(MIN)}$ value (see Applications section). This utilizes the differential mode operation of the A/D.

The zero error of the A/D converter relates to the location of the first riser of the transfer function and can be measured by grounding the V (−) input and applying a small magnitude positive voltage to the V (+) input. Zero error is the difference between the actual DC input voltage which is necessary to just cause an output digital code transition from 0000 0000 to 0000 0001 and the ideal 1/2 LSB value (1/2 LSB = 9.8 mV for $V_{REF}/2$ = 2.500 V_{DC}).

2.5.2 Full-Scale

The full-scale adjustment can be made by applying a differential input voltage which is 1-1/2 LSB down from the desired analog full-scale voltage range and then adjusting the magnitude of the $V_{REF}/2$ input (pin 9 or the V_{CC} supply if pin 9 is not used) for a digital output code which is just changing from 1111 1110 to 1111 1111.

2.5.3 Adjusting for an Arbitrary Analog Input Voltage Range

If the analog zero voltage of the A/D is shifted away from ground (for example, to accommodate an analog input signal which does not go to ground) this new zero reference should be properly adjusted first: A $V_{IN}(+)$ voltage which equals this desired zero reference plus 1/2 LSB (where the LSB is calculated for the desired analog span, 1 LSB = analog span/256) is applied to pin 6 and the zero reference voltage at pin 7 should then be adjusted to just obtain the 00_{HEX} to 01_{HEX} code transition.

The full-scale adjustment should then be made (with the proper $V_{IN}(-)$ voltage applied) by forcing a voltage to the $V_{IN}(+)$ input which is given by:

$$V_{IN}(+) \text{ fs adj} = V_{MAX} - 1.5 \left[\frac{(V_{MAX} - V_{MIN})}{256} \right]$$

where:

V_{MAX} = The high end of the analog input range

and

V_{MIN} = the low end (the offset zero) of the analog range. (Both are ground referenced.)

The $V_{REF}/2$ (or V_{CC}) voltage is then adjusted to provide a code change from FE_{HEX} to FF_{HEX}. This completes the adjustment procedure.

2.6 Clocking Option

The clock for the A/D can be derived from the CPU clock or an external RC can be added to provide self-clocking. The CLK IN (pin 4) makes use of a Schmitt trigger as shown in *Figure 6*.

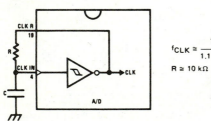

$$f_{CLK} \cong \frac{1}{1.1 \, RC}$$

$$R \cong 10 \, k\Omega$$

FIGURE 6. Self-Clocking the A/D

Heavy capacitive or DC loading of the clock R pin should be avoided as this will disturb normal converter operation. Loads less than 50 pF, such as driving up to 7 A/D converter clock inputs from a single clock R pin of 1 converter, are allowed. For larger clock line loading, a CMOS or low power T^2L buffer or PNP input logic should be used to minimize the loading on the clock R pin (do not use a standard T^2L buffer).

2.7 Restart During a Conversion

If the A/D is restarted ($\overline{CS}$ and $\overline{WR}$ go low and return high) during a conversion, the converter is reset and a new conversion is started. The output data latch is not updated if the conversion in process is not allowed to be completed, therefore the data of the previous conversion remains in this latch. The $\overline{INTR}$ output also simply remains at the "1" level.

2.8 Continuous Conversions

For operation in the free-running mode an initializing pulse should be used, following power-up, to insure circuit operation. In this application, the $\overline{CS}$ input is grounded and the $\overline{WR}$ input is tied to the $\overline{INTR}$ output. This $\overline{WR}$ and $\overline{INTR}$ node should be momentarily forced to logic low following a power-up cycle to guarantee operation.

2.9 Driving the Data Bus

This MOS A/D, like MOS microprocessors and memories, will require a bus driver when the total capacitance of the data bus gets large. Other circuitry, which is tied to the data bus, will add to the total capacitive loading, even in TRI-STATE (high impedance mode). Backplane bussing also greatly adds to the stray capacitance of the data bus.

There are some alternatives available to the designer to handle this problem. Basically, the capacitive loading of the data bus slows down the response time, even though DC specifications are still met. For systems operating with a relatively slow CPU clock frequency, more time is available in which to establish proper logic levels on the bus and therefore higher capacitive loads can be driven (see typical characteristics curves).

At higher CPU clock frequencies time can be extended for I/O reads (and/or writes) by inserting wait states (8080) or using clock extending circuits (6800).

Finally, if time is short and capacitive loading is high, external bus drivers must be used. These can be TRI-STATE buffers (low power Schottky is recommended such as the DM74LS240 series) or special higher drive current products which are designed as bus drivers. High current bipolar bus drivers with PNP inputs are recommended.

2.10 Power Supplies

Noise spikes on the V_{CC} supply line can cause conversion errors as the comparator will respond to this noise. A low inductance tantalum filter capacitor should be used close to the converter V_{CC} pin and values of 1 μF or greater are recommended. If an unregulated voltage is available in the system, a separate LM340LAZ-5.0, TO-92, 5V voltage regulator for the converter (and other analog circuitry) will greatly reduce digital noise on the V_{CC} supply.

2.11 Wiring and Hook-Up Precautions

Standard digital wire wrap sockets are not satisfactory for breadboarding this A/D converter. Sockets on PC boards can be used and all logic signal wires and leads should be grouped and kept as far away as possible from the analog signal leads. Exposed leads to the analog inputs can cause undesired digital noise and hum pickup, therefore shielded leads may be necessary in many applications.

A single point analog ground should be used which is separate from the logic ground points. The power supply bypass capacitor and the self-clocking capacitor (if used) should both be returned to digital ground. Any $V_{REF}/2$ bypass capacitors, analog input filter capacitors, or input signal shielding should be returned to the analog ground point. A test for proper grounding is to measure the zero error of the A/D converter. Zero errors in excess of 1/4 LSB can usually be traced to improper board layout and wiring (see section 2.5.1 for measuring the zero error).

3.0 TESTING THE A/D CONVERTER

There are many degrees of complexity associated with testing an A/D converter. One of the simplest tests is to apply a known analog input voltage to the converter and use LEDs to display the resulting digital output code as shown in *Figure 7*.

For ease of testing, the $V_{REF}/2$ (pin 9) should be supplied with 2.560 V_{DC} and a V_{CC} supply voltage of 5.12 V_{DC} should be used. This provides an LSB value of 20 mV.

If a full-scale adjustment is to be made, an analog input voltage of 5.090 V_{DC} (5.120 − 1 1/2 LSB) should be applied to the $V_{IN}(+)$ pin with the $V_{IN}(-)$ pin grounded. The value of the $V_{REF}/2$ input voltage should then be adjusted until the digital output code is just changing from 1111 1110 to 1111 1111. This value of $V_{REF}/2$ should then be used for all the tests.

The digital output LED display can be decoded by dividing the 8 bits into 2 hex characters, the 4 most significant (MS) and the 4 least significant (LS). Table I shows the fractional binary equivalent of these two 4-bit groups. By adding the decoded voltages which are obtained from the column: Input voltage value for a 2.560 $V_{REF}/2$ of both the MS and the LS groups, the value of

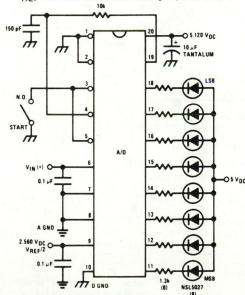

FIGURE 7. Basic A/D Tester

the digital display can be determined. For example, for an output LED display of 1011 0110 or B6 (in hex), the voltage values from the table are 3.520 + 0.120 or 3.640 V_{DC}. These voltage values represent the center-values of a perfect A/D converter. The effects of quantization error have to be accounted for in the interpretation of the test results.

For a higher speed test system, or to obtain plotted data, a digital-to-analog converter is needed for the test set-up. An accurate 10-bit DAC can serve as the precision voltage source for the A/D. Errors of the A/D under test can be provided as either analog voltages or differences in 2 digital words.

A basic A/D tester which uses a DAC and provides the error as an analog output voltage is shown in *Figure 8*. The 2 op amps can be eliminated if a lab DVM with a numerical subtraction feature is available to directly readout the difference voltage, "A−C". The analog input voltage can be supplied by a low frequency ramp generator and an X-Y plotter can be used to provide analog error (Y axis) versus analog input (X axis). The construction details of a tester of this type are provided in the NSC application note AN-179, "Analog-to-Digital Converter Testing".

For operation with a microprocessor or a computer-based test system, it is more convenient to present the errors digitally. This can be done with the circuit of *Figure 9*, where the output code transitions can be detected as the 10-bit DAC is incremented. This provides 1/4 LSB steps for the 8-bit A/D under test. If the results of this test are automatically plotted with the analog input on the X axis and the error (in LSB's) as the Y axis, a useful transfer function of the A/D under test results. For acceptance testing, the plot is not necessary and the testing speed can be increased by establishing internal limits on the allowed error for each code.

4.0 MICROPROCESSOR INTERFACING

To discuss the interface with 8080A and 6800 microprocessors, a common sample subroutine structure is used. The microprocessor starts the A/D, reads and stores the results of 16 successive conversions, then returns to the user's program. The 16 data bytes are stored in 16 successive memory locations. All Data and Addresses will be given in hexadecimal form. Software and hardware details are provided separately for each type of microprocessor.

4.1 Interfacing 8080 Microprocessor Derivatives (8048, 8085)

This converter has been designed to directly interface with derivatives of the 8080 microprocessor. The A/D can be mapped into memory space (using standard memory address decoding for $\overline{CS}$ and the $\overline{MEMR}$ and $\overline{MEMW}$ strobes) or it can be controlled as an I/O device by using the $\overline{I/O\ R}$ and $\overline{I/O\ W}$ strobes and decoding the address bits A0 → A7 (or address bits A8 → A15 as they will contain the same 8-bit address information) to obtain the $\overline{CS}$ input. Using the I/O space provides 256 additional addresses and may allow a simpler 8-bit address decoder but the data can only be input to the accumulator. To make use of the additional memory reference instructions, the A/D should be mapped into memory space. An example of an A/D in I/O space is shown in *Figure 10*.

National Semiconductor

A to D, D to A

DAC0808, DAC0807, DAC0806 8-Bit D/A Converters

General Description

The DAC0808 series is an 8-bit monolithic digital-to-analog converter (DAC) featuring a full scale output current settling time of 150 ns while dissipating only 33 mW with ±5V supplies. No reference current (I_{REF}) trimming is required for most applications since the full scale output current is typically ±1 LSB of 255 I_{REF}/256. Relative accuracies of better than ±0.19% assure 8-bit monotonicity and linearity while zero level output current of less than 4 μA provides 8-bit zero accuracy for $I_{REF} \geq 2$ mA. The power supply currents of the DAC0808 series are independent of bit codes, and exhibits essentially constant device characteristics over the entire supply voltage range.

The DAC0808 will interface directly with popular TTL, DTL or CMOS logic levels, and is a direct replacement for the MC1508/MC1408. For higher speed applications, see DAC0800 data sheet.

Features

- Relative accuracy: ±0.19% error maximum (DAC0808)
- Full scale current match: ±1 LSB typ
- 7 and 6-bit accuracy available (DAC0807, DAC0806)
- Fast settling time: 150 ns typ
- Noninverting digital inputs are TTL and CMOS compatible
- High speed multiplying input slew rate: 8 mA/μs
- Power supply voltage range: ±4.5V to ±18V
- Low power consumption: 33 mW @ ±5V

Block and Connection Diagrams

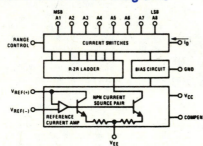

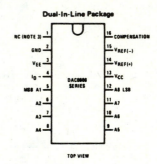

Typical Application

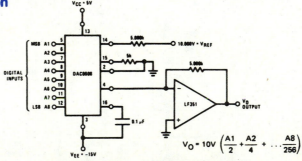

$$V_O = 10V \left(\frac{A1}{2} + \frac{A2}{4} + \cdots \frac{A8}{256} \right)$$

FIGURE 1. +10V Output Digital to Analog Converter

Ordering Information

ACCURACY	OPERATING TEMPERATURE RANGE	ORDER NUMBERS*					
		D PACKAGE (D16C)		J PACKAGE (J16A)		N PACKAGE (N16A)	
8-bit	−55°C ≤ T_A ≤ +125°C	DAC0808LD	MC1508L8				
8-bit	0°C ≤ T_A ≤ +75°C			DAC0808LCJ	MC1408L8	DAC0808LCN	MC1408P8
7-bit	0°C ≤ T_A ≤ +75°C			DAC0807LCJ	MC1408L7	DAC0807LCN	MC1408P7
6-bit	0°C ≤ T_A ≤ +75°C			DAC0806LCJ	MC1408L6	DAC0806LCN	MC1408P6

*Note. Devices may be ordered by using either order number.

Absolute Maximum Ratings

Power Supply Voltage		Power Dissipation (Package Limitation)	1000 mW
V_{CC}	$+18\ V_{DC}$	Derate above $T_A = 25°C$	$6.7\ mW/°C$
V_{EE}	$-18\ V_{DC}$	Operating Temperature Range	
Digital Input Voltage, V5–V12	$-10\ V_{DC}$ to $+18\ V_{DC}$	DAC0808L	$-55°C \leq T_A \leq +125°C$
Applied Output Voltage, V_O	$-11\ V_{DC}$ to $+18\ V_{DC}$	DAC0808LC Series	$0 \leq T_A \leq +75°C$
Reference Current, I_{14}	5 mA	Storage Temperature Range	$-65°C$ to $+150°C$
Reference Amplifier Inputs, V14, V15	V_{CC}, V_{EE}		

Electrical Characteristics

($V_{CC} = 5V$, $V_{EE} = -15\ V_{DC}$, $V_{REF}/R14 = 2$ mA, DAC0808: $T_A = -55°C$ to $+125°C$, DAC0808C, DAC0807C, DAC0806C, $T_A = 0°C$ to $+75°C$, and all digital inputs at high logic level unless otherwise noted.)

	PARAMETER	CONDITIONS	MIN	TYP	MAX	UNITS
E_r	Relative Accuracy (Error Relative to Full Scale I_O)	(Figure 4)				%
	DAC0808L (LM1508-8),				±0.19	%
	DAC0808LC (LM1408-8)					
	DAC0807LC (LM1408-7), (Note 1)				±0.39	%
	DAC0806LC (LM1408-6), (Note 1)				±0.78	%
	Settling Time to Within 1/2 LSB (Includes t_{PLH})	$T_A = 25°C$ (Note 2), (Figure 5)		150		ns
t_{PLH}, t_{PHL}	Propagation Delay Time	$T_A = 25°C$, (Figure 5)		30	100	ns
TCI_O	Output Full Scale Current Drift			±20		ppm/°C
MSB	Digital Input Logic Levels	(Figure 3)				
V_{IH}	High Level, Logic "1"		2			V_{DC}
V_{IL}	Low Level, Logic "0"				0.8	V_{DC}
MSB	Digital Input Current	(Figure 3)				
	High Level	$V_{IH} = 5V$		0	0.040	mA
	Low Level	$V_{IL} = 0.8V$		-0.003	-0.8	mA
I_{15}	Reference Input Bias Current	(Figure 3)		-1	-3	μA
	Output Current Range	(Figure 3)				
		$V_{EE} = -5V$	0	2.0	2.1	mA
		$V_{EE} = -15V$, $T_A = 25°C$	0	2.0	4.2	mA
I_O	Output Current	$V_{REF} = 2.000V$, R14 = 1000Ω, (Figure 3)	1.9	1.99	2.1	mA
	Output Current, All Bits Low	(Figure 3)		0	4	μA
	Output Voltage Compliance	$E_r \leq 0.19\%$, $T_A = 25°C$				
	Pin 1 Grounded,				-0.55, +0.4	V_{DC}
	V_{EE} Below -10V				-5.0, +0.4	V_{DC}
SRI_{REF}	Reference Current Slew Rate	(Figure 6)	4	8		mA/μs
	Output Current Power Supply Sensitivity	$-5V \leq V_{EE} \leq -16.5V$		0.05	2.7	μA/V
	Power Supply Current (All Bits Low)	(Figure 3)				
I_{CC}				2.3	22	mA
I_{EE}				-4.3	-13	mA
	Power Supply Voltage Range	$T_A = 25°C$, (Figure 3)				
V_{CC}			4.5	5.0	5.5	V_{DC}
V_{EE}			-4.5	-15	-16.5	V_{DC}
	Power Dissipation					
	All Bits Low	$V_{CC} = 5V$, $V_{EE} = -5V$		33	170	mW
		$V_{CC} = 5V$, $V_{EE} = -15V$		106	305	mW
	All Bits High	$V_{CC} = 15V$, $V_{EE} = -5V$		90		mW
		$V_{CC} = 15V$, $V_{EE} = -15V$		160		mW

Note 1: All current switches are tested to guarantee at least 50% of rated current.
Note 2: All bits switched.
Note 3: Range control is not required.

Test Circuits

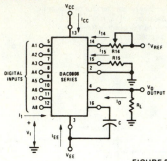

V_I and I_1 apply to inputs A1–A8.

The resistor tied to pin 15 is to temperature compensate the bias current and may not be necessary for all applications.

$$I_O = K \left(\frac{A1}{2} + \frac{A2}{4} + \frac{A4}{16} + \frac{A5}{32} + \frac{A6}{64} + \frac{A7}{128} + \frac{A8}{256} \right)$$

where $K \cong \dfrac{V_{REF}}{R14}$

and A_N = "1" if A_N is at high level

A_N = "0" if A_N is at low level

FIGURE 3. Notation Definitions Test Circuit

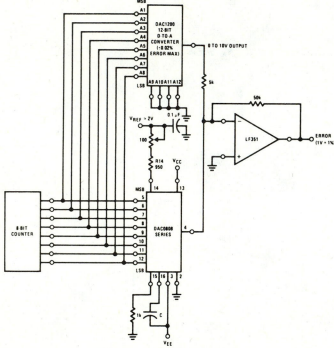

FIGURE 4. Relative Accuracy Test Circuit

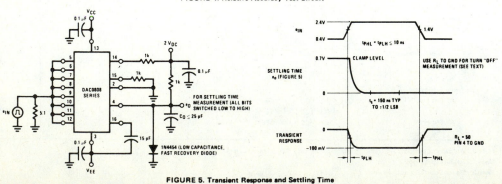

FIGURE 5. Transient Response and Settling Time

Test Circuits (Continued)

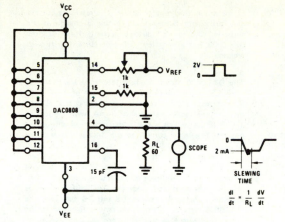

FIGURE 6. Reference Current Slew Rate Measurement

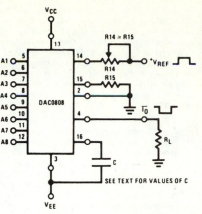

FIGURE 7. Positive V_REF

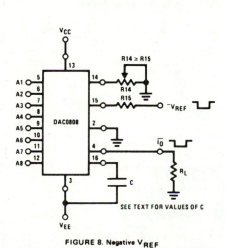

FIGURE 8. Negative V_REF

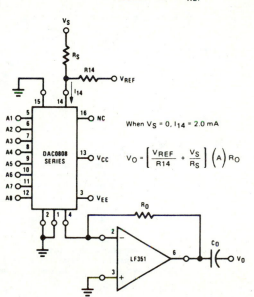

When $V_S = 0$, $I_{14} = 2.0$ mA

$$V_O = \left[\frac{V_{REF}}{R14} + \frac{V_S}{R_S} \right] (A) R_O$$

FIGURE 9. Programmable Gain Amplifier or Digital Attenuator Circuit

Application Hints

REFERENCE AMPLIFIER DRIVE AND COMPENSATION

The reference amplifier provides a voltage at pin 14 for converting the reference voltage to a current, and a turn-around circuit or current mirror for feeding the ladder. The reference amplifier input current, I_{14}, must always flow into pin 14, regardless of the set-up method or reference voltage polarity.

Connections for a positive voltage are shown in *Figure 7*. The reference voltage source supplies the full current

I_{14}. For bipolar reference signals, as in the multiplying mode, R15 can be tied to a negative voltage corresponding to the minimum input level. It is possible to eliminate R15 with only a small sacrifice in accuracy and temperature drift.

The compensation capacitor value must be increased with increases in R14 to maintain proper phase margin; for R14 values of 1, 2.5 and 5 kΩ, minimum capacitor values are 15, 37 and 75 pF. The capacitor may be tied to either V_{EE} or ground, but using V_{EE} increases negative supply rejection.

Application Hints (Continued)

A negative reference voltage may be used if R14 is grounded and the reference voltage is applied to R15 as shown in *Figure 8*. A high input impedance is the main advantage of this method. Compensation involves a capacitor to V_{EE} on pin 16, using the values of the previous paragraph. The negative reference voltage must be at least 4V above the V_{EE} supply. Bipolar input signals may be handled by connecting R14 to a positive reference voltage equal to the peak positive input level at pin 15.

When a DC reference voltage is used, capacitive bypass to ground is recommended. The 5V logic supply is not recommended as a reference voltage. If a well regulated 5V supply which drives logic is to be used as the reference, R14 should be decoupled by connecting it to 5V through another resistor and bypassing the junction of the 2 resistors with 0.1 μF to ground. For reference voltages greater than 5V, a clamp diode is recommended between pin 14 and ground.

If pin 14 is driven by a high impedance such as a transistor current source, none of the above compensation methods apply and the amplifier must be heavily compensated, decreasing the overall bandwidth.

OUTPUT VOLTAGE RANGE

The voltage on pin 4 is restricted to a range of −0.6 to 0.5V when V_{EE} = −5V due to the current switching methods employed in the DAC0808.

The negative output voltage compliance of the DAC0808 is extended to −5V where the negative supply voltage is more negative than −10V. Using a full-scale current of 1.992 mA and load resistor of 2.5 kΩ between pin 4 and ground will yield a voltage output of 256 levels between 0 and −4.980V. Floating pin 1 does not affect the converter speed or power dissipation. However, the value of the load resistor determines the switching time due to increased voltage swing. Values of R_L up to 500Ω do not significantly affect performance, but a 2.5 kΩ load increases worst-case settling time to 1.2 μs (when all bits are switched ON). Refer to the subsequent text section on Settling Time for more details on output loading.

OUTPUT CURRENT RANGE

The output current maximum rating of 4.2 mA may be used only for negative supply voltages more negative than −7V, due to the increased voltage drop across the resistors in the reference current amplifier.

ACCURACY

Absolute accuracy is the measure of each output current level with respect to its intended value, and is dependent upon relative accuracy and full-scale current drift. Relative accuracy is the measure of each output current level as a fraction of the full-scale current. The relative accuracy of the DAC0808 is essentially constant with temperature due to the excellent temperature tracking

of the monolithic resistor ladder. The reference current may drift with temperature, causing a change in the absolute accuracy of output current. However, the DAC0808 has a very low full-scale current drift with temperature.

The DAC0808 series is guaranteed accurate to within ±1/2 LSB at a full-scale output current of 1.992 mA. This corresponds to a reference amplifier output current drive to the ladder network of 2 mA, with the loss of 1 LSB (8 μA) which is the ladder remainder shunted to ground. The input current to pin 14 has a guaranteed value of between 1.9 and 2.1 mA, allowing some mismatch in the NPN current source pair. The accuracy test circuit is shown in *Figure 4*. The 12-bit converter is calibrated for a full-scale output current of 1.992 mA. This is an optional step since the DAC0808 accuracy is essentially the same between 1.5 and 2.5 mA. Then the DAC0808 circuits' full-scale current is trimmed to the same value with R14 so that a zero value appears at the error amplifier output. The counter is activated and the error band may be displayed on an oscilloscope, detected by comparators, or stored in a peak detector.

Two 8-bit D-to-A converters may not be used to construct a 16-bit accuracy D-to-A converter. 16-bit accuracy implies a total error of ±1/2 of one part in 65,536, or ±0.00076%, which is much more accurate than the ±0.019% specification provided by the DAC0808.

MULTIPLYING ACCURACY

The DAC0808 may be used in the multiplying mode with 8-bit accuracy when the reference current is varied over a range of 256:1. If the reference current in the multiplying mode ranges from 16 μA to 4 mA, the additional error contributions are less than 1.6 μA. This is well within 8-bit accuracy when referred to full-scale.

A monotonic converter is one which supplies an increase in current for each increment in the binary word. Typically, the DAC0808 is monotonic for all values of reference current above 0.5 mA. The recommended range for operation with a DC reference current is 0.5 to 4 mA.

SETTLING TIME

The worst-case switching condition occurs when all bits are switched ON, which corresponds to a low-to-high transition for all bits. This time is typically 150 ns for settling to within ±1/2 LSB, for 8-bit accuracy, and 100 ns to 1/2 LSB for 7 and 6-bit accuracy. The turn OFF is typically under 100 ns. These times apply when $R_L \leq 500\Omega$ and $C_O \leq 25$ pF.

Extra care must be taken in board layout since this is usually the dominant factor in satisfactoy test results when measuring settling time. Short leads, 100 μF supply bypassing for low frequencies, and minimum scope lead length are all mandatory.

Appendixes

APPENDIX A: MISCELLANEOUS CHARTS AND TABLES

TABLE A.1 Binary Bit Weights or Resolution

BIT	2^{-n}	$1/2^n$ (Fraction)	dB	$1/2^n$ (Decimal)	%	ppm
FS	2^0	1	0	1.0	100	1,000,000
MSB	2^{-1}	1/2	-6	0.5	50	500,000
2	2^{-2}	1/4	-12	0.25	25	250,000
3	2^{-3}	1/8	-18.1	0.125	12.5	125,000
4	2^{-4}	1/16	-24.1	0.0625	6.2	62,500
5	2^{-5}	1/32	-30.1	0.03125	3.1	31,250
6	2^{-6}	1/64	-36.1	0.015625	1.6	15,625
7	2^{-7}	1/128	-42.1	0.007812	0.8	7,812
8	2^{-8}	1/256	-48.2	0.003906	0.4	3,906
9	2^{-9}	1/512	-54.2	0.001953	0.2	1,953
10	2^{-10}	1/1,024	-60.2	0.0009766	0.1	977
11	2^{-11}	1/2,048	-66.2	0.00048828	0.05	488
12	2^{-12}	1/4,096	-72.2	0.00024414	0.024	244
13	2^{-13}	1/8,192	-78.3	0.00012207	0.012	122
14	2^{-14}	1/16,384	-84.3	0.000061035	0.006	61
15	2^{-15}	1/32,768	-90.3	0.0000305176	0.003	31
16	2^{-16}	1/65,536	-96.3	0.0000152588	0.0015	15
17	2^{-17}	1/131,072	-102.3	0.00000762939	0.0008	7.6
18	2^{-18}	1/262,144	-108.4	0.000003814697	0.0004	3.8
19	2^{-19}	1/524,288	-114.4	0.000001907349	0.0002	1.9
20	2^{-20}	1/1,048,576	-120.4	0.0000009536743	0.0001	0.95

TABLE A.2 Vibration and the Likely Cause

Frequency — CPM or RPM	Cause	Amplitude	Phase	Remarks
Approximately ½ × rotating speed	Oil whip or oil whirl	Often very severe	Erratic	Occurs only on high speed machines where pressure lubricated bearings are used.
1 × rpm	Unbalance	Proportional to unbalance	Single reference mark	Most common cause. If high in vertical direction check for loose mounting.
2 × rpm	Mechanical looseness	Erratic	Two marks	Usually high in vertical direction. Causes: loose mountings, worn bearing housings.
2 × rpm	Misalignment or bent shaft	Should be ½ that at 1 × rpm. Large in axial direction.	Two marks	Use dial indicator for positive diagnosis.
1, 2, 3, 4 × rpm	Bad drive belts	Erratic	One, two, three or four. Unsteady	Use strobe light to freeze faulty belt.
Synchronous or 2 × synchronous	Electrical	Usually low	Single or double rotating mark	If vibration amplitude drops instantly when power is turned off, cause is electrical.
Many times rpm 8,000 to 25,000 cpm	Bad bearings (antifriction)	Erratic	Erratic — many reference marks	If amplitude exceeds .25 mils, suspect faulty bearings.
RPM × number of gear teeth	Gear noise	Usually low	Many reference marks	
RPM × number of blades on fan or pump	Aerodynamic or hydraulic			Uncommon cause

TABLE A.3   Significance of Motor Locked-Rotor Indicating Code Letter

Note: Section 430-7 of the *National Electrical Code* requires that all AC motors rated 1/2 hp or more (other than polyphase wound rotor motors) be marked with a locked-rotor code letter in accordance with the table below. Current drawn by the motor under stall conditions can be calculated by using the values given in this table. Current drawn by the motor under stalled conditions must be considered in selecting the motor protection and starting package and in coordinating power system protective devices. However, current drawn by the motor upon starting will be higher than locked-rotor current because of transient inrush. In some cases, it might be necessary or more feasible to select a motor with a different locked-rotor code letter.

Code Letter	Kilovolt-Amperes per Horsepower with Locked Rotor
A	0–3.14
B	3.15–3.54
C	3.55–3.99
D	4.0–4.49
E	4.5–4.99
F	5.0–5.59
G	5.6–6.29
H	6.3–7.09
J	7.1–7.99
K	8.0–8.99
L	9.0–9.99
M	10.0–11.19
N	11.2–12.49
P	12.5–13.99
R	14.0–15.99
S	16.0–17.99
T	18.0–19.99
U	20.0–22.39
V	22.4–and up

TABLE A.4   NEMA Design Letter Designations

Designation	Explanation
NEMA Design A	Design A covers a wide variety of motors similar to Design B except that their maximum torque and starting current are higher. Motors of this design are not regularly offered but are built to order for special applications.
NEMA Design B	These motors are the standard general purpose design. They have low starting current, normal torque, and normal slip. Their field of application is broad and includes fans, blowers, pumps, and machine tools.
NEMA Design C	Design C motors have high breakaway torque, low starting current, and normal slip. The higher breakaway torque makes this motor advantageous for "hard-to-start" applications such as plunger pumps, conveyors, and compressors
NEMA Design D	These motors have a high breakaway torque combined with high slip. Breakaway torque for 4-, 6-, and 8-pole motors is 275 percent or more of full load torque. The 5 to 8 percent and 8 to 13 percent slip motors are recommended for punch presses, shears, and other high inertia machinery. In these applications it is frequently desirable to make use of the energy stored in the flywheel under heavy, fluctuating load conditions. These high slip motors are also used for multimotor conveyor drives where motors operate in mechanical parallel. The 13 percent, or greater, Design D motors also have high starting torque but are limited to short-time duty. By developing a high breakaway torque and high running horsepower in the smallest possible frame size, these motors find use on cranes, hoists, elevators, and for auxiliary movement of machine tools.

TABLE A.5   NEMA Standard M61-2.01: Nominal Voltages

Voltage Required at Motor	Minimum Voltage at Generator
110-single phase	120
208	
220	240
440	480
550	600
2200	2400
2300	2500

TABLE A.6   Insulation Class, Operating Temperature, and Service Factor*

Note: Motor insulation class, permissible operating temperature, and service factor are inextricably intertwined. The insulation class defines the temperature that the insulation can be subjected to without suffering damage. The actual temperature that the winding is subjected to is, in turn, determined by the ambient temperature and the temperature rise resulting from motor loading. Service factor is a multiplier indicating the extent to which a motor can be overloaded under specified conditions.

Class	Explanation
Class A	Now obsolescent insofar as industrial motors are concerned, Class A was once the most common of motor insulations, especially for small motors. Class A comprises materials or combinations of materials such as cotton or paper, when suitably impregnated or coated, or other materials capable of operation at the temperature rise assigned for Class A insulation for the particular machine.
Class B	The predominant class of insulation used in motor manufacturing and rewinding today, this class is the basic standard of the industry. It comprises materials such as mica, glass fiber, polyester and aramid laminates, etc., with suitable bonding substances, or other materials, not necessarily inorganic, capable of operation at the temperature rise assigned for Class B insulation for the particular machine.
Class F	Class F incorporates materials that are similar to those in Class B but capable of operation at the temperature rise assigned for Class F for the particular machine.
Class H	Class H insulation systems comprise materials or combinations of materials such as silicone elastomer, mica, glass fiber, polyester and aramid laminates, etc., with suitable bonding substances such as silicone resins, or other materials capable of operation at the temperature rise assigned for Class H insulation for the particular machine.

*In the past, motor nameplates gave temperature rise (based on 40° C ambient), service factor, and insulation class (if other than Class A). Maximum permissible operating temperatures were also identified for each class. These criteria were difficult for the user to understand or apply, because the actual operating temperature is a function of both ambient temperature and temperature rise, and temperature rise is virtually impossible for the user to determine.

TABLE A.7   Industrial Programmable Controller Languages—Advantages and Disadvantages

LANGUAGE	ISSUE	ADVANTAGES	DISADVANTAGES
Ladder	Sequential process	Simulates actual devices Solves many applications	Slow scan time Lack of extensibility
	Continuous process	Widespread acceptance and support	Hard to use Longer design time
	Mathematics	Simple functions	Lack of advanced functions
	Report generation		Manufacturer dependent
	PID		Limited capability
	Networking		Manufacturer dependent
	Plant support	Extensive knowledge	
	Diagnostics	Power flow, force I/O	Limited capability
	Portability	Easy for basic functions	Rewrite for advanced functions
Flowchart	Sequential process	Displays logic flow	Not a common format
	Continuous process	Design engineer's tool Easy to understand	
	Mathematics	Included as function block	
	Report generation	Included as function block	
	PID	Good relation to original design Displays algorithm	
	Networking		Hard to represent
	Plant Support	Depends on prior experience	
	Diagnostics	Shows overall flow	Plant support varies
	Portability	Good	Not common for PLCs
Boolean	Sequential process	Represent logical flow Faster execution	Not for complex application Limited to small PLCs
	Continous process	Language familiar to designers	Hard to understand
	Mathematics		CPU dependent
	Report generation	Depends on supplier	
	PID	Depends on supplier	
	Networking		Requires special instructions
	Plant support	Depends on plant experience	Harder to support
	Diagnostics		Limited to simple control
	Portability	Good for basic functions	Can be CPU dependent
Graphics	Sequential process	Display total process Show logical flow	Not well known
	Continuous process	Display total process Show logical flow	Not well known
	Mathematics	Include as function block	
	Report generation		No special support
	PID	Include as function block	
	Networking		No special support
	Plant support	Easy to understand	Not in common use Training required
	Diagnostics	Display failed process	
	Portability	Good at top levels	Not offered by many suppliers

TABLE A.8 Relay Contact Forms

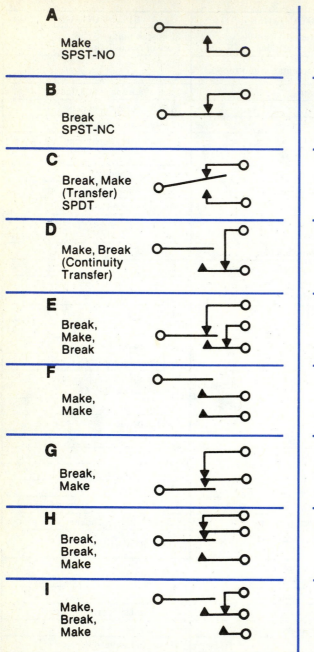

A
Make
SPST-NO

B
Break
SPST-NC

C
Break, Make
(Transfer)
SPDT

D
Make, Break
(Continuity
Transfer)

E
Break,
Make,
Break

F
Make,
Make

G
Break,
Make

H
Break,
Break,
Make

I
Make,
Break,
Make

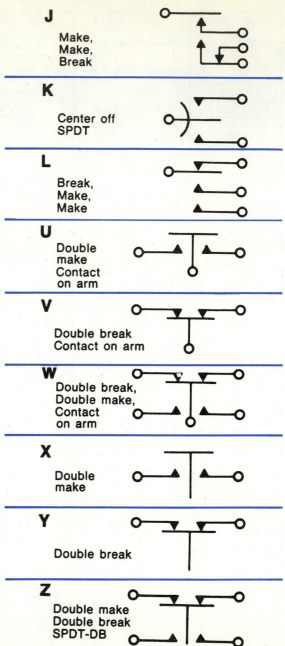

J
Make,
Make,
Break

K
Center off
SPDT

L
Break,
Make,
Make

U
Double
make
Contact
on arm

V
Double break
Contact on arm

W
Double break,
Double make,
Contact
on arm

X
Double
make

Y
Double break

Z
Double make
Double break
SPDT-DB

APPENDIX B: ASCII CHARACTER CODES

							HEX LSD / HEX MSD	0	1	2	3	4	5	6	7
b7								0	0	0	0	1	1	1	1
	b6							0	0	1	1	0	0	1	1
		b5						0	1	0	1	0	1	0	1
b7	b6	b5	b4	b3	b2	b1		0	1	2	3	4	5	6	7
			0	0	0	0	0	NUL	DLE	SP	0	@	P	'	p
			0	0	0	1	1	SOH	DC1	!	1	A	Q	a	q
			0	0	1	0	2	STX	DC2	"	2	B	R	b	r
			0	0	1	1	3	ETX	DC3	#	3	C	S	c	s
			0	1	0	0	4	EOT	DC4	$	4	D	T	d	t
			0	1	0	1	5	ENQ	NAK	%	5	E	U	e	u
			0	1	1	0	6	ACK	SYN	&	6	F	V	f	v
			1	1	1	1	7	BEL	ETB	'	7	G	W	g	w
			1	0	0	0	8	BS	CAN	(	8	H	X	h	x
			1	0	0	1	9	HT	EM	)	9	I	Y	i	y
			1	0	1	0	A(10)	LF	SUB	*	:	J	Z	j	z
			1	0	1	1	B(11)	VT	ESC	+	;	K	[	k	{
			1	1	0	0	C(12)	FF	FS	,	<	L	\	l	¦
			1	1	0	1	D(13)	CR	GS	-	=	M	]	m	}
			1	1	1	0	E(14)	SO	RS	.	>	N	ˆ	n	~
			1	1	1	1	F(15)	SI	US	/	?	O	—	o	DEL

NUL Null
SOH Start of heading
STN Start of text
ETX End of text
EOT End of transmission
ENQ Enquiry
ACK Acknowledge
BEL Bell, or alarm
BS   Backspace

HT   Horizontal tabulation
LF   Line feed
VT   Vertical tabulation
FF   Form feed
CR   Carriage return
SO   Shift out
SI   Shift in
DLE Data link escape
DC1 Device control 1

DC2 Device control 2
DC3 Device control 3
DC4 Device control 4
NAK Negative acknowledge
SYN Synchronous idle
ETB End of transmission block
CAN Cancel
EM   End of medium
SUB Substitute

ESC Escape
FS   File separator
GS   Group separator
RS   Record separator
US   Unit separator
SP   Space
DEL Delete

APPENDIX C: EBCDIC CODE

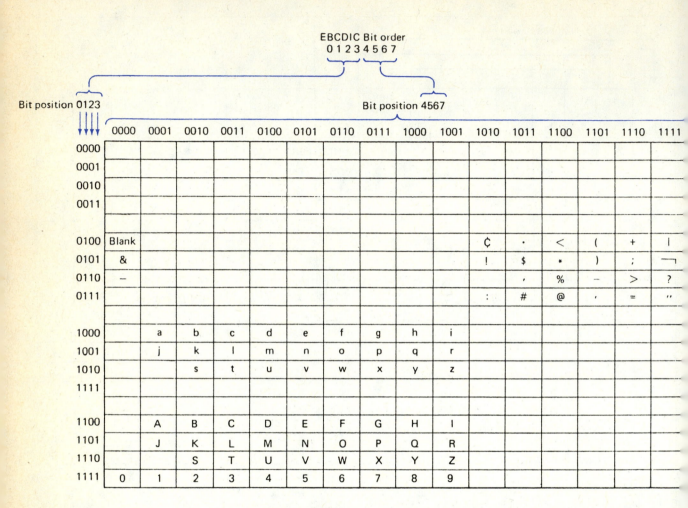

EBCDIC Bit order
0 1 2 3 4 5 6 7

Bit position 0123

Bit position 4567

0123 \ 4567	0000	0001	0010	0011	0100	0101	0110	0111	1000	1001	1010	1011	1100	1101	1110	1111
0000																
0001																
0010																
0011																
0100	Blank										¢	•	<	(	+	|
0101	&										!	$	*	)	;	¬
0110	−											,	%	−	>	?
0111											:	#	@	'	=	"
1000		a	b	c	d	e	f	g	h	i						
1001		j	k	l	m	n	o	p	q	r						
1010			s	t	u	v	w	x	y	z						
1111																
1100		A	B	C	D	E	F	G	H	I						
1101		J	K	L	M	N	O	P	Q	R						
1110			S	T	U	V	W	X	Y	Z						
1111	0	1	2	3	4	5	6	7	8	9						

APPENDIX D: OPERATIONAL AMPLIFIERS

Introduction

The *operational amplifier* (*op amp*) has a short but eventful history. The concept of the op amp dates back to the late 1940s when they were used in analog computers. They performed the mathematical operations of addition, subtraction, integration, and differentiation—hence, the name *operational* amplifier. The original op amp was constructed mainly of vacuum tubes. Those early op amps consumed a lot of power and were costly and bulky.

The integrated circuit op amp as we know it today was developed in 1965 by Robert Widlar. Widlar was then working for Fairchild Semiconductor, which marketed the first integrated circuit op amp, the µA709. The 709 was so well designed that it is still in use today. In the intervening years, that one design has been further developed and expanded. Today, it is estimated that there are several thousand types of op amps. Furthermore, over one-third of all linear integrated circuits (ICs) are op amps. In fact, the op amp is one of the most popular active linear devices on the market today. The reason for its popularity is threefold. First, many different kinds of op amps are commercially available, and they are low in cost. Second, the op amp displays enormous versatility in application. Third, op amps are easy to use in designing and developing prototypes.

Just what is an op amp? We can describe an op amp very simply as a high-gain, direct-coupled amplifier. Characteristics such as circuit gain and frequency response are established by external components.

Ideal Op Amp

If asked to describe the perfect or ideal amplifier, what characteristics would we choose? For instance, what should the gain, frequency response, and input and output impedances be? If small signals are to be amplified, certainly we would want the voltage gain to be high, ideally infinite ($A_V = \infty$). (Note that we use A to symbolize gain, although G is sometimes used in other textbooks.) Again, if the input signal is very small, we do not want the amplifier to load

down (reduce) the signal. So, ideally, the input impedance should be infinite ($Z_i = \infty$). What about frequency response? If an amplifier is to be versatile, it should amplify (ideally) any signal from 0 Hz (DC) to infinity (bandwidth $BW = \infty$). Finally, what is the ideal output impedance? There are basically two conditions we need to think about in choosing an output impedance: maximum transfer of power and maximum transfer of voltage. Since the op amp is not a power device, we do not need to worry about matching impedances for maximum power transfer. If the output impedance of the op amp is zero, we have maximum transfer of voltage, which is what we want. Also, if the output impedance is high, the gain of the amplifier is reduced. The output impedance forms a voltage divider with the load and feedback resistors. Ideally, then, the amplifier's output impedance should be zero ($Z_o = 0\ \Omega$).

We do not normally need to be concerned with the internal workings of the op amp. However, a brief consideration of the op amp's internal circuitry is helpful in understanding how the ideal op amp parameters are so nearly achieved. It also gives some insight into the limitations in testing and measuring op amp performance.

Inside the Op Amp

Basically, the op amp can be broken down internally into three stages: input, intermediate, and output.

Input Stage. The *input stage* gives the op amp its high Z_i and contributes to its high A_V. The basic component in this stage, and the heart of the op amp, is the *differential amplifier*, illustrated in Figure D.1. Note that V_{i1} and V_{i2} are the two input voltages to the differential amplifier, and V_{o1} and V_{o2} are the two outputs. If V_{o2} is taken as a reference, then a signal applied at V_{i1} is inverted at the V_{o1} output terminal. This input terminal is usually identified with a minus sign. If a signal is applied at V_{i2}, the signal is unchanged in phase at V_{o1}. This input terminal is called the *noninverting terminal*. It is usually identified with a plus sign.

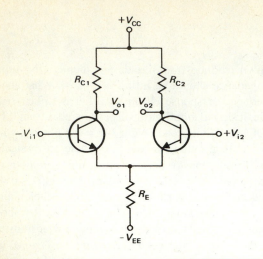

FIGURE D.1   Differential Amplifier

Recall that the differential amplifier exhibits a very high input impedance, mathematically approximated by the following equation:

$$Z_i \approx 2\beta r'_e \qquad \textbf{(D.1)}$$

where

ß = current gain of transistor (h_{fe})

r'_e = AC resistance of base-emitter junction, approximately 25 mV/I_E

I_E = emitter current

Equation D.1 is derived from considering the differential amp as a common emitter cascaded to a common base. The common emitter's input impedance is the current gain times r'_e. The common base's input (r'_e) is in series with r'_e from the common emitter. Since r'_e is relatively large owing to small emitter currents, Z_i approaches the range 1–2 MΩ.

Another interesting feature of the differential amplifier is its supply voltages. Note that we must have both a positive and a negative supply. This requirement will be especially important later when we discuss level shifters.

A useful property of the differential amplifier is its ability to reject common-mode signals. *Common-*

mode rejection is defined as the amplifier's ability to reject identical signals when they occur on both inputs at the same time. Why is this property so important? Many industrial electronics applications exhibit high noise levels. This unwanted interference is fed equally to both the inputs of the op amp. Since the amplitude and phase of the noise are equal at any instant, the differential amp amplifies the common-mode signals equally. The difference voltage at the output terminals due to the common-mode signal is then zero. The desired signal, however, uses the difference-mode amplifying capabilities of the differential amp. Here, the input signal is the difference between the input terminals, and the amplifier amplifies this difference.

Well-designed differential amplifiers are rarely seen in the form shown in Figure D.1. One reason is that R_E is typically a large-value resistor and is not well suited to IC fabrication techniques. Transistors generally replace large-value resistors in IC design. Thus, differential amplifiers are more often seen in the form shown in Figure D.2.

What is the difference between Figure D.1 and Figure D.2? Notice that the resistor R_E in Figure D.1 has been replaced by a *constant-current source*—that

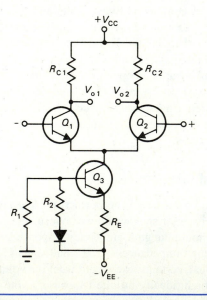

FIGURE D.2   Differential Amplifier with Constant-Current Source

is, a transistor Q_3 with constant bias. This procedure improves common-mode rejection characteristics. Good common-mode rejection characteristics come about when the common-mode gain is low. Since the impedance of the constant-current source is high and is in the denominator of the gain equation, the common-mode gain is low.

Sometimes, other components and other configurations are used to increase the input impedance of the differential amplifier. Two of the most common configurations are illustrated in Figure D.3.

The circuit shown in Figure D.3A has a *Darlington pair input*. The input impedance of the Darlington configuration is approximately $3ß^2 r'_e$. This value increases the input impedance of the differential amplifier by a factor of 1.5ß.

Note that in the circuit in Figure D.3B, the input devices are *field effect transistors* (FETs), not bipolar junction transistors (BJTs). The advantage of this configuration is the high input impedance provided by the FETs.

Intermediate Stage. The *intermediate stage* serves two functions. First, it increases the current and voltage gain of the op amp. Second, it provides level shifting. These functions are described next.

A problem encountered in the differential amplifier stage is overcome here in the intermediate stages. The gain of the common emitter is inversely related to input impedance. As the designer seeks to raise input impedance by lowering emitter current, gain is also lowered. The intermediate stage provides the additional voltage gain needed to provide high overall amplifier gain.

Level shifting is a change in DC bias levels owing to direct-coupled cascaded amplifiers. Level shifting is necessary to combat the phenomenon of voltage buildup. This phenomenon occurs in any direct-coupled amplifier using cascaded common-emitter stages. Thus, each succeeding stage requires a higher collector supply to keep the collector at a higher potential than that of the base.

Look at Figure D.2 again. If the inputs are grounded, there is no difference in potential between the inputs, and we expect 0 V out. We can see that if the bases of Q_1 and Q_2 are grounded, their collectors must be at a higher potential in order to prevent saturation. Thus, the output is not zero, as we desired. But suppose we couple the outputs of the circuit in Figure D.2 to the inputs of the circuit of Figure D.4A. In this situation, the collectors of Figure D.4A are at a lower potential (and closer to 0 V).

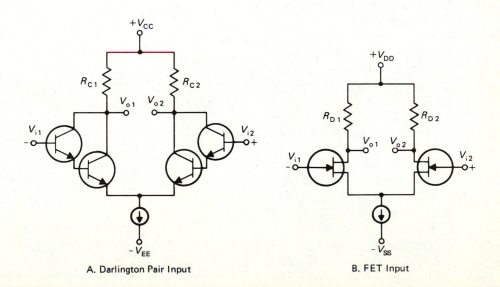

A. Darlington Pair Input   B. FET Input

FIGURE D.3 Differential Amplifiers

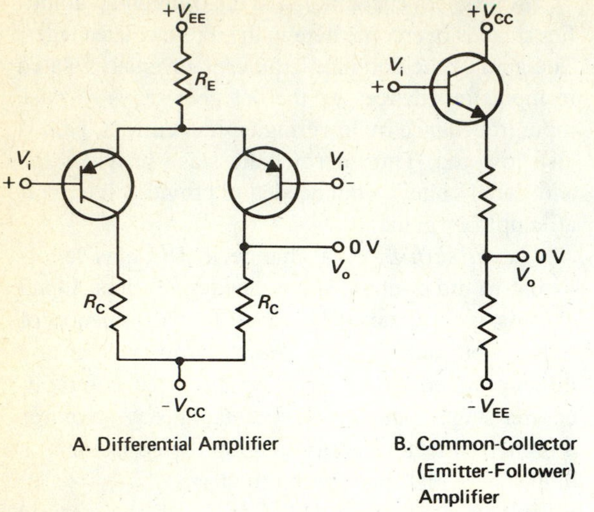

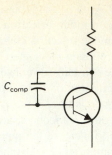

A. Differential Amplifier

B. Common-Collector
(Emitter-Follower)
Amplifier

FIGURE D.4 Level Shifting

FIGURE D.5 Intermediate-Stage Circuit
with Compensating Capacitor

Level shifting and additional voltage and current gain are provided in the circuit shown in Figure D.4A. Another method for solving the voltage buildup problem is the circuit of Figure D.4B. The common-collector amplifier in Figure D.4B only gives level shifting and current gain.

Another circuit generally employed in the inter-mediate stage is the one shown in Figure D.5. This stage contains an *internal compensating capacitor* (C_{comp}) or provides for an external capacitor connection. Among other things, as discussed later, the capacitor C_{comp} helps prevent high-frequency oscillations and instability.

Output Stage. The *output stage* provides the power necessary to drive the load while allowing maximum output signal voltage swing and minimum output impedance. The simplest circuit that satisfies these requirements is the *emitter-follower*. A disadvantage of the emitter-follower is that its emitter resistor consumes too much power at high current levels. A more practical substitute for the emitter-follower is the simplified *complementary symmetry power amplifier* shown in Figure D.6. This amplifier can supply large currents with good power gain and low output impedance.

Figure D.7 shows a simplified but complete op amp schematic. See whether you can identify the components that make up the three stages discussed in this section. Note that all resistor values are typical.

Op Amp Schematic Symbol

The op amp schematic symbol is shown in Figure D.8. Terminals 2 and 3 are the differential input terminals of the device. Terminal 2 is the inverting

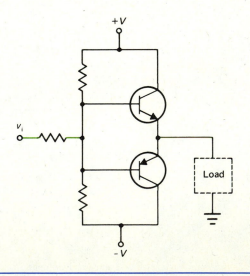

FIGURE D.6 Output Stage of
Complementary Symmetry Power Amplifier

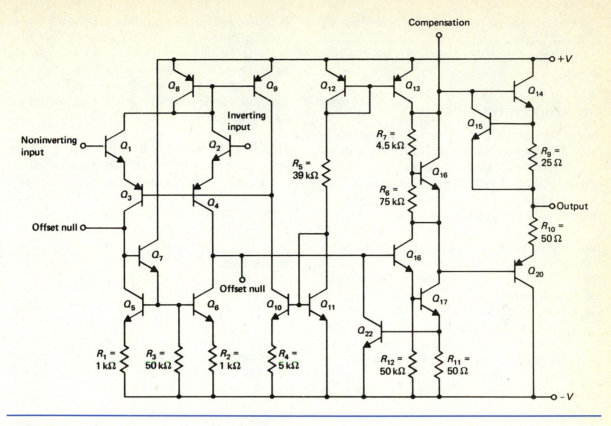

FIGURE D.7 Integrated Circuit Op Amp Schematic with Provision for External Compensation

terminal. Any signal connected to this terminal will experience a 180° phase shift at the output. Terminal 3 is the noninverting terminal. No phase inversion occurs between this input and the output.

Terminals 1 and 4 are the power supply terminals. Normally, a positive voltage is connected to terminal 1 and a negative voltage to terminal 4. The op amp may have either one of these terminals grounded. In such a case, the terminal shows a ground schematic symbol attached to the proper terminal. Be cautious of operating the op amp with either power input grounded. This mode is not the normal mode of operation, and special consideration must be taken. For example, the output cannot go to 0 V, and the op amp can be destroyed if the input signal is of the incorrect polarity. Another word of caution: No matter what the power supply voltage is, within its specified range, the input signal should not exceed it. If no power supply connections are shown on the schematic, you can assume that positive and negative voltages are connected as shown.

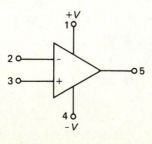

FIGURE D.8 Op Amp Schematic Symbol

Terminal 5 is the output terminal. Notice that the output is single-ended. In other words, it has one output terminal that carries the signal with respect to ground.

Op Amp Parameters

Technicians need to be familiar with op amp parameters. The op amp parameters are contained in the data sheets supplied by the device manufacturer. The specifications for the μA741 are contained in the Data Sheets. Here, we will discuss only the more frequently used op amp parameters.

Input Offset Voltage. In the ideal op amp, the output voltage is 0 V when the input voltage is 0 V. However, it is impossible to construct a perfectly balanced differential amplifier. Therefore, there is an output with no input voltage. The voltage needed at the input to adjust the amplifier for 0 V out is called the *input offset voltage* (V_{io}). It ranges in value from a few millivolts to microvolts. The popular 741 op amp's input offset voltage is typically 1 mV. In general, the lower the input offset voltage, the better the device is.

Input Offset Current. The ideal op amp has equal input currents when the output is 0 V. However, the input currents are not always equal. The difference between the input currents when the output voltage is 0 V is called the *input offset current* (I_{os}). Typically, the input offset current for the 741 is about 20 nA.

Input Bias Current. We have noted that the input currents are not necessarily equal. The *input bias current* is the average of the bias currents flowing into or out of both inputs. This parameter specifies what the approximate input current is. In general, the smaller the input bias current, the better the op amp is. Op amps that use FET front ends generally have much smaller input currents than those with BJT front ends. The 741 has a bias current of approximately 80 nA, while FET op amps have input bias currents that extend down into the picoampere range. Input bias current is symbolized as I_b.

Frequency Response Parameters. There are several parameters that involve the op amp's frequency response. Of all the ideal op amp characteristics we have discussed, frequency response does not approach the ideal very closely. It is important, therefore, that you have a thorough understanding of the op amp's frequency limitations.

Slew Rate: The slew rate limitation is probably the most important frequency limitation since it affects the amplifier's large-signal performance. Basically, the *slew rate parameter* indicates how well an amplifier follows a rapidly changing input signal. The slew rate (SR) limitation is defined mathematically as the change in output voltage (ΔV_o) over a change in time (Δt):

$$SR = \frac{\Delta V_o}{\Delta t} \qquad \text{(D.2)}$$

The slew rate parameter is measured in volts per microsecond. For example, the 741 has a typical slew rate of 0.5 V/μs. Thus, during 1 μs, the output voltage cannot change more than 0.5 V.

Slew rate limiting is caused by the presence of internal or external capacitances, the largest being the compensating capacitor. As we know, it takes a finite amount of time to charge and discharge a capacitor with a fixed amount of current available. The op amp has a number of internal constant-current sources that fix the amount of current the op amp can provide to charge the compensating capacitor. In the 741, this current is about 15 μA, and the compensating capacitor is about 30 pF. Thus, the slew rate is as follows:

$$SR = \frac{\Delta V_o}{\Delta t} = \frac{I_{max}R_L}{R_L C_{comp}} = \frac{I_{max}}{C_{comp}}$$

$$= \frac{15\ \mu A}{30\ pF}$$

$$= 0.5\ V/\mu s \qquad \text{(D.3)}$$

where

I_{max} = maximum charging current for C_{comp}
R_L = load resistance

We see from Equations D.2 and D.3 that several things can be done to prevent slew rate limiting:

1. Decrease the input signal amplitude (which decreases the transistor's output current requirement).

2. Decrease the input signal frequency (which reduces the slope of the input and thus the output).

3. Increase the current the op amp provides to charge the compensating capacitor or decrease the size of the compensating capacitor.

The most difficult of these solutions is increasing the current the op amp provides. To do so, we must substitute another op amp.

You may be required to measure the slew rate characteristic of an op amp. Slew rate measurements require a function generator and an oscilloscope. A square wave input produces an output like that shown in Figure D.9A if the frequency is high enough to cause slew rate limiting. Notice that the output signal has risen 5 V in 10 μs. The slew rate is the slope of the line from point A to point B. Therefore, the slew rate is rise divided by run, or 5 V divided by 10 μs, which equals 0.5 V/μs.

Now, look at Figure D.9B. Notice that the shape of the output waveform is triangular for a sine wave input. It must be a pronounced triangle wave before slew rate limiting can be measured. The slew rate, as before, is the slope of this distorted waveform. In this case, we again choose points A and B for our measurements. So, the calculation is rise divided by run, or 10 V divided by 20 μs, which equals 0.5 V/μs.

It should be obvious from the previous discussion that the two important factors in slew rate limiting are input frequency and amplitude. A useful equation is derived from an application of the calculus. The equation enables us to calculate the op amp's maximum frequency or amplitude limitations (for sine waves only). The equation is as follows:

$$f_{max} = \frac{SR}{2\pi \times V_{pk}} \qquad \textbf{(D.4)}$$

where

V_{pk} = peak output voltage of sine wave
f_{max} = maximum undistorted frequency (in hertz) before slewing occurs

The term f_{max} is sometimes referred to as the *power bandwidth*.

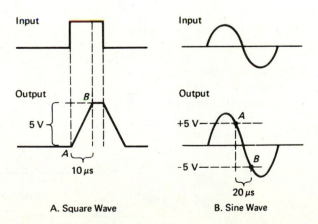

A. Square Wave   B. Sine Wave

FIGURE D.9 Waveforms for Slew Rate Limiting

Gain-Bandwidth Product: We have seen that the op amp's frequency response is limited in large-signal applications by the slew rate. In small-signal applications, the op amp's response again centers around circuit capacitances, of which the compensating capacitor is the largest.

Most amplifiers at high frequencies have a problem with feedback. As we know, an amplifier may be made to oscillate if enough positive feedback is applied. All amplifiers experience a significant phase shift between input and output at some high frequency. This result is especially true in multistage amplifiers. Suppose a sufficient number of stages contribute a portion of the phase shift. Then, eventually, some of the output signal will be fed back in phase with the input, and oscillations may occur.

For prevention of these oscillations, some op amps have a built-in capacitor that causes the gain to decrease at high frequencies. The phase shift may still be there. But the decrease in gain ensures that the amount of regenerative feedback will never be enough to sustain oscillations. If the op amp has an internal capacitor, the amplifier is said to be *internally compensated*. If the capacitor is added on externally, the op amp is said to be *externally compensated*.

Recall, from basic AC circuit theory, that the reactance X_C of a capacitor varies inversely with frequency f. The equation that expresses this relationship is as follows:

$$X_C = \frac{1}{2\pi fC} \qquad \textbf{(D.5)}$$

As frequency increases, the capacitor's reactance decreases, and vice versa. If frequency is increased by a factor of 10 (a decade), then reactance decreases by a factor of 10. Since the compensating capacitor is in the collector load of the differential amplifier, the gain is as follows:

$$A_V = \frac{r_C \parallel X_{C(comp)}}{r'_e} \qquad \textbf{(D.6)}$$

where

r_C = AC collector resistance
$\parallel$ indicates a parallel circuit

According to the equation, as X_C decreases by a factor of 10, the op amp gain also decreases by 10, or –20 dB, where 1 dB = 20 $\log_{10} V_o/V_i$. This decrease in gain as frequency increases is called *roll-off*. Roll-off is expressed in either of two ways: a number of decibels per decade or a number of decibels per octave. A decade is a tenfold change in frequency, while an octave is a twofold change. The change may be either an increase, expressed as a positive number, or a decrease, expressed as a negative number. The roll-off of a 741 is –20 dB per decade or –6 dB per octave. That is, as the frequency increases by 10, voltage gain decreases by 20 dB. This concept is graphically illustrated in Figure D.10, which plots gain versus frequency response. This diagram is called a *Bode diagram*, which is a plot of gain and phase angle versus frequency. In Figure D.10, only the gain-frequency relationship is shown.

From this graph, we see that frequency and gain are inversely related and by factors of 10. Notice that when the gain is 1 (0 dB), the frequency limit is 1 MHz. Data books often call this frequency the small-signal, *unity-gain frequency* (f_T). Sometimes, you will not be able to find this parameter in the data books. You may, instead, see a parameter called *transient-response rise time* (T_R). Rise time is defined as

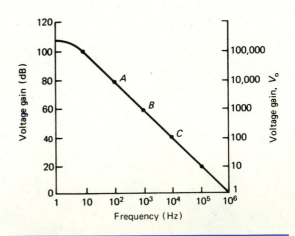

FIGURE D.10 Bode Diagram of Frequency Response Characteristics of a Compensated Op Amp (μA741)

the time it takes a waveform to go from 10% to 90% of its final amplitude. Measurement of rise time is shown in Figure D.11.

Unity-gain frequency bandwidth (BW) can be calculated from the rise time T_R by the following equation:

$$BW = \frac{0.35}{T_R} \qquad \text{(D.7)}$$

A factor intimately related to frequency response is the *gain-bandwidth product* (GBP). The GBP is defined mathematically by the following equation:

$$GBP = BW \times A_V \qquad \text{(D.8)}$$

where

A_V = closed-loop voltage gain of circuit

This equation is used to find the bandwidth of an op amp at a specific gain. Suppose you are working with an amplifier whose gain is 100 and whose GBP is 1 MHz. What is the bandwidth of this op amp? The bandwidth is calculated as follows:

$$BW = \frac{GBP}{A_V} = \frac{1\ MHz}{100}$$
$$= 10\ kHz \qquad \text{(D.9)}$$

One interesting observation about the GBP is that it is always constant. From the graph in Figure D.10, we see that as gain goes up by 10, frequency

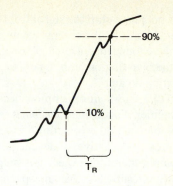

FIGURE D.11 Rise Time Measurement

goes down by the same amount. Thus, the product is always the same. In practice, it is difficult to measure the GBP for low-gain settings. This limitation is due, in part, to the small voltages and currents involved.

For purposes of comparison, Table D.1 lists some common parameters for several common types of op amps.

Inverting Op Amp

Op amps are used in two basic configurations: the inverting and the noninverting amplifier. An *inverting op amp* is one that takes a signal at the input and phase-shifts it 180° at the output. The *noninverting*

TABLE D.1 Parameters of Various Op Amps

Parameters	Op Amp					
	709	201	301	741	791	Ideal
Input bias current (nA)	1500	500	250	500	500	0
Input offset voltage (mV)	7.5	7.5	7.5	6.0	6.0	0
Input offset current (nA)	500	200	50	200	200	0
GBP (MHz)	1.0	1	1	1	0.2	∞
Slew rate (V/µs)	3.0	2.0	2.0	0.5	6.0	∞
Input impedance (MΩ)	0.7	4.0	2.0	2.0	2.0	∞

op amp does not phase-shift the signal from the input to the output. Almost all the circuits we discuss later are built on these foundational circuits. In this section, we discuss the inverting op amp. Later, we consider the noninverting op amp.

Figure D.12 shows an op amp connected as an inverting amplifier. Notice that a *feedback resistor* R_F connects the output to the inverting terminal. This connection provides negative, or degenerative, feedback. We would therefore expect this amplifier to have a gain significantly less than the op amp's open-loop gain. *Open-loop gain* is the gain of the op amp when there is no feedback path. For the ideal op amp, the open-loop gain is infinity. However, the practical op amp's gain is somewhat less than infinity. For example, the 741 has an open-loop gain of about 200,000.

Also, notice in Figure D.12 that the noninverting terminal (*C*) is grounded and that R_i connects the input (*B*) to the inverting terminal (*A*).

Virtual Ground. The concept of the virtual ground is an important one in understanding the behavior of inverting op amps.

Figure D.13 shows a basic op amp with no feedback. Now, let us suppose that we have a power supply of ±12 V powering the op amp. Let us also suppose that the maximum output voltage (saturation voltage) can go no higher than +10 V and no lower than −10 V. In this case, the output voltage of +10 or −10 V represents output saturation voltage, V_{sat}. *Output saturation voltage* occurs when the value of the output voltage of the op amp can get no closer to the value of its supply voltage because a portion of the supply voltage is dropped across the saturated output transistors.

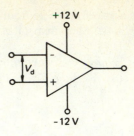

FIGURE D.13 Op Amp Circuit with No Feedback

With a gain of 200,000, what is the input voltage that gives an output of 10 V? The voltage gain equation is as follows:

$$A_V = \frac{V_o}{V_i} \tag{D.10}$$

Rearranging this equation, we get the following:

$$V_i = \frac{V_o}{A_V} = \frac{10 \text{ V}}{200,000} \tag{D.11}$$

$$= 50 \text{ μV}$$

We see, then, that the maximum difference (without signal distortion) in potential (V_d) between the inverting and noninverting terminals is 50 μV. This important fact brings us to the first rule in understanding op amps.

- *Rule 1*: The difference in potential between the two input terminals of an op amp is approximately zero.

Since the input voltage is almost always significantly higher than V_d, we can assume that V_d is approximately equal to zero.

Now, if the noninverting terminal is grounded, what is the approximate difference in potential between the inverting terminal and ground? Look at the circuit in Figure D.14. The noninverting terminal is at ground potential, and there is essentially no

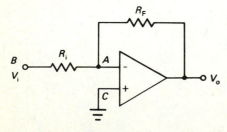

FIGURE D.12 Basic Inverting Op Amp Circuit

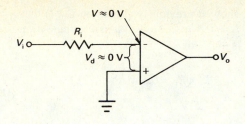

FIGURE D.14 Op Amp with Noninverting Terminal Grounded

difference in potential between the input terminals. Thus, the inverting input can be considered to be 0 V. Or we can say it is *virtually grounded*, from which comes the name virtual ground.

Input Current. The second rule we need for analyzing the inverting amplifier concerns the input current. The input impedance of a practical op amp is very high; for instance, the 741 input impedance is about 2 MΩ. We then find the maximum input current by using the 50 µV input voltage V_i, the input impedance Z_i, and Ohm's law:

$$I_i = \frac{V_i}{Z_i} = \frac{50\,\mu V}{2\,M\Omega}$$
$$= 25\,pA$$

(D.12)

Notice how small the input current is. This result brings us to our second rule.

• *Rule 2*: Essentially no current flows into or out of the input terminals of the op amp.

Since the input current is in the order of 10^{-12} A, we can safely neglect it.

Analyzing the Inverting Amplifier. Armed with these two rules, we can develop equations to help us understand this amplifier. Refer to the schematic in Figure D.12. If the potential at point A is 0 V, we can say that all of V_i is dropped across R_i. Using

Ohm's law, we can then find the current flowing through R_i:

$$I_i = \frac{V_i}{R_i}$$

(D.13)

Since no appreciable current flows into or from the inverting terminal, all the current flowing through R_i also flows through R_F. So, we have the following equation:

$$I_i = I_F$$

(D.14)

Now, also because of the virtual ground, V_o is dropped across R_F. From Ohm's law, V_o can be expressed as follows:

$$V_o = -I_F R_F$$

(D.15)

where
I_F = DC feedback current

And from Equation D.13, V_i is as follows:

$$V_i = I_i R_i$$

(D.16)

Therefore, we have the following relationship:

$$\frac{V_o}{V_i} = -\frac{R_F}{R_i} = A_V$$

(D.17)

We now have an equation that defines the gain of an inverting op amp circuit. Note that, due to the op amp's high open-loop gain, the circuit gain is entirely determined by components external to the op amp. The negative sign indicates that a 180° phase shift occurs between input and output. Or in the case of a DC input, a change in polarity between input and output occurs. The polarities shown in Figure D.12 support this statement.

Input Impedance. One interesting feature of the inverting op amp configuration is its input impedance. Look at the op amp circuit in Figure D.12 again. If point A is always at 0 V, what is the input impedance at point B? Obviously, if all the input voltage is dropped across R_i, the input impedance at point B is simply the resistance of R_i. Within limits, then, the input impedance of this inverting amplifier is changed by changing R_i. It is recommended that R_i be no less than 1 kΩ in order to limit the possible current drain on the op amp.

Output Current. As a technician, you may be called upon to evaluate an op amp's output current. We will calculate the output current for the op amp circuit in Figure D.15. Notice that the current (I_o) flowing in the output terminal is a combination of the currents flowing through R_F (I_F) and R_L (I_L). Thus, we can equate the output current to the sum of those currents:

$$I_o = I_F + I_L \qquad \text{(D.18)}$$

Both I_F and I_L can be easily calculated.

Suppose we are using the circuit in Figure D.15. All the input voltage (5 V) is dropped across R_i. The current through R_i, then, is as follows:

$$I_i = \frac{V_i}{R_i} = \frac{5 \text{ V}}{10 \text{ k}\Omega}$$

$$= 0.5 \text{ mA} \qquad \text{(D.19)}$$

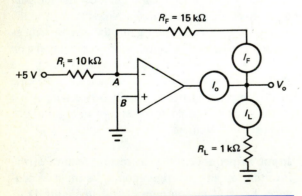

FIGURE D.15 Circuit for Output Current Calculation

We showed previously (Equation D.14) that $I_i = I_F$. So, $I_F = 0.5$ mA. Now, V_F, the voltage drop across R_F, is as follows:

$$V_F = I_F \times R_F = 0.5 \text{ mA} \times 15 \text{ k}\Omega$$

$$= 7.5 \text{ V} \qquad \text{(D.20)}$$

Notice that since A and B are at the same potential (ground), 7.5 V must be dropped across R_L as well as R_F. Therefore, we can find I_L as follows:

$$I_L = \frac{V_L}{R_L} = \frac{7.5 \text{ V}}{1 \text{ k}\Omega}$$

$$= 7.5 \text{ mA} \qquad \text{(D.21)}$$

The total output current is, then, as follows:

$$I_o = I_L + I_F = 7.5 \text{ mA} + 0.5 \text{ mA}$$

$$= 8 \text{ mA} \qquad \text{(D.22)}$$

The maximum output current that an op amp can deliver to a load is found in its data sheets. In no case should the output current demanded by the circuit and the load exceed the maximum allowable output current. Some op amps, including the 741, are protected from excess currents by circuitry that limits the output current to a maximum value. Some op amps, though, are not so protected. If the maximum allowable output current is exceeded, the gain equation (D.17) is no longer valid since the op amp is operating outside its normal range. If the load requirements exceed the op amp's capabilities (25 mA for the 741), a transistor circuit can be added to the op amp output to boost its current output.

AC Input Signals. Thus far, we have only considered DC input voltages in our examples. The inverting op amp is not limited to DC applications, though. Let us now consider an AC input to the op amp.

The op amp circuit in Figure D.16A has a gain of −10 with an input signal of ±1 V peak. By looking at the output in Figure D.16B, we see that the signal has been amplified by a factor of 10, with 180° phase inversion.

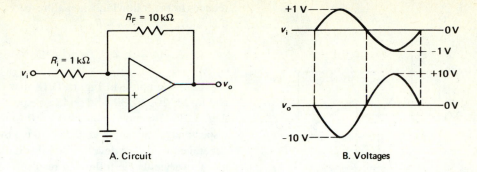

FIGURE D.16   AC Output Voltage Calculation Diagrams for Inverting Amp

AC and DC Input Signals. The op amp circuits you will encounter may have AC signals riding on DC levels. Let's see how the op amp treats such a signal. Figure D.17A depicts an inverting op amp; the input signal is shown in Figure D.17B. Notice that the input signal is shifted in phase at the output by 180°. The input DC voltage level is translated from +3 V to –6 V.

Algebraically, this change can be calculated by knowing that the gain is –2. We use the following equations:

$$A_V = -2 \quad \text{and} \quad A_V = \frac{V_o}{V_i} \qquad \text{(D.23)}$$

So, if V_i is +3 V, then V_o is +3 V multiplied by –2. The output DC voltage is then –6 V, as shown in Figure D.17B. By the same token, the AC output signal is calculated by knowing that the gain is –2 and the input signal is ±2 V peak.

Summing Op Amp

The *summing op amp* is a special application of the inverting amplifier. The summing op amp performs the mathematical operation of addition. It can be considered as an inverting op amp with several inputs. A diagram of the summing amp configuration is shown in Figure D.18.

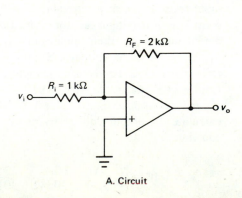

A. Circuit

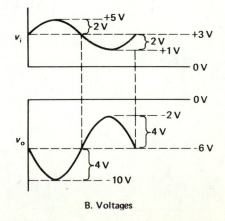

B. Voltages

FIGURE D.17   AC and DC Output Voltage Calculation Diagrams for Inverting Amp

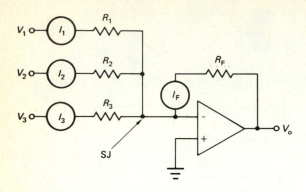

FIGURE D.18 Inverting Summing Amp

An important point to notice in this circuit is that each applied voltage is dropped across its respective resistor. Thus, V_1 is dropped across R_1, V_2 across R_2, and V_3 across R_3. Why? Because the voltage at the summing junction (SJ) is 0 V due to the virtual ground. Note also that the currents join at SJ to form the feedback current. (Recall that no current flows into the inverting terminal.) So, we have the following equation:

$$I_1 + I_2 + I_3 = I_F \qquad \text{(D.24)}$$

Also,

$$I_F = -\frac{V_o}{R_F} \qquad \text{(D.25)}$$

Substituting voltage and resistance for current, we have the following equation:

$$\frac{V_1}{R_1} + \frac{V_2}{R_2} + \frac{V_3}{R_3} = -\frac{V_o}{R_F} \qquad \text{(D.26)}$$

Solving for V_o, we obtain the next equation:

$$V_o = -R_F\left(\frac{V_1}{R_1} + \frac{V_2}{R_2} + \frac{V_3}{R_3}\right) \qquad \text{(D.27)}$$

This equation is the general equation used to find the output voltage of a summing amp. This equation

reduces to the following when all resistors are equal; that is, $R_F = R_1 = R_2 = R_3$:

$$V_o = -(V_1 + V_2 + V_3) \qquad \text{(D.28)}$$

Let us suppose in Figure D.18 that R_1, R_2, R_3, and R_F are 10, 20, 30, and 60 kΩ, respectively. Further, let V_1, V_2, and V_3 equal +0.1, +0.2, and +0.3 V, respectively. The output voltage is found by using the general equation we developed. Substituting the values above yields the following result:

$$V_o = -60\text{ k}\Omega\left(\frac{0.1\text{ V}}{10\text{ k}\Omega} + \frac{0.2\text{ V}}{20\text{ k}\Omega} + \frac{0.3\text{ V}}{30\text{ k}\Omega}\right)$$

$$= -1.8\text{ V} \qquad \text{(D.29)}$$

The summing amp is used for AC applications as well as DC applications. In fact, the audio microphone mixer shown in Figure D.19 is an excellent and useful AC application. The advantage of this circuit arises from the fact that each input voltage is dropped across its own input resistor. Therefore, there is no interaction between microphones. Also, provision is made to vary the gain of each input individually (R_1, R_2, R_3) or of all channels at the same time (R_F).

Averaging Op Amp

The *averaging op amp* is a special case of the summing amp. Consider Figure D.18 again. If all the input resistors are of equal value, and R equals the resistance of one input resistor divided by the number of inputs (N), then this circuit arithmetically averages the input voltages. Suppose each input resistor is 30 kΩ. To average the input voltages, the feedback resistor would have to be 10 kΩ. With V_1, V_2, and V_3 of +3, +5, and +2 V, respectively, we can calculate the output voltage by using the general equation:

$$V_o = -10\text{ k}\Omega\left(\frac{3\text{ V}}{30\text{ k}\Omega} + \frac{4\text{ V}}{30\text{ k}\Omega} + \frac{2\text{ V}}{30\text{ k}\Omega}\right)$$

$$= -3\text{ V} \qquad \text{(D.30)}$$

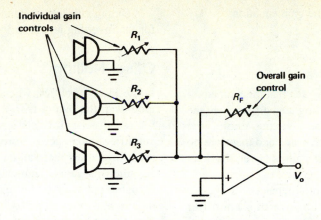

FIGURE D.19 Summing Amp Used as a Microphone Mixer

You can see from this example that this circuit does indeed average voltages. Like the summing amp, it can also be used to average AC input signals. You may also have noticed that the output is inverted. This problem can easily be solved by using another inverting amplifier with a gain of 1 in series with the averaging amp.

Noninverting Op Amp

Up to this time, we have concentrated on inverting op amps. As stated previously, the noninverting amp makes up the second major division of op amp circuits.

Figure D.20 illustrates the basic noninverting op amp configuration. Note the difference between this amplifier and the inverting amplifier of Figure D.12. In the noninverting amplifier, the noninverting terminal is connected to the input rather than ground, while R_i is grounded. Going back to op amp Rule 1, we can state that V_i appears across R_i because the voltage difference between the inverting and noninverting terminals is essentially zero. Therefore, we have the following equations:

$$I_i = \frac{V_i}{R_i} \quad \text{and} \quad V_F = I_i R_F \qquad \text{(D.31)}$$

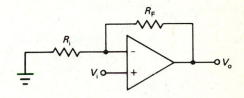

FIGURE D.20 Basic Noninverting Op Amp Circuit

Substituting, we get the following:

$$V_F = \frac{R_F}{R_i} V_i \qquad \text{(D.32)}$$

Looking at the circuit in Figure D.20, we also see that the output voltage (V_o) is equal to V_F added to V_i: $V_o = V_F + V_i$. Substituting the known factors into this equation gives us the following equation:

$$V_o = \frac{R_F}{R_i} V_i + V_i \qquad \text{(D.33)}$$

Factoring this equation, we get the following:

$$V_F = \left(\frac{R_F}{R_i} + 1 \right) V_i \qquad \text{(D.34)}$$

Rearranging, we arrive at the gain equation:

$$A_V = \frac{V_o}{V_i} = \frac{R_F}{R_i} + 1 \qquad \textbf{(D.35)}$$

From this derivation, we see that the gain equation (D.35) for the noninverting amp differs from the gain equation (D.17) for the inverting amp in two ways: (1) There is no negative sign and therefore no phase inversion, and (2) the gain is never less than unity.

There are other important differences. Recall that the input impedance of the inverting amplifier equaled the input resistor. As we can see in the noninverting amplifier, the input impedance is that of the device itself—that is, very high. The input impedance of the noninverting amp does not depend on external circuit characteristics.

Let us apply an input to the noninverting amp, as shown in Figure D.21. The input signal is a 1 V peak sine wave riding on a +2 V DC level, with an input resistor of 10 kΩ and a feedback resistor of 20 kΩ. The gain of the amplifier is as follows:

$$A_V = \frac{R_F}{R_i} + 1 = \frac{20\text{ k}\Omega}{10\text{ k}\Omega} + 1$$

$$\qquad \textbf{(D.36)}$$

$$= 3$$

Therefore, the +2 V DC level becomes +6 V, while the 1 V peak sine wave becomes 3 V peak.

Output current is calculated in the same way as for the inverting amplifier.

Offset Nulling

In the ideal amplifier, when the input (or inputs) is grounded, the output is 0 V. This result is not true for the practical amplifier owing to differential amplifier imbalances. Normally, such imbalances produce outputs that can be neglected. However, if the amplifier gain is high, or if it is necessary for the output to approach 0 V as closely as possible, offset nulling may be in order. *Offset nulling* is the procedure by which the output of the op amp circuit is adjusted for 0 V with 0 V input.

The best way to null an op amp circuit uses the built-in circuit provided by the manufacturer for that purpose. The 741 nulling circuit is shown in Figure D.22.

The nulling procedure for the 741 is as follows:

1. Ground the input (V_i, not pin 2 of the op amp).

2. Place a voltmeter from the output to ground.

3. Adjust the offset, or adjust until 0 V is reached at the output.

4. Unground the input—do not move the offset-adjust potentiometer once it is set.

Note that the amplifier is nulled with feedback. It is virtually impossible to null the amplifier without

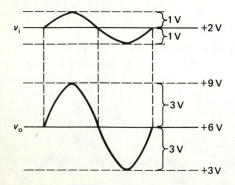

FIGURE D.21 AC and DC Input Voltage to Noninverting Amp

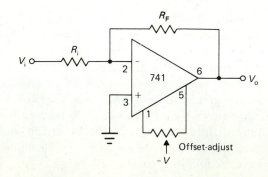

FIGURE D.22 Offset-Nulling Circuit for 741 Op Amp

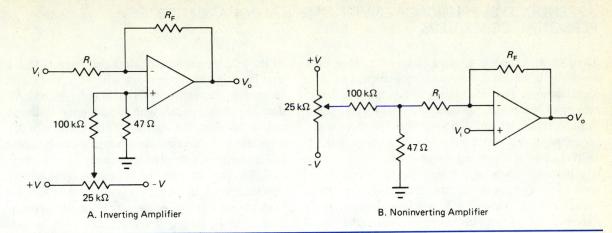

FIGURE D.23 Offset-Nulling Circuits

DC feedback because of capacitance charging associated with high open-loop gain.

If an op amp does not have provisions for nulling, external circuits may be used. Examples of these circuit configurations are shown in Figures D.23A and D.23B. In each case, the nulling procedure is the same as previously stated.

Another configuration you may see associated with an inverting amp is illustrated in Figure D.24. Note the resistor R_b in series with the noninverting

terminal. The resistor R_b compensates for the error caused by bias currents of the op amp's input differential amplifier. For best results, the value usually selected for R_b is as follows:

$$R_b = R_i \parallel R_F \qquad \text{(D.37)}$$

(The $\parallel$ symbol indicates that the resistors are electrically connected in parallel.)

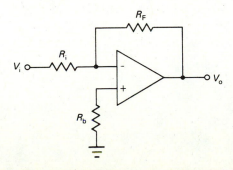

FIGURE D.24 Bias Current Compensation

APPENDIX E: SUPPLIERS OF ENVIRONMENTALLY HARDENED PERSONAL COMPUTERS

The following table is a listing of suppliers of environmentally hardened personal computers and a few specifications. The information in this table is current as of November 1987 and does not include the most recent models and data available. Therefore, we recommend that you contact the companies listed individually for additional information.

In the table headings, "CPU" means central processing unit and is the microprocessor used in the system. "RAM" means random-access memory and indicates the amount of memory available. The category of "Disks" shows two types: "F" for floppy disk drive and "H" for hard disk drive. The column headed "I/O Ports" shows the number of serial ports on the left and the number of parallel ports on the right—(none), 1, or 2. The "Free Slots" column lists on the left, the number of slots that have IBM XT slots (an edge card connector with 62 contacts); in the middle, the number of slots that have IBM AT slots (the XT connector with an additional 36 lines); and on the right, the number of short slots (the XT connector with the component part of the board about half as long as a regular XT board). In the "Keyboard/Monitor" column, "O" represents optional, "Y" represents yes (it is part of the system), and CGA and EGA are the types of adapters present, which determine the type of monitor used. The CGA (color-graphics adapter) is used with a color monitor only; the EGA (enhanced-graphics adapter) can be connected to a monochrome display, a color display, or an enhanced-color display.

Vendor	Model (Price)	CPU	RAM	Disks	I/O Ports Serial/Parallel	Free Slots XT/AT/short	Keyboard/ Monitor
Action Instruments 8601 Aero Dr San Diego, CA 92123 (619) 279-5726	BC-6 ($1995)	8088	640k	—	—/—	5/-/-	O/O
	BC-10+ ($3295)	8088	256k	—	1/1	8/-/-	O/O
	BC-12 ($3395)	8088	256k	—	1/1	6/-/-	O/O
	BC-16 ($3595)	80286	1M	—	1/1	-/4/-	O/O
	BC-20 ($3995)	80286	1M	—	1/1	-/8/-	O/O
	BC-22 ($5995)	80286	1M	—	1/1	-/6/-	O/O
ADAC Corp 70 Tower Office Park Woburn, MA 01801 (617) 935-6668	4000AT ($5850)	80286	1.2M	F (360k) H (22M)	2/2	—/1/4	Y / EGA
Allen Bradley Industrial Computer Div 747 Alpha Dr Highland Hts, OH 44143 (216) 449-6700	6120COX ($4333)	8088	640k	F (360k) H (10M)	1/1	3/-/-	Y/O
	6121-COF ($6425)	80286	640k	F (1.2M) H (20M)	—/—	2/5/-	Y/O
	6122COP ($9371)	80286	512k	H (10M)	-/-	-/5/-	O/O
Analog Devices Inc Three Technology Way Norwood, MA 02062 (617) 329-4700	Macsym 120 ($7750)	8088	256k	F (360k) H (10M)	1/1	4/-/1	Y/O
Comark Corp Box 474 Medfield, MA 02052 (617) 359-4700	Expert-XT ($5495)	8088	640k	—	1/1	6/-/-	Y / CGA
	Expert-AT ($6595)	80286	640k	—	1/1	—/6/—	Y / CGA
	Expert Diskless ($6995)	80286	512k (sys) 256k (user)	256k ROM	1/1	—/6/—	Y / EGA
Diversified Technologies Corp 112 E State St Ridgeland, MS 39158 (601) 856-4121	CHI/C10 ($2100)	80286	640k	—	2/1	3/5/—	O/O
Electro Design Inc 690 Rancheros Dr San Marcos, CA 92069 (619) 471-0680	IMP-12X ($2835)	8088	—	—	—	11/—/—	Y/O
	80286 ($2115)	512k	—	—	—	—/11/—	Y/O
GE Fanuc Automation Box 8106 Charlottesville, VA 22906 (804) 978-5000	Workmaster ($4800)	8088	640k	F (720k, 3.5")	1/1	3/—/2	Y / CGA (amber)
	Cimstar I ($6500)	80286	1M	—	2/1	1/5/—	O/O

Vendor	Model (Price)	CPU	RAM	Disks	I/O Ports Serial/Parallel	Free Slots XT/AT/short	Keyboard/ Monitor
Heath/Zenith Hilltop Rd St Joseph, MI 49085 (616) 982-3200	SW-3000 ($4500)	80286	512k	F (1.2M) H (20M)	1/1	1/5/-	Y/O
IBI Systems Inc 6842 NW 20th Ave Ft Lauderdale, FL 33309 (305) 978-9225	ST-1000EGA ($5900)	80286	1M	F (1.2M) H (20M)	1/1	-/4/-	Y/EGA
	ST-3000EGA ($9500)	80286	1M	F (1.2M) H (40M)	1/1	-/5/-	Y/EGA
IBM Corp Industrial Systems Products Box 1328 Boca Raton, FL 33432 (800) 447-4700	7531 ($6425)	80286	512k H (20M)	F (2.1M)	—/—	2/5/—	Y/O
	7552 (Gearbox, $6770)	80286	512k	—	—/—	—/7/—	O/O
I-Bus 5780 Chesapeake Ct San Diego, CA 92123 (619) 569-0646	F286/12 slot ($2640)	80286	512k	—	—	11/-/-	O/O
ICS Corp 5466 Complex St, Ste 208 San Diego, CA 92123 (619) 279-0084	9531 ($3725)	80286	1M	F (360k) H (20M)	—/—	6/4/—	O/O
	8531-20EGA/AT ($7850)	80286	1M	F (360) H (20M)	—	4/3/—	Light pen/EGA
IndTech Inc 1275 Hammerwood Ave Sunnyvale, CA 94089 (408) 743-4300	6170 ($4167)	80286	512k	F (1.2M)	1/1	3/5/1	K/O
Materials Development Corp 21541 Nordhoff St Chatsworth, CA 91311 (818) 700-8290	RM-1600 ($2195)	8088	512k	2F (360k)	1/1	4/-/-	Y/Y
	RM-1600-AT ($3595)	80286	512k	F (1.2M)	1/1	2/2/-	Y/Y
Pragmatic Computers Inc 807 Aldo Ave, Ste 110 Santa Clara, CA 95054 (408) 986-9350	BPC-EG ($6150)	8088	512k	F (360k) H (20M)	2/2	7/-/-	Light pen/EGA
	BAT-EG ($7250)	80286	640k	F(1.2M) H (20M)	-/-	-/7/-	Light pen/EGA
Qualogy 2241 Lundy Ave San Jose, CA 95131 (408) 434-5200	QPC-8121 ($9095)	80286	1M	2F (360k, 1.2M or 720k, 3.5") H (40M)	1/1	5/4/—	Y/EGA
Texas Microsystems Inc 10618 Rockley Rd Houston, TX 77099 (713) 933-8050	2001X ($6175)	8088	640k	F (360k) H (20M)	1/1	6/—/—	Y/CGA
	2001A ($6770)	80286	1M	F (360k) F (1.2M) H (20M)	1/1	4/2/—	Y/CGA

APPENDIX F: PC MANUFACTURERS' PRODUCTS

The following table is a listing of the known products of PC (programmable controller) manufacturers and some of the more important PC specifications, as of January 1988. These devices meet the historical definition of a conventional PC. There are many factors that cannot be grouped into a table, and some of these factors may alter your perception about which system will best suit your needs; therefore, you should check with the manufacturers for specific data.

The manufacturers caution that scan rate data is, at most, an *estimate* of typical performance. For any given program, the scan rate will vary depending on the types of actions called for by the program.

Most systems can accept any combination of local or remote I/Os. However, some do have dedicated quantities of inputs versus outputs, such as 20 inputs and 12 outputs. Only the totals are shown.

Where high-level language usage is indicated, both the high-level languages, such as TINY BASIC, and additional commands beyond normal ladder logic diagramming are available.

The category "Documentation" is not to be confused with manufacturer-generated documentation such as owners' manuals and service manuals. The term "Documentation" in the table means the ability of the system to provide a hard copy of its configuration by means of a built-in or external printer.

The symbol † indicates that information can be found in the "Comments" column.

The column heading "Type of interface" refers to the type of connection to the programmable controller *system*, not the I/O interface discussed in Chapter 14. For more information on the system interface connections, refer to the section on "Data Communication" in Chapter 12.

Manufacturer	Model	Total system I/O	Max discrete I/O	Max analog I/O	Relay ladder logic	High level language	PID capabilities	Motion control	Documentation	Diagnostics	Type of interface	Scan rate/1K	Type of memory	Size of memory	Country of origin	Comments
ASEA IND'L. SYS.	MP51	64	64						Y	Y	A	<5ms	CMOS		Sweden/US	
(New Berlin, WI)	MP120	128	128	128	Y				Y	Y	A	†	CMOS	16K	Sweden/US	†User defined
	MP140	128	128	128	Y				Y	Y	A	†	CMOS	16K	Sweden/US	†User defined
	MP160	128	128	128	Y	Y	Y		Y	Y	A	†	CMOS	16K	Sweden/US	†User defined
	MP150T	128	128	128	Y	Y			Y	Y	A	†	CMOS	16K	Sweden/US	†User defined
	MP170T	128	128	128	Y	Y			Y	Y	A	†	CMOS	16K	Sweden/US	†User defined
	MP220	5500	5500		Y				Y	Y	A	†	CMOS	1M	Sweden/US	†User defined
	MP240	1300	1300	600	Y	Y	Y		Y	Y	A	†	CMOS	1M	Sweden/US	†User defined
	MP260	1300	1300	600	Y	Y	Y		Y	Y	A	†	CMOS	1M	Sweden/US	†User defined
	MP280	1300	1300	600	Y	Y	Y	Y	Y	Y	A	†	EEPROM, CMOS	1M	Sweden/US	†User defined
	MP290/1	5000	5000	2000	Y	Y	Y	Y	Y	Y	B	†	CMOS	4M	Sweden	
ASC COMPUTER SYST.	PC/86		512	128	Y	Y	Y	Y	Y	Y	B	20ms	EPROM, RAM	256K	U.S.A.	AB Data Highway
(St. Clair Shores, MI)	PC/88	512	512	64	Y	Y	Y	Y	Y	Y	B, C	20ms	EPROM, RAM	1M	U.S.A.	LAN/Data Highway
	PC/188		1000	128	Y	Y	Y	Y	Y	Y	B, C	10ms	RAM, EPROM, EEPROM	1M	U.S.A.	
ADATEK, INC.	System 10	1272	1176	96		Y	Y		Y	Y	A	10ms	EPROM, CMOS RAM	48K	U.S.A.	422 multidrop network
(Sandpoint, ID)																
ALLEN-BRADLEY	PLC-3	8192	8192	4096	Y	Y	Y	Y	Y	Y	A	2.5ms	RAM EDC	2M	U.S.A.	
(Milwaukee, WI)	PLC-3/10	4098	4098	2048	Y	Y	Y	Y	Y	Y	A	2.5ms	RAM EDC	64K	U.S.A.	
	SLC 100	112	112	24	Y				Y	Y	J	†	EEPROM, CMOS RAM	885	U.S.A./Japan	†15ms/500 words
	SLC 150	112	112	24	Y				Y	Y	J	†	EEPROM, CMOS RAM	1200	U.S.A Japan	†2ms 500 words
	PLC-2/02	512	512	256	Y		Y	Y	Y	Y	A	12.5ms	RAM, EEPROM	1K	U.S.A	
	PLC-2/16	512	512	256	Y		Y	Y	Y	Y	A	12.5ms	RAM, EEPROM	3K	U.S.A	Net. is Data Highway
	PLC-2/17	512	512	256	Y		Y	Y	Y	Y	A	12.5ms	RAM, EEPROM	6K	U.S.A	for all entries
	PLC-2/30	2688	2688	896	Y	Y	Y	Y	Y	Y	A	5ms	RAM	16K	U.S.A	except SLC 100
	PLC-5/12	512	512	256	Y	Y	Y	Y	Y	Y	A	8ms	RAM, EEPROM	6K	U.S.A	and SLC 150
	PLC-5/15	1024	1024	512	Y	Y	Y	Y	Y	Y	A	8ms	RAM, EEPROM	14K	U.S.A	
	PLC-5/25	2048	2048	1024	Y	Y	Y	Y	Y	Y	A	8ms	RAM, EEPROM	21K	U.S.A	
	PLC-5/VME	512	512	256	Y	Y	Y	Y	Y	Y	†	8ms	RAM, EEPROM	14K	U.S.A	†Backplane
ANDERSON CORNELIUS	APC-3	†	†	†	Y	Y	Y	Y	Y	Y	B	18ms	PROM, RAM	80K	U.S.A.	†240/controller
(Eden Prairie, MN)	MCU-250	†	†	†	Y	Y	Y	Y	Y	Y	B	40ms	PROM, RAM	80K	U.S.A.	†240/controller

Manufacturer / Model	Total system I/O	Max discrete I/O	Max analog I/O	Relay ladder logic	High level language	PID capabilities	Motion control	Documentation	Diagnostics	Type of interface	Scan rate/1K	Type of memory	Size of memory	Country of origin	Comments
AUTOMATIC TIMING & CONTROLS (King of Prussia, PA) ATCOM 64	72	64	8		Y	Y		Y	Y	A	7-20ms	CMOS RAM, EPROM	8K	U.S.A.	
AUTOMATION SYSTEMS (Eldridge, IA) PAC-5	1024	1024	512	Y	Y	Y	Y	Y	Y	B	1.1ms	RAM, EPROM	32K	U.S.A.	12 commun. ports
B & R INDUSTRIAL AUTOMATION CORP. (Stone Mountain, GA) Minicontrol	96	96	8	Y	Y	Y	Y	Y	Y	B, F	4ms	RAM, EEPROM	16K	Austria	All units use
Midicontrol	128	128	64	Y	Y	Y	Y	Y	Y	B, F	4ms	RAM, EEPROM	16K	Austria	Ring network
Multicontrol CP40	1024	1024	128	Y	Y	Y	Y	Y	Y	B, F	4ms	RAM, EEPROM	16K	Austria	
Multicontrol CP80	1024	1024	128	Y	Y	Y	Y	Y	Y	B, F	2.5ms	RAM, EEPROM	74K	Austria	
PROVICON 75200	275	200	75	Y	Y	Y	Y	Y	Y	B, F	2.5ms	RAM, EEPROM	74K	Austria	
BAILEY CONTROLS CO. (Wickliffe, OH) MPC01	1024	1024	960	Y	Y	Y		Y	Y	A	2ms	BBRAM, ROM	276K	U.S.A.	All units use
LMM02	512	512		Y	Y			Y	Y		15ms	BBRAM, ROM	80K	U.S.A.	proprietary network
CSC01	28	28		Y	Y			Y	Y		2ms	BBRAM, ROM	272K	U.S.A.	
MFC03	5000	5000	5000	Y	Y	Y		Y	Y	A	1ms	BBRAM, ROM	848K	U.S.A.	
BARBER-COLMAN CO. IND'L. INST. DIV. (Loves Park, IL) Network 8000	5000	5000	5000	Y	Y	Y		Y	Y	A, D	5ms	RAM, EPROM, EEPROM	512K	U.S.A.	
BONAR AUGUST SYST. (Tigard, OR) Trigard/CS330S	7000	7000	3000	Y	Y	Y		Y	Y	B	15μs	RAM, PROM	750K	U.S.A.	Uses peer-to-peer net.
BRITISH BROWN-BOVARI (Telford, England) K200	128	128		Y	Y			Y	Y		5ms	EPROM	2K	FRG	
Procontic B	1024	768	256	Y	Y	Y	Y	Y	Y	B	2.5ms	RAM, EPROM	512	FRG	
DP800	4000	3000	1000	Y	Y	Y	Y	Y	Y	B	1ms	RAM, EPROM	inf	UK	
P214	1164	1100†	64†	Y	Y	Y	Y	Y	Y	B, I	1ms	RAM, EPROM	700K	UK	†per station
CINCINNATI MILACRON, ELECTR. SYST. (Lebanon, OH) APC105	64	64		Y					Y		5ms	CMOS EEPROM	1K	FRG	
APC500 RELAY		512	64	Y				Y	Y	B	5ms	CMOS RAM	8K	U.S.A.	
APC500 MCL		2048	128		Y	Y	Y	Y	Y	B	4.5ms	CMOS RAM	8K	U.S.A.	
CONTROL SYST. INTL. (Dallas, TX) 6400	76	48	36	Y	Y		Y	Y	Y	D		EPROM, RAM, NOVRAM	24K	U.S.A.	Uses token passing net.
CONTROL TECHNOLOGY (Hopkinton, MA) 2200	80	80		Y		Y	Y	Y	Y	A		BBRAM	64K	U.S.A.	
2800iEA	2208	2048	128	Y	Y	Y	Y	Y	Y	B		BBRAM	256K	U.S.A.	Uses multidrop net.
2800iE	416	256	128	Y	Y	Y	Y	Y	Y	B		BBRAM	256K	U.S.A.	Uses multidrop net.
2400iE	416	256	128	Y	Y	Y	Y	Y	Y	B		BBRAM	256K	U.S.A.	Uses multidrop net.
CROUZET CONTROLS (Schaumburg, IL) CMP-31	32	32	8	Y			Y	Y	Y		7ms	EPROM	2K	France	
CMP-34	256	256	64	Y			Y	Y	Y	propri.	7ms	EPROM	4K	France	Uses C-line net.
CMP-340	512	512		Y			Y	Y	Y	propri.	7ms	EPROM	8K	France	Uses C-line net.
DATEM LTD. (Ottawa, Canada) DCX538	48	48		Y	Y	Y	Y	Y	Y	D	1ms	RAM, EPROM	64K	Canada	All units use
DCX535	19	16	3	Y	Y	Y	Y	Y	Y	D	1ms	RAM, EPROM	64K	Canada	Bitbus network
DCX537	24	24		Y	Y	Y	Y	Y	Y	D	1ms	RAM, EPROM	64K	Canada	
DCX536	38	32	6	Y	Y	Y	Y	Y	Y	D	1ms	RAM, EPROM	64K	Canada	
DCX530	11	8	3	Y	Y	Y	Y	Y	Y	D	1ms	RAM, EPROM	64K	Canada	
DCX531	22	16	6	Y	Y	Y	Y	Y	Y	D	1ms	RAM, EPROM	64K	Canada	
DCX2000	160	†	†	Y	Y	Y	Y	Y	Y	B, D	10ms	RAM	64K	Canada	†Expandable
DI-AN MICRO SYSTEMS (Stockport, England) DMS563	256	128	128	Y	Y	Y		Y	Y	B	5ms	RAM	48K	UK	
DIVELBISS CORP. (Fredericktown, OH) BB-04	249	249	20	Y				Y	Y		5ms	EPROM	4K	U.S.A.	
BB-40	249	249	20	Y				Y	Y		5ms	EPROM	4K	U.S.A.	
BB-15	249	249	20	Y				Y	Y		2ms	EPROM	16K	U.S.A.	
DYNAGE CONTROLS (Cromwell, CT) SAFE 8000	2176	2048	128	Y	Y			Y	Y	B		CMOS RAM	40K	U.S.A.	
EAGLE SIGNAL CONTROLS (Austin, TX) EPTAK 100	16	16		Y				Y	Y		20ms	EEPROM	250†	Japan	†Statements
EPTAK 120	66	66		Y				Y	Y		20ms	ROM, EPROM	520†	Korea	†Statements
EPTAK 225	128	128		Y				Y	Y	B	39ms	CMOS	2800†	U.S.A.	†Statements
EPTAK 245	128	128	32	Y		Y		Y	Y	B	46ms	CMOS	2800†	U.S.A.	†Statements
EPTAK 700	2048	2048	1000			Y		Y	Y	B	1.5ms	CMOS	48K	U.S.A.	
EATON CORP. CUTLER-HAMMER (Milwaukee, WI) MPC1	128	128	8	Y				Y	Y	A	4ms	RAM, EEPROM	2K	U.S.A.	
D100															All D100 units use
CRA28	28	28		Y				Y	Y	B	7ms	RAM, EEPROM	1K	Japan	RS 422 multi-drop
CR20	40	40		Y				Y	Y	B	7ms	RAM, EEPROM	1K	Japan	
CRA40	80	80		Y				Y	Y	B	7ms	RAM, EEPROM	1K	Japan	
CAA40H	120	120		Y				Y	Y	B	4ms	RAM, EEPROM	1K	Japan	
CRA14	34	34	2	Y				Y	Y	B	6ms	RAM, EEPROM	1K	Japan	
CR20A	40	40	2	Y				Y	Y	B	6ms	RAM, EEPROM	1K	Japan	
CRA40A	80	80	2	Y				Y	Y	B	6ms	RAM, EEPROM	1K	Japan	
D500															All D500 units use
CPU20	224	224	28	Y		Y	Y	Y	Y	B	1.5μs	RAM, EEPROM	4K	Japan	Easynet parity line
CPU25	256	256	32	Y		Y	Y	Y	Y	B	1.5μs	RAM, EEPROM	4K	Japan	
CPU50	512	512	64	Y		Y	Y	Y	Y	B	0.9μs	RAM, EEPROM	8K	Japan	
ELECTROMATIC CONTROLS CORP. (Hoffman Estates, IL) 230816	56	56		Y				Y	Y		30ms	CMOS RAM	1.5K	Denmark	
330816	56	56		Y			Y	Y	Y		30ms	CMOS RAM	1.5K	Denmark	
300606	12	12		Y				Y	Y		30ms	CMOS RAM	1.5K	Denmark	
PLCF 223232	64	64	4	Y				Y	Y		30ms	CMOS RAM	1.5K	Denmark	Programming: built in
PLCF 323232	64	64	4	Y				Y	Y		30ms	CMOS RAM	1.5K	Denmark	Programming: external
ENCODER PROD. CO. (Sandpoint, ID) 7152	408	408	232		Y		Y		Y	B		RAM, EEPROM	52K	U.S.A.	Uses BASIC
7252	264	264	88		Y		Y		Y	B		EPROM, EEPROM	32K	U.S.A.	Uses BASIC
Synergy	6000	6000	2000	Y	Y	Y	Y		Y	A, D		RAM, disk	32K†	U.S.A.	†Per board

Manufacturer	Model	Total system I/O	Max discrete I/O	Max analog I/O	Relay ladder logic	High level language	PID capabilities	Motion control	Documentation	Diagnostics	Type of Interface	Scan rate / 1K	Type of memory	Size of memory	Country of origin	Comments
ENG'RING TOOLS (Naperville, IL)	86-Ladder-PC	2048	2048		Y	Y			Y	Y	B	5ms	RAM	8K	U.S.A.	
	86-Ladder-MBI	2048	2048		Y				Y	Y	B	5ms	BBRAM	8K	U.S.A.	
	86-Ladder-MBII	2048	2048		Y				Y	Y	B	5ms	BBRAM	8K	U.S.A.	
ENTERTRON INDUSTR. (Gasport, NY)	SK-1600	64	64		Y				Y	Y	†		EPROM	2K	U.S.A.	†15-30ms
	SK-1600R	56	56		Y				Y	Y	†		EPROM	4K	U.S.A.	†15-30ms
	SK-1800	88	88		Y				Y	Y	†		EPROM	8K	U.S.A.	†0.8-1.6ms
FF-ELEKTRONIIKKA FREDRIKSSON KY (Helsinki, Finland)	AL32	16	16	6		Y			Y	Y	A	25ms	EEPROM	2000	Finland	All units use Modbus network
	AL36	20	36	2		Y			Y	Y	A	25ms	EEPROM	2000	Finland	
	AL64		64	32	Y	Y	Y		Y	Y	A	12ms	EEPROM	2000	Finland	
	AL1024		512	64		Y			Y	Y	A	7ms	EEPROM	6000	Finland	
FOXBORO CO. (Foxboro, MA)	3PC-3A	64†	512	128	Y		Y	Y	Y	Y	D	1ms	EEPROM, RAM	32K	U.S.A.	†Modules
	3PC-4A	256†	2048	128	Y	Y	Y	Y	Y	Y	D	1ms	EEPROM, RAM	48K	U.S.A.	†Modules
FURNAS ELECTRIC (Batavia, IL)	PC/96	256	256	56	Y		Y		Y	Y	J	5ms	CMOS RAM	2K	U.S.A.	
	PC/96 Plus	480	480	56	Y		Y		Y	Y	J	5ms	CMOS RAM	5K	U.S.A.	
	96HM12XX	32	32		Y			Y	Y	Y	J	0.05ms†	CMOS RAM, EPROM	320	Japan	†Per step
	96HM20XX	40	40		Y			Y	Y	Y	J	0.05ms†	CMOS RAM, EPROM	320	Japan	†Per step
	96JM40XX	80	80		Y			Y	Y	Y	J	7µs†	CMOS RAM, EPROM	1000	Japan	†Per step
	96JM60XX	120	120		Y			Y	Y	Y	J	7µs†	CMOS RAM, EPROM	1000	Japan	†Per step
GEC AUTOMATION PROJECTS (Southfield, MI)	MICROGEM	12	24		Y				Y	Y	F	5ms	BBRAM, EEPROM	1K	U.S.A./UK	All units use current loop network
	MINIGEM	64	64	10	Y		Y	Y	Y	Y	F	5ms	RAM, EPROM	500	U.S.A./UK	
	GEM 80/100	512	512	32	Y				Y	Y	F	20ms	BBRAM, EPROM	12K	U.S.A./UK	
	GEM 80/130	512	512	32	Y		Y		Y	Y	F	20ms	BBRAM, EPROM	12K	U.S.A./UK	
	GEM 80/141	512	512	32	Y		Y	Y	Y	Y	J	1.25ms	BBRAM, EPROM	256K	U.S.A./UK	
	GEM 80/142	2048	2048	64	Y		Y	Y	Y	Y	J	1.25ms	BBRAM, EPROM	256K	U.S.A./UK	
	GEM 80/300	8192	8192	256	Y		Y	Y	Y	Y	J	1.25ms	BBRAM, EPROM	1.6M	U.S.A./UK	
	GEM 80/700	8192	8192	448	Y		Y	Y	Y	Y	J	1.25ms	BBRAM, EPROM	1.6M	U.S.A./UK	
GE FANUC AUTOMATION (Charlottesville, VA)	SERIES ONE JR	96	96		Y				Y	Y	J	40ms	RAM, EPROM	700		All units use GENET network
	SERIES ONE	112	112		Y				Y	Y	J	40ms	RAM, EPROM	1.7K		
	SERIES ONE/E	112	112	24	Y				Y	Y	J	12ms	RAM, EPROM	1.7K		
	SERIES ONE PLUS	168	168	24	Y				Y	Y	J	12ms	RAM, EPROM	3.7K		
	SERIES THREE	400	400	24	Y				Y	Y	J	12ms	RAM, EPROM	4K		
	SERIES FIVE	2048	2048	512	Y		Y	Y	Y	Y	B	1ms	RAM, EPROM, EEPROM	16K		
	SERIES SIX PLUS	8000	8000	992	Y	Y	Y	Y	Y	Y	B	1ms	CMOS RAM	32K		
	SERIES SIX PLUS/II	8000	8000	992	Y	Y	Y	Y	Y	Y	B	0.8ms	CMOS RAM	64K		
GENERAL NUMERIC (Elk Grove Village, IL)	100U	256	256	16	Y	Y		Y	Y	Y	F	7ms	RAM/EPROM	2K	FRG	All units use Sinec L1 network
	101U	64	64		Y	Y			Y	Y	F	75ms	RAM, EPROM, EEPROM	1K	FGR	
	115U	2048	2048	64	Y	Y	Y	Y	Y	Y	K	18ms	RAM, EPROM, EEPROM	24K	FRG	
	135U	8192	8192	384	Y	Y	Y	Y	Y	Y	K	8ms	RAM, EPROM, EEPROM	32K	FRG	
	150U	8192	8192	384	Y	Y	Y	Y	Y	Y	F	3.75ms	RAM, EPROM, EEPROM	128K	FRG	
G & L ELECTRONICS (Fond du Lac, WI)	PiC4.9	43	40	40	Y	Y	Y	Y	Y	Y	B	0.8ms	CMOS BBRAM	68K	U.S.A.	
	PiC49	240	232	232	Y	Y	Y	Y	Y	Y	B	0.8ms	CMOS BBRAM	40K	U.S.A.	
	PiC409	2040	2032	2032	Y	Y	Y	Y	Y	Y	B	0.8ms	CMOS BBRAM	288K	U.S.A.	
GOULD, INDUST. AUTOMATION DIV. (N. Andover, MA)	0085	120	120		Y				Y	Y	J	6ms	CMOS BBRAM, EPROM	1K	Japan	
	0185	512	512	†		Y			Y	Y	J	2ms	CMOS BBRAM, EPROM	3.5K	Japan	†Based on data memory addressing
	MICRO 84	112	112	12	Y				Y	Y	A	40ms	EAROM, PROM	504	U.S.A.	
	484	512	512	32	Y		Y		Y	Y	A	10ms	CMOS BBRAM	8K	U.S.A.	
	584L	8192	8192	512	Y	Y	Y	Y	Y	Y	A	1.5ms	CMOS BBRAM	128K	U.S.A.	
	884	1024	1280	1040	Y		Y		Y	Y	A	25ms	CMOS BBRAM	8K	U.S.A.	Network: Modbus or Modway for all models
	984A	2048	1024	1920	Y	Y	Y		Y	Y	A	0.75ms	CMOS BBRAM	16K	U.S.A.	
	984B	64K	8192	2048	Y		Y	Y	Y	Y	A	0.75ms	CMOS BBRAM	128K	U.S.A.	
	984X	8192	1048	448	Y		Y	Y	Y	Y	A	0.75ms	CMOS BBRAM	8K	U.S.A.	
	984-380	1024	256	64	Y		Y		Y	Y	A	5ms	CMOS BBRAM	6K	U.S.A.	
	984-480	7168	1024	448	Y		Y	Y	Y	Y	A	5ms	CMOS BBRAM	10K	U.S.A.	
	984-680	7168	2048	448	Y		Y	Y	Y	Y	A	3ms	CMOS BBRAM	10K	U.S.A.	
GUARDIAN/HITACHI (Chicago, IL)	J16	128	16		Y				Y	Y	B	20ms	EEPROM	2K	Japan	
HAWKER SIDDELEY LTD. (Welwyn Gardens City, UK)	SEQUEL	4640	4096	544		Y	Y		Y	Y	A		PROM, RAM	56K	UK	
	SM-PLC	25	19	6		Y	Y		Y		A		PROM, RAM	8K	UK	
HONEYWELL INDUST. CONTROLS DIV. (York, PA)	620-10	512	512		Y	Y		Y	Y	Y	B	10ms	CMOS, EPROM	4K	U.S.A.	All units use peer-to-peer net.
	620-15	512	512	512	Y	Y		Y	Y	Y	B	10ms	CMOS, EPROM	4K	U.S.A.	
	620-20	512	512	512	Y	Y	Y	Y	Y	Y	B	3ms	CMOS	8K	U.S.A.	
	620-25	2048	2048	2048	Y	Y	Y	Y	Y	Y	B	2.3ms	CMOS	32K	U.S.A.	
	620-35	2048	2048	2048	Y	Y	Y	Y	Y	Y	B	2.3ms	CMOS	32K	U.S.A.	
IDEC SYSTEMS & CONTROLS CORP. (Sunnyvale, CA)	FA-1	256	256	56	Y			Y	Y	Y	A	32ms	RAM, EPROM, EEPROM	4K	Japan	
	FA-1J	256	256	31	Y		Y	Y	Y	Y	A	32ms	RAM, EPROM, EEPROM	4K	Japan	
	MACH 1	32	32		Y					Y		40ms	RAM	0.7K	Japan	
INDUST. INDEXING SYST. (Victor, NY)	MM-10+	18	10	8			Y	Y	Y	Y	A, F	3ms	CMOS RAM		U.S.A.	
	MM-20	18	10	8			Y	Y	Y	Y	A, F	3ms	CMOS RAM		U.S.A.	
	MSC-100	40	40			Y	Y	Y	Y	Y	A, F	<3ms	CMOS RAM	12K	U.S.A.	
	MSC-800	168	72	96		Y	Y	Y	Y	Y	A, F	<3ms	CMOS RAM	16K	U.S.A.	
KLOCKNER-MOELLER CORP. (Natick, MA)	PS3	84	68	16	Y	Y	Y	Y	Y	Y	D	<5ms	RAM, EEPROM	3.6K	FRG	
	PS32	2176	2048	128	Y	Y	Y	Y	Y	Y	B, D	2.5ms	RAM, EPROM	32K	FRG	

Manufacturer / Model	Total system I/O	Max discrete I/O	Max analog I/O	Relay ladder logic	High level language	PID capabilities	Motion control	Documentation	Diagnostics	Type of interface	Scan rate / 1K	Type of memory	Size of memory	Country of origin	Comments
LEHIGH FLUID POWER (Lambertville, NJ)															
TPC-20	20	20		Y					Y		20ms	CMOS RAM	570	Japan	
MTS SYSTEMS CORP. (Minn., MN)															
TDC/EDC	22	22			Y	Y	Y	Y	A		EPROM	500	U.S.A.	Multidrop RS-485 net.
470 SERIES	†32				Y	Y	Y	Y	A		EPROM	32K	U.S.A.	†Modules
MAXITRON CORP. (Corte Madera, CA)															
PLC 47Jr	80	80	16	Y			Y	Y	B, F	2ms	CMOS RAM, EPROM	32K	US/France	High lev. languages:
PLC 47-20	256	256	16	Y		Y	Y	Y	B, D, F	2ms	CMOS RAM, EPROM	32K	US/France	Grafcet & literal
PLC 67-30	512	512	64	Y	Y	Y	Y	Y	B, D, F	0.45ms	CMOS RAM, EPROM	32K	US/France	for all units
PLC 87-10	992	992	120	Y	Y	Y	Y	Y	B, D, F	0.45ms	CMOS RAM, EPROM	128K	US/France	
PLC 87-30	2048	2048	256	Y	Y	Y	Y	Y	B, D, F	0.45ms	CMOS RAM, EPROM	128K	US/France	Maxinet One is used
DPC 87-10	976	976	120	Y	Y	Y	Y	y	B, D, F	0.45ms	CMOS RAM, EPROM	2.3M	US/France	for all units
DPC 87-30	2032	2032	124	Y	Y	Y	Y	Y	B, D, F	0.45ms	CMOS RAM, EPROM	4.9M	US/France	
AC 107-30	2032	2032	124	Y	Y	Y	Y	Y	B, D, F	0.45ms	CMOS RAM, EPROM	17.1M	US/France	
McGILL MFG. CO. ELEC. DIV. (Valparaiso, IN)															
1701-2000	472	472	16	Y		Y		Y	Y	A	3-4ms	RAM, EPROM	4K	U.S.A.	
1701-7000	512	512	32	Y		Y		Y	Y	A	3-4ms	RAM, EPROM	8K	U.S.A.	
MILLER FLUID POWER (Bensenville, IL)															
Epic 24	48	48		Y				Y	Y		50ms	RAM, EPROM, EEPROM	640	Japan	
Epic 40	120	120		Y				Y	Y		50ms	RAM, EPROM, EEPROM	1000	Japan	
Epic 24K	104	104		Y				Y	Y		50ms	RAM, EPROM, EEPROM	1000	Japan	
Epic 40K	120	120		Y				Y	Y		50ms	RAM, EPROM, EEPROM	1000	Japan	
MINARIK ELECTRIC CO. (Los Angeles, CA)															
LS 1000	177	121	56	Y	Y	Y	Y	Y	Y	B	7.5ms	RAM, ROM	3K	U.S.A.	
WP6200	12	12		Y				Y				RAM	79	U.S.A.	
WP6300	20	20		Y				Y				RAM	79	U.S.A.	
MITSUBISHI ELECT. SALES AMERICA (Mt. Prospect, IL)															
A1CPU	256			Y	Y	Y	Y	Y	B	1.25ms	RAM, EPROM, EEPROM	6K	Japan	All units use
A2CPU	512			Y	Y	Y	Y	Y	B	1.25ms	RAM, EPROM	14K	Japan	proprietary network
A3CPU	2048			Y	Y	Y	Y	Y	B	1.25ms	RAM, EPROM	60K	Japan	
F1-12	32	32		Y		Y	Y	Y	J	12ms	RAM, EPROM, EEPROM	1K	Japan	
F1-20M	40	40	6	Y		Y	Y	Y	J	12ms	RAM, EPROM, EEPROM	1K	Japan	
F1-30M	70	70	6	Y		Y	Y	Y	J	12ms	RAM, EPROM, EEPROM	1K	Japan	
F1-40M	80	80	12	Y		Y	Y	Y	J	12ms	RAM, EPROM, EEPROM	1K	Japan	
F1-60M	120	120	18	Y		Y	Y	Y	J	12ms	RAM, EPROM, EEPROM	1K	Japan	
F2-40M	80	80	12	Y		Y	Y	Y	J	10ms	RAM, EPROM, EEPROM	2K	Japan	
F2-60M	120	120	18	Y		Y	Y	Y	J	10ms	RAM, EPROM, EEPROM	2K	Japan	
NAVCOM, INC. (Huron, OH)															
F10A	10	10		Y		Y	Y	Y	B, D	40ms	EPROM, RAM	32K	U.S.A.	
F20A	20	20		Y		Y	Y	Y	B, D	40ms	EPROM, RAM	32K	U.S.A.	
F26A	26	26	16	Y		Y	Y	y	B, D	40ms	EPROM, RAM	32K	U.S.A.	
F36B	36	36	16	Y		Y	Y	Y	B, D	40ms	EPROM, RAM	32K	U.S.A.	
F40A	40	40		Y		Y	Y	Y	B, D	40ms	EPROM, RAM	32K	U.S.A.	
OMRON ELECTRONICS (Schaumburg, IL)															
S6	64	64					Y	Y		10ms	RAM, EPROM	1024	Japan	
C20	140	140		Y			Y	Y	H	10ms	RAM, EPROM	1194	Japan	
C20K	84	84		Y			Y	Y	H	10ms	RAM, EPROM	1K	Japan	
C120	256	256	22	Y			Y	Y	B, H	10ms	RAM, EPROM	2.6K	Japan	
C500	512	512	64	Y		Y	Y	Y	B, H	5ms	RAM, EPROM	8K	Japan	
C200H	1024	1024		Y	Y	Y	Y	Y	B, H	0.75ms	RAM, EPROM	7K	Japan	
C1000H	1024	1024	128	Y	Y	Y	Y	Y	B, H	0.4ms	RAM, EPROM	32K	Japan	
C2000H	2048	2048	128	Y	Y	Y	Y	Y	B, H	0.4ms	RAM, EPROM	32K	Japan	
PHILIPS B. V. (Eindhoven, The Netherlands)															
MC30	120	120		Y	Y		Y	Y	B, D	2ms	RAM, EPROM, EEPROM	2K	NL	All units use
PC20	2000	2000	300	Y	Y	Y		Y	Y	B, D	1ms	RAM, EPROM, EEPROM	16K	NL	a proprietary network
PHOENIX DIGITAL CORP. (Phoenix, AZ)															
DPAC	832	320	512	Y	Y	Y		Y	Y	B, H	Var.	CMOS RAM, UV PROM	180K	U.S.A.	Fib. optic/RS-232C nets
RELIANCE ELECTRIC (Euclid, OH)															
AutoMate 15	64	64		Y			Y	Y	A	4ms	RAM, EEPROM, NVRAM	1K	U.S.A.	Uses R-NET
AutoMate 20	256	256		Y	Y		Y	Y	A	10ms	RAM, EEPROM, NVRAM	2K	U.S.A.	Uses R-NET
AutoMate 30	512	512	128	Y	Y	Y	Y	Y	A	2ms	RAM, EEPROM, NVRAM	8K	U.S.A.	Uses R-NET
AutoMate 40	8192	8192	2048	Y	Y	Y	Y	Y	A	0.8ms	RAM, EEPROM, NVRAM	104K	U.S.A.	Uses R-NET
Shark	60	60	4				Y	Y	A	5ms	EPROM, EEPROM	2K	Japan	
Shark XL	160	160	16				Y	Y	A	5ms	EPROM, EEPROM	2K	Japan	
DCS 5000	12K+	12K+	4K+	Y	Y	Y	Y	Y	A		RAM, EPROM	80K	U.S.A.	Uses DCS NET
SELECTRON (Lyss, Switzerland)															
Euroline	32	32	4				Y	Y		10ms	EPROM	1K	Switzerland	
Picoline	28	28	4				Y	Y		10ms	EPROM	1K	Switzerland	
128 PEX	98	98	12				Y	Y		5ms	EPROM	2K	Switzerland	
256 PEX	112	112	12				Y	Y		5ms	EPROM	2K	Switzerland	
PMC 20	144	144	9				Y	Y	B, D	3.5ms	RAM, EPROM, EEPROM	16K	Switzerland	Selecontrol net
PMC 30	256	256	128				Y	Y	B, D	3.5ms	RAM, EPROM, EEPROM	32K	Switzerland	Selecontrol net
SIEMENS ENERGY & AUTOMATION (Peabody, MA)															
S5-100U†	128	128	8	Y						70ms	RAM, EPROM, EEPROM	1K	FRG	†Uses CPU100
S5-100U†	256	256	16	Y				D		7ms	RAM, EPROM, EEPROM	2K	FRG	†Uses CPU102
S5-101U	64	64		Y	Y			Y		30ms	RAM, EPROM, EEPROM	1K	FRG	
S5-101R	32	32		Y	Y			Y		2.5ms	RAM, EPROM, EEPROM	384	FRG	Most units use Sinec
S5-105R	128	128		Y	Y			Y		5ms	RAM, EPROM, EEPROM	1000	FRG	
S5-115U/942	2048	2048	128	Y	Y	Y	Y	Y	A	15ms	RAM, EPROM, EEPROM	42K	FRG	
S5-115U/943	2048	2048	128	Y	Y	Y	Y	Y	A	10ms	RAM, EPROM, EEPROM	48K	FRG	
S5-135U-R	8192	8192	384	Y	Y	Y	Y	Y	A	†	RAM, EPROM	128K	FRG	†20ms/8 loops
S5-135U-S	8192	8192	384	Y	Y	Y	Y	Y	A	2ms	RAM, EPROM, EEPROM	128K	FRG	
S5-150U	38K	38K	384	Y	Y	Y	Y	Y	A	1.5ms	RAM, EPROM	112K	FRG	

Manufacturer	Model	Total system I/O	Max discrete I/O	Max analog I/O	Relay ladder logic	High level language	PID capabilities	Motion control	Documentation	Diagnostics	Type of interface	Scan rate/1K	Type of memory	Size of memory	Country of origin	Comments
SOLID CONTROLS, INC. (Minneapolis, MN)	EPIC 1	520	512	8			Y	Y	Y			2.5ms	EPROM	16K	U.S.A.	
	SYSTEM 10	136	128	8			Y	Y	Y			2.5ms	EPROM	16K	U.S.A.	
	EPIC 8B	584	384	200		Y	Y	Y	Y		B, D	1.5ms	EPROM, RAM	392K	U.S.A.	
SPRECHER & SCHUH (Aarau, Switzerland)	SESTEP 390	160	160	120	Y			Y	Y	Y	A	20ms	EEPROM	2K	Jap./Swiss	
	SESTEP 490	2080	2080	108	Y	Y		Y	Y	Y	B	9.6ms	RAM, EPROM	7.6K	Jap./Swiss	
	SESTEP 590	2432	2432	240	Y	Y		Y	Y	Y	B	9.6ms	RAM, EPROM	15.7K	Jap./Swiss	
	SESTEP 690	3840	3840	768	Y	Y		Y	Y	Y	B	5.1ms	RAM, EPROM	48.5K	Jap./Swiss	
	SESTEP 300	128	128		Y				Y	Y	A	5ms	RAM, EPROM	2K	Japan	
	SESTEP 430	144	144	32	Y				Y	Y	A	50ms	RAM, EPROM	16K	Switzerland	
	SESTEP 530	1024	1024	64	Y				Y	Y	A	45ms	RAM, EPROM	16K	Switzerland	
SQUARE D CO. (Milwaukee, WI)	SY/MAX 50	256	256	32	Y					Y	J	7ms	RAM, EPROM, EEPROM	4K	Japan	All units use Time Token Passing network except SY/MAX 50
	SY/MAX 100	40	40		Y					Y	J	10ms	RAM, UVPROM	420	UK	
	SY/MAX 300	256	256	112	Y	Y	Y	Y	Y	Y	J	30ms	RAM, UVPROM	2K	U.S.A.	
	SY/MAX 500	2000	2000	1792	Y	Y	Y	Y	Y	Y	J	2.6ms	RAM, UVPROM	8K	U.S.A.	
	SY/MAX 700	14K	14K	3584	Y	Y	Y	Y	Y	Y	J	1.3ms	RAM, Bubble	64K	U.S.A.	
	SY/MAX LC	80		28	Y	Y	Y	Y	Y	Y	J	200ms†	RAM	256	UK	†Per 4 loops
TELEMECANIQUE (Westminster, MD)	TSX 27	80	80		Y	Y			Y	Y	F	2ms	RAM, EPROM	32K	France	All units except TSX 27 and MPC-007 use peer-to-peer net.
	MPC-007	256	256	32	Y				Y	Y		32ms	RAM, EPROM	4K	Japan	
	TSX 17	120	120	12	Y	Y	Y	Y	Y	Y	B	10ms	RAM, EPROM	24K	France	
	TSX 47 Jr	80	80	22	Y	Y	Y	Y	Y	Y	B	2ms	RAM, EPROM	32K	France	
	TSX 47	256	256	44	Y	Y	Y	Y	Y	Y	B	2ms	RAM, EPROM	32K	France	
	TSX 47-30	256	256	64	Y	Y	Y	Y	Y	Y	B	0.5ms	RAM, EPROM	32K	France	
	TSX 67-30	512	512	64	Y	Y	Y	Y	Y	Y	B	0.5ms	RAM, EPROM	64K	France	
	TSX 87-10	1024	1024	128	Y	Y	Y	Y	Y	Y	B	0.5ms	RAM, EPROM	128K	France	
	TSX 87-30	2048	2048	256	Y	Y	Y	Y	Y	Y	B	0.5ms	CMOS RAM, EPROM	128K	France	
TEMPATRON, LTD (Reading, England)	TPC 9000	252	252	60		Y	Y	Y	Y	Y	B	10ms	RAM, EPROM	32K	UK	
TENOR CO. (New Berlin, WI)	100	252	252	15	Y	Y	Y	Y	Y	Y	B, D	10ms	RAM, EPROM	32K	UK	T-NET network
	PSC 763	96	96			Y							EPROM	128	U.S.A.	
TEXAS INSTRUMENTS INDUSTRIAL SYST. (Johnson City, TN)	5TI	512	512		Y			Y	Y	Y	L	8.2ms	RAM, EPROM	4K	U.S.A.	All units use TIWAY1 network
	510	40	40		Y			Y	Y	Y	A, L		RAM, EPROM	256	U.S.A.	
	TI100	128	128		Y			Y	Y	Y	L	5ms	RAM, EPROM	1K	Japan	
	TI160	24	18	6				Y	Y	Y	A		RAM, NOVRAM	762	U.S.A.	
	520C-1102	512	512	512	Y	Y	Y	Y	Y	Y	B, I	4ms	RAM, EPROM	3.5K	U.S.A.	
	530C-1104	1023	1023	1023	Y	Y	Y	Y	Y	Y	B, I	4ms	RAM, EPROM	8K	U.S.A.	
	530C-1108	1023	1023	1023	Y	Y	Y	Y	Y	Y	B, I	4ms	RAM, EPROM	15K	U.S.A.	
	530C-1112	1023	1023	1023	Y	Y	Y	Y	Y	Y	B, I	4ms	RAM, EPROM	20K	U.S.A.	
	525-1102	512	512	64	Y	Y	Y	Y	Y	Y	A, I	3.7ms	RAM, EPROM, EEPROM	5K	U.S.A.	
	525-1104	1023	1023	1023	Y	Y	Y	Y	Y	Y	B, I	3.7ms	RAM, EPROM, EEPROM	8K	U.S.A.	
	525-1208	1023	1023	1023	Y	Y	Y	Y	Y	Y	B, I	3.7ms	RAM, EPROM, EEPROM	15K	U.S.A.	
	525-1212	1023	1023	1023	Y	Y	Y	Y	Y	Y	B, I	3.7ms	RAM, EPROM, EEPROM	20K	U.S.A.	
	535-1204	1023	1023	1023	Y	Y	Y	Y	Y	Y	B, I	0.83ms	RAM, EPROM, EEPROM	8K	U.S.A.	
	535-1212	1023	1023	1023	Y	Y	Y	Y	Y	Y	B, I	0.83ms	RAM, EPROM, EEPROM	20K	U.S.A.	
	PM550C	640	512	128	Y	Y	Y	Y	Y	Y	B	8ms	RAM	7K	U.S.A.	
	560/565	8192	8192	8192	Y	Y	Y	Y	Y	Y	B, I	2.2ms	RAM/RAM EPROM	256K	U.S.A.	
	8640	46	32	14	Y	Y	Y		Y	Y	A		RAM, EPROM	120	U.S.A.	
	8641	248	248	148	Y	Y	Y		Y	Y	A		RAM, EPROM	256K	U.S.A.	
	8642	248	248	248	Y	Y	Y		Y	Y	A		RAM, EPROM	256K	U.S.A.	
	8650	24	16	8	Y	Y	Y		Y	Y	A		RAM, EPROM	128K	U.S.A.	
THESAURUS (Huntsville, AL)	CBPC-1	256	256	256	Y	Y	Y	Y	Y	Y	B, C	0.5ms	RAM	500K	U.S.A.	
	CBPC-2	512	512	512	Y	Y	Y	Y	Y	Y	B, C	0.2ms	RAM	1M	U.S.A.	
	CBPC-3	1024	1024	1024	Y	Y	Y	Y	Y	Y	B, C	0.1ms	RAM	2M	U.S.A.	
	CBPC-4	2048	2048	2048	Y	Y	Y	Y	Y	Y	B, C	0.01ms	RAM	16M	U.S.A.	
TOSHIBA (Houston, TX)	EX200	240	224	16	Y		Y	Y	Y	Y	B	9ms	CMOS RAM	4K	Japan	Tosline-30 Data Hwy
	EX250	240	256	16	Y		Y	Y	Y	Y	B	7ms	CMOS RAM	4K	Japan	Tosline-30 Data Hwy
	EX500	544	512	32	Y		Y	Y	Y	Y	B	5ms	CMOS RAM	8K	Japan	Tosline-30 Data Hwy
	EX14B	34	34		Y			Y	Y	Y	B	60ms	CMOS RAM	1K	Japan	
	EX20-PLUS	40	40	2	Y			Y	Y	Y	B	60ms	CMOS RAM	1K	Japan	
	EX28B	28	28		Y			Y	Y	Y	B	60ms	CMOS RAM	1K	Japan	
	EX40-PLUS	80	80	2	Y			Y	Y	Y	B	60ms	CMOS RAM	1K	Japan	
TRICONEX (Irvine, CA)	TRICON	2208	2208	2208	Y	Y	Y		Y	Y	A	2.9ms	RAM, PROM	378K	U.S.A.	Modbus net.
TRIPLEX (Torrance, CA)	REGENT	2560	2560	2560	Y	Y	Y		Y	y	A	1ms	CMOS RAM	512K	U.S.A.	Modbus net.
TURNBULL CONTROLS (Reston, VA)	6433	32	32	32		Y	Y		Y	Y	J		RAM	8K	UK	ANSI X3.28 net.
UTICOR TECHNOLOGY (Bettendorf, IA)	DIR. ONE	128	128		Y			Y	Y	Y	A	20ms	RAM	1970	Japan	All units use RS-422 net.
	DIR. 4001	384	384	128	Y		Y	Y	Y	Y	B	10ms	RAM, EEPROM	6K	U.S.A.	
	DIR. 4002	64	64	64	Y		Y	Y	Y	Y	B	10ms	RAM, EEPROM	6K	U.S.A.	
VEEDER-ROOT CO. (Hartford, CT)	V-12	120	120	15	Y			Y	Y			40ms	CMOS RAM, EPROM	944	Japan	Standard unit
	V-12 EXP	80	80	8	Y			Y	Y			45ms	CMOS RAM, EPROM	832	Japan	Expanded CPU
WESTINGHOUSE ELECTRIC CO., (Pittsburgh, PA)	PC-100	30	30								†	8ms	CMOS RAM, EPROM	320	Japan	†Special
	PC-110	112	112								†	8ms	CMOS RAM, EPROM	1K	Japan	†Special
	PC-1100	144	128	16	Y		Y	Y	Y	Y	A	8ms	CMOS RAM	3.5K	U.S.A.	All units use WESTNET II except PC-100 & PC-110
	PC-900	288	256	32	Y	Y	Y	Y	Y	Y	A	20ms	CMOS RAM, EPROM	2.5K	U.S.A.	
	PC-700	576	512	64	Y	Y	Y	Y	Y	Y	A	0ms	CMOS RAM	8K	U.S.A.	
	HPPC	8192	8192	8192	Y	Y	Y	Y	Y	Y	A	0.8ms	CMOS RAM	224K	U.S.A.	

Bibliography

Chapter 1: Operational Amplifiers for Industrial Applications

*Berlin, H. W. *Design of Active Filters with Experiments*. Indianapolis: Sams, 1977.

_____. *Design of Op-Amp Circuits with Experiments*. Indianapolis: Sams, 1977.

*Boyce, J. *Operational Amplifiers for Technicians*. North Scituate, Mass.: Breton, 1983.

Burr-Brown Research Corporation. *Handbook of Operational Amplifier Applications*. Tucson, 1963.

*Cirovic, M. M. *Integrated Circuits: A User's Handbook*. Reston, Va.: Reston, 1977.

*Coughlin, R. F., and Driscoll, F. F. *Operational Amplifiers and Linear Integrated Circuits*. 2nd ed. Englewood Cliffs, N.J.: Prentice-Hall, 1982.

*Deboo, G. J., and Burrous, C. N. *Integrated Circuits and Semiconductor Devices*. New York: McGraw-Hill, 1977.

*Dungan, F. R. *Linear Integrated Circuits for Technicians*. North Scituate, Mass.: Breton, 1984.

Faulkenberry, L. M. *An Introduction to Operational Amplifiers*. 2nd ed. New York: Wiley, 1984.

Garrett, P. H. *Analog I/O Design*. Reston, Va.: Reston, 1981.

Gayawkwad, R. A. *Op-Amps and Linear Integrated Circuits*. 2nd ed. Englewood Cliffs, N.J.: Prentice-Hall, 1988.

Graeme, J. G. *Designing with Operational Amplifiers*. New York: McGraw-Hill, 1977.

Hnatek, E. R. *Applications of Linear Integrated Circuits*. New York: Wiley, 1975.

Jacob, J. M. *Applications and Designs with Analog Integrated Circuits*. Reston, Va.: Reston, 1982.

---

*Books that have a technical rather than an engineering emphasis.

*Jung, W. C. *IC Op Amp Cookbook*. Indianapolis: Sams, 1974.

*Lancaster, D. *Active Filter Cookbook*. Indianapolis: Sams, 1974.

*Malcolm, D. R. *Fundamentals of Electronics*. 2nd ed. North Scituate, Mass.: Breton, 1983.

National Semiconductor Corporation. *Linear Databook*. Vols. 1, 2, and 3. Santa Clara, Calif., 1988.

_____. *Application Note AN–79*. Santa Clara, Calif., 1980.

*Pasahow, E. *Principles of Integrated Circuits*. North Scituate, Mass.: Breton, 1982.

*Rutkowski, G. B. *Handbook of Integrated Circuit Operational Amplifiers*. Englewood Cliffs, N.J.: Prentice-Hall, 1982.

Wait, J. V. *Introduction to Operational Amplifier Theory and Applications*. New York: McGraw-Hill, 1975.

Chapter 2: Linear Integrated Circuits for Industrial Applications

*Cirovic, M. M. *Integrated Circuits: A User's Handbook*. Reston, Va.: Reston, 1977.

*Coughlin, R. F., and Driscoll, F. F. *Operational Amplifiers and Linear Integrated Circuits*. 2nd ed. Englewood Cliffs, N.J.: Prentice-Hall, 1982.

*Dungan, F. R. *Linear Integrated Circuits for Technicians*. North Scituate, Mass.: Breton, 1984.

Garrett, P. H. *Analog I/O Design*. Reston, Va.: Reston, 1981.

Gayawkwad, R. A. *Op-Amps and Linear Integrated Circuits*. 2nd ed. Englewood Cliffs, N.J.: Prentice-Hall, 1988.

Hnatek, E. R. *Applications of Linear Integrated Circuits*. New York: Wiley, 1975.

Jacob, J. M. *Applications and Designs with Analog Integrated Circuits*. Reston, Va.: Reston, 1982.

*Malcolm, D. R. *Fundamentals of Electronics*. 2nd ed. North Scituate, Mass.: Breton, 1983.

National Semiconductor Corporation. *Linear Databook*. Vols. 1, 2, and 3. Santa Clara, Calif., 1988.

*Pasahow, E. *Principles of Integrated Circuits*. North Scituate, Mass.: Breton, 1982.

Signetics Corporation. *Analog Data Manual*. Sunnyvale, Calif., 1985.

_____. *Signetics Linear LSI Data and Applications Manual*. Vol. 1. Sunnyvale, Calif., 1985.

_____. *Signetics Linear LSI Data and Applications Manual*. Vol. 2. Sunnyvale, Calif., 1985.

Chapter 3: Wound-Field DC Motors and Generators

Anderson, L. R. *Electric Machines and Transformers*. Reston, Va.: Reston, 1981.

Department of the Army, Headquarters. *Electric Motor and Generator Repair*. Washington, D.C.: Government Printing Office, 1972.

*Emmanuel, P. *Motors, Generators, Transformers and Energy*. Englewood Cliffs, N.J.: Prentice-Hall, 1985.

Fisher, F. "Convenient Comparison Charts Aid in Motor Selection." *EDN*, 23 (August 5, 1978): 97-99.

Gingrich, H. W. *Electrical Machinery, Transformers, and Controls*. Englewood Cliffs, N.J.: Prentice-Hall, 1979.

*Humphries, James T. *Motors and Controls*. Columbus, Ohio: Merrill, 1988.

Kosow, I. L. *Electric Machinery and Transformers*. Englewood Cliffs, N.J.: Prentice-Hall, 1972.

*Lister, E. C. *Electric Circuits and Machines*. 6th ed. New York: McGraw-Hill, 1984.

Lloyd, T. C. *Electric Motors and Their Applications*. New York: Wiley, 1969.

*Naval Education and Training Support Command. *Electrician's Mate 3 & 2*. Washington, D.C.: Government Printing Office, 1972.

Richardson, D. V. *Rotating Electric Machinery and Transformer Technology*. Reston, Va.: Reston, 1978.

Rosenblatt, J., and Friedman, M. H. *Direct and Alternating Current Machinery*. 2nd ed. Columbus, Ohio: Merrill, 1984.

Wildi, T. *Electric Power Technology*. New York: Wiley, 1981.

Chapter 4: Brushless and Stepper DC Motors

Airpax Corporation. *Stepper Motor Handbook*. Cheshire, Conn., 1985.

Electro-craft Corporation. *DC Motors, Speed Controls and Servo Systems*. 6th ed. Hopkins, Minn., 1982.

*Emmanuel, P. *Motors, Generators, Transformers and Energy*. Englewood Cliffs, N.J.: Prentice-Hall, 1985.

*Humphries, James T. *Motors and Controls*. Columbus, Ohio: Merrill, 1988.

International Rectifier Corporation. *HEXFET Power MOSFET Databook*. El Segundo, Calif., 1985.

Kuo, B. C. *Theory and Applications of Stepper Motors*. St. Paul, Minn.: West, 1974.

Kuo, B. C., and Tal, J. *DC Motors and Control Systems*. Champaign, Ill.: SRL, 1978.

Oriental Motor U.S.A. Corporation. *Technical Information on Stepping Motors*. Torrance, Calif.: 1983.

Siliconix Corporation. *MOSPOWER Design Catalog*. Santa Clara, Calif., 1983.

Chapter 5: AC Motors

Anderson, L. R. *Electric Machines and Transformers*. Reston, Va.: Reston, 1981.

Emmanuel, P. *Motors, Generators, Transformers and Energy*. Englewood Cliffs, N.J.: Prentice-Hall, 1985.

Gingrich, H. W. *Electrical Machinery, Transformers, and Controls*. Englewood Cliffs, N.J.: Prentice-Hall, 1979.

*Humphries, James T. *Motors and Controls*. Columbus, Ohio: Merrill, 1988.

Kosow, I. L. *Electric Machinery and Transformers*. Englewood Cliffs, N.J.: Prentice-Hall, 1972.

*Lister, E. C. *Electric Circuits and Machines*. 6th ed. New York: McGraw-Hill, 1984.

Lloyd, T. C. *Electric Motors and Their Applications*. New York: Wiley, 1969.

*Naval Education and Training Support Command. *Electrician's Mate 3 & 2*. Washington, D.C.: Government Printing Office, 1972.

Richardson, D. V. *Rotating Electric Machinery and Transformer Technology*. Reston, Va.: Reston, 1978.

Rosenblatt, J., and Friedman, M. H. *Direct and Alternating Current Machinery*. 2nd ed. Columbus, Ohio: Merrill, 1984.

*Wildi, T. *Electric Power Technology*. New York: Wiley, 1981.

Chapter 6: Industrial Control Devices

*Bell, D. A. *Electronic Devices and Circuits*. 2nd ed. Reston, Va.: Reston, 1978.

*Coughlin, R. F., and Driscoll, F. F. *Operational Amplifiers and Linear Integrated Circuits*. 2nd ed. Englewood Cliffs, N.J.: Prentice-Hall, 1982.

Datta, S. K. *Power Electronics and Controls*. Reston, Va.: Reston, 1985.

*Deboo, G. J., and Burrous, C. N. *Integrated Circuits and Semiconductor Devices*. New York: McGraw-Hill, 1977.

General Electric Company. *SCR Manual*. 2nd ed. Auburn, N.Y., 1969.

_____. *Transistor Manual*. 2nd ed. Auburn, N.Y., 1969.

Gottlieb, I. M. *Solid-State Power Electronics*. Indianapolis: Sams, 1977.

International Rectifier Corporation. *HEXFET Power MOSFET Databook*. El Segundo, Calif., 1985.

_____. *SCR Applications Handbook*. El Segundo, Calif., 1985.

Maloney, T. J. *Industrial Solid-State Electronics*. 2nd ed. Englewood Cliffs, N.J.: Prentice-Hall, 1986.

Malvino, A. P. *Transistor Circuit Approximations*. 3rd ed. New York: McGraw-Hill, 1980.

Motorola Inc. *Thyristor Data*. 1985.

Newman, M. *Industrial Electronics and Controls*. New York: Wiley, 1986.

RCA Corporation. *Thyristor and Rectifier Manual*. Somerville, N.J.: 1975.

Signetics Corporation. *Analog Data Manual*. Sunnyvale, Calif., 1985.

_____. *Signetics Linear LSI Data and Applications Manual*. Vol. 1. Sunnyvale, Calif., 1985.

_____. *Signetics Linear LSI Data and Applications Manual*. Vol. 2. Sunnyvale, Calif., 1985.

Chapter 7: Power Control Circuits

Airpax Corporation. *Stepper Motor Handbook*. Cheshire, Conn., 1985.

*Coughlin, R. F., and Driscoll, F. F. *Operational Amplifiers and Linear Integrated Circuits*. 2nd ed. Englewood Cliffs, N.J.: Prentice-Hall, 1982.

Datta, S. *Power Electronics and Controls*. Reston, Va.: Reston, 1985.

Deboo, G. J., and Burrous, C. N. *Integrated Circuits and Semiconductor Devices*. New York: McGraw-Hill, 1977.

Electro-craft Corporation. *DC Motors, Speed Controls and Servo Systems*. 6th ed. Hopkins, Minn., 1982.

General Electric Company. *SCR Manual*. 2nd ed. Auburn, N.Y., 1979.

_____. *Transistor Manual*. 2nd ed. Auburn, N.Y., 1969.

*Gottlieb, I. M. *Solid-State Power Electronics*. Indianapolis: Sams, 1977.

International Rectifier Corporation. *HEXFET Power MOSFET Databook*. El Segundo, Calif., 1985.

_____. *SCR Applications Handbook*. El Segundo, Calif., 1985.

Kuo, B. C. *Theory and Applications of Stepper Motors*. St. Paul, Minn.: West, 1974.

Kuo, B. C., and Tal, J. *DC Motors and Control Systems*. Champaign, Ill.: SRL, 1978.

Kusko, A. *Solid-State DC Motor Drives*. Cambridge, Mass.: MIT Press, 1969.

*Maloney, T. J. *Industrial Solid-State Electronics.* 2nd ed. Englewood Cliffs, N.J.: Prentice-Hall, 1986.

Mazda, F. F. *Thyristor Control.* London: Butterworth, 1973.

Motorola Inc. *Thyristor Data.* 1985.

Murphy, J. M. *Thyristor Control of AC Motors.* Oxford: Pergamon Press, 1973.

Newman, M. *Industrial Electronics and Controls.* New York: Wiley, 1986.

Oriental Motor U.S.A. Corporation. *Technical Information of Stepping Motors.* Torrance, Calif., 1983.

RCA Corporation. *Thyristor and Rectifier Manual.* Somerville, N.J., 1975.

Signetics Corporation. *Analog Data Manual.* Sunnyvale, Calif., 1985.

_____. *Signetics Linear LSI Data and Applications Manual.* Vol. 1. Sunnyvale, Calif., 1985.

_____. *Signetics Linear LSI Data Applications Manual.* Vol. 2. Sunnyvale, Calif., 1985.

Siliconix Corporation. *MOSPOWER Design Catalog.* Santa Clara, Calif., 1983.

Traister, J. E. *Complete Handbook of Electric Motor Controls.* Englewood Cliffs, N.J.: Prentice-Hall, 1986.

Herceg, E. E. *Schaevitz Handbook of Measurement and Control.* Camden, N.J.: Schaevitz Engineering, 1976.

*Johnson, C. D. *Process Control Instrumentation Technology.* 2nd ed. New York: Wiley, 1982.

*Lenk, J. D. *Handbook of Controls and Instrumentation.* Englewood Cliffs, N.J.: Prentice-Hall, 1980.

Mansfield, P. H. *Electrical Transducers for Industrial Measurement.* London: Butterworth, 1973.

Micro Switch. *Solid State Sensors.* Freeport, Ill.: Honeywell.

Minnar, E. J., ed. *ISA Transducer Compendium.* Pittsburgh: Instrumentation Society of America, 1963.

Moore, R. L., ed. *Basic Instrumentation Lecture Notes and Study Guide.* 2nd ed. Pittsburgh: Instrument Society of America, 1976.

National Semiconductor Corporation. *Pressure Transducer Handbook.* Santa Clara, Calif., 1980.

*Norton, H. M. *Handbook of Transducers for Electronic Measuring Systems.* Englewood Cliffs, N.J.: Prentice-Hall, 1969.

O'Higgins, P. J. *Basic Instrumentation.* New York: McGraw-Hill, 1966.

Sheingold, D. H. *Transducer Interfacing Handbook.* Norwood, Mass.: Analog Devices, 1980.

Chapter 8: Transducers

Allocca, J. A. *Transducers: Theory and Applications.* Reston, Va.: Reston, 1984.

Anderson, N. A. *Instrumentation for Process Measurement and Control.* 2nd ed. Radnor, Pa.: Chilton, 1972.

Andrew, W. G., and Williams, H. B. *Applied Instrumentation in the Process Industries.* 2nd ed. Houston: Gulf, 1979.

Deboo, G. J., and Burrous, C. N. *Integrated Circuits and Semiconductor Devices.* New York: McGraw-Hill, 1977.

*Driscoll, F. F. *Industrial Electronics: Devices, Circuits and Applications.* Chicago: American Technical Society, 1976.

*Fribance, A. E. *Industrial Instrumentation Fundamentals.* New York: McGraw-Hill, 1962.

Chapter 9: Optoelectronics

*Center for Occupational Research and Development. *Introduction to Lasers.* Waco, Texas, 1984.

Hewlett-Packard. *LED Indicators and Display Application Handbook.* Palo Alto, Calif., 1986.

*Hitz, C. B. *Understanding Laser Technology.* Tulsa, OK: PennWell Publishing Co., 1987.

Luxon, J. T. *Industrial Lasers and Their Applications.* Englewood Cliffs, N.J.: Prentice-Hall, 1985.

Chapter 10: Industrial Process Control

Brewer, J. W. *Control Systems: Analysis, Design, and Simulation.* Englewood Cliffs, N.J.: Prentice-Hall, 1974.

Hougen, J. O. *Measurements and Control Applications.* Research Park Triangle, N.C.: Instrument Society of America, 1979.

*Hunter, R. P. *Automatic Process Control Systems: Concepts and Hardware.* Englewood Cliffs, N.J.: Prentice-Hall, 1978.

*Johnson, C. D. *Process Control Instrumentation Technology.* 2nd ed. New York: Wiley, 1982.

Kuo, B. C. *Automatic Control Systems.* Englewood Cliffs, N.J.: Prentice-Hall, 1978.

Needler, M. A., and Baker, D. E. *Digital and Analog Controls.* Reston, Va.: Reston, 1985.

Newman, M. *Industrial Electronics and Controls.* New York: Wiley, 1986.

Pearman, R. A. *Solid-State Industrial Electronics.* Reston, Va.: Reston, 1984.

Pericles, E., and Leff, E. *Introduction to Feedback Control Systems.* New York: McGraw-Hill, 1979.

Weyrick, R. C. *Fundamentals of Automatic Control.* New York: McGraw-Hill, 1975; Reston, Va.: Reston, 1978.

Chapter 11: Pulse Modulation

*Hnatek, E. R. *Applications of Linear Integrated Circuits.* New York: Wiley, 1975.

Journal Ministere Russie Defense 47, section 7, 25 (1845).

*Kennedy, G. *Electronic Communication Systems.* 2nd ed. New York: McGraw-Hill, 1977.

*Miller, G. M. *Modern Electronic Communication.* 2nd ed. Englewood Cliffs, N.J.: Prentice-Hall, 1983.

*National Semiconductor Corporation. *Special Functions Databook.* Santa Clara, Calif., 1979.

Shannon, C. E., and Weaver, W. *The Mathematical Theory of Communications.* Urbana: University of Illinois Press, 1949.

Taub, H., and Schilling, D. L. *Principles of Communication Systems.* New York: McGraw-Hill, 1971.

*Tomasi, W. *Fundamentals of Electronic Communications Systems.* Englewood Cliffs, N.J.: Prentice-Hall, 1988.

Chapter 12: Industrial Telemetry and Data Communication

*Campbell, J. *The RS–232 Solution.* Berkeley, Calif.: Sybex, 1984.

Fisher, H. F. *Telemetry Transducer Handbook.* Technical Report No. WADD–TR–61–67, Vol. I, Rev. I. Wright-Patterson Air Force Base, Ohio: Air Force Systems Command, 1963.

*Gruenberg, E. L., ed. *Handbook of Telemetry and Remote Control.* New York: McGraw-Hill, 1967.

Inter-Range Instrumentation Group. *Telemetry Standards.* Document No. 106–73, rev. White Sands Missile Range, N.M.: Secretariat, Range Commanders Council, 1973.

*Kennedy, G. *Electronic Communication Systems.* 2nd ed. New York: McGraw-Hill, 1977.

Martin, J. *Telecommunications and the Computer.* Englewood Cliffs, N.J.: Prentice-Hall, 1969.

*Miller, G. M. *Modern Electronic Communication.* Englewood Cliffs, N.J.: Prentice-Hall, 1978.

*Tomasi, W. *Fundamentals of Electronic Communications Systems.* Englewood Cliffs, N.J.: Prentice-Hall, 1988.

*Zanger, H. *Electronic Systems Theory and Applications.* Englewood Cliffs, N.J.: Prentice-Hall, 1977.

Chapter 13: Sequential Process Control

Brewer, J. W. *Control Systems: Analysis, Design, and Simulation.* Englewood Cliffs, N.J.: Prentice-Hall, 1974.

Hougen, J. O. *Measurements and Control Applications.* Research Park Triangle, N. C.: Instrument Society of America, 1979.

*Hunter, R. P. *Automatic Process Control Systems: Concepts and Hardware*. Englewood Cliffs, N.J.: Prentice-Hall, 1978.

*Johnson, C. D. *Process Control Instrumentation Technology*. 2nd ed. New York: Wiley, 1982.

Kuo, B. C. *Automatic Control Systems*. Englewood Cliffs, N.J.: Prentice-Hall, 1978.

*Maloney, T. J. *Industrial Solid-State Electronics*. 2nd ed. Englewood Cliffs, N.J.: Prentice-Hall, 1986.

Needler, M. A., and Baker, D. E. *Digital and Analog Controls*. Reston, Va.: Reston, 1985.

Pearman, R. A. *Solid-State Industrial Electronics*. Reston, Va.: Reston, 1984.

Pericles, E., and Leff, E. *Introduction to Feedback Control Systems*. New York: McGraw-Hill, 1979.

Weyrick, R. C. *Fundamentals of Automatic Control*. New York: McGraw-Hill, 1975; Reston, Va: Reston, 1978.

Chapter 14: Programmable Controllers

*Allen-Bradley Company. *Bulletin 1770: Industrial Terminal Systems User's Manual*. Cleveland.
_____. *Bulletin 1772: Mini–PLC-2 Programmable Controller*. Cleveland.

Andrew, W. G., and Williams, H. B. *Applied Instrumentation in the Process Industries*. Houston: Gulf, 1979.

Deltano, D. "Programming Your PC." *Instruments & Control Systems* 53 (July 1980): 37–40.

General Electric Company. *Series Six Plus Programmable Controller Manual*. Charlottesville, Va.: GE Fanuc Automation, 1986.

Hickey, J. "Programmable Controller Roundup." *Instruments & Control Systems* 54 (July 1981): 57–64.

Jannotta, K. "What Is a PC?" *Instruments & Control Systems* 53 (February 1980): 21–25.

Webb, J. W. *Programmable Controllers—Principles and Applications*. Columbus, Ohio: Merrill, 1988.

Chapter 15: Introduction to Robotics

*Goetsch, D. L. *Fundamentals of CIM Technology*. Albany, N.Y.: Delmar Publishing Co., 1988.

Hoekstra, R. L. *Robotics and Automated Systems*. Cincinnati, Ohio: Southwestern Publishing Co., 1986.

*Malcolm, D. R. *Robotics, An Introduction*. Albany, N.Y.: Delmar Publishing Co., 1985.

Answers to Odd-Numbered Problems

Chapter 1

1. 10 V
3. $R_A = 50\ \Omega$
   $A_V = 400$
5. $A_V = 20$
   (a) $V_o = 1.1$ V
   (b) $R_9 = \Delta\ 18.17\ \Omega$
7. $C = 6666\ \mu F$
9. ±1.24 V
11. 5 mA
13. Supply voltage = 5 V
15. (a) $\dfrac{V_o}{V_i} = \dfrac{R_3}{R_1} = \dfrac{3}{4}$

    (b) 2.5 V
17. $I_C = \dfrac{3\ V}{100\ \Omega} = 30\ mA$

    $I_B = \dfrac{I_C}{\beta} = \dfrac{30\ mA}{50} = 0.6\ mA$
19. $I_{ref} = 60\ \mu A$
    Upper trip voltage = 7.5 V
    Lower trip voltage = 6 V
    Window = 1.5 V
21. 1.59 μF
23. $V_{ut} = 8.4$ V
    $V_{lt} = 6$ V
25. $I_{L1} = 20$ mA
    $I_{L2} = 10$ mA
27. $I_b = 2.4\ \mu A$
    $V_o = 4.8$ V
    $A_V = 13.3$
    $V_o = 2.66\ V_{pk}$
29. $V_{ut} = 9.99$ V
    $V_{lt} = 4.99$ V
    $I_{ut} = 6.66\ \mu A$
    $I_{lt} = 3.33\ \mu A$
31. $g_m = 1650\ \mu S$
33. $I_{ABC} = 80\ \mu A$

Chapter 2

1. (a) $f_o = 1739$ Hz
   (b) $V_{pin\ 5} = 10.17$ V
   (c) $f_o = 2033$ Hz
   (d) $V_i = 10.27$ V and 10.07 V
       $f_o = 1922$ Hz
       $f_o = 2144$ Hz
   (e) 0.18 V
3. (a) $R_1 = 808.8\ \Omega$
   (b) $\Delta f_o = 8241$ Hz
       So, f_o will be 15,000 Hz ± 8244 Hz, or about ±55% deviation.
5. (a) $f_o = 1111$ Hz
   (b) $f_c = 83.47$ Hz
   (c) $f_l = 740.7$ Hz
7. (a) $f_c = 31.35$ Hz
   (b) $f_l = 266.9$ Hz
   (c) $f_o = 667.3$ Hz
9. (a) $V_e = 7.5$ V
   (b) $K_o = 294$ Hz/V
   (c) The upper lock-range is 295 Hz above f_o. Since the K_o is 294 Hz/V, the output voltage will decrease 1 V (from 7.5 to 6.5) when f_i goes from 740 Hz to 1034 Hz. The output voltage will increase 1 V (from 7.5 to 8.5) when f_i goes from 740 Hz to 446 Hz. The frequency change for ½ V would be ½ of 294 or 147 Hz. The output voltage will increase ½ V (from 7.5 to 8) when f_i goes from 740 Hz to 593 Hz (740 Hz − 147 Hz).
   (d) $\Delta V = 0.5$ V
       $\Delta f = 147$ Hz
       At the upper limit of the lock range (1034 Hz), the error voltage goes to 6.5; at the lower edge of the lock range (446 Hz), the voltage goes from 7.5 to 8.5 V.
11. (a) $f_o = 6060$ Hz
    (b) $BW = 121.5$ Hz

13. (a) $f_o = 1289$ Hz
  (b) $BW = 332$ Hz
15. (a) $V_o = 2.4$ V
  (b) $f_{i(max)} = 1250$ Hz
  (c) $V_{r(p-p)} = 0.107$ V
17. Resolution (in percentage) = 0.0244%
19. $V_o = -3$ V
21. $I_{ref} = 2$ mA
  $I_o = 1.05$ mA
23. $C_1 = 0.375$ μF
  $R_1 = 8000$ Ω
  $C_2 = 10.9$ μF
25. (a) $V_o = 1.875$ V
  (b) $V_o = 3.125$ V
  (c) $R_F = 4.4$ kΩ

Chapter 3

1. $I_A = 0.3$ A
  cemf = 195.5 V
  $n_A = 2095$ r/min
3. cemf = 80 V
  $I_{field} = 0.013$ A
  $R_{field} = 972$ Ω
5. $I_{field} = 0.2$ A
  $I_A = 5$ A
  $T = 0.5$ Nm
  $P = 500$ W
7. emf = 157.5 V
  $n_A = 2287$ r/min
9. $I_A = 20.66$ A

Chapter 4

1. $M = 4.84$
  $E_{max} = 0.629$ or 62.9%
  $P_i = 116.1$ W
  $P_o = 76.3$ W
  $N_m = 6636$ r/min
  $T_m = 108.7$ mNm
3. $f = 12$ cycles
  $n_{max} = 1667$ r/min

Chapter 5

1. (a) $V_L = 830.4$ V AC
  (b) The current in any line in a wye-connected generator will be equal to the phase current, 20 A.
3. (a) $I_L = 1.73I_P = 1.73(150$ A$) = 295.5$ A
    The phase voltage is equal to the line voltage, 240 V AC.
  (b) $P_{kW} = 91.58$ kW
  (c) $P_{kVA} = 107.74$ kVA
5. $s = 0.111$ or 11%
7. (a) $T_A = 108.5$ Nm
  (b) $V_A = 251.7$ V AC
9. % speed regulation = 2.94%
11. $P_o = 425$ W or 0.425 kW
13. (a) $n_{st} = 1800$ r/min
  (b) $s = 0.0277$ or 2.77%
  (c) $T_{rated} = 225$ lb-ft
  (d) $P_{i(kVA)} = 70.7$ kVA
  (e) $P_{i(W)} = 61.5$ kW
  (f) $I_{FL} = 187.5$ A
15. $p = 4$

Chapter 6

1. Inrush current = 4.167 A
  Sealed current = 1.04 A
3. (a) $T_J = 143.5°C$
  (b) $R_{\theta CA} = 9.33°C/W$
  (c) $P_J = 390$ mW
  (d) $T_C = 51.19°C$
5. $T_J = 120°C$
7. $T_C = 74.5°C$
9. $V_T = 1.8$ V
  $\theta_{fire} = 0.6°$
11. $T = 2.29$ ms
  $f = 436$ Hz
13. $I_{max} = 16.7$ W
15. $V_{pk} = 170$ V
  $V_{L(av)} = 31.7$ V
17. $\eta = 0.627$
  $T = 2.16$ ms
  $V_{fire} = 16.3$ V

19. (a) 230 V AC; 60 Hz; 25° firing angle
    $V_L = 98.6$ V
    (b) $\theta_{cond} = 155°$
    (c) $V_{L(av)} = 98.6$ V
    $I_{L(av)} = 0.986$ A
21. $T = 10$ ms
    $R_1 = 109.17$ kΩ
23. $T = 0.407$ s
    $f = 2.45$ Hz
25. $T = 1.27$ ms
    $f = .78$ Hz
27. $\eta = 0.286$
    $T = 68$ ms
    $f = 14.7$ Hz
29. $R_1 = 1.8$ MΩ

Chapter 7

1. $V_A = 103.6 \, V_{av}$
3. $P_{(hp)} = 1.341$ hp
   $n_{st} = 1200$ r/min
   $n_{rt} = 1164$ r/min
   $T = 6.05$ ft-lb
   $T_{(Nm)} = 8.2$ Nm
   (a) V/Hz ratio = 3.83 V/Hz
   (b) $n_{st} = 1030.9$ r/min
       $f = 51.55$ Hz
       $V_{st} = 197.4$ V AC
   (c) $f = 51.55$ Hz
   (d) $P_{o(W)} = 975$ W
   (e) % power decrease = 2.5%

5. $V_{13} = \dfrac{R_2}{R_2 + R_1} (6.5 \text{ V})$

   $= \dfrac{2000}{2000 + 200} (6.5 \text{ V}) = 5.9$ V

   $t_{off} = 100°C$

Chapter 8

1. (a) Using Table 8.2, a 6.907 mV potential is given by a type J thermocouple at 130°C.
   (b) Using Table 8.2, we find a voltage of 1.277 mV produced at 25°C. Adding this to

6.907 mV we get 8.184 mV. This 8.184 mV is the potential we would expect to measure if the reference were 0°C. Using the table again, we see that an 8.184 mV potential at a reference of zero is produced by a temperature between 153° and 154°C.

3. $y = 0.859$ cm
5. $y = 0.0162$ cm
7. Using the type J table in the text, we see that the measurement junction reading of 5.485 mV corresponds to a temperature of 104°C.
9. A type J thermocouple with a reference junction at 85°C will have a reference junction output of 4.455 mV. To find the output voltage of the measuring junction at 0°C, we add the value of the reference junction output at 85°C to the measuring junction output voltage at 85°C.
   $V_{meas \, @ \, 0°C \, ref} = 19.052$ mV
   We may now look at the type J table for a measuring junction temperature of 19.052 mV with a reference junction temperature of 0°C. The closest output voltage is 19.033 mV, which corresponds to a temperature of 349°C. Actually, the temperature will be slightly less than this.
11. Using the type J table in the text, we see that the measuring junction at 450°C will have an output of 24.607 mV.
13. $G = 10$
15. $T = 96.3°C$
17. $T = 46.4°C$
    $T = 64°C$
19. $V_o = -2$ V
21. $V_{BE} = 0.43125$ V
23. $R_{45°C} = 117.6$ Ω
    The total output voltage is the sum of both contributions, 2.975 V + (−2.657 V) = 0.318 V.
25. $V_o = 3.58$ V
27. (a) 3.28 V
    (b) 3.48 V
29. $\sigma = E\varepsilon = (1 \times 10^8 \text{ kPa})(2 \times 10^{-6}) = 200$ kPa
31. $\Delta R = 0.6$ Ω
33. $V_o = 8.64 \, \mu V$
35. $p = 2047$ lb/ft^2
    $h = 32.8$ ft
37. $h = 49.45$ ft

Chapter 9

1. 565 nm
3. 1.243 μm
5. 200 MHz
7. 130 μW/sr
9. 0.5622 m
11. Peak power = 62.5 kW
    Average power = 75 W
13. I_{photo} = 1.083 μA
    V_o = 27 mV

Chapter 10

1. Gain = 1
   Gain = 2
3. Controller gain = 10
   Proportional band = 10%
5. (a) V_o = 3.33 V
   (b) Reference voltage should equal 3.33 V.
7. R_2 = 281 kΩ
   R_5 = 5.02 MΩ
9. An 18 mA signal across 250 Ω = −4.5 V developed across R_4.
   With R_1 adjusted to +1 V, the output of 1IC would be +3.5 V.
   With R_8 set to 5 kΩ, the output voltage of 2IC would be +8.75 V.
11. t = 6.67 min
13. UTP = 3.73 V
    LTP = 3.47 V
    A voltage of 3.73 V corresponds to a temperature of 373 K or 100°C.
    A voltage of 3.47 V corresponds to a temperature of 347 K or 74°C.
    The differential gap in temperature is 100°C − 74°C or 26°C.
    The differential gap in voltage is 3.73 V − 3.47 V = 0.26 V.
15. e = 2.5%
    C_o = 56%

17. T_{osc} = 0.094 s or 94 ms
    Duty cycle = 40.88%
    P_{load} = 61.3 W

Chapter 11

1. Sampling rate is two times highest frequency to be sampled: 2×20 kHz = 40 kHz
3. f_o = 675 Hz
5. f_o = 41.39 kHz
7. The digital bit values for the two data words shown in the upper left of the photo, given in the order of their transmission are: First word transmitted is 01010101. The second word transmitted is 01010100.
9. R_{max} = 16 MΩ
11. R_{max} = 8 MΩ
    C_1 = 0.0028 μF
13. One bit = 0.00122 V or 1.22 mV

Chapter 12

1. Frequency deviation = 100 kHz
3. Data BW = 11 Hz
   BW_{ch3} = 132 Hz
   Lower limit = 664 Hz
   Upper limit = 796 Hz
   t_r = 0.0318 s or 31.8 ms
5. t_p = 500 μs
   T_f = 125 μs
   Frame rate (f_f) = 8 kHz
   T_s = 13.88 μs
   BW = 36 kHz
   The composite BW will be 2(36 kHz) = 72 kHz
7. f_f = 8 kHz
   T_f = 125 μs
   T_s = 12.5 μs
   $t_{r\,max}$ = 1% of 12.5 μs or 0.125 μs
   BW = 4 MHz
   The composite BW will be 2(4 MHz) = 8 MHz

Chapter 13

1. Problem 1 designs can vary widely.

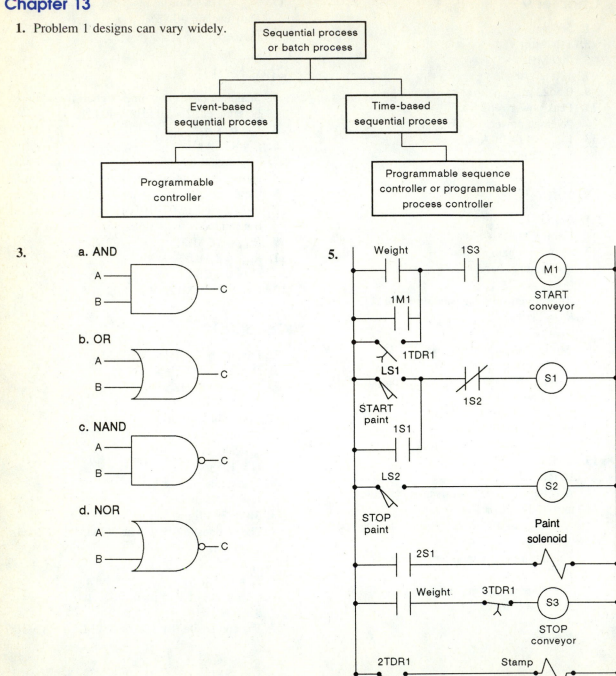

3.

a. AND

b. OR

c. NAND

d. NOR

5.

7.

A.

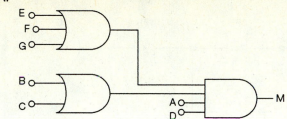

B.

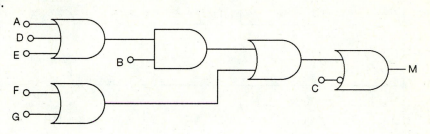

9.

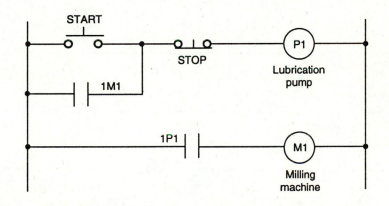

Chapter 14

1.

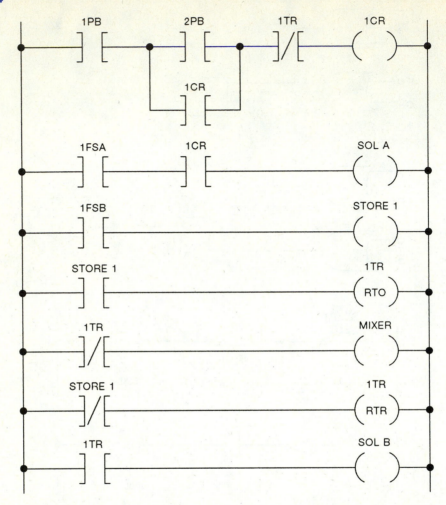

Notice that 1TR is used both in the examine on and examine off condition. If 1TR contacts represent one memory location, then examine on and examine off represent opposite conditions for the same function—which is necessary sometimes.

3.

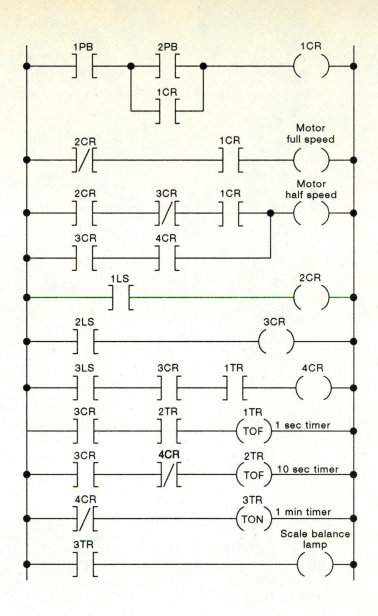

5.

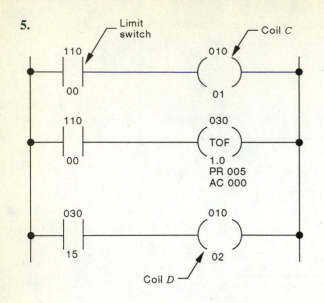

7.

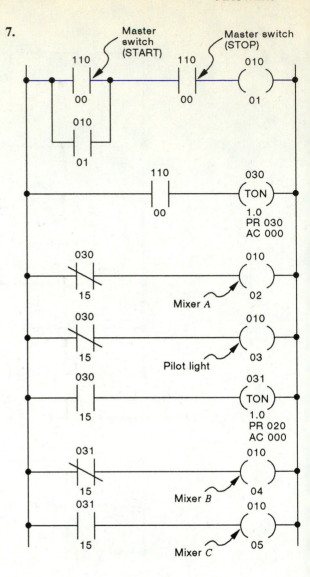

Index